Prentice Hall Advanced Reference Series

Engineering

ANTOGNETTI AND MILUTINOVIC, EDS. *Neural Networks: Concepts, Applications, and Implementations, Vols. I-VI*
DENNO *Power System Design and Applications for Alternative Energy Sources*
ESOGBUE, ED. *Dynamic Programming for Optimal Water Resources Systems Analysis*
FERRY, AKERS, AND GREENEICH *Ultra Large Scale Integrated Microelectronics*
GNADT AND LAWLER, EDS. *Automating Electric Utility Distribution Systems: The Athens Automation and Control Experiment*
HAYKIN, ED. *Advances in Spectrum Analysis and Array Processing, Vol. I*
HAYKIN, ED. *Advances in Spectrum Analysis and Array Processing, Vol. II*
HAYKIN, ED. *Advances in Spectrum Analysis and Array Processing, Vol. III*
HAYKIN, ED. *Selected Topics in Signal Processing*
HENSEL *Inverse Theory and Applications for Engineers*
JOHNSON *Lectures on Adaptive Parameter Estimation*
MILUTINOVIC, ED. *Microprocessor Design for GaAs Technology*
QUACKENBUSH, BARNWELL III, AND CLEMENTS *Objective Measures of Speech Quality*
ROFFEL, VERMEER, AND CHIN *Simulation and Implementation of Self-Tuning Controllers*
SASTRY AND BODSON *Adaptive Control: Stability, Convergence, and Robustness*

Prentice Hall Signal Processing Series

Alan V. Oppenheim, Editor

ANDREWS & HUNT *Digital Image Restoration*
BRACEWELL *Two Dimensional Imaging*
BRIGHAM *The Fast Fourier Transform and Its Applications*
BURDIC *Underwater Acoustic System Analysis 2/E*
CASTLEMAN *Digital Image Processing*
COHEN *Time-Frequency Analysis*
CROCHIERE & RABINER *Multirate Digital Signal Processing*
DUDGEON & MERSEREAU *Multidimensional Digital Signal Processing*
HAYKIN *Advances in Spectrum Analysis and Array Processing Vols. I, II & III*
HAYKIN, ED. *Array Signal Processing*
JOHNSON & DUDGEON *Array Signal Processing*
KAY *Fundamentals of Statistical Signal Processing*
KAY *Modern Spectral Estimation*
KINO *Acoustic Waves: Devices, Imaging, and Analog Signal Processing*
LIM *Two-Dimensional Signal and Image Processing*
LIM, ED. *Speech Enhancement*
LIM & OPPENHEIM, EDS., *Advanced Topics in Signal Processing*
MARPLE *Digital Spectral Analysis with Applications*
MCCLELLAN & RADER *Number Theory in Digital Signal Processing*
MENDEL *Lessons in Estimation Theory for Signal Processing Communications and Control 2/E*
NIKIAS & PETROPULU *Higher-Order Spectra Analysis*
OPPENHEIM & NAWAB *Symbolic and Knowledge-Based Signal Processing*
OPPENHEIM, WILLSKY, WITH YOUNG *Signals and Systems*
OPPENHEIM & SCHAFER *Digital Signal Processing*
OPPENHEIM & SCHAFER *Discrete-Time Signal Processing*
PHILLIPS & NAGLE *Digital Control Systems Analysis and Design, 3/E*
PICINBONO *Random Signals and Systems*
RABINER & GOLD *Theory and Applications of Digital Signal Processing*
RABINER & SCHAFER *Digital Processing of Speech Signals*
RABINER & JUANG *Fundamentals of Speech Recognition*
ROBINSON & TREITEL *Geophysical Signal Analysis*
STEARNS & DAVID *Signal Processing Algorithms in Fortran and C*
THERRIEN *Discrete Random Signals and Statistical Signal Processing*
TRIBOLET *Seismic Applications of Homomorphic Signal Processing*
VAIDYANATHAN *Multirate Systems and Filter Banks*
VETTERLI & KOVACEVIC *Wavelets and Subband Coding*
WIDROW & STEARNS *Adaptive Signal Processing*

Advances in Spectrum Analysis and Array Processing

Volume III

Simon Haykin, editor
McMaster University

Prentice Hall PTR, Upper Saddle River, New Jersey 07458

Library of Congress Cataloging-in-Publication Data
(Revised for vol. 3)

Advances in spectrum analysis and array processing.

(Prentice Hall advanced references series. Engineering)
(Prentice Hall signal processing series)
Includes bibliographical references and indexes.
1. Signal processing. 2. Signal theory (Telecommunication).
I. Haykin, Simon S., 1931 . II. Series.
TK5102.5.A334 1991 621.382'23 89-26459
ISBN 0-13-007444-6 (v. 1)
ISBN 0-13-008574-X (v. 2)
ISBN 0-13-061540-4 (v. 3)

Editorial production: *bookworks*
Acquisitions editior: *Karen Gettman*
Cover designer: *Karen Salzbach*
Manufacturing manager: *Alexis R. Heydt*

Prentice-Hall. Inc.
A Pearson Education Company
Upper Saddle River, NJ 07458

Printed and bound by Antony Rowe Ltd, Eastbourne

Transferred to digital print on demand, 2002

ISBN 0-13-061540-4

Prentice-Hall International (UK) Limited,London
Prentice-Hall of Australia Pty. Limited, Sydney
Prentice-Hall Canada Inc., Toronto
Prentice-Hall Hispanoamericana, S.A., Mexico
Prentice-Hall of India Private Limited, New Delhi
Prentice-Hall of Japan, Inc., Tokyo
Pearson Education Asia Pte. Ltd., Singapore
Editora Prentice-Hall do Brasil, Ltda., Rio de Janeiro

Contents

Contributors

KEVIN M. BUCKLEY, University of Minnesota, Department of Electrical Engineering, 40174 Electrical Engineering/Computer Science Building, 200 Union Street SE, Minneapolis, Minn. 55455

RONALD A. DELAP, United States Air Force, Air Force Institute of Technology, Dayton, OH 45433

MITHAT C. DOGAN, Signal and Imaging Processing Institute, Department of Electrical Engineering-Systems, University of Southern California, Los Angeles, CA 90089-2564

R. HAMZA, University of Minnesota, Department of Electrical Engineering, 40174 Electrical Engineering/Computer Science Building, 200 Union Street SE, Minneapolis, Minn. 55455

SIMON HAYKIN, Communications Research Laboratory, McMaster University, 1280 Main Street West, Hamilton, Ontario, L8S 4K1, Canada

BRAND HEDSTROM, Electrical and Computer Engineering, University of Victoria, P.O. Box 1700, Victoria B.C., V8W 3P6

ALFRED O. HERO, III, Department of EECS, The University of Michigan, Ann Arbor, MI, 48109-2122

JACK JACHNER, Atlantic Aerospace Electronics Corporation, 470 Totten Pond Road, Waltham, MA 02154

R. LYNN KIRLIN, Electrical and Computer Engineering, University of Victoria, P.O. Box 1700, Victoria, B.C. V8W 3P6

HARRY B. LEE, Atlantic Aerospace Electronics Corporation, Atlantic Aerospace Electronics Corporation, 470 Totten Pond Road, Waltham, MA 02154

HENRY LEUNG, Radar Division, Defense Research Establishment Ottawa, Ontario K1A OK2, Canada

TITUS LO, Communications Research Laboratory, McMaster University, 1280 Main Street West, Hamilton, Ontario, L8S 4K1, Canada

CHERIAN P. MATHEWS, School of Electrical Engineering, 1285 Electrical Engineering Building, Purdue University, West Lafayette, IN 47907-1285

JERRY M. MENDEL, Signal and Image Processing Institute, Department of Electrical Engineering-Systems, University of Southern California, University Park, Los Angeles, CA 90089-2564

W. RADICH, University of Minnesota, Department of Electrical Engineering, 40174 Electrical Engineering/Computer Science Building, 200 Union Street S.E., Minneapolis, Minn. 55455

JACOB SHEINVALD, Center for Signal Processing, Rafael 83, Box 2250, Haifa, 31021, Israel

P. STOICA, Department of Automatic Control, The Polytechnic Institute of Bucharest, Bucharest, Romania

EMILY SU, Electrical and computer Engineering, University of Victoria, P.O. Box 1700, Victoria, B.C. V8W 3P6

MATI WAX, Center for Signal Processing, Rafael 83, Box 2250, Haifa 31021, Israel

K. M. WONG, Communications Research Laboratory, McMaster University, 1280 Main Street West, Hamilton, Ontario, L8S 4K1, Canada

Q. WU, Communications Research Laboratory, McMaster University, 1280 Main Street West, Hamilton, Ontario, L8S 4K1, Canada

MICHAEL D. ZOLTOWSKI, School of Electrical Engineering, 1285 Electrical Engineering Building, Purdue University, West Lafayette, IN 47907-1285

Preface

This is the third and the final volume in a series of books which I have had the honor of editing on *Advances in Spectrum Analysis and Array Processing*. The first two volumes were published in 1991.

The present volume is organized in 10 chapters. Chapter 1 by Mati Wax presents optimal and suboptimal model-based processing techniques for the detection, localization, and beamforming of multiple narrowband sources by means of passive arrays. Chapter 2 by Jack Jachner and Harry Lee utilizes the Cramér Rao lower bound on parameter estimation variance to identify fundamental limitations of unbiased direction-finding algorithms for closely spaced source scenarios. Chapter 3 by W. Radich, R. Hamza, and Kevin Buckley addresses another fundamental issue, namely, the robustness of subspace-based direction-finding algorithms. Chapter 4 by Lynn Kirlin, Emily Su, and Brad Hedstrom uses analogy with phase-back loop solutions to shed further light on the array processing problem.

Chapter 5 by Cherian Mathews and Michael Zoltowski focuses on the special case of uniform circular arrays, describing subspace-based methods for two-dimensional angle estimation. Chapter 6 by Max Wong, Qiang Wu, and Peter Stoica describes the development of a new technique called generalized correlation decomposition and its application to the array processing problem in an unknown noise background. Chapter 7 by Jacob Sheinvald and Mati Wax presents a new technique for solving the array processing problem, which permits sampling arbitrary subarrays sequentially in a computationally efficient manner.

Chapter 8 by Alfred Hero III and Ronald DeLap develops task-specific criteria for adaptive beamforming, with the aim of optimizing the best achievable signal detection or parameter estimation at the output of the beamformer.

Chapter 9 by Mithat Dogan and Jerry Mendel exploits cumulants as a tool for extracting more phase information than is possible by using only second-order statistics as is ordinarily the case. By so doing, significant improvements in the performance of array processing systems are realized.

Finally, Chapter 10 by Henry Leung and Titus Lo describes the use of another new tool, neural networks, as the basis for using array data to solve the inverse problem of source location.

Much of the material presented in this book has not appeared in book form before. It has been my distinct pleasure to have worked with these fine researchers in editing the book.

Simon Haykin

1

Model-Based Processing in Sensor Arrays

Mati Wax

1.1 INTRODUCTION

The topic of signal processing in sensor arrays arises in a variety of fields ranging from radar, sonar, oceanography, and seismology to medical imaging and radio-astronomy, and has been the subject of numerous papers and books [13, 19, 12, 16].

Sensor arrays are characterized by their spatial geometry and by the directional patterns of the sensors. If the geometry of the array and the directional patterns are regular, the array is referred to as structured; otherwise it is referred to as unstructured. Although structured arrays present the possibility of sophisticated processing algorithoms that exploit the structure, mutual coupling between the sensors and geometrical constraints in many cases preclude the possibility of implementing such arrays in practice.

The signals impinging on sensor arrays are characterized by their bandwidth and by their mutual correlation. If the bandwidth of the impinging signal is much smaller than the reciprocal of the propagation time across the array, the signal is referred to as narrowband; otherwise it is referred to as wideband. Since wideband signals in many cases have narrowband components, the core of the research in sensor arrays has been centered on narrowband signals. The impinging signals can be uncorrelated as it happens when the signals are radiated from two independent sources, or can be partially or fully correlated as it happens in multipath propagation.

The scenario we address is that of narrowband and arbitrarily correlated sources impinging on an arbitrarily structured array. For this rather general scenario we present optimal and suboptimal solutions to the problems of detection, localization, and beamforming, which are the key problems in adaptive processing in sensor arrays. In section 1.1 we present the mathematical and statistical models for the signals and the noise and formulate the problems of interest. In section 1.2 we present the key concepts upon which the solutions are based. Then, in section 1.3 we present the conditions that guarantee the uniqueness of the solutions. In sections 1.4 and 1.5 we present, respectively, suboptimal non-model-based and model-based solutions. Then, in sections 1.6 and 1.7 we present the optimal solutions to the case of spatially white noise. Finally, in section 1.8 we present a suboptimal solution to the case of spatially colored noise.

The Mathematical Model

Consider an array composed of p sensors with arbitrary locations and arbitrary directional characteristics. Assume that q narrowband sources, centered around a known frequency, say ω_0, impinge on the array from distinct locations $\theta_1, \ldots, \theta_q$. For simplicity, assume that the sources and the sensors are coplanar and that the sources are in the far field of the array. In this case the only parameter that characterizes the source location is its direction-of-arrival θ.

Using complex envelope representation, the $p \times 1$ vector received by the array can be expressed by

$$\mathbf{x}(t) = \sum_{k=1}^{q} \mathbf{a}(\theta_k) s_k(t) + \mathbf{n}(t) \tag{1.1}$$

where $\mathbf{a}(\theta)$ is the $p \times 1$ "steering vector" of the array toward direction θ

$$\mathbf{a}(\theta) = \left[a_1(\theta)e^{-j\omega_0\tau_1(\theta)}, \ldots, a_p(\theta)e^{-j\omega_0\tau_p(\theta)}\right]^T \tag{1.2}$$

$a_k(\theta)$ denotes the amplitude response of the k-th sensor toward direction θ; $\tau_k(\theta)$ denotes the proapagation delay between the reference point and the k-th sensor to a wavefront impinging from direction (θ); and $s_k(t)$ denotes the signal of the k-th source as received at the reference point, and $\mathbf{n}(t)$ denotes the $p \times 1$ vector of the noise at the sensors.

In matrix notation this becomes

$$\mathbf{x}(t) = \mathbf{A}(\mathbf{\Theta})\mathbf{s}(t) + \mathbf{n}(t) \tag{1.3}$$

where $\mathbf{A}(\mathbf{\Theta})$ is the $p \times q$ matrix of the steering vectors

$$\mathbf{A}(\mathbf{\Theta}) = \left[\mathbf{a}(\theta_1), \ldots, \mathbf{a}(\theta_q)\right] \tag{1.4}$$

Suppose that the received vector $\mathbf{x}(t)$ is sampled M times, at $t_1, \ldots, t_M$. From

(1.1), the sampled data can be expressed as

$$\mathbf{X} = \mathbf{A}(\boldsymbol{\Theta})\mathbf{S} + \mathbf{N} \tag{1.5}$$

where $\mathbf{X}$ and $\mathbf{N}$ are the $p \times M$ matrices

$$\mathbf{X} = [\mathbf{x}(t_1), \ldots, \mathbf{x}(t_M)] \tag{1.6}$$

$$\mathbf{N} = [\mathbf{n}(t_1), \ldots, \mathbf{n}(t_M)] \tag{1.7}$$

and $\mathbf{S}$ is the $q \times M$ signal matrix

$$\mathbf{S} = [\mathbf{s}(t_1), \ldots, \mathbf{s}(t_M)] \tag{1.8}$$

The Statistical Models

There are two common statistical models for the noise. One is referred to as the White Noise (WN) model and the other as the Colored Noice (CN) model.

WN: The noise samples $\{\mathbf{n}(t_i)\}$ are statistically independent Gaussian random vectors with zero mean and covariance matrix $\sigma^2\mathbf{I}$.

CN: The noise samples $\{\mathbf{n}(t_i)\}$ are statistically independent Gaussain random vectors with zero mean and covariance matrix $\mathbf{Q}$.

Similarly, there are two common statistical models for the signals. One is referred to as the Deterministic Signals (DS) model and the other as the Stochastic Signals (SS) model.

DS: The signals $\{\mathbf{s}(t_i)\}$ are regarded as unknown deterministic constants.

SS: The signal samples $\{\mathbf{s}(t_i)\}$ are independent and identical Gaussain random vectors with zero mean and covariance matrix $\mathbf{P}$.

In both models, the shape of the signals and the correlation among them can be arbitrary. Specifically, the signals can be fully correlated as it happens, for example, in specular multipath propagation.

Problem Formulation

The problem can now be stated as follows. Given the sampled data $\mathbf{X}$, determine:

(i) the number of sources q
(ii) the directions-of-arrival $\theta_1, \ldots, \theta_q$
(iii) the signal waveforms $\mathbf{s}(t_1), \ldots, \mathbf{s}(t_M)$.

We shall refer to (i) as the detection problem, to (ii) as the localization problem, and to (iii) as the beamforming problem.

Mathematically Equivalent Problems

The mathematical model of the sensor array problem is identical to the model that arises in other key problems in signal processing. The solutions to the sensor array problems that we shall describe are therefore applicable to these problems as well.

Harmonic Retrieval. Consider a signal $x(t)$ composed of q sinusoids with unknown parameters embedded in additive noise:

$$x(t) = \sum_{k=1}^{q} a_k \cos(\omega_k t + \phi_k) + n(t) \tag{1.9}$$

where ω_k, a_k, and ϕ_k are the frequency, amplitude, and phase, respectively, of the k-th sinusoid.

Let a tapped-delay-line with p equally spaced taps D delay units apart be used to sample the signal. The signal at the $(i+1)$-th tap is given by

$$x(t - iD) = \sum_{k=1}^{q} a_k \cos(\omega_k(t - iD)) + \phi_k) + n_i(t) \tag{1.10}$$

Using complex representation we can express the $p \times 1$ vector of the tapped-delay-line outputs as

$$\mathbf{x}(t) = \sum_{k=1}^{q} \mathbf{a}(\omega_k) s_k(t) + \mathbf{n}(t) \tag{1.11}$$

where $s_k(t)$ is the complex sinusoid

$$s_k(t) = a_k e^{j(\omega_k t + \phi_k)} \tag{1.12}$$

and $\mathbf{a}(\omega_k)$ is the $p \times 1$ vector

$$\mathbf{a}(\omega_k) = \left[1, e^{-j\omega_k D}, \ldots, e^{-j\omega_k (p-1)D}\right]^T \tag{1.13}$$

which is exactly in the form of (1.1).

Pole Retrieval. Consider next a linear system that is excited by an impulse. Assuming that the system has q unknown and distinct poles in the complex plane, the response of the system can be expressed as

$$x(t) = \sum_{k=1}^{q} a_k e^{\alpha_k} \cos(\omega_k t + \phi_k) + n(t) \tag{1.14}$$

where $s_k = \alpha_k + j\omega_k$ is the location of the k-th pole in the complex plane and a_k and ϕ_k are its residue and phase, respectively.

Let a tapped-delay-line with p equally spaced taps D delay units apart be used to sample the signal. Using complex representation we can express the $p \times 1$ vector of the tapped-delay-line outputs as

$$\mathbf{x}(t) = \sum_{k=1}^{q} \mathbf{a}(s_k) h_k(t) + \mathbf{n}(t) \tag{1.15}$$

where $h_k(t)$ is the response to the k-th pole

$$h_k(t) = a_k e^{s_k t + \phi_k} \tag{1.16}$$

and $\mathbf{a}(s_k)$ is the $p \times 1$ vector

$$\mathbf{a}(s_k) = \left[1, e^{-s_k D}, \ldots, e^{-s_k(p-1)D}\right]^T \tag{1.17}$$

which is exactly in the form of (1.1).

Echo Retrieval. Consider a radar or sonar system that transmits a *known* pulse $s(t)$ and receives a backscattered signal. Assume that the backscattered signal can be modeled as a superposition of q scaled and delayed echoes embedded in additive noise:

$$x(t) = \sum_{k=1}^{q} m_k s(t - \tau_k) + n(t) \tag{1.18}$$

where m_k and τ_k are the amplitude and delay of the k-th echo, respectively.

Let a tapped-delay-line with p equally spaced taps D delay units apart be used to sample the signal. The $p \times 1$ vector of the tapped-delay-line outputs can be expressed as

$$\mathbf{x}(t) = \sum_{k=1}^{q} \mathbf{a}_k(\tau_k) m_k + \mathbf{n}(t) \tag{1.19}$$

where $\mathbf{a}(\tau_k)$ is the $p \times 1$ vector

$$\mathbf{a}(\tau_k) = [s(t - \tau_k), \ldots, s(t - (p-1)D - \tau_k)]^T \tag{1.20}$$

which is exactly in the form of (1.1).

1.2 PRELIMINARIES

In this section we present the fundamental concepts upon which the solutions will be based.

The Signal Subspace

To gain some insight into the nature of the problems at hand, we cast them in a geometrical frameworks first presented by Schmidt [28, 29].

Observe that the p-dimensional complex vector $\mathbf{x}(t)$ can be regarded as a point in $\mathbf{C}^p$. The location of $\mathbf{x}(t)$ in $\mathbf{C}^p$ is determined by the constraints imposed by (1.1). To see the nature of these constraints we consider two simple cases.

Consider first the case of a single source with no noise present. In this case (1.1) becomes

$$\mathbf{x}(t) = \mathbf{a}(\theta_1) s_1(t) \tag{1.21}$$

implying that $\mathbf{x}(t)$ is confined to a one-dimensional subspace spanned by $\mathbf{a}(\theta_1)$.

Similarly, for the case of two sources with no noise present, (1.1) becomes

$$\mathbf{x}(t) = \mathbf{a}(\theta_1)s_1(t) + \mathbf{a}(\theta_2)s_2(t) \tag{1.22}$$

implying that $\mathbf{x}(t)$ is confined to two-dimensional subspace spanned by $\mathbf{a}(\theta_1)$ and $\mathbf{a}(\theta_2)$. The path traced by $\mathbf{x}(t)$ in this plane depends on the values of $s_1(t)$ and $s_2(t)$ and on their statistical correlation. If $s_1(t)$ and $s_2(t)$ are uncorrelated then $\mathbf{x}(t)$ traces a random pattern in the plane. If, on the other hand, $s_1(t)$ and $s_2(t)$ are fully correlated, that is, $s_1(t) = cs_2(t)$, as it happens in specular multipath propagation, the path traced by $\mathbf{x}(t)$ is confined to the line $\mathbf{a}(\theta_1) + c\mathbf{a}(\theta_2)$ in this plane.

In general, in the case of q sources with no noise present, (1.1) becomes

$$\mathbf{x}(t) = \sum_{i=1}^{q} \mathbf{a}(\theta_i)s_i(t) \tag{1.23}$$

that is, $\mathbf{x}(t)$ is confined to the q-dimensional subspace span $\{\mathbf{a}(\theta_1), \ldots, \mathbf{a}(\theta_q)\}$. For obvious reasons, this subspace is referred to as the "signal subspace," and its complement is referred to as the "noise subspace." If the signals are uncorrelated, then $\mathbf{x}(t)$ spans the whole signal subspace, while if part of the signals or all of them are fully correlated, then $\mathbf{x}(t)$ spans only part of the signal subspace.

The solution of the localization problem in the absence of noise amounts therefore to finding the *smallest* number q and the locations $\theta_1, \ldots, \theta_q$ for which the signal subspace, span $\{\mathbf{a}(\theta_1), \ldots, \mathbf{a}(\theta_q)\}$, *contains* the sampled data $\{\mathbf{x}(t_i)\}$.

In the presence of noise, the vectors $\mathbf{x}(t)$ are no longer confined to the signal subspace and instead are scattered around this subspace. The solution of the localization problem under these more realistic conditions is therefore much more subtle. Intuitively, it amounts to finding the *smallest* number q and the locations $\theta_1, \ldots, \theta_q$ for which span$\{\mathbf{a}(\theta_1), \ldots, \mathbf{a}(\theta_q)\}$ "best fits" the sampled data $\{\mathbf{x}(t_i)\}$.

The MDL Model Selection Criterion

The measure for the "goodness-of-fit" between the potential signal subspace and the observed data will be derived from information theoretic criteria for model selection pioneered by Akaike [1, 2] and further developed by Schwartz [30] and Rissanen [24, 25, 26]. Despite their very similar nature and form, these criteria are derived from quite different principles. Since we believe that Rissanen's approach is the most powerful operationally and conceptually, we base our derivations on his Minimum Description Length (MDL) principle.

The MDL principle asserts the following: given a data set and a family of competing statistical models, the best model is the one that yields the *minimal description length* of the data. The rationale is that since a good model is judged by its ability to "summarize" the data, then the shorter is the "summary," the better is the corresponding model.

The description length of the data, that is, the number of bits it takes to encode the data, can be evaluated quantitatively. Indeed, suppose we want to encode the data set $\mathbf{X} = \{\mathbf{x}(t_1), \ldots, \mathbf{x}(t_M)\}$ using the probabilistic model $P(\mathbf{X}|\mu)$, where μ is an *unknown* (vector) parameter. One possible encoding scheme, referred to as "two-part coding", is to first encode an estimate of the parameter vector μ and then encode the data set $\mathbf{X}$ using the value of the encoded parameter. The code length in this scheme is therefore given by

$$L(\mathbf{X}) = L(\tilde{\mu}) + L(\mathbf{X}|\tilde{\mu}) \tag{1.24}$$

where $L(\tilde{\mu})$ is the code length of the parameter vector and $L(\mathbf{X}|\tilde{\mu})$ is the code length of the data.

To encode the parameters we use a simple coding scheme. Suppose the value of the estimated parameter vector is $\hat{\mu}$ and suppose we encode the parameters by truncating them to some precision, say $\tilde{\mu}_i$, to the precision $\delta_i = 2^{-q_i}$, where q_i is the number of fractional binary digits taken in the truncation. In this case

$$L(\tilde{\mu}) = \sum_{i=1}^{\nu} -\log\delta_i \tag{1.25}$$

where ν denotes the number of free parameters in μ.

It has been shown by Shannon [32] that if the parameter μ is known then the optimal code length for the data is given by $-\log P(\mathbf{X}|\mu)$. Since the value of μ is unknown, it is natural to look for the value of μ for which the code length is minimized. This yields

$$\min_{\mu}\{-\log P(\mathbf{X}|\mu)\} = \max_{\mu} \log P(\mathbf{X}|\mu) \tag{1.26}$$

which is the maximum likelihood criterion, that is

$$L(\mathbf{X}|\tilde{\mu}) = -\log P(\mathbf{X}|\tilde{\mu}) \tag{1.27}$$

where $\tilde{\mu}$ is the truncated value of the maximum likelihood estimator $\tilde{\mu}$.

Combining the two code lengths, we get

$$L(\mathbf{X}) = -\log P(\mathbf{X}|\tilde{\mu}) + \sum_{i=1}^{\nu} -\log\delta_i \tag{1.28}$$

Not surprisingly, this code length depends on the precision selected for the parameters. While the last term increases as the precision increases since the $\{\delta_i\}$ are smaller, the first term decreases since the truncated parameters better approximate the optimal value $\hat{\mu}$. Consequently, there is a certain precision for which the code length is minimized.

To find this optimal precision, we expand the first term in a Taylor series around the optimal value $\hat{\mu}$. Since the first derivative vanishes at the optimal value, we get

$$L(\mathbf{X}|\delta) = -\log P(\mathbf{X}|\hat{\mu}) + \frac{1}{2}\delta^T\Sigma\delta + \sum_{i=1}^{\nu} -\log\delta_i \tag{1.29}$$

where $\delta = [\delta_1, \ldots, \delta_\nu]^T$ and Σ is the Hessian matrix of $-\log P(\mathbf{X}|\hat{\mu})$.

Differentiating the right-hand side with respect to δ, we find that the minimum is achieved for the value of δ given by the equation

$$\Sigma\delta = \delta^{(-1)} \tag{1.30}$$

where

$$\delta^{(-1)} = \left[\frac{1}{\delta_1}, \ldots, \frac{1}{\delta_\nu}\right]^T \tag{1.31}$$

Substituting the optimizing precision in the expression for the code length, we get

$$L(\mathbf{X}|\hat{\delta}) = -\log P(\mathbf{X}|\hat{\mu}) + \frac{1}{2}\nu + \sum_{i=1}^{\nu} -\log \hat{\delta}_i \tag{1.32}$$

To further simplify this expression, we consider its asymptotic behavior. Assuming that $-\log P(\mathbf{X}|\mu)$ grows proportionally to the number of samples M, as is normally the case, the elements of $\frac{1}{M}\sum$ are of order 1, regardless of M. Equation (1.30) then implies

$$\hat{\delta}_i = \frac{c_i(M)}{\sqrt{(M)}} \tag{1.33}$$

where $\{c_i(\mathbf{M})\}$ are some bounded parameters, and hence the description length of the data is asymptotically given by

$$L(\mathbf{X}) = -\log P(\mathbf{X}|\hat{\mu}) + \frac{1}{2}\nu \log M + O(\nu) \tag{1.34}$$

where

$$O(\nu) = \frac{\nu}{2} + \sum_{i=1}^{\nu} \log c_i(M) \tag{1.35}$$

Assuming M is large and $\frac{\nu}{\log M} \to 0$, the Minimum Description Length (MDL) is given by

$$\text{MDL} = -\log P(\mathbf{X}|\hat{\mu}) + \frac{1}{2}\nu \log M. \tag{1.36}$$

Notice that since the asymptotic analysis is valid only if $\frac{\nu}{\log M} \to 0$, the MDL model selection criterion (1.36) is not applicable to the pathological case where the number of free parameters grows linearly with the number of samples.

It is appropriate at this point to compare MDL model selection criterion with the AIC. This criterion, based on an asymptotic analysis of the Kulback-Liebler distance between the "true" and estimated models, is given by

$$\text{AIC} = -\log P(\mathbf{X}|\hat{\mu}) + \nu \tag{1.37}$$

Notice that the only difference between the two criteria is in the coefficient multiplying the number of free parameters in the second term; it is $\frac{1}{2}\log M$ in the MDL and 1 for the AIC.

It turns out that this seemingly minor difference has important implications in terms of the asymptotic behavior of the two criteria. Indeed, in all the analyzed models the MDL has been proven to be consistent, that is, it selects the true model with probability one when the number of samples grows to infinity, while the AIC has been proven to be inconsistent, with a tendency to select more complex models then the true one.

It should be pointed out that the consistency property is shared by a larger family of Information Criteria. Indeed, any Information Criterion with the following generic structure

$$IC = -\log(\mathbf{X}|\hat{\mu}) + \alpha(M)\log M \tag{1.38}$$

can be shown to be consistent if $\alpha(M) \to \infty$ and $\frac{\alpha(M)}{M} \to 0$ as $M \to \infty$ [43].

1.3 CONDITIONS FOR UNIQUENESS

In this section we consider the uniqueness of the solution. Our presentation follows Wax and Ziskind [40] with a slight simplification due to Nehorai, Starer, and Stoica (1990).

By its nature, the uniqueness problem is decoupled from the estimation problem. Thus, we may ignore the noise in the following analysis and rewrite (1.5) as

$$\mathbf{X} = \mathbf{A}(\Theta)\mathbf{S} \tag{1.39}$$

Expression (1.39) can be regarded as a set of equations in which the left-hand side contains the given data while the right-hand side contains the unknowns. Our goal is to specify the conditions under which the solution $(\Theta, \mathbf{S})$ of this set of equations is unique.

Referring to the set $\{\mathbf{a}(\theta), \theta \in \Omega\}$, where Ω denotes the field-of-view, as the "array manifold," the conditions we impose on the array are given by:

A1: The array manifold is known.

A2: Any subset of p distinct steering vectors from the array manifold is linearly independent.

A1 can be fulfilled by either computing the array manifold analytically, if possible, or, alternatively, by measuring it in situ. A2 imposes mild constraints on the array geometry which can easily be fulfilled.

To specify the conditions on the sources, let η denote the rank of the $q \times M$ signal matrix $\mathbf{S}$

$$\eta = \text{rank } \mathbf{S} \tag{1.40}$$

Note that

$$\eta = \text{rank } \left[\mathbf{S}\mathbf{S}^H\right] = \text{rank } \left[\sum_{i=1}^{M} \mathbf{s}(t_i)\mathbf{s}(t_i)^H\right] \tag{1.41}$$

where H denotes the Hermitian (conjugate) transpose, which implies that η is the rank of the sample covariance of the signals

$$\eta = \text{rank}\left[\frac{1}{M}\sum_{i=1}^{M}\mathbf{s}(t_i)\mathbf{s}(t_i)^H\right] \tag{1.42}$$

To point out the implication of $\mathbf{S}$ being of rank η, we rewrite (1.39) as

$$[\mathbf{X}_1, \mathbf{X}_2] = \mathbf{A}(\mathbf{\Theta})\,[\mathbf{S}_1, \mathbf{S}_2] \tag{1.43}$$

where $\mathbf{X}_1$ and $\mathbf{S}_1$ are $p \times \eta$ matrices, and $\mathbf{X}_2$ and $\mathbf{S}_2$ are $p \times (M - \eta)$ matrices. Now, since the matrix $\mathbf{S}$ is of rank η we can assume, without loss of generality, that $\mathbf{S}_1$ is full column rank and that

$$\mathbf{S}_2 = \mathbf{S}_1\mathbf{B} \tag{1.44}$$

where $\mathbf{B}$ is the matrix expressing the linear dependence. From (1.43) and (1.44) we get

$$\mathbf{X}_2 = \mathbf{A}(\mathbf{\Theta})\mathbf{S}_1\mathbf{B} = \mathbf{X}_1\mathbf{B}, \tag{1.45}$$

and hence the system of equations (1.43) is equivalent to the reduced system

$$\mathbf{X}_1 = \mathbf{A}(\mathbf{\Theta})\mathbf{S}_1 \tag{1.46}$$

so that we can confine the analysis to $p \times \eta$ matrices $\mathbf{S}$ of full column rank.

We shall consider two types of uniqueness: (i) uniqueness for every batch $\mathbf{X}$, (ii) uniqueness for almost every batch $\mathbf{X}$, that is, with the exception of a set of batches $\mathbf{X}$ of measure zero.

We first state the conditions that guarantee uniqueness for every batch $\mathbf{X}$.

Theorem 1.1. An array satisfying conditions A1 and A2 can always, that is for every batch $\mathbf{X}$, uniquely localize q sources provided that

$$q < \frac{p+\eta}{2} \tag{1.47}$$

Proof. We shall show that if (1.47) holds true then for every $\mathbf{X}$

$$\mathbf{X} = \mathbf{A}(\mathbf{\Theta})\mathbf{S} \neq \mathbf{A}(\mathbf{\Theta}')\mathbf{S}' \tag{1.48}$$

for any locations $\mathbf{\Theta}' \neq \mathbf{\Theta}$ and any set of signals $\mathbf{S}'$.

To this end, suppose $\theta_i = \theta'_j$ for d pairs of i and $j\,(i, j = 1, \ldots, q)$. Observe that the case $\theta_i \neq \theta'_j$ for all i and $j\,(i, j = 1, \ldots, q)$ corresponds to $d = 0$.

In this case we can rewrite (1.48) as

$$\left[\mathbf{A}(\mathbf{\Theta}), \tilde{\mathbf{A}}(\mathbf{\Theta}')\right]\begin{bmatrix}\tilde{\mathbf{S}}\\ -\tilde{\mathbf{S}}'\end{bmatrix} \neq \mathbf{0} \tag{1.49}$$

where $\tilde{\mathbf{A}}(\mathbf{\Theta}')$ denotes the $p \times (q - d)$ matrix obtained from $\mathbf{A}(\mathbf{\Theta}')$ by deleting those d columns which are identical to the columns of $\mathbf{A}(\mathbf{\Theta})$, $\tilde{\mathbf{S}}'$ denotes the $(q - d) \times \eta$ matrix obtained from $\mathbf{S}'$ by deleting the d rows corresponding to the deleted columns in $\mathbf{A}(\mathbf{\Theta}')$, and $\tilde{\mathbf{S}}$ denotes the $q \times \eta$ matrix obtained from $\mathbf{S}$ by substracting the deleted rows of $\mathbf{S}'$ from the corresponding ones in $\mathbf{S}$. Denote

$$\zeta = \text{null}\left[\mathbf{A}(\mathbf{\Theta}), \tilde{\mathbf{A}}(\mathbf{\Theta}')\right] = (2q - d) - \text{rank}\left[\mathbf{A}(\mathbf{\Theta}), \tilde{\mathbf{A}}(\mathbf{\Theta}')\right] \tag{1.50}$$

and

$$\nu = \text{rank} \begin{bmatrix} \tilde{\mathbf{S}} \\ -\tilde{\mathbf{S}}' \end{bmatrix} \tag{1.51}$$

It follows from (1.49) that to prove the theorem it suffices to show that

$$\zeta < \nu \tag{1.52}$$

To this end, first note that since rank $\mathbf{S} = \eta$ and since $\tilde{\mathbf{S}}$ is obtained from $\mathbf{S}$ by deleting d rows then rank $\tilde{\mathbf{S}} \geq \eta - d$, and hence we have

$$\text{rank} \begin{bmatrix} \tilde{\mathbf{S}} \\ -\tilde{\mathbf{S}}' \end{bmatrix} \geq \eta - d \geq 1 \tag{1.53}$$

Next, by A2 we have

$$\text{rank} \left[\mathbf{A}(\Theta), \tilde{\mathbf{A}}(\Theta') \right] = \min\{p, 2q - d\} \tag{1.54}$$

Now if $d < 2q - p$ then (1.54) implies that $\zeta = 2q - p - d$ and hence since from (1.47) $2q - p - d < \eta - d$ and from (1.53) $\nu \geq \eta - d$, it follows that $\zeta < \nu$.

Similarly, if $d \geq 2q - p$, then it follows from (1.54) that $\zeta = 0$ and since from (1.53) $\nu \geq 1$, we get that $\zeta < \nu$. This completes the proof. ■

It is of interest to examine condition (1.47) for some special cases. One important case is $\eta = q$, occurring when the sources are uncorrelated. In this case (1.47) implies that a unique solution is guaranteed for every batch of data if $q < p$. Another important case is $\eta = 1$, occurring when the sources are fully correlated (as it happens in specular multipath propagation), and in the case of a single snapshot, that is, when $M = 1$. In this case (1.47) implies that a unique solution is guaranteed for every batch of data if $q < (p + 1)/2$.

Theorem 1.1 states the conditions required to guarantee uniqueness for *every* batch $\mathbf{X}$. However, condition (1.47) is rather demanding and cannot be fulfilled in many applications. It is of interest therefore whether this condition can be weakened so as to guarantee uniqueness for *almost every* batch $\mathbf{X}$, that is, with the exception of a set of batches of measure zero.

To formalize this approach we introduce the notion of "topological dimension" of a set. This notion is rigorously defined in Hurewicz and Whallman [14]. However, in order not to clutter the discussion with inessential technical details, we adopt the following definition which is sufficient for our problem. A set S is said to be m-dimensional if m is the least number of real parameters needed to describe its points.

Lemma 1.1. Let $S(\Theta)$ be a m-dimensional set, with Θ denoting a $n \times 1$ parameter vector in R^n, and let S denote the union of all possible $S(\Theta)$'s

$$S = \cup_{\Theta} S(\Theta) \tag{1.55}$$

Then

$$\dim S \leq m + n \tag{1.56}$$

Proof. In order to prove the lemma it suffices to show that the set S can be described with no more than $m+n$ real parameters. Indeed, for each point P in the set S, n real parameters are needed to specify Θ and hence the subset $S(\Theta)$ in which P lies, and m real parameters are needed to specify the coordinates of P inside $S(\Theta)$. This amounts to a total of $m+n$ real parameters and hence the dimension of S is less than or equal to $m+n$.

Having established the above result from dimension theory, we now state the conditions that guarantee the uniqueness for almost every batch **X**. ■

Theorem 1.2. Let (Θ) be fixed and let **S** be a $q \times \eta$ random matrix, drawn from the set of all rank-η matrices whose elements are jointly distributed according to some absolutely continuous distribution on $R^{2q\eta}$. An array satisfying A1 and A2 can then, with probability one, uniquely localize q sources provided that

$$q < \frac{2\eta}{2\eta+1} p \tag{1.57}$$

Proof. Let the set of all matrices **S** for which Θ and Θ' are two ambiguous solutions, where Θ' is arbitrary, be denoted by $S(\Theta, \Theta')$

$$S(\Theta, \Theta') = \{\mathbf{S}, \mathbf{A}(\Theta)\mathbf{S} = \mathbf{A}(\Theta')\mathbf{S}'\} \tag{1.58}$$

Here $\mathbf{S}'$ denotes some $q \times \eta$ matrix. Finally, let $S(\Theta)$ denote the union of the sets $S(\Theta, \Theta')$, that is, the set of all matrices **S** for which ambiguous solutions exist

$$S(\Theta) = \cup_{\Theta'} S(\Theta, \Theta') \tag{1.59}$$

Our proof is based on comparing the topological dimension of the set S to that of the set $S(\Theta)$. To this end, observe that $q\eta$ complex parameters are required to describe a (rank-η) matrix **S** in S. Therefore

$$\dim S = 2q\eta \tag{1.60}$$

To derive the topological dimension of the set $S(\Theta)$, we have to evaluate the topological dimension of the set $S(\Theta, \Theta')$. To this end, as in the proof of Theorem 1.1, suppose that $\theta_i = \theta'_j$ for d pairs of i and $j (i, j = 1, \ldots, q)$.

In this case $\text{null}\big[\mathbf{A}(\Theta), \mathbf{A}(\Theta')\big]$ is given by (1.50). This implies that $\zeta\eta$ complex parameters are required to describe a (rank-η) matrix **S** in $S(\Theta, \Theta')$ and therefore

$$\dim S(\Theta, \Theta') = 2\zeta\eta \tag{1.61}$$

Now, if $d < 2q - p$ we get from (1.54) that

$$\dim S(\Theta, \Theta') = 2(2q-p)\eta \tag{1.62}$$

Using Lemma 1.1, we then have

$$\dim S(\Theta) \leq 2(2q-p)\eta + q \tag{1.63}$$

Now if (1.57) holds true, then $2(2q-p)\eta + q < 2q\eta$. This then implies, by (1.65) and (1.60), that

$$\dim S(\Theta) < \dim S \tag{1.64}$$

Similarly, if $d \geq 2q - p$ it follows from (1.54) that

$$\dim S(\Theta, \Theta') = 2d\eta \tag{1.65}$$

Using Lemma 1.1 this yields

$$\dim S(\Theta) \leq 2d\eta + (q - d) \tag{1.66}$$

Now evidently $2(q - d)\eta > (q - d)$ and hence $2d\eta + q - d < 2q\eta$. This then implies, by (1.66) and (1.60), that

$$\dim S(\Theta) < \dim S \tag{1.67}$$

We have thus shown that the topological dimension of the set $S(\Theta)$ is smaller than that of S. That is, $S(\Theta)$ is a *proper lower dimensional subset* of S. Now since our stochastic setting defines an absolutely continuous measure on the set S, the total measure induced on a proper lower dimensional subset of S is zero. Consequently, the probability of an ambiguous solution is zero. This completes the proof. ■

Observe that Theorem 1.2 does not exclude the possibility of ambiguous solutions. In fact, if $(p + \eta)/2 \leq q < (2\eta p)/(2\eta + 1)$ then for every location vector Θ there exist some matrices **S**, and hence some batches **X**, for which the solution is ambiguous. What Theorem 2 assures is that the measure of this set of batches within the set of all possible batches, when **S** is drawn at random from the set of rank-η $q \times \eta$ matrices, is zero.

Observe that in the special case of coherent sources, $\eta = 1$, condition (1.57) becomes $q < 2p/3$. It should be pointed out that Bresler and Macovski [8] proved that for the case of a uniform linear array this condition is $q \leq 2p/3$, which slightly less demanding.

Theorem 1.1 and 1.2 specify sufficient conditions for uniqueness. It turns out that a slightly weaker version of (1.57) can be shown to be necessary. Indeed, we now show that a necessary condition for uniqueness for every batch **X** is

$$q \leq \frac{2\eta}{2\eta + 1} p \tag{1.68}$$

Suppose that (1.68) does not hold, namely

$$2p\eta < 2q\eta + q \tag{1.69}$$

The right-hand side of (1.69) represents the number of unknowns, while the left-hand side represents the number of independent equations. Thus, (1.69) implies [that Eq. (3.2) has more unknowns than equations] and hence, by the implicit function theorem (see, e.g., [27]), assuming that the conditions for its existence hold, an infinite number of ambiguous solutions exist in the proximity of the solution $(\Theta, \mathbf{S})$. This establishes the necessity of (1.68) for uniqueness.

In Wax and Ziskind [40] it is conjectured that the necessary condition (1.68) is aslo sufficient to establish uniqueness with probability one. In fact, for the special case of a uniform linear array and $\eta = 1$, the sufficiency of (1.68) has been established already by Bresler and Macovski [8].

1.4 SUBOPTIMAL NON-MODEL-BASED SOLUTIONS

In this section we present the "classical" solutions that do not exploit the detailed nature of the model of $\mathbf{x}(t)$ nor the statistical models of the signals and noise.

Delay-and-Sum

The oldest and still one of the most common solutions in use today is the "delay-and-sum" technique, also referred to as the classical beamformer. To motivate this technique, suppose that a wavefront impinges on the array from direction θ and we want to coherently sum the received signals at the different sensors. To this end we have to steer a beam toward direction θ, namely first properly delay the received signals at the different sensors and then sum them. This is carried out by

$$\mathbf{w}(\theta)^H \mathbf{x}(t) \tag{1.70}$$

where $\mathbf{w}(\theta)$ denotes the vector of delays (phase shifts) causing coherent summation from direction θ

$$\mathbf{w}(\theta) = \left[e^{-j\omega_0\tau_1(\theta)}, \dots, e^{-j\omega_0\tau_p(\theta)}\right]^T \tag{1.71}$$

The average power at the delay-and-sum processor is given by

$$B(\theta) = \frac{1}{M} \sum_{i=1}^{M} \|\mathbf{w}(\theta)^H \mathbf{x}(t_i)\|^2 \tag{1.72}$$

which can be rewritten as

$$B(\theta) = \mathbf{w}(\theta)^H \hat{\mathbf{R}} \mathbf{w}(\theta) \tag{1.73}$$

where $\hat{\mathbf{R}}$ denotes the sample covariance matrix

$$\hat{\mathbf{R}} = \frac{1}{M} \sum_{i=1}^{M} \mathbf{x}(t_i)\mathbf{x}(t_i)^H \tag{1.74}$$

In order to solve the stated problem first we have to compute $B(\boldsymbol{\Theta})$ for every θ, then determine the locations of the sources as the $\hat{q}$ peaks of $B(\theta)$ that pass a certain threshold, and finally steer beams to these $\hat{q}$ directions to estimate the signal waveforms.

Evidently, the delay-and-sum technique is suited for a *single* source; in this case $B(\theta)$ will have, asymptotically, a single maximum at the correct direction. However, if more then one source is present, the peaks may be at the wrong directions or may not be resolved at all because of the relatively poor resolution properties of this technique.

Minimum Variance

The minimum variance technique was proposed by Capon [9] to overcome the resolution problems of the delay-and-sum solution. Capon realized that the poor resolution of the delay-and-sum technique can be attributed to the fact that the power of the delay-and-sum processor at a given direction does not depend only on the power of the source at that direction, but also on undesireable contributions from other sources. To improve the resolution, he proposed to modify the delay-and-sum technique so as to minimize these interferences.

His approach was as follows. Suppose we want to receive a wavefront from direction θ and carry it out by a linear combination of the received signals, that is by

$$\mathbf{w}(\theta)^H \mathbf{x}(t) \tag{1.75}$$

where $\mathbf{w}(\theta)$ is a $p \times 1$ complex vector to be determined. It follows from (1.1) that to sum the wavefront from direction θ, the vector $\mathbf{w}$ must obey the constraint

$$\mathbf{w}(\theta)^H \mathbf{a}(\theta) = 1 \tag{1.76}$$

The output power of the linear combiner is given by

$$C(\theta) = \frac{1}{M} \sum_{i=1}^{M} \|\mathbf{w}(\theta)^H \mathbf{x}(t_i)\|^2 \tag{1.77}$$

which can be rewritten as

$$C(\theta) = \mathbf{w}(\theta)^H \hat{\mathbf{R}} \mathbf{w}(\theta) \tag{1.78}$$

where $\hat{\mathbf{R}}$ denotes the sample covariance matrix.

Now in order to minimize the contribution to the output power of other sources at directions different from θ, Capon proposed to select the vector $\mathbf{w}(\theta)$ so as minimize the output power, that is

$$\min_{w(\theta)^H a(\theta)=1} \mathbf{w}(\theta)^H \hat{\mathbf{R}} \mathbf{w}(\theta) \tag{1.79}$$

The solution of this minimization problem, obtained easily by the Langrange-multiplier technique, is given by

$$\hat{\mathbf{w}}(\theta) = \alpha \hat{\mathbf{R}}^{-1} \mathbf{a}(\theta) \tag{1.80}$$

where α is a positive scalar given by

$$\alpha = \frac{1}{\mathbf{a}(\theta)^H \hat{\mathbf{R}}^{-1} \mathbf{a}(\theta)} \tag{1.81}$$

The minimization of the power output, dictated by (1.79), may create serious difficulties when the interfering signal are *correlated* with the desired signals. In this case, the resulting vector $\hat{\mathbf{w}}$ will operate on the interfering signals so as to partially or fully cancel the desired signals, as dictated by the minimization requirement. Note that constraint on the vector $\mathbf{w}$ does not prevent this phenomenon since although the desired signal is summed coherently, the interfering signals are phase-shifted and scaled by $\mathbf{w}$ so as to partially or fully cancel the desired signal. This phenomenon is referred to as signal cancellation by Widrow et al. [42].

The average power at the output of the linear combiner obtained by using the optimal vector $\hat{\mathbf{w}}$ is

$$C(\theta) = \frac{1}{\mathbf{a}(\theta)^H \hat{\mathbf{R}}^{-1} \mathbf{a}(\theta)} \tag{1.82}$$

Thus, to solve our problems we have first to compute $C(\theta)$ for every θ, then determine the locations of the sources as the $\hat{q}$ peaks of $C(\theta)$ that pass a certain threshold, and finally steer beams to these $\hat{q}$ directions to estimate the signal waveforms.

Maximum Signal-to-Interference Interpretation. We shall now show that in the case where the interfering signals are uncorrelated with the desired signals, the minimum variance technique can be interpreted also as maximizing the signal-to-interference ratio at the beamformer output.

Indeed, let us first rewrite (1.1) as

$$\mathbf{x}(t) = \mathbf{a}(\theta)s(t) + \mathbf{i}(t) \tag{1.83}$$

where $\mathbf{i}(t)$ represents the interference due to the other impinging signals and the thermal noise.

The average of the power of the desired signal at the beamformer output is given by

$$E\,\|\mathbf{w}(\theta)^H \mathbf{a}(\theta)s(t)\|^2 = p\mathbf{w}(\theta)^H \mathbf{a}(\theta)\mathbf{a}(\theta)^H \mathbf{w}(\theta) \tag{1.84}$$

where p denotes the average power of the signal $s(t)$. Similarly, the average power output of the interference is given by

$$E\,\|\mathbf{w}(\theta)^H \mathbf{i}(t)\|^2 = \mathbf{w}(\theta)^H \mathbf{R}_I \mathbf{w}(\theta) \tag{1.85}$$

where $\mathbf{R}_I$ denotes the covariance matrix of the interference $\mathbf{i}(t)$.

The signal-to-interference at the beamformer output is therefore given by

$$\frac{S}{I} = \frac{p\mathbf{w}(\theta)^H \mathbf{a}(\theta)\mathbf{a}(\theta)^H \mathbf{w}(\theta)}{\mathbf{w}(\theta)^H \mathbf{R}_I \mathbf{w}(\theta)} \tag{1.86}$$

Denoting the inner product between two vectors $\mathbf{u}$ and $\mathbf{v}$ as

$$(\mathbf{u}, \mathbf{v}) = \mathbf{u}^H \mathbf{R}_I \mathbf{v} \tag{1.87}$$

we can rewrite (1.86) as

$$\frac{S}{I} = p\frac{\|(\mathbf{w}(\theta), \mathbf{R}_I^{-1}\mathbf{a}(\theta))\|^2}{(\mathbf{w}(\theta), \mathbf{w}(\theta))} \tag{1.88}$$

Now by the well-known Cauchy-Schwartz inequality, this ratio is maximized for $\tilde{\mathbf{w}}(\theta)$ given by

$$\hat{\mathbf{w}}(\theta) = \beta \mathbf{R}_I^{-1} \mathbf{a}(\theta) \tag{1.89}$$

where β is some complex scalar.

To see the relation between (1.89) and (1.80), observe that if the signal and the interference are *uncorrelated*, then the covariance matrix of $\mathbf{x}(t)$ is given by

$$\mathbf{R} = p\mathbf{a}(\theta)\mathbf{a}(\theta)^H + \mathbf{R}_I \tag{1.90}$$

Hence, by the well-known matrix inversion lemma

$$\mathbf{R}^{-1} = \mathbf{R}_I^{-1} - \mathbf{R}_I^{-1}\mathbf{a}(\theta)\left(\mathbf{a}(\theta)^H\mathbf{R}_I^{-1}\mathbf{a}(\theta) + \frac{1}{p}\right)^{-1}\mathbf{a}(\theta)^H\mathbf{R}_I^{-1} \tag{1.91}$$

which implies, by multiplying (1.91) by $\mathbf{a}(\theta)$, that

$$\mathbf{R}^{-1}\mathbf{a}(\theta) = \alpha\mathbf{R}_I^{-1}\mathbf{a}(\theta) \tag{1.92}$$

where α is a complex scalar given by

$$\alpha = 1 - \left(\mathbf{a}(\theta)^H\mathbf{R}_I^{-1}\mathbf{a}(\theta) + \frac{1}{p}\right)^{-1}\mathbf{a}(\theta)^H\mathbf{R}_I^{-1}\mathbf{a}(\theta) \tag{1.93}$$

From (1.93) and (1.89) we get

$$\tilde{\mathbf{w}}(\theta) = \gamma\mathbf{R}^{-1}\mathbf{a}(\theta) \tag{1.94}$$

where γ is a scalar constant.

Hence, since $\hat{\mathbf{R}} \to \mathbf{R}$, the equivalence between the minimum variance weight vector and the maximal signal-to-interference weight vector, for the case of uncorrelated interferences, is established.

Adapted Angular Response

An interesting variant of the minimum variance solution has been proposed by Borgiotti and Kaplan [7]. Their starting point is the expression (1.80) for the optimal weight vector in the minimum variance solution. They are looking for a weight vector of the form

$$\mathbf{w}(\theta) = \mu\tilde{\mathbf{R}}^{-1}\mathbf{a}(\theta) \tag{1.95}$$

However, unlike Capon, which determined μ so as to satisfy the constraint (1.76), they proposed to determine it so as to satisfy

$$\mathbf{w}(\theta)^H\mathbf{w}(\theta) = 1 \tag{1.96}$$

This modification leads to the desireable property that the contribution of the noise at the output of the linear combiner is, on the average, identical for every direction.

Combining (1.95) and (1.96) we get

$$\hat{\mu} = \frac{1}{\mathbf{a}(\theta)^H\hat{\mathbf{R}}^{-2}\mathbf{a}(\theta)} \tag{1.97}$$

implying that the power output of the linear combiner, when steered toward direction (θ), is given by

$$A(\theta) = \frac{\mathbf{a}(\theta)^H\hat{\mathbf{R}}^{-1}\mathbf{a}(\theta)}{\mathbf{a}(\theta)^H\hat{\mathbf{R}}^{-2}\mathbf{a}(\theta)} \tag{1.98}$$

Thus, to solve our problem we have to first compute $A(\theta)$ for every θ, then determine the locations of the sources as the $\hat{q}$ peaks of $A(\theta)$ that pass a certain threshold, and finally steer beams to these directions to estimate the signal waveforms.

1.5 SUBOPTIMAL MODEL-BASED SOLUTIONS

Unlike the classical solutions presented in the previous section, in this section we present suboptimal solutions that exploit the structural and statistical models of the signals and noise. Specifically, the solutions will heavily exploit the WN model. As for the signal model, though the derivation will be based on the SS model, the results are applicable also to the DS model. The only requirement is that the signals will not be coherent, that is, that the covariance matrix of the signal will be *nonsingular*. Thus, the solutions we present are applicable to Noncoherent Signals and White Noise, which we refer to as NCSWN.

It should be pointed out that in the case of uniform linear array, the requirement for nonsingularity of the covariance matrix can be overcome by a preprocessing technique referred to as spatial smoothing [10, 31].

Note that from (1.3) and the SS and WN models it follows that $\mathbf{x}(t)$ is a complex Gaussian vector with zero mean and covariance matrix given by

$$\mathbf{R} = \mathbf{A}(\Theta)\mathbf{C}\mathbf{A}(\Theta)^H + \sigma^2\mathbf{I} \tag{1.99}$$

This matrix has interesting properties provided that the covariance matrix of the sources, $\mathbf{C}$, is *nonsingular*. Indeed, since $\mathbf{A}(\Theta)$ is full column-rank, the nonsingularity of $\mathbf{C}$ implies that the $p \times p$ matrix $\mathbf{A}(\Theta)\mathbf{C}\mathbf{A}(\Theta)^H$ has rank q and hende that $p-q$ of its eigenvalues are zero. Consequently, denoting by $\lambda_1 \geq \ldots \geq \lambda_p$ and $\mathbf{v}_1, \ldots, \mathbf{v}_p$ the eingevalues and the corresponding eigenvectors of $\mathbf{R}$, it follows from (1.99) that the $p-q$ smallest eigenvalues of $\mathbf{R}$ are all equal to σ^2, that is

$$\lambda_{q+1} = \ldots = \lambda_p = \sigma^2 \tag{1.100}$$

and hence that the corresponding eigenvectors satisfy

$$\mathbf{R}\mathbf{v}_i = \sigma^2\mathbf{v}_i \quad i = q+1, \ldots, p \tag{1.101}$$

This implies, using (1.99), that

$$\mathbf{A}(\Theta)\mathbf{C}\mathbf{A}(\Theta)^H\mathbf{v}_i = \mathbf{0} \quad i = q+1, \ldots, p \tag{1.102}$$

and since $\mathbf{A}(\Theta)$ is full column rank and $\mathbf{C}$ is by assumption nonsingular, it then follows that

$$\mathbf{A}(\Theta)^H\mathbf{v}_i = \mathbf{0} \quad i = q+1, \ldots, p \tag{1.103}$$

or alternatively

$$\{\mathbf{a}(\theta_1), \ldots, \mathbf{a}(\theta_q)\} \perp \{\mathbf{v}_{q+1}, \ldots, \mathbf{v}_p\} \tag{1.104}$$

Relations (1.100) and (1.104) are the key to the suboptimal solutions. Indeed, (1.100) implies that the number of sources can be determined from the multiplicity of the smallest eigenvalue, while (1.104) implies that the locations of the sources can be determined by searching over the array manifold for those steering vectors that are orthogonal to the eigensubspace of the smallest eigenvalue.

The problem is that in practice we do not have the eigenvalues and eigenvectors of the matrix $\mathbf{R}$ but only an estimate of them formed from the data. With probability one, the estimates do not obey the relations (1.100) and (1.104). Instead, the eigenvalues corresponding to the smallest eigenvalue are "spread" around some value, and their corresponding eigenvectors are only "nearly orthogonal" to steering vectors of the sources.

As a result, to solve the detection and estimation problems we have to resort to more sophisticated approaches. To this end, it will be useful to have a compact representation of $\mathbf{R}$ in terms of its eigenstructure.

From (1.100) and the well-known spectral representation theorem of matrix algebra, we get

$$\mathbf{R} = \sum_{i=1}^{q} \lambda_i \mathbf{v}_i \mathbf{v}_i^H + \sum_{i=q+1}^{p} \sigma^2 \mathbf{v}_i \mathbf{v}_i^H \tag{1.105}$$

Now since the eigenvectors from an orthonormal basis in $\mathbf{C}^P$, we have

$$\mathbf{I} = \sum_{i=1}^{q} \mathbf{v}_i \mathbf{v}_i^H + \sum_{i=q+1}^{p} \mathbf{v}_i \mathbf{v}_i^H \tag{1.106}$$

which, when substituted into (1.105), yields

$$\mathbf{R} = \sum_{i=1}^{q} (\lambda_i - \sigma^2) \mathbf{v}_i \mathbf{v}_i^H + \sigma^2 \mathbf{I} \tag{1.107}$$

or in matrix notation

$$\mathbf{R} = \mathbf{V}_S(\mathbf{\Lambda}_\mathbf{S} - \sigma^2 \mathbf{I}) \mathbf{V}_S^H + \sigma^2 \mathbf{I} \tag{1.108}$$

where

$$\mathbf{V}_S = \left[\mathbf{v}_1, \ldots, \mathbf{v}_q\right] \tag{1.109}$$

and

$$\mathbf{\Lambda}_\mathbf{S} = \text{diag}\,(\lambda_1, \ldots, \lambda_q) \tag{1.110}$$

Estimator of the Eigensystem

Having established the key role played by the eigenstructure of $\mathbf{R}$, the first step is to estimate the eigensystem from the sampled data. To this end we derive the maximum likelihood estimator. The derivation follows Anderson [3].

Relation (1.108) expresses $\mathbf{R}$ in terms of the eigensystem. The ML estimators of these parameters are the values that maximize the log-likelihood function given by

$$L(\mathbf{X}|\{\lambda_i\}, \sigma^2, \{\mathbf{v}_i\}) = -M \log |\mathbf{R}| - M \text{ tr} \left[\mathbf{R}^{-1}\hat{\mathbf{R}}\right] \tag{1.111}$$

where $\hat{\mathbf{R}}$ is the sample-covariance matrix.

By applying the well-known matrix inversion lemma to (1.108) we get

$$\mathbf{R}^{-1} = \frac{1}{\sigma^2}\mathbf{I} - \mathbf{V}_S \mathbf{\Gamma}_\mathbf{S} \mathbf{V}_S^H \tag{1.112}$$

where

$$\boldsymbol{\Gamma}_S = \text{diag}\,(\gamma_1, \ldots, \gamma_q) = \left[\left(\frac{1}{\sigma^2}\boldsymbol{\Lambda}_S - \sigma^2 \mathbf{I} \right)^{-1} + \sigma^2 \mathbf{I} \right] \tag{1.113}$$

Also, by the well-known spectral representation

$$\hat{\mathbf{R}} = \mathbf{U}\mathbf{L}\mathbf{U}^H \tag{1.114}$$

where

$$\mathbf{U} = \left[\mathbf{u}_1, \ldots, \mathbf{u}_p \right] \tag{1.115}$$

and

$$\mathbf{L} = \text{diag}\,(l_1, \ldots, l_p) \tag{1.116}$$

Now, substituting (1.112) and (1.114) into (1.111), and using the fact that the determinant of a matrix is given by the product of its eigenvalues, we obtain

$$L(\mathbf{X}|\{\lambda_i\}, \sigma^2, \{\mathbf{v}_i\}) = -M \left(\sum_{i=1}^{q} \log \lambda_i - (p-q)\log \sigma^2 - \frac{M}{\sigma^2}\,\text{tr}\left[\hat{\mathbf{R}}\right] + \text{tr}\left[\boldsymbol{\Gamma}_S \mathbf{Q}^H \mathbf{L} \mathbf{Q}\right] \right) \tag{1.117}$$

where $\mathbf{Q}$ is the $p \times q$ matrix

$$\mathbf{Q} = \mathbf{U}^H \mathbf{V}_S \tag{1.118}$$

Note that in order to maximize this expression with respect to $\{\mathbf{v}_i\}$, it suffices to maximize with respect to all matrices $\mathbf{Q}$ such that $\mathbf{Q}^H\mathbf{Q} = \mathbf{I}$. To this end, note that since $\gamma_1 > \ldots > \gamma_q$ and $l_1 > \ldots l_p$, it can be readily verified that

$$\text{tr}\left[\boldsymbol{\Gamma}_S \mathbf{Q}^H \mathbf{L} \mathbf{Q}\right] \leq \sum_{i=1}^{q} \lambda_i l_i \tag{1.119}$$

with equality if and only if $\mathbf{Q}$ is given by

$$\hat{\mathbf{Q}} = \begin{pmatrix} \pm 1 & & 0 \\ & \ddots & \\ 0 & & \pm 1 \\ \hline 0 & & 0 \end{pmatrix} \tag{1.120}$$

Thus, the maximum of (1.117) will be achieved for $\mathbf{Q}$ given by (1.120). The ML estimate of the eigenvector matrix $\mathbf{V}_S$ is therefore given by

$$\hat{\mathbf{V}}_S = \mathbf{U}\hat{\mathbf{Q}} \tag{1.121}$$

which implies that, up to a sign, the ML estimates of $\{\mathbf{v}_i\}$ are given by the corresponding eigenvectors of the sample-covariance matrix

$$\hat{\mathbf{v}}_i = \mathbf{u}_i \quad i = 1, \ldots, q \tag{1.122}$$

Substituting the maximizing value of $\hat{\mathbf{Q}}$, from (1.120), into (1.117) and using the well-known results that the determinant and the trace of a matrix are given by the product and sum, respectively, of its eigenvalues, we get

$$L(\mathbf{X}|\{\lambda_i\}, \sigma^2) = -M\left(\sum_{i=1}^{q} \log\lambda_i + (p-q)\log\sigma^2 - \frac{1}{\sigma^2}\sum_{i=1}^{p} l_i + \sum_{i=1}^{q}\left(\frac{l_i}{\sigma^2} - \frac{l_i}{\lambda_i}\right)\right)$$

$$= -M\left(\sum_{i=1}^{q} \log\lambda_i + (p-q)\log\sigma^2 - \frac{1}{\sigma^2}\sum_{i=q+1}^{p} l_i - \sum_{i=1}^{q}\frac{l_i}{\lambda_i}\right) \quad (1.123)$$

Now, by straightforward differentiation with respect to σ^2, we find that the ML estimator of this parameter is given by

$$\hat{\sigma}^2 = \frac{1}{p-q}\sum_{i=q+1}^{p} l_i \quad (1.124)$$

This estimator is intuitively very pleasing. As one would expect form (1.100), the estimator of σ^2 is given by the average of the $p-q$ smallest eigenvalues of sample-covariance matrix.

Similarly, straightforward differentiation with respect to $\{\lambda_i\}$ yields

$$\hat{\lambda}_i = l_i \quad i = 1, \ldots, q \quad (1.125)$$

That is, the estimators of the large eigenvalues and their associated eigenvectors are given by their corresponding values in the sample-covariance matrix.

It should be noted that since the eigenvectors corresponding to the smallest eigenvalue do not appear in the likelihood function, it follows that the only requirement on their ML estimator is that they span the orthogonal complement of the subspace spanned by the eigenvectors corresponding to the large eigenvalues. Therefore, up to a unitary transformation

$$\hat{v}_i = \mathbf{u}_i \quad i = q+1, \ldots, p \quad (1.126)$$

Detection of the Number of Sources

With the estimates of the eigenvalues at hand, we next address the problem of detecting the number of sources from the "multiplicity" of the smallest eigenvalue. The solution follows Wax and Kailath [39] and is based on information-theoretic criteria for model selection.

Let k denote the hypothesized number of signals, $k \in \{0, 1, \ldots, p-1\}$. With this notation, our model for the covariance matrix, from (1.108), can be rewritten as

$$\mathbf{R} = \sum_{i=1}^{k}(\lambda_i - \sigma^2)\mathbf{v}_i\mathbf{v}_i^H + \sigma^2\mathbf{I} \quad (1.127)$$

Every k represents a different multiplicity and hence a different model. The problem is to determine the value of k that best fits the sampled data. We shall solve the problem by the MDL criterion (1.24).

Let $\phi^{(k)}$ denote the parameter vector of the k-th model

$$\phi^{(k)} = (\lambda_1, \ldots, \lambda_k, \sigma^2, \mathbf{v}_1^T, \ldots, \mathbf{v}_k^T)^T \tag{1.128}$$

To compute the first term in (1.24) we sostitute the ML estimator of these parameters into the log-likelihhod function (1.123), and add $M \sum_{i=1}^{p} \log l_i$, which is independent of k. This yields

$$L(\mathbf{X}|\hat{\phi}^{(k)}) = M \sum_{i=k+1}^{p} \log l_i - M(p-k)\log(\frac{1}{p-k}\sum_{i=k+1}^{p} l_i) \tag{1.129}$$

or alternatively

$$L(\mathbf{X}|\hat{\phi}^{(k)}) = -M(p-k)\log\left(\frac{\frac{1}{p-k}\sum_{i=k+1}^{p} l_i}{\prod_{i=k+1}^{p} l_i^{\frac{1}{(p-k)}}}\right) \tag{1.130}$$

We next compute the number of free parameters in $\phi^{(k)}$. Recall that a complex covariance matrix has real eigenvalues but complex eigenvectors. Thus, the number of parameters in $\phi^{(k)}$ is $k + 1 + 2pk$. However, since the eigenvectors are constrained to be mutualluy orthogonal and to have unit norm, their parameters are not independent. The mutual othogonality constraints result in a reduction of $k(k-1)$ degrees of freedom, while the unit norm constraint results in a further reduction of k degrees of freedom. The number of degrees of freedom in $\phi^{(k)}$ is therefore $k + 1 + 2pk - k - k(k-1) = k(2p - k + 1) + 1$.

Substituting (1.130) and the number of degrees-of-freedom into (1.24), the MDL criterion becomes

$$\hat{q} = \arg \min_{k \in \{0,\ldots,p-1\}} MDL_{NCSWN}(k) \tag{1.131}$$

where

$$MDL_{NCSWN}(k) = M(p-k)\log\left(\frac{\frac{1}{p-k}\sum_{i=k+1}^{p} l_i}{\left(\prod_{i=k+1}^{p} l_i\right)^{\frac{1}{(p-k)}}}\right) + \frac{1}{2}k(2p-k+1)\log M \tag{1.132}$$

Analogous to the MDL criterion, we can write down the AIC

$$\hat{q} = \arg \min_{k \in \{0,\ldots,p-1\}} AIC_{NCSWN}(k) \tag{1.133}$$

where

$$AIC_{NCSWN}(k) = M(p-k)\log\left(\frac{\frac{1}{p-k}\sum_{i=k+1}^{p} l_i}{\left(\prod_{i=k+1}^{p} l_i\right)^{\frac{1}{(p-k)}}}\right) + k(2p-k+1) \qquad (1.134)$$

Estimation of the Locations

In the previous section, we saw how to estimate the number of sources from the eigenvalues of the sample-covariance matrix. Using this estimator, we now show how to estimate the locations of the sources from the eigenvectors of the sample-covariance matrix. We describe two different estimators. One is referred to as MUltiple SIgnal Classification (MUSIC) estimator and the other as Minimum Norm estimator.

The MUSIC Estimator. This estimator was proposed by Schmidt [28] and Bienvenu and Kopp [4, 5].

As we have seen in (1.104), the steering vectors of the sources are orthogonal to the subspace spanned by the eigenvectors corresponding to the smallest eigenvalue of the covariance matrix. Thus, if these eigenvectors were known, the locations of the sources could have been determined simply by searching over the array manifold for those steering vectors that are orthogonal to the subspace they span. However, in practice only an estimate of this subspace is available, given by the $\hat{q}$ eigenvectors corresponding to the smallest eigenvalues. Unfortunately, the problem is that this estimated subspace is not orthogonal, with probability one, to the steering vectors, thus rendering this simple approach inapplicable.

A simple solution to this problem is to introduce a "measure of orthogonality" and use it to select the $\hat{q}$ steering vectors which are "most nearly orthogonal" to the estimated subspace. A natural measure of orthogonality between a vector and a subspace is the squared cosine of the angle between this vector and the subspace. Since the eigenvectors $\mathbf{u}_{\hat{q}+1}, \ldots, \mathbf{u}_p$ are orthogonal, the squared cosine of the angle between the vector $\mathbf{a}(\theta)$ and subspace span $\{\mathbf{u}_{\hat{q}+1}, \ldots, \mathbf{u}_p\}$, and hence the orthogonality measure, is given by

$$g(\theta) = \sum_{i=q+1}^{p} \frac{\|\mathbf{a}(\theta)^H \mathbf{v}_i\|^2}{\mathbf{a}(\theta)^H \mathbf{a}(\theta)} \qquad (1.135)$$

The estimates of the locations of the sources are obtained by evaluating (1.135) for every θ and selecting the $\hat{q}$ values of θ that yield the lowest minima.

The asymptotic performance of this estimator, referred to as MUltiple SIgnal Classification (MUSIC), has been the topic of extensive investigation by Sharman et al. [33], Kaveh and Barabell [17], Porat and Friedlander [22], and Stoica and Nehorai [35]; also see Chapter 2 of this book. The analysis shows that the estimator is consistent but that it is statistically efficient only in the special case that the covariance matrix of the sources $\mathbf{C}$ is diagonal. That is, when the sources are correlated, the MUSIC estimator does not achieve the Cramer-Rao lower bound.

The Minimum Norm Estimator. This estimator has been proposed by Kumaresan and Tufts (1984). In the special case of a uniform linear array this estimator coincides with the estimator proposed by Reddi [23].

Instead of measuring the orthogonality between a vector and a subspace by the squared cosine of the angle between the vector and the subspace, as was done in MUSIC, one can use the squared cosine of the angle between this vector and *some* vector $\mathbf{d}$ in this subspace. The only question is how to select $\mathbf{d}$.

To this end, observe that since $\mathbf{d}$ is in span $\{\mathbf{u}_{\hat{q}+1}, \ldots, \mathbf{u}_p\}$, it is orthogonal to the complement subspace, that is

$$\mathbf{U}_S^H \mathbf{d} = 0 \tag{1.136}$$

where $\mathbf{U}_S$ is the $p \times \hat{q}$ matrix of the signal subspace eigenvectors

$$\mathbf{U}_S = \left[\mathbf{u}_1, \ldots, \mathbf{u}_{\hat{q}}\right] \tag{1.137}$$

To determine $\mathbf{d}$ we can therefore fix the first component of $\mathbf{d}$ to be one and then solve the following set of linear equations:

$$\mathbf{G}^H \tilde{\mathbf{d}} = \mathbf{g}^* \tag{1.138}$$

where $*$ denotes the complex conjugate and $\mathbf{G}$, $\mathbf{g}$, and $\tilde{\mathbf{d}}$ are defined by

$$\mathbf{U}_S = \begin{bmatrix} \mathbf{g}^T \\ \mathbf{G} \end{bmatrix} \tag{1.139}$$

and

$$\mathbf{d} = \begin{bmatrix} 1 \\ \tilde{\mathbf{d}} \end{bmatrix} \tag{1.140}$$

Since this set of equations contains more unknown than equations, it does not have a unique solution. A question then arises as to which of all the possible solutions should be preferred. Motivated by the interpretation of the norm of $\tilde{\mathbf{d}}$ in the case of a uniform linear array, Kumaresan and Tufts proposed to select the minimum norm solution, given by

$$\tilde{\mathbf{d}} = (\mathbf{G}^H \mathbf{G})^{-1} \mathbf{G}^H \mathbf{g}^* \tag{1.141}$$

A simpler form of the estimator can be obtained by exploiting the unitarity of the matrix $\mathbf{U}_S$. Indeed, we have

$$[\mathbf{g}^*, \mathbf{G}^H] \begin{bmatrix} \mathbf{g}^T \\ \mathbf{G} \end{bmatrix} = \mathbf{I} \tag{1.142}$$

which implies that

$$\mathbf{g}^* \mathbf{g}^T + \mathbf{G}^H \mathbf{G} = \mathbf{I} \tag{1.143}$$

and hence, by the matrix inversion lemma

$$(\mathbf{G}^H\mathbf{G})^{-1} = (\mathbf{I} - \mathbf{g}^*\mathbf{g}^T)^{-1} = \mathbf{I} - \mathbf{g}^*(1 - \mathbf{g}^T\mathbf{g}^*)^{-1}\mathbf{g}^T \tag{1.144}$$

Substituting this result into (1.143) and performing some algebraic manipulations, we get

$$\tilde{\mathbf{d}} = \frac{1}{1 - \mathbf{g}^H\mathbf{g}}\mathbf{G}\mathbf{g}^* \tag{1.145}$$

We can also express $\tilde{\mathbf{d}}$ in terms of the noise subspace eigenvectors. To this end, let $\mathbf{U}_N$ denote the matrix of the noise subspace eigenvectors

$$\mathbf{U}^N = \left[\mathbf{u}_{\hat{q}+1}, \ldots, \mathbf{u}_p\right] \tag{1.146}$$

and let it be partioned as

$$\mathbf{U}_N = \begin{bmatrix} h^T \\ \mathbf{H} \end{bmatrix} \tag{1.147}$$

Now since $[\mathbf{U}_S, \mathbf{U}_N]$ is an orthonormal basis in $\mathbf{C}^P$, it follows that

$$[\mathbf{U}_S, \mathbf{U}_N][\mathbf{U}_S, \mathbf{U}_N]^H = \mathbf{I} \tag{1.148}$$

and hence

$$\begin{bmatrix} \mathbf{g}^T & \mathbf{h}^T \\ \mathbf{G} & \mathbf{H} \end{bmatrix} \begin{bmatrix} \mathbf{g}^* & \mathbf{G}^H \\ \mathbf{h}^* & \mathbf{H}^H \end{bmatrix} = \mathbf{I} \tag{1.149}$$

Using this identity, the expression (1.145) becomes

$$\tilde{\mathbf{d}} = \frac{1}{\mathbf{h}^H\mathbf{h}}\mathbf{H}\mathbf{g}^* \tag{1.150}$$

The vector $\mathbf{d}$, with $\tilde{\mathbf{d}}$ given by (1.145) or (1.150), "represents" the signal subspace. The squared cosine between $\mathbf{a}(\theta)$ and $\mathbf{d}$, which we use as a measure of orthogonality between $\mathbf{a}(\Theta)$ and the signal subspace, is given by

$$g(\theta) = \frac{\|\mathbf{a}(\theta)^H\mathbf{d}\|^2}{\mathbf{a}(\theta)^H\mathbf{a}(\theta)} \tag{1.151}$$

The estimates of the locations of the sources are obtained by evaluating (1.151) for every θ and selecting the $\tilde{q}$ values of θ that yield the lowest minima.

1.6 OPTIMAL SOLUTION FOR THE DSWN MODEL

In this section we present the optimal solution to the Deterministic Signals (DS) and White Noise (WN), which we refer to as the DSWN model.

Before we present this solution, we derive the maximum likelihood estimator which plays a key role in the solution.

The Maximum Likelihood Estimator

The derivation of the Maximum Likelihood (ML) estimator follows Wax [36].

From (1.3) and the DS and WN models, the joint density function of the sampled data is given by

$$f(\mathbf{X}) = \prod_{i=1}^{M} \frac{1}{\pi|\sigma^2 \mathbf{I}|} \exp(-\frac{1}{\sigma^2}\|\mathbf{x}(t_i) - \mathbf{A}(\Theta)\mathbf{s}(t_i)\|^2) \tag{1.152}$$

In this statistical model, the number of parameters grows linearly with the number of samples since $\{\mathbf{s}(t_i)\}$ represent $2qM$ free real parameters. This rather pathological behavoir of the DS model has serious implications on the performance of the ML estimator since, as is well known, the more unknowns there are, the worse is the performance.

From (1.152), the log-likelihood of the sampled data, ignoring constant terms, is given by

$$L(\mathbf{X}|\Theta, \{\mathbf{s}(t_i)\}, \sigma^2) = -Mp\log\sigma^2 - \frac{1}{\sigma^2}\sum_{i=1}^{M}\|\mathbf{x}(t_i) - \mathbf{A}(\Theta)\mathbf{s}(t_i)\|^2 \tag{1.153}$$

To compute the ML estimator we have to maximize the log-likelihood with respect to the unknown parameters. We carry out this maximization in three steps: (i) maximize with respect to σ^2 with $\{\mathbf{s}(t_i)\}$ and Θ being fixed; then (ii) substitute the resulting estimate of σ^2 expressed as a function of $\{\mathbf{s}(t_i)\}$ and Θ, back into the log-likelihood function and maximize with respect to parameter $\{\mathbf{s}(t_i)\}$, with Θ fixed; and (iii) substitute the resulting estimate of $\{\mathbf{s}(t_i)\}$, expressed as a function of Θ, back into the log-likelihood function and obtain a function to be maximized over Θ only.

To carry out the first step, we fix Θ and $\{\mathbf{s}(t_i)\}$, and then maximize with respect to σ^2. This yields

$$\tilde{\sigma}^2(\Theta), \{\mathbf{s}(t_i)\}) = \frac{1}{Mp}\sum_{i=1}^{M}\|\mathbf{x}(t_i) - \mathbf{A}(\Theta)\mathbf{s}(t_i)\|^2 \tag{1.154}$$

Substituting this result into (1.153), the log-likelihood becomes

$$L(\mathbf{X}|\Theta, \{\mathbf{s}(t_i)\} = -Mp\log\left(\sum_{i=1}^{M}\|\mathbf{x}(t_i) - \mathbf{A}(\Theta)\mathbf{s}(t_i)\|^2\right) \tag{1.155}$$

Notice that since the logarithm is a monotonic function, the values of Θ and $\{\mathbf{s}(t_i)\}$ that maximize this expression are in fact the least-squares estimators and hence are meaningful even if the WN model is not valid.

To carry out the second step, we fix Θ and maximize this expression with respect to $\mathbf{s}(t_i)$. This yields

$$\hat{\mathbf{s}}(t_i) = \left[\mathbf{A}(\Theta)^H\mathbf{A}(\Theta)\right]^{-1}\mathbf{A}(\Theta)^H\mathbf{x}(t_i) \tag{1.156}$$

Substituting this expression into the log-likelihood function, ignoring constant terms, we get

$$L(\mathbf{X}|\boldsymbol{\Theta}) = -Mp \log \left(\sum_{i=1}^{M} \|\mathbf{x}(t_i) - \mathbf{P}_{\mathbf{A}(\boldsymbol{\Theta})}\mathbf{x}(t_i)\|^2 \right) \tag{1.157}$$

where $\mathbf{P}_{\mathbf{A}(\boldsymbol{\Theta})}$ is the orthogonal projection onto the space spanned by the columns of the matrix $\mathbf{A}(\boldsymbol{\Theta})$

$$\mathbf{P}_{\mathbf{A}(\boldsymbol{\Theta})} = \mathbf{A}(\boldsymbol{\Theta}) \left[\mathbf{A}(\boldsymbol{\Theta})^H \mathbf{A}(\boldsymbol{\Theta})\right]^{-1} \mathbf{A}(\boldsymbol{\Theta})^H \tag{1.158}$$

Alternatively, we can rewrite this as

$$L(\mathbf{X}|\boldsymbol{\Theta}) = -Mp \log \left(\sum_{i=1}^{M} \|\mathbf{P}^{\perp}_{\mathbf{A}(\boldsymbol{\Theta})}\mathbf{x}(t_i)\|^2 \right) \tag{1.159}$$

where $\mathbf{P}^{\perp}_{\mathbf{A}(\boldsymbol{\Theta})}$ is the orthogonal projection onto the orthogonal complement of the space spanned by the columns of the matrix $\mathbf{A}(\boldsymbol{\Theta})$

$$\mathbf{P}^{\perp}_{\mathbf{A}(\boldsymbol{\Theta})} = \mathbf{I} - \mathbf{P}_{\mathbf{A}(\boldsymbol{\Theta})} \tag{1.160}$$

Thus, the ML estimator $\hat{\boldsymbol{\Theta}}$ is the solution of the following maximization problem:

$$\hat{\boldsymbol{\Theta}} = \arg\max_{\boldsymbol{\Theta}} \sum_{i=1}^{M} \|\mathbf{P}_{\mathbf{A}(\boldsymbol{\Theta})}\mathbf{x}(t_i)\|^2 \tag{1.161}$$

or, alternatively, the solution of the following minimization problem:

$$\hat{\boldsymbol{\Theta}} = \arg\min_{\boldsymbol{\Theta}} \sum_{i=1}^{M} \|\mathbf{P}^{\perp}_{\mathbf{A}(\boldsymbol{\Theta})}\mathbf{x}(t_i)\|^2 \tag{1.162}$$

This estimator has an appealing geometric interpretation. Recall that in the absence of noise the vector $\mathbf{x}(t)$ is confined to the q-dimensional space spanned by the columns of the matrix $\mathbf{A}(\boldsymbol{\Theta})$, referred to as the signal subspace. Expression (1.161) implies that the ML estimator selects the signal subspace which is "closest" to the sampled vectors $\{\mathbf{x}(t_i)\}$, where closeness is measured by the sum of squares of the projections of $\{\mathbf{x}(t_i)\}$ onto this subspace. Similarly, expression (1.162) implies that the ML estimator selects the noise subspace which is "mostly orthogonal" to the sampled vectors $\{\mathbf{x}(t_i)\}$, where the orthogonality is measured by the sum of squares of the projections of $\{\mathbf{x}(t_i)\}$ onto this subspace.

Since the projection operator is independent, that is, $\mathbf{P}_{\mathbf{A}(\boldsymbol{\Theta})}\mathbf{P}_{\mathbf{A}(\boldsymbol{\Theta})} = \mathbf{P}_{\mathbf{A}(\boldsymbol{\Theta})}$, we have

$$\begin{aligned} \sum_{i=1}^{M} \|\mathbf{P}_{\mathbf{A}(\boldsymbol{\Theta})}\mathbf{x}(t_i)\|^2 &= \sum_{i=1}^{M} \mathbf{x}(t_i)^H \mathbf{P}_{\mathbf{A}(\boldsymbol{\Theta})}\mathbf{x}(t_i) \\ &= \sum_{i=1}^{M} \operatorname{tr}\left[\mathbf{P}_{\mathbf{A}(\boldsymbol{\Theta})}\mathbf{x}(t_i)\mathbf{x}(t_i)^H\right] \end{aligned} \tag{1.163}$$

where tr [ ] denotes the trace operator. Substituting this result into (1.161), we can recast the ML estimator as the solution to the following problems:

$$\max_{\Theta} \operatorname{tr}\left[\mathbf{P}_{\mathbf{A}(\Theta)}\hat{\mathbf{R}}\right] \tag{1.164}$$

or, alternatively

$$\min_{\Theta} \operatorname{tr}\left[\mathbf{P}^{\perp}_{\mathbf{A}(\Theta)}\hat{\mathbf{R}}\right] \tag{1.165}$$

where $\hat{\mathbf{R}}$ is the sample covariance matrix

$$\hat{\mathbf{R}} = \frac{1}{M}\sum_{i=1}^{M}\mathbf{x}(t_i)\mathbf{x}(t_i)^H \tag{1.166}$$

This expression shows the central role played by the sample-covariance matrix in this problem. Indeed, the sample-covariance is the *sufficient statistic* for the problem; all the relevant information in the data $\{\mathbf{x}(t_i)\}$ is captured by $\mathbf{R}$.

An interesting interpretation of the ML estimator is obtained by expressing (1.165) in terms of the eigenstructure of $\hat{\mathbf{R}}$. Let $l_1 \geq \ldots \geq l_p$ and $\mathbf{u}_1, \ldots, \mathbf{u}_p$ denote the eigenvalues and eigenvectors, respectively, of $\hat{\mathbf{R}}$. From the spectral representation theorem of matrix theory, we can express $\hat{\mathbf{R}}$ as

$$\hat{\mathbf{R}} = \sum_{i=1}^{p} l_i \mathbf{u}_i \mathbf{u}_i^H \tag{1.167}$$

Using this representation and the properties of the projection and trace operators, we can rewrite (1.164) as

$$\hat{\Theta} = \arg\max_{\Theta} \sum_{i=1}^{p} l_i \|\mathbf{P}_{\mathbf{A}(\Theta)}\mathbf{u}_i\|^2 \tag{1.168}$$

This expression shows that the measure of "closeness" between the observed data, as summarized by the sufficient statistic $\hat{\mathbf{R}}$, and the potential signal subspace can be also expressed in terms of the projections of the eigenvectors of $\hat{\mathbf{R}}$ onto this subspace; the larger the eigenvalue the more important it is that the projection of the corresponding eigenvector onto the signal subspace be maximized.

It is of interest to examine the performance of the ML estimator as $M \to \infty$. Unfortunately, the classical results on the consistency and the statistical efficiency of the ML estimator cannot be used in this problem because the regularity conditions required for their applicability do not hold. This "pathological" behavior stems from the linear growth of the number of unknown parameters with the number of samples we have pointed out above.

Following Stoica and Nehorai [34], we first establish the consistency of the ML estimator. Assuming $\hat{\mathbf{R}} \to \mathbf{R}$ and $\frac{1}{M}\sum_{i=1}^{M}\mathbf{s}(t_i)\mathbf{s}(t_i)^H \to \mathbf{P}$ as $M \to \infty$, we find that as $M \to \infty$ the function to be minimized becomes

$$\operatorname{tr}\left[\mathbf{P}^{\perp}_{\mathbf{A}(\hat{\Theta})}\hat{\mathbf{R}}\right] = \operatorname{tr}\left[\mathbf{P}^{\perp}_{\mathbf{A}(\hat{\Theta})}(\mathbf{A}(\Theta)\mathbf{P}\mathbf{A}(\Theta)\right] + \sigma^2\mathbf{I}) \geq (p-q)\sigma^2 \tag{1.169}$$

which is clearly minimized for $\hat{\Theta} = \Theta$. The consistency of $\hat{\Theta}$ does not imply, however, the consistency of the other parameters of the model. Indeed, as $M \to \infty$ we get

$$\hat{\sigma}^2 \to \frac{1}{p} \operatorname{tr}\left[\mathbf{P}_{\mathbf{A}(\Theta)}^{\perp}\mathbf{R}\right] = \frac{(p-q)}{p}\sigma^2 \tag{1.170}$$

and

$$\hat{\mathbf{s}}(t_i) \to \left[\mathbf{A}(\Theta)^H\mathbf{A}(\Theta)\right]^{-1}\mathbf{A}(\Theta)^H\mathbf{x}(t_i) =$$

$$\mathbf{s}(t_i) + \left[\mathbf{A}(\Theta)^H\mathbf{A}(\Theta)\right]^{-1}\mathbf{A}(\Theta)^H\mathbf{n}(t_i) \tag{1.171}$$

The "pathological" behavior of the ML estimator in the DS model is not reflected only in the inconsistency of the above estimators. As shown by Stoica and Nehorai [34], the ML estimator of Θ is statistically inefficient, namely, it does not achieve the Cramer-Rao lower bound as $M \to \infty$.

Simultaneous Detection and Localization

We shall solve the detection and localization problem simultaneously via the MDL principle for model selection. The derivation follows Wax and Ziskind [41].

Denoting by k the unknown number of sources, our model (1.3) becomes

$$\mathbf{x}(t) = \mathbf{A}(\Theta^{(k)})\mathbf{s}(t) + \mathbf{n}(t) \tag{1.172}$$

where $\mathbf{A}(\Theta^{(k)})$ is a $p \times k$ matrix of the steering vectors of the sources

$$\mathbf{A}(\Theta^{(k)}) = [\mathbf{a}(\theta_1), \ldots, \mathbf{a}(\theta_k)] \tag{1.173}$$

Note that every different k represents a potential and competing model. Our problem is to choose the model in this family that best fits the sampled data.

Unfortunately, this family of models does not allow a straightforward application of the MDL principle for model selection since the number of parameters in this family grows with the number of samples. Indeed, as noted in the derivation of the maximum likelihood estimator, the number of parameters in this family is $1 + q + 2kM$.

To overcome this problem, we shall recast the problem so as to eliminate the $2kM$ parameters of the signals $\{\mathbf{s}(t_i)\}$, which are nuisance parameters in the detection and localization problems.

To this end, consider the partition of the vector $\mathbf{x}(t)$ into its component in the signal subspace, span $\{\mathbf{A}(\Theta^{(k)})\}$, and its complement, the noise subspace. We have

$$\mathbf{x}(t) = \mathbf{G}(\Theta^{(k)})\begin{bmatrix}\mathbf{x}_S(t)\\ \mathbf{x}_N(t)\end{bmatrix} \tag{1.174}$$

where $\mathbf{x}_S(t)$ denotes the $k \times 1$ component in the signal subspace, $\mathbf{x}_N(t)$ denotes the $(p-k) \times 1$ component in the noise subspace, and $\mathbf{G}(\Theta^{(k)})$ denotes a unitary coordinate transformation matrix determined so as to align the signal subspace with the first k coordinates, that is

$$\mathbf{P}_{\mathbf{A}(\Theta^{(k)})}\mathbf{x}(t) = \mathbf{G}(\Theta^{(k)})\begin{bmatrix}\mathbf{x}^S(t)\\ \mathbf{0}\end{bmatrix} \tag{1.175}$$

Consequently, this transformation aligns the last $p-k$ coordinates in the noise subspace, that is

$$\mathbf{P}^{\perp}_{\mathbf{A}(\mathbf{\Theta}^{(k)})}\mathbf{x}(t) = \mathbf{G}(\mathbf{\Theta}^{(k)}) \begin{bmatrix} \mathbf{0} \\ \mathbf{x}_N(t) \end{bmatrix} \tag{1.176}$$

The representation (1.174) defines a family of competing models for the sampled vector $\mathbf{x}(t)$. To apply the MDL principle to this family, we must compute the code length required to encode the data $\{\mathbf{x}(t_i)\}$ using each of the competing models. Now since $\mathbf{x}(t)$ is expressed in terms of its components in the signal and noise subspaces, the code length of $\{\mathbf{x}(t_i)\}$ should be expressed in these terms.

We start with the noise subspace components $\{\mathbf{x}_N(t_i)\}$. According to (1.24), the code length required to encode them depends on the model for their probability distribution. To this end, observe that from (1.172) and (1.176) we get

$$\mathbf{P}^{\perp}_{\mathbf{A}(\mathbf{\Theta}^{(k)})}\mathbf{n}(t) = \mathbf{G}(\mathbf{\Theta}^{(k)}) \begin{bmatrix} \mathbf{0} \\ \mathbf{x}_N(t) \end{bmatrix} \tag{1.177}$$

Thus, since the noise $\mathbf{n}(t)$ is a complex Gaussian random vector with zero mean and covariance matrix $\sigma^2\mathbf{I}$, it follows that given $\mathbf{\Theta}^{(k)}$, $\mathbf{x}_N(t)$ is a $(p-k)$-dimensional complex Gaussian random vector with zero mean and covariance matrix $\sigma^2\mathbf{I}$, as shown by

$$\mathbf{x}_N(t)|\mathbf{\Theta}^{(k)} \sim N_{p-k}(\mathbf{0}, \sigma^2\mathbf{I}) \tag{1.178}$$

Hence, the probabilistic model for the noise subspace component is given by

$$f(\{\mathbf{x}_N(t_i)\}|\mathbf{\Theta}^{(k)}) = \prod_{i=1}^{M} \frac{1}{|\pi\sigma^2\mathbf{I}|} \exp\{-\mathbf{x}_N(t_i)^H \sigma^{-2} \mathbf{x}_N(t_i)\} \tag{1.179}$$

which can be rewritten as

$$f(\{\mathbf{x}_N(t_i)\}|\mathbf{\Theta}^{(k)}) = |\pi\sigma^2\mathbf{I}|^{-M} \exp\{-\sigma^{-2}M \operatorname{tr}\left[\hat{\mathbf{R}}_{NN}(\mathbf{\Theta}^{(k)})\right]\} \tag{1.180}$$

where $\hat{\mathbf{R}}_{NN}(\mathbf{\Theta}^{(k)})$ denotes the $(p-k) \times (p-k)$sample-covariance matrix of $\mathbf{x}_N(t)$

$$\hat{\mathbf{R}}_{NN}(\mathbf{\Theta}^{(k)}) = \frac{1}{M} \sum_{i=1}^{M} \mathbf{x}_N(t_i)\mathbf{x}_N(t_i)^H \tag{1.181}$$

The probabilistic model (1.180) has only a single parameter, σ^2, whose ML estimator is given by

$$\hat{\sigma}^2(\mathbf{\Theta}^{(k)}) = \frac{1}{(p-k)} \operatorname{tr}\left[\hat{\mathbf{R}}_{NN}(\mathbf{\Theta}^{(k)})\right] \tag{1.182}$$

Thus, it follows from (1.24) that the code length required to encode the noise subspace components, ignoring constant terms that are indipendent of k, is given by

$$L\{\mathbf{x}_N(t_i)|\mathbf{\Theta}^{(k)}\} = M \log|\hat{\sigma}^2(\mathbf{\Theta}^{(k)})\mathbf{I}| + M(p-k) + \frac{1}{2}\log M \tag{1.183}$$

Next we evaluate the code length required to encode the signal subspace components $\{\mathbf{x}_S(t_i)\}$. To this end we have to select a probabilistic model for the conditional distribution of $\mathbf{x}_S(t)$ given $\mathbf{x}_N(t)$ and $\mathbf{\Theta}^{(k)}$. A mathematically convenient choice is

$$\mathbf{x}_S(t)|\mathbf{x}_N(t), \mathbf{\Theta}^{(k)} \sim N_k(\mathbf{B}\mathbf{x}_N(t), \mathbf{V}) \tag{1.184}$$

where $\mathbf{B}$ and $\mathbf{V}$ are unknown complex matrices of dimension $k \times (p-k)$ and $k \times k$, respectively. The probabilistic model for the signal subspace components is therefore given by

$$f(\{\mathbf{x}_S(t_i)\}|\mathbf{\Theta}_{(k)}, \{\mathbf{x}_N(t_i)\}) =$$
$$\prod_{i=1}^{M} \frac{1}{|\pi \mathbf{V}|} \exp\{-(\mathbf{x}_S(t_i) - \mathbf{B}\mathbf{x}_N(t_i))^H \mathbf{V}^{-1}(\mathbf{x}_S(t_i) - \mathbf{B}\mathbf{x}_N(t_i))\} \tag{1.185}$$

It can be easily verified that the number of free parameters in the matrices $\mathbf{B}$ and $\mathbf{V}$ is given by $2(p-k)k$ and k^2, respectively, and that their maximum likelihood estimators are given by

$$\hat{\mathbf{B}} = \hat{\mathbf{R}}_{SN}\hat{\mathbf{R}}_{NN}^{-1} \tag{1.186}$$

$$\hat{\mathbf{V}} = \hat{\mathbf{R}}_{SS} - \hat{\mathbf{R}}_{SN}\hat{\mathbf{R}}_{NN}^{-1}\hat{\mathbf{R}}_{NS} \tag{1.187}$$

where

$$\hat{\mathbf{R}}_{SS} = \frac{1}{M}\sum_{i=1}^{M} \mathbf{x}_S(t_i)\mathbf{x}_S(t_i)^H \tag{1.188}$$

and

$$\hat{\mathbf{R}}_{SN} = \frac{1}{M}\sum_{i=1}^{M} \mathbf{x}_S(t_i)\mathbf{x}_N(t_i)^H \tag{1.189}$$

From (1.24), the code length required to encode the signal subspace components is given by

$$L(\{\mathbf{x}_S(t_i)\}|\mathbf{\Theta}^{(k)}, \{\mathbf{x}_N(t_i)\}) = M\log|\hat{\mathbf{V}}| + Mk + \frac{1}{2}(k^2 + 2k(p-k))\log M \tag{1.190}$$

Combining (1.184) and (1.190), the total code length required to encode the signal and noise subspace components is given by

$$L(\{\mathbf{x}_N(t_i)\}, \{\mathbf{x}_S(t_i)\}|\mathbf{\Theta}^{(k)}) = M\log\left(\frac{1}{(p-k)}\operatorname{tr}\hat{\mathbf{R}}_{NN}\right)^{p-k}$$
$$+M\log|\hat{\mathbf{R}}_{SS} - \hat{\mathbf{R}}_{SN}\hat{\mathbf{R}}_{NN}^{-1}\hat{\mathbf{R}}_{NS}| + \frac{1}{2}(k^2 + 2k(p-k) + 1)\log M \tag{1.191}$$

In order to simplify this expression, we rewrite the sample-covariance matrix $\hat{\mathbf{R}}$, from (1.174), as

$$\hat{\mathbf{R}} = \mathbf{G}(\mathbf{\Theta}^{(k)})\begin{pmatrix}\hat{\mathbf{R}}_{SS} & \hat{\mathbf{R}}_{SN}\\ \hat{\mathbf{R}}_{NS} & \hat{\mathbf{R}}_{NN}\end{pmatrix}\mathbf{G}(\mathbf{\Theta}^{(k)})^H \tag{1.192}$$

Taking the determinant of both sides and recalling that $\mathbf{G}(\boldsymbol{\Theta}^{(k)})$ is unitary, we get

$$|\hat{\mathbf{R}}| = \begin{vmatrix} \hat{\mathbf{R}}_{SS} & \hat{\mathbf{R}}_{SN} \\ \hat{\mathbf{R}}_{NS} & \hat{\mathbf{R}}_{NN} \end{vmatrix} = |\hat{\mathbf{R}}_{NN}||\hat{\mathbf{R}}_{SS} - \hat{\mathbf{R}}_{SN}\hat{\mathbf{R}}_{NN}^{-1}\hat{\mathbf{R}}_{NS}| \tag{1.193}$$

When substituted into (1.191), and ignoring terms that are independent of k, this yields

$$L\{\mathbf{x}_N(t_i), \mathbf{x}_S(t_i)|\boldsymbol{\Theta}^{(k)}\} = M(p-k)\log\left(\frac{\frac{1}{(p-k)}\operatorname{tr}\hat{\mathbf{R}}_{NN}(\boldsymbol{\Theta}^{(k)})}{|\hat{\mathbf{R}}_{NN}(\boldsymbol{\Theta}^{(k)})|^{\frac{1}{p-k}}}\right) + \frac{1}{2}k(2p-k)\log M \tag{1.194}$$

Here we have changed the notation from $\hat{\mathbf{R}}_{NN}$ to $\hat{\mathbf{R}}_{NN}(\boldsymbol{\Theta}^{(k)})$ to emphasize that $\hat{\mathbf{R}}_{NN}$ depends on $\boldsymbol{\Theta}^{(k)}$. To make the dependence on $\boldsymbol{\Theta}$ more explicit, observe that from (1.176) we get

$$\mathbf{P}^{\perp}_{\mathbf{A}(\boldsymbol{\Theta}^{(k)})}\hat{\mathbf{R}}\mathbf{P}^{\perp}_{\mathbf{A}(\boldsymbol{\Theta}^{(k)})} = \mathbf{G}(\boldsymbol{\Theta}^{(k)})\begin{pmatrix} \mathbf{0} & \mathbf{0} \\ \mathbf{0} & \hat{\mathbf{R}}_{NN}(\boldsymbol{\Theta}^{(k)}) \end{pmatrix}\mathbf{G}(\boldsymbol{\Theta}^{(k)})^H \tag{1.195}$$

Hence, by the invariance of the trace and the determinant under the transformation $\mathbf{F} \rightarrow \mathbf{G}\mathbf{F}\mathbf{G}^H$ when $\mathbf{G}$ is unitary

$$\operatorname{tr}\hat{\mathbf{R}}_{NN}(\boldsymbol{\Theta}^{(k)}) = \sum_{i=1}^{p-k} l_i(\boldsymbol{\Theta}^{(k)}) \tag{1.196}$$

and

$$\det\hat{\mathbf{R}}_{NN}(\boldsymbol{\Theta}^{(k)}) = \prod_{i=1}^{p-k} l_i(\boldsymbol{\Theta}^{(k)}) \tag{1.197}$$

where $l_i(\boldsymbol{\Theta}^{(k)}) \geq \ldots \geq l_{(p-k)}(\boldsymbol{\Theta}^{(k)})$ denote the nonzero eigenvalues of the rank-$(p-k)$ matrix $\mathbf{P}^{\perp}_{\mathbf{A}(\boldsymbol{\Theta}^{(k)})}\hat{\mathbf{R}}\mathbf{P}^{\perp}_{\mathbf{A}(\boldsymbol{\Theta}^{(k)})}$.

Substituting (1.196) and (1.197) into (1.194), we get

$$L\{\mathbf{x}_N(t_i), \mathbf{x}_S(t_i)|\boldsymbol{\Theta}^{(k)}\} = M(p-k)\log\left(\frac{\frac{1}{(p-k)}\sum_{i=1}^{p-k} l_i(\hat{\boldsymbol{\Theta}}^{(k)})}{\left(\prod_{i=1}^{p-k} l_i(\hat{\boldsymbol{\Theta}}^{(k)})\right)^{\frac{1}{(p-k)}}}\right) + \frac{1}{2}k(2p-k)\log M \tag{1.198}$$

The code length (1.198) is conditioned on knowing the $k \times 1$ location vector $\boldsymbol{\Theta}^{(k)}$. However, this value is unknown and hence it must be estimated from the data and encoded as a preamble to the code string. A natural choice for the estimator is the ML estimator derived above. Thus, substituting for $\boldsymbol{\Theta}$ the ML estimator and increasing by k the number of free parameters in (1.193), the code length becomes

$$L(\mathbf{X}|k) = M(p-k)\log\left(\frac{\frac{1}{(p-k)}\sum_{i=1}^{p-k} l_i(\hat{\boldsymbol{\Theta}}^{(k)})}{\prod_{i=1}^{p-k} l_i(\hat{\boldsymbol{\Theta}}^{(k)})^{\frac{1}{(p-k)}}}\right) + \frac{1}{2}k(2p-k+1)\log M \tag{1.199}$$

with $\hat{\boldsymbol{\Theta}}^{(k)}$ being the ML estimator.

The MDL estimator of the number of sources is given by the value of k that minimizes the code length. Hence

$$\hat{k} = \arg\min_{k\in\{0,\ldots,p-1)\}} MDL_{DSWN}(k), \tag{1.200}$$

where

$$MDL_{DSWN}(k) = M(p-k)\log\left(\frac{\frac{1}{(p-k)}\sum_{i=1}^{p-k} l_i(\hat{\boldsymbol{\Theta}}^{(k)})}{\prod_{i=1}^{p-k} l_i(\hat{\boldsymbol{\Theta}}^{(k)})^{\frac{1}{(p-k)}}}\right) + \frac{1}{2}k(2p-k+1)\log M \tag{1.201}$$

Analogously, the AIC criterion is given by

$$\hat{k} = \arg\min_{k\in\{0,\ldots,p-1)\}} AIC_{DSWN}(k) \tag{1.202}$$

where

$$AIC_{DSWN}(k) = M(p-k)\log\left(\frac{\frac{1}{(p-k)}\sum_{i=1}^{p-k} l_i(\hat{\boldsymbol{\Theta}}^{(k)})}{\prod_{i=1}^{p-k} l_i(\hat{\boldsymbol{\Theta}}^{(k)})^{\frac{1}{(p-k)}}}\right) + k(2p-k+1) \tag{1.203}$$

Beamforming

Having solved the detection and localization problem, the solution to the beamforming problem is striaghtforward. Indeed, from (1.156) we get

$$\hat{\mathbf{s}}(t_i) = \left[\mathbf{A}(\hat{\boldsymbol{\Theta}}^{(\hat{q})})^H \mathbf{A}(\hat{\boldsymbol{\Theta}}^{(\hat{q})})\right]^{-1} \mathbf{A}(\boldsymbol{\Theta}^{(\hat{q})})^H \mathbf{x}(t_i) \tag{1.204}$$

1.7 OPTIMAL SOLUTION FOR THE SSWN MODEL

In this section we present the optimal solution for the Stochastic Signals (SS) and White Noise (WN), which we refer to as the SSWN model.

Analogous to the DSWN model, the optimal solution involves a *simultaneous* solution to the detection and estimation problems and is based on the maximum likelihood estimator.

The Maximum Likelihood Estimator

The derivation of the ML estimator follows Bohme [6] (also see [15]).

Note that from (1.3) and the SS and WN models it follows that $\mathbf{x}(t)$ is a complex Gaussian vector with zero mean and covariance matrix

$$\mathbf{R} = \mathbf{A}(\boldsymbol{\Theta})\mathbf{C}\mathbf{A}(\boldsymbol{\Theta})^H + \sigma^2\mathbf{I} \tag{1.205}$$

Hence, the joint density function of the sampled data is given by

$$f(\mathbf{X}) = \prod_{i=1}^{M} \frac{1}{\pi \det[\mathbf{R}]} \exp\{(-\mathbf{x}(t_i)^H\mathbf{R}^{-1}\mathbf{x}(t_i))\} \tag{1.206}$$

Observe that unlike in the DSWN model, here the number of parameters does not grow with the number of samples. Indeed, the number of free parameters in σ^2, $\mathbf{C}$, and $\boldsymbol{\Theta}$ is given by 1, q^2, and q, respectively.

The log-likelihood of the sampled data, ignoring constant terms, is given by

$$L(\mathbf{X}|\boldsymbol{\Theta}, \mathbf{C}, \sigma^2) = -M \log|\mathbf{R}| - \sum_{i=1}^{M} \mathbf{x}(t_i)^H\mathbf{R}^{-1}\mathbf{x}(t_i) \tag{1.207}$$

or more compactly

$$L(\mathbf{X}|\boldsymbol{\Theta}, \mathbf{C}, \sigma^2) = -M \log|\mathbf{R}| - M \operatorname{tr}\left[\mathbf{R}^{-1}\hat{\mathbf{R}}\right] \tag{1.208}$$

where $\hat{\mathbf{R}}$ is the sample-covariance matrix.

To compute the ML estimator we have to maximize the log-likelihood with respect to the unknown parameters.We shall carry out this maximization in three steps: (i) maximize with respect to $\mathbf{C}$ with σ^2 and $\boldsymbol{\Theta}$ being fixed; then (ii) substitute the resulting estimate of $\mathbf{C}$, expressed as a function of σ^2 and $\boldsymbol{\Theta}$, back into the log-likelihood function and maximize with respect to parameter σ^2, with $\boldsymbol{\Theta}$ fixed; and (iii) substitute the resulting estimate of σ^2, expressed as a function of $\boldsymbol{\Theta}$, back into the log-likelihood function and obtain a function to be maximized over $\boldsymbol{\Theta}$ only.

To carry out the first step we have to differentiate (1.207) with respect to the matrix $\mathbf{C}$. Using the well-known rules of matrix algebra [11], and after some rather lengthy algebraic manipulations, we find that the maximizing value of $\mathbf{C}$ expressed in terms of $\boldsymbol{\Theta}$ and σ^2 is given by

$$\hat{\mathbf{C}}(\boldsymbol{\Theta}, \sigma^2) = \left[\mathbf{A}(\boldsymbol{\Theta})^H\mathbf{A}(\boldsymbol{\Theta})\right]^{-1}\mathbf{A}(\boldsymbol{\Theta})^H(\hat{\mathbf{R}} - \sigma^2\mathbf{I})\mathbf{A}(\boldsymbol{\Theta})\left[\mathbf{A}(\boldsymbol{\Theta})^H\mathbf{A}(\boldsymbol{\Theta})\right]^{-1} \tag{1.209}$$

Substituting this expression into (1.205), we get

$$\hat{\mathbf{R}}(\Theta, \sigma^2) = \mathbf{A}(\Theta)\left[\mathbf{A}(\Theta)^H\mathbf{A}(\Theta)\right]^{-1}\mathbf{A}(\Theta)^H\hat{\mathbf{R}}\mathbf{A}(\Theta)\left[\mathbf{A}(\Theta)^H\mathbf{A}(\Theta)\right]^{-1}\mathbf{A}(\Theta)^H + \sigma^2\mathbf{I}$$
$$- \sigma^2\left[\mathbf{A}(\Theta)^H\mathbf{A}(\Theta)\right]^{-1}\mathbf{A}(\Theta)^H\mathbf{A}(\Theta)\left[\mathbf{A}(\Theta)^H\mathbf{A}(\Theta)\right]^{-1} \tag{1.210}$$

or more compactly, using the projection operator notation (1.158)

$$\hat{\mathbf{R}}(\Theta, \sigma^2) = \mathbf{P}_{\mathbf{A}(\Theta)}\hat{\mathbf{R}}\mathbf{P}_{\mathbf{A}(\Theta)} + \sigma^2\mathbf{P}^{\perp}_{\mathbf{A}(\Theta)} \tag{1.211}$$

To carry out the second step in the computation of the ML estimator, we substitute (1.211) into (1.208) and maximize it with respect to σ^2. After some straightforward algebraic manipulations, we get

$$\hat{\sigma}^2(\theta) = \frac{1}{p-q}\,\mathrm{tr}\left[\mathbf{P}^{\perp}_{\mathbf{A}(\Theta)}\hat{\mathbf{R}}\right] \tag{1.212}$$

Substituting this expression into (1.208), the log-likelihood becomes

$$L(\mathbf{X}|\Theta) = -M\log|\hat{\mathbf{R}}(\Theta)| - Mp \tag{1.213}$$

where

$$\hat{\mathbf{R}}(\Theta) = \mathbf{P}_{\mathbf{A}(\Theta)}\hat{\mathbf{R}}\mathbf{P}_{\mathbf{A}(\Theta)} + \frac{1}{p-q}\,\mathrm{tr}\left[\mathbf{P}^{\perp}_{\mathbf{A}(\Theta)}\hat{\mathbf{R}}\right]\mathbf{P}^{\perp}_{\mathbf{A}(\Theta)} \tag{1.214}$$

Hence, since the logarithm is a monotonic function, the ML estimator of Θ is given by the solution of the following minimization problem:

$$\hat{\Theta} = \arg\min_{\Theta}|\hat{\mathbf{R}}(\Theta)| \tag{1.215}$$

Though the intuitive interpretation of this estimator, in its present form, is rather obscure, we shall, in the sequel, present an intuitively pleasing interpretation of this estimator and a simple proof of its consistency.

Using the ML estimate of Θ, the ML estimates of σ^2 and $\mathbf{C}$ become

$$\hat{\sigma}^2 = \frac{1}{p-q}\,\mathrm{tr}\left[\mathbf{P}^{\perp}_{\mathbf{A}(\hat{\Theta})}\hat{\mathbf{R}}\right] \tag{1.216}$$

and

$$\hat{\mathbf{C}} = \left[\mathbf{A}(\hat{\Theta})^H\mathbf{A}(\hat{\Theta})\right]^{-1}\mathbf{A}(\hat{\Theta})^H\left[\hat{\mathbf{R}} - \hat{\sigma}^2\mathbf{I}\right]\mathbf{A}(\hat{\Theta})\left[\mathbf{A}(\hat{\Theta})^H\mathbf{A}(\hat{\Theta}\right]^{-1} \tag{1.217}$$

It can be readly verified that unlike in the DSWN model, the ML estimators of σ^2 and $\mathbf{C}$ are consistent. Indeed, assuming that $(\hat{\Theta})$ is consistent, as $M \to \infty$ we get

$$\hat{\sigma}^2 \to \frac{1}{p-q}\,\mathrm{tr}\left[\mathbf{P}^{\perp}_{\mathbf{A}(\Theta)}\mathbf{R}\right] = \frac{\sigma^2}{p-q}\,\mathrm{tr}\left[\mathbf{P}^{\perp}_{\mathbf{A}}\right] = \sigma^2 \tag{1.218}$$

and

$$\hat{\mathbf{C}} \to \left[\mathbf{A}(\Theta)^H\mathbf{A}(\Theta)\right]^{-1}\mathbf{A}(\Theta)^H\left[\mathbf{R} - \sigma^2\mathbf{I}\right]\mathbf{A}(\Theta)\left[\mathbf{A}(\Theta)^H\mathbf{A}(\Theta)\right]^{-1} = \mathbf{C} \tag{1.219}$$

Observe that $\hat{\mathbf{C}}$ is a "sensible" estimate of $\mathbf{C}$. Indeed, $\hat{\mathbf{C}}$ is obtained by simply solving (1.205) for $\mathbf{C}$, with the unknown quantities replaced by their estimates from the data. In fact, Schmidt [28] proposed this estimator on intuitive grounds, only in his approach $\hat{\mathbf{\Theta}}$ and $\hat{\sigma}^2$ were not the ML estimators. Interestingly, note that $\hat{\mathbf{C}}$ is not guaranteed to be non-negative definite for finite sample sizes.

Simultaneous Detection and Localization

Since the number of parameters in the SSWN model does not grow with the number of samples, the simultaneous solution to the detection and localization problems can be derived by straightforwardly applying the MDL criterion for model selection in conjunction with the ML estimator derived above. Nevertheless, we present a different approach, based also on the MDL principle, which exploits the structure of the signal and noise subspaces and yields an intuitively pleasing interpretation of the ML estimator. The derivation follows Wax [36].

Analogous to the DSWN model, we shall evaluate the description length in three steps: (i) compute the description length of the noise subspace components $\{\mathbf{x}_N(t_i)\}$ assuming that $\mathbf{\Theta}^{(k)}$ is given; (ii) compute the description length of the signal subspace components $\{\mathbf{x}_S(t_i)\}$ assuming that the noise subspace components $\{\mathbf{x}_N(t_i)\}$ and $\mathbf{\Theta}^{(k)}$ are given; and (iii) compute the description length of $\mathbf{\Theta}^{(k)}$.

Observe that since the noise model is identical to that used in the DSWN model, the description length of the noise subspace components is identical to that given in (1.183). Nevertheless, we rederive this expression here for the sake of completeness.

The WN model and (1.177) imply that $\mathbf{x}_N(t)$ is a $(p-k)\times 1$ complex Gaussian random vector with zero mean and covariance matrix $\sigma^2\mathbf{I}$, that is

$$\mathbf{x}_N(t)|\mathbf{\Theta}_{(k)} \sim N_{p-k}(\mathbf{0}, \sigma^2\mathbf{I}) \tag{1.220}$$

Hence, the probabilistic model for the noise subspace components is given by

$$f(\{\mathbf{x}_N(t_i)\}|\mathbf{\Theta}^{(k)}) = \prod_{i=1}^{M} \frac{1}{|\pi\sigma^2\mathbf{I}|} \exp\{-\mathbf{x}_N^H(t_i)\sigma^{-2}\mathbf{x}_N(t_i)\} \tag{1.221}$$

which can be rewritten as

$$f(\{\mathbf{x}_N(t_i)\}|\mathbf{\Theta}^{(k)}) = |\pi\sigma^2\mathbf{I}|^{-M} \exp\{-\sigma^{-2}M \operatorname{tr}\left[\hat{\mathbf{R}}_{NN}(\mathbf{\Theta}^{(k)})\right]\} \tag{1.222}$$

where $\hat{\mathbf{R}}_{NN}(\mathbf{\Theta}^{(k)})$ denotes the $(p-k)\times(p-k)$ sample-covariance matrix of the $\mathbf{x}_N(t)$

$$\hat{\mathbf{R}}_{NN}(\mathbf{\Theta}^{(k)}) = \frac{1}{M}\sum_{i=1}^{M} \mathbf{x}_N(t_i)\mathbf{x}_N^H(t_i) \tag{1.223}$$

As can be readily verified, the probabilistic model (1.222) has only a single parameter, σ^2, and its ML estimator is given by

$$\hat{\sigma}^2(\mathbf{\Theta}^{(k)}) = \frac{1}{(p-k)} \operatorname{tr}\left[\hat{\mathbf{R}}_{NN}(\mathbf{\Theta}^{(k)})\right] \tag{1.224}$$

It then follows from (1.36) that the code length required to encode the noise subspace components, ignoring constant terms which are independent of k, is given by

$$L\{\mathbf{x}_N(t_i)|\Theta^{(k)}\} = M \log|\hat{\sigma}^2(\Theta^{(k)})\mathbf{I}| + M(p-k) + \frac{1}{2}\log M \tag{1.225}$$

Next, we evaluate the description length of the signal subspace components $\{\mathbf{x}_S(t_i)\}$. Note that the SS model and (1.175) imply that $\mathbf{x}_S(t)$ is a $k \times 1$ complex Gaussian random vector with zero mean. Denoting by $\mathbf{R}_{SS}(\Theta^{(k)})$ its covariance matrix, we have

$$\mathbf{x}_S(t)|\Theta^{(k)} \sim N_k(\mathbf{0}, \mathbf{R}_{SS}(\Theta^{(k)})) \tag{1.226}$$

The probabilistic model for the signal subspace components is therefore given by

$$f(\{\mathbf{x}_S(t_i)\} \mid \Theta^{(k)}) = \prod_{i=1}^{M} \frac{1}{|\pi\mathbf{R}_{SS}(\Theta^{(k)})|}\exp\{-\mathbf{x}_S^H(t_i)\mathbf{R}_{SS}^{-1}(\Theta^{(k)})\mathbf{x}_S(t_i)\} \tag{1.227}$$

which can be rewritten as

$$f(\{\mathbf{x}_S(t_i)\} \mid \Theta^{(k)}) = |\pi\mathbf{R}_{SS}(\Theta^{(k)})|^{-M}\exp\{-M \text{ tr}\left[\mathbf{R}_{SS}^{-1}(\Theta^{(k)})\hat{\mathbf{R}}_{SS}(\Theta^{(k)})\right]\} \tag{1.228}$$

where

$$\hat{\mathbf{R}}_{SS}(\Theta^{(k)}) = \frac{1}{M}\sum_{i=1}^{M}\mathbf{x}_S(t_i)\mathbf{x}_S^H(t_i) \tag{1.229}$$

As can be readly verified, the number of free real parameters in $\mathbf{R}_{SS}(\Theta^{(k)})$ is k^2 and its maximum likelihood estimator is given by the sample covariance matrix $\hat{\mathbf{R}}_{SS}(\Theta^{(k)})$. Consequently, from (1.36), the description length of the signal subspace components, ignoring constant terms that are independent of k, is given by

$$L\{\mathbf{x}_S(t_i)|\Theta^{(k)}\} = M\log|\hat{\mathbf{R}}_{SS}(\Theta^{(k)})| + Mk + \frac{1}{2}k^2\log M \tag{1.230}$$

Thus, summing up (1.225) and (1.230), again ignoring constant terms that are independent of k, the total description length of the signal and noise subspace component is given by

$$L\{\mathbf{x}_N(t_i), \mathbf{x}_S(t_i)|\Theta_{(k)}\} = M\log(|\hat{\mathbf{R}}_{SS}(\Theta^{(k)})||\hat{\sigma}^2(\Theta^{(k)})\mathbf{I}|) + \frac{1}{2}(k^2+1)\log M \tag{1.231}$$

The computed description length was conditioned on knowing the $k \times 1$ location vector $\Theta^{(k)}$. However, this value is unknown and hence it must be estimated from the data and encoded as a preamble to the code string. Now, since our goal is to obtain the shortest code length, which we claim assures the best detection performance, the optimal estimate is obtained by minimizing (1.231), that is

$$\hat{\Theta}^{(k)} = \arg\min_{\Theta^{(k)}}\log(|\hat{\mathbf{R}}_{SS}(\Theta^{(k)})||\hat{\sigma}^2(\Theta^{(k)})\mathbf{I}|) \tag{1.232}$$

This estimator has an interesting geometric interpretation. Indeed, note that since $\mathbf{R}_{SS}(\theta^{(k)}) = \frac{1}{M}\mathbf{X}_S\mathbf{X}_S^H$, where $\mathbf{X}_S = [\mathbf{x}_S(t_1), \ldots, \mathbf{x}_S(t_M)]$, it follows that $|\hat{\mathbf{R}}_{SS}(\theta^{(k)})|$ is a Grammian and as such it represents the volume occupied by the signal subspace components. Note also that since $\sigma^2(\Theta^{(k)})$ represents the variance of each of the $(p-k)$ noise subspace components, it follows that $|\sigma^2(\Theta^{(k)})\mathbf{I}|$ represents the volume of the noise subspace components under the spherical WN model. We can therefore interpret the estimator (1.232) as the value of $\boldsymbol{\Theta}^{(k)}$ that minimizes the volume occupied by the data in *both* the signal and the noise subspaces. For comparison, the ML estimator of the deterministic signals model, (1.165), minimizes the value of $\hat{\sigma}^2 = \frac{1}{p-k} \operatorname{tr} \left[\mathbf{P}^{\perp}_{\mathbf{A}(\hat{\boldsymbol{\Theta}})} \hat{\mathbf{R}} \right]$ which is equivalent to minimizing the volume $|\hat{\sigma}^2(\boldsymbol{\Theta}^{(k)})\mathbf{I}|$.

As we shall now show, the estimator (1.232) coincides with the ML estimator (1.215). To establish this, notice that from (1.175) and (1.176), suppressing the index k for notational compactness, we get

$$\mathbf{P}_{\mathbf{A}(\boldsymbol{\Theta})} \hat{\mathbf{R}} \mathbf{P}_{\mathbf{A}(\boldsymbol{\Theta})} = \mathbf{G}(\boldsymbol{\Theta}) \begin{pmatrix} \hat{\mathbf{R}}_{SS}(\boldsymbol{\Theta}) & \mathbf{0} \\ \mathbf{0} & \mathbf{0} \end{pmatrix} \mathbf{G}^H(\boldsymbol{\Theta}) \tag{1.233}$$

and

$$\mathbf{P}^{\perp}_{\mathbf{A}(\boldsymbol{\Theta})} \hat{\mathbf{R}} \mathbf{P}^{\perp}_{\mathbf{A}(\boldsymbol{\Theta})} = \mathbf{G}(\boldsymbol{\Theta}) \begin{pmatrix} \mathbf{0} & \mathbf{0} \\ \mathbf{0} & \hat{\mathbf{R}}_{NN}(\boldsymbol{\Theta}) \end{pmatrix} \mathbf{G}^H(\boldsymbol{\Theta}) \tag{1.234}$$

where $\hat{\mathbf{R}}$ denotes the sample-covariance matrix, and also

$$\mathbf{P}^{\perp}_{\mathbf{A}(\boldsymbol{\Theta})} \hat{\sigma}^2(\boldsymbol{\Theta}) \mathbf{P}^{\perp}_{\mathbf{A}(\boldsymbol{\Theta})} = \mathbf{G}(\boldsymbol{\Theta}) \begin{pmatrix} \mathbf{0} & \mathbf{0} \\ \mathbf{0} & \hat{\sigma}^2(\boldsymbol{\Theta})\mathbf{I} \end{pmatrix} \mathbf{G}^H(\boldsymbol{\Theta}) \tag{1.235}$$

Now, taking the trace of both sides of (1.234), recalling that $\mathbf{G}(\boldsymbol{\Theta}^{(k)})$ is unitary, we have

$$\operatorname{tr} \left[\hat{\mathbf{R}}_{NN}(\boldsymbol{\Theta}) \right] = \operatorname{tr} \left[\mathbf{P}^{\perp}_{\mathbf{A}(\boldsymbol{\Theta})} \hat{\mathbf{R}} \right] \tag{1.236}$$

and hence, by (1.224)

$$\hat{\sigma}^2(\boldsymbol{\Theta}^{(k)}) = \frac{1}{p-k} \operatorname{tr} \left[\mathbf{P}^{\perp}_{\mathbf{A}(\boldsymbol{\Theta}^{(k)})} \hat{\mathbf{R}} \right] \tag{1.237}$$

Also, summing up (1.233) and (1.235) we get

$$\mathbf{P}_{\mathbf{A}(\boldsymbol{\Theta})} \hat{\mathbf{R}} \mathbf{P}_{\mathbf{A}(\boldsymbol{\Theta})} + \mathbf{P}^{\perp}_{\mathbf{A}(\boldsymbol{\Theta})} \hat{\sigma}^2(\boldsymbol{\Theta}) = \mathbf{G}(\boldsymbol{\Theta}) \begin{pmatrix} \hat{\mathbf{R}}_{SS}(\boldsymbol{\Theta}) & \mathbf{0} \\ \mathbf{0} & \hat{\sigma}^2(\boldsymbol{\Theta})\mathbf{I} \end{pmatrix} \mathbf{G}^H(\boldsymbol{\Theta}) \tag{1.238}$$

Taking the determinant of both sides, recalling that $\mathbf{G}(\boldsymbol{\Theta})$ is unitary, we get

$$|\mathbf{P}_{\mathbf{A}(\boldsymbol{\Theta})} \hat{\mathbf{R}} \mathbf{P}_{\mathbf{A}(\boldsymbol{\Theta})} + \mathbf{P}^{\perp}_{\mathbf{A}(\boldsymbol{\Theta})} \hat{\sigma}^2(\boldsymbol{\Theta})| = |\hat{\mathbf{R}}_{SS}(\boldsymbol{\Theta})||\hat{\sigma}^2(\boldsymbol{\Theta})\mathbf{I}| \tag{1.239}$$

and therefore, by (1.232)

$$\hat{\boldsymbol{\Theta}}^{(k)} = \arg\min_{\boldsymbol{\Theta}^{(k)}} |\mathbf{P}_{\mathbf{A}(\boldsymbol{\Theta}^{(k)})} \hat{\mathbf{R}} \mathbf{P}_{\mathbf{A}(\boldsymbol{\Theta}^{(k)})} + \frac{1}{p-k} \operatorname{tr} \left[\mathbf{P}^{\perp}_{\mathbf{A}(\boldsymbol{\Theta}^{(k)})} \hat{\mathbf{R}} \right] \mathbf{P}^{\perp}_{\mathbf{A}(\boldsymbol{\Theta}^{(k)})}| \tag{1.240}$$

which coincides with the ML estimator (1.215).

Having shown that the estimator (1.232) coincides with the ML estimator, we shall now prove the consistency of the ML estimator using this more convenient form.

From (1.174), suppressing the index k, we get

$$\hat{\mathbf{R}} = \mathbf{G}(\Theta) \begin{pmatrix} \hat{\mathbf{R}}_{SS}(\Theta) & \hat{\mathbf{R}}_{SN}(\Theta) \\ \hat{\mathbf{R}}_{NS}(\Theta) & \hat{\mathbf{R}}_{NN}(\Theta) \end{pmatrix} \mathbf{G}^H(\Theta) \tag{1.241}$$

where $\hat{\mathbf{R}}_{SS}(\Theta)$ and $\hat{\mathbf{R}}_{NN}(\Theta)$ are given by (1.229) and (1.223), respectively, and

$$\hat{\mathbf{R}}_{SN}(\Theta) = \frac{1}{M} \sum_{i=1}^{M} \mathbf{x}_S(t_i)\mathbf{x}_S^H(t_i) \tag{1.242}$$

Taking the determinant of both sides, recalling that $\mathbf{G}(\Theta)$ is unitary, we get

$$|\hat{\mathbf{R}}| = \begin{vmatrix} \hat{\mathbf{R}}_{SS}(\Theta) & \hat{\mathbf{R}}_{SN}(\Theta) \\ \hat{\mathbf{R}}_{NS}(\Theta) & \hat{\mathbf{R}}_{NN}(\Theta) \end{vmatrix} = |\hat{\mathbf{R}}_{NN}(\Theta)||\hat{\mathbf{R}}_{SS}(\Theta) - \hat{\mathbf{R}}_{SN}(\Theta)\hat{\mathbf{R}}_{NN}^{-1}(\Theta)\hat{\mathbf{R}}_{NS}(\Theta)| \tag{1.243}$$

which, since $\hat{\mathbf{R}}_{SN}(\Theta)\hat{\mathbf{R}}_{NN}^{-1}(\Theta)\hat{\mathbf{R}}_{NS}(\Theta)$ is non-negative definite, implies that

$$|\hat{\mathbf{R}}| \leq |\hat{\mathbf{R}}_{NN}(\Theta)||\hat{\mathbf{R}}_{SS}(\Theta)| \tag{1.244}$$

Now

$$|\hat{\mathbf{R}}_{NN}(\Theta)| = \prod_{i=1}^{p-q} l_i^N(\Theta) \tag{1.245}$$

and by (1.237)

$$|\hat{\sigma}^2(\Theta)\mathbf{I}| = \left(\frac{1}{p-q} \sum_{i=1}^{p-q} l_i^N(\Theta) \right)^{p-q} \tag{1.246}$$

where $l_i^N(\Theta) \geq \ldots \geq l_{p-q}^N(\Theta)$ denote the nonzero eigenvalues of the rank-$(p-q)$ matrix $\mathbf{P}_{\mathbf{A}(\Theta)}^{\perp}\hat{\mathbf{R}}\mathbf{P}_{\mathbf{A}(\Theta)}^{\perp}$.

By the inequality of the arithmetic and geometric means, we have

$$\frac{1}{p-q} \sum_{i=1}^{p-q} l_i^N(\Theta) \geq \left(\prod_{i=1}^{p-q} l_i^N(\Theta) \right)^{\frac{1}{p-q}} \tag{1.247}$$

with equality if and only if $l_i^N(\Theta) = \ldots = l_{p-q}^N(\Theta)$. Hence, using (1.245) and (1.246), we obtain

$$|\hat{\mathbf{R}}_{NN}(\Theta)| \leq |\hat{\sigma}^2(\Theta)\mathbf{I}| \tag{1.248}$$

and therefore, by (1.244)

$$|\hat{\mathbf{R}}| \leq |\hat{\mathbf{R}}_{SS}(\Theta)||\hat{\sigma}^2(\Theta)\mathbf{I}| \tag{1.249}$$

with equality if and only if $\hat{\mathbf{R}}_{NS}(\Theta) = 0$ and $l_i^N(\Theta) = \ldots = l_{p-q}^N(\Theta)$.

Since both conditions are satisfied for the true Θ as $M \to \infty$, it follows that the minimum of (1.232) is obtained at the true Θ. This establishes the consistency of the ML estimator $\hat{\Theta}$.

Yet another form of the ML estimator can be obtained by casting it in terms of the eigenvalues of the matrices involved. Indeed, from (1.233) and (1.234), by the well-known properties of the trace and the determinant and their invariance under the transformation $\mathbf{F} \rightarrow \mathbf{GFG}^H$ when $\mathbf{G}$ is unitary, we get

$$|\hat{\mathbf{R}}_{SS}(\mathbf{\Theta}^{(k)})| = \prod_{i=1}^{k} l_i^S(\mathbf{\Theta}^{(k)}) \tag{1.250}$$

and

$$\mathrm{tr}\left[\hat{\mathbf{R}}_{NN}(\mathbf{\Theta}^{(k)})\right] = \sum_{i=1}^{p-k} l_i^N(\mathbf{\Theta}^{(k)}) \tag{1.251}$$

where $l_1^S(\mathbf{\Theta}^{(k)}) \geq \ldots \geq l_k^S(\mathbf{\Theta}^{(k)})$ denote the nonzero eigenvalues of the rank-k matrix $\mathbf{P}_{\mathbf{A}(\mathbf{\Theta}^{(k)})}\hat{\mathbf{R}}\mathbf{P}_{\mathbf{A}(\mathbf{\Theta}^{(k)})}$ and $l_1^N(\mathbf{\Theta}^{(k)}) \geq \ldots \geq l_{p-k}^N(\mathbf{\Theta}^{(k)})$ denote the nonzero eigenvalues of the rank-$(p-k)$ matrix $\mathbf{P}^{\perp}_{\mathbf{A}(\mathbf{\Theta}^{(k)})}\hat{\mathbf{R}}\mathbf{P}^{\perp}_{\mathbf{A}(\mathbf{\Theta}^{(k)})}$.

From (1.224) we therefore obtain

$$\hat{\sigma}^2(\mathbf{\Theta}) = \frac{1}{p-k}\sum_{i=1}^{p-k} l_i^N(\mathbf{\Theta}^{(k)}) \tag{1.252}$$

and hence, substituting (1.252) and (1.250) into (1.232), we get

$$\hat{\mathbf{\Theta}}^{(k)} = \arg\min_{\mathbf{\Theta}^{(k)}} \left(\prod_{i=1}^{k} l_i^S(\mathbf{\Theta}^{(k)})\right)\left(\frac{1}{p-k}\sum_{i=1}^{p-k} l_i^N(\mathbf{\Theta}^{(k)})\right)^{p-k} \tag{1.253}$$

Substituting the ML estimator $\hat{\mathbf{\Theta}}^{(k)}$ into (1.232) and adding $\frac{1}{2}k \log M$ for the description length of the k real parameters, we find that the MDL estimator of the number of sources is given by the value of k that minimizes the following criterion:

$$\hat{k} = \arg\min_{k} MDL_{SSWN}(k) \tag{1.254}$$

where

$$\begin{aligned} MDL_{SSWN}(k) &= M \log\left(\prod_{i=1}^{k} l_i^S(\hat{\mathbf{\Theta}}^{(k)}) \left(\frac{1}{p-k}\sum_{i=1}^{p-k} l_i^N(\hat{\mathbf{\Theta}}^{(k)})\right)^{p-k}\right) \\ &\quad + \frac{1}{2}k(k+1)\log M \end{aligned} \tag{1.255}$$

with $\hat{\mathbf{\Theta}}^{(k)}$ denoting the ML estimator (1.232).

Similarly, the AIC estimator is given by

$$\hat{k} = \arg\min_{k} AIC_{SSWN}(k) \tag{1.256}$$

where

$$AIC_{SSWN}(k) = M \log \left(\prod_{i=1}^{k} l_i^S(\hat{\Theta}^{(k)}) \left(\frac{1}{p-k} \sum_{i=1}^{p-k} l_i^N(\hat{\Theta}^{(k)}) \right)^{p-k} \right) + k(k+1). \quad (1.257)$$

Beamforming

Unlike in the DSWN model, in the SSWN model the signals $\{\mathbf{s}(t_i)\}$ are not parameters of the model and hence the solution of the beamforming problem is less striagthforward. The presentation follows Wax [39].

Note that from (1.3) and the SS model it follows that if $\boldsymbol{\Theta}$ is given, then $\mathbf{s}(t_i)$ is independent of $\mathbf{x}(t_j)$ for $j \neq i$. Consequently, having an estimate of $\boldsymbol{\Theta}$ at hand, it is natural to base the estimate of $\mathbf{s}(t_i)$ solely on $\mathbf{x}(t_i)$.

To derive the estimate, observe that the SSWN model implies that $\mathbf{s}(t_i)$ and $\mathbf{x}(t_i)$ are jointly Gaussian. This, in turn, implies that the best estimator of $\mathbf{s}(t_i)$ given $\mathbf{x}(t_i)$ is the conditional mean $E[\mathbf{s}(t_i)|\mathbf{x}(t_i)]$, given by the well-known expression

$$\hat{\mathbf{s}}(t_i) = E\left[\mathbf{s}(t_i)\mathbf{x}(t_i)^H\right] E\left[\mathbf{x}(t_i)\mathbf{x}(t_i)^H\right]^{-1} \mathbf{x}(t_i) \quad (1.258)$$

where E is the expectation operator. Now, from (1.3), we have

$$E\left[\mathbf{s}(t_i)\mathbf{x}(t_i)^H\right] = E\left[\mathbf{s}(t_i)\mathbf{s}(t_i)^H\right]\mathbf{A}(\boldsymbol{\Theta})^H = \mathbf{C}\mathbf{A}(\boldsymbol{\Theta})^H \quad (1.259)$$

so that we can rewrite (1.258) as

$$\hat{\mathbf{s}}(t_i) = \mathbf{C}\mathbf{A}(\boldsymbol{\Theta})^H \mathbf{R}^{-1}\mathbf{x}(t_i) \quad (1.260)$$

Replacing the above quantities by their estimates from the data, we get

$$\hat{\mathbf{s}}(t_i) = \hat{\mathbf{C}}^{(\hat{q})}\mathbf{A}(\hat{\boldsymbol{\Theta}}^{(\hat{q})})^H \hat{\mathbf{R}}^{-1}\mathbf{x}(t_i) \quad (1.261)$$

Since $\mathbf{R}$ is given by (1.205), and since the following easily verified matrix identity holds

$$\mathbf{AC}(\mathbf{ACA}^H + \sigma^2\mathbf{I})^{-1} = (\mathbf{A}^H\mathbf{A} + \sigma^2\mathbf{C}^{-1})^{-1}\mathbf{A}^H \quad (1.262)$$

we get

$$\hat{\mathbf{s}}(t_i) = \left(\mathbf{A}(\hat{\boldsymbol{\Theta}}^{(\hat{q})})^H \mathbf{A}(\hat{\boldsymbol{\Theta}}^{(\hat{q})}) + \hat{\sigma}^2(\hat{\mathbf{C}}^{(\hat{q})})^{-1}\right)^{-1} \mathbf{A}(\hat{\boldsymbol{\Theta}}^{(\hat{q})})^H \mathbf{x}(t_i) \quad (1.263)$$

The asymptotic behavior of (1.263) can be easily derived and compared with DSWN beamformer. Indeed, when the signal-to-noise ratio is high, namely when norm $\left[\mathbf{A}(\hat{\boldsymbol{\Theta}}^{(\hat{q})})^H \mathbf{A}(\hat{\boldsymbol{\Theta}}^{(\hat{q})})\right] \gg$ norm$\left[\hat{\sigma}^2(\hat{\mathbf{C}}^{(\hat{q})})^{-1}\right]$ (any sensible matrix norm will do), the above estimator reduces to

$$\hat{\mathbf{s}}(t_i) = (\mathbf{A}(\hat{\boldsymbol{\Theta}}^{(\hat{q})})^H \mathbf{A}(\hat{\boldsymbol{\Theta}}^{(\hat{q})})^{-1} \mathbf{A}(\hat{\boldsymbol{\Theta}}^{(\hat{q})})^H \mathbf{x}(t_i) \quad (1.264)$$

which is identical in form to the optimal beamformer in the DSWN model (1.205).

It is instructive to compare this beamformer also with the minimum variance beamformer. To this end, observe that when the signals are uncorrelated, that is, when $\mathbf{C} = \text{diag}\{p_1, \ldots, p_q\}$, it follows from (1.261) that the estimator of the k-th signal is given by

$$\hat{s}_k(t_i) = \hat{p}_k \mathbf{a}(\hat{\theta}_k)^H \hat{\mathbf{R}}^{-1} \mathbf{x}(t_i) \tag{1.265}$$

which is identical, up to a scalar factor, to the minimum variance beamformer.

However, when the signals are correlated, that is, when the matrix $\mathbf{C}$ is nondiagonal, the two estimators differ drastically. Unlike the minimum variance estimator, which suffers severe degradation when the signals are partially correlated and fails completely when the signals are fully correlated, this estimator exploits the correlaction to improve the performance.

1.8 SUBOPTIMAL SOLUTION TO THE SSCN MODEL

The solutions presented in the previous sections, both the optimal and suboptimal, were for the White Noise (WN) model. In this section we address the much more complex Colored Noise (CN) model. Specifically, we present a suboptimal solution for the Stochastic Signals (SS) and Colored Noise (CN) model.

Simultaneous Detection and Localization

As in the previous sections, our approach is based on a simultaneous solution of the detection and localization problems via the Minimum Description Length (MDL) principle for model selection. The derivation follows Wax [38].

A straigthforward application of the MDL principle to our problem is computationally very unattractive because of the large number of unknown parameters in the SSCN model. Indeed, from (1.3) and the SS and CN model it follows that $\mathbf{x}(t)$ is a complex Gaussian random vector with zero mean and covariance matrix given by

$$\mathbf{R} = \mathbf{A}(\mathbf{\Theta})\mathbf{P}\mathbf{A}^H(\theta) + \mathbf{Q} \tag{1.266}$$

Thus, since the vector $\mathbf{\Theta}$ is real and the matrices $\mathbf{P}$ and $\mathbf{Q}$ are Hermitian, it follows that the number of unknown real parameters in $\mathbf{\Theta}$, $\mathbf{P}$ and $\mathbf{Q}$, are q, q^2, and p^2, respectively, which amount to a total of $q + q^2 + p^2$ parameters. The computation of the maximum likelihood estimator therefore calls for the solution of a $(q + q^2 + p^2)$ dimensional nonlinear maximization problem, thus rendering a straigthforward application of the MDL model selection criterion very unattractive.

To circumvent this difficulty, we recast the problem in the signal subspace framework and compute the description length by models based on the decomposition of $\mathbf{x}(t)$ into its components in the signal and noise subspaces. These models, though suboptimal, simplify comsiderably the computational load.

As in the optimal solutions for the WN model, we compute the description length by summing three terms: (i) the description length of the noise subspace componenets $\{\mathbf{x}_N(t_i)\}$, assuming that $\mathbf{\Theta}^{(k)}$ is given; (ii) the description length of the signal subspace componenets $\{\mathbf{x}_S(t_i)\}$, assuming that the noise subspace componenets $\{\mathbf{x}_N(t_i)\}$ and $\mathbf{\Theta}^{(k)}$ are given; and (iii) the description length of $\mathbf{\Theta}^{(k)}$.

To compute the first term we need a probabilistic model for the noise subspace components. Note that from the CN model and (1.177) it follows that $\mathbf{x}_N(t)$ is a $(p-k) \times 1$ complex Gaussian random vector with zero mean. Denoting its covariance matrix by $\mathbf{R}_{NN}(\Theta^{(k)})$, we have

$$\mathbf{x}_N(t)|\Theta^{(k)} \sim N_{p-k}(\mathbf{0}, \mathbf{R}_{NN}(\Theta^{(k)})) \tag{1.267}$$

Hence

$$f(\{\mathbf{x}_N(t_i)\}|\Theta_{(k)}) = \prod_{i=1}^{M} \frac{1}{|\pi \mathbf{R}_{NN}(\Theta_{(k)})|} \exp(-\mathbf{x}_N^H(t_i)\mathbf{R}_{NN}^{-1}(\Theta^{(k)})\mathbf{x}_N(t_i)) \tag{1.268}$$

or more compactly

$$f(\{\mathbf{x}_N(t_i)\} \mid \Theta_{(k)}) = |\pi \mathbf{R}_{NN}(\Theta_{(k)})|^{-M} \exp\left(-M \text{ tr} \left[\mathbf{R}_{NN}^{-1}(\Theta^{(k)})\hat{\mathbf{R}}_{NN}(\Theta^{(k)})\right]\right) \tag{1.269}$$

where $\hat{\mathbf{R}}_{NN}(\Theta^{(k)})$ denotes the $(p-k) \times (p-k)$ sample-covariance matrix of $\mathbf{x}_N(t)$,

$$\hat{\mathbf{R}}_{NN}(\Theta^{(k)}) = \frac{1}{M}\sum_{i=1}^{M} \mathbf{x}_N(t_i)\mathbf{x}_N^H(t_i) \tag{1.270}$$

As can be readily verified, the maximum likelihood estimator of $\mathbf{R}_{NN}(\Theta^{(k)})$ is given by $\hat{\mathbf{R}}_{NN}(\Theta^{(k)})$. Hence, since $\mathbf{R}_{NN}(\Theta^{(k)})$ containes $(p-k)^2$ real parameters, it follows from (1.24) that the code length required to encode the noise subspace components, ignoring constant terms that are independent of k, is given by

$$L\{\mathbf{x}_N(t_i) \mid \Theta^{(k)}\} = M \log |\hat{\mathbf{R}}_{NN}(\Theta^{(k)})| + M(p-k) + \frac{1}{2}(p-k)^2 \log M \tag{1.271}$$

To compute the second term, we construct a probabilistic model for the signal subspace components given $\Theta^{(k)}$. The dependence on the noise subspace componets will be ignored to simplify the model. To this end, observe that from the SSCN model and (1.175) it follows that, given $\Theta^{(k)}$, $\mathbf{x}_S(t)$ is a $k \times 1$ complex Gaussian random vector with zero mean. Denoting its covariance matrix by $\mathbf{R}_{SS}(\Theta^{(k)})$, we have

$$\mathbf{x}_S(t) \mid \Theta^{(k)} \sim N_k(\mathbf{0}, \mathbf{R}_{SS}(\Theta^{(k)})) \tag{1.272}$$

Hence, our model is

$$f(\{\mathbf{x}_S(t_i)\} \mid \Theta_{(k)}) = \prod_{i=1}^{M} \frac{1}{|\pi \mathbf{R}_{SS}(\Theta_{(k)})|} \exp(-\mathbf{x}_S^H(t_i)\mathbf{R}_{SS}^{-1}(\Theta^{(k)})\mathbf{x}_S(t_i)) \tag{1.273}$$

or more compactly

$$f(\{\mathbf{x}_S(t_i)\} \mid \Theta_{(k)}) = |\pi \mathbf{R}_{SS}(\Theta_{(k)})|^{-M} \exp\left\{-M \text{ tr} \left[\mathbf{R}_{SS}^{-1}(\Theta^{(k)})\hat{\mathbf{R}}_{SS}(\Theta^{(k)})\right]\right\} \tag{1.274}$$

where

$$\hat{\mathbf{R}}_{SS}(\Theta^{(k)}) = \frac{1}{M}\sum_{i=1}^{M} \mathbf{x}_S(t_i)\mathbf{x}_S^H(t_i) \tag{1.275}$$

The maximum likelihood estimator of $\mathbf{R}_{SS}(\Theta^{(k)})$ is given by the sample covariance matrix $\hat{\mathbf{R}}_{SS}(\Theta^{(k)})$. Thus, since the number of free parameters in $\mathbf{R}_{SS}(\Theta^{(k)})$ is k^2, it follows from (1.36) that the description length of the signal subspace components, ignoring constant terms that are independent of k, is given by

$$L\{\mathbf{x}_S(t_i) \mid \Theta^{(k)}\} = M \log|\hat{\mathbf{R}}_{SS}(\Theta^{(k)})| + Mk + \frac{1}{2}k^2 \log M \tag{1.276}$$

Combining (1.272) and (1.276) and ignoring, again, constant terms that are independent of k, the total description length of the signal and noise subspace components is given by

$$\begin{aligned} L\{\mathbf{x}_N(t_i), \mathbf{x}_S(t_i) \mid \Theta^{(k)}\} &= M \log|\hat{\mathbf{R}}_{SS}(\Theta^{(k)})||\hat{\mathbf{R}}_{NN}(\Theta^{(k)})| \\ &\quad + \frac{1}{2}((p-k)^2 + k^2)\log M \end{aligned} \tag{1.277}$$

To compute the third term, we first have to decide on the algorithm used for the estimation of $\Theta^{(k)}$. Indeed, since this value is unknown, it must be estimated from the data and encoded as a preamble to the code string. To derive this estimator, recall that our goal is to obtain the shortest code length, which we claim assures the best detection performance. Hence, the optimal estimate of $\Theta^{(k)}$ is obtained by minimizing the description length of the data as given by (1.277)

$$\hat{\Theta}^{(k)} = \arg\min_{\Theta^{(k)}} \log(|\hat{\mathbf{R}}_{SS}(\Theta^{(k)})||\hat{\mathbf{R}}_{NN}(\Theta^{(k)})|) \tag{1.278}$$

This estimator has an interesting and intuitively appealing geometric interpretation. Observe that $|\hat{\mathbf{R}}_{SS}(\Theta^{(k)})|$ represents the volume occupied by the data in the signal subspace while $|\hat{\mathbf{R}}_{NN}(\Theta^{(k)})|$ represents the volume occupied by the data in the noise subspace. We can therefore interpret the estimator (1.278) as the value of $\Theta^{(k)}$ that minimizes the volume occupied by the data in *both* the signal and the noise subspaces. This interpretation is intuitively pleasing since the minimization of the volume of the projections onto the noise subspaces guarantees good fit of the signal subspace while the minimization of the volume of the projections onto the signal subspace guarantees the exploitation of the stochastic model of the signals, or more specifically, their being zero mean vectors.

Another form of this estimator can be obtained by casting it in terms of the eigenvalues of the matrices involved. Indeed, observe that from (1.233) and (1.234), by the well-known invariance property of the determinant under the transformation $\mathbf{F} \rightarrow \mathbf{G}\mathbf{F}\mathbf{G}^H$ when $\mathbf{G}$ is unitary, we get

$$|\hat{\mathbf{R}}_{SS}(\Theta^{(k)})| = \prod_{i=1}^{k} l_i^S(\Theta^{(k)}) \tag{1.279}$$

and

$$|\hat{\mathbf{R}}_{NN}(\Theta^{(k)})| = \prod_{i=1}^{p-k} l_i^N(\Theta^{(k)}) \tag{1.280}$$

where $l_1^S(\Theta^{(k)}) \geq \ldots \geq l_k^S(\Theta^{(k)})$ denote the nonzero eigenvalues of the rank-k matrix $\mathbf{P}_{\mathbf{A}(\Theta^{(k)})}\hat{\mathbf{R}}\mathbf{P}_{\mathbf{A}(\Theta^{(k)})}$ and $l_1^N(\Theta^{(k)}) \geq \ldots \geq l_{p-k}^N(\Theta^{(k)})$ denote the nozero eigenvalues of the rank-$(p-k)$ matrix $\mathbf{P}^{\perp}_{\mathbf{A}(\Theta^{(k)})}\hat{\mathbf{R}}\mathbf{P}^{\perp}_{\mathbf{A}(\Theta^{(k)})}$.

Hence, substituting (1.279) and (1.280) into (1.278), we get

$$\hat{\Theta}^{(k)} = \arg\min_{\Theta^{(k)}} \left(\prod_{i=1}^{k} l_i^S(\Theta^{(k)}) \prod_{i=1}^{p-k} l_i^N(\Theta^{(k)}) \right) \tag{1.281}$$

Using this estimator in (1.277) and adding $\frac{1}{2}k \log M$ for the description length of its k real parameters, the total number of free parameters becomes $k^2 + (p-k)^2 + k = p^2 - k(2p - 2k - 1)$. Hence, dropping terms that do not depend on k, the MDL criterion is given by

$$\hat{k} = \arg\min_{k} MDL_{SSCN}(k) \tag{1.282}$$

where

$$MDL_{SSCN}(k) = M \log \left(\prod_{i=1}^{k} l_i^S(\hat{\Theta}^{(k)}) \prod_{i=1}^{p-k} l_i^N(\hat{\Theta}^{(k)}) \right) - \frac{1}{2}k(2p-2k-1)\log M \tag{1.283}$$

with $\hat{\Theta}^{(k)}$ denoting the MDL estimator given by (1.281).

By analogy, the AIC [1] is given by

$$AIC_{SSCN}(k) = M \log \left(\prod_{i=1}^{k} l_i^S(\hat{\Theta}^{(k)}) \prod_{i=1}^{p-k} l_i^N(\hat{\Theta}^{(k)}) \right) - k(2p - 2k - 1) \tag{1.284}$$

Observe that the resulting form of the MDL criterion is rather unconventional. First, the second term, referred to as the complexity term, is *negative* and is not a monotonic function of k. Rather, it has a parabolic behavior with the peak being at $p/2$. Second, which is to be expected in light of the behavior of the second term, the first term is also not a monotonic function of k. The exact behavior of this term, however, is very difficult to analyze since it also depends on the data.

The computational load is relatively modest since only a k-dimensional minimization is involved and is similar to that involved in the optimal techniques for the WN model described in the previous sections.

REFERENCES

1. H. Akaike, "Information Theory and an Extension of the Maximum Likelihood Principle," *In Proc. 2nd Int. Symp. Inform. Theory*, B. N. Petrov and F. Caski eds., pp. 267–81, 1973.
2. H. Akaike, "A New Look at the Statistical Model Identification," *IEEE Trans. on AC*, vol. 19, pp. 716–23, 1974.

3. T. W. ANDERSON, "Asymptotic Theory for Principal Components Analysis," *Ann. of Math. Stat.*, vol. 34, pp. 122–48, 1963.

4. J. BIENVENU and L. KOPP, "Principle de la Goniometric Passive Adaptive," *Proc. 7'eme Collque GRESTI*, (Nice, France), pp. 106/1–106/10, 1979.

5. J. BIENVENU and L. KOPP, "Adaptivity to Background Noise Spatial Coherence for High Resolution Passive Methods," *ICASSP 80* (Denver, CO), pp. 307–10, 1980.

6. J. F. BOHME, "Estimation of Spectral Parameters of Correlated Signals in Wavefields," *Signal Processing*, vol. 11, pp. 329–37, 1986.

7. G. V. BORGIOTTI and L. J. KAPLAN, "Superresolution of Uncorrelated Interference Sources by Using Adaptive Array Techniques," *IEEE Trans. on AP*, vol 27, pp. 842–45, 1979.

8. Y. BRESLER and A. MACOVSKI, "On the Number of Signals Resolvable by a Uniform Linear Array," *IEEE Trans. on ASSP*, vol. 34, pp. 1361–75, 1986.

9. J. CAPON, "High Resolution Frequency Wave Number Spectrum Analysis," *Proc. IEEE*, vol. 57, pp. 1408–18, 1969.

10. J. E. EVANS, J. R. JOHNSON and D. F. SUN, "Application of Advanced Signal Processing Angle-of-Arrival Estimation in ATC Navigation and Surveillance Systems," MIT Lincoln Lab, Lexington, MA, Rep. 582, 1982.

11. A. GRAHAM, *Kroneker Products and Matrix Calculus with Applications*, Chichester, UK: Elis Horwood Ltd., 1981.

12. S. HAYKIN et al., editors, Array Signal Processing, Englewood Cliffs, NJ, Prentice Hall, 1985.

13. J. E. HUDSON, *Adaptive Array Processing*, Peter Peregrinus, 1981.

14. W. HUREWICZ and H. WALLMAN, *Dimension Theory*, Princeton NJ: Princeton University Press, 1948.

15. A. G. JAFFER, "Maximum Likelihood Direction Finding of Stochastic Sources: A Separable Solution," *ICASSP 88*, pp. 2296–2893,1988.

16. D. JOHNSON and D. E. DUDGEON, *Array Signal Processing: Concepts and Techniques*, Englewood Cliffs, NJ, Prentice Hall, 1993.

17. M. KAVEH and A. J. BARABELL, "The Statistical Performace of the MUSIC and Minimum Norm Algorithms in Resolving Plane Waves in Noise," *IEEE Trans. on ASSP*, vol. 34, pp. 331–41, 1986.

18. R. KUMARESAN and D. W. TUFTS, "Estimating the Angle-of-Arrival of Multiple Plane Waves," *IEEE Trans. on AES*, vol. 19, pp. 134–39, 1983.

19. R. A. MONZINGO and T. W. MILLER, *Introduction to Adaptive Arrays*, New York, NY: Wiley-Interscience, 1980.

20. A. NEHORAI D. STARER and P. STOICA, "Consistency of Direction-of-Arrival Estimation with Multipath and Few Snapshots," *ICASSP 90*, pp. 2819–22, 1990.

21. B. OTTERSTEN and L. LJUNG, "Asymptotic Results for Sensor Array Processing," *ICASSP 89*, pp. 2266–69, 1989.

22. B. PORAT and B. FRIEDLANDER, "Analysis of the Asymptotic Relative Efficiency of the MUSIC Algorithm," *IEEE Trans, on ASSP*, vol. 36, pp. 532–44, 1988.

23. S. S. REDDI, "Multiple Source Location—A Digital Approach," *IEEE Trans. on AES*, vol. 15, pp. 95–105, 1979.

24. J. RISSANEN, "Modeling by the Shortest Description," *Automatica*, vol. 14, pp. 465–71, 1978.

25. J. RISSANEN, "A Universal Prior for the Integers and Estimation by Minimum Description Length," *Ann. of Stat.*, vol. 11, pp. 416–31, 1983.

26. J. RISSANEN, "Stochastic Complexity in Statistical Inquiry," *World Scientific*, Series in Computer Science, vol. 15, 1989.

27. W. RUDIN, *Principles of Mathematical Analysis*, New York, NY: McGraw-Hill, 1976.

28. R. O. SCHMIDT, "Multiple Emitter Location an Signal Parameter Estimation," *Proc. RADC Spectrum Estimation Workshop*, (Griffis AFB, N.Y.), pp. 243–58, 1979.

29. R. O. SCHMIDT, "A Signal Subspace Approach to Multiple Emitter Location and Spectral Estimation," Ph.D dissertation, Stanford University, CA, 1981.

30. G. SCHWARTZ, "Estimating the Dimension of the Model," *Ann. Stat.*, vol. 6, pp. 461–64, 1978.

31. T. J. SHAN, M. WAX, and T. KAILATH, "On Spatial Smoothing for Direction-of-Arrival Estimation of Coherent Signals," *IEEE Trans. on ASSP*, vol. 33, pp. 806–11, 1985.

32. C. E. SHANNON, "The Mathematical Theory of Communication," *Bell Syst. Tech. J.*, vol. 46. pp. 497–511, 1948.

33. K. SHARMAN, T. S. DURRANI, M. WAX, and T. KAilath, "Asymptotic Performance of Eigenstructure Spectral Analysis Methods," *ICASSP 84*, pp. 45.5.1–45.5.4, 1984.

34. P. STOICA, and A. NEHORAI, "MUSIC, Maximum Likelihood and the Cramer-Rao Bound," *IEEE Trans. on ASSP*, vol. 37, pp. 720–43, 1989.

35. P. STOICA, and A. NEHORAI, "MUSIC, Maximum Likelihood and the Cramer-Rao bound: Further Results and Comparisons," *IEEE SP*, vol. 38, pp. 2140–50, 1990.

36. M. WAX), "Detection and Estimation of Superimposed Signals," Ph.D Dissertation, Stanford University, CA, 1985.

37. M. WAX, "Detection of Coherent and Noncoherent Signals via the Stochastic Signals Model," *IEEE Trans. on SP*, vol. 39, pp. 2450–56, 1991.

38. M. WAX, "Detection and Localization of Multiple Source in Spatially Colored Noise," *IEEE Trans on SP*, vol. 40, pp. 245–49, 1992.

39. M. WAX, and T. KAILATH, "Detection of Signals by Information Theoretic Criteria," *IEEE Trans. on ASSP*, vol. 33, pp. 387–92, 1985.

40. M. WAX, and I. ZISKIND, "On Unique Localization of Multiple Sources in Passive Sensor Arrays," *IEEE Trans. on ASSP*, vol. 37, pp. 996–1000, 1989.

41. M. WAX, and I. ZISKIND, "Detection of the Number of Coherent and Noncoherent Signals by the MDL Principle," *IEEE Trans. on ASSP*, vol. 37, pp. 1190–96, 1989.

42. B. WIDROW, K. M. DUVALL, R. P. GOOCH, and W. C. NEWMAN, "Signal Cancellation Phenomena in Adaptive Antennas: Causes and Cures," *IEEE Trans. on AP*, vol. 30, pp. 469–78, 1982.

43. L. C. ZHAO, P. R. KRISHNAIAH, and Z. D. BAI, "On Detection of the Number of Signals in the Presence of White Noise," *J. Multivariate Anal.*, vol. 20, pp. 1–20, 1986.

44. I. ZISKIND, and M. WAX, "Maximum Likelihood Localization of Multiple Sources by Alternating Projection," *IEEE Trans. on ASSP*, vol. 36, pp. 1553–60, 1988.

2

Fundamental Limitations on Direction Finding Performance for Closely Spaced Sources

Jack Jachner and Harry B. Lee

2.1 INTRODUCTION

Determining the direction of propagating signals incident upon a sensor array, in the challenging case when separation between signal sources is small, has been a topic of active interest over the last two decades. Numerous high resolution direction-finding techniques have been proposed to resolve closely spaced sources and estimate their directions [1]–[5]. The Cramér-Rao (CR) lower bound on the variance of unbiased estimates is commonly used as a yardstick to assess the estimation accuracy of candidate algorithms. The CR bound provides fundamental insight into performance limitations, since it is not algorithm specific, but rather characterizes the optimum performance of any unbiased algorithm.

This chapter develops simple analytical expressions that describe the behavior of the CR bound for closely spaced sources. The expressions are directly useful in that they indicate the feasibility of accurate direction estimation. These expressions are also utilized to identify the minimum signal-to-noise ratio (SNR) and data sample size for which one can expect an optimum DF algorithm to resolve multiple closely spaced sources. Due to their analytical basis, the results apply to a broad class of algorithms and scenarios.

Expressions are developed both for 1-D scenarios, with source direction specified by a single direction parameter (e.g., azimuth), and also for multi-D scenarios, with multiple source-location parameters to be estimated for each source [e.g., azimuth and elevation (2-D), or azimuth, elevation, and range (3-D)]. The results provide insight into the individual performance impact of scenario parameters such as SNR, data sample size, sensor array geometry, source powers and correlations, source configuration, and maximum source spacing $\delta\omega$. A striking feature of the results is their strong dependence on the source-separation factor $\delta\omega$. An example result is that for *any* one-dimensional scenario involving $M = 3$ sources, uniformly reducing the source separations by a factor of 10 requires that source powers be increased by $2(M-1)10$ dB = 40 dB for an optimum unbiased estimator to maintain constant frequency standard deviation!

The results also elucidate the impact of dimensionality upon the performance of unbiased direction estimators. Comparison of the CR bounds for 1-D and multi-D scenarios shows that the multi-D results in some ways parallel the 1-D results, but also differ in important ways. For a given number M of closely spaced sources, we find for multi-D scenarios, in relation to 1-D scenarios, that the Cramér-Rao variance lower bounds are much lower (more favorable) so that direction estimation is more promising, and the source resolution problems are easier. For example, for a typical two-dimensional scenario involving $M = 3$ sources, uniformly reducing the source separations by a factor of 10, requires only that source powers be increased by $2(2-1)10$ dB = 20 dB for an optimum unbiased estimator to maintain constant frequency standard deviation.

The form of the CR bound depends on the model for the data vectors. One common model assumes that the sequence of vectors of signal complex amplitudes is fixed for all ensemble realizations [6–8]. A second common model is that the vector sequence is a (Gaussian) random variable across the ensemble [9–12]. Following Stoica and Nehorai, we refer to the former model as the Conditional Model, and to the latter model as the Unconditional Model [11, 12]. The designations Deterministic Model and Stochastic Model are also used frequently [9, 10, 13]. The results herein identify the small-source-separation behavior of CR bound for the conditional model. It is well known that the CR bound for the conditional model lower bounds that for the unconditional model (with suitable interpretations) [11, 12]. Therefore, the results also illuminate the behavior of the CR bound for the unconditional model.

Evaluation of the CR bound requires inverting the applicable Fisher Information matrix $\mathcal{F}$. The dimension of $\mathcal{F}$ equals the number of real parameters to be estimated. For the problem of estimating the spatial frequency ω (i.e., 1-D source location parameter) of a single source in white Gaussian noise, the matrix $\mathcal{F}$ can be inverted analytically. As a result, simple explicit expressions are available for the CR frequency bounds in this case [5, 6, 7]. The bound for the conditional model takes the following form:

$$\mathrm{Var}\{\hat{\omega}\} \geq \left(\frac{3}{2\pi^2\|\vec{a}(\omega)\|^2}\right)\frac{\Omega^2}{N \cdot \mathrm{SNR}} \tag{2.1}$$

where Ω denotes an applicable resolution cell size, SNR the (point) signal-to-noise ratio, N the number of data samples, and $\vec{a}(\omega)$ is the signal vector.

For applications involving M signals, inversion of the relevant $\mathcal{F}$ matrix rapidly becomes intractable with increasing M. As a result, for $M > 1$, CR bounds are typically computed numerically rather than analytically [14, 15]. Stoica and Nehorai [8] have provided a useful expression for calculating CR bounds applicable to frequency and direction estimation problems for 1-D scenarios and the conditional model. The expression requires calculating and inverting an $M \times M$ matrix function of the signal vectors, their derivatives, and the covariance matrix for the signal complex amplitudes. Yau and Bresler [16] extended the result [8] to multi-D scenarios, for which calculation and inversion of an $M\mathcal{D} \times M\mathcal{D}$ matrix is required, where $\mathcal{D}$ denotes the scenario dimensionality.

The formulations in [8, 16] are very useful for both analytical and numerical work in that they bypass the tedious computation associated with calculating and inverting the complete (large) Fisher Information Matrix. However, a shortcoming to the formulations is that the dependence of CR bounds upon scenario parameters such as array geometry, source configuration, source powers, and correlations remains implicit. This chapter builds upon the Stoica and Nehorai 1-D formulation, and the Yau and Bresler multi-D formulation, to develop simple approximate expressions that make explicit the dependence of the CR bound upon such parameters for the closely spaced source problem. The results herein provide insight into the dependence of the CR bound upon scenario elements, and facilitate derivation of fundamental performance metrics such as the minimum (threshold) SNR, and data set size, at which closely spaced sources are resolvable.

2.2 SUMMARY OF RESULTS

The main result for 1-D scenarios is the following:

$$\operatorname{Var}\{\hat{\omega}_j\} \geq c_{M,j} \frac{\Omega^2}{N \cdot \mathrm{SNR}_j} \frac{1}{(\delta\omega/\Omega)^{2(M-1)}} + \mathcal{O}\left(\delta\omega^{-2(M-1)+1}\right) \tag{2.2}$$

for small $\delta\omega$, where $\hat{\omega}_j$ denotes the estimate of spatial frequency for the jth source, Ω denotes an applicable resolution cell size, $\delta\omega$ denotes the spatial frequency difference between the two most widely spaced sources, SNR_j denotes the signal-to-noise ratio for the jth signal, and M denotes the number of sources. The $c_{M,j}$ denote constants that depend on the other scenario parameters of sensor array geometry, source powers and correlations, and source configuration. The main result for multi-D scenarios is

$$\operatorname{Var}\{\hat{\omega}_{ij}\} \geq c_{\chi,i,j} \frac{\Omega_i^2}{N \cdot \mathrm{SNR}_j} \frac{1}{(\delta\omega/\Omega_i)^{2(\chi-1)}} + \mathcal{O}\left(\delta\omega^{-2(\chi-1)+1}\right) \tag{2.3}$$

where $\hat{\omega}_{ij}$ denotes the estimate of the ith component of spatial frequency vector $\vec{\omega}_j$ for the jth source, Ω_i denotes the array resolution cell size along the ith spectral coordinate, parameter χ is defined in section 2.10 and importantly satisfies $\chi < M$ for $M > 2$ in typical multi-D scenarios, and the $c_{\chi,i,j}$ denote suitable constants that depend on the other scenario parameters. Clearly, (2.2) and (2.3) represent a generalization of (2.1). The primary impact of the additional signals is to introduce the factor $1/(\delta\omega/\Omega)^{2(M-1)}$ in 1-D scenarios and the factor $1/(\delta\omega/\Omega_i)^{2(\chi-1)}$ in multi-D scenarios. The additional factor accounts for the sensitivity of estimation results for closely spaced signals. Significantly, since $\chi < M$ for typical multi-D scenarios, the CR lower variance bound is typically much smaller (more favorable) in multi-D than in 1-D for scenarios with $M > 2$ closely spaced sources.

Based on (2.2) and (2.3), it is argued that the threshold (minimum) SNR at which any unbiased estimator can reliably resolve M sources is proportional to $\delta\omega^{-2M}$ in 1-D scenarios and to $\delta\omega^{-2\chi}$ in multi-D scenarios. It is further argued that the minimum data set size N for reliable resolution is also proportional to $\delta\omega^{-2M}$ in 1-D scenarios and to $\delta\omega^{-2\chi}$ in multi-D scenarios. That is

$$\mathcal{E}_R \simeq \frac{K_R}{N \cdot \delta\omega_\chi^2} \tag{2.4}$$

$$\mathcal{N}_R \simeq \frac{K_R}{\text{SNR} \cdot \delta\omega_\chi^2} \tag{2.5}$$

for large N and small $\delta\omega$, where $\mathcal{E}_R$ denotes the SNR resolution threshold, $\mathcal{N}_R$ the data set size N resolution threshold, K_R is a suitable constant, and $\chi = M$ for 1-D scenarios. Since $\chi < M$ for typical multi-D scenarios, the threshold SNR and data set size required to resolve $M > 2$ closely spaced sources are typically much smaller (more favorable) in multi-D than in 1-D scenarios.

The chapter is organized as follows. Sections 2.3 to 2.7 develop and discuss the result (2.2) for 1-D. Section 2.8 illustrates that the result (2.2) accurately models the exact CR bound in 1-D for small signal separations. Section 2.9 addresses the problem of resolving multiple closely spaced signals in 1-D to develop results (2.4) and (2.5) for 1-D. Sections 2.10 and 2.11 develop the result (2.3) for multi-D for a convenient set of analysis assumptions. Section 2.12 illustrates the accuracy of the multi-D result (2.3) and section 2.13 addresses the resolution thresholds for multi-D. Section 2.14 considers (degenerate) multi-D scenarios for which the basic analysis assumptions are not satisfied. Section 2.15 is a brief summary.

2.3 ASSUMPTIONS AND NOTATION

The problem addressed is that of estimating parameter vectors $\vec{\omega}_1 \cdots \vec{\omega}_M$ of the conditional model described in [8, 11, 12]:

$$\vec{y}(t) = A \cdot \vec{x}(t) + \vec{e}(t) \qquad t = 1 \cdots N \tag{2.6}$$

$\vec{y}(t)$ is a noisy (complex) data vector observed for the sample index t; $\vec{y}(t)$ is assumed to be $W \times 1$. A is a constant $W \times M$ matrix of M signal vectors having the special form

$$A = \left[\vec{a}(\vec{\omega}_1) \cdots \vec{a}(\vec{\omega}_M)\right] \tag{2.7}$$

$\vec{a}(\vec{\omega})$ is a known generic signal vector function of a source-location parameter vector $\vec{\omega}$. $\vec{\omega}_1 \cdots \vec{\omega}_M$ are the source parameter vectors to be estimated, and are assumed to be $\mathcal{D} \times 1$. For 1-D problems, $\mathcal{D} = 1$ and $\vec{\omega}_j = \omega_j$ are scalar parameters. $\vec{x}(t)$ is an unknown $M \times 1$ vector of complex amplitudes which can change from one value of sample index to the next. $\vec{e}(t)$ is a $W \times 1$ vector representing additive complex noise for sample index t; $\vec{e}(t)$ is modeled as a zero-mean Gaussian random vector with covariance $\sigma^2 I$. The vectors $\vec{a}(\vec{\omega})$, $\vec{x}(t)$, and $\vec{e}(t)$ are assumed to satisfy additional conditions detailed in Appendix 2A for 1-D scenarios and in Appendix 2B for multi-D scenarios. As noted in [8], this model is applicable to a variety of frequency and direction estimation problems. The central estimation problem associated with the model (2.6) is to estimate the source parameter vectors $\vec{\omega}_1, \cdots, \vec{\omega}_M$ given a sequence of vector data samples (or snapshots) $\vec{y}(1), \cdots, \vec{y}(N)$, and knowledge of the array response $\vec{a}(\vec{\omega})$ for all $\vec{\omega}$.

For direction-finding applications, the number of sources is M, of sensors is W, of data vector samples is N, of dimensions in the parameter vector $\vec{\omega}$ is $\mathcal{D}$.

We utilize the notation that $E\{\ \}$ denotes expectation and that superscript ^{T} denotes the transpose, * the conjugate, ^{H} the Hermitian (conjugate) transpose, $^{-1}$ the matrix inverse, and $^+$ the pseudo-inverse.

Projection matrices play a fundamental role in our results. To simplify the discussion, we introduce the following additional notation.

$Q_{[Z]} \triangleq ZZ^+$	projection matrix onto the column space of Z
$Q_{[Z^H]} \triangleq Z^H(Z^H)^+ = Z^+Z$	projection matrix onto the row space of Z
$P_{[Z]} \triangleq I - Q_{[Z]}$	projection matrix onto the column null space of Z
$P_{[Z^H]} \triangleq I - Q_{[Z^H]}$	projection matrix onto the row null space of Z

2.4 CRAMÉR-RAO BOUNDS

Evaluation of the CR bound generally requires inverting the applicable Fischer Information matrix $\mathcal{F}$ of dimension equal to the number of unknown (real and imaginary) model parameters. In the conditional model the unknown parameters are not only the source spatial frequency vectors $\vec{\omega}_1, \cdots, \vec{\omega}_M$ of interest, but also the noise variance σ^2 and the complex signal amplitude vector sequence $\vec{x}(1), \cdots, \vec{x}(N)$. These latter unknowns are essentially nuisance parameters for the DF problem, which significantly enlarge $\mathcal{F}$ and make direct calculation of $\mathcal{F}^{-1}$ exceedingly cumbersome.

The CR bound of present interest is that on the covariance of the spatial frequency vectors. This bound is given by a (small) submatrix of $\mathcal{F}^{-1}$. Since only a submatrix of $\mathcal{F}^{-1}$ is required, it is useful for both analytical and numerical work to have available an explicit formulation for the applicable submatrix of $\mathcal{F}^{-1}$. Such formulations have been developed by Stoica and Nehorai for 1-D scenarios [8], and extended by Yau and Bresler to multi-D scenarios [16].

For 1-D scenarios, the CR bound on sample frequency covariances takes the form

$$E\left\{\left(\hat{\vec{\Upsilon}} - \vec{\Upsilon}\right)\left(\hat{\vec{\Upsilon}} - \vec{\Upsilon}\right)^H\right\} \geq B_C \qquad (2.8)$$

where $A \geq B$ means that the matrix $A - B$ is non-negative definite, and

$$\vec{\Upsilon} \triangleq [\omega_1, \omega_2, \cdots, \omega_M]^T$$
$$\hat{\vec{\Upsilon}} \triangleq \left[\hat{\omega}_1, \hat{\omega}_2, \cdots, \hat{\omega}_M\right]^T \tag{2.9}$$

$\hat{\omega}_j$ denotes an unbiased estimate of the spatial frequency ω_j for the jth source. The matrix B_C is

$$B_C = \frac{\sigma^2}{2N}\left[\mathrm{Re}\{H \odot \hat{P}^T\}\right]^{-1} \tag{2.10}$$

where $\odot$ denotes the Hadamard (element-by-element) matrix product, and

$$H \triangleq D^H\left[I - A(A^H A)^{-1}A^H\right]D \qquad M \times M \tag{2.11}$$

$$D \triangleq \left[\vec{d}(\omega_1), \vec{d}(\omega_2), \cdots, \vec{d}(\omega_M)\right] \tag{2.12}$$

$$\vec{d}(\omega_j) \triangleq \left.\frac{d\vec{a}(\omega)}{d\omega}\right|_{\omega=\omega_j} \tag{2.13}$$

$$\hat{P} \triangleq \frac{1}{N}\sum_{t=1}^{N}\vec{x}(t)\vec{x}(t)^H \tag{2.14}$$

Vector $\vec{a}(\omega)$ is the generic arrival vector for (scalar) spatial frequency ω, matrix A is the source arrival matrix (2.7), and $\hat{P}$ is the sample source amplitude covariance matrix. The result (2.8), (2.10) is valid for 1-D scenarios under the conditional model assumptions **X1**, **X2**, and **E1** of Appendix 2A. The result is due to Stoica and Nehorai [8].

Yau and Bresler [16] have extended the result (2.10)–(2.13) to multi-D scenarios with parameter vectors $\vec{\omega}_1 \cdots \vec{\omega}_M$, again under the conditional model assumptions **X1**, **X2**, and **E1** of Appendix 2A. For multi-D scenarios, the CR bound applicable to the parameter vectors $\vec{\omega}_1 \cdots \vec{\omega}_M$ also takes the form (2.8), this time with

$$\vec{\Upsilon} \triangleq [\omega_{11} \cdots \omega_{\mathcal{D}1} \cdots \omega_{1M} \cdots \omega_{\mathcal{D}M}]^T$$
$$\hat{\vec{\Upsilon}} \triangleq \left[\hat{\omega}_{11} \cdots \hat{\omega}_{\mathcal{D}1} \cdots \hat{\omega}_{1M} \cdots \hat{\omega}_{\mathcal{D}M}\right]^T \tag{2.15}$$

$\hat{\omega}_{ij}$ denotes an unbiased estimate of the ith element of $\vec{\omega}_j$ ($i = 1 \cdots \mathcal{D}$, $j = 1 \cdots M$). In the multi-D case, B_C is an $M\mathcal{D} \times M\mathcal{D}$ matrix with the compact form [16]

$$B_C = \frac{\sigma^2}{2N}\left[\mathrm{Re}\left\{H \odot \hat{P}_+^T\right\}\right]^{-1} \tag{2.16}$$

with

$$H \triangleq D^H\left[I - A(A^H A)^{-1}A^H\right]D \qquad M\mathcal{D} \times M\mathcal{D} \tag{2.17}$$

$$D \triangleq \left[\dot{D}(\vec{\omega}_1), \cdots \dot{D}(\vec{\omega}_M)\right] \tag{2.18}$$

$$\dot{D}(\vec{\omega}_j) \triangleq \left[\frac{\partial}{\partial\omega_1}\vec{a}(\vec{\omega}), \cdots \frac{\partial}{\partial\omega_{\mathcal{D}}}\vec{a}(\vec{\omega})\right]_{\vec{\omega}=\vec{\omega}_j} \tag{2.19}$$

$$\hat{P}_+ \triangleq \hat{P} \otimes 1_{\mathcal{D}\times\mathcal{D}}$$
$$= \begin{bmatrix} 1_{\mathcal{D}\times\mathcal{D}} \cdot \hat{p}_{11} & \cdots & 1_{\mathcal{D}\times\mathcal{D}} \cdot \hat{p}_{M1} \\ \vdots & & \vdots \\ 1_{\mathcal{D}\times\mathcal{D}} \cdot \hat{p}_{1M} & \cdots & 1_{\mathcal{D}\times\mathcal{D}} \cdot \hat{p}_{MM} \end{bmatrix} \quad M\mathcal{D} \times M\mathcal{D} \tag{2.20}$$

where $\hat{p}_{ij}$ is the i, j^{th} scalar element of the sample source amplitude covariance matrix $\hat{P}$ of (2.14), $\otimes$ denotes Kronecker product, and $1_{\mathcal{D}\times\mathcal{D}}$ denotes the $\mathcal{D} \times \mathcal{D}$ matrix of ones.

A shortcoming of the B_C expressions of (2.10) and (2.16) is that the dependence of B_C on physical scenario parameters such as sensor array geometry, source configuration, source powers, and correlations remains implicit. A major contribution of this chapter is to make explicit the dependence of B_C upon such scenario parameters for the closely spaced source problem. Simple explicit expressions in terms of the maximum source separation $\delta\omega$ and the foregoing scenario parameters are developed in section 2.6 for 1-D scenarios, and in section 2.10 for multi-D scenarios. These expressions clarify the dependence of the CR bound upon scenario elements, and facilitate derivation of fundamental performance metrics such as the minimum (threshold) SNR, and data set size, at which closely spaced sources are resolvable.

Relationship to the MUSIC Null Spectrum

Insight into the formulations (2.10) and (2.16) for the 1-D and multi-D CR bound on $\text{Cov}\{\hat{\vec{\Upsilon}}\}$ can be gained by reference to the MUSIC null spectrum.

The null spectrum $\Delta(\vec{\omega})$ for the MUSIC algorithm for an arbitrary direction $\vec{\omega}$ is defined as follows:

$$\Delta(\vec{\omega}) \triangleq \frac{\vec{a}(\vec{\omega})^H \left[I - A(A^H A)^{-1} A^H\right] \vec{a}(\vec{\omega})}{\|\vec{a}(\vec{\omega})\|^2} \tag{2.21}$$

where A denotes the matrix (2.7) of source arrival vectors $\vec{a}(\vec{\omega}_j)$, $j = 1 \cdots M$. This scalar function has the useful property that it equals zero whenever $\vec{\omega}$ coincides with a source spatial frequency $\vec{\omega}_j$. The MUSIC algorithm uses this property as a basis for estimating source directions [5].

For 1-D scenarios, straightforward differentiation of (2.21) with respect to (scalar) ω shows that the second derivative of $\Delta(\omega)$ at $\omega = \omega_j$ is

$$\Delta^{(2)}(\omega_j) = \frac{2 \cdot \text{Re}\left\{\vec{d}(\omega_j)^H \left[I - A(A^H A)^{-1} A^H\right] \vec{d}(\omega_j)\right\}}{\|\vec{a}(\omega_j)\|^2}$$
$$= \frac{2 \cdot (H)_{jj}}{\|\vec{a}(\omega_j)\|^2} \tag{2.22}$$

where

$$\Delta^{(2)}(\omega) \Delta \frac{d^2}{d\omega} [\Delta(\omega)] \tag{2.23}$$

$\vec{d}(\omega_j)$ is defined in (2.13), and $(H)_{jj}$ is the (real) diagonal element of (Hermitian) H in (2.11) corresponding to the jth source.

For multi-D scenarios, straightforward differentiation of (2.21) with respect to $\vec{\omega}$ shows that the Hessian matrix $\mathcal{H}_j$ of $\Delta(\vec{\omega})$ at $\vec{\omega} = \vec{\omega}_j$ is

$$\begin{aligned} \mathcal{H}_j &= \frac{2 \cdot \text{Re}\left\{\dot{D}(\vec{\omega}_j)^H \left[I - A(A^H A)^{-1} A^H\right] \{\dot{D}(\vec{\omega}_j)\}\right.}{\|\vec{a}(\vec{\omega}_j)\|^2} \\ &= \frac{2 \cdot \text{Re}\{(H)_{[jj]}\}}{\|\vec{a}(\vec{\omega}_j)\|^2} \end{aligned} \tag{2.24}$$

where $\dot{D}(\vec{\omega})$ is defined in (2.19), and $(H)_{[jj]}$ is the $\mathcal{D} \times \mathcal{D}$ diagonal block of H in (2.17) corresponding to the jth source.

Consequently, the diagonal block of B_C^{-1} corresponding the jth source is

$$(B_C^{-1})_{[jj]} = \begin{cases} \dfrac{N \cdot \hat{p}_{jj} \cdot \|\vec{a}(\omega_j)\|^2}{\sigma^2} \Delta^{(2)}(\omega_j) & \text{1-D} \\ \dfrac{N \cdot \hat{p}_{jj} \cdot \|\vec{a}(\vec{\omega}_j)\|^2}{\sigma^2} \mathcal{H}_j & \text{multi-D} \end{cases} \tag{2.25}$$

where $\hat{p}_{jj}$ denotes the sample power of the jth source.

For uncorrelated sources and large N, matrix $\hat{P}$ essentially is diagonal, hence matrix B_C is diagonal for 1-D and block diagonal for multi-D. Therefore, from (2.25), the block of B_C corresponding to the jth source is

$$(B_C)_{[jj]} = \begin{cases} \dfrac{\sigma^2}{N \cdot \hat{p}_{jj} \cdot \|\vec{a}(\omega_j)\|^2} \dfrac{1}{\Delta^{(2)}(\omega_j)} & \text{1-D} \\ \dfrac{\sigma^2}{N \cdot \hat{p}_{jj} \cdot \|\vec{a}(\vec{\omega}_j)\|^2} \mathcal{H}_j^{-1} & \text{multi-D} \end{cases} \tag{2.26}$$

For correlated sources, a lower limit on the $(B_C)_{[jj]}$ can be established as follows. From the identity for the inverse of any partitioned positive definite matrix Z

$$\left(Z^{-1}\right)_{[jj]} \geq \left[Z_{[jj]}\right]^{-1} \tag{2.27}$$

Use of (2.27) with $Z = B_C^{-1}$, and (2.25) gives

$$(B_C)_{[jj]} \geq \begin{cases} \dfrac{\sigma^2}{N \cdot \hat{p}_{jj} \cdot \|\vec{a}(\omega_j)\|^2} \dfrac{1}{\Delta^{(2)}(\omega_j)} & \text{1-D} \\ \dfrac{\sigma^2}{N \cdot \hat{p}_{jj} \cdot \|\vec{a}(\vec{\omega}_j)\|^2} \mathcal{H}_j^{-1} & \text{multi-D} \end{cases} \tag{2.28}$$

We note from (2.26) and (2.28) that for given source powers, the Cramér-Rao variance bound for correlated sources is lower bounded by the CR bound for uncorrelated sources.

Equations (2.26), (2.28) indicate for 1-D that the Cramér-Rao bound $(B_C)_{jj}$ on $\text{Var}\{\hat{\omega}_j\}$ is proportional to the radius of curvature of the MUSIC null spectrum at ω_j. It is well known that the Hessian describes the curvature of a quadratic surface.

Accordingly, equations (2.26), (2.28) indicate for multi-D that the CR bound can be expected to be small for the ith element of $\vec{\omega}_j$ corresponding to a large i^{th} diagonal element of $\mathcal{H}_j$, or, equivalently, corresponding to small radius of curvature of the spectrum $\Delta(\vec{\omega})$ along the ith spatial frequency coordinate. Similarly, the CR bound can be expected to be unfavorable for the ith element of $\vec{\omega}_j$ corresponding to a large radius of curvature of the spectrum $\Delta(\vec{\omega})$ along the ith coordinate at $\vec{\omega}_j$.

2.5 ANALYSIS APPROACH

The main results of our analysis are obtained by identifying expressions for the CR bound B_C as source spacing becomes small. We illustrate the analysis approach for the multi-D scenario case; the approach for the 1-D case is a straightforward scalar analog.

To facilitate the analysis, we express the spatial frequency vector for the jth source as

$$\vec{\omega}_j = \vec{\omega}_0 + \delta\omega \cdot \vec{q}_j \tag{2.29}$$

$j = 1, \cdots M$, where $\vec{\omega}_0$ is a nearby fixed reference vector, $\delta\omega$ is a *variable real scale factor*, and

$$\vec{q}_j = \left[q_{1j}, \cdots q_{\mathcal{D}j}\right]^T \tag{2.30}$$

is a *normalized offset vector with constant real elements.* The $\vec{q}_j$ are normalized so that $\delta\omega$ equals the maximum separation $\|\vec{\omega}_j - \vec{\omega}_k\|$ between pairs of vectors $\vec{\omega}_1, \cdots \vec{\omega}_M$. That is,

$$\delta\omega = \max_{j,k} \|\vec{\omega}_j - \vec{\omega}_k\| \tag{2.31}$$

The analysis strategy is to examine the Cramér-Rao bound B_C as scaling factor $\delta\omega \to 0$, while the $\vec{q}_j$ are held constant. The leverage in the representation (2.29) is that it replaces the M variable spatial vectors $\vec{\omega}_1 \cdots \vec{\omega}_M$ by a single variable scalar parameter $\delta\omega$, thereby greatly simplifying analysis. The condition $\delta\omega \to 0$ corresponds to the coalescing of all source parameter vectors to the reference vector $\vec{\omega}_0$, while the relative (normalized) source configuration remains unchanged.

Example 2.1

To illustrate the analysis strategy, consider the 2-D problem of estimating a 2-element spatial frequency vector $\vec{\omega}_j = [\omega_{xj}, \omega_{yj}]^T$ for each source ($j = 1, 2, 3$) in a cluster of 3 far-field sources in the triangular configuration illustrated in Figure 2.1A. To implement the analysis approach, we express each source spatial frequency vector as in (2.29). We define a reference vector $\vec{\omega}_0$ in the vicinity of the source spatial frequency vectors, and a scalar parameter $\delta\omega$ to be the maximum source separation, which in Fig. 2.1A are

$$\begin{aligned} \vec{\omega}_0 &= [0, 0]^T \\ \delta\omega &= \|\vec{\omega}_1 - \vec{\omega}_3\| \end{aligned} \tag{2.32}$$

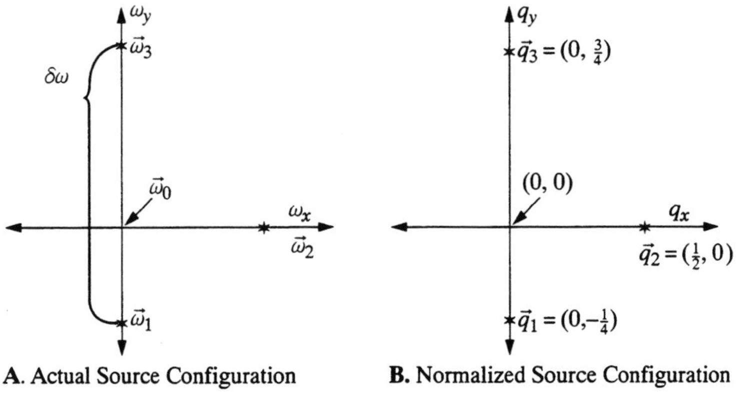

Figure 2.1 Actual and normalized source configuration for a 2-D DF scenario.

Finally, we define normalized offset vectors $\vec{q}_1, \vec{q}_2, \vec{q}_3$ to satisfy (2.29). The normalized source configuration is illustrated in Figure 2.1B. As $\delta\omega \to 0$, the actual source configuration in Figure 2.1A coalesces to the reference direction $\vec{\omega}_0$, but the normalized configuration in Figure 2.1B remains fixed.

2.6 ASYMPTOTIC CR BOUND FOR 1-D

We first analyze the behavior of the CR bound (2.10) on directional covariance for M closely spaced sources in one-dimensional space. The analysis is performed subject to assumptions presented in Appendix 2A. These assumptions typically are satisfied in direction-finding problems. The primary assumptions are that derivatives of $\vec{a}(\omega)$ with respect to ω exist for all orders, and that $\vec{a}(\omega)$ and its first M derivatives are linearly independent.

The signal spatial frequencies are assumed to be scalars $\omega_1 \cdots \omega_M$. The representation (2.29) here takes the form:

$$\omega_j = \omega_0 + q_j \delta\omega \tag{2.33}$$

$j = 1 \cdots M$. Here ω_0 denotes a fixed reference frequency, the q_j are normalized scalar offsets such that $q_1 < q_2 < \cdots < q_M$ with $q_M - q_1 = 1$, and $\delta\omega$ is a variable scale parameter corresponding to the separation of the extreme frequencies.

Taylor Series for Matrices A and D

Taylor series representations are central to the analysis. To facilitate identification of the small $\delta\omega$ structure of the CR bound B_C in (2.10) for 1-D scenarios, we construct the Taylor series of the generic arrival vector $\vec{a}(\omega)$, of the generic arrival vector derivative $\vec{d}(\omega)$ in (2.13), and of matrix A of (2.7). The generic spatial frequency parameter ω is

$$\omega = \omega_0 + q\delta\omega \tag{2.34}$$

where q is a *normalized generic spatial frequency*. To explicitly denote dependence on the elements of (2.34), we express the Taylor series of the generic arrival vector $\vec{a}(\omega)$ about the reference ω_0 as

$$\vec{a}(\omega) = \sum_{p=0}^{\infty} \delta\omega^p \dot{\vec{a}}_p \cdot \gamma_p(q) \tag{2.35}$$

where $\dot{\vec{a}}_p$ is the pth order derivative of $\vec{a}(\omega)$ with respect to ω at ω_0. That is

$$\dot{\vec{a}}_p \triangleq \left[\frac{d^p\vec{a}(\omega)}{d\omega^p}\right]_{\omega=\omega_0} \qquad (W \times 1) \tag{2.36}$$

For 1-D, scalar $\gamma_p(q)$ depends on the normalized direction offset q as

$$\gamma_p(q) = q^p/p! \tag{2.37}$$

The $\dot{\vec{a}}_p$ and $\gamma_p(q)$ are constant with $\delta\omega$; $\dot{\vec{a}}_p$ is typically complex, while $\gamma_p(q)$ is always real.

It follows from (2.35) that matrix A in (2.7) has Taylor series of the form

$$\begin{aligned} A &= \left[\vec{a}(\omega_1), \cdots \vec{a}(\omega_M)\right] \\ &= \sum_{p=0}^{\infty} \delta\omega^p \dot{\vec{a}}_p \Gamma_p \end{aligned} \tag{2.38}$$

where matrix $\dot{\vec{a}}_p$ is as in (2.36), and Γ_p is a constant real $1 \times M$ matrix of the form

$$\Gamma_p \triangleq \left[\gamma_p(q_1), \cdots \gamma_p(q_M)\right] \tag{2.39}$$

For 1-D scenarios,

$$\begin{aligned} \Gamma_0 &= [1, & \cdots & \quad 1] \\ \Gamma_1 &= [q_1, & \cdots & \quad q_M] \\ \Gamma_2 &= [q_1^2/2, & \cdots & \quad q_M^2/2] \\ \Gamma_3 &= [q_1^3/6, & \cdots & \quad q_M^3/6] \end{aligned} \tag{2.40}$$

As defined in (2.12), (2.13), each column $\vec{d}(\omega_j)$ of matrix D is the derivative of the generic arrival vector $\vec{a}(\omega)$ at ω_j; hence

$$\vec{d}(\omega_j) = \left[\frac{d}{d\omega}\vec{a}(\omega)\right]_{\omega=\omega_j} = \frac{1}{\delta\omega}\left[\frac{d}{dq}\vec{a}(\omega_0 + \delta\omega q)\right]_{q=q_j} \tag{2.41}$$

Hence

$$D = \delta\omega^{-1} \sum_{p=0}^{\infty} \delta\omega^p \dot{\vec{a}}_p \dot{\Gamma}_p \tag{2.42}$$

where

$$\dot{\Gamma}_p \triangleq \left[\left.\frac{d\gamma_p(q)}{dq}\right|_{q=q_1}, \left.\frac{d\gamma_p(q)}{dq}\right|_{q=q_2}, \ldots, \left.\frac{d\gamma_p(q)}{dq}\right|_{q=q_M} \right] \tag{2.43}$$

and $\dot{\vec{a}}_p$, $\gamma_p(q)$ are as in (2.36), (2.37). Note that $\dot{\Gamma}_0 = [0, \cdots 0]$ since $\gamma_0(q) = 1$, and that the $\dot{\vec{a}}_p$ and $\dot{\Gamma}_p$ are constant with $\delta\omega$, so that (2.42) is a Taylor series in $\delta\omega$.

Behavior of $\left[I - A(A^H A)^{-1} A^H\right] D$ for Small $\delta\omega$

To elucidate the CR bound B_C in (2.10), consider the matrix factor H given by (2.11). The factor $\left[I - A(A^H A)^{-1} A^H\right]$ is a projection matrix onto column null space of $W \times M$ matrix A, has constant rank $W - M$, and has $W - M$ unit eigenvalues for all $\delta\omega \neq 0$ from Assumptions **A1, A3**. Therefore, $\left[I - A(A^H A)^{-1} A^H\right]$ does not approach zero as $\delta\omega \to 0$. Similarly, the typical column $\vec{d}(\omega_j)$ of D approaches the constant vector $\vec{d}(\omega_0)$ as $\delta\omega \to 0$. However, numerical examples show that the product $\left[I - A(A^H A)^{-1} A^H\right] D \to 0$ as $\delta\omega \to 0$. Consequently, the essential properties of H and therefore of B_C, for $\delta\omega \to 0$ derive from the *interaction* of the factors $\left[I - A(A^H A)^{-1} A^H\right]$ and D.

To clarify the small $\delta\omega$ behavior of H, it is desirable to obtain appropriate small $\delta\omega$ behavior of the product $\left[I - A(A^H A)^{-1} A^H\right] D$. Analysis based on the Taylor series shows that

$$A(A^H A)^{-1} A^H = \dot{A}(\dot{A}^H \dot{A})^{-1} \dot{A}^H + \mathcal{O}(\delta\omega) \tag{2.44}$$

for small $\delta\omega$, where

$$\dot{A} \triangleq \left[\vec{a}(\omega_0), \dot{\vec{a}}_1, \cdots, \dot{\vec{a}}_{M-1}\right] \tag{2.45}$$

and $\dot{\vec{a}}_p$ are defined in (2.36). (See Appendix C of [17].) Equation (2.44) asserts that the limiting vector subspace spanned by the M arrival vectors $\vec{a}(\vec{\omega}_1), \cdots, \vec{a}(\vec{\omega}_M)$ as $\delta\omega \to 0$ is identical to that spanned by the first M vector terms of the Taylor series (2.35), not a surprising result. From (2.44) we have

$$\left[I - A(A^H A)^{-1} A^H\right] = \left[I - \dot{A}(\dot{A}^H \dot{A})^{-1} \dot{A}^H\right] + \mathcal{O}(\delta\omega) \tag{2.46}$$

Equation (2.46) suggests that, for small $\delta\omega$, pre-factor $\left[I - A(A^H A)^{-1} A^H\right]$ in the product $\left[I - A(A^H A)^{-1} A^H\right] D$ acts to annihilate all components of the columns of D which lie in the subspace defined by the columns of $\dot{A}$. Analysis presented in Appendices 2A, 2B of [17] shows that this indeed is the case. The essence of the analysis is as follows.

First, the derivative vector $\vec{d}(\omega_j)$ in (2.41) is expressed as a linear combination of the arrival vectors represented by the columns of A, and a remainder vector $\vec{f}_j$:

$$\vec{d}(\omega_j) = \delta\omega^{-1}\left[A\vec{\beta}_j + \vec{f}_j\right] \tag{2.47}$$

The rationale for utilizing the formulation (2.47) for $\vec{d}(\omega_j)$ is to concentrate in the term $\delta\omega^{-1} A\vec{\beta}_j$ as much of the vector $\vec{d}(\omega_j)$ as practicable that will be annihilated when $\vec{d}(\omega_j)$ is premultiplied by the factor $\left[I - A(A^H A)^{-1} A^H\right]$ in the product $\left[I - A(A^H A)^{-1} A^H\right] D$. This concentrates in the remainder vector $\delta\omega^{-1} \vec{f}_j$ components that will survive in the product.

Vector $\vec{f}_j$ can be regarded as the error in the approximation $\delta\omega \cdot \vec{d}(\omega_j) \simeq A\vec{\beta}_j$. For each candidate weight vector $\vec{\beta}_j$, the remainder vector $\vec{f}_j$ has some order μ in $\delta\omega$; that is, $\vec{f}_j = \mathcal{O}(\delta\omega^{\mu})$. The weight vector $\vec{\beta}_j$ is selected to maximize μ; this serves to minimize the approximation error $\vec{f}_j$ as $\delta\omega \to 0$. Maximization of μ is accomplished by selecting $\vec{\beta}_j$ to zero as many leading terms as possible of the Taylor series of $\vec{f}_j$. This leads to a unique vector $\vec{\beta}_j$; for this $\vec{\beta}_j$,

$$\mu = M \tag{2.48}$$

and

$$\vec{f}_j = \delta\omega^M \dot{\vec{a}}_M \psi(q_j) + \mathcal{O}\left(\delta\omega^{M+1}\right) \tag{2.49}$$

where

$$\dot{\psi}(q) \triangleq \frac{d\psi(q)}{dq} \tag{2.50}$$

$$\psi(q) \triangleq \frac{1}{M!}\prod_{l=1}^{M}(q - q_l) \tag{2.51}$$

(See Appendices 2A and 2B of [17].) Substitution of (2.47) and (2.49) in the product $\left[I - A(A^H A)^{-1}A^H\right] D$ followed by use of (2.46) shows that

$$\left[I - A(A^H A)^{-1}A^H\right] D = \delta\omega^{M-1}\vec{\varepsilon}_M \left[1, \cdots, 1\right] \dot{\Psi} + \mathcal{O}(\delta\omega^M) \tag{2.52}$$

where

$$\vec{\varepsilon}_M \triangleq \left[I - \dot{A}(\dot{A}^H \dot{A})^{-1}\dot{A}^H\right] \dot{\vec{a}}_M \tag{2.53}$$

$$\dot{\Psi} \triangleq \text{Diag}\left[\dot{\psi}(q_1), \dot{\psi}(q_2), \cdots \dot{\psi}(q_M)\right] \tag{2.54}$$

The factors in (2.52) have the following interpretations. The vector $\vec{\varepsilon}_M$ is simply the Mth order derivative of $\vec{a}(\omega)$ at ω_0, less its projection onto the column space spanned by the derivatives of lower order. As a consequence of Assumption **A6** of Appendix 2A, $\vec{\varepsilon}_M \neq 0$. The jth diagonal element of $\dot{\Psi}$ is the derivative of polynomial function $\psi(q)$ at the zero crossing $q = q_j$. Since the q_j are by assumption distinct, the derivatives of $\psi(q)$ at zero-crossings are nonzero.

Substitution of (2.52) in (2.11) for small $\delta\omega$ gives

$$H = \delta\omega^{2(M-1)}(\vec{\varepsilon}_M^H \vec{\varepsilon}_M)\dot{\Psi} 1_{M\times M}\dot{\Psi} + \mathcal{O}(\delta\omega^{2(M-1)+1}) \tag{2.55}$$

where $1_{M\times M}$ is an $M \times M$ matrix of ones. From (2.55) and the diagonal form of (2.54) we find

$$H \odot \hat{P}^T = \delta\omega^{2(M-1)}(\vec{\varepsilon}_M^H \vec{\varepsilon}_M)\dot{\Psi}\hat{P}^T\dot{\Psi} + \mathcal{O}(\delta\omega^{2(M-1)+1}) \tag{2.56}$$

CR Bound B_C for Small $\delta\omega$

Substitution of Eq. (2.56) into (2.10) and use of the real property of $\dot{\Psi}$ gives our main result for 1-D scenarios:

$$B_C = \delta\omega^{-2(M-1)} K + \mathcal{O}(\delta\omega^{-2(M-1)+1}) \tag{2.57}$$

where

$$K = \frac{\sigma^2}{2N} \frac{1}{\vec{\varepsilon}_M^H \vec{\varepsilon}_M} \dot{\Psi}^{-1} \left[\mathrm{Re}\,\{\hat{P}\}\right]^{-1} \dot{\Psi}^{-1} \tag{2.58}$$

Assumption **X2** of Appendix 2A ensures that $\hat{P}$ is positive definite, which in turn ensures that $\mathrm{Re}\,\{\hat{P}\}$ is positive definite, so that $[\mathrm{Re}\,\{\hat{P}\}]^{-1}$ exists. Note that $\mathrm{Re}\,\{\hat{P}\}$ is symmetric since $\hat{P}$ is Hermitian by assumption **X2** of Appendix 2A.

2.7 NOTES ON 1-D EXPRESSION FOR B_C

The result (2.57), (2.58) is quite useful in that it is applicable to a broad range of scenarios, and it makes explicit trade-offs among physical scenario parameters such as source separations, signal powers and correlations, and the sampling grid. Specifically, these quantities are represented in (2.57), (2.58) as follows:

$$\text{source separations} \iff \dot{\psi}(q_j) \cdot (\delta\omega)^{M-1}$$

$$\text{signal powers and correlations} \iff \left[\mathrm{Re}\,\{\hat{P}\}\right]^{-1}$$

$$\text{sampling grid} \iff \|\vec{\varepsilon}_M\|^2$$

Thus, for example, it is immediately clear from (2.57) for *any* one-dimensional scenario that reducing the source separation factor $\delta\omega$ by a factor of 10 in a $M = 3$ signal scenario requires that the source powers be increased by $2(M-1) \cdot 10\,\mathrm{dB} = 40\,\mathrm{dB}$ for an unbiased estimator to maintain the same direction estimate standard deviation.

Given the interpretations of the factors $\vec{\varepsilon}_M$ and $\dot{\Psi}$, we obtain the following insight on the small $\delta\omega$ behavior of the CR bound. The bound B_C will have a small norm (i.e., be favorable) if

1. the Mth order derivative of the generic arrival vector $\vec{a}(\omega)$ is large and well separated from the vector space spanned by the derivatives of lower order (i.e., $\vec{\varepsilon}_M$ has large norm); and
2. the function $\psi(q)$ has steep slope at the zero crossings $q = q_j$ (i.e., the diagonal elements of $\dot{\Psi}$ have large norm).

We note that 1 depends only on the sensor array, while 2 depends only on the source configuration.

CR Bound on $\mathrm{Var}\,\{\hat{\omega}_j\}$

The corresponding bound on the variance of $\hat{\omega}_j$, the estimate of the jth source parameter ω_j, is given by the diagonal entries of B_C. Therefore

$$\text{Var}\{\hat{\omega}_j\} \geq (B_C)_{jj} = \frac{b_{M,j}}{N}\left[\frac{\delta\omega^{-2(M-1)}}{\text{SNR}_j}\right] + \mathcal{O}(\delta\omega^{-2(M-1)+1}) \tag{2.59}$$

for small $\delta\omega$, where

$$b_{M,j} = \frac{1}{2\left[\dot{\psi}(q_j)\right]^2 \vec{\varepsilon}_M^H \vec{\varepsilon}_M}([\text{Re}\{\rho\}]^{-1})_{jj} \tag{2.60}$$

SNR $_j$ denotes the signal-to-noise ratio for the jth signal

$$\text{SNR}_j = (\hat{P})_{jj}/\sigma^2 \tag{2.61}$$

and ρ denotes the matrix of (complex) signal correlation coefficients

$$\rho = \hat{P}_D^{-1/2}\hat{P}\hat{P}_D^{-1/2} \tag{2.62}$$

$$\hat{P}_D = \text{Diag}\left[(\hat{P})_{11}, (\hat{P})_{22}, \cdots, (\hat{P})_{MM}\right] \tag{2.63}$$

In direction-finding analyses, results frequently are expressed in terms of the Array-Signal-to-Noise Ratio (ASNR) defined as

$$\text{ASNR}_j = \frac{(\hat{P})_{jj}}{\sigma^2}\|\vec{a}(\omega_j)\|^2 = \text{SNR}_j \cdot \|\vec{a}(\omega_j)\|^2 \tag{2.64}$$

Use of (2.64) in (2.59) together with the relationship

$$\|\vec{a}(\omega_j)\|^2 = \|\vec{a}(\omega_0)\|^2 + \mathcal{O}(\delta\omega) \tag{2.65}$$

gives the alternative expression

$$(B_C)_{jj} = \frac{b'_{M,j}}{N}\left[\frac{\delta\omega^{-2(M-1)}}{\text{ASNR}_j}\right] + \mathcal{O}(\delta\omega^{-2(M-1)+1}) \tag{2.66}$$

where

$$b'_{M,j} = b_{M,j}\|\vec{a}(\omega_0)\|^2 \tag{2.67}$$

The results (2.59) and (2.66) show for arbitrary M in I-D scenarios that *the CR bound B_C on Var$\{\hat{\omega}_j\}$ always is proportional to $1/(\delta\omega)^{2(M-1)}$ for small signal separations $\delta\omega$*. Clearly for $M > 1$, the bound increases rapidly as $\delta\omega \to 0$.

The result (2.59)–(2.63) can be simplified considerably by referencing the MUSIC null spectrum. Specifically (2.59), (2.60) can be rewritten

$$(B_C)_{jj} = \frac{1}{N \cdot \text{SNR}_j \cdot 2(H)_{jj}}([\text{Re}\{\rho\}]^{-1})_{jj} + \mathcal{O}(\delta\omega^{-2(M-1)+1}) \tag{2.68}$$

Use of (2.22) in (2.68) to eliminate the factor $2(H)_{jj}$ gives the alternative formulation

$$(B_C)_{jj} = \frac{1}{N \cdot \text{SNR}_j \cdot \|\vec{a}(\omega_j)\|^2 \Delta^{(2)}(\omega_j)}([\text{Re}\{\rho\}]^{-1})_{jj} + \mathcal{O}(\delta\omega^{-2(M-1)+1}) \tag{2.69}$$

The counterpart of (2.66) is

$$(B_C)_{jj} = \frac{1}{N \cdot \text{ASNR}_j \cdot \Delta^{(2)}(\omega_j)}([\text{Re}\{\rho\}]^{-1})_{jj} + \mathcal{O}(\delta\omega^{-2(M-1)+1}) \tag{2.70}$$

Thus, for 1-D scenarios with closely spaced sources the Cramér-Rao lower bound $(B_C)_{jj}$ on $\mathrm{Var}\{\hat{\omega}_j\}$ is proportional to the radius of curvature of the MUSIC null spectrum in direction ω_j, for both correlated and uncorrelated source signals.

Alternate Form of CR Bound

It is useful to recast the results to this point in terms of a resolution cell Ω defined as follows:

$$\Omega^2 = \frac{\pi^2}{3} \frac{\|\vec{a}(\omega_0)\|^2}{\|\vec{\alpha}_1\|^2} \tag{2.71}$$

where

$$\vec{\alpha}_1 \triangleq \left[I - \frac{1}{\|\vec{a}(\omega_0)\|^2} \vec{a}(\omega_0)\vec{a}(\omega_0)^H \right] \frac{d\vec{a}(\omega)}{d\omega}\bigg|_{\omega=\omega_0} \tag{2.72}$$

Note that Ω has the dimensions of the parameter(s) ω_j to be estimated. The quantity Ω is identical to the "Cramér-Rao Angle" defined by Schmidt, except for a multiplicative factor [5]. Whereas Schmidt's angle differs from the Rayleigh beamwidth by a factor approximating π, the foregoing measure closely approximates the Rayleigh beamwidth/bandwidth.

Use of (2.71) in (2.59) and (2.66) gives the following alternative expressions for the conditional model CR bounds:

$$(B_C)_{jj} = c_{M,j} \frac{\Omega^2}{N \cdot \mathrm{SNR}_j} \frac{1}{(\delta\omega/\Omega)^{2(M-1)}} + \mathcal{O}(\delta\omega^{-2(M-1)+1}) \tag{2.73}$$

$$= c'_{M,j} \frac{\Omega^2}{N \cdot \mathrm{ASNR}_j} \frac{1}{(\delta\omega/\Omega)^{2(M-1)}} + \mathcal{O}(\delta\omega^{-2(M-1)+1}) \tag{2.74}$$

where $c_{M,j}$ and $c'_{M,j}$ denote the dimensionless parameters

$$c_{M,j} = b_{M,j} \left[\frac{3\|\vec{\alpha}_1\|^2}{\pi^2 \|\vec{a}(\omega_0)\|^2} \right]^M \tag{2.75}$$

$$c'_{M,j} = c_{M,j} \|\vec{a}(\omega_0)\|^2 \tag{2.76}$$

Clearly, (2.73) and (2.74) represent generalizations of the familiar result (2.1) for estimation problems involving one signal. *The primary impact of the additional signals is to introduce the factor* $(\delta\omega/\Omega)^{2(M-1)}$ *into the denominator of the bound.* This factor accounts for the sensitivity of estimation results in the case of multiple closely spaced signals.

2.8 NUMERICAL EXAMPLE FOR 1-D

The foregoing analysis indicates for $\delta\omega/\Omega \ll 1$ that the CR bounds (2.59) approximate straight lines on a log-log plot of $(B_C)_{jj}$ versus $\delta\omega$. The following example illustrates this point.

Example 2.2

Consider the case of $M = 2, 3,$ and 4 signals having the generic signal vector

$$\vec{a}(\omega) = g\left[1, e^{jh(\omega)}, e^{j2h(\omega)}, \cdots e^{j(M-1)h(\omega)}\right]^T \tag{2.77}$$

with $W = 10$, and uncorrelated equal-power sources. Assume that the signal frequencies are equally spaced in the parameter ω, and that $\omega_1 = 0$.

Straightforward calculation shows that $\|\vec{\varepsilon}_M\|^2 / \left|g \cdot h'(\omega_0)^M\right|^2$ has the following values

	$M=1$	$M=2$	$M=3$	$M=4$
$\|\vec{\varepsilon}_M\|^2 / \left\|g \cdot h'(\omega_0)^M\right\|^2$	82.5	528	3, 089	16, 474

Use of $\|\vec{a}(\omega_0)\|^2 = 10|g|^2$, and $\|\vec{\alpha}_1\|^2 = 82.5$ in (2.71) gives $\Omega = 0.631/|h'(\omega_0)|$. The solid curves in Figures 2.2–2.4 depict the exact CR bounds (2.10) for the frequencies

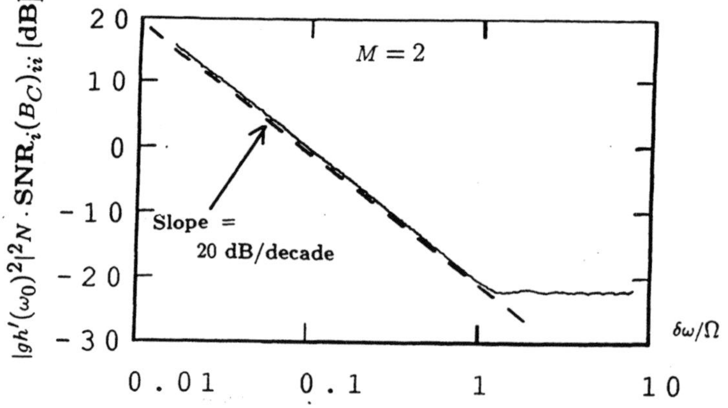

Figure 2.2 Normalized CR bound on variance of 1-D frequency estimates $\hat{\omega}_i$ for $M = 2$ uncorrelated signals having frequency separation $\delta\omega$, observed by a uniform linear array of 10 sensors.

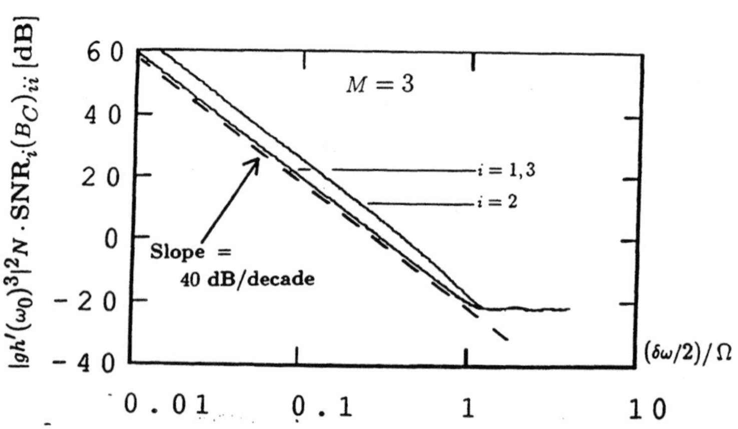

Figure 2.3 Normalized CR bound on variance of 1-D frequency estimates $\hat{\omega}_i$ for $M = 3$ uncorrelated signals having frequency separation $\delta\omega$, observed by a uniform linear array of 10 sensors.

ω_j. SNR denotes the common value of the SNR_j. The horizontal coordinate $[\delta\omega/(M-1)]/\Omega$ denotes the frequency separation between adjacent signals in units of resolution cells. The lower curve in Figures 2.3 and 2.4 applies to the outermost signals; the upper curve applies to the centermost signals. Note the increase in the vertical scale from Figure 2.2 through Figure 2.4. Figure 2.5 presents all curves for $j = 1$ on a common scale to emphasize the dependence of slope upon M.

The slanted dashed lines in Figures 2.2–2.4 depict the asymptotic bound (2.73) for the outermost signal. Clearly the straight lines capture the essence of the multiple-signal bound.

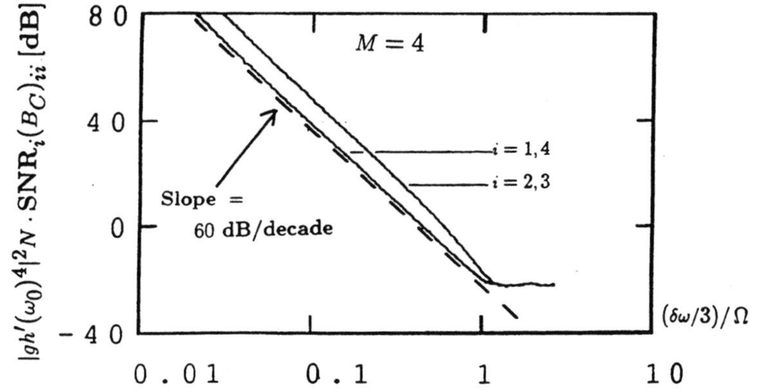

Figure 2.4 Normalized CR bound on variance of 1-D frequency estimates $\hat{\omega}_i$ for $M = 4$ uncorrelated signals having frequency separation $\delta\omega$, observed by a uniform linear array of 10 sensors.

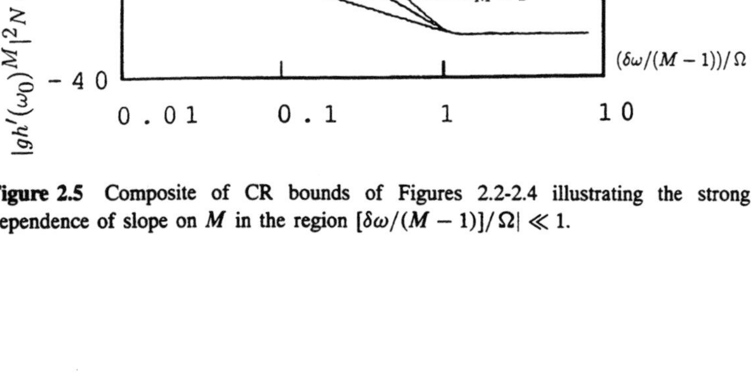

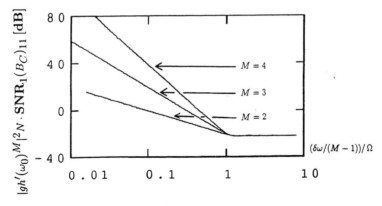

Figure 2.5 Composite of CR bounds of Figures 2.2-2.4 illustrating the strong dependence of slope on M in the region $[\delta\omega/(M-1)]/\Omega| \ll 1$.

2.9 RESOLUTION THRESHOLDS IN 1-D

An important aspect of Direction Finding (DF) algorithm performance is the ability to provide useful direction estimates for each of M closely spaced sources; if successful, the algorithm is said to have *resolved* the sources.

One useful measure of the "resolving power" of a DF algorithm is the signal-to-noise ratio (SNR) threshold $\mathcal{E}_R$ at which the algorithm can reliably resolve M sources for a given source-array configuration, and a given number N of data snapshots. An alternative performance measure is the data set size N threshold $\mathcal{N}_R$ at which the algorithm can reliably resolve the M sources for a given source-array configuration, and a given SNR. These threshold values also can be regarded respectively as the minimum SNR and N at which "one can see" each of the M sources present.

For the two-signal problem, Stoica and Nehorai argued that the MUSIC algorithm is unlikely to resolve closely spaced signals if the standard deviation of the $\hat{\omega}_j$ exceeds $\delta\omega/8$ [8]. We now apply this logic to the asymptotic CR bound (2.59) for M signals.

Thus, consider M closely spaced signals, and define $\delta\omega_j$ to denote the minimum spacing between the jth and any other source. That is

$$\delta\omega_j = \delta\omega \cdot \min_{k\neq j} \|q_j - q_k\| \tag{2.78}$$

where $q_1, \cdots q_M$ are the normalized spectral frequency offsets. If the root CR bound $\sqrt{(B_C)_{jj}}$ at the jth source is small compared to $\delta\omega_j$ for all $j = 1 \cdots M$, then there is a basis for seeking an unbiased estimator for resolving the signals. On the other hand, if one or more of the $\sqrt{(B_C)_{jj}}$ are large compared to the associated $\delta\omega_j$, then it is unlikely that there exists an unbiased estimator which can resolve the signals with high probability. Accordingly, one strongly suspects that a necessary condition for the existence of an unbiased estimator capable of resolving M closely spaced sources with high probability is that

$$\sqrt{(B_C)_{jj}} \leq f \cdot \delta\omega_j \tag{2.79}$$

$j = 1 \cdots M$, where f is a suitable fraction (e.g., $f = 1/8$).

To define the resolution thresholds, we represent the source amplitude correlation matrix $\hat{P}$ as follows:

$$\hat{P} = p\hat{P}_0 \tag{2.80}$$

where $\hat{P}_0$ is a constant matrix the eigenvalues of which sum to unity, and p is a variable scale factor. Note that representation (2.80) retains the correlations between the source powers. We define the signal SNR to be the ratio of the scale factor p to the noise power σ^2. That is

$$\text{SNR} = p/\sigma^2 \tag{2.81}$$

We deem satisfactory resolution performance to be possible whenever condition (2.79) is satisfied for all j (i.e., for all signals). We define the resolution threshold power to be the smallest value $p_{\min}$ of p for which (2.79) is satisfied for a fixed N, and define the resolution threshold SNR to be

$$\mathcal{E}_R = p_{\min}/\sigma^2 \tag{2.82}$$

Analogously, we define the data set size resolution threshold $\mathcal{N}_R$ to be the smallest value of N for which (2.79) is satisfied for a fixed power factor p.

Squaring of (2.79), substitution of (2.59), followed by rearrangement gives the equivalent condition

$$\text{SNR} \cdot N \geq K_{R,j} \delta\omega^{-2M} \tag{2.83}$$

where

$$K_{R,j} \triangleq \frac{1}{2f^2 \cdot \min_{k \neq j} \|q_j - q_k\|^2} \frac{\left(\left[\text{Re}\{\hat{P}_0\}\right]^{-1}\right)_{jj}}{\left[\psi(q_j)\right]^2 \|\vec{\varepsilon}_M\|^2} \tag{2.84}$$

$j = 1 \cdots M$.

We note that an equivalent condition for (2.83) to be satisfied for all $j = 1 \cdots M$ is

$$\text{SNR} \cdot N \geq K_R \delta\omega^{-2M} \tag{2.85}$$

where we define

$$K_R = \max_j \{K_{R,j}\} \tag{2.86}$$

It follows from (2.85) that

$$\text{SNR} \geq \mathcal{E}_R \tag{2.87}$$

where $\mathcal{E}_R$ denotes the SNR resolution threshold:

$$\mathcal{E}_R = \frac{K_R}{N} \delta\omega^{-2M} \tag{2.88}$$

It likewise follows from (2.85) that

$$N \geq \mathcal{N}_R \tag{2.89}$$

where $\mathcal{N}_R$ denotes the data set size resolution threshold:

$$\mathcal{N}_R = \frac{K_R}{\text{SNR}} \delta\omega^{-2M} \tag{2.90}$$

Equation (2.88) implies that the minimum (threshold) value $\mathcal{E}_R$ of SNR at which multiple closely spaced signals are resolvable by *any unbiased algorithm* is proportional to $(\delta\omega)^{-2M}/N$ for the conditional model, and at least proportional to $(\delta\omega)^{-2M}/N$ for the unconditional model. This (apparent) limitation is consistent with a sizable (albeit finite) body of evidence. For example, Cox has shown that resolution of two closely spaced signals using Capon's MLM technique requires SNR proportional to $(\delta\omega)^{-4}$ [18]. Simulations by Gabriel for $M = 2$ sources with spacings between 0.1 and one beamwidth for a variety of algorithms indicate that SNR proportional to $(\delta\omega)^{-3.3}$ is required to resolve the sources [19]. More recently, Kaveh and Barabell have shown for $M = 2$ uncorrelated equal-power sources, a uniform linear sampling grid, the unconditional model, and a selected resolution criterion, that the threshold SNR for MUSIC is proportional to $(\delta\omega)^{-4}/N$ for large SNR [20]. Lee and Wengrovitz have extended this result to $M = 2$ uncorrelated sources of any relative powers, to arbitrary planar sampling grids for both Conventional MUSIC, and also Beamspace MUSIC [21]. Finally, the very high threshold SNRs implied by (2.88) in the case of $M \geq 2$ sources (signals) are consistent with simulation results of Gatlin using Gabriel's Thermal Noise Algorithm [22].

Since $\hat{P}_0$ is positive definite, $\text{Re}\{\hat{P}_0\}$ also is positive definite. From the identity (2.27) for the inverse of any partitioned positive definite matrix Z, we obtain

$$([\text{Re}\{\hat{P}_0\}]^{-1})_{jj} \geq 1/(\text{Re}\{\hat{P}_0\})_{jj} = 1/(\hat{P}_0)_{jj} \\ = 1 \tag{2.91}$$

Use of (2.91) in (2.84), (2.88), and (2.90) indicates that correlation among signals typically increases (never decreases) the threshold values of SNR and N necessary to resolve signals relative to the uncorrelated case.

2.10 ASYMPTOTIC CR BOUND FOR MULTI-D

We next analyze the CR bound (2.16) on directional covariance for M closely spaced sources in a $\mathfrak{D}$-dimensional space. The analysis is performed subject to assumptions presented in Appendix 2B. These conditions typically (but not always) are satisfied in direction-finding problems.

Explicit expressions for the multi-D CR bound are identified by extending the approach used above for 1-D CR bound.

Taylor Series for Matrices A and $\dot{D}(\vec{\omega})$

In the multi-D case, the generic spatial frequency parameter $\vec{\omega}$ is

$$\vec{\omega} = \vec{\omega}_0 + \delta\omega\,\vec{q} \tag{2.92}$$

where

$$\vec{q} = [q_1, \cdots q_{\mathfrak{D}}]^T \tag{2.93}$$

is a *normalized $\mathfrak{D}$-dimensional generic spatial frequency vector*. The generic arrival vector $\vec{a}(\vec{\omega})$ has a Taylor series about $\delta\omega = 0$ of the form

$$\vec{a}(\vec{\omega}) = \sum_{p=0}^{\infty} \delta\omega^p \dot{A}_p \vec{\gamma}_p(\vec{q}) \qquad (W \times 1) \tag{2.94}$$

where the columns of $\dot{A}_p$ are the pth order spatial derivatives of $\vec{a}(\vec{\omega})$ at $\vec{\omega}_0$,

$$\dot{A}_p \triangleq \left[\frac{\partial^p \vec{a}(\vec{\omega})}{\partial\omega_1^p}, \frac{\partial^p \vec{a}(\vec{\omega})}{\partial\omega_1^{p-1}\partial\omega_2}, \cdots, \frac{\partial^p \vec{a}(\vec{\omega})}{\partial\omega_{\mathfrak{D}}^p} \right]_{\vec{\omega}=\vec{\omega}_0} \qquad (W \times \bar{n}_p) \tag{2.95}$$

where $\bar{n}_p$ is the number of pth order spatial derivatives. Note that the number of columns of $\dot{A}_p$ *increases* with p. As we will see, this circumstance complicates analysis of the multi-D CR bound. For 2-D applications

$$\bar{n}_p = p + 1 \tag{2.96}$$

(i.e., $\bar{n}_0 = 1, \bar{n}_1 = 2$, etc.). Vector $\vec{\gamma}_p(\vec{q})$ is $\bar{n}_p \times 1$, real, constant with $\delta\omega$, and depends only on the normalized spatial frequency offset vector $\vec{q}$. For 2-D applications with $\vec{q} = \left[q_x, q_y\right]^T$, the vectors $\vec{\gamma}_p(\vec{q})$ are

$$\vec{\gamma}_0(\vec{q}) = [1], \vec{\gamma}_1(\vec{q}) = \begin{bmatrix} q_x \\ q_y \end{bmatrix}, \vec{\gamma}_2(\vec{q}) = \begin{bmatrix} q_x^2/2 \\ q_x q_y \\ q_y^2/2 \end{bmatrix}, \vec{\gamma}_3(\vec{q}) = \begin{bmatrix} q_x^3/6 \\ q_x^2 q_y/2 \\ q_x q_y^2/2 \\ q_y^3/6 \end{bmatrix} \tag{2.97}$$

The general expression for $\vec{\gamma}_p(\vec{q})$ for 2-D scenarios is

$$\vec{\gamma}_p(\vec{q}) \triangleq \left[c_{0,p} q_x^p, c_{1,p} q_x^{p-1} q_y^1, \cdots, c_{p,p} q_y^p\right]^T \tag{2.98}$$

where

$$c_{i,p} \triangleq \binom{p}{i} \frac{1}{p!} \tag{2.99}$$

$i = 0, \cdots p$, and the first factor of (2.99) is the binomial coefficient.

Expressions analogous to (2.95), (2.98) can be written for Taylor series of any dimensionality.

It follows from (2.94) that matrix A in (2.7) has Taylor series of the form

$$\begin{aligned} A &= \left[\vec{a}(\vec{\omega}_1), \cdots \vec{a}(\vec{\omega}_M)\right] \\ &= \sum_{p=0}^{\infty} \delta\omega^p \dot{A}_p \Gamma_p \end{aligned} \tag{2.100}$$

where matrix $\dot{A}_p$ is as in (2.95), and Γ_p is a constant real $\bar{n}_p \times M$ matrix of the form

$$\Gamma_p \triangleq \left[\vec{\gamma}_p(\vec{q}_1), \cdots \vec{\gamma}_p(\vec{q}_M)\right] \tag{2.101}$$

As defined in (2.19), each column of $\dot{D}(\vec{\omega}_j)$ is a partial derivative of the generic arrival vector $\vec{a}(\vec{\omega})$ at $\vec{\omega}_j$ of the form

$$\left[\frac{\partial}{\partial \omega_i} \vec{a}(\vec{\omega})\right]_{\vec{\omega}=\vec{\omega}_j} = \frac{1}{\delta\omega} \left[\frac{\partial}{\partial q_i} \vec{a}(\vec{\omega}_0 + \delta\omega\vec{q})\right]_{\vec{q}=\vec{q}_i} \tag{2.102}$$

Hence

$$\dot{D}(\vec{\omega}_j) = \delta\omega^{-1} \sum_{p=0}^{\infty} \delta\omega^p \dot{A}_p \dot{\Gamma}_p(\vec{q}_j) \tag{2.103}$$

where

$$\dot{\Gamma}_p(\vec{q}_j) \triangleq \left[\frac{\partial \vec{\gamma}_p(\vec{q})}{\partial q_1}, \frac{\partial \vec{\gamma}_p(\vec{q})}{\partial q_2}, \cdots, \frac{\partial \vec{\gamma}_p(\vec{q})}{\partial q_{\mathcal{D}}}\right]_{\vec{q}=\vec{q}_j} \tag{2.104}$$

and $\dot{A}_p, \vec{\gamma}_p(\vec{q})$ are as in (2.95), (2.97). Note that $\dot{\Gamma}_0(\vec{q}_j) = 0$ since $\vec{\gamma}_0(\vec{q}) = 1$, and that the $\dot{A}_p$ and $\dot{\Gamma}_p(\vec{q}_j)$ are constant with $\delta\omega$, so that (2.103) is a Taylor series in $\delta\omega$.

Behavior of $\left[I - A(A^H A)^{-1} A^H\right]$ for Small $\delta\omega$

As in the 1-D case, the essential properties of H in (2.17), and therefore of B_C, for $\delta\omega \to 0$ derive from the *interaction* of the factors $\left[I - A(A^H A)^{-1} A^H\right]$ and $\dot{D}(\vec{\omega}_j)$. Accordingly, it is desirable to identify the small $\delta\omega$ behavior of $\left[I - A(A^H A)^{-1} A^H\right]$, or equivalently of $A(A^H A)^{-1} A^H$.

The limiting behavior of $A(A^H A)^{-1} A^H$ as $\delta\omega \to 0$ is analyzed in section 9.1 of [23] for the multi-D case. It is shown there that the dominant term of $A(A^H A)^{-1} A^H$ is determined by the first partial sum of the Taylor series (2.100) for A which has full rank. This circumstance parallels the situation for 1-D. However, the detailed multi-D result is more complicated since the ranks of successive terms of (2.100) are not constant at unity, but rather increase with term order.

We designate the first partial sum of the Taylor series (2.100) with full rank as $\tilde{A}$. Matrix $\tilde{A}$ has the form

$$\begin{aligned}\tilde{A} &= \sum_{p=0}^{r} \delta\omega^p \dot{A}_p \Gamma_p \\ &= \dot{A} \nabla \Gamma\end{aligned} \tag{2.105}$$

where

$$\dot{A} \triangleq \left[\dot{A}_0, \dot{A}_1, \cdots, \dot{A}_r\right], \Gamma \triangleq \begin{bmatrix} \Gamma_0 \\ \vdots \\ \Gamma_{r-1} \\ \Gamma_r \end{bmatrix} \tag{2.106}$$

and ∇ is a diagonal matrix with diagonal elements equal to powers of $\delta\omega$. By definition

$$\text{Rank}\,\{\tilde{A}\} = M \tag{2.107}$$

But from (2.105)

$$\begin{aligned}\text{Rank}\,\{\tilde{A}\} &= \text{Rank}\,\{\dot{A} \nabla \Gamma\} \\ &= \text{Rank}\,\{\dot{A}\Gamma\}\end{aligned} \tag{2.108}$$

since ∇ is nonsingular for $\delta\omega \neq 0$. Therefore, r is the smallest integer for which

$$\text{Rank}\,\{\dot{A}\Gamma\} = M \tag{2.109}$$

To further articulate the value of r, it is convenient to make the following:

Definition of $\bar{n}_{\{0\cdots k\}}$. Integer $\bar{n}_{\{0\cdots k\}}$ is the number of columns in the aggregate matrix $\left[\dot{A}_0, \dot{A}_1, \cdots \dot{A}_k\right]$. That is

$$\bar{n}_{\{0\cdots k\}} \triangleq \sum_{p=0}^{k} \bar{n}_p \tag{2.110}$$

To facilitate analysis of B_C for the multi-D scenarios, we make assumptions which include the following.

C1. The $W \times \bar{n}_{\{0\cdots r-1\}}$ submatrix $\left[\dot{A}_0, \dot{A}_1, \cdots \dot{A}_{r-1}\right]$ has full column rank.

C2. The $\bar{n}_{\{0\cdots r-1\}} \times M$ submatrix $\left[\Gamma_0^H, \Gamma_1^H, \cdots \Gamma_{r-1}^H\right]^H$ has full row rank.

(See Conditions **C1** and **C2** of Appendix 2B.)

Given satisfaction of the foregoing conditions, it is evident from (2.109) that r is the integer which satisfies the double inequality

$$\bar{n}_{\{0\cdots r-1\}} < M \leq \bar{n}_{\{0\cdots r\}} \tag{2.111}$$

Analysis in [23] for scenarios that satisfy the assumptions of Appendix 2B shows that

$$A(A^H A)^{-1} A^H = \bar{A}\bar{A}^+ + \mathcal{O}(\delta\omega) \tag{2.112}$$

where

$$\bar{A} \triangleq \left[\dot{A}_0, \cdots, \dot{A}_{r-1}, \dot{A}_r T_r\right] \qquad W \times \bar{n}_{\{0\cdots r\}} \tag{2.113}$$

$$T_r \triangleq \Gamma_r P_{[\Gamma_0^H,\cdots,\Gamma_{r-1}^H]} \left(\Gamma_r P_{[\Gamma_0^H,\cdots,\Gamma_{r-1}^H]}\right)^+ \qquad \bar{n}_r \times \bar{n}_r \tag{2.114}$$

$$P_{[\Gamma_0^H,\cdots,\Gamma_{r-1}^H]} \triangleq I - \left[\Gamma_0^H, \cdots, \Gamma_{r-1}^H\right]\left[\Gamma_0^H, \cdots, \Gamma_{r-1}^H\right]^+ \qquad M \times M \tag{2.115}$$

Equation (2.112) generalizes (2.44) to multi-D scenarios that satisfy the assumptions of Appendix 2B. Equation (2.112) asserts that the limiting column space of A as $\delta\omega \to 0$ is contained in that of $\dot{A}$, and is identical to that of $\bar{A}$. From (2.112), we have

$$\left[I - A(A^H A)^{-1} A^H\right] = \left[I - \bar{A}\bar{A}^+\right] + \mathcal{O}(\delta\omega) \tag{2.116}$$

Subject to condition **C1**, all the column spaces of submatrices $\dot{A}_0, \cdots \dot{A}_{r-1}$ of $\bar{A}$ are linearly independent, and make a rank contribution of $\bar{n}_{\{0,\cdots r-1\}}$ to (2.113). The final matrix $\dot{A}_r T_r$ contributes additional independent columns to complete the column space of $\bar{A}$. The role of the post-factor T_r of $\dot{A}_r$ in (2.113) is to select a subspace of that defined by the columns of $\dot{A}_r$ to produce the result

$$\text{Rank}\{\bar{A}\} = \text{Rank}\{A\} = M \tag{2.117}$$

Subject to conditions **C1–C3** of Appendix 2B

$$\text{Rank}\{T_r\} = M - \bar{n}_{\{0,\cdots r-1\}} \tag{2.118}$$

(See [23].)

From (2.114) it is clear that T_r is defined by $\vec{q}_1, \cdots, \vec{q}_M$ and, therefore, by the relative geometry of the sources. We note specifically that projection T_r is independent of $\delta\omega$ or of the sensor array geometry.

Note that in the special case where the matrix $\bar{A}$ has

$$\bar{n}_{\{0,\cdots r\}} = M \tag{2.119}$$

we have from (2.118)

$$\begin{aligned} \text{Rank}\{T_r\} &= \bar{n}_{\{0,\cdots r\}} - \bar{n}_{\{0,\cdots r-1\}} \\ &= \bar{n}_r \end{aligned} \tag{2.120}$$

That is, projection matrix T_r has full rank so that

$$T_r = I \tag{2.121}$$

and

$$\bar{A} = \dot{A} \tag{2.122}$$

Distinct Cases

Reference to (2.45) shows that for 1-D scenarios the number of columns of $\dot{A}$ equals the number M of sources. By contrast, it is evident from (2.106) and (2.111) that the number $\bar{n}_{\{0,\cdots r\}}$ of columns of $\dot{A}$ for multi-D scenarios equals or *exceeds* M. This is a consequence of the fact that successive partial derivative matrices $\dot{A}_0, \dot{A}_1, \cdots \dot{A}_r$ have increasing numbers of columns. As we will see, this circumstance introduces a new ingredient into the multi-D problem. Specifically, subsequent analysis shows that the CR bound B_C depends on $\delta\omega$ in three fundamentally different ways according to the difference

$$\nu \triangleq \bar{n}_{\{0,\cdots r\}} - M \tag{2.123}$$

which equals the number of "extra columns" in $\dot{A}$. Distinct behaviors are exhibited in the following three cases:

Case I. No "extra columns"; that is

$$\nu = 0 \tag{2.124}$$

Case II. $\mathcal{D}$ or more "extra columns"; that is

$$\nu \geq \mathcal{D} \tag{2.125}$$

Case III. Some, but fewer than $\mathcal{D}$ "extra columns"; that is

$$0 < \nu < \mathcal{D} \tag{2.126}$$

Behavior of $\left[I - A(A^H A)^{-1} A^H\right] \dot{D}(\vec{\omega}_j)$ for Small $\delta\omega$

Case I Behavior of $\left[I - A(A^H A)^{-1} A^H\right] \dot{D}(\vec{\omega}_j)$ for Small $\delta\omega$. Paralleling the 1-D case analysis outlined in section 2.6, the matrix $\dot{D}(\vec{\omega}_j)$ of partial first derivatives is expressed as a linear transformation of the columns of A plus a remainder matrix F_j

$$\dot{D}(\vec{\omega}_j) = \delta\omega^{-1}\left[AB + F_j\right] \qquad (W \times \mathcal{D}) \tag{2.127}$$

where $M \times \mathcal{D}$ matrix B contains $\mathcal{D}$ weight vectors, and

$$F_j = \mathcal{O}(\delta\omega^{\mu}) \qquad (W \times \mathcal{D}) \tag{2.128}$$

The weight vectors contained in B are selected to maximize μ. Once again, this is accomplished by selecting B to zero out as many leading terms as possible in the Taylor series for F_j. Only in Case I where $\dot{A}$ has $\nu = 0$ excess columns does this process uniquely determine B and F_j. In Cases II and III where $\dot{A}$ has $\nu > 0$ extra columns, an infinity of choices for B and F_j exist which maximize μ.

In Case I, the resulting values for μ and F_j are

$$\mu = r + 1 \tag{2.129}$$

$$F_j = \sum_{p=r+1}^{\infty} \delta\omega^{p} \dot{A}_p \dot{\Psi}_p(\vec{q}_j) \tag{2.130}$$

where

$$\dot{\Psi}_p(\vec{q}_j) \triangleq \left[\dot{\Gamma}_p(\vec{q}_j), \Gamma_p\right]\begin{bmatrix} I \\ -\Gamma^{+}\dot{\Gamma}(\vec{q}_j) \end{bmatrix} \tag{2.131}$$

$$\Gamma \triangleq \begin{bmatrix} \Gamma_0 \\ \vdots \\ \Gamma_{r-1} \\ \Gamma_r \end{bmatrix} \qquad \bar{n}_{\{0,\cdots r\}} \times M \tag{2.132}$$

$$\dot{\Gamma}(\vec{q}) \triangleq \begin{bmatrix} \dot{\Gamma}_0(\vec{q}) \\ \vdots \\ \dot{\Gamma}_{r-1}(\vec{q}) \\ \dot{\Gamma}_r(\vec{q}) \end{bmatrix} \qquad \bar{n}_{\{0,\cdots r\}} \times \mathcal{D} \tag{2.133}$$

See [23] for details.

To interpret $\dot{\Psi}_p(\vec{q}_j)$, we express (2.131) as

$$\begin{aligned} \dot{\Psi}_p(\vec{q}_j) &= \dot{\Gamma}_p(\vec{q}_j) - \Gamma_p\Gamma^{+}\dot{\Gamma}(\vec{q}_j) \\ &= \left[\frac{\partial\vec{\psi}_p(\vec{q})}{\partial q_1}, \frac{\partial\vec{\psi}_p(\vec{q})}{\partial q_2}, \ldots, \frac{\partial\vec{\psi}_p(\vec{q})}{\partial q_{\mathcal{D}}}\right]_{\vec{q}=\vec{q}_j} \end{aligned} \tag{2.134}$$

where $\vec{\psi}_p(\vec{q})$ denotes the vector polynomial in $\vec{q}$:

$$\vec{\psi}_p(\vec{q}) \triangleq \left[\vec{\gamma}_p(\vec{q}) - \Gamma_p\Gamma^{+}\vec{\gamma}(\vec{q})\right] \tag{2.135}$$

and vector $\vec{\gamma}(\vec{q})$ is defined as

$$\vec{\gamma}(\vec{q}) \triangleq \begin{bmatrix} \vec{\gamma}_0(\vec{q}) \\ \vdots \\ \vec{\gamma}_{r-1}(\vec{q}) \\ \vec{\gamma}_r(\vec{q}) \end{bmatrix} \tag{2.136}$$

The vector polynomial $\vec{\psi}_p(\vec{q})$ has zeros at each of the normalized source directions $\vec{q} = \vec{q}_j,\ j = 1 \cdots M$, since $\vec{\gamma}(\vec{q}_j)$ is the jth column of Γ, and matrix Γ has full column rank M subject to Conditions **C1–C3**, so that

$$\Gamma^+ \vec{\gamma}(q_j) = \vec{u}_j \tag{2.137}$$

where $\vec{u}_j$ is a vector of zeros except for element j which is 1.

Substitution of (2.127), (2.130) and (2.116), (2.122) in the product $[I - A(A^H A)^{-1} A^H] \dot{D}(\vec{\omega}_j)$ gives

$$\left[I - A(A^H A)^{-1} A^H\right] \dot{D}(\vec{\omega}_j) = \delta\omega^r \mathcal{E}_{r+1} \dot{\Psi}_{r+1}(\vec{q}_j) + \mathcal{O}(\delta\omega^{r+1}) \tag{2.138}$$

where

$$\mathcal{E}_{r+1} = \left[I - \dot{A}\dot{A}^+\right] \dot{A}_{r+1} \tag{2.139}$$

Note that matrix $\mathcal{E}_{r+1}$ consists of the $(r+1)^{th}$ order spatial derivatives of $\vec{a}(\vec{\omega})$ at $\vec{\omega}_0$, less their projection onto the column space spanned by all the spatial derivatives of lower order. Matrix $\mathcal{E}_{r+1}$ specializes for 1-D to the vector $\vec{\varepsilon}_M$ of (2.53).

Subject to conditions **C1–C3** and **CR1–CR3** of Appendix 2B, it can be shown that matrix $\mathcal{E}_{r+1} \dot{\Psi}_{r+1}(\vec{q}_j)$ in the leading term of (2.138) has full rank $(= \mathcal{D})$. See [23].

Cases II and III Behavior of $\left[I - A(A^H A)^{-1} A^H\right] \dot{D}(\vec{\omega}_j)$ for Small $\delta\omega$. In Cases II and III, selection of B in (2.127), (2.128) to minimize μ produces the results

$$\mu = r \tag{2.140}$$

$$F_j = \delta\omega^r \dot{A}_r C_j + \mathcal{O}(\delta\omega^{r+1}) \tag{2.141}$$

with an infinity of possible choices for matrix C_j. Therefore, the remainder term F_j is not uniquely defined. An additional criterion is needed to usefully specify F_j.

The rationale for developing an expression of the form (2.127) is to concentrate in the term $\delta\omega^{-1} AB$ as much of the submatrix $\dot{D}(\vec{\omega}_j)$ as practicable which will be annihilated when $\dot{D}(\vec{\omega}_j)$ is premultiplied by $[I - A(A^H A)^{-1} A^H]$ in the product

$$\left[I - A(A^H A)^{-1} A^H\right] \dot{D}(\vec{\omega}_j) = \delta\omega^{-1} \left[I - A(A^H A)^{-1} A^H\right] F_j \tag{2.142}$$

The key to further useful specialization of matrix B in (2.127) for Cases II and III is provided by the result (2.112). Note that as $\delta\omega \to 0$ the limiting value of the prefactor $[I - A(A^H A)^{-1} A^H]$ eliminates all vector components of F_j in the column space of $\dot{A}$. Specifically, the limiting prefactor removes all vector components in the column space of $\dot{A}_r T_r$. Accordingly, to determine B and F_j in (2.127), we require not only that

- B maximize μ, which produces (2.140) and (2.141), but also that
- B be such that the leading term of (2.141) has columns that are confined to the column space of $\dot{A}_r\,[I - T_r]$.

The resulting value for F_j is

$$F_j = \sum_{p=r}^{\infty} \delta\omega^p \dot{A}_p \dot{\Psi}_p(\vec{q}_j) \tag{2.143}$$

where

$$\dot{\Psi}_p(\vec{q}_j) \triangleq \left[\dot{\Gamma}_p(\vec{q}_j), \Gamma_p\right] \begin{bmatrix} I \\ -\Gamma^+\dot{\Gamma}(\vec{q}_j) \end{bmatrix} \tag{2.144}$$

$$\Gamma \triangleq \begin{bmatrix} \Gamma_0 \\ \vdots \\ \Gamma_{r-1} \\ T_r\Gamma_r \end{bmatrix} \qquad \bar{n}_{\{0,\cdots r\}} \times M \tag{2.145}$$

$$\dot{\Gamma}(\vec{q}) \triangleq \begin{bmatrix} \dot{\Gamma}_0(\vec{q}) \\ \vdots \\ \dot{\Gamma}_{r-1}(\vec{q}) \\ T_r\dot{\Gamma}_r(\vec{q}) \end{bmatrix} \qquad \bar{n}_{\{0,\cdots r\}} \times \mathcal{D} \tag{2.146}$$

See [23] for details.

As in Case I, to interpret $\dot{\Psi}_p(\vec{q}_j)$, we express (2.144) as

$$\begin{aligned} \dot{\Psi}_p(\vec{q}_j) &= \dot{\Gamma}_p(\vec{q}_j) - \Gamma_p\Gamma^+\dot{\Gamma}(\vec{q}_j) \\ &= \left[\frac{\partial\vec{\psi}_p(\vec{q})}{\partial q_1}, \frac{\partial\vec{\psi}_p(\vec{q})}{\partial q_2}, \ldots, \frac{\partial\vec{\psi}_p(\vec{q})}{\partial q_{\mathcal{D}}}\right]_{\vec{q}=\vec{q}_j} \end{aligned} \tag{2.147}$$

where $\vec{\psi}_p(\vec{q})$ denotes the vector polynomial

$$\vec{\psi}_p(\vec{q}) \triangleq \left[\vec{\gamma}_p(\vec{q}) - \Gamma_p\Gamma^+\vec{\gamma}(\vec{q})\right] \tag{2.148}$$

with vector $\vec{\gamma}(\vec{q})$ now defined as

$$\vec{\gamma}(\vec{q}) \triangleq \begin{bmatrix} \vec{\gamma}_0(\vec{q}) \\ \vdots \\ \vec{\gamma}_{r-1}(\vec{q}) \\ T_r\vec{\gamma}_r(\vec{q}) \end{bmatrix} \tag{2.149}$$

As in Case I, the vector polynomial $\vec{\psi}_p(\vec{q})$ has zeros at each of the normalized source directions $\vec{q} = \vec{q}_j$, $j = 1 \cdots M$.

Substitution of (2.143) in (2.142) and use of (2.116) gives

$$\left[I - A(A^H A)^{-1}A^H\right]\dot{D}(\vec{\omega}_j) = \delta\omega^{r-1}\mathcal{E}_r\dot{\Psi}_r(\vec{q}_j) + \mathcal{O}(\delta\omega^r) \tag{2.150}$$

where

$$\mathcal{E}_r = \left[I - \bar{A}\bar{A}^+\right]\dot{A}_r \tag{2.151}$$

Matrix $\mathcal{E}_r$ consists of the rth order spatial derivatives of $\vec{a}(\vec{\omega})$ at $\vec{\omega}_0$, less their projection onto the limiting column space of A as $\delta\omega \to 0$.

Subject to conditions **C1–C3** and **CR1–CR3** of Appendix 2B, it can be shown that matrices in (2.150) have rank properties as follows:

Case II. Matrix $\mathcal{E}_r\dot{\Psi}_r(\vec{q}_j)$ has full rank $\mathcal{D}$,

Case III. Matrix $\delta\omega^{r-1}\mathcal{E}_r\dot{\Psi}_r(\vec{q}_j)$ has incomplete rank $v < \mathcal{D}$.

Thus in Case II the leading term of (2.150) is a full rank representation of $[I - A(A^HA)^{-1}A^H]\dot{D}(\vec{\omega}_j)$ valid for small $\delta\omega$. By contrast, in Case III, the leading term of (2.150) is only a partial rank representation of $[I - A(A^HA)^{-1}A^H]\dot{D}(\vec{\omega}_j)$; in order to obtain a full rank representation, the remainder term $\mathcal{O}(\delta\omega^r)$ must be further detailed.

To identify the full rank representation of the dominant behavior of the product $[I - A(A^HA)^{-1}A^H]\dot{D}(\vec{\omega}_j)$ for Case III, we express

$$\left[I - A(A^HA)^{-1}A^H\right]\dot{D}(\vec{\omega}_j) = \delta\omega^{r-1}\tilde{Z}_r(\vec{q}_j) + \delta\omega^r\tilde{Z}_{r+1}(\vec{q}_j) \tag{2.152}$$

where $\tilde{Z}_r(\vec{q}_j)$ and $\tilde{Z}_{r+1}(\vec{q}_j)$ by construction have orthogonal columns

$$\tilde{Z}_r(\vec{q}_j)^H\tilde{Z}_{r+1}(\vec{q}_j) = 0 \tag{2.153}$$

and for small $\delta\omega$ have the form

$$\begin{aligned}\tilde{Z}_r(\vec{q}_j) &= \mathcal{E}_r\dot{\Psi}_r(\vec{q}_j) + \mathcal{O}(\delta\omega)\\ \tilde{Z}_{r+1}(\vec{q}_j) &= \mathcal{E}_{r+1}\dot{\Psi}_{r+1}(\vec{q}_j) + \mathcal{O}(\delta\omega)\end{aligned} \tag{2.154}$$

See [23] for details.

Behavior of H for Small $\delta\omega$. Use of (2.138), (2.150), and (2.152) in (2.17), and use of the orthogonal property (2.153) for Case III, gives

$$H = \begin{cases}\delta\omega^{2r}H_{2(r+1)} + \mathcal{O}(\delta\omega^{2r+1}) & \text{Case I}\\ \delta\omega^{2(r-1)}H_{2r} + \mathcal{O}(\delta\omega^{2(r-1)+1}) & \text{Case II}\\ \delta\omega^{2(r-1)}H_{2r}(\delta\omega) + \delta\omega^{2r}H_{2(r+1)}(\delta\omega) & \text{Case III}\end{cases} \tag{2.155}$$

where

$$\begin{aligned}H_{2p}(\delta\omega) &= \left[\tilde{Z}_p(\vec{q}_1), \cdots \tilde{Z}_p(\vec{q}_M)\right]^H\left[\tilde{Z}_p(\vec{q}_1), \cdots \tilde{Z}_p(\vec{q}_M)\right]\\ &= H_{2p} + \mathcal{O}(\delta\omega)\end{aligned} \tag{2.156}$$

$$H_{2p} = \dot{\Psi}_p^H(1_{M\times M}\otimes\mathcal{E}_p^H\mathcal{E}_p)\dot{\Psi}_p \tag{2.157}$$

$$\dot{\Psi}_p = \text{Block Diag}\,\{\dot{\Psi}_p(\vec{q}_1), \dot{\Psi}_p(\vec{q}_2), \cdots \dot{\Psi}_p(\vec{q}_M)\} \tag{2.158}$$

for $p = r, r + 1$, where $\otimes$ denotes Kronecker product [as in (2.20)].

The block diagonal matrix $\dot{\Psi}_p$ generalizes to multi-D the diagonal matrix $\dot{\Psi}$ in (2.54) used in the 1-D small $\delta\omega$ CR bound expressions.

CR Bound B_C for Small $\delta\omega$

Substitution of Eq. (2.155) into (2.16) gives

$$B_C = \begin{cases} \frac{\sigma^2}{2N}\left[\delta\omega^{2r}\mathrm{Re}\{H_{2(r+1)} \odot \hat{P}_+^T\} + \mathcal{O}(\delta\omega^{2r+1})\right]^{-1} & \text{Case I} \\ \frac{\sigma^2}{2N}\left[\delta\omega^{2(r-1)}\mathrm{Re}\{H_{2r} \odot \hat{P}_+^T\} + \mathcal{O}(\delta\omega^{2(r-1)+1})\right]^{-1} & \text{Case II} \\ \frac{\sigma^2}{2N}\left[\delta\omega^{2(r-1)}\mathrm{Re}\{H_{2r}(\delta\omega) \odot \hat{P}_+^t\} \right. & \\ \qquad \left. + \delta\omega^{2r}\mathrm{Re}\{H_{2(r+1)}(\delta\omega) \odot \hat{P}_+^T\}\right]^{-1} & \text{Case III} \end{cases} \quad (2.159)$$

It can be shown [23] that the constant matrices in (2.159) have rank properties as follows:

Case I. The leading term $\mathrm{Re}\{H_{2(r+1)} \odot \hat{P}_+^T\}$ has full rank $M\mathcal{D}$.

Case II. The leading term $\mathrm{Re}\{H_{2r} \odot \hat{P}_+^T\}$ has full rank $M\mathcal{D}$.

Case III. The leading term $\mathrm{Re}\{H_{2r}(\delta\omega) \odot \hat{P}_+^T\}$ has incomplete rank $M\nu$; however, the two terms together have full rank $M\mathcal{D}$.

To identify the dominant term of B_C in (2.159), we apply well-known matrix inverse properties, to obtain the general result

$$B_C = \delta\omega^{-2(\chi-1)}K_\chi + \mathcal{O}(\delta\omega^{-2(\chi-1)+1}) \quad (2.160)$$

where parameter χ determines the order of the dominant term of B_C, and satisfies

$$\chi = \begin{cases} r+1 & \text{Case I} \\ r & \text{Case II} \\ r+1 & \text{Case III} \end{cases} \quad (2.161)$$

For 1-D scenarios, $\chi = M$. Importantly for Case III, since the leading term of (2.159) is not full rank, the order of the dominant term of B_C is determined by the second term in (2.159).

The constant matrix K_χ is of the form

$$K_\chi = \begin{cases} \frac{\sigma^2}{2N}\left[\dot{\Psi}_{r+1}^T\mathrm{Re}\{\hat{P}^T \otimes \mathcal{E}_{r+1}^H\mathcal{E}_{r+1}\}\dot{\Psi}_{r+1}\right]^{-1} & \text{Case I} \\ \frac{\sigma^2}{2N}\left[\dot{\Psi}_r^T\mathrm{Re}\{\hat{P}^T \otimes \mathcal{E}_r^H\mathcal{E}_r\}\dot{\Psi}_r\right]^{-1} & \text{Case II} \\ \frac{\sigma^2}{2N}\left[P_{[\dot{\Psi}_r^T]}\dot{\Psi}_{r+1}^T\mathrm{Re}\{\hat{P}^T \otimes \mathcal{E}_{r+1}^H\mathcal{E}_{r+1}\}\dot{\Psi}_{r+1}P_{[\dot{\Psi}_r^T]}\right]^{+} & \text{Case III} \end{cases} \quad (2.162)$$

where use has been made of (2.155)–(2.157), of the real and block diagonal property of $\dot{\Psi}_p$, and the block constant property of P_+.

2.11 NOTES ON MULTI-D EXPRESSION FOR B_C

It is shown in Appendix 2C for multi-D scenarios that

$$\chi < M \text{ for } M > 2 \tag{2.163}$$

and

$$\chi = M \text{ for } M = 2 \tag{2.164}$$

Comparison of B_C expressions (2.160) and (2.57) and reference to (2.163) shows that *for $M \geq 3$, B_C depends much less strongly on the source separation factor $\delta\omega$ in multi-D scenarios than in 1-D scenarios.* This circumstance indicates that accurate direction estimation for $M \geq 3$ closely spaced sources is more promising in multi-D scenarios than in 1-D scenarios. Comparison of (2.160) and (2.57) and reference to (2.164) shows that for $M = 2$ sources B_C has the same $\delta\omega$ dependence in both multi-D and 1-D scenarios. The reason for this circumstance is discussed in section 2.14.

Table 2.1 presents example χ values for $\mathcal{D} = 1, 2,$ and 3. The table entries show that inequality (2.163) tends to become stronger with increasing M or $\mathcal{D}$.

TABLE 2.1 χ values for $\mathcal{D} = 1, 2, 3$

$M =$	1	2	3	4	5	6	7	8
$\mathcal{D} = 1$	1	2	3	4	5	6	7	8
$\mathcal{D} = 2$	1	2	2	2	3	3	3	3
$\mathcal{D} = 3$	1	2	2	2	2	2	2	3

The result (2.160)–(2.162) is quite useful for multi-D applications in that it makes explicit trade-offs among scenario parameters such as source separations, signal powers and correlations, and the sampling grid. Specifically, these quantities are represented in (2.160)–(2.162) as follows:

$$\text{source separations} \Longleftrightarrow \dot{\Psi}_\chi(q_j), \ (\delta\omega)^{-2(\chi-1)}$$

$$\text{signal powers and correlations} \Longleftrightarrow \hat{P}$$

$$\text{sampling grid} \Longleftrightarrow \mathcal{E}_\chi$$

Thus, for example, it is immediately clear from (2.160) for a two-dimensional scenario that reducing the frequency separation factor $\delta\omega$ by a factor of 10 in an $M = 3, (\chi = 2)$ signal scenario requires that the source powers be increased by $2(\chi - 1) \cdot 10\,\text{dB} = 20\,\text{dB}$ for an unbiased estimator to maintain the same frequency standard deviation. Note that the same conditions require a 40 dB SNR increase in 1-D scenarios.

Given the interpretations of the factors $\mathcal{E}_p$ and $\dot{\Psi}_p$, we obtain the following insight on the small $\delta\omega$ behavior of the multi-D CR bound. The bound B_C will have a small norm (i.e., be favorable) if

1. the χth order partial derivatives of the generic arrival vector $\vec{a}(\vec{\omega})$ are large and well-separated from the vector space spanned by the partial derivatives of lower order (i.e., $\mathcal{E}_\chi$ has large norm), and
2. the scalar functions that make up the vector function $\vec{\psi}_\chi(\vec{q})$ have steep slope at the zero crossings $\vec{q} = \vec{q}_j$ (i.e., $\dot{\Psi}_\chi(\vec{q}_j)$ has large norm).

We note that in multi-D, as in 1-D, condition 1 depends only on the sensor array, while condition 2 depends only on the source configuration.

CR Bound on $\text{Var}\{\hat{\omega}_{ij}\}$

The corresponding bound on the variance of $\hat{\omega}_{ij}$, the estimate of the ith component of the jth source parameter vector $\vec{\omega}_j$, is given by the diagonal entries of B_C. Therefore

$$\text{Var}\{\hat{\omega}_{ij}\} \geq (B_C)_{ll} = \frac{b_{\chi,i,j}}{N}\left[\frac{\delta\omega^{-2(\chi-1)}}{\text{SNR}_j}\right] + \mathcal{O}(\delta\omega^{-2(\chi-1)+1}) \tag{2.165}$$

where $l = \mathcal{D}(j-1) + i$ and

$$b_{\chi,i,j} \triangleq \begin{cases} \frac{1}{2}\left(\left[\dot{\Psi}_{r+1}^T \text{Re}\{\rho^T \otimes \mathcal{E}_{r+1}^H \mathcal{E}_{r+1}\}\dot{\Psi}_{r+1}\right]^{-1}\right)_{ll} & \text{Case I} \\ \frac{1}{2}\left(\left[\dot{\Psi}_r^T \text{Re}\{\rho^T \otimes \mathcal{E}_r^H \mathcal{E}_r\}\dot{\Psi}_r\right]^{-1}\right)_{ll} & \text{Case II} \\ \frac{1}{2}\left(\left[P_{[\dot{\Psi}_r^T]}\dot{\Psi}_{r+1}^T \text{Re}\{\rho^T \otimes \mathcal{E}_{r+1}^H \mathcal{E}_{r+1}\}\dot{\Psi}_{r+1} P_{[\dot{\Psi}_r^T]}\right]^{+}\right)_{ll} & \text{Case III} \end{cases} \tag{2.166}$$

SNR_j denotes the signal-to-noise ratio for the jth source defined in (2.61) and ρ denotes the matrix of (complex) signal correlation coefficients in (2.62). Use of the multi-D counterpart of the Array-Signal-to-Noise Ratio (ASNR) definition (2.64) gives the alternative expression

$$\text{Var}\{\hat{\omega}_{ij}\} \geq \frac{b'_{\chi,i,j}}{N}\left[\frac{\delta\omega^{-2(\chi-1)}}{\text{ASNR}_j}\right] + \mathcal{O}(\delta\omega^{-2(\chi-1)+1}) \tag{2.167}$$

where

$$b'_{\chi,i,j} \triangleq b_{\chi,i,j}\|\vec{a}(\vec{\omega}_0)\|^2 \tag{2.168}$$

Finally, we define the multi-D counterpart of a resolution cell along the ith coordinate direction to be

$$\Omega_i^2 = \frac{\pi^2}{3}\frac{\|\vec{a}(\vec{\omega}_0)\|^2}{\|\vec{\alpha}_{1,i}\|^2} \tag{2.169}$$

where

$$\vec{\alpha}_{1,i} \triangleq \left[I - \frac{1}{\|\vec{a}(\vec{\omega}_0)\|^2}\vec{a}(\vec{\omega}_0)\vec{a}(\vec{\omega}_0)^H \right] \left.\frac{\partial \vec{a}(\vec{\omega})}{\partial \omega_i}\right|_{\vec{\omega}=\vec{\omega}_0} \tag{2.170}$$

Use of (2.169) in (2.165) and (2.167) gives the following alternative expressions for the multi-D Conditional-Model CR bounds:

$$(B_C)_{ll} = c_{\chi,i,j} \frac{\Omega_i^2}{N \cdot \mathrm{SNR}_j} \frac{1}{(\delta\omega/\Omega_i)^{2(\chi-1)}} + \mathcal{O}(\delta\omega^{-2(\chi-1)+1}) \tag{2.171}$$

$$= c'_{\chi,i,j} \frac{\Omega_i^2}{N \cdot \mathrm{ASNR}_j} \frac{1}{(\delta\omega/\Omega_i)^{2(\chi-1)}} + \mathcal{O}(\delta\omega^{-2(\chi-1)+1}) \tag{2.172}$$

where $c_{\chi,i,j}$ and $c'_{\chi,i,j}$ denote the dimensionless parameters

$$c_{\chi,i,j} = b_{\chi,i,j} \left[\frac{3\|\vec{\alpha}_{1,i}\|^2}{\pi^2\|\vec{a}(\vec{\omega}_0)\|^2} \right]^{\chi} \tag{2.173}$$

$$c'_{\chi,i,j} = c_{\chi,i,j}\|\vec{a}(\vec{\omega}_0)\|^2 \tag{2.174}$$

CR Bound in Preferred Directions

Expression (2.160) showed that for Case III scenarios the dominant term of B_C for small $\delta\omega$ is $\delta\omega^{-2r}K_{r+1}$. Matrix K_{r+1} in (2.162) can be shown to have only partial rank ($= M(\mathcal{D} - \nu)$) for Case III scenarios. Consequently, in Case III scenarios, there exist coordinate directions for which the coefficient $b_{\chi,i,j}$ of the $\delta\omega^{-2r}$ term of the variance bound (2.165) vanishes, and for which the small $\delta\omega$ CR bound is more favorable (smaller). We designate such coordinates as lying in *preferred directions.*

Analysis in [23] shows that if Q_{Pref} is defined as a projection onto the preferred coordinate directions, then

$$Q_{\mathrm{Pref}} B_C Q_{\mathrm{Pref}} = \delta\omega^{-2(r-1)} \frac{\sigma^2}{2N} \left[\dot{\Psi}_r^T \mathrm{Re}\{\hat{P}^T \otimes \mathcal{E}_r^H \mathcal{E}_r\} \dot{\Psi}_r \right]^+ + \mathcal{O}(\delta\omega^{-2(r-1)+1}) \tag{2.175}$$

and that for small $\delta\omega$

$$Q_{\mathrm{Pref}} = \dot{\Psi}_r^+ \dot{\Psi}_r + \mathcal{O}(\delta\omega) \tag{2.176}$$

Since $\dot{\Psi}_r$ has block diagonal structure, it follows that preferred coordinate directions at the jth source for small $\delta\omega$ are specified by coordinate vectors $\hat{i}_{\mathrm{Pref}}$ that lie in the row space of $\dot{\Psi}_r(\vec{q}_j)$.

Equation (2.175) shows that the CR bound along a preferred coordinate for small $\delta\omega$ is proportional to $\delta\omega^{-2(r-1)}$, which is more favorable than the typical $\delta\omega^{-2r}$ dependence along nonpreferred directions for Case III scenarios.

A geometric interpretation of preferred coordinate directions in terms of source configuration can be derived straightforwardly for 2-D scenarios. The results are

Geometric Interpretation of Preferred Directions in 2-D: For small $\delta\omega$, the preferred directions are normal to the unique r^{th} order polynomial curve specified by the M source locations $\vec{q}_1, \cdots \vec{q}_M$.

Two simple examples of Case III scenarios are

1. $M = 2$ source 2-D scenario for which the preferred direction is normal to line between the two source locations specified by $\vec{q}_1, \vec{q}_2$.
2. $M = 5$ source 2-D scenario for which the preferred directions are normal to the unique conic section curve specified by the $\vec{q}_1, \cdots \vec{q}_5$ (e.g., circle, ellipse, parabola, hyperbola, etc.).

The latter example is illustrated in the simulation examples in the following section.

The practical effect of preferred directions in Case III is that for these types of multi-D scenarios, the direction estimation ability of any unbiased DF algorithm is likely to be severely degraded in certain spectral directions. In a 2-D example with $M = 5$ sources in a circular configuration, it is likely to be much more difficult to accurately estimate the tangential than the radial spectral parameter of each source. Note that preferred directions do not arise in 1-D scenarios, for which there is only one spectral coordinate.

2.12 NUMERICAL EXAMPLES FOR 2-D

To illustrate the accuracy of the foregoing limiting theoretical expressions for directional CR bounds as $\delta\omega \to 0$, we compare the small $\delta\omega$ representations to the exact CR bounds for four 2-D direction-finding scenarios. The scenarios differ only in the number M of sources present, and illustrate the occurrence of Cases I, II, and III for $M = 3, 4, 5, 6$.

Each example involves a planar array of $W = 16$ unit-gain, isotropic sensors, and far-field sources clustered near to the array broadside.

The generic arrival vector takes the form

$$\vec{a}(\vec{\omega}) = \left[e^{j\vec{r}_1^T\vec{\omega}}, e^{j\vec{r}_2^T\vec{\omega}}, \ldots, e^{j\vec{r}_W^T\vec{\omega}}\right]^T \tag{2.177}$$

where $\vec{r}_i = [r_{xi}, r_{yi}]^T$ is the location of the ith sensor in sensor plane. The reference parameter vector $\vec{\omega}_0$ is taken to be at array broadside, defined as

$$\vec{\omega}_0 = [0, 0]^T \tag{2.178}$$

From (2.177), (2.178), and (2.95) we find that for this scenario the matrices $\dot{A}_0, \dot{A}_1, \dot{A}_2$, and $\dot{A}_3$ are

$$\dot{A}_0 = \begin{bmatrix} 1 \\ \vdots \\ 1 \end{bmatrix}, \dot{A}_1 = j \cdot \begin{bmatrix} r_{x1} & r_{y1} \\ \vdots & \vdots \\ r_{xW} & r_{yW} \end{bmatrix}, \dot{A}_2 = -1 \cdot \begin{bmatrix} r_{x1}^2 & r_{x1}r_{y1} & r_{y1}^2 \\ \vdots & \vdots & \vdots \\ r_{xW}^2 & r_{xW}r_{yW} & r_{yW}^2 \end{bmatrix}$$

$$\dot{A}_3 = -j \cdot \begin{bmatrix} r_{x1}^3 & r_{x1}^2 r_{y1} & r_{x1} r_{y1}^2 & r_{y1}^3 \\ \vdots & \vdots & \vdots & \vdots \\ r_{xW}^3 & r_{xW}^2 r_{yW} & r_{xW} r_{yW}^2 & r_{yW}^3 \end{bmatrix} \tag{2.179}$$

From (2.40), (2.97), (2.101) we find that

$$\Gamma_0 = \begin{bmatrix} 1, & \cdots & 1 \end{bmatrix}$$

$$\Gamma_1 = \begin{bmatrix} q_{x1}, & \cdots & q_{xM} \\ q_{y1}, & \cdots & q_{yM} \end{bmatrix}$$

$$\Gamma_2 = \begin{bmatrix} q_{x1}^2/2, & \cdots & q_{xM}^2/2 \\ q_{x1}q_{y1}, & \cdots & q_{xM}q_{yM} \\ q_{y1}^2/2, & \cdots & q_{yM}^2/2 \end{bmatrix}$$

$$\Gamma_3 = \begin{bmatrix} q_{x1}^3/6, & \cdots & q_{xM}^3/6 \\ q_{x1}^2 q_{y1}/2, & \cdots & q_{xM}^2 q_{yM}/2 \\ q_{x1} q_{y1}^2/2, & \cdots & q_{xM} q_{yM}^2/2 \\ q_{y1}^3/6, & \cdots & q_{yM}^3/6 \end{bmatrix} \tag{2.180}$$

The matrices $\dot{\Gamma}_p(\vec{q}_j)$ in (2.104) with $\vec{q}_j = [q_{xj}, q_{yj}]$ take the form

$$\dot{\Gamma}_0(\vec{q}_j) = \begin{bmatrix} 0 \; , \; 0 \end{bmatrix}$$

$$\dot{\Gamma}_1(\vec{q}_j) = \begin{bmatrix} 1 \; , & 0 \\ 0 \; , & 1 \end{bmatrix}$$

$$\dot{\Gamma}_2(\vec{q}_j) = \begin{bmatrix} q_{xj} \; , & 0 \\ q_{yj} \; , & q_{xj} \\ 0 \; , & q_{yj} \end{bmatrix}$$

$$\dot{\Gamma}_3(\vec{q}_j) = \begin{bmatrix} q_{xj}^2/2 \; , & 0 \\ q_{xj}q_{yj} \; , & q_{xj}^2/2 \\ q_{yj}^2/2 \; , & q_{xj}q_{yj} \\ 0 \; , & q_{yj}^2/2 \end{bmatrix} \tag{2.181}$$

We assume that the sources are *correlated* and equal power. The source cross-power matrix P is taken to be of the form

$$P = 1/M \cdot \begin{bmatrix} 1 & p_{12} & \cdots & p_{12}^{n-1} \\ p_{12}^* & 1 & p_{12} & \vdots \\ \vdots & \ddots & \ddots & \ddots \\ (p_{12}^{n-1})^* & \cdots & p_{12}^* & 1 \end{bmatrix} \tag{2.182}$$

with $p_{12} = 0.4 + j0.6$.

The exact CR bounds computed using (2.16) are compared to the asymptotic values for small $\delta\omega$ predicted by the result (2.160) of our analysis in the following numerical example.

Example 2.3

Consider the scenarios defined by $M = 2, 3, 4, 5$ sources taken from the leading subsets of set {SC1, SC2, SC3, SC4, SC5, SC6} in the source configuration of Figure 2.6. Let the sensor array be as in Figure 2.6.

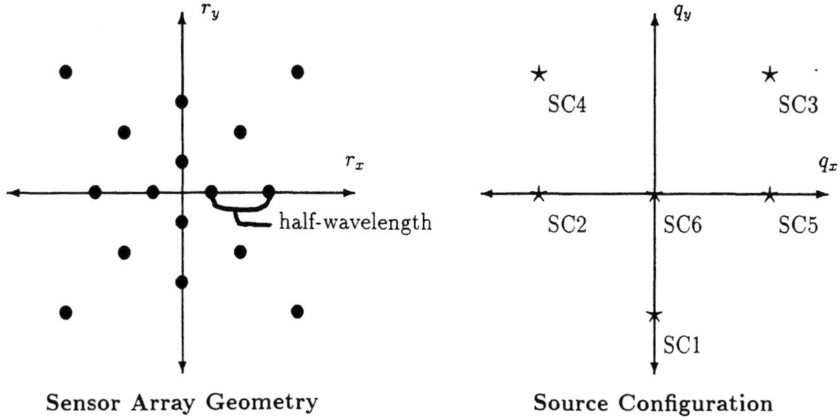

Figure 2.6 Planar sensor array geometry and normalized source configuration for multi-D CR bound examples.

It can be verified that these scenarios satisfy conditions **C1–C3** and **CR1–CR3**, with values of r, χ, and ν as in Table 2.2.

TABLE 2.2 Values of r, χ, and ν for $M = 3, 4, 5, 6$

	$M=3$	$M=4$	$M=5$	$M=6$
r	1	2	2	2
χ	2	2	3	3
ν	0	2	1	0
Case	I	II	III	I

For each of the four scenarios, Figures 2.7–2.10 show the values of the CR bounds for parameter estimates along the x and y spectral frequency axes for one of the sources, specifically SC1 in Figure 2.6. The solid curves depict the exact CR

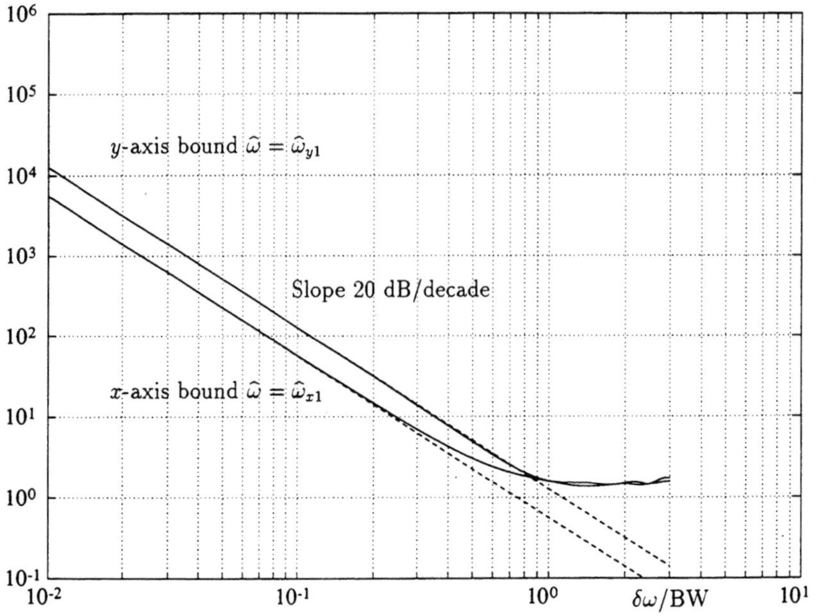

Figure 2.7 Limiting CR bounds for source SC1 in $M = 3$ source scenario.

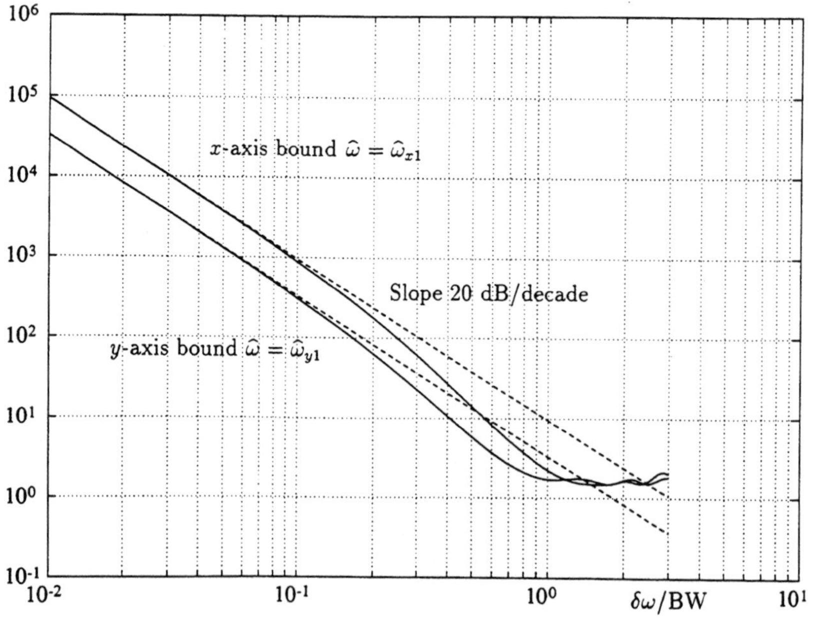

Figure 2.8 Limiting CR bounds for source SC1 in $M = 4$ source scenario.

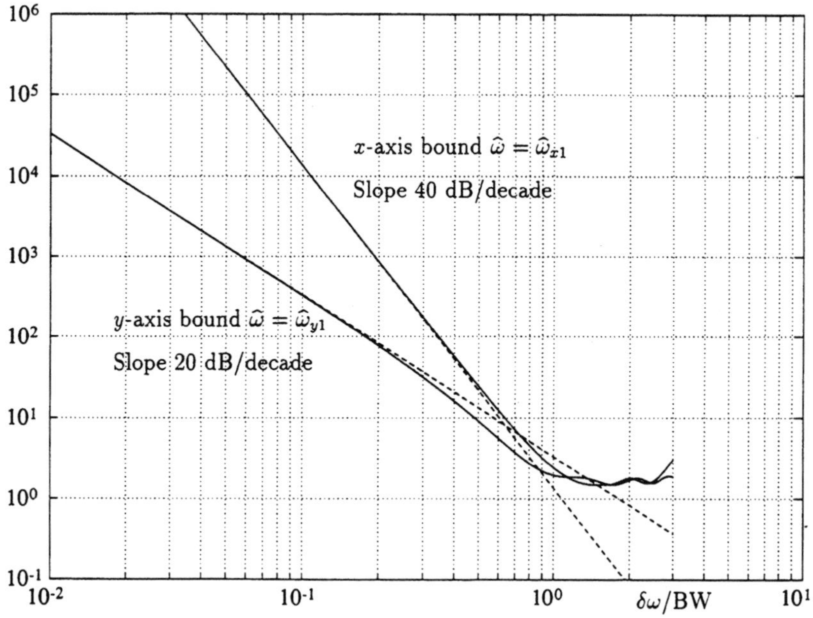

Figure 2.9 Limiting CR bounds for source SC1 in $M = 5$ source scenario.

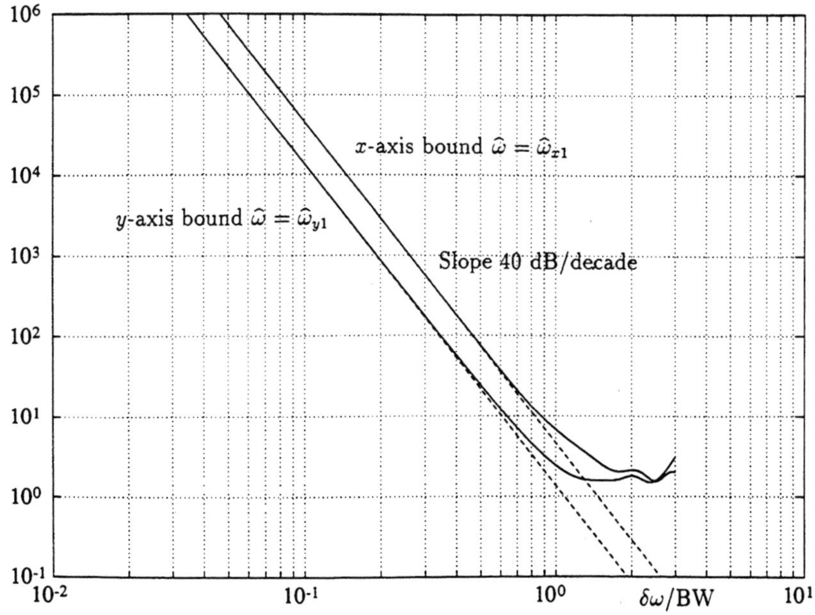

Figure 2.10 Limiting CR bounds for source SC1 in $M = 6$ source scenario.

bounds; the dashed lines depict the asymptotic behavior predicted by Eq. (2.165). The horizontal scale denotes spatial frequency separation $\delta\omega$ normalized by the array beamwidth BW, so that unity on the horizontal scale of the graph corresponds to maximum source separation of one beamwidth. The vertical scale depicts the value of the normalized bound

$$N \cdot \mathrm{SNR}_1 \cdot (B_C)_{ll} \tag{2.183}$$

where $(B_C)_{ll}$ is as in (2.165).

Clearly, the simplified asymptotic expressions capture the essence of the bounds for source separation less than one beamwidth. As predicted, the CR bounds exhibit a $\delta\omega^{-2(\chi-1)}$ behavior for small $\delta\omega$, with a slope of 20 dB/decade in Figures 2.7 and 2.8, and 40 dB/decade in Figure 2.10. In Figure 2.9, the 5 sources lie on an ellipse centered at coordinate origin. As predicted, the CR bounds exhibit a preferred $\delta\omega^{-2}$ behavior for small $\delta\omega$, with a slope of 20 dB/decade, along the preferred direction normal to the ellipse curve (y-axis at SC1), and exhibit a $\delta\omega^{-4}$ behavior, with a slope of 40 dB/decade, along the tangential direction (x-axis at SC1).

Thus, the theoretical expressions accurately predict the CR bounds for small separations $\delta\omega$ for these scenarios.

2.13 RESOLUTION THRESHOLDS IN MULTI-D

To obtain resolution threshold expressions for both SNR and N in multi-D scenarios, we extend the 1-D arguments of section 2.9 that any unbiased estimator can successfully resolve M closely spaced sources only if the standard deviation of the directional estimates is substantially less than the minimum spacing between any two sources. The SNR and N values required to satisfy the above condition are deemed to be the resolution threshold values.

Thus, consider M closely spaced signals, and define $\delta\omega_j$ to denote the minimum spacing between the jth and any other source. That is

$$\delta\omega_j = \delta\omega \cdot \min_{k \neq j} \|\vec{q}_j - \vec{q}_k\| \tag{2.184}$$

where $\vec{q}_1, \cdots \vec{q}_M$ are the normalized spectral frequency offsets. If the root CR bounds $\sqrt{(B_C)_{ll}}$ for all coordinate directions at the jth source ($l = \mathfrak{D}(j-1) + i$ and $i = 1 \cdots \mathfrak{D}$) are small compared to $\delta\omega_j$ for each source $j = 1 \cdots M$, then there is a basis for seeking an unbiased estimator for resolving the signals. On the other hand, if one or more of the $\sqrt{(B_C)_{ll}}$ are large compared to the associated $\delta\omega_j$, then it is unlikely that there exists an unbiased estimator which can resolve the signals with high probability. Accordingly, one strongly suspects that a necessary condition for the existence of an unbiased estimator capable of resolving M closely spaced sources with high probability is that

$$\sqrt{(B_C)_{ll}} \leq f \cdot \delta\omega_j \tag{2.185}$$

for all $i = 1 \cdots \mathfrak{D}$, $j = 1 \cdots M$, with $l = \mathfrak{D}(j-1) + i$ and where f is a suitable fraction (e.g., $f = 1/8$).

To define the resolution thresholds, we represent the source amplitude correlation matrix $\hat{P}$ as in (2.80). We define the signal SNR to be the ratio as in (2.81), and define the solution threshold SNR $\mathcal{E}_R$ and $\mathcal{N}_R$ as in 1-D. We deem satisfactory resolution performance to be possible whenever condition (2.185) is satisfied for all i, j (i.e., for all signals and for all coordinate directions).

To elucidate the necessary conditions for resolution, we square (2.185) and substitute expression (2.160) to obtain

$$\delta\omega^{-2(\chi-1)} \frac{\sigma^2}{2N} (K_\chi)_{ll} \le f^2 \cdot \delta\omega^2 \cdot \min_{k \ne j} \|\vec{q}_j - \vec{q}_k\|^2 \tag{2.186}$$

for small $\delta\omega$. Rearrangement of (2.186) and use of (2.80) gives the equivalent condition

$$\mathrm{SNR} \cdot N \ge K'_{R,i,j} \delta\omega^{-2\chi} \tag{2.187}$$

where we define

$$K'_{R,i,j} \triangleq \begin{cases} \dfrac{\sigma^2}{2N} \dfrac{1}{2f^2 \cdot \min_{k \ne j} \|\vec{q}_j - \vec{q}_k\|^2} \left(\left[\dot{\Psi}^T_{r+1} \mathrm{Re}\,\{\hat{P}^T_0 \otimes \mathcal{E}^H_{r+1} \mathcal{E}_{r+1}\} \dot{\Psi}_{r+1} \right]^{-1} \right)_{ll} & \text{Case I} \\ \dfrac{\sigma^2}{2N} \dfrac{1}{2f^2 \cdot \min_{k \ne j} \|\vec{q}_j - \vec{q}_k\|^2} \left(\left[\dot{\Psi}^T_r \mathrm{Re}\,\{\hat{P}^T_0 \otimes \mathcal{E}^H_r \mathcal{E}_r\} \dot{\Psi}_r \right]^{-1} \right)_{ll} & \text{Case II} \\ \dfrac{\sigma^2}{2N} \dfrac{1}{2f^2 \cdot \min_{k \ne j} \|\vec{q}_j - \vec{q}_k\|^2} \left(\left[P_{[\dot{\Psi}^T_r]} \dot{\Psi}^T_{r+1} \mathrm{Re}\,\{\hat{P}^T_0 \otimes \mathcal{E}^H_{r+1} \mathcal{E}_{r+1}\} \dot{\Psi}_{r+1} P_{[\dot{\Psi}^T_r]} \right]^{+} \right)_{ll} & \text{Case III} \end{cases} \tag{2.188}$$

where $l = \mathcal{D}(j-1) + i$.

We note that an equivalent condition for (2.187) to be satisfied for all $j = 1 \cdots M, i = 1 \cdots \mathcal{D}$ is the condition

$$\mathrm{SNR} \cdot N \ge K'_R \delta\omega^{-2\chi} \tag{2.189}$$

where we define

$$K'_R = \max_{i,j} \{K'_{R,i,j}\} \tag{2.190}$$

Thus, we deem satisfactory resolution performance to be possible whenever condition (2.189) is satisfied.

Identification of the SNR resolution threshold $\mathcal{E}_R$ gives

$$\text{SNR} \geq \mathcal{E}_R \tag{2.191}$$

where

$$\mathcal{E}_R = \frac{K'_R}{N}\delta\omega^{-2\chi} \tag{2.192}$$

Similarly, identification of the data set size resolution threshold N_R gives

$$N \geq \mathcal{N}_R \tag{2.193}$$

where

$$\mathcal{N}_R = \frac{K'_R}{\text{SNR}}\delta\omega^{-2\chi} \tag{2.194}$$

The threshold expressions (2.192) and (2.194) are important since they provide explicit expressions for the minimum SNR and data set size N required to satisfy the "necessary condition" (2.185) for resolution of M closely spaced sources using any unbiased spectral estimation algorithm. The threshold expressions can be used to generate model resolution curves for any given scenario, since the constant K'_R can be calculated explicitly given the array geometry, sensor directional response, source configuration, and source correlations.

The threshold expressions (2.192) and (2.194) also clarify the trade-off between SNR, N and maximum source spacing $\delta\omega$ required to maintain resolution performance in multi-D scenarios. For example, if noise power is doubled in a given scenario, then the size of the data set must increase by a factor of 2 to maintain resolution performance. If, on the other hand, the maximum source spacing $\delta\omega$ is decreased by a factor of 2 in a typical 2-D scenario with $M = 6$ sources (with $\chi = 3$), then to maintain resolution performance with a fixed data set size N, the SNR must increase by a factor of $2^6 = 64$! Alternately if $\delta\omega$ is halved while the SNR remains fixed, then the data set size N must increase by a factor of $2^6 = 64$!

By way of comparison, if the maximum source spacing $\delta\omega$ is decreased by a factor of 2 in a 1-D scenario with $M = 6$ sources, then to maintain resolution performance with a fixed data set size N (or fixed SNR), the SNR (or N) must increase by a factor of $2^{12} = 4096$!!! For small $\delta\omega$ and a given number of sources M, the resolution thresholds are typically much smaller (more favorable) in multi-D than in 1-D scenarios.

2.14 DEGENERATE SCENARIOS

The foregoing analysis presumes that the scenario of interest satisfies the assumptions of Appendix 2B, which includes Assumptions **C1** and **C2** articulated in section 2.10. We note that those assumptions *typically* are satisfied in multi-D scenarios, but are *not always* satisfied.

For example, consider a 2-D scenario in which the sensor array consists of a planar array of identical omni-directional sensors, and $M = 6$ closely spaced sources illuminate the array. Applicable expressions for $\dot{A}_0, \dot{A}_1, \dot{A}_2$ are given by Eq. (2.179);

applicable values of $\Gamma_0, \Gamma_1, \Gamma_2$ are given by Eq. (2.180). Note that if the array sensors lie on a conic section defined by the 2-D polynomial in r_x and r_y:

$$0 = 1 \cdot \alpha_0 + (r_x \cdot \alpha_x + r_y \cdot \alpha_y) + (r_x^2 \cdot \alpha_{xx} + r_x r_y \cdot \alpha_{xy} + r_y^2 \cdot \alpha_{yy}) \tag{2.195}$$

with constants $\alpha_0, \alpha_x, \alpha_y, \alpha_{xx}, \alpha_{xy}$, and α_{yy}, then the columns of $\dot{A}_0, \dot{A}_1, \dot{A}_2$ defined by (2.179) are linearly dependent, which violates Assumption **C1**. More generally, Assumption **C1** is satisfied if and only if the sensors do not lie upon a generalized conic section defined by the equation

$$0 = 1 \cdot \alpha_0 + (r_x \cdot \alpha_x + r_y \cdot \alpha_y) + \cdots + (r_x^{r-1} \cdot \alpha_{x\cdots x} + r_x r_y^{r-2} \cdot \alpha_{xy\cdots y} + r_y^{r-1} \cdot \alpha_{y\cdots y}) \tag{2.196}$$

with r defined by (2.111).

Dually, if $M = 6$ sources lie upon a conic section defined by the 2-D polynomial in q_x and q_y:

$$0 = 1 \cdot \zeta_0 + (q_x \cdot \zeta_x + q_y \cdot \zeta_y) + (q_x^2 \cdot \zeta_{xx} + q_x q_y \cdot \zeta_{xy} + q_y^2 \cdot \zeta_{yy}) \tag{2.197}$$

with constants $\zeta_0, \zeta_x, \zeta_y, \zeta_{xx}, \zeta_{xy}$, and ζ_{yy}, then the rows of $\Gamma_0, \Gamma_1, \Gamma_2$ defined by (2.180) are linearly dependent, which violates Assumption **C2**. More generally, Assumption **C2** is satisfied if and only in the sources do not lie upon a generalized conic section defined by the equation

$$0 = 1 \cdot \zeta_0 + (q_x \cdot \zeta_x + q_y \cdot \zeta_y) + \cdots + (q_x^{r-1} \cdot \zeta_{x\cdots x} + q_x q_y^{r-2} \cdot \zeta_{xy\cdots y} + q_y^{r-1} \cdot \zeta_{y\cdots y}) \tag{2.198}$$

with r defined by (2.111).

Therefore, the results on the behavior of B_c derived in this chapter apply to scenarios with "random" sensor locations which do not satisfy (2.196), and with "random" source locations which do not satisfy (2.198). Specifically, for 2-D scenarios the results *do not* apply to scenarios in which the receiving array consists of omni-directional sensors confined to a curve of the form (2.196). Likewise, the results do not apply to scenarios in which the sources lie upon a curve of the form (2.198). These observations extend in the obvious manner to multi-D scenarios with $\mathfrak{D} > 2$.

We designate scenarios for which either Assumption **C1** or **C2** (but not both) are violated as *partially degenerate scenarios*. We designate scenarios for which both Assumptions **C1** and **C2** are violated as (fully) *degenerate scenarios*. The analysis methodology presented here can be extended both to partially degenerate, and to fully degenerate scenarios. However, the analysis becomes more complicated in that the assumptions can be violated in many ways and, therefore, there are many cases to consider. See [24] for a discussion of partially degenerate scenarios with co-linear sources, and [23] for a discussion of arbitrary partially degenerate scenarios.

The thrust of the analysis of partially and fully degenerate cases is that both

r and χ tend to assume larger values than those defined by (2.111) and (2.161), and source direction estimation, or source resolution, becomes correspondingly more difficult. It follows that satisfaction of Condition **C1** can be exploited as a criterion in the design of an "efficient" sensor array that maximizes source direction estimation accuracy and source resolution ability. Such a criterion leads to the (trivial) conclusion that a co-linear array is "bad" for source direction estimation in 2-D, but also to the nonobvious conclusion that a *circular array* (or any other conic section) of omni-directional sensors is "bad" for direction estimation of $M \geq 6$ closely spaced sources in 2-D.

Dually, violation of Condition **C2** can be exploited as a design criterion for *covert source configurations* that make closely spaced sources difficult to locate accurately, or resolve reliably. Specifically, a co-linear configuration of $M > 3$ sources presents a much more difficult source localization problem than a typical "random" configuration (see [24]). Similarly, a conic section (e.g., circular) configuration of $M \geq 6$ sources is much more challenging than a typical and other than co-linear configuration.

2.15 CONCLUSIONS

The ability of a candidate direction-finding (DF) algorithm to resolve multiple closely spaced sources and to accurately estimate source-location parameters has become a standard test of DF algorithm effectiveness. This chapter has utilized the Cramér-Rao (CR) lower bound on parameter estimation variance to identify fundamental limitations of unbiased DF algorithms for the closely spaced source scenario. The CR bound has been used as the analysis vehicle since it is not algorithm specific, but rather characterizes the optimum performance of any unbiased algorithm. The behavior of the bound has been identified for one-dimensional (1-D) and multi-dimensional (multi-D) scenarios.

The main results are Eqs. (2.57) and (2.160) for the CR bound on directional covariance for 1-D scenarios, and for multi-D scenarios that satisfy the assumptions of Appendix 2B. These expressions provide great insight into achievable direction-finding performance in that they make explicit the individual impact of physical scenario parameters such as maximum source separation $\delta\omega$, relative source configuration, source powers and correlations, and sensor array geometry. A striking feature of the results is their strong dependence on the source separation factor $\delta\omega$. This dependence accounts for the sensitivity of the estimation results in the case of multiple closely spaced sources. Comparison of the multi-D result (2.160) with the 1-D result (2.57) indicates that effective direction estimation is more promising in multi-D than in 1-D as $\delta\omega \rightarrow 0$, as a consequence of (2.163).

For multi-D applications which satisfy the analysis assumptions of Appendix 2B, it has been shown that the CR bound can depend on the source separation factor $\delta\omega$ in three fundamentally different ways. Distinct behaviors are exhibited in Cases designated as Cases I–III (see section 2.10). In Case III "preferred directions" exist in which directional variances for the preferred directions are much smaller than for other directions. For example, in a 2-D scenario with $M = 5$ sources located on a circle, the radial direction is preferred; that is, the results indicate that one can estimate source location much more accurately in the radial coordinate than in the tangential coordinate.

A variety of 1-D and multi-D numerical examples have been presented which confirm the analysis results.

Arguing that all direction estimation standard deviations must be smaller than source separations in order for an unbiased estimator to reliably resolve the sources, we have exploited the CR bound results to derive approximate expressions for the minimum SNR $\mathcal{E}_R$ and data set size $\mathcal{N}_R$ for source resolution. These expressions are given by (2.88), (2.90) for 1-D, and by (2.192), (2.194) for multi-D. The 1-D results are consistent with a sizable (albeit finite) set of specialized analytical and empirical results.

Multi-D scenarios which do not satisfy the assumptions of Appendix 2B are designated as degenerate. It is noted that the CR bound on directional variance for degenerate scenarios typically is more sensitive to the source-separation scale factor $\delta\omega$ than scenarios which satisfy the analysis assumptions of Appendix 2B. Satisfaction of the analysis assumptions may be exploited in the design of "efficient" sensor arrays that maximize source direction estimation accuracy and source resolution ability, or dually, in the design of "covert" source configurations that make closely spaced sources difficult to locate accurately, or resolve reliably.

The results provide useful analytical benchmarks for direction finding in 1-D and multi-D scenarios. It is hoped that the results will facilitate the assessment of candidate DF techniques, and provide insight helpful to the development of improved DF algorithms.

APPENDIX 2A ASSUMPTIONS FOR 1-D DIRECTION-FINDING SCENARIOS

The vectors $\vec{a}(\vec{\omega})$, $\vec{x}(t)$, and $\vec{e}(t)$ are assumed to satisfy the following conditions:

$\vec{a}(\vec{\omega})$:

A1.
$$W \geq M + \mathcal{D} \tag{2A.1}$$

A2. Matrix A is a $W \times M$ matrix of the form (2.7), with columns $\vec{a}(\vec{\omega}_1), \cdots \vec{a}(\vec{\omega}_M)$.

A3. Matrix A has linearly independent columns, provided $\vec{\omega}_i \neq \vec{\omega}_j$ for $i \neq j$.

A4. The elements of $\vec{a}(\vec{\omega})$ are bounded and possess partial derivatives of all orders with respect to the elements of $\vec{\omega}$, within a convex region $\mathcal{R}$ of $\vec{\omega}$ space that includes all source spatial frequency vectors $\vec{\omega}_1, \cdots, \vec{\omega}_M$.

A5. The $\mathcal{D}$ first-order partial derivatives of $\vec{a}(\vec{\omega})$ with respect to the elements of $\vec{\omega}$, at each source direction $\vec{\omega} = \vec{\omega}_j$, $(j = 1 \cdots M)$ are linearly independent from the source arrival vectors $\vec{a}(\vec{\omega}_1), \cdots \vec{a}(\vec{\omega}_M)$.

A6. For 1-D scenarios, the vector $\vec{a}(\omega)$ and its first M derivatives are linearly independent for ω in $\mathcal{R}$.

$\vec{x}(t)$:

X1. The sequence of vectors $\vec{x}(t)$, $t = 1 \cdots N$ is fixed for all realizations of the data sequence $\vec{y}(t)$.

X2. The matrix

$$\hat{P} \triangleq \frac{1}{N} \sum_{t=1}^{N} \vec{x}(t)\vec{x}(t)^H \tag{2A.2}$$

is positive definite.

$\vec{e}(t)$:

E1. The vector $\vec{e}(t)$ varies randomly across the ensemble of data vectors $\vec{y}(t)$. Specifically, the $\vec{e}(t)$ are samples of a zero-mean complex Gaussian random process with

$$E\{\vec{e}(t)\vec{e}(s)^H\} = \begin{cases} \sigma^2 I & t = s \\ 0 & t \neq s \end{cases} \tag{2A.3}$$

$$E\{\vec{e}(t)\vec{e}(s)^T\} = 0 \tag{2A.4}$$

APPENDIX 2B ASSUMPTIONS FOR 2-D DIRECTION-FINDING SCENARIOS

For multi-D scenarios, all assumptions of Appendix 2A are retained except for assumption **A6**. Analysis of multi-D scenarios is simplified by identification of additional structural conditions satisfied in typical (i.e., *nondegenerate*) DF scenarios. This appendix defines these conditions.

Recall the Taylor series (2.100) of matrix A for closely spaced sources in multi-D. Reference to (2.95) shows that $\dot{A}_0$ has rank of unity, and successive $\dot{A}_p$ have small and increasing ranks. As a consequence, successive terms of (2.100) are of low and slowly increasing rank, and a number of such terms typically must be included in a partial sum to obtain a full-rank approximation of A. To characterize the minimum number of such terms we define integer parameter r as follows:

Definition of r. Integer r is the smallest number such that the partial Taylor sum formed by the successive terms $p = 0, \cdots r$ of (2.100) has full rank.

Provided Conditions **C1–C3** (detailed subsequently) are satisfied, r is determined by the relationship

$$\sum_{p=0}^{r-1} \bar{n}_p < M \leq \sum_{p=0}^{r} \bar{n}_p \tag{2B.1}$$

where $\bar{n}_p$ is the number of columns in $\dot{A}_p$. If Conditions **C1–C3** are not all satisfied, r may be greater than that determined by (2B.1).

Conditions **C1–C3** sufficient for (2B.1) to determine r are the following:

C1. The generic arrival vector $\vec{a}(\vec{\omega})$ and its partial derivatives at $\vec{\omega} = \vec{\omega}_0$ up to order $r-1$ with respect to the elements of $\vec{\omega}$ are linearly independent. That is, matrices $\dot{A}_p$ have full rank $\bar{n}_p$ for $p = 0 \cdots r-1$, and the columns of $\dot{A}_p$ are linearly independent from the vector space spanned by the columns of the sequence $\dot{A}_0, \cdots, \dot{A}_{p-1}$ for $p = 1 \cdots r-1$. Specifically

$$\begin{aligned} \text{Rank}\,\{\dot{A}_0\} &= \bar{n}_0 \\ \text{Rank}\,\{P_{[\dot{A}_0, \cdots \dot{A}_{p-1}]}\dot{A}_p\} &= \bar{n}_p \qquad \text{for}\, p = 1 \cdots r-1 \end{aligned} \tag{2B.2}$$

where $P_{[Z]}$ is the notation for the projection onto the null space of the columns of Z as defined in section 2.3.

C2. The matrices Γ_p have full rank $\bar{n}_p$ for $p = 0 \cdots r-1$, and the rows of Γ_p are linearly independent from the space spanned by the rows of the sequence $\Gamma_0, \cdots, \Gamma_{p-1}$ for $p = 1 \cdots r-1$. Specifically

$$\begin{aligned} \text{Rank}\,\{\Gamma_0\} &= \bar{n}_0 \\ \text{Rank}\,\{\Gamma_p P_{[\Gamma_0^H, \cdots \Gamma_{p-1}^H]}\} &= \bar{n}_p \qquad \text{for}\, p = 1 \cdots r-1 \end{aligned} \tag{2B.3}$$

where $P_{[Z^H]}$ is the notation for the projection onto the null space of the rows of Z as defined in section 2.3.

C3. For $p = r$, the component of the product $\dot{A}_r\Gamma_r$, which has columns orthogonal to those of the sequence $\dot{A}_0, \cdots, \dot{A}_{r-1}$ *and* has rows orthogonal to those of the sequence $\Gamma_0, \cdots, \Gamma_{r-1}$ has sufficient rank to complete the rank of A. That is,

$$\text{Rank}\,\{P_{[\dot{A}_0, \cdots \dot{A}_{r-1}]}\dot{A}_r\Gamma_m P_{[\Gamma_0^H, \cdots, \Gamma_{r-1}^H]}\} = M - \sum_{p=0}^{r-1} \bar{n}_p \tag{2B.4}$$

Conditions **C1–C3** generalize to multi-D Assumption **A6** defined for 1-D scenarios. Conditions **C1–C3** are sufficient to guarantee that r determined by (2B.1) is such that the partial Taylor series consisting of terms of order $p = 0$ through $p = r$, does in fact have full rank M. For 1-D scenarios, $r = M - 1$.

Condition **C1** depends on the array geometry and sensor directional response, and is independent of source configuration or source powers and correlations. Condition **C2** depends only on the normalized source coordinates $\vec{q}_j (j = 1 \cdots M)$. Thus, **C2** depends only on normalized source configuration, and is independent of the array geometry, sensor directional response, or source powers and correlations.

Note that Conditions **C1–C3** assume (require) that matrix A have more columns than rows ($W > M$), and that matrix A be full rank ($= M$). These prerequisites are satisfied under our Assumptions **A1–A3**.

Analysis shows that the behavior of the CR bound for small $\delta\omega$ depends fundamentally on the interaction between the columns of matrix A and of the matrix $\dot{D}(\vec{\omega}_j)$ of first spatial derivatives of $\vec{a}(\vec{\omega})$ at $\vec{\omega}_j$, for each $j = 1 \cdots M$. For convenience, we define the *augmented matrices*

$$\Xi_j \triangleq \left[\delta\omega \cdot \dot{D}(\vec{\omega}_j), A\right] \qquad W \times (M + \mathcal{D}) \tag{2B.5}$$

for $j = 1 \cdots M$, which aggregate the columns of $\dot{D}(\vec{\omega}_j)$, scaled by $\delta\omega$, and the columns of A. We note that Ξ_j has full rank $M + \mathcal{D}$ by Assumption **A5**.

Since the constituent matrices have Taylor series (2.103) and (2.100), each Ξ_j also has Taylor series of the form

$$\Xi_j = \sum_{p=0}^{\infty} \delta\omega^p \dot{A}_p \Gamma'_p(\vec{q}_j) \tag{2B.6}$$

where $\dot{A}_p$ is as in (2.95) and

$$\Gamma'_p(\vec{q}_j) \triangleq \left[\dot{\Gamma}_p(\vec{q}_j), \Gamma_p\right] \qquad \bar{n}_p \times (M + \mathcal{D}) \tag{2B.7}$$

with $\dot{\Gamma}_p(\vec{q}_j)$ as in (2.104) and Γ_p as in (2.101).

As in the Taylor series of A, the successive terms of (2B.6) are of low and slowly increasing rank. To characterize the minimum number of terms that must be included in a partial sum to obtain a full-rank approximation of Ξ_j, we define integer parameter χ_j as follows:

Definition of χ_j. Integer χ_j is the smallest number such that the partial Taylor sum formed by the successive terms $p = 0, \cdots \chi_j$ of Taylor series (2B.6) of Ξ_j has full rank.

Provided Conditions **CR1–CR3** (detailed subsequently) are satisfied, each $\chi_j = \chi$ where χ is determined by the relationship

$$\sum_{p=0}^{\chi-1} \bar{n}_p < M + \mathcal{D} \leq \sum_{p=0}^{\chi} \bar{n}_p \tag{2B.8}$$

Note that $M + \mathcal{D}$ is the number of columns in augmented matrix Ξ_j. If Conditions **CR1–CR3** are not all satisfied, χ may be greater than that determined by (2B.8).

Conditions **CR1–CR3** sufficient for (2B.8) to determine all χ_j are the following:

CR1.

$$\begin{aligned} \text{Rank}\,\{\dot{A}_0\} &= \bar{n}_0 \quad && \text{for } p = 0 \\ \text{Rank}\,\{P_{[\dot{A}_0, \cdots \dot{A}_{p-1}]} \dot{A}_p\} &= \bar{n}_p \quad && \text{for } p = 1, \cdots \chi - 1 \end{aligned} \tag{2B.9}$$

CR2. For $j = 1 \cdots M$,

$$\begin{aligned} \text{Rank}\,\{\Gamma'_0(\vec{q}_j)\} &= \bar{n}_0 \quad && \text{for } p = 0 \\ \text{Rank}\,\{\Gamma'_p(\vec{q}_j) P_{[\Gamma'_0(\vec{q}_j)^H, \cdots, \Gamma'_{p-1}(\vec{q}_j)^H]}\} &= \bar{n}_p \quad && \text{for } p = 1, \cdots \chi - 1 \end{aligned} \tag{2B.10}$$

CR3. For $j = 1 \cdots M$,

$$\text{Rank}\,\{P_{[\dot{A}_0, \cdots \dot{A}_{\chi-1}]} \dot{A}_\chi \Gamma'_\chi(\vec{q}_j) P_{[\Gamma'_0(\vec{q}_j)^H, \cdots, \Gamma'_{\chi-1}(\vec{q}_j)^H]}\} = M + \mathcal{D} - \sum_{p=0}^{\chi-1} \bar{n}_p \tag{2B.11}$$

Note that Conditions **CR1–CR3** parallel for the augmented matrices Ξ_j the Conditions **C1–C3** defined for matrix A.

APPENDIX 2C PROOF OF (2.163) AND (2.164)

From (2.111) we have

$$\sum_{p=0}^{r-1} \bar{n}_p < M \tag{2C.1}$$

or

$$r + \sum_{p=0}^{r-1} (\bar{n}_p - 1) < M \tag{2C.2}$$

Since $\bar{n}_0 = 1$, and $\bar{n}_p \geq 2$ for $p = 1, 2, \cdots$ in multi-D scenarios, it follows from (2C.2) that

$$r + 1 < M \qquad \text{for } r \geq 2 \tag{2C.3}$$

We note from (2.111) that $r \geq 2$ for $M > \mathcal{D} + 1$.

For $r = 1$, possible M values are $M = 2, 3, \cdots \mathcal{D} + 1$. Clearly (2C.3) is satisfied for $M = 3 \cdots \mathcal{D} + 1$. Therefore, the range of validity of (2C.3) can be extended to:

$$r + 1 < M \qquad \text{for } M \geq 3 \tag{2C.4}$$

For $r = 1$ and $M = 2$ we have

$$r + 1 = M \tag{2C.5}$$

Inequality (2C.4) together with (2.161) establishes (2.163) for each of Cases I, II, and III. Equation (2.164) follows from (2C.5), and the observation that for $M = 2$ and $\mathcal{D} \geq 2$ is Case III.

REFERENCES

1. J. P. Burg, "Maximum Entropy Spectral Analysis," Proc. 37th Annual Meeting of the Society of Exploration Geophysicists, Oklahoma City, Oct. 31, 1967.
2. J. Capon, "High Resolution Frequency-Wavenumber Spectrum Analysis," *Proc. IEEE*, vol. 57, pp. 1408–18, Aug. 1969.
3. S. S. Reddi, "Multiple Source Location—A Digital Approach," *IEEE Trans. Aerosp. Electron. Syst. AES-15*, 1, pp. 95–105, Jan. 1979.
4. R. Kumaresan and D. W. Tufts, "Estimating the Angles of Arrival of Multiple Plane Waves," *IEEE Trans. Aerosp. Electron. Syst.*, vol. AES-19, no. 1, pp. 134–39, Jan. 1983.
5. R. O. Schmidt, "Multiple Emitter Location and Signal Parameter Estimation," *Proc. RADC Spectral Est. Workshop*, Grifiss AFB, NY, 1979.
6. L. E. Brennan, "Angular Accuracy of a Phased Array Radar," *IRE Trans. on Antennas and Propagation*, vol. AP-9, pp. 268–75, May 1961.
7. D. C. Rife and R. R. Boorstyn, "Single-tone Parameter Estimation from Discrete-time Observations," *IEEE Trans. Inform. Theory*, vol. IT-20, pp. 591–98, Sept. 1974.

8. P. Stoica, and A. Nehorai, "MUSIC, Maximum Likelihood, and Cramér-Rao Bound," *IEEE Trans. on Acoustics, Speech and Signal Processing*, vol. 37, no. 5, May 1989.
9. M. I. Miller and D. R. Fuhrmann, "Maximum Likelihood Direction-of-Arrival Estimation for Multiple Narrowband Signals in Noise," *The 1987 Conf. on Information Sciences and Systems*, Baltimore, MD, pp. 710–12, March 1987.
10. A. G. Jaffer, "Maximum Likelihood Direction Finding of Stochastic Sources: A Separable Solution," ICASSP 88, pp. 2893–96.
11. P. Stoica and A. Nehorai, "Mode, Maximum Likelihood and Cramér-Rao Bound: Conditional and Unconditional Results," pp. 2715–18, ICASSP 90.
12. P. Stoica and A. Nehorai, "Performance Study of Conditional and Unconditional Direction of Arrival Estimation," *IEEE Trans. on Acoust., Speech, Signal Processing*, October 1990.
13. H. Messer and Y. Adar, "New Lower Bounds of Frequency Estimation of a Multitone Random Signal in Noise," *Signal Processing*, Elsevier Science Publishers, pp. 413–24, 1989.
14. J. R. Sklar and F. C. Schweppe, "The Angular Resolution of Multiple Targets," Lincoln Lab, Group Rep. 1964, Jan. 1964.
15. D. C. Rife and R. R. Boorstyn, "Multiple Tone Parameter Estimation from Discrete-time Observations," *Bell Syst. Tech. J.*, vol. 55, pp. 1389–1410, Nov. 1976.
16. S. F. Yau and Y. Bresler, "A Compact Cramér-Rao Bound Expression for Parametric Estimation of Superimposed Signals," *IEEE Trans. on Signal Processing*, vol. 40, no. 5, May 1992.
17. H. B. Lee, "The Cramér-Rao Bound on Frequency Estimates of Signals Closely Spaced in Frequency," *IEEE Trans. on SP*, June 1992.
18. H. Cox, "Resolving Power and Sensitivity to Mismatch of Optimum Array Processors," *Journal of Acoustical Society of America*, vol. 54, no. 3, 1973.
19. W. Gabriel, "Spectral Analysis and Adaptive Array Superresolution Techniques," *IEEE Proceedings*, June 1980.
20. M. Kaveh, and A. J. Barabell, "The Statistical Performance of the MUSIC and the Minimum-Norm Algorithms in Resolving Plane Waves in Noise," *IEEE Trans. ASSP*, ASSP-34, no. 2, pp. 331–41, 1986; corrections in *IEEE Trans. Acoust., Speech, Signal Processing*, vol. ASSP-34, p. 633, June 1986.
21. H. B. Lee, and M. Wengrovitz, "Resolution Threshold of Beamspace MUSIC for Two Closely Spaced Emitters," *IEEE Trans. on Acoust., Speech, Signal Processing*, Sept. 1990.
22. B. Gatlin, "Superresolution of Multiple Noise Sources in Antenna Beam," *IEEE Trans. on Antennas and Propagation*, May 1983.
23. J. Jachner, "High-Resolution Direction Finding in Multi-Dimensional Scenarios," MIT Ph.D. Thesis, September 1993.
24. J. Jachner, and H. Lee, "Cramér Rao Bounds on Direction Estimates for Closely Spaced Emitters in Multi-Dimensional Applications," *ICASSP-92*, vol. II, pp. 513–16, March 1992.

3

Robustness and Sensitivity Analysis for Eigenspace Localization Methods

W. Radich, R. Hamza, and K. Buckley

3.1 INTRODUCTION

As a community of signal processors investigating the possibilities of determining locations of narrowband, cofrequency sources from array data, we have strived for higher and higher performance as measured by location estimate accuracy and by resolvability of closely positioned sources. Concerning higher resolution, one of the topics that has received intense consideration is eigenspace-based spectral methods (e.g., [1–6]). Spectral methods are attractive because, compared to multidimensional search methods (e.g., those based on least squares, maximum likelihood, and MAP estimation [7–11]), they are computationally inexpensive. Ideally, eigenspace spectral methods provide unbiased estimates, and thus compared to more traditional spatial filter-based methods [12–14], they can provide higher resolution. In this chapter we focus for the most part [1] on eigenspace-based spectral methods. [2]

[1] For background and completeness, several beamformer-based spatial-spectrum estimators will also be reviewed.

[2] Rotational invariance [15, 16] and polynomial rooting [17] eigenspace methods have been developed for specific array configurations which are roughly similar to their spectral counterparts in terms of performance

To implement eigenspace-based spectral approaches to source localization, the following are required: (1) a well-designed (unambiguous) array; (2) noncoherent sources of number less than the number of sensors; (3) the exact model of sources are observed across the outputs of the array elements; and (4) a sufficient second-order characterization of the array data for given signal and noise assumptions.[3] We know how to specify unambiguous arrays, and we have special techniques for handling coherent sources. It is the impracticality of the latter two requirements that justifies our continued investigation into eigenspace spectral techniques. In this chapter we concentrate on the performance and enhancement of eigenspace-based spatial-spectral estimators under the condition that the source observation model cannot be exactly characterized, and given only an estimate of the output data (observation) covariance matrix. We assume that the noise covariance is known.[4]

Issues associated with observation modeling fall under the categories of array calibration (the estimation of the source observation model), sensitivity analysis, and robust algorithm development. We assume that a nominal source observation model is given, obtained either analytically or via calibration. We thus concentrate on robust algorithm development and sensitivity analysis.

Resolution relates to ability to distinguish between closely located sources. Robustness herein refers to a lack of sensitivity to inaccuracies in the model of the observation of the source across the array. Resolution and robustness go hand-in-hand. On the one hand, we desire algorithms that are highly sensitive to source location (primary) parameters. On the other hand, we need estimators with a certain degree of insensitivity to secondary parameters associated with the source (e.g., frequency, polarization, range), the propagation channel (e.g., conductivities, boundaries) and the array (e.g., sensor positions and responses). Resolution and robustness are seemingly conflicting objectives, and thus one might strive to achieve a proper balance between them. One might also ask if we can affect both enhanced resolution and improved robustness. As shown in this chapter, the answer to this is a qualified yes. In some cases only, is it possible to achieve both. In this chapter we aim to provide an understanding of and tools for achieving a balance between spectral resolution and robustness, and for pursuing both objectives simultaneously. The understanding presented here is geometric, being based on weighting in a linear observation space, and on principal component subspaces.

Section 3.2 of this chapter is a standard description of the array observation of a narrowband source, and a basic review of high-resolution spatial-spectrum estimation. We describe popular beamformer- and eigenspace-based estimators. An approach to spectral resolution enhancement is described which is based on weighting in the observation space. In section 3.3, source modeling errors are discussed. A mathematical model is presented. Several physical sources of such errors are

and computational requirements. Although we explicitly consider only spectral methods here, much of this discussion pertains to these other eigenspace methods.

[3] We assume additive noise which is uncorrelated with source-generated signals, so that the exact covariance matrices of the array output data and noise vectors are required.

[4] Several spectrum estimation approaches discussed herein do not require knowledge of the noise covariance. Concerning those that do, see, for example, [18, 19] for sensitivity analysis and robust algorithm development paralleling that presented in section 3.5.

described which were selected to exemplify model error structure in the observation space. In section 3.4 several robust spatial-spectrum estimators are described. These are modifications of basic methods described in section 3.3, designed to reduce sensitivity to model error (generally at the expense of resolution). Several also incorporate weighting to provide enhanced resolution. In section 3.5 model error sensitivity is determined for a general class of weighted eigenspace-based estimators, and robust estimators are derived by identifying the weighting which minimizes this sensitivity. Numerical examples are then presented in section 3.6, and in section 3.7 the chapter is summarized.

Notation

Generally we adhere to the following notational convention: matrices are represented in uppercase bold or italic; column vectors in lowercase bold or italic with over arrows; scalar variables in lowercase italic; and spatial-spectral functions in script uppercase. For scalar constants (e.g., sensor, sources) we try to conform to notation common in the literature.

Matrix and Vector Notation

$$\mathbf{X}^H = \text{Hermitian transpose of matrix } \mathbf{X}$$

$$\mathbf{X}^T = \text{Transpose of } \mathbf{X}$$

$$\mathbf{X}^* = \text{Complex conjugate of } \mathbf{X}$$

$$\hat{x} = \text{an estimate of } x$$

$$\dot{x}_i = \frac{\delta x}{\partial \theta_i}; \text{ the first partial derivative of } x \text{ with respect to } \theta_i$$

$$\ddot{x}_{ij} = \frac{\partial^2 x}{\partial \theta_j \partial \theta_i}; \text{ the second partial derivative of } x \text{ with respect to } \theta_i \text{ and } \theta_j$$

$$\mathbf{X}^+ = \text{Left pseudoinverse of } \mathbf{X}$$

$$\Delta x = \hat{x} - x$$

$$\mathcal{R}\{x\} = \text{the real part of the complex variable } x$$

$$Tr\{\mathbf{X}\} = \text{the trace of the matrix } \mathbf{X}$$

$$E[x] = \text{the expected value of } x$$

Principal Symbols

$$M, d, N = \text{the number of sensors, sources, and snapshots}$$

$$\vec{x}(t) = \text{the } M \times 1 \text{ array data vector at time } t$$

$$\vec{w} = \text{the } M \times 1 \text{ coefficient vector of an array spatial filter}$$

θ = the source arrival angle relative to broadside of a linear array

$\mathcal{D}(\theta)$ = the null (inverse) spatial spectrum

$\omega_i(\theta) = 2\pi l_i \sin\theta, i = 1, \ldots, M$

l_i = the spacing between the ith sensor element and the reference point

$\vec{\mathbf{a}}(\theta) = [1, e^{-j\omega_1(\theta)}, \ldots, e^{-j\omega(M-1)^{(\theta)}}]^T$; the $M \times 1$ steering vector at direction θ

$\vec{\mathbf{b}}(\theta) = \vec{\mathbf{a}}(\theta)/\|\vec{\mathbf{a}}(\theta)\|$

$\vec{d}(\theta_i) = \frac{\partial}{\partial\theta_i}\vec{\mathbf{a}}(\theta_i)$, for all $i = 1, \cdots, d$

$A = [\vec{\mathbf{a}}(\theta_1, \cdots, \vec{\mathbf{a}}(\theta_d)]$

$D = [\vec{d}(\theta_1), \vec{d}(\theta_2), \cdots, \vec{d}(\theta_d)]$

$R_x = E\left[\vec{x}(t)\vec{x}^H(t)\right]$; the data covariance matrix

λ_i = the ith ordered eigenvalue of R_x

$\vec{e}_i$ = the ith normalized eigenvector of R_x

E_s, E_n = matrices of the signal and noise subspace eigenvectors

Ψ, Γ = the signal and noise subspaces

$P = P_A$ the signal subspace projection operator; $P = AA^H$

$P^{\perp} = P_A^{\perp} = I_m - P = E_n E_n^H.$

3.2 EIGENSPACE SPATIAL-SPECTRUM ESTIMATION

This section provides an overview of eigenspace DOA estimation under the assumption that the employed source observation model is accurate. We first describe this model and identify its associated eigenstructure property which forms the basis of eigenspace-based DOA estimation. We then present a general class of eigenspace spatial spectrum estimation which can be tuned both to enhance resolution and to combat model pertubation errors. We close this section by presenting results of finite data analysis which are pertinent to the sensitivity analysis discussed in section 3.4.

The Narrowband Observation

Consider, as a function of temporal index t, an M-dimensional narrowband spatial observation vector $\vec{x}(t)$ which could be the output of either M sensor-array elements or M channels of an array preprocessor (e.g., a beamspace preprocessor). Let θ represent the direction-of-arrival (DOA) of an impinging source, and $\vec{a}(\theta)$ denote an M-dimensional array response vector—the standard narrowband source observation model which is composed of the relative magnitudes and phases across the observation of a signal impinging form θ. Assume additive noise uncorrelated with the source signals, and $d < M$ sources impinging from locations $\theta_i; i = 1, 2, \ldots, d$, with $\vec{\theta} = [\theta_1, \theta_2, \ldots, \theta_d]^T$. Let $\vec{x}(t)$ denote an array observation vector

$$\vec{x}(t) = A(\vec{\theta})\vec{s}(t) + \vec{n}(t) \tag{3.1}$$

where $A(\vec{\theta})$ is the $M \times d$ matrix whose linearly independent columns are the array response vectors $\vec{\mathbf{a}}(\theta_i)$, for $i = 1, 2, \ldots d$. $\vec{s}(t)$ is the d vector of source time-series as observed at the array phase center, and $\vec{n}(t)$ is additive noise. The covariance matrix of $\vec{x}(t)$ is

$$\mathbf{R}_x = \mathbf{A}(\vec{\theta})\mathbf{P}_s\mathbf{A}^H(\vec{\theta}) + \sigma_n^2\mathbf{R}_n \tag{3.2}$$

where the $d \times d$ source covariance matrix $\mathbf{P}_s$ is full rank only if the sources are noncoherent. $\mathbf{R}_n$, the normalized (with trace equal to M) noise spatial covariance matrix, is the noise model commonly used. $\mathbf{R}_n$ represents sufficient information for the estimation approaches considered here and is statistically sufficient if the noise is zero-mean Gaussian. Note that we assume that the noise is uncorrelated with the source signals.

Algorithms considered here exploit properties of the eigenstructure of $\mathbf{R}_x$. For noncoherent sources, assuming the $\mathbf{R}_n$ is known and prewhitened if necessary, the eigenstructure representation is

$$\mathbf{R}_x = \mathbf{A}(\vec{\theta})\mathbf{P}_s\mathbf{A}^H(\vec{\theta}) + \sigma_n^2\mathbf{I}_M = \sum_{i=1}^{M} \lambda_i\vec{e}_i\vec{e}_i^H = \mathbf{E}\Lambda\mathbf{E}^H \tag{3.3}$$

where $\mathbf{I}_M$ is the M-dimensional identity matrix. The eigenstructure representation of $\mathbf{R}_x$ in (3.3) leads to the following eigenspace decomposition:

$$\mathbf{R}_x = \mathbf{E}_s\Lambda_s\mathbf{E}_s^H + \sigma_n^2\mathbf{E}_n\mathbf{E}_n^H, \tag{3.4}$$

where $\mathbf{E} = [\mathbf{E}_s \mid \mathbf{E}_n]$, and the columns of $\mathbf{E}_s$ and $\mathbf{E}_n$ are, respectively, the d signal and the $M - d$ noise eigenvectors. The column space of $\mathbf{E}_s$ is that of $\mathbf{A}(\vec{\theta})$. The column spaces of $\mathbf{E}_s$ (denoted Ψ) and $\mathbf{E}_n$ (denoted Γ), termed the signal and noise subspaces respectively, are orthogonal complements. $\Lambda_s = \mathrm{Diag}\{\lambda_1, \lambda_2, \ldots, \lambda_d\}$ where $\lambda_i > \sigma_n^2$, for $i = 1, 2, \ldots, d$. It is the orthogonal complement property of Ψ and Γ relative to the range of $\mathbf{A}(\vec{\theta})$, whose columns are the array response vectors of the d impinging sources, which is central to both the high-resolution algorithms considered here and their robust extensions.

The finite sample effects discussed in section 3.1 are a result of estimating $\mathbf{R}_x$ from available data. Let $\hat{\mathbf{R}}_x$ denote the sample covariance matrix estimate of $\mathbf{R}_x$, obtained by averaging the outer products of N observation vectors:

$$\hat{\mathbf{R}}_x = \frac{1}{N}\sum_{n=1}^{N} \vec{x}(t_n)\vec{x}^H(t_n) \tag{3.5}$$

Let $\hat{\lambda}_i; i = 1, 2, \ldots, M; \hat{\lambda} \geq \hat{\lambda}_{i+1}$ and $\hat{\vec{e}}_i; i = 1, 2, \ldots, M$ be the ordered eigenvalues and corresponding normalized eigenvectors of $\hat{\mathbf{R}}_x$. The estimated signal subspace $\hat{\Psi}$ and noise subspace $\hat{\Gamma}$ are, respectively, the column spaces of the matrices

$$\hat{\mathbf{E}}_s = \left[\hat{\vec{e}}_1, \hat{\vec{e}}_2, \ldots, \hat{\vec{e}}_{\hat{d}}\right], \hat{\mathbf{E}}_n = \left[\hat{\vec{e}}_{\hat{d}+1}, \hat{\vec{e}}_{\hat{d}+2} \ldots, \hat{\vec{e}}_M\right] \tag{3.6}$$

It is often assumed in the analysis of eigenspace-based estimators that $\hat{d} = d$, even though estimators of d are very sensitive to eigenstructure characteristics of $\mathbf{R}_n$.

Spatial-Spectrum Estimation

The well-known MUSIC algorithm [4, 20] forms a spatial-spectral estimate based on projections onto the entire noise-only subspace Γ. Thus, given the ideal covariance matrix $\mathbf{R}_x$, the true source locations can be found by identifying the vectors within the array manifold which are orthogonal to the noise subspace. That is, by finding the zeros of the null spectrum

$$\mathcal{D}_{mu}(\theta, \mathbf{E}_s) = \frac{\vec{\mathbf{a}}^H(\theta)(\mathbf{I}_M - \mathbf{E}_s\mathbf{E}_s^H)\vec{\mathbf{a}}(\theta)}{\vec{\mathbf{a}}^H(\theta)\vec{\mathbf{a}}(\theta)} \tag{3.7}$$

In practice, the true noise subspace is unknown, so the DOA estimates are given by the d minima of

$$\mathcal{D}_{mu}(\theta, \hat{\mathbf{E}}_s) = \frac{\vec{\mathbf{a}}^H(\theta)(\mathbf{I}_M - \hat{\mathbf{E}}_s\hat{\mathbf{E}}_s^H)\vec{\mathbf{a}}(\theta)}{\vec{\mathbf{a}}^H(\theta)\vec{\mathbf{a}}(\theta)} \tag{3.8}$$

Resolution-enhanced, eigenspace-based spectrum methods can be achieved by emphasizing certain subspaces of the observation space [21]. The idea is to project onto a subspace of $\hat{\Gamma}$ that is in some sense *closest* to the array manifold over a given source sector. A particular class of CLOSEST spectral estimators is described as follows. Let Θ represent a location sector of interest, and form the source covariance matrix

$$\mathbf{R}_\Theta = \int_\Theta \vec{\mathbf{a}}_\theta \vec{\mathbf{a}}_\theta^H \, d\theta \tag{3.9}$$

Let $(\eta_i; i = 1, 2, \ldots, M; \eta_i \geq \eta_{i+1})$ and $(\vec{u}_i; i = 1, 2, \ldots, M)$ be the ordered eigenvalues and corresponding eigenvectors of $\mathbf{R}_\Theta$. The range of the $M \times D_\Theta$ matrix

$$\mathbf{U}_\Theta = \left[\vec{u}_1, \vec{u}_2, \ldots, \vec{u}_{D_\Theta}\right] \tag{3.10}$$

is termed the *source representation subspace*. The CLOSEST spectrum estimator considered here is described by the null spectrum

$$\begin{aligned} \mathcal{D}_c(\theta) &= \frac{\vec{\mathbf{a}}^H(\theta)\hat{\mathbf{P}}^\perp \mathbf{U}_\Theta \mathbf{U}_\Theta^H \hat{\mathbf{P}}^\perp \vec{\mathbf{a}}(\theta)}{\vec{\mathbf{a}}^H(\theta)\vec{\mathbf{a}}(\theta)} \\ &= \frac{|\vec{\mathbf{a}}^H(\theta)\hat{\mathbf{P}}^\perp \mathbf{U}_\Theta|^2}{\vec{\mathbf{a}}^H(\theta)\vec{\mathbf{a}}(\theta)} \end{aligned} \tag{3.11}$$

where $\hat{\mathbf{P}}^\perp = \hat{\mathbf{E}}_n\hat{\mathbf{E}}_n^H$. In [21] the resolution enhancement properties of CLOSEST are identified with the fact that the columns of the weighting matrix $\mathbf{U}_\Theta$ essentially span the subspace represented by the manifold over the DOA sector Θ.[5]

[5] Besides MUSIC and the CLOSEST estimator described by (3.11), in this chapter we are interested in what have been termed the MIN-NORM [6] and FINE [21] estimators. Both of these are of the (3.11) form,

Spatial filter (beamformer) techniques predate high-resolution subspace methods for source location estimation. In general, for each DOA θ on a dense grid, a beamformer with weighting vector $\vec{\omega}(\theta)$ is designed. A spatial power spectrum is then constructed as an estimate of

$$\mathcal{P}_{\text{BEAM}}(\theta) = E\{|\vec{w}^H(\theta)\vec{x}(t)|^2\} = \vec{w}^H(\theta)\mathbf{R}_x\vec{w}(\theta) \tag{3.12}$$

In a conventional fixed beamformer approach each $\vec{w}(\theta)$ is designed, independent of the data statistics, so as to provide a good spatial filter response.

The minimum variance distortionless response (MVDR) beamformer provides a power spectrum estimate as a function of θ by minimizing the output power of the spatial filter $y(t) = \vec{w}^H(\theta)\vec{x}(t)$ subject to the constraint that the resulting filter passes signals from the specified source location θ without attenuation. Consider the constrained minimization problem.

$$\min_{\vec{w}} \quad \vec{w}^H\mathbf{R}_x\vec{w} \tag{3.13}$$
$$\text{subj. to} \quad \vec{\mathbf{a}}^H(\theta)\vec{w} = 1$$

The optimal MVDR weight is given by

$$w_{\text{MVDR}}(\theta) = \frac{\mathbf{R}_x^{-1}\vec{\mathbf{a}}(\theta)}{\vec{\mathbf{a}}^H(\theta)\mathbf{R}_x^{-1}\vec{\mathbf{a}}(\theta)} \tag{3.14}$$

and thus results in the power density estimate $\mathcal{P}_{\text{MVDR}}(\theta) = \mathcal{D}_{\text{MVDR}}^{-1}(\theta)$, with the corresponding MVDR null spectrum

$$\mathcal{D}_{\text{MVDR}}(\theta) = \vec{\mathbf{a}}^H(\theta)\mathbf{R}_x^{-1}\vec{\mathbf{a}}(\theta) \tag{3.15}$$

As reported in [22], the relationship of the MVDR null spectrum to eigenspace methods becomes apparent by expressing (3.15) in terms of the eigenstructure of $\mathbf{R}_x$:

$$\mathcal{D}_{\text{MVDR}}(\theta) = \vec{\mathbf{a}}^H(\theta)\,[\mathbf{E}_s \mid \mathbf{E}_n] \begin{bmatrix} \frac{1}{\lambda_1} & & & & & 0 \\ & \ddots & & & & \\ & & \frac{1}{\lambda_d} & & & \\ & & & \frac{1}{\sigma_n^2} & & \\ & 0 & & & \ddots & \\ & & & & & \frac{1}{\sigma_n^2} \end{bmatrix} [\mathbf{E}_s \mid \mathbf{E}_n]^H\,\vec{\mathbf{a}}(\theta) \tag{3.16}$$
$$= \vec{\mathbf{a}}^H(\theta)\left(\mathbf{E}_s\boldsymbol{\Lambda}_s^{-1}\mathbf{E}_s^H + \frac{1}{\sigma_n^2}\mathbf{E}_n\mathbf{E}_n^H\right)\vec{\mathbf{a}}(\theta)$$

Note that at high SNR ($\lambda_i \gg \sigma_n^2$ for $i = 1, \ldots, d$), (3.16) reduces to MUSIC: $\mathcal{D}_{\text{MVDR}}(\theta) \approx \frac{1}{\sigma_n^2}\vec{\mathbf{a}}^H(\theta)\mathbf{E}_n\mathbf{E}_n^H\vec{\mathbf{a}}(\theta)$.

The MVDR weight vector obtained in (3.14) is the result of a single DOA-dependent linear constraint. In [23] the MVDR spatial filter is extended to a full

where $\mathbf{U}_\Theta$ is a column vector. For FINE, which is a specific CLOSEST estimator, $\mathbf{U}_\Theta$ is $\vec{u}_1$ from (3.10). For MIN-NORM, $\mathbf{U}_\Theta$ is the first standard basis vector $[1, 0, \ldots, 0]^T$.

3-dimensional space to localize the internal sources of EEG activity. Because an EEG source is represented by a rank-3 subspace model, three rank-1 spatial filters corresponding to mutually orthogonal dipole orientations must be constructed to pass the desired signal without distortion.

Consider the more general constraint specification of the linearly constrained minimum variance (LCMV) problem:

$$\begin{array}{ll} \min\limits_{\vec{w}} & \vec{w}^H \mathbf{R}_x \vec{w} \\ \text{subj. to} & \mathbf{C}^H(\theta)\vec{w} = \vec{f} \end{array} \tag{3.17}$$

where $\vec{f}(\theta)$ is the $(J \times 1)$ desired response vector and $\mathbf{C}(\theta)$ is an $(M \times J)$ constraint matrix. The optimum solution is [24]

$$\vec{w}_{\text{LCMV}}(\theta) = \mathbf{R}_x^{-1}\mathbf{C}(\theta)(\mathbf{C}^H(\theta)\mathbf{R}_x^{-1}\mathbf{C}(\theta))^{-1}\vec{f}(\theta) \tag{3.18}$$

and results in the null spectrum

$$\mathcal{D}_{\text{LCMV}}(\theta) = \frac{1}{\vec{f}^H(\theta)(\mathbf{C}^H(\theta)\mathbf{R}_x^{-1}\mathbf{C}(\theta))^{-1}\vec{f}(\theta)}. \tag{3.19}$$

The more general form of the constraint equation in (3.17) allows the null spectrum given above to incorporate a broad number of different desired characteristics. In [25] eigenvector constraints were used for broadband spatial filtering to pass signals from a single source location θ, over a range of temporal frequencies. In addition to eigenvector constraints, point and derivative constraints have been used to control response level and mainlobe shape at desired steer locations, and also to force nulls in other directions.

As will be discussed in section 3.4, the linear constraint formulation can be extended to include quadratic constraints [26]. The quadratic constraint provides an effective means to combat the problem of signal cancellation due to mismatch between actual and assumed array response.

Relative Performance Characteristics

Above we have presented several of the most popular spatial spectrum estimators. In order of improving resolvability, these are: the conventional fixed beamformer, the LCMV, the MVDR, the MUSIC, the MIN-NORM, and the CLOSEST estimator. The resolution of the fixed beamformer approach is limited by the array aperture (the array spatial extent, measured in propagation wavelengths). For LCMV, as the number of linear constraints approaches the number of sensors, performance ranges from that of MVDR to that of a conventional fixed beamformer-based spatial spectrum estimator. Asymptotically (as the number of observation goes to infinity) eigenspace estimators have perfect resolution. As noted above, CLOSEST is a subspace approach which is resolution enhanced (relative to MUSIC, for a finite number of observations). Resolvability of MVDR depends on SNR. At low SNR, MVDR behaves as does the conventional fixed beamformer approach. As noted above, as SNR goes to infinity, MVDR reduces to MUSIC in which case it will provide infinite resolution (asymptotically).

Relative performance as measured by DOA estimate variance depends to a great extent on the particular situation. Although for selected situations each of these estimators will achieve Cramér-Rao bounds, in general none are asymptotically efficient. Of a general weighted eigenspace-based spectral class which includes the eigenspace methods mentioned above, it has been shown [38, 39] that MUSIC is optimum. However, in [45] it is established that CLOSEST methods provide DOA variance comparable to MUSIC (sufficient conditions on U_Θ for equivalent variance are identified), whereas MIN-NORM provides substantially higher variance.

Either DOA estimate bias or variance can limit resolvability. For beamformer-based methods, bias plays a predominant role. In [44, 45] it is shown that resolution advantage of CLOSEST over MUSIC is due to reduced DOA estimate bias.

For the previous discussion on relative resolution and DOA estimate variance and bias, it is assumed that the array response vectors are known. Generally speaking, concerning modeling errors and the algorithms considered to this point, the better the resolvability, the more sensitive to modeling errors. However, relative sensitivity will depend on the specifics of the given situation.[6] In sections 3.4 and 3.5 we in effect exploit this dependence. We describe approaches which can, for some situations, provide both enhanced resolution and reduced sensitivity to modeling errors.

3.3 MODELING NARROWBAND SOURCE OBSERVATION ERRORS

In this section we present a quantitative description of the source observation uncertainty. That is, we model the source modeling error. We assume that this error is described by a stochastic model. Therefore, for a given scenario, each source observation vector $\vec{\mathbf{a}}(\theta)$ is viewed as one realization from a distribution over the observation space.

In array processing, source observation model errors result from a wide variety of physical uncertainties involving the array, the channel, and the actual sources. Errors may be, for example, additive or multiplicative. More generally, they are either linear or nonlinear. Thus, there is no single best model for source modeling errors. Our approach here, to model the uncertainty as linear, is standard. Nonlinear errors are linearized. Thus, the model is valid for either large or small additive errors, but otherwise only small errors are considered.

In this section we first formulate the additive model of source observation uncertainty. We then base this model on reality by relating it to several physical causes of source modeling uncertainty. The section closes with an observation-space (geometric) discussion of characteristics of modeling errors that might be exploited in the development of robust algorithms.

---

[6] For example, in [28] it is illustrated that to a commonly considered type of model perturbation (moderate complex perturbation of array response vectors) FINE is no more sensitive than MUSIC.

Additive Modeling of Source Model Errors

As described in section 3.2, our model of the array observation of a source is the array response vector $\vec{\mathbf{a}}(\theta)$. Since we assume that this vector only approximates the actual array response to a source impinging from θ, we also call this model vector the ideal or the nominal vector. We start by considering the following simple additive model of error:

$$\hat{\vec{\mathbf{a}}}(\theta) = \vec{\mathbf{a}}(\theta) + \Delta\vec{\mathbf{a}}(\theta) \tag{3.20}$$

The vector $\Delta\vec{\mathbf{a}}(\theta)$ is the additive perturbation from the nominal value. In the development of robust algorithms we will sometimes exploit characteristics of the $\Delta\vec{\mathbf{a}}(\theta)$. At these times we will assume that it is a zero-mean vector with covariance $\mathcal{B}(\theta)$.[7] The eigenstructure of this matrix describes the distribution of error energy in the observation space along principal component directions.

Thinking of the actual array response vector in terms of a perturbation from a nominal value is in many situations appropriate, in particular when our uncertainty is limited so that our prior identification of a nominal observation vector is fairly accurate. However, in some practical cases we can encounter uncertainty that results in a substantial variation in the direction of an array response vector. Of course, if this variation is distributed substantially throughout the entire observation space, locating sources may be futile. But as illustrated below, situations arise where the orientation of the array response vector in the observation space, while substantially uncertain, is effectively limited to a low-rank subspace. For these structured situations, consider the following model:

$$\vec{\mathbf{a}}(\theta) = \mathbf{G}(\theta)\vec{m}(\theta) \tag{3.21}$$

where the $p < K$ columns of the $K \times p$-dimensional deterministic matrix $\mathbf{G}(\theta) = \left[\vec{g}_1(\theta), \vec{g}_2(\theta), \ldots, \vec{g}_p(\theta)\right]$ identify for each θ a p-dimensional subspace in which $\vec{\mathbf{a}}(\theta)$ is known to be restricted. The stochastic, unknown vector $\vec{m}(\theta)$ determines where $\vec{\mathbf{a}}(\theta)$ is in the range of $G(\theta)$.

Whereas (3.20) directly models the source modeling error as linear, (3.21) models the source model as linear in the modes $\vec{g}_i(\theta)$; $i = 1, 2, \ldots, p$. Although (3.20) generally pertains to both large and small modeling errors, we use it below when considering small-error sensitivity analysis and robust algorithms. Discussions related to (3.21) relate to both large and small structured errors. Note that (3.20) also represents structured modeling errors when $\mathcal{B}(\theta)$ is rank deficient or has large eigenvalue spread.

Examples of Modeling Source Modeling Errors as Additive

This section begins with a general discussion of small array response vector gain and phase errors which can be modeled as additive perturbations as in (3.20). Then

[7] This matrix $\mathcal{B}(\theta)$ is central to the observation space weighting approach to robustness that is discussed in this chapter. As shown below, in some important cases this matrix is independent of DOA and can be represented as $\mathcal{B}$.

several structured modeling error examples of both the (3.20) and (3.21) type are presented. Included are examples of uncertainty associated with the array platform, with the propagation channel and with source parameter variation.

Small Phase and Gain Errors. The following model is general in the sense that many types of small modeling errors can be described in terms of gain and phase uncertainties. Such errors may be due to changes in the surrounding environment, undesirable channel interacts, sensor misplacements, and reverberation in the antenna platform. For example see [27] for details.

Let $g_k(\theta)$ and $\tau_k(\theta)$ be respectively the nominal gain and phase response of the kth sensor to a signal impinging from the angle θ, that is, the kth sensor response

$$a_k(\theta) = g_k(\theta)e^{j\tau_k(\theta)} \tag{3.22}$$

For the random antenna element imperfection model, each sensor response deviates from the exact value of (3.22) due to drift in gain and phase response. Denote the additive deviation error in the gain component $g_k(\theta)$ by $\Delta g_k(\theta)$ and the additive deviation error in the phase $\tau_k(\theta)$ by $\Delta\tau_k(\theta)$, then the resulting perturbed k^{th} sensor response may be written as

$$\begin{aligned}\hat{a}_k(\theta) &= (g_k(\theta) + \Delta g_k(\theta))e^{j[\tau_k(\theta)+\Delta\tau_k(\theta)]} \\ &= g_k(\theta)e^{j\tau_k(\theta)}\left(1 + \frac{\Delta g_k(\theta)}{g_k(\theta)}\right)e^{j\Delta\tau_k(\theta)} \end{aligned}\tag{3.23}$$

The perturbed k^{th} sensor response can always be expressed as a sum of the nominal value of the kth sensor response and some additive error term, that is, $\hat{a}^k(\theta) = a_k(\theta) + \Delta a_k(\theta)$. This error term can be determined from (3.23) to be

$$\begin{aligned}\Delta a_k(\theta) &= a_k(\theta)\left[\left(1 + \frac{\Delta g_k(\theta)}{g_k(\theta)}\right)e^{j\Delta\tau_k(\theta)} - 1\right] \\ &= a^k(\theta)\gamma_k(\theta) \end{aligned}\tag{3.24}$$

The perturbation of the array response vector (steered to angle θ_η) can thus be described by the equation

$$\Delta\vec{\mathbf{a}}(\theta_\eta) = \mathbf{\Sigma}_\eta\vec{\mathbf{a}}(\theta_\eta) \tag{3.25}$$

where $\mathbf{\Sigma}_\eta = \text{diag}\{\gamma_1(\theta_\eta), \cdots, \gamma_M(\theta_\eta)\}$. If there exists uncalibrated mutual coupling between the elements of the array, the matrix $\mathbf{\Sigma}_\eta$ is no longer a diagonal matrix, that is, it may have non-zero off-diagonal components.

A simple model that has often been used (e.g., [27]) for performance analysis is to assume that the pertubation error has a white Gaussian distribution of the following form:

$$E\left[\Delta\vec{\mathbf{a}}(\theta_\eta)\Delta\vec{\mathbf{a}}^H(\theta_\zeta)\right] = \delta_{\eta\zeta}\mathcal{B} \tag{3.26}$$

$$E\left[\Delta\vec{\mathbf{a}}(\theta_\eta)\Delta\vec{\mathbf{a}}^T(\theta_\zeta)\right] = 0 \tag{3.27}$$

where $\delta_{\eta\zeta} = 0$ for $\eta \neq \zeta$. If the perturbation errors are not correlated from sensor to sensor, $\mathcal{B}$ will be simply a diagonal matrix.

Flexible Array Platforms. Examples of flexible arrays are: towed SONAR arrays [29]; aircraft wing mounted arrays [30]; and space-deployed arrays [31]. These arrays have configurations which are random but physically constrained, and there is evidence that sensor position errors can be represented as linear combinations of modes. As shown below, these modal sensor position errors can correspond to additive, structured modeling errors.

Sensor position errors for large space-deployed arrays were analyzed by Shaw [31]. Consider the (3.20) model error type. Assume, for example, a line array and arrival angle θ measured relative to array broadside. Consider the kth sensor positioned, nominally, at location x_k on the array axis. Let $z(x_k)$ be the position error (restricted to the array axis for simplicity). The actual sensor response is

$$\hat{a}_k(\theta) = e^{j\omega(x_k \sin\theta + z(x_k)\cos\theta)} \tag{3.28}$$

where ω is the spatial frequency. For small position errors

$$\hat{a}_k(\theta) = a_k(\theta) + j\omega\cos\theta z(x_k)e^{j\omega x_k \sin\theta} \tag{3.29}$$

where

$$a_k(\theta) = e^{j\omega x_k \sin\theta} \tag{3.30}$$

Modal position errors are of the form

$$z(x_k) = \sum_{n=1}^{N} a_n \sin(n\pi x_k/L) + b_n\left[1 - \cos(n\pi x_k/L)\right] \tag{3.31}$$

where L is the array aperture. For one mode (e.g., $n = 1$), the actual array response vector is

$$\hat{\vec{\mathbf{a}}}(\theta) = \vec{\mathbf{a}}(\theta) + j\omega a_1 \cos\theta \vec{g}_1(\theta) + j\omega b_1 \cos\theta \vec{g}_2(\theta) \tag{3.32}$$

where a_1 and b_1 are random variables, and $\vec{g}_1(\theta)$ and $\vec{g}_2(\theta)$ are deterministic vectors. Thus, the actual array response vector is modeled as a structured, additive perturbation on the nominal one.

In [30] distortion of an aircraft wing mounted array is modeled by the parabolic function

$$z(x_k) = \alpha(x_k)^2 \tag{3.33}$$

where α is an unknown distortion parameter. For a uniform linear array $x_k = \frac{\lambda}{2}(k-1)$ and

$$z(x_k) = \alpha\frac{\lambda^2}{4}(k-1)^2 \tag{3.34}$$

so that for small errors

$$\hat{a}_k(\theta) = a_k(\theta) + j\pi\gamma\cos\theta(k-1)^2 e^{j\pi(k-1)\sin\theta} \tag{3.35}$$

where the new constant γ is defined by $\gamma = \alpha\frac{\lambda}{2}$. Equation (3.35) can be written in the vector form given by (3.20) with

$$\Delta\vec{\mathbf{a}}(\theta) = j\pi\gamma\cos\theta \begin{bmatrix} 0 \\ 1 \\ 4 \\ \vdots \\ (M-1)^2 \end{bmatrix} \odot \vec{\mathbf{a}}(\theta) \tag{3.36}$$

It should be noted that the approximation given in (3.35) is not valid unless $\gamma(k-1)^2$ is close to zero. As a result, for a large array the approximation may be poor for the outer elements (because $(M-1)^2$ may be large compared to γ).

If γ is modeled as a zero-mean random variable, then from (3.36) we see that $\mathcal{B}(\theta)$ is a rank-1 matrix. Furthermore, for θ close to zero, the eigenvector corresponding to the nonzero eigenvalue of $\mathcal{B}$ is simply a constant times $[0\,1\,4\cdots(k-1)^2\cdots(M-1)^2]^T$.

Matched-Field Processing Underwater Multipath Propagation. In many applications, the effects of multipath propagation and refraction make the simple plane wave model of array observation a poor approximation. If the propagation characteristics are known, a propagation model for the signal response from a source at any array input can be formulated. Matched field processing is a generalization of the plane wave processing, which exploits this propagation model for the reception and localization of sources. Because of the complicated and variable nature of matched-field array response vectors, robust methods are required. However, arrays with good matched-field beam patterns are typically impossible to design, in which case high-resolution methods are necessary.

For example, Krolik [32] describes a structured underwater multipath propagation model, of Eq. (3.21) form, which accounts for variation in temperature profile. To summarize this, consider underwater source location parameters range and depth denoted by θ, and consider a vertical array of M hydrophones. Krolik showed that an observation is of the form

$$\vec{\mathbf{a}}(\theta) = \mathbf{P}(\theta)\vec{s}(\theta) \tag{3.37}$$

where $\mathbf{P}(\theta)$ is the $M \times N$ matrix with kn^{th} element corresponds to the attenuation and propagation delay of the nth (of N) modal signal component at the kth element, and the elements of $\vec{s}(\theta)$ correspond to (scaled) modal phase perturbations which depend on the temperature profile. For reasonable ranges of ocean temperature profiles, Krolik showed that $\mathbf{P}(\theta)\vec{s}(\theta)$ was low rank (i.e., rank much less than either M and N), so that (3.37) can be replaced with (3.21) with small p.

Source Parameter Variation in EEG and MEG. Recent research in EEG (electroencephalogram) and MEG (magnetoencephalogram) signal processing has been directed toward high-resolution localization of evoked brain activity regions. In [33], a modified MUSIC estimator has been used successfully for this application. The modified algorithm used is the one considered by Schmidt [4] and Buckley [34]. As with any application, modeling error sensitivity is of concern. Source observation modeling is based on a dipole source model, which is generally considered to only approximate a true source.

For a single dipole source, denote the gain matrix that describes the transfer function from 3-d source location $\vec{L} = [L_x, L_y, L_z]^T$ to the external sensors as the $(M \times 3)$ matrix $\mathbf{G}(\vec{L})$. The observed EEG/MEG measurement is described by

$$\vec{x}(t) = \mathbf{G}(\vec{L})\vec{q}(t) + \vec{n}(t) \tag{3.38}$$

where $\vec{q}(t) = [q_x(t), q_y(t), q_z(t)]^T$ is the dipole source moment vector. Now consider the case where the orientation of $\vec{q}(t)$ remains fixed over the observation interval. This can be modeled as $\vec{q}(t) = \vec{m}s(t)$, where $\vec{m}$ is the unit-norm, fixed-orientation vector, and $s(t)$ represents a time-varying scalar. For this case we can write.

$$\mathbf{G}(\vec{L})\vec{q}(t) = \left[\mathbf{G}(\vec{L})\vec{m}\right] s(t) \tag{3.39}$$

Equation (3.39) explicitly shows that the dipole source observation is restricted to a one-dimensional subspace. However, if the moment orientation $\vec{m}$ is assumed unknown, we can only indentify this rank-1 observation as existing somewhere within the subspace spanned by the columns of $\mathbf{G}(\vec{L})$.

Discussion

Given either (3.20) or (3.21), the descriptions in section 3.2 of the observation covariance matrix and its eigenstructure, Eqs. (3.3) (3.4), still apply. We just don't know the exact function of $\vec{\mathbf{a}}(\theta)$. Of course, one should consider obtaining a better estimate of $\vec{\mathbf{a}}(\theta)$ (i.e., calibration), but there will be practical limits to this approach, so that in the end we are left with some uncertainty. Our problem starts with (3.20) or (3.21) and ends with estimates of $\theta_i; i = 1, 2, \ldots, D$. As stated in the introduction, we are interested in controlling sensitivity to modeling errors, but we still want high resolution. How can this be done?

Before getting to this question in the next two sections let's consider, qualitatively, the effect that additive modeling errors have on the eigenspaces of the observation covariance matrix $\mathbf{R}_x$, and therefore on eigenspace-based spectrum estimators. Under (3.20), Eq. (3.3) becomes:

$$\mathbf{R}_x = (\mathbf{A}(\vec{\theta}) + \Delta\mathbf{A}(\vec{\theta}))\mathbf{P}_s(\mathbf{A}(\vec{\theta}) + \Delta\mathbf{A}(\vec{\theta}))^H + \sigma_n^2\mathbf{I}_k \tag{3.40}$$

where $\Delta\mathbf{A}(\vec{\theta}) = [\Delta\vec{\mathbf{a}}(\theta_1), \Delta\vec{\mathbf{a}}(\theta_2), \ldots, \Delta\vec{\mathbf{a}}(\theta_D)]$. These perturbations result in the following perturbation of the eigenstructure shown in (3.3):[8]

$$\mathbf{R}_x = (\mathbf{E}_s + \Delta\mathbf{E}_s)(\mathbf{\Lambda}_s + \Delta\mathbf{\Lambda}_s)(\mathbf{E}_s + \Delta\mathbf{E}_s)^H + \sigma_n^2(\mathbf{E}_n + \Delta\mathbf{E}_n)(\mathbf{E}_n + \Delta\mathbf{E}_n)^H \tag{3.41}$$

Let $\hat{\Psi}$ and $\hat{\Sigma}$ denote the actual signal and noise subspaces, from (3.41). Recall that eigenspace-based estimators exploit orthogonality properties of these observation space subspaces, that is, the $\vec{\mathbf{a}}(\theta_i); i = 1, 2, \ldots, D$ are orthogonal to $\hat{\Sigma}$. Unfortunately, the nominal vectors $\vec{\mathbf{a}}(\theta_i); i = 1, 2, \ldots, D$ that we use are

[8] As in section 3.2, we assume that the sources are noncoherent and the columns of $(\mathbf{A}(\vec{\theta}) + \Delta A(\vec{\theta}))$ are linearly independent. Thus, Λ_s is $D \times D$. Also we assume that the number of sources D is known. Given the first two of these assumptions, the last is no more restrictive here that it was under the idealistic situation considered in section 3.2 since the noise covariance is assumed known. (Perturbation of the noise covariance model would degrade our ability to estimate D from available observations.)

not. Consider processing in a subspace, with associated projection matrix $\mathbf{P}_{B^\circ}$, for which $\mathbf{P}_{B^\circ}, \Delta\vec{\mathbf{a}}(\theta_i) = \vec{0}; i = 1, 2, \ldots, D$. Processing in this subspace would be advantageous since then the $\mathbf{P}_{B^\circ}\vec{\mathbf{a}}(\theta_i); i = 1, 2, \ldots, D$ are orthogonal to $\hat{\Sigma}$. Such a subspace would include the union of the null space of $\mathcal{B}(\theta_i); i = 1, 2, \ldots, D$. Thus, it exists only if the $\mathcal{B}(\theta_i)$ are rank deficient, and design of the projector would require prior knowledge of the $\mathcal{B}(\theta_i)$. The point is not that such a subspace exists, although it might for some structured modeling errors. However, it does suggest a basic qualitative objective that can be quantitatively formalized. That objective is to stay away from observation subspaces of relatively high model error energy.

3.4 ROBUST HIGH-RESOLUTION SPATIAL SPECTRUM ESTIMATION

In this section we describe algorithms that incorporate high resolution with robustness to model errors. In doing so, utility is made at the end of the section of the additive model for perturbations in the array response. First, some previously developed, yet related, techniques are reviewed.

It should be mentioned that a number of other techniques not reviewed here exist if one is willing to increase the dimensionality of the search space. A straightforward Bayesian method for dealing with probabilistic array uncertainty is the MAP joint estimate of DOA's and random model parameters [35]. This method has the benefit of not requiring an additive model for the array uncertainty. In addition, recent work by Viberg and Swindlehurst [19] discusses algorithms that combat the combined variance due to finite sample and model error effects in a weighted-subspace (i.e., multidimensional search) framework.

Robust Beamforming Methods

Quadratic Constraints and Noise Injection. As discussed in [26], mismatch between the actual and assumed array response can lead to significant cancellation of signals from the desired steer direction, and result in an uncontrolled growth of the optimum filter weight vector for beamforming methods such as MVDR and LCMV. A very general formulation to incorporate robustness is thus to consider the combination of linear and quadratic constraints in the minimization problem:

$$\min_{\vec{w}} \quad \vec{w}^H \mathbf{R}_x \vec{w} \tag{3.42}$$

$$\text{subj. to} \quad \mathbf{C}^H(\theta)\vec{w} = \vec{f}(\theta)$$

$$\text{and} \quad \vec{w}^H Q \vec{w} \leq \delta^{-2}$$

where $\mathbf{Q}^{-1}$ is a positive, semidefinite, Hermitian matrix selected to control the norm of $\vec{w}(\theta)$ with variable emphasis in the observation space.

It can be shown through the use of Lagrange multipliers [26] that the optimum filter weight for (3.42) is of the form

$$\vec{w}_{\mathrm{LQCMV}}(\theta) = [\mathbf{R}_x + \varepsilon\mathbf{I}]^{-1}\,\mathbf{C}(\theta)(\mathbf{C}^H(\theta)\,[\mathbf{R}_x + \varepsilon\mathbf{I}]^{-1}\,\mathbf{C}(\theta))^{-1}\vec{f}(\theta) \tag{3.43}$$

for some value of ε that depends on δ in a nonstraightforward manner. When $\mathbf{Q} = \mathbf{I}$, the weight vector is being constrained in the l^2 norm, and this quadratic constraint is seen by (3.43) to correspond to the early ad hoc technique of incorporating robustness by adding a small constant to the diagonal of $\mathbf{R}_x$, otherwise known as *white noise injection.*

Note that this technique is not based explicitly on any of the models for array perturbation as discussed in section 3.3.

An Expanded Source Model

Another method that does not require the additive perturbation model is described in [4] and [34]. The approach is applicable when source observations, known to be rank-1, can only be identified as existing somewhere in a low-rank subspace. A low-rank source model is developed for representations of source locations over a variation of some uncertain secondary parameters (for instance, temporal frequency or source range [34], or polarized sources [4]).

Assume, for example, that the array manifold uncertainty is parameterized by the random scalar ϕ, so that the steering vector can be denoted $\vec{\mathbf{a}}\ (\theta, \phi)$, with ϕ varying over some range Φ. Form the covariance matrix

$$\mathbf{R}_\phi = \int_\Phi p(\phi)\vec{\mathbf{a}}(\theta, \phi)\vec{\mathbf{a}}(\theta, \phi)^H d\phi \tag{3.44}$$

where $p(\phi)$ is the density or weighting function for ϕ. Let $\vec{v}_{j,\theta,j} = 1, \ldots, M$ represent the eigenvectors corresponding to decreasing eigenvalues of $\mathbf{R}_\phi$. By assuming a small number D_ϕ of significant eigenvalues, the set $\{\vec{v}_{j,\theta};\ j = 1, 2, \ldots, D_\phi\}$ is an efficient orthogonal basis for the source representation space.

For a particular ϕ, the actual steering can be approximated as

$$\vec{\mathbf{a}}(\theta, \phi) = \sum_{j=1}^{D_\phi} b_{j,\theta}\vec{v}_{j,\theta} \tag{3.45}$$

where the $b_{j,\theta}$'s represent unknown scalars. This can be written in the vector/matrix form

$$\vec{\mathbf{a}}(\theta, \phi) = \mathbf{V}_\theta\vec{b}_\theta \tag{3.46}$$

with $\mathbf{V}_\theta = \left[\vec{v}_{1,\theta}, \cdots, \vec{v}_{D_\phi,}\theta\right]$ and $\vec{b}_\theta = \left[b_{1,\theta}, \ldots, b_{D_\phi\theta}\right]^T$.

Now consider this expanded model for the MUSIC DOA estimator. The MUSIC null spectrum $\mathcal{D}(\theta) = 1 - \vec{\mathbf{a}}^H(\theta, \phi)\hat{\mathbf{E}}_s\hat{\mathbf{E}}_s^H\vec{\mathbf{a}}(\theta, \phi)$ becomes

$$\mathcal{D}(\theta) = 1 - \mathcal{S}(\theta) \tag{3.47}$$

where

$$\mathcal{S}(\theta) = \vec{b}_\theta^H(\mathbf{V}_\theta^H\hat{\mathbf{E}}_s\hat{\mathbf{E}}_s^H\mathbf{V}_\theta)\vec{b}_\theta \tag{3.48}$$

For each θ select the $\vec{b}_\theta$ that maximizes $\mathbf{S}(\theta)$. The desired vector is simply the eigenvector of $\mathbf{F}_\theta = \mathbf{V}_\theta^H\hat{\mathbf{E}}_s\hat{\mathbf{E}}_s^H\mathbf{V}_\theta$ corresponding to the largest eigenvalue, denoted

$\lambda_{\theta,\max}$. It is straightforward to show that $\lambda_{\theta\max}$ is the $\cos^2$ of the minimum (principal) angle between the ranges of $\hat{\mathbf{E}}_s$ and $\mathbf{V}_\theta$.

This technique is an appropriate choice for situations in which the steering vector pertubations cannot be modeled as additive, but the actual (perturbed) source observation space is still limited to a low-rank subspace. In addition to the example identified in [4, 34], specific examples are given in section 3.3 for the case of matched-field processing in an underwater multipath propagation environment and MEG and EEG dipole sources.

In [36] an essentially equivalent method is used to estimate source locations when the intensity coefficients α_{mn} (real parameters associated with the directional patterns and relative locations of the nth source and the mth sensor) are assumed unknown. However, for this problem, using an expanded source model also introduces ambiguity complications.

Subspace Approaches for Robustness and Enhanced Resolution

Enhanced MVDR. Recall the observation made in section 3.2 from (3.16); namely, that at high SNR the MVDR null spectrum approaches the MUSIC null spectrum. This observation led Owsley to the more general family of *Enhanced* MVDR spectra [22]

$$\mathcal{D}_{\mathrm{EMVDR}}(\theta, c) = \vec{\mathbf{a}}^H(\theta)\,[\mathbf{E}_s \mid \mathbf{E}_n] \begin{bmatrix} \frac{1}{c\lambda_1} & & & & & \\ & \ddots & & & 0 & \\ & & \frac{1}{c\lambda_d} & & & \\ & & & \frac{1}{\sigma_n^2} & & \\ & 0 & & & \ddots & \\ & & & & & \frac{1}{\sigma_n^2} \end{bmatrix} [\mathbf{E}_s \mid \mathbf{E}_n]^H\,\vec{\mathbf{a}}(\theta) \quad (3.49)$$

where the scalar c is a variable "robustness" parameter; $c = 1$ results in MVDR location estimates, and $c \to \infty$ gives MUSIC location estimates.

By choosing smaller values of c, the EMVDR estimates are generally made more robust (for instance, to the white noise assumption) at the cost of higher resolution. However, varying the single scalar does not provide a straightforward way for incorporating prior knowledge of model uncertainties.

Perturbation Injection. Consider the general null spectrum form:

$$\mathcal{D}(\theta) = \frac{|\vec{\mathbf{a}}^H(\theta)\mathbf{E}\,|^2}{\vec{\mathbf{a}}^H(\theta)\vec{\mathbf{a}}(\theta)} \quad (3.50)$$

To achieve asymptotic infinite resolution it is desired that the range of the $M \times I$ matrix E be a subspace of the noise subspace. As discussed in section 3.2 enhanced finite-sample resolution can then be incorporated by constraining $\mathbf{E}$ further such that Range $\{\mathbf{E}\}$ is, in some sense, close to the space spanned by the significant eigenvectors R_Θ. To incorporate robustness to model errors we would, in addition, like Range{E} to contain *few* of the significant eigenvectors of the error covariance $\mathcal{B}$.

Now assume the case $I = 1$, so that $\mathbf{E}$ is a vector and is therefore denoted $\vec{e}$. Consider the constrained minimization

$$\min_{\vec{e}} \frac{\vec{e}^H(\mathbf{I} - U_\Theta U_\Theta^H + \varepsilon\mathcal{B})\vec{e}}{\vec{e}^H\vec{e}} \tag{3.51}$$

$$\text{subj. to } \mathbf{E}_s^H\vec{e} = 0$$

In this formulation ε is an adjustable parameter. Not that $\varepsilon = 0$ (and $D_\Theta = 1$) results in the nonrobust FINE vector $\vec{e}_f = \mathbf{P}_n\vec{u}_1$, from (3.10) and (3.11) in section 3.2. As ε gets larger, $\vec{e}$ gets penalized more for residing in the perturbation subspace defined by the significant eigenvectors of $\mathcal{B}$. The form of (3.51) is reminiscent of the "noise injection" method for robust beamforming described earlier, and so in this chapter the solution to (3.51) will be referred to as the "perturbation injection" method. The solution follows from the equivalent unconstrained minimization

$$\min_{\vec{e}_n} \frac{\vec{e}_n^H\mathbf{R}_n\vec{e}_n}{\vec{e}_n^H\vec{e}_n} \tag{3.52}$$

where $\vec{e}_n = \mathbf{E}_n^H\vec{e}$ and $\mathbf{R}_n = \mathbf{E}_n^H(\mathbf{I} - \mathbf{U}_\Theta\mathbf{U}_\Theta^H + \varepsilon\mathcal{B})\mathbf{E}_n$. A solution of (3.52) is

$$\vec{e}_{pi} = \mathbf{E}_n\vec{v} \tag{3.53}$$

where $\vec{v}$ represents the eigenvector corresponding to the smallest eigenvalue of the $(M-d)\times(M-d)$ matrix $\mathbf{R}_n$.

It should be noted that the "perturbation injection" method described here differs fundamentally from the "noise injection" method. In particular, the constrained minimization seeks a vector in the noise subspace, *not* a filter weight vector $\vec{w}$. In addition, this technique does make explicit use of the additive model for perturbations by seeking a vector in the noise subspace that is simultaneously close to the source representation subspace yet away from the range space defined by the significant eigenvectors of $\mathcal{B}$.

3.5 SENSITIVITY ANALYSIS AND ROBUSTNESS

Through statistical analysis, accurate expression of estimator performance (e.g., estimator variance and bias) can be derived. With these expressions, estimators can be efficiently evaluated for a broad range of processing situations. Additionally, by deriving general performance expressions for a parameterized class of estimators, optimum estimators (within the class) can be identified by selecting the parameter set which optimizes to performance expression.

The field of statistical analysis on eigenspace-based spectrum estimators has been extensively explored. Concerning finite sample performance, many of the results that have been presented are based on statistics, presented by Kaveh and Barabell [37], of the sample covariance matrix eigenvectors. Based on these statistics, Stoica and Nehorai [38, 39] performed some of the more important research on

DOA estimate variance (other pertinent references include [40–45]), and Xu and Buckley [44, 45] derived DOA bias expressions. Concerning analysis of sensitivity to modeling errors, Swindlehurst and Kailath [18, 27] and Hamza and Buckley [46] derive variance and bias expressions, respectively.

Viberg and Swindlehurst [19] have recently presented a combined finite sample and model error sensitivity analysis for variance of a general weighted subspace fitting class of eigenspace DOA estimators. In this section we parallel their presentation, considering spectral methods specifically. We analyze the variance of the DOA estimates using the weighted eigenspace methods with limited number of snapshots and under model perturbation, and we will discuss optimum weighting. The results here are based on a *first-order* statistical analysis. The analysis approach is for general additive modeling errors. However, it should be noted that, as in [19], specific results are restricted to the Eqs. (3.26, 3.27) cases.

Background and Assumptions

Since the parameter estimates of these methods are direction matrices $\mathbf{A}$ and $\mathbf{E}_s$ dependent, we restate the general null spectrum function from section 3.2 as

$$\mathcal{D}(\theta, \hat{\mathbf{A}}, \hat{\mathbf{E}}_s) = \vec{\hat{b}}^H(\theta)(\mathbf{I}_M - \hat{\mathbf{E}}_s\hat{\mathbf{E}}_s^H)\mathbf{W}(\mathbf{I}_M - \hat{\mathbf{E}}_s\hat{\mathbf{E}}_s^H)\vec{\hat{b}}(\theta) \tag{3.54}$$

where $\hat{\mathbf{P}}_A^{\perp}$ is replaced by $\mathbf{I}_M - \hat{\mathbf{E}}_s\hat{\mathbf{E}}_s^H$. The vector $\vec{\hat{b}}^H(\theta)$ is the perturbation of the normalized array response vector $\vec{b}(\theta)$. In (3.54), we redefine the spectrum function so that perturbation in $\mathbf{E}_s$ is due only to finite data effects. The model errors are included as calibration errors in the direction vectors $\vec{b}(\theta)$. The perturbation errors in the steering vectors (columns of A) are equivalent, in a *first-order* sense, to calibration measurement errors [18]. In (3.54), $\mathcal{D}(\theta, \mathbf{A}, \mathbf{E}_s)$ denotes the null spectrum function and satisfies, for all types of weighted spectral methods,

$$\mathcal{D}(\theta, \mathbf{A}, \mathbf{E}_s) > 0 \; s.t. \; \theta \neq \theta_\zeta \tag{3.55}$$

and

$$\mathcal{D}(\theta_\zeta, \mathbf{A}, \mathbf{E}_s) = 0 \text{ for all } \zeta = 1, 2, \cdots, d \tag{3.56}$$

Let $\hat{\theta}_o$ denote the estimate of θ_o. By definition, the perturbed null spectrum has a minimum at $\hat{\theta}_o$. Two observations are very essential to the analysis that follows and can be expressed mathematically:

$$\dot{\mathcal{D}}(\hat{\theta}_o, \hat{\mathbf{A}}, \hat{\mathbf{E}}_s) = 0$$

and

$$\dot{\mathcal{D}}(\theta_o, \mathbf{A}_o, \mathbf{E}_{s_o}) = 0 \tag{3.57}$$

where $\mathbf{A}_o$ represents an exact presentation of $\mathbf{A}$ and $\dot{\mathcal{D}}(.,..,.)$ is the first partial derivative of $\mathcal{D}(.,..,.)$ with respect to θ. By assumption, $\dot{\mathcal{D}}(\theta, \mathbf{A}, \mathbf{E}_s)$ is analytical in $\mathbf{A}$ and $\mathbf{E}_s$ in the neighborhood of the exact point $(\theta_o, \mathbf{A}_o, \mathbf{E}_{s_o})$.

Variance Analysis

To relate the statistics of the errors in $\mathbf{A}$ and $\mathbf{E}_s$ to those of the DOA estimates, it will be assumed that the array errors are sufficiently small, in a first-order sense, so that an expression for $\Delta\theta_o$ may be obtained via the following first-order Taylor expansion. For notation simplicity, we introduce $\vec{g}_k(\theta, \mathbf{A}, \mathbf{E}_s)$ to denote the gradient of $\dot{\mathcal{D}}(\theta, \mathbf{A}, \mathbf{E}_s)$ with respect to $\vec{\mathbf{a}}_k$ ($\vec{\mathbf{a}}_k$ is the kth column of $\mathbf{A}$). The gradient of $\dot{\mathcal{D}}(\theta, \mathbf{A}, \mathbf{E}_s)$ with respect to $\vec{e}_k$ ($\vec{e}_k$ is the kth column of $\mathbf{E}_s$) is denoted by $\vec{q}_k(\theta, \mathbf{A}, \mathbf{E}_s)$.

$$\dot{\mathcal{D}}(\hat{\theta}_o, \hat{\mathbf{A}}, \hat{\mathbf{E}}_s) = \dot{\mathcal{D}}(\theta_o, \mathbf{A}_o, \mathbf{E}_{s_o}) + \ddot{\mathcal{D}}(\theta_o, \mathbf{A}_o, \mathbf{E}_{s_o})\Delta\theta_\circ$$
$$+ 2\sum_{i=1}^{d} \mathcal{R}[\vec{g}_i^T(\theta_o, \mathbf{A}_o, \mathbf{E}_{s_o})\Delta\vec{\mathbf{a}}_i + \vec{q}_i^T(\theta_o, \mathbf{A}_o, \mathbf{E}_{s_o})\Delta\vec{e}_i] + r \quad (3.58)$$

where r includes *second- and higher-order* terms. Note that all terms on the right side of Eq. (3.58) are evaluated at the exact point $(\theta_o, \mathbf{A}_o, \mathbf{E}_{s_o})$. The term of interest, $\Delta\theta_\circ$, appears in the second term on the right of Eq. (3.58). Our approach is to derive an expression for $\Delta\theta_o$ from (3.58), and to evaluate its mean and covariance. Considering (3.57), a unified expression for the error in the parameter estimate can be defined as

$$\Delta\theta_o = -\frac{\sum_i^d (\vec{g}_i^T(\theta_o, \mathbf{A}_o, \mathbf{E}_{s_o})\Delta\vec{\mathbf{a}}_i + \vec{g}_i^H(\theta_o, \mathbf{A}_o, \mathbf{E}_{s_o})\Delta\vec{\mathbf{a}}_i^*)}{\ddot{\mathcal{D}}(\theta_o, \mathbf{A}_o, \mathbf{E}_{s_o})}$$
$$-\frac{\sum_i^d (\vec{q}_i^T(\theta_o, \mathbf{A}_o, \mathbf{E}_{s_o})\Delta\vec{e}_i + \vec{q}_i^H(\theta_o, \mathbf{A}_o, \mathbf{E}_{s_o})\Delta\vec{e}_i^*)}{\ddot{\mathcal{D}}(\theta_o, \mathbf{A}_o, \mathbf{E}_{s_o})} - \tilde{r} \quad (3.59)$$

where $\tilde{r} = \frac{r}{\ddot{\mathcal{D}}(\theta_o, \mathbf{A}_o, \mathbf{E}_{s_o})}$. Since the remainder term is of higher order, it will be neglected and therefore $\Delta\theta_o$ has the same statistics as the first two terms of (3.59). Noting that these terms are linear functions of the errors in columns of $\mathbf{A}$ and $\mathbf{E}_s$, it is straightforward to obtain the statistical distribution of the errors in the desired DOA $\hat{\theta}_o$:

$$E\left[\Delta\theta_o\right] \simeq -\frac{2\sum_i^d \mathcal{R}\left(\vec{g}_i^T\left(\theta_o, \mathbf{A}_o, \mathbf{E}_{s_o}\right) E\left[\Delta\vec{\mathbf{a}}_i\right] + \vec{q}_i^T(\theta_o, \mathbf{A}_o, \mathbf{E}_{s_o}) E\left[\Delta\vec{e}_i\right]\right)}{\ddot{\mathcal{D}}(\theta_o, \mathbf{A}_o, \mathbf{E}_{s_o})} = 0 \quad (3.60)$$

for zero-mean perturbation, $E\left[\Delta\vec{\mathbf{a}}_i\right] = 0$. $E\left[\Delta\vec{e}_i\right] = 0$ (see [37]). A reasonable model for certain types of array errors was defined in (3.26). This particular model simplifies searching for an optimum matrix weighting, as we shall see in the following. Let $\mathcal{H}_{kl} = E\left[\Delta\vec{e}_k\Delta\vec{e}_l^H\right]$ and $\breve{\mathcal{H}}_{kl} = E\left[\Delta\vec{e}_k\Delta\vec{e}_l^T\right]$, and assuming model (3.26), the asymptotic variance is

$$E\left[\Delta\theta_o^2\right] \approx \frac{2}{\ddot{\mathcal{D}}^2(\theta_o, \mathbf{A}_o, \mathbf{E}_{s_o})} \sum_{k,l=1}^{d} \mathcal{R}\{\vec{g}_k^T(\theta_o, \mathbf{A}_o, \mathbf{E}_{s_o})\mathcal{B}\vec{g}_k^*(\theta_o, \mathbf{A}_o, \mathbf{E}_{s_o})\}$$
$$+ \mathcal{R}\{\vec{q}_k^T(\theta_o, \mathbf{A}_o, \mathbf{E}_{s_o})\mathcal{H}_{kl}\vec{q}_l^*(\theta_o, \mathbf{A}_o, \mathbf{E}_{s_o})\}$$
$$+ \mathcal{R}\{\vec{q}_k^T(\theta_o, \mathbf{A}_o, \mathbf{E}_{s_o})\breve{\mathcal{H}}_{kl}\vec{q}_l(\theta_o, \mathbf{A}_o, \mathbf{E}_{s_o})\} \quad (3.61)$$

It is obvious that the covariance between the DOA estimates are easily reducible from (3.61). Since the errors due to the perturbation of nominal values are independent from the errors due to finite data effects, the expected values of their cross terms vanish and do not appear in (3.61).

To obtain specific expressions for each algorithm, expressions for $\ddot{\mathcal{D}}(\theta_o, \mathbf{A}_o, \mathbf{E}_{s_o})$ and the gradients must be evaluated for any particular algorithm. An expression for the second derivative of the cost function can be shown to be

$$\ddot{\mathcal{D}}(\theta_o, \mathbf{A}_o, \mathbf{E}_{s_o}) = 2\dot{\vec{b}}_o^H(\theta_o)(\mathbf{P}_A^\perp \mathbf{W} \mathbf{P}_A^\perp)\dot{\vec{b}}_o(\theta_o). \tag{3.62}$$

Concise expressions for the gradient vector $\vec{g}_h(\theta_o, \mathbf{A}_o, \mathbf{E}_{s_o})$ of any weighted eigenspace method were derived in [46] under perturbation model using an infinite record of data. Similarly, expressions for the gradient $\vec{q}_h(\theta_o, \mathbf{A}_o, \mathbf{E}_{s_o})$ were obtained for MUSIC in [44] and for MIN-NORM and FINE in [45] under finite data effects with no model perturbations. These expressions are still valid here since both sources of errors are independent and their effects are treated independently in our framework. Using the results of [46, 45], we can easily evaluate the asymptotic variance for the combined finite and model error effects by

$$E\left[\Delta\theta_o^2\right] \approx A\mathrm{Var}_s(\Delta\theta_o^2) + A\mathrm{Var}_e(\Delta\theta_o^2) \tag{3.63}$$

where

$$A\mathrm{Var}_s(\Delta\theta_o^2) \approx \frac{\sigma_n^2}{N\ddot{\mathcal{D}}(\theta_o, \mathbf{A}_o, \mathbf{E}_{s_o})} Tr\{\mathbf{P}^\perp \mathbf{W}\}\dot{\vec{b}}(\theta_o)^H \mathbf{E}_{s_o} \tilde{\mathbf{\Lambda}}_{s_o}^{-2} \mathbf{\Lambda}_{s_o} \mathbf{E}_{s_o}^H \dot{\vec{b}}(\theta_o) \tag{3.64}$$

and

$$A\mathrm{Var}_e(\Delta\theta_o^2) \approx \frac{2}{\ddot{\mathcal{D}}^2(\theta_o, \mathbf{A}_o, \mathbf{E}_{s_o})} \dot{\vec{b}}^H(\theta_o)(\mathbf{P}^\perp \mathbf{W} \mathbf{P}^\perp)\mathcal{B}(\mathbf{P}^\perp \mathbf{W} \mathbf{P}^\perp)^H \dot{\vec{b}}(\theta_o) \tag{3.65}$$

for limited number of snapshots N. $\mathbf{\Lambda}_{s_o}$ represents the exact value of $\mathbf{\Lambda}_s$ and $\tilde{\mathbf{\Lambda}}_{s_o} = \mathbf{\Lambda}_{s_o} - \sigma_n^2\mathbf{I}$. These expressions have been shown to be very accurate when either source of errors is present. In section 3.6, we shall illustrate that these expressions are valid even for the combined sources of errors.

Robustness

In real systems, many issues related to the sensitivity and the sampling effects are encountered that preclude good DOA estimators. Based on the results of (3.63), one can consider determining an optimal weighting that lowers the variance of the estimates that is due to combined finite sample and model error effects. The variance expression in (3.63) was interpreted as a summation of two independent terms (each due to a different source of errors). This makes the optimization of the spectral function a challenging problem. No overall optimal choice of weighting matrices has yet been proposed.

In [19], Viberg and Swindlehurst consider a weighted subspace fitting for a class of multidimensional search methods, and suggest one particular weighting that leads to improved performance under both model perturbation and sampling effects. They concede that their proposed weighting is only optimum for the extreme cases (where the degradations in the DOA's are due mostly to one particular source of errors). Below we parallel their approach, considering the weighted subspace class of spectral methods.

At a sufficiently large SNR, one might argue that the model perturbation has a larger impact than the sampling effects in degrading the DOA. In addition, the errors due to the imperfection of the model are almost unavoidable, whereas the data sampling effects are significant only if the number of snapshots is considerably small. For the moment, we ignore the effect of the finite data and consider only array perturbation effects. In this case, the variance expression simplifies to (3.65):

$$E\left[\Delta\theta_o^2\right] \approx \frac{2\vec{b}^H(\theta_o)(\mathbf{P}^\perp\mathbf{W}\mathbf{P}^\perp)\mathcal{B}(\mathbf{P}^\perp\mathbf{W}\mathbf{P}^\perp)^H\vec{b}(\theta_o)}{\ddot{\mathcal{D}}(\theta_o, \mathbf{A}_o, \mathbf{E}_{s_o})}. \tag{3.66}$$

Minimization of (3.66) can easily be shown to be achieved with

$$\mathbf{W}_{\text{APOSM}} = (\mathbf{P}^\perp\mathcal{B}\mathbf{P}^\perp)^+ \tag{3.67}$$

This is, effectively, the result presented by Swindlehurst and Kailath [27], who expect that therein the weighting matrix is applied to the noise subspace (as opposed to the whole observation space). In this chapter, the resulting optimal method is referred to as *Array Perturbation Optimal Spectrum Method (APSOM)*. Its variance is expressed as

$$A\,\text{Var}_{\text{APOSM}} \approx \frac{1}{2\vec{b}^H(\theta_o)\mathbf{P}^\perp(\mathbf{P}^\perp\mathcal{B}\mathbf{P}^\perp)^+\mathbf{P}^\perp\vec{b}(\theta_o)}. \tag{3.68}$$

At a moderate SNR, signal resolvability becomes very critical to some eigenspace spectrum methods. CLOSEST (e.g., FINE) has been found to be very promising in resolving closely spaced signals at even low SNR with limited number of snapshots. Based on this method and the results of (3.67), we propose a choice of weighting matrices that normally leads to improved performance at a wide range of SNR and model uncertainties. The proposed *Robust Enhanced-resolution Method (REM)* has a weighting of the form:

$$\mathbf{W}_{\text{REM}} = \mu(\mathbf{P}^\perp\mathcal{B}\mathbf{P}^\perp)^+ + \frac{1}{N}\mathbf{U}_\Theta\mathbf{U}_\Theta^H \tag{3.69}$$

where $\mathbf{U}_\Theta$ is defined in (3.10) and μ is a scaling that controls the relative contribution of the resolution enhancement (due to $\mathbf{U}_\Theta$) and the array perturbation weightings. It is important to note that this is not an overall optimal weighting for a general case of both finite sample and model errors. However, it can improve the resolution performance while controlling sensitivity due to modeling errors, as we shall illustrate in section 3.6.

The weighting in (3.67) has intuitive appeal in that it deemphasizes subspaces of high perturbation error power, which, as argued at the end of section 3.3, is advantageous. Note, from (3.67), that if the perturbation error is evenly distributed within the observation space, no preferential weighting is applied to the observation space, so that incorporated robustness is not possible. Finally, again note that the shortcoming of this approach is that general optimum weighting for both modeling error and finite sample effects has not been derived. Thus, for providing both robustness and resolution enhancement, the resulting approach is similar to the perturbation injection approach in that a parameter must be selected to trade off the two objectives.

3.6 NUMERICAL STUDIES

In this section, Monte Carlo simulations of the spatial-spectrum estimators described in sections 3.2 and 3.4 are presented, as are evaluations of and studies employing the performance equations of section 3.5. In these studies, probability-of-resolution (of closely positioned sources) and direction-of-arrival (DOA) estimator variance are used as comparative performance measures. DOA estimator bias, significant for some of the estimators considered here (e.g., MUSIC), is implicitly considered here in that it affects resolvability.

We are mainly concerned with the possibility of enhancing spectral resolution while providing robustness to source modeling errors. Eigenspace algorithms described in sections 3.2, 3.4, and 3.5 are evaluated, and several of the source model error examples discussed in section 3.3 are considered. Six examples are presented, illustrating relative performance of the estimators in sections 3.2 and 3.4 for: modal sensor position errors, source parameter uncertainty with an EEG electrode array, and a contrived example. Then the analytical variance expressions presented in section 3.5 are employed, along with Monte Carlo simulation results, to study the robust estimators described therein. Model phase errors (discussed in general terms in section 3.3) are simulated.

Simulated Low-Rank Perturbations

The first simulation example demonstrates the "perturbation injection" method for a rank-1, angle-independent, contrived perturbation. An $M = 5$ element uniform linear array was simulated with $d = 2$ equi-powered and uncorrelated sources, $\theta_1 = 19°$, and $\theta_1 = 21°$. The source representation subspace was chosen with $D_\Theta = 1$ (i.e., $U_\Theta = \vec{u}_1$) and $\Theta = [16°, 24°]$. The number of snapshots was $N = 100$, and $\text{SNR} = 30\,\text{dB}$. The rank-1 perturbation, $\vec{b}$, was modeled such that $\cos^2$ of the angle between $\vec{b}$ and $\vec{u}_1$ is equal to 0.2055. Hence, most of the perturbation energy is restricted to a subspace of the noise subspace. All results are based on Monte Carlo simulation with 400 trials.

Figure 3.1 shows the probability of resolution of the first source as a function of perturbation power, σ_γ^2, for the following algorithms: MUSIC, FINE, perturbation injection, and a robust variant of MUSIC, called here "Robust MUSIC."[9] The

[9] For a general perturbation, the robust MUSIC algorithm is given in [27]. The optimum observation space weighting for the general spatial null-spectrum estimator of Eq. (3.54) is given by Eq. (3.67). For low-rank modeling errors such as considered here, Eqs. (3.54, 3.67) indicate that the robust MUSIC algorithm

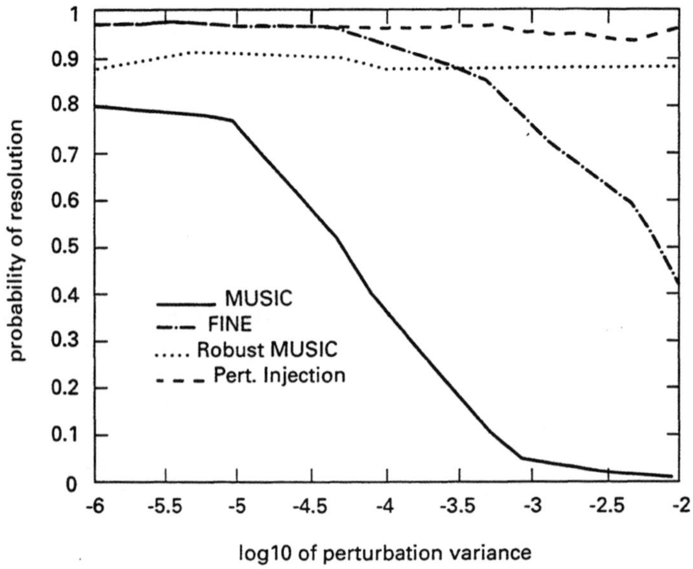

Figure 3.1 Empirical probability of resolution versus perturbation power for simulated angle independent perturbation.

perturbation variance σ_γ^2 and the normalized vector $\vec{b}$ determine the perturbation covariance $\mathfrak{B}$, and are therefore considered prior knowledge. The perturbation injection algorithm thus utilized a varying scalar ε that increased as σ_γ^2 increased. However, no effort was made to search for an "optimum" ε for each perturbation variance. Figure 3.2 is a plot of the variance for source 1 corresponding to each algorithm as a function of σ_γ^2. Figure 3.3 is a composite spectral plot of the four algorithms for a single trial with $\sigma_\gamma^2 = 10^{-3}$.

As expected, the nonrobust algorithms (MUSIC and FINE) both demonstrate increasing variance as a function of increasing perturbation power. The robust MUSIC and perturbation injection algorithms take advantage of the known pertubation structure of this contrived example to eliminate the effect of the model error on DOA estimator variance. Both process in the subspace orthogonal to the low-rank pertubation. Pertubation injection, by additionally emphasizing processing in a subspace where higher spectral resolution can be achieved, provides a higher probability of resolution. Note that this example was constructed so that both the high resolution and the robustness objectives can be realized simultaneously (the source representation and model perturbation subspaces are well separated).

---

simply projects the array manifold onto the orthogonal complement of the subspaces spanned by the columns of both $\hat{\mathbf{E}}_s$ and $\vec{b}$. In effect, processing is restricted to the subspace orthogonal to the low-rank perturbation.

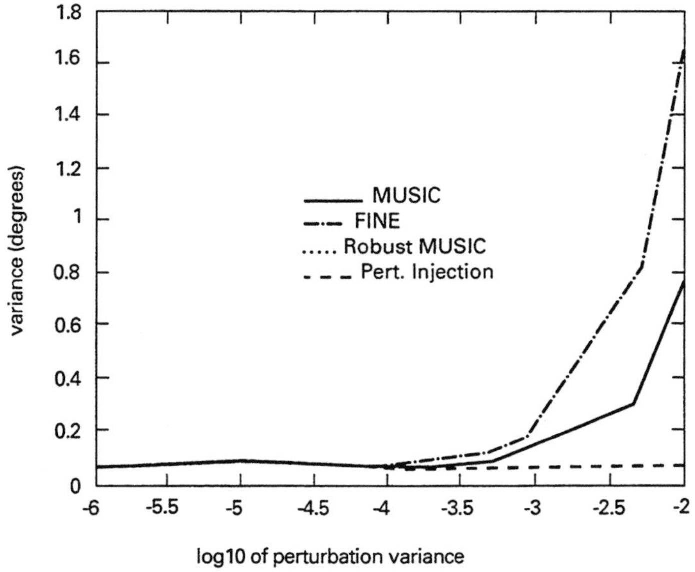

Figure 3.2 Variance of DOA estimates perturbation power for simulated angle independent perturbation.

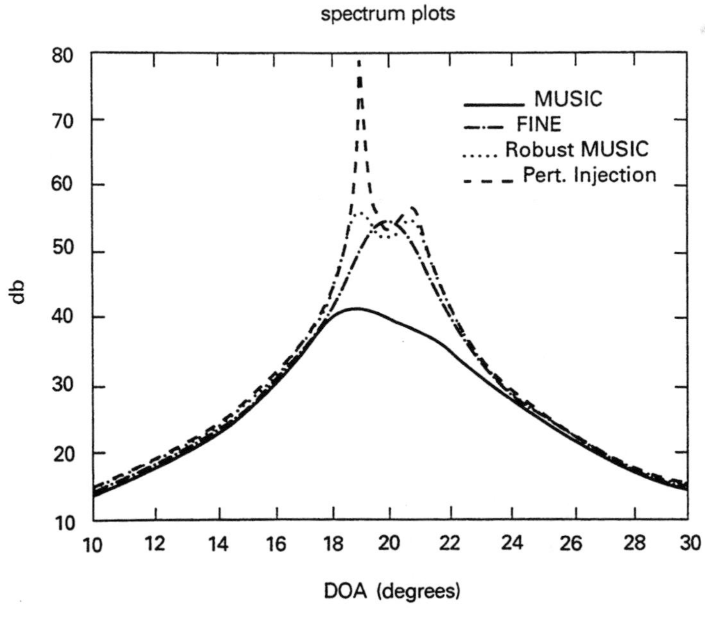

Figure 3.3 Single trial spectral plots for simulated angle independent perturbation.

Modal Perturbations of the Array Response

The second example in this section is for the case of a simulated modal position error. Consider the following special case (second mode) corresponding to equation (3.29), with

$$z(x_k) = a \sin\left(\frac{2\pi x_k}{L}\right) \tag{3.70}$$

where a is a random scalar. For a uniform linear array of M elements, $x_k = \frac{\lambda}{2}(k-1)$ and $L = (M-1)\frac{\lambda}{2}$. Therefore

$$\Delta\vec{\mathbf{a}}(\theta) = j\gamma \cos\theta\{\vec{\mathbf{U}} \odot \vec{\mathbf{a}}(\theta)\} \tag{3.71}$$

where $\vec{\mathbf{U}} = \left[0, \sin\left(\frac{2\pi}{(M-1)}\right), \sin\left(\frac{4\pi}{(M-1)}\right), \ldots, \sin(2\pi)\right]^T$ and the random parameter γ is defined by $\gamma = \left(\frac{2\pi a}{\lambda}\right)$. For sources close to broadside, $\Delta\vec{\mathbf{a}}(\theta)$ is approximately angle-independent

$$\Delta\vec{\mathbf{a}} \approx j\gamma\vec{\mathbf{U}} \tag{3.72}$$

The simulation results are again for $M = 5$, and $d = 2$ equi-powered, uncorrelated sources. The DOA's are $\theta_1 = -2°$ and $\theta_2 = 2°$. The source representation subspace was chosen with $D_\Theta = 1$ (i.e., $\mathbf{U}_\Theta = \vec{u}_1$) and $\Theta = [-4°, 4°]$. Also, the number of snapshots was $N = 50$, and SNR $= 20$ dB. In this example, the perturbations were incorporated by simulating the modal perturbations directly on the array manifold.

Figure 3.4 is a plot of probability of resolution for θ_1 as a function of

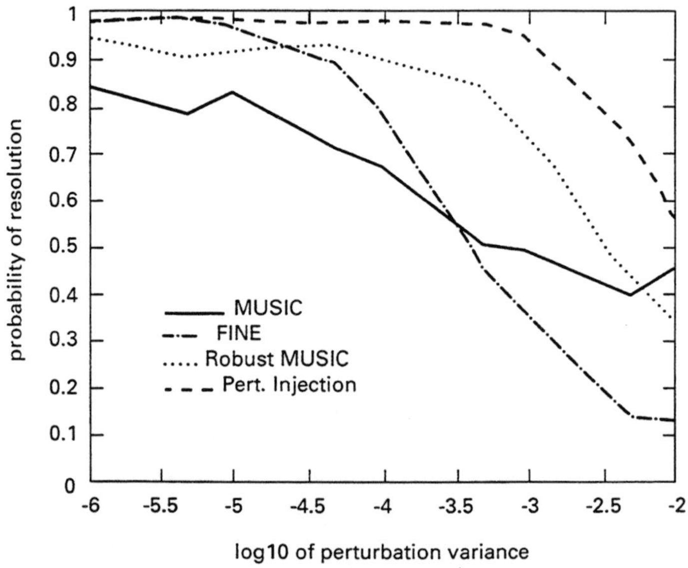

Figure 3.4 Empirical probability of resolution versus perturbation power for simulated modal perturbation.

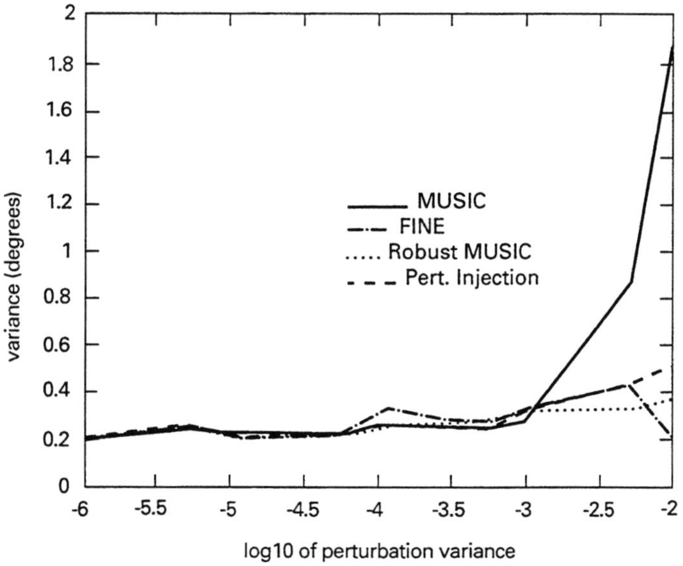

Figure 3.5 Variance of DOA estimates perturbation power for simulated modal perturbation.

perturbation power, σ_γ^2, for the same algorithms as displayed in Figures 3.1–3.3. Figure 3.5 is the corresponding plot of DOA estimate variance.

Again the perturbation-injection algorithm is seen to exhibit the best empirical probability of resolution, at variance levels comparable to the robust MUSIC algorithm. The variance of FINE is low even at relatively high perturbation power. This can be explained by the facts that: (1) for this example, the additive perturbation subspace is nearly orthogonal to the source representation subspace, and thus one would expect FINE to be as relatively insensitive to perturbations as possible; and (2) as noted in section 3.2 it has been illustrated in [28] that FINE is no more sensitive than MUSIC to moderate levels of model errors. However, as can be seen in Figure 3.4, the resolution of FINE still breaks down for large pertubations, likely due to its strong dependence on an accurately known array manifold.

Simulated Parabolic Distortion of a Line Array

In the preceding simulations, the possibility of combining robustness and high resolution within the framework of spectral algorithms was demonstrated. However, in each case, the principal perturbation subspace was well separated from the source representation subspace. As a result, the FINE vector (corresponding to perturbation injection with $\varepsilon = 0$) and the maximally perturbation-insensitive vector (corresponding to perturbation injection with $\varepsilon = \infty$) were, in a geometric sense, somewhat close. This is a very favorable condition, and not always likely to occur in practice. The third simulation example in this section demonstrates the difficulty of incorporating both high resolution and robustness for a more unfavorable situation.

Recall, from section 3.3, the low-rank approximation for $\Delta\vec{\mathbf{a}}(\theta)$ for parabolic distortion of a line array:

$$\Delta\vec{\mathbf{a}}(\theta) = j\pi\gamma\cos\theta \begin{bmatrix} 0 \\ 1 \\ 4 \\ \vdots \\ (M-1)^2 \end{bmatrix} \odot \vec{\mathbf{a}}(\theta) \tag{3.73}$$

For θ sufficiently close to broadside, the principal eigenvector of $\mathcal{B}$ is a constant times $\vec{\mathbf{U}} = [0, 1, 4, \cdots, (k-1)^2, \cdots, (M-1)^2]^T$. Thus, the perturbation-injection vector $\vec{e}_{pi}(\varepsilon)$ can be computed from the approximation $\mathcal{B} = \sigma_\gamma^2 \vec{\mathbf{U}}\vec{\mathbf{U}}^H$, where it is assumed that normalizing constants can be absorbed into σ_γ^2. As an example, consider a uniform linear array with $M = 5$ elements, and two equi-powered, uncorrelated sources from $\theta_1 = 19°$ and $\theta_1 = 21°$ as in the first example. The source representation subspace was chosen with $D_\Theta = 1$ (i.e., $U_\Theta = \vec{u}_1$) and $\Theta = [16°, 24°]$. The number of snapshots was $N = 100$, and SNR $= 30$ dB. The perturbation variance, σ_γ^2, was kept constant at 2×10^{-5} so that effects of varying ε could be observed for perturbation injection,

Figure 3.6 is a plot of the empirical probability of resolution (corresponding to θ_1) for the family of perturbation-injection algorithms, as ε was increased from 10^{-5} to 0.1. Although $\varepsilon = 0.1$ is certainly much smaller than ∞, this value is sufficiently large for this case to make $\vec{e}_{pi}(\varepsilon)$ essentially orthogonal to $\vec{u}_1$. This figure also includes the MUSIC algorithm for comparison. Of course MUSIC does not depend

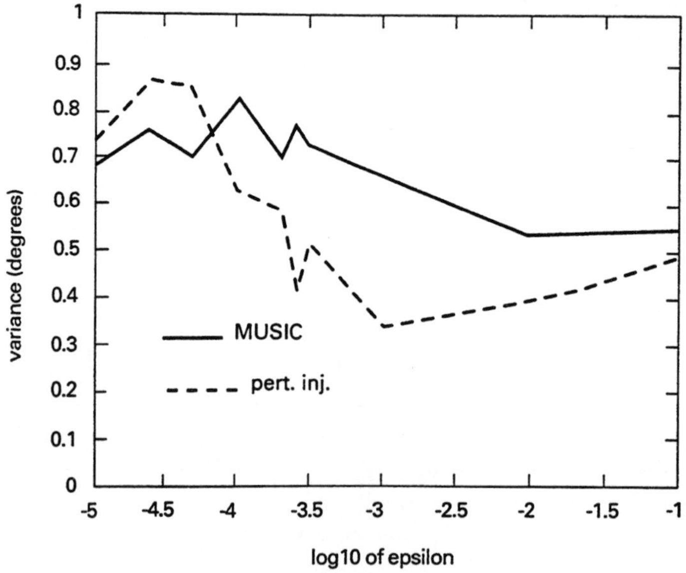

Figure 3.6 Empirical probability of resolution versus perturbation injection constant for simulated parabolic distortion of line array.

on ε, and therefore any small variations are due solely to Monte Carlo effects. It is clear from this figure that increasing ε merely degrades the performance of the perturbation injection algorithm. This penalty in resolution is most likely due to the fact that the model perturbation variance is relatively low, and therefore, straying from the source representation subspace (i.e., increasing ε) has a detrimental effect on the performance contribution due to finite samples only.

Figure 3.7 is a plot of the variance for θ_1. It appears from the figure that lower variance can be achieved (relative to MUSIC) for sufficiently high ε. However, we see from Figure 3.6 that ε's in this region correspond to resolution performance that is *worse* than nonrobust MUSIC.

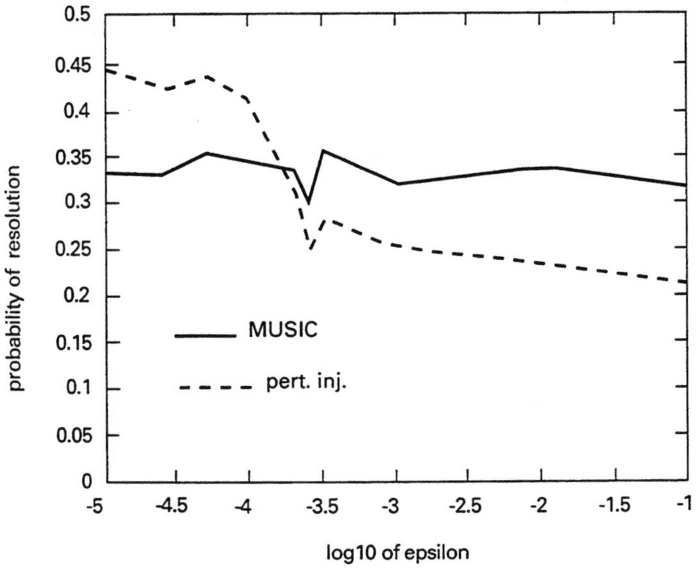

Figure 3.7 Variance of DOA estimates versus perturbation injection constant for simulated parabolic distortion of line array.

This example illustrates the limitation of the subspace weighting approach to combining, for the spectral class of DOA estimators, resolution enhancement and robustness. In this case both objectives cannot be realized simultaneously because, as translated to subspace weighting, the objectives are contradictory. It should be noted that the Expanded Source Model approach, outlined in section 3.4, is also not effective in this case. Specifically, it is shown in Eqs. (3.20), (3.36) that for small array errors a rank-2 model can be used to represent a source observation. However, in this case Eqs. (3.47), and (3.48) the spectrum estimator is prone to false peaks associated with low-rank model ambiguity. We conclude that, in terms of the joint objectives of robustness and enhanced spectral resolution, this is a very difficult case.

EEG Electrode Array with Dipole Orientation Uncertainty

The fourth simulation example in this section demonstrates the expanded source model technique of section 3.4 for the case of EEG dipole localization. Consider the scalp measurements due to two fixed-orientation current dipoles (see section 3.3):

$$\vec{x}(t) = \left[\mathbf{G}(\vec{L}_1)\vec{m}_1\right] s_1(t) + \left[\mathbf{G}(\vec{L}_2)\vec{m}_2\right] s_2(t) + \vec{n}(t) \tag{3.74}$$

where we assume that $\vec{m}_1$ and $\vec{m}_2$ are unknown. The primary goal is to establish the source locations $\vec{L}_1$ and $\vec{L}_2$. As in [33], the locations and moments can be chosen by minimizing the following cost function over $\vec{L}$ and $\vec{m}$:

$$D(\vec{L}, \vec{m}) = \frac{\vec{m}^T \mathbf{G}^T(\vec{L}) \hat{\mathbf{P}}^{\perp} \mathbf{G}(\vec{L}) \vec{m}}{\vec{m}^T \mathbf{G}^T(\vec{L}) \mathbf{G}(\vec{L}) \vec{m}} \tag{3.75}$$

subject to the constraint that $\vec{m}^T \vec{m} = 1$. From section 3.4 and [33] we know that the same location estimates can be obtained by searching for the 3-d "peaks" in the spectrum

$$\frac{1}{D(\vec{L})} = \frac{1}{\lambda_{\min}\{\mathbf{U}_{\vec{L}}^T \hat{\mathbf{P}}^{\perp} \mathbf{U}_{\vec{L}}\}} \tag{3.76}$$

where the SVD is utilized to write $\mathbf{G}(\vec{L}) = \mathbf{U}\Sigma\mathbf{V}^T$, and $\mathbf{U}_L$ denotes the columns of $\mathbf{U}$ corresponding to the 3 nonzero singular values.

For the simulation example an array of $M = 127$ EEG sensors was distributed about the entire upper hemisphere of a spherical surface of radius equal to 8.8 cm. The 4-sphere head model [47] was used to determine $\mathbf{G}(\vec{L})$. The two moment vectors $\vec{m}_1$ and $\vec{m}_2$ were oriented orthogonal to each other, and the scalar time series for each source, $s_1(t)$ and $s_2(t)$, is plotted in Figure 3.8. Case 1 is for 2 deep sources, with locations given by $\vec{L}_1 = [0, 0, 5]$ cm, and $\vec{L}_2 = [1, 1, 5]$ cm (in cartesian coordinates). The source estimates are based on $n = 50$ distinct temporal snapshots, each snapshot corresponding to an entire vector of spatial EEG data. The measurement noise is zero-mean and Gaussian, with standard deviation equal to 0.10 times the peak noiseless external measurement. For the subspace method, a 3-dimensional space must be searched for the 2 maxima. Figure 3.9 shows the subspace spectrum at the $z = 5$ cm plane as a function of x and y. This figure also shows the subspace spectrum corresponding to the second case, characterized by dipole source locations $\vec{L}_1 = [0, 5, 5]$ cm, and $\vec{L}_2 = [1, 6, 5]$ cm. The location estimates of the two dipoles are clearly indicated by the maxima of these spectrum plots.

Analytical Result Verification

In this example we use two computer simulations to verify the statistical analysis of weighted spectrum methods presented in section 3.5. In this example, MUSIC, MIN-NORM, and CLOSEST were used to compare the analytical and the empirical results. Two far-field narrowband emitters were impinging from 12° and 18° on an

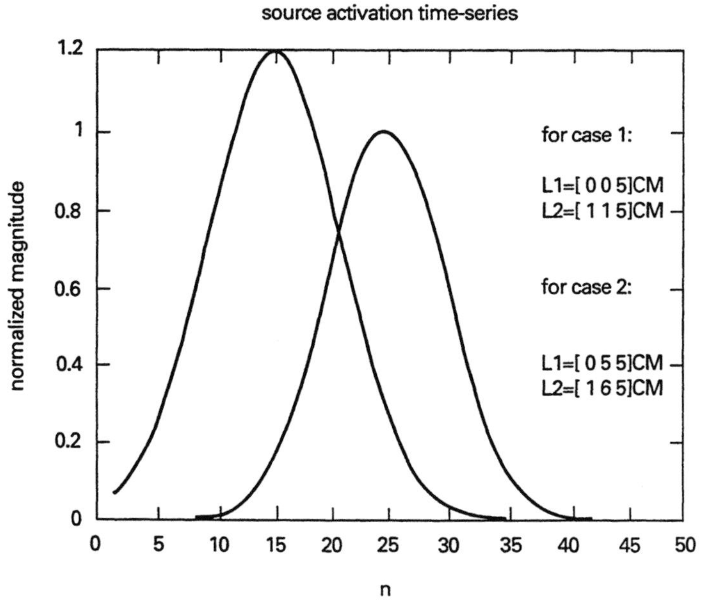

Figure 3.8 Activation profiles for dipole sources, $s_1(t)$ and $s_2(t)$.

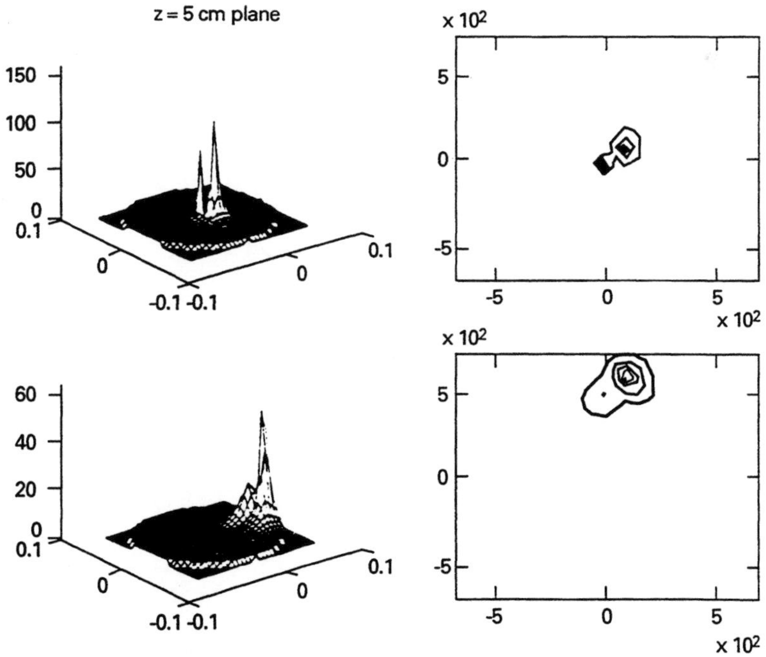

Figure 3.9 Subspace spectrum plots. Case 1: top, Case 2: bottom.

array of 10 omnidirectional sensors. The sensors were matched with half wavelength spacing. Fifty independent snapshots were employed. For CLOSEST, the sector of focusing, Θ, was chosen such that its center was 15° and its width was two beamwidths, with $D_\Theta = 2$. 500 trials were conducted to form the perturbed covariance matrix. The sensor errors were simulated only in the phase. For simplicity, the perturbation was assumed to be *angle-sensor* independent. The phase error of any sensor was zero-mean uniform random variable in an interval $\left[-\sqrt{3\sigma_\gamma^2}, \sqrt{3\sigma_\gamma^2}\right]$ (where σ_γ^2 is the phase error variance). The definition of empirical spectral resolution in [21] was employed in obtaining the probability of resolution of the methods. Algorithms were evaluated in terms of variance only if they can be resolved with probability close to one.

Figure 3.10 shows the statistical performance (in terms of the standard deviation) of the DOA estimates of the three methods versus the signal-to-noise ratio. The continuous lines represent the theory predictions and the symbols indicate the empirical results. Figure 3.11 illustrates the statistical performance of these three methods versus the error magnitude variations. The phase error STD was allowed to range from .06° to 29°. Figures 3.10 and 3.11 clearly show that our analytical variance expressions of DOA estimates are accurate and valid over a wide range of error variations.

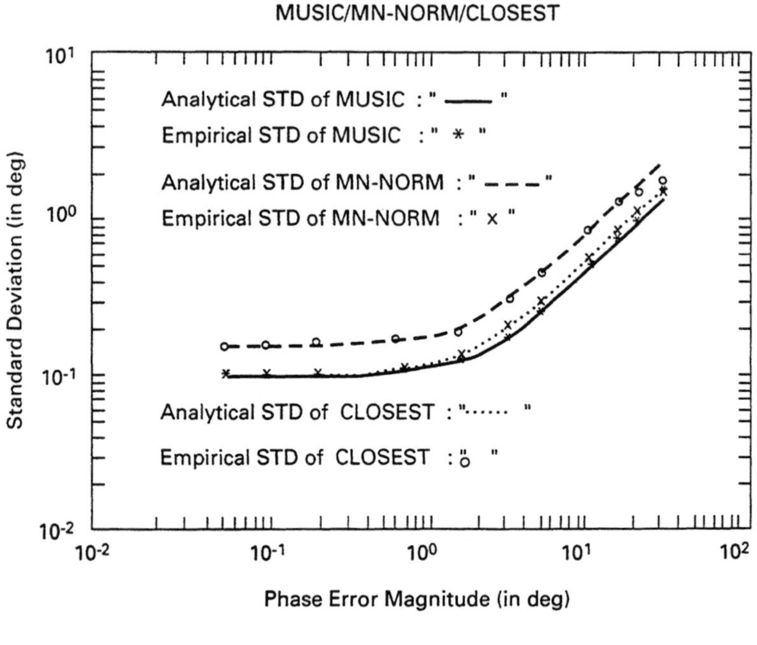

Figure 3.10 Standard deviation versus phase error.

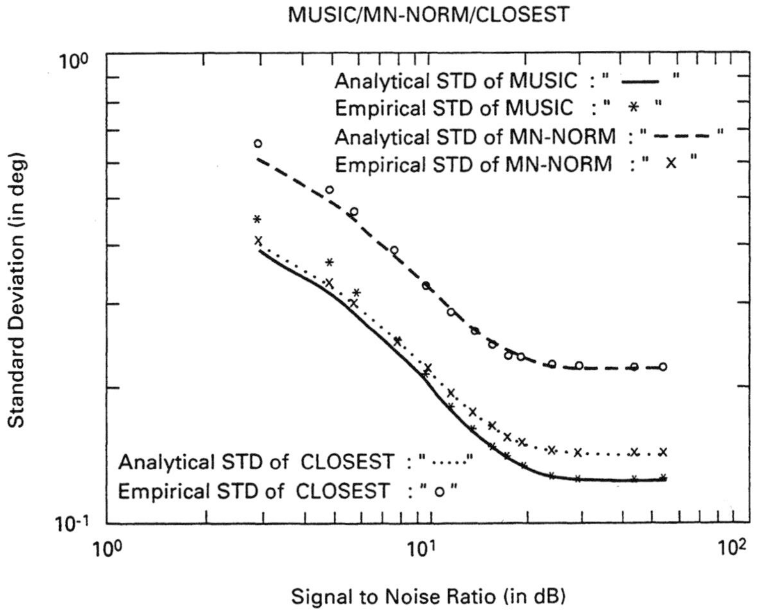

Figure 3.11 Standard deviation versus SNR.

Analytic Performance Comparison

In this example we illustrate the resolution and the asymptotic performance of the Robust Enhanced-resolution Method (REM). We compare the performance of REM with that of the conventional algorithms, for example, MUSIC, CLOSEST. The scenario is identical to the previous case, except the array model errors is assumed to have Gaussian distribution according to the model of (3.26, 3.27), with

$$\mathcal{B} = \sigma_\gamma^2 \text{diag}\,\{0, .7, 0, .6, .5, 1, .3, .8, .2, 0\} \tag{3.77}$$

The scaling $\mu = 0.4$. The array error STD σ_γ was allowed to vary from .01 to .5.

Figure 3.12 shows the probability of resolution versus the error magnitude. We notice that the probability of resolution of CLOSEST decreases exponentially whereas that of REM maintains its high resolution throughout the whole error range. Both methods outperform MUSIC. Figure 3.13 displays the predicted asymptotic performance of MUSIC, CLOSEST, APOSM (is REM with $\mu = 0$), and REM. The results illustrated in Figure 3.13 are based on a high signal-to-noise ratio (SNR = 15 dB) (this is done to guarantee a probability of one for all methods). It is clear that OSM outperforms both CLOSEST and MUSIC in terms of the estimation root mean squared error. However, the lowest estimation error is achieved using APOSM. This is expected since, at high signal-to-noise ratio, the array perturbation has a greater effect on the degradation of the DOA estimates than the sampling effects. On the other hand, APOSM weighting does not always have the resolution capability of REM. We have experienced significant variation in the resolution performance of APOSM. On the other hand, REM is more robust to changes in the

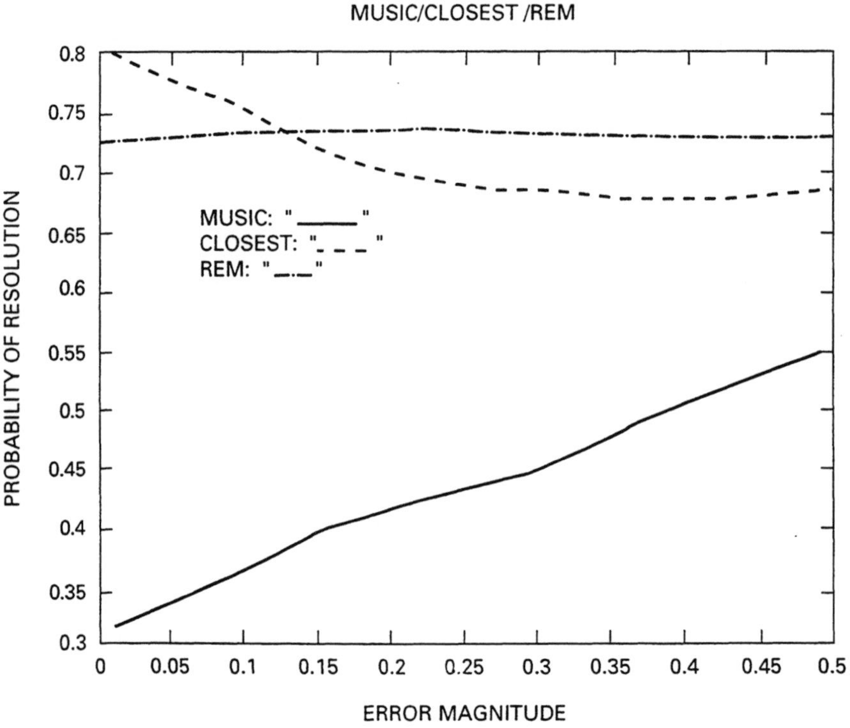

Figure 3.12 Probability of resolution versus error magnitude.

number of samples or the signal power level and maintained a comparable asymptotic performance to APOSM.

3.7 CONCLUDING REMARKS

Resolution relates to the ability to distinguish between located sources. Robustness, as used herein, refers to a lack of sensitivity to inaccuracies in the model of the observation of the source across the array. Resolution and robustness go hand-in-hand. On the one hand, we desire algorithms which are highly sensitive to source location (primary) parameters. On the other, we need estimators with a certain degree of insensitivity to secondary parameters associated with the sources (e.g., frequency, polarization, range), the propagation channel (e.g., conductivities, boundaries), and the array (e.g., sensor positions and responses).

In this chapter, spatial-spectrum based DOA estimators are considered with principal focus on resolution enhancement and robustness to model errors. Tools and an understanding are presented for achieving a balance between spectral-resolution and robustness, and for pursuing both objectives simultaneously. The tools are:

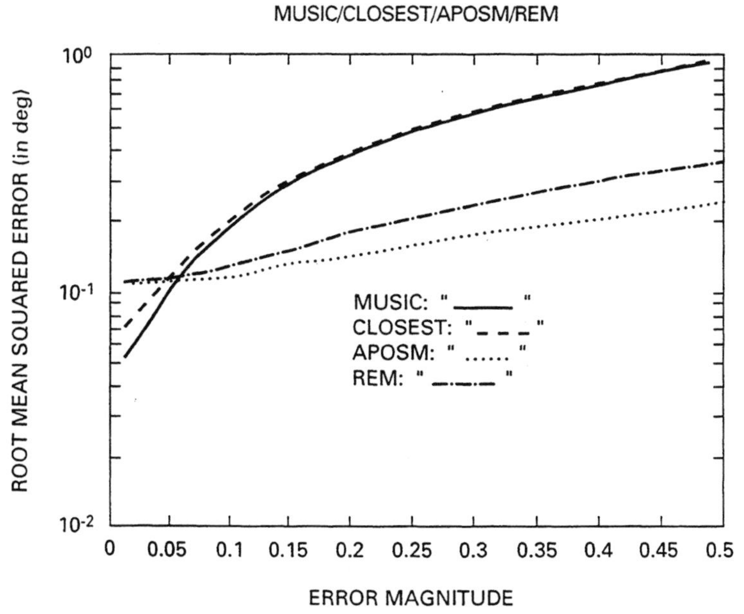

Figure 3.13 Asymptotic RMS error versus error magnitude.

1. an observation space-weighting approach, based on a geometric interpretation of the enhanced resolution and robustness objectives;
2. a broad range of beamformer and eigenspace based spatial spectrum estimators; and
3. results from a finite sample and model error analysis of a general weighted eigenspace spectrum class of DOA estimators.

The understanding presented here is geometric, being based on a linear observation space, on weighting in this space, and on principal component subspaces. Concepts and performance of several of the estimators presented are illustrated using several application specific examples of modeling errors.

Can we enhance both spectral-resolution and improved robustness? The answer to this is a qualified yes. We can if the model error is structured in the observation space, with principal components well separated from a subspace where high resolution can be achieved.

REFERENCES

1. V. F. Pisarenko, "The Retrieval of Harmonics from a Covariance Function," *Geo. J. Royal Astron. Soc.*, vol. 33, pp. 347–66, 1973.
2. N. L. Owsley, "Adaptive Data Orthogonal," *Proc. ICASSP'78*, pp. 109–12, Apr. 1978.

3. G. Bienvenu, "Influence of the Spatial Coherence of the Background Noise on High Resolution Passive Methods," *Proc. ICASSP'79*, pp. 306–09, April 1979.

4. R. O. Schmidt, "A Signal Subspace Approach to Multiple Emitter Location," Ph.D. dissertation, Stanford University, 1980.

5. S. S. Reddi, "Multiple Source Location—A Digital Approach," *IEEE Trans, on AES,* vol. 15, no. 1, pp. 95–105, Jan. 1979.

6. R. Kumaresan, and D. W. Tufts, "Estimating the Angle of Arrival of Multiple Plane Waves," *IEEE Trans. on AES,* vol. 19, pp. 134–39, Jan 1983.

7. F. C. Schweppe, "Sensor-Array Data Processing for Multiple-Signal Sources," *IEEE Trans. on Information Theory*, vol. IT-14, pp. 294–305, March 1968.

8. J. F. Bohme, "Estimation of Source Parameters by Maximum Likelihood and Nonlinear Regression," *Proc. IEEE ICASSP-84*, pp. 7.3.1.–7.3.4, May 1984.

9. W. J. Bangs, and P. M. Schultheiss, "Space-time Processing for Optimum Parameter Estimation," in *Signal Processing: Proc. NATO Advanced Study Institute on Signal Processing–Underwater Acoustics*, ed. J.W.R. Griffiths et al., New York Academic Press, pp. 577–90, 1973.

10. D. C. Rife, and R. R. Boorstyn, "Multiple Tone Parameter Estimation from Discrete-time Observation, *Bell Sys. Tech Jour.*, pp. 1389–1410, Nov. 1976.

11. S. Haykin, J. P. Reilly, V. Kezys, and E. Vertatschitsh, "Some Aspects of Array Signal Processing," *Proc. IEE Part F,* vol. 139, pp. 1–26, Feb. 1992.

12. J. Capon, "High-resolution Frequency-wavenumber Spectrum Analysis," *Proceedings of The IEEE,* vol. 57, pp. 1408–18, Aug. 1969.

13. S. H. Nawab, F. U. Dowla, and R. T. Lacoss, "Direction Determination of Wideband Signals," *IEEE Trans. on ASSP,* ASSP-33, no. 4, pp. 1114–22, Oct. 1985.

14. A. B. Baggeroer, W. A. Kuperman, and H. Schmidt, "Matched Field Processing: Source Localization in Correlated Noise as an Optimum Parameter Estimation Problem," *Jour. of the Acoustical Society of America,* vol., 83, pp. 571–87, Feb. 1988.

15. A. Paulraj, R. Roy, and T. Kailath, "Estimation of Signal Parameters via Rotational Invariance Techniques–ESPRIT," *Asilomar Conference on Circuits, Systems and Computers,* Nov. 1985.

16. R. Roy, and T. Kailath, "Total Least Squares ESPRIT," *Asilomar Conference on Circuits, Systems and Computers,* pp. 297–301, Nov. 1988.

17. A. J. Barabell, "Improving the Resolution Performance of Eigenstructure-based Direction-finding Algorithms, *Proc. ICASSP'83,* pp. 336–39, April 1983.

18. A. L. Swindlehurst, Applications of Subspace Fitting to Estimation and Identification, Ph.D. dissertation, Stanford University, pp. 71–110, June 1991.

19. M. Viberg and A. L. Swindlehurst, *Analysis of the Combined Effects of Finite Samples and Model Errors on Array Processing Performance*, submitted to *IEEE Trans. on SP*, Feb. 93.

20. R. O. Schmidt, "Multiple Emitter Location and Signal Parameter Estimation," *IEEE Trans. Antenn. Prop.,* vol. AP-34, pp. 276–80, March 1986.

21. X. L. Xu and K. M. Buckley, "Spatial-spectrum Estimation in a Location Sector," *IEEE Trans. ASSP,* vol. 38, No. 11, pp. 1829–41, Nov. 1990.

22. N. Owsley, "Enhanced minimum variance beamforming," *Underwater Acoustic Data Processing,* ed. Y. Chan, NATO ASI Series, Klewer, 1989, pp. 285–93.

23. B. VAN VEEN, J. JOSEPH, and K. HECOX, "Localization of Intra-cerebral Sources of Electrical Activity via Linearly Constrained Minimum Variance Spatial Filtering," *Proc. Sixth SSAP Workshop on Statistical Signal Array Processing*, pp. 526–29, Oct. 1992.

24. B. D. VAN VEEN and K. M. BUCKLEY, "Beamforming: A Versatile Approach to Spatial Filtering," *IEEE ASSP*, pp. 4–23, April 1988.

25. K. M. BUCKLEY, "Spatial/Spectral Filtering with Linearly Constrained Minimum Variance Beamformers," *IEEE Trans. ASSP*, vol. 35, no. 3, pp. 249–65, March 1987.

26. H. COX, R. M. ZESKIND, and M. M. OWEN, "Robust adaptive beamforming," *IEEE Trans. ASSP*, vol. 35, no. 10, pp. 1365–76, Oct. 1987.

27. A. L. SWINDLEHURST and T. KAILATH, "A Performance Analysis of Subspace-based Methods in the Presence of Model Errors, Part I: The MUSIC Algorithm," *IEEE Trans. on Signal Processing*, vol. 40, no. 7, pp. 1758–74, July 1992.

28. J. A. MARKS, and K. M. BUCKLEY, "Improved Rooting Method for Source Localization," *IEE Proc.—F*, vol. 139, no. 5, pp. 321–26, Oct. 1992.

29. E. C. van BALLEGOOIJEN, C. W. M. van MIERLO, C. van SCHOONEVELD, P. P. M. van der ZALM, A. T. PARSONS, and N. H. FIELD, "Measurement of Towed Array Position, Shape and Attitude," *IEEE Tour. on Ocean Eng.*, vol. 14, no. 4, pp. 375–83, Oct. 1989.

30. F. HABER, and Q. SHI, "Direction Finding in Array Geometry Uncertainty," *Proc. IEEE AP-S International Sys.*, pp. 258–61, June 1988.

31. G. M. SHAW, "The Effect of Model Element Position Errors on the Radiation Pattern of Large Space-deployable Arrays," *Proc. IEEE AP-S International Sys.*, pp. 23–26, June 1986.

32. J. L. KROLIK, "Matched Field Minimum Variance Beamforming in a Random Ocean Channel," *Jour. of Acoustical Society of America*, accepted April 1992.

33. J. C. MOSCHER, P.S. LEWIS, and R. M. LEAHY, "Multiple Dipole Modeling and Localization from Spatio-temporal MEG Data," *IEEE Trans. on Bio. Eng.*, vol. 39, no. 6, pp. 541–57, June 1992.

34. K. M. BUCKLEY, "Incorporated Robustness in Narrow-band Signal Subspace Spatial Spectral Estimators," *PROC. ICASSP 87*, pp. 53–56, April 1987.

35. B. WAHLBERG. B. OTTERSTEN, and M. VIBERG, "Robust Signal Parameter Estimation in the Presence of Array Perturbations," *Proc. ICASSP 91*, Toronto, Canada, May 1991.

36. A. J. WEISS, A. S. WILLSKY, and B. C. LEVY, "Eigenstructure Approach for Array Processing with Unknown Intensity Coefficients," *IEEE Trans. ASSP*, vol. 36, no. 10, pp. 1613–17, Oct. 1988.

37. M. KAVEH, and A. J. BARABELL, "The Statistical Performance of the MUSIC and Minimum-norm Algorithms in Resolving Plane Waves in Noise," *IEEE Trans. Acoust., Speech, Signal Proc.*, vol. ASSP-34, pp. 331–41, April 1986.

38. P. STOICA, and A. NEHORAJ, "MUSIC, Maximum Likelihood, and Cramer-Rao Bound," *IEEE Trans. Acoust., Speech, Signal Proc.*, vol. ASSP-37, no. 5, 720–41, May 1989.

39. P. STOICA, and A. NEHORAI, "MUSIC, Maximum Likelihood, and Cramer-Rao Bound: Further Results and Comparisons," *IEEE Trans. Acoust., Speech, Signal Proc.*, vol Assp-38, no. 12, pp. 2140–50, Dec. 1990.

40. K. SHARMAN. T. S. DURRANI, M. WAX, and T. KAILATH, "Asymptotic Performance of Eigenstructure Spectral Analysis Methods," *Proc. ICASSP'83*. pp. 45.5.1.–45.5.4, April 1983.

41. B. PORAT, and B. FRIEDLANDER, "On the Asymptotic Relative Efficiency of the MUSIC Algorithm," *IEEE Trans. Acoust., Speech, Signal Proc.*, vol. ASSP-36, no. 4, pp. 532–44, April 1988.

42. B. D. RAO, and K. V. S. HARI, "Performance Analysis of the MIN-NORM Method," *IEE Proc.*, vol. 136, Part F, no. 3, Jan. 1989.

43. F. LI, and R. VACCARO, "Analysis of Min-Norm and MUSIC with Arbitrary Array Geometry," *IEEE Trans. on AES*, vol. 26, no. 6, pp. 976–85, Nov. 1990.

44. X.-L. XU, and K. M. BUCKLEY, "Bias Analysis of the MUSIC Location Estimator," *IEEE Trans. Signal Proc.*, vol. 40, no. 10, pp. 2559–69, Oct., 1992.

45. X.-L. XU, *A New Eigenspace Based Approach to Source Localization and Its Performance Analysis*, Ph.D. dissertation, University of Minnesota, 1991.

46. R. M. HAMZA and K. M. BUCKLEY, *An Analysis of Weighted Eigenspace Methods in the Presence of Sensor Errors*, submitted to *IEEE Trans. on SP* for publication, Feb. 1993.

47. B. N. CUFFIN, and D. COHEN, "Comparison of the Magnetoencephalogram and Electroencephalogram," *Electroencephalography and Clinical Neurophysiology*, pp. 132–46, 1979.

4

The Scorefunction Approach to Bearing Estimation: Phase-Lock Loops and Eigenstructure

R. Lynn Kirlin, Emily Su, and Brad Hedstrom

4.1 INTRODUCTION

In recent years many scientists have addressed aspects of beamforming, bearing estimation, and spectral estimation utilizing the methodology of sample covariance (of sensor-array output time-slice vector samples) eigenstructure, or subspace processing. In particular the MUSIC (multiple signal classification) algorithm [1, 2] has received notoriety. That algorithm and a variation, the Minimum Norm (MN) [3], have received considerable attention and were analyzed for error variance, and resolution capabilities and threshold effects by Kaveh and Barabell [3]. Many similar subspace method performance analyses have since been published, for example [11]. Closely comparing the MU and MN algorithms and other spectral estimators, which utilize eigenstructure, leads one to note that the differences are simply in the selection of eigenvectors' coefficients, a scalar multiplier, and in the placement of magnitude-squared operation with respect to each or all of the inner products of direction vector estimates with noise-space eigenvectors [4, 5, 6]. Such comparison are detailed in [4] and [11].

We relate the orthogonality of solution signal-space vectors with noise-space eigenvectors (as emphasized in MUSIC) to the orthogonality of received and locally generated signals as utilized in phased-locked loops (PLLs) [8, 9]. The new scorefunction form of maximum likelihood (ML) source location algorithm yields a relation of ML to MU and MN. Weightings on the eigenvectors are optimal at the solution values with respect to maximizing the ML gradient slope in the vicinity of the true (single-source) bearing, thus minimizing error variance per Cramér-Rao theory. Using expressions given in [11], we show that MU and ML have the same mean-squared error.

Although the derived ML scorefunction algorithm may be considered more instructive than useful (due mostly to the ambiguous nulls in the scorefunction), the insights gained lead to a better understanding of the minimizing functions in the various algorithms. For example, most subspace spectral or bearing estimation algorithms contain a denominator function which nulls at the solution point when SNR is infinite. The reciprocal of the nulling function appears as a "spectral" peak. We show that only the ML scorefunction can give an expected true null at SNR less than infinity; the structure of other estimators generally minimizes but does not null when any noise is present.

In section 4.3 we use the initial steps in the derivation of the eigenstructure version of the scorefunction solution and produce an iterative, nonsubspace method of solution both for bearing and the temporal phase of an isolated narrowband, angle-modulated carrier source. The resulting estimator is realized as a cross-coupled pair of phase-lock loops (CPLL), one tracking bearing and one tracking phase. An analysis of loop filter and gain parameters follows, leading to design equations.

For multiple sources, multiple cross-coupled PLLs are required. A thorough analysis of such a system is beyond our scope here, although we have made progress on the companion problem of separating two signals having identical carrier frequency, arising from sources either at the same bearing or sensed by a single sensor. In section 4.4 we show that separation of such signals is impossible only if the carriers are exactly in phase *and* the state transition matrices for the two angle modulating processes are identical. We quantify our results through the means provided by the observability matrix.

Thus, the link between orthogonal subspace processing and PLL tracking is established in this chapter. Both schemes incorporate cross-correlation of the incoming signal (direction vector) with a local copy of that signal's temporal derivative (vector's angular spatial derivative). Cross-correlation with the signal implies peak-seeking, while cross-correlation with the derivative implies null-seeking. Both correlations are ML solutions.

The fundamental difference between the eigenstructure method and the PLL method is that the former gathers all data into the covariance matrix before searching while the other is an iterative gradient method, storing past information with the time constants of the PLL's transfer function.

In our gradient (scorefunction) version of the subspace method, the incoming direction vector (signal) and its local copy derivative are sought by search of bearing angle. Assuming the covariance has been correctly analyzed for signal and noise subspaces, the direction vector's derivative must lie in the noise subspace. Thus, the dot product of the trial direction vector with its derivative's projection onto the noise subspace will ideally null when the trial vector is a correct solution. When it is not correct, the trial vector contains a noise-subspace component and the product is not null.

4.2 THE MAXIMUM LIKELIHOOD METHOD

The maximum likelihood (ML) approach to the problem of multiple-source location with sensor arrays has already been addressed in the literature. Ziskind and Wax [7] establish the density of the data as a function of deterministic source signals and bearings and the noise statistics:

$$f(\mathbf{x}(i)|\mathbf{A},\mathbf{s}) = \prod_{i=1}^{N} \frac{1}{\pi M \sigma^2} \exp\left\{-\frac{1}{\sigma^2}|\mathbf{x}(i) - \mathbf{A}\mathbf{s}(i)|^2\right\} \tag{4.1}$$

where M is the number of sensors, N is the number of vector sensor-output samples $\mathbf{x}(i)$,

$$\mathbf{x}(i) = (x_1(i), x_2(i), \ldots x_M(i))^T \tag{4.2a}$$

$$= \mathbf{A}\mathbf{s}(i) + \mathbf{n}(i) \tag{4.2b}$$

where $\mathbf{s}$ is the vector of the D deterministic source signals $s_j(i)$

$$\mathbf{s} = \mathbf{s}(i) = (s_1(i), s_2(i), \ldots, s_D(i))^T \tag{4.3}$$

$\mathbf{A} = \mathbf{A}(\theta)$ is the source direction matrix with k^{th} column the direction vector of the kth source (using the array center for reference)

$$\mathbf{a}_k(e^{-(M-1)j\theta_k/2} \ldots e^{(M-1)j\theta_k/2}/\sqrt{M} \tag{4.4}$$

and $\mathbf{n}(i)$ is the vector of additive noise samples. In the above, $(\cdot)^T$ denotes transpose. In the following, $(\cdot)^H$ denotes conjugate transpose and $(\cdot)^*$ denotes conjugation. For a linear equally spaced sensor array, $\theta_k = w_k \Delta \sin \alpha_k / c$, where α_k is the kth source direction, w_k is the kth source frequency, Δ is the sensor spacing, and c is the velocity of the wave.

When the noise samples are stationary, Gaussian, independent in time and space, and have equal variance σ^2 at each sensor, it is shown [8] that the ML estimate of σ^2, fixing bearing and signal variables, is

$$\hat{\sigma}^2 = \frac{1}{MN} \sum_{i=1}^{N} \|\mathbf{x}(i) - \mathbf{A}\mathbf{s}(i)\|^2 \tag{4.5}$$

Inserting $\hat{\sigma}^2$ into the log-likelihood expression, the exponent in Eq. (4.1), gives the quantity to be minimized over all θ_k and s_k as

$$F = \sum_{i=1}^{N} \|\mathbf{x}(i) - \mathbf{A}\mathbf{s}(i)\|^2 \tag{4.6}$$

Ziskind and Wax find the least squares solution for deterministic $\mathbf{s}(i)$ in terms of

$\mathbf{x}(i)$ and assumed-known $\mathbf{A}$ and substitute it back into (4.6). They next produce a sequential method of repeatedly iterating the θ_k in $\boldsymbol{\theta} = (\theta_1, \theta_2 \ldots, \theta_D)$ to maximization.

We diverge from [7] at Eq. (4.6) and assume that the signals $\mathbf{s}(i)$ are stationary, white, independent zero-mean Gaussian, with diagonal covariance matrix $\mathbf{P}_s$. Thus, Eq. (4.1) is actually $f(\mathbf{x}|\mathbf{A}, \sigma^2, \mathbf{s})$ for this situation. σ^2 and $\mathbf{A}$ are deterministic unknown parameters, but $\mathbf{s} = (s(1)^T s(2)^T \ldots s(N)^T)^T$ is a random vector process with block-diagonal covariance having N identical diagonal subblocks $\mathbf{P}_s$ of size D.

To find the ML $\mathbf{A}$ in the case of nondeterministic $\mathbf{s}$, we would maximize over $\mathbf{A}(\theta)$

$$L = \log f(\mathbf{x}, \mathbf{s}|\mathbf{A}) = \log f(\mathbf{x}|\mathbf{s}, \mathbf{A}) + \log f(\mathbf{s}) \tag{4.7}$$

Note that the first RH term in (4.7) is the same as the log of (4.1) except that $\mathbf{s}$ is considered a random vector, so the ML solution for σ^2 in (4.5) still holds. Further, the second RH term in (4.7) is not a function of $\mathbf{A}$ because $\mathbf{s}$ is independent of $\mathbf{A}$. Thus, in order to proceed to the ML solution of $\mathbf{A}$ from $\log f(\mathbf{x}|\mathbf{s}, \mathbf{A})$, knowledge of $\mathbf{s}$ is needed. A maximum likelihood estimate of $\mathbf{s}$ from $f(\mathbf{s}|\mathbf{x}, \mathbf{A}) = f(\mathbf{x}, \mathbf{s}|\mathbf{A})/f(\mathbf{x}|\mathbf{A})$ is found in Appendix 4A where $\mathbf{P}_s$ and $E\{\mathbf{x}\mathbf{s}^H\}$ are assumed also to be known. Although these statistics are not truly known, they are implicity estimated with large sample size N, as will be shown. An estimator for each $\mathbf{s}(i)$ is found in the next section using the orthogonality principle; the result is identical to the ML estimator.

Assuming we have an estimate $\hat{\mathbf{s}}(i)$ of $\mathbf{s}(i)$ at each step, the likelihood function in (4.7) again reduces to F in (4.6), but with $\hat{\mathbf{s}}(i)$ replacing $\mathbf{s}(i)$.

Writing out the magnitude-squared summand in (4.6) gives three terms if the sources are uncorrelated, only the sum of the two cross-terms equaling 2 $\mathrm{Re}\{\mathbf{x}^H(i)\mathbf{A}\hat{\mathbf{s}}(i)\}$ is a function of θ. The first term is $\mathbf{x}^H(i)\mathbf{x}(i)$ and depends only on the data. The third term $\mathbf{s}^H(i)\mathbf{A}^H\mathbf{A}\mathbf{s}(i)$ has subterms $\mathbf{s}_d^*(i)\mathbf{a}_d^H\mathbf{a}_e\mathbf{s}_e(i)$ with expected value zero unless $d = e$. Thus, summation over large N effectively nulls this term as if it were noise, except for the subterm with $d = e$; when $d = e$, $\mathbf{a}_d^H\mathbf{a}_e|_{d=e} = 1$ and this subterm is not a function of θ. Therefore, minimizing F for uncorrelated sources is equivalent to maximizing

$$J_\theta = \mathrm{Re}\left\{\sum_{i=1}^{N} \mathbf{x}^H(i)\mathbf{A}\mathbf{s}(i)\right\} \tag{4.8}$$

In many communications problems $s(i)$ is deterministic and perhaps periodic, as in typical PLL applications, but for the current considerations we assume $\mathbf{s}(i)$ is nondeterministic. Although J_θ is periodic in θ, its gradient will have a null at any peak or valley point in J_θ. Because of the ambiguity, we must subsequently restrict the method we are proposing to applications where a priori estimates have been obtained.

An Estimate of $\mathbf{A}\mathbf{s}(i)$

Recognizing that $\mathbf{x}(i)$ is a noisy version of $\mathbf{A}\mathbf{s}(i)$ and holding $\mathbf{A} = \hat{\mathbf{A}}$, we can replace the vector $\mathbf{A}\mathbf{s}(i)$ with its minimum mean-squared error (mmse) estimate $\mathbf{G}\mathbf{x}(i)$ given through the orthogonality principle (the alternate ML approach is given in Appendix 4A); thus

$$E\left\{\left[\hat{\mathbf{A}}\mathbf{s}(i) - \mathbf{G}\mathbf{x}(i)\right]\mathbf{x}^H(i)\right\} = 0 \tag{4.9}$$

Solving (4.9) for $\mathbf{G}\mathbf{x}(i)$, using $E\{\mathbf{s}(i)\mathbf{x}^H(i)\} = \mathbf{P}_s\mathbf{A}^H$, gives $\mathbf{G} = \hat{\mathbf{A}}\mathbf{P}_s\mathbf{A}^H\mathbf{R}_x^{-1}$, so that

$$\mathbf{G}\mathbf{x}(i) = \hat{\mathbf{A}}\mathbf{P}_s\mathbf{A}^H\mathbf{R}_x^{-1}\mathbf{x}(i) \tag{4.10}$$

where $\mathbf{P}_s = E\{\mathbf{s}(i)\mathbf{s}^H(i)\} = \operatorname{diag}(\sigma_d^2)$, and $\sigma_d^2 = E\{|s_d|^2\}$. Note in (4.10) that both the assumed $\hat{\mathbf{A}}$ and the true $\mathbf{A}$ appear. We deal with this eventually by letting $\hat{\mathbf{A}}$ correspond to trial values $\hat{\theta}$, and constraining $\mathbf{A}$ to be made up of signal subspace basis vectors.

We recognize in $\mathbf{G}\mathbf{x}(i) = \hat{\mathbf{A}}\mathbf{P}_s\mathbf{A}^H\mathbf{R}_x^{-1}\mathbf{x}(i)$ that for any estimate $\hat{\mathbf{A}}$, $\hat{\mathbf{A}}\mathbf{P}_s\mathbf{A}^H\mathbf{R}_x^{-1}\mathbf{x}(i) = \hat{\mathbf{A}}\hat{\mathbf{s}}(i)$, where we must have

$$\hat{\mathbf{s}}(i) = \mathbf{P}_s\mathbf{A}^H\mathbf{R}_x^{-1}\mathbf{x}(i) \tag{4.11}$$

Replacing $\hat{\mathbf{A}}$ in (4.8) with $\mathbf{A}$ as we did in (4.9), and replacing $\mathbf{s}(i)$ with $\hat{\mathbf{s}}(i)$ from (4.11), gives

$$\begin{aligned} J_\theta &= \sum_i \operatorname{Re}\left\{\mathbf{x}^H(i)\hat{\mathbf{A}}\mathbf{P}_s\mathbf{A}^H\mathbf{R}_x^{-1}\mathbf{x}(i)\right\} \\ &= \sum_i \operatorname{Re}\left\{\sum_d \mathbf{x}^H(i)\hat{\mathbf{a}}_d\sigma_d^2\mathbf{a}_d^H\mathbf{R}_x^{-1}\mathbf{x}(i)\right\} \\ &= \sum_d \sigma_d^2 \operatorname{Re}\left\{\mathbf{a}_d^H\sum_i \mathbf{x}(i)\mathbf{x}^H(i)\mathbf{R}_x^{-1}\mathbf{a}_d\right\} \end{aligned} \tag{4.12}$$

We next reduce (4.12) by using the large snapshot assumption (analogous to MUSIC), as derived in [2], and then replace $\mathbf{a}_d$ with its signal subspace eigenvector expansion. The large snapshot assumption gives that $\sum_i \mathbf{x}(i)\mathbf{x}^H(i) = \mathbf{R}_x$, we thus have the covariance matrix,

$$\hat{\mathbf{R}}_x = \frac{1}{N}\sum_{i=1}^{N}\mathbf{x}(i)\mathbf{x}^H(i) = \mathbf{R}_x + \mathbf{E}_x(N) \tag{4.13}$$

where $\mathbf{E}_x$ is an error matrix,

$$\mathbf{R}_x = \mathbf{V}\mathbf{\Lambda}\mathbf{V}^H \tag{4.14}$$

$$= \mathbf{V}_D\mathbf{\Lambda}_D\mathbf{V}_D^H + \mathbf{V}_{D'}\mathbf{\Lambda}_{D'}\mathbf{V}_{D'}^H \tag{4.15}$$

and $\mathbf{V}_D$ and $\mathbf{V}_{D'}$ are matrices whose columns are the eigenvectors of the signal and noise space respectively, and $\mathbf{\Lambda}_D$ and $\mathbf{\Lambda}_{D'}$ are the corresponding diagonal matrices of eigenvalues. Then, for now, considering that $\mathbf{E}_x$ is negligible and expanding $\mathbf{a}_d$ in the eigenvectors $\mathbf{V}_D$, (4.12) becomes

$$J_\theta = \sum_d \sigma_d^2 \operatorname{Re}\left\{\hat{\mathbf{a}}_d^H\mathbf{V}_D\mathbf{V}_D^H\mathbf{a}_d\right\} \tag{4.16}$$

The Scorefunction

Taking the derivative of J_θ with respect to $\hat{\theta}_k$, assuming $\mathbf{V}_D$ and $\mathbf{a}_k$ are known, gives the scorefunction

$$J'_\theta = \text{Re}\left\{(\mathbf{B}\hat{\mathbf{a}}_k)^H \mathbf{V}_D \mathbf{V}_D^H \mathbf{a}_k\right\} = \text{Re}\left\{\mathbf{a}_k^H \mathbf{V}_D \mathbf{V}_D^H \mathbf{B}\hat{\mathbf{a}}_k\right\} \tag{4.17}$$

where, using the phase of the signal at the array's center sensor as reference,

$$\mathbf{B} = j \text{ diag}\left\{-\frac{M-1}{2} - \frac{M-3}{2} \cdots \frac{M-1}{2}\right\} \tag{4.18}$$

If the $\mathbf{a}_k$ were known, of course, there would be nothing to estimate. However, the exact $\mathbf{a}_k$ in Eqs. (4.12) through (4.17) arises from assuming that the expected value of $\mathbf{s}(i)\mathbf{x}^H(i)$ in Eq. (4.9) is known $(\mathbf{P}_s\mathbf{A}^H)$. We now propose that $\mathbf{a}_k$ in Eq. (4.17) simply be replaced by $\hat{\mathbf{a}}_k$, using what information we have through $\hat{\mathbf{R}}_x$ to estimate $\mathbf{V}_D$ and $\mathbf{V}_{D'}$. Then as $\hat{\theta}_k \to \hat{\theta}_k$, $\hat{\mathbf{a}}_k \to \mathbf{a}_k$ and $\mathbf{B}\hat{\mathbf{a}}_k^H \mathbf{V}_D \mathbf{V}_D^H \hat{\mathbf{a}}_k \to 0$.

In Appendix 4B it is shown that when $\mathbf{V}_D$ is not an approximation, $\hat{\mathbf{a}}^H \mathbf{V}_D \mathbf{V}_D^H \mathbf{B}\hat{\mathbf{a}}$ is real for all $\hat{\theta}$, and the Re$\{\cdot\}$ operation in (4.17) may be discarded. Thus, the scorefunction becomes

$$J'_\theta = \hat{\mathbf{a}}^H \mathbf{V}_D \mathbf{V}_D^H \mathbf{B}\hat{\mathbf{a}} \tag{4.19}$$

Because $\mathbf{I} = \mathbf{V}_D \mathbf{V}_D^H + \mathbf{V}_{D'} \mathbf{V}_{D'}^H$ and $< \hat{a}^H, \mathbf{B}\hat{\mathbf{a}} >= 0$ (from Appendix 4C), we can also write

$$J'_\theta = -\hat{\mathbf{a}}^H \mathbf{V}_{D'} \mathbf{V}_{D'}^H \mathbf{B}\hat{\mathbf{a}} \tag{4.20}$$

This indicates that for the ML nulling solution, any inner product of $\hat{\mathbf{a}}$ with noise subspace eigenvectors should have weights on those eigenvectors which are proportional to their respective expansion coefficients for $\mathbf{B}\mathbf{a}$; that is,

$$J'_\theta = \hat{\mathbf{a}}^H \mathbf{V}_{D'} \boldsymbol{\beta} \tag{4.21}$$

where the weight vector

$$\boldsymbol{\beta} = \mathbf{V}_{D'}^H \hat{\mathbf{a}}' \tag{4.22}$$

achieves $\boldsymbol{\beta} = \mathbf{V}_{D'}^H \mathbf{B}\mathbf{a}$ when $\hat{\theta}_k = \theta_k$ for any $k = 1, 2, \ldots, D$. Appendix 4D shows that this weighting is optimal from the point of view of maximizing the slope of the gradient (scorefunction) at $\hat{\theta}_k = \theta_k$. This maximization of gradient slope is required by the Cramér-Rao bound. It is clear that J'_θ in (4.21) is a cross-correlation of a trial direction vector with the noise subspace projection of its derivative. At solution the trial direction vector lies in the signal subspace, and its angular spatial derivative lies totally in the noise subspace. Otherwise J'_θ gives a nonzero inner product of $\hat{\mathbf{a}}$ and $\mathbf{B}\hat{\mathbf{a}}$ components in the noise subspace.

Multiple Sources

The effect of multiple sources on J'_θ is examined in the following.

In Appendix 4B, Eq. (4B.4) is

$$J'_\theta = \begin{pmatrix} <\hat{\mathbf{a}}, \mathbf{a}_1> \\ <\hat{\mathbf{a}}, \mathbf{a}_2> \\ \vdots \\ <\hat{\mathbf{a}}, \mathbf{a}_D> \end{pmatrix}^T \mathbf{G} \begin{pmatrix} <\mathbf{a}_1, \hat{\mathbf{a}}'> \\ <\mathbf{a}_2, \hat{\mathbf{a}}'> \\ \vdots \\ <\mathbf{a}_D, \hat{\mathbf{a}}'> \end{pmatrix} \tag{4.23}$$

$$= \sum_i \sum_k <\hat{\mathbf{a}}, \mathbf{a}_i> G_{ik} <a_k, \hat{a}'> \tag{4.24}$$

where from Appendix 4C, $\mathbf{G} = (\mathbf{A}^H\mathbf{A})^{-1}$ and for M odd, for example,

$$<\mathbf{a}_i, \mathbf{a}_k> = 2M^{-1}\frac{\sin(M\theta_{ik}/2)\cos((M-2)\theta_{ik}/2)}{\sin(\theta_{ik}/2)} \tag{4.25}$$

$$<\hat{\mathbf{a}}, \mathbf{a}_i> = 2M^{-1}\frac{\sin(M\delta_i/2)\cos((M-2)\delta_i/2)}{\sin(\delta_i/2)} \tag{4.26}$$

$$\begin{aligned} <\mathbf{a}_k, \hat{\mathbf{a}}'> = {} & \frac{1}{2M\sin^2(\delta_k/2)}[(M-1)\sin(\delta_k/2)\cos((M-1)\delta_k/2) \\ & -\sin((M-1)\delta_k/2)\cos(\delta_k/2)] \end{aligned} \tag{4.27}$$

$\theta_{ki} = \theta_k - \theta_i$ and $\delta_i = \hat{\theta}_i - \theta_i$.

Equations (4.24)-(4.27) show that *for only one source*

$$\begin{aligned} J'_\theta &= <\hat{\mathbf{a}}, \mathbf{a}_1><\mathbf{a}_1, \hat{\mathbf{a}}'> \\ &= (M^2\sin^3(\delta_1/2))^{-1}[\sin(M\delta_1/2)\cos((M-2)\delta_1/2)] \\ &\quad \cdot [(M-1)\sin(\delta_1/2)\cos((M-1)\delta_1/2) \\ &\quad -\sin((M-1)\delta_1/2)\cos(\delta_1/2)] \end{aligned} \tag{4.28}$$

A plot of Eq. (4.28) for $M = 10$ is shown in Figure 4.1. Aside from $\delta_1 = 0$, the next closest zeros of this function are where either of the two bracketed factors in (4.28) equals zero. For the first factor this occurs when $\cos((M-2)\delta_1/2) = 0$ or $\delta_1 = \pm\pi/(M-2)$. The last factor can be written $b = \cos((M-1)\delta_1/2 + \delta)$ where $b = [(M-1)^2\sin(\delta_1/2) + \cos^2(\delta_1/2)]^{1/2}$ and $\gamma = \tan^{-1}[\cos(\delta_1/2)/((M-1)\sin(\delta_1/2))] \simeq ((M-1)\delta_1/2)^{-1}$. This factor approximately equals zero when $(M-1)\delta_1/2 + ((M-1)\delta_1/2)^{-1} = \pm\pi/2$ or $\delta_1 \simeq \pm\pi/(M-1)$. Because both factors are zero at $\delta_1 \simeq \pm\pi/(M-1)$, it will be required to know a priori the bearing angle to within half this region; that is, $|\delta_1| \leq \pi/(2(M-1))$. This is essentially the Fourier resolution, and Fourier techniques are often first used to roughly determine target bearings before high-resolution methods are applied.

To determine the ideal response of J'_θ of *two close sources*, let G be 2×2. Then

$$\begin{aligned} \|\mathbf{A}^H\mathbf{A}\|J'_\theta = {} & <\hat{\mathbf{a}}, \mathbf{a}_1><\mathbf{a}_2, \mathbf{a}_2><\mathbf{a}_1, \hat{\mathbf{a}}'> \\ & - <\hat{\mathbf{a}}, \mathbf{a}_1><\mathbf{a}_2, \mathbf{a}_1><\mathbf{a}_2, \hat{\mathbf{a}}'> \\ & - <\hat{\mathbf{a}}, \mathbf{a}_2><\mathbf{a}_1, \mathbf{a}_2><\mathbf{a}_1, \hat{\mathbf{a}}'> \\ & + <\hat{\mathbf{a}}, \mathbf{a}_2><\mathbf{a}_1, \mathbf{a}_1><\mathbf{a}_2, \hat{\mathbf{a}}'> \end{aligned} \tag{4.29}$$

The inner products $<\mathbf{a}_1, \hat{\mathbf{a}}'>$ null properly when $\hat{\mathbf{a}} = \mathbf{a}_1$, and the other two terms exactly cancel. When $\hat{\mathbf{a}} = \mathbf{a}_2$ the situation is analogous, giving a null. There

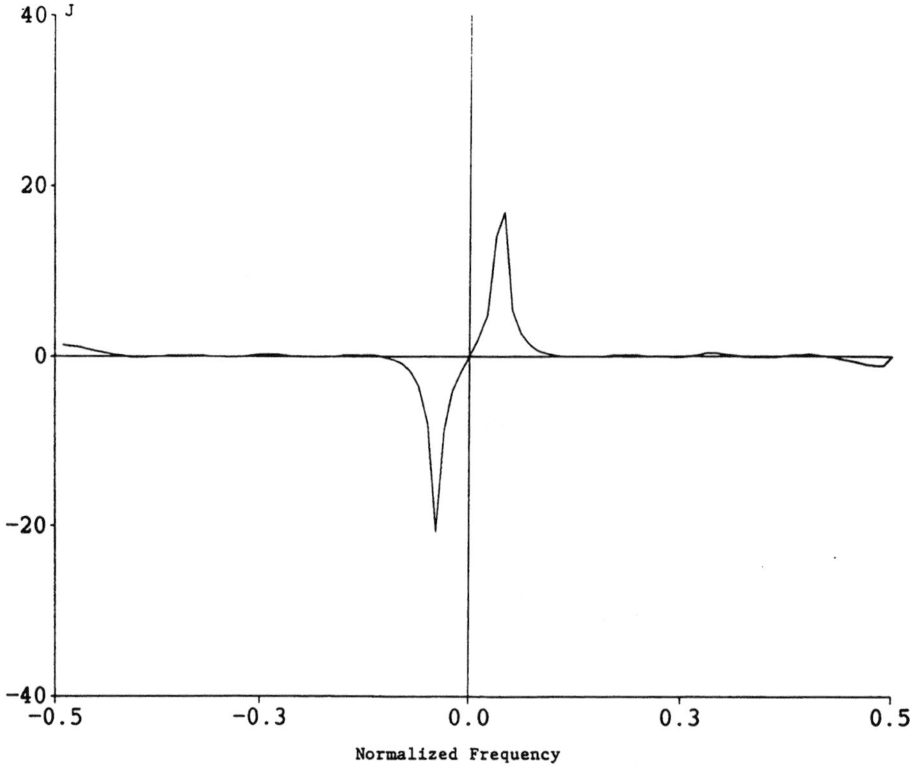

Figure 4.1 Zero crossings of J'_θ for a single source at 0 degrees. SNR $= \infty$, $M = 10$.

is a concern with two sources, however, because when $\hat{\theta} = (\theta_2 + \theta_1)/2$, another null will ideally appear, corresponding to a minimum between the two maxima of J_θ. The merging of two nulling functions with odd symmetry around their zeros give obvious difficulties for resolution.

The actual response of J'_θ to sources in noise will be shown by simulation in Results, Section IV.

It is clear that MUSIC, with minimizing function $J_{MU} = \hat{a}^H V_{D'} V_{D'} \hat{a} = M - \hat{a}^H V_D V_D^H \hat{a}$, is basically a signal subspace version of the ML solution, except that the peak of the correlation function for $\hat{\theta}$ is subtracted from the appropriate constant (M) to give a minimum which is ideally zero. The term $\hat{a}^H V_D V_D^H \hat{a}$ could be derived to maximize the log-likelihood in the same way that we derived $\hat{a} V_{D'} V_{D'}^H B \hat{a}$ to null the scorefunction.

Figure 4.1 shows real (J'_θ) for one source at $\theta = 0$, SNR$= \infty$, and number of sensors $M = 10$. An example is shown in Figure 4.2 for two sources at normalized bearing (frequency) 0.2 and 0.25.

Figure 4.3 shows the effect on a single run when noise is present, SNR$= 10$ dB, $N = 119$ snapshots, $M = 10$. The zero crossings between the two frequencies and at $f = 0.22$ have disappeared, but a minimum of $|J'_\theta|$ near 0.22 still exists.

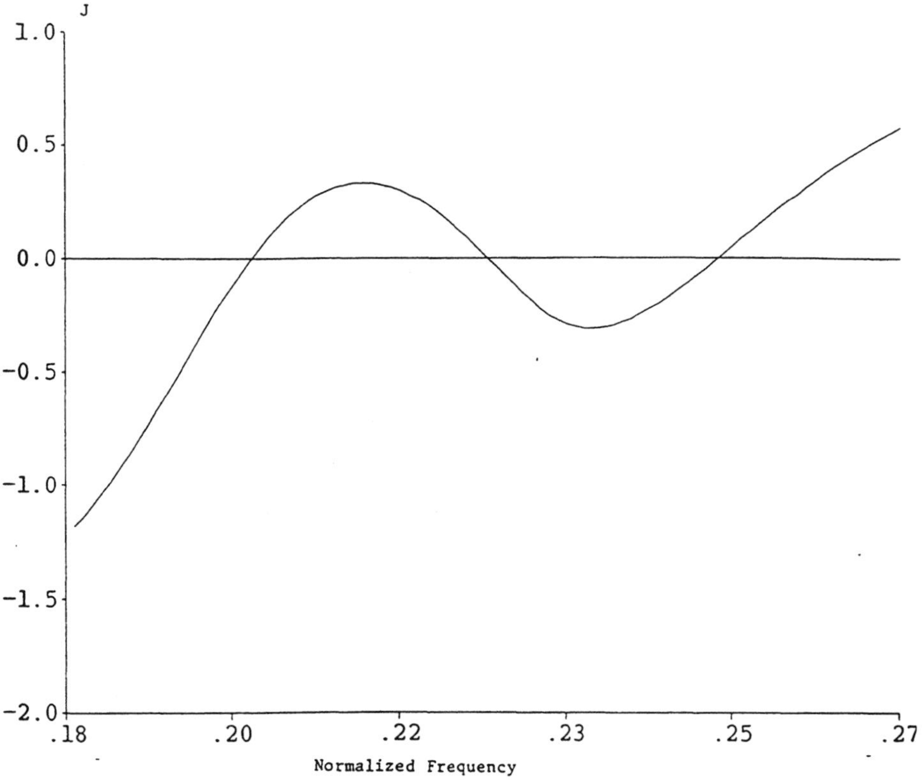

Figure 4.2 J'_θ for two sources at $f = 0.20, 0.25$, SNR $= \infty$, $M = 10$.

Discussion and Performance

The scorefunction estimator has been derived from maximum likelihood principles, solving for values of bearing parameters for which the partial of the likelihood function with respect to the bearing parameter (the scorefunction) is zero. As derived, it is only applicable in regions which are known a priori to contain sources, because any ambiguous peaks and valleys in the likelihood function also produce zeros in the scorefunction J'_θ. The new estimator is shown to give optimal weights to the noise-space eigenvectors from the point of view of maximizing the slope of the scorefunction at the solution point.

Near solution the scorefunction linearizes. As will be seen in the following sections, the linearization allows noise analysis. Following [11], we may assume that rather than the true eigenstructure, we have an estimate $\hat{\mathbf{R}}_x$, yielding $\hat{\mathbf{V}}_D = \mathbf{V}_D + \Delta\mathbf{V}_D$ and $\hat{\mathbf{V}}_{D'} = \mathbf{V}_{D'} + \Delta\mathbf{V}_{D'}$ where $\Delta\mathbf{V}_D$ lies in the noise space and $\Delta\mathbf{V}_{D'}$ lies in the signal space. It has been shown that in terms of the data matrix, $\mathbf{Y}$ and its perturbation $\Delta\mathbf{Y}$ due to noise, that is,

$$\Delta\mathbf{V}_{D'} = -\mathbf{V}_D\mathbf{\Lambda}_D^{-1}\mathbf{V}_D^H\Delta\mathbf{Y}^H\mathbf{V}_{D'} \tag{4.30}$$

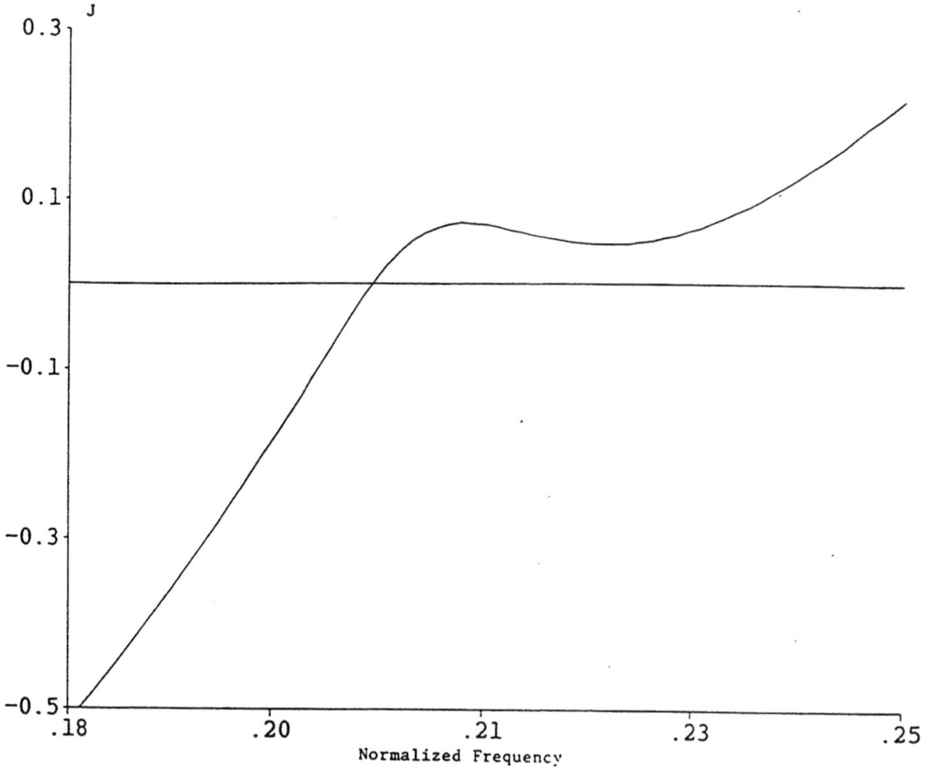

Figure 4.3 Example of J'_θ for SNR = 10 dB, $M = 10$, 119 samples $f_1 = 0.2$, $f_2 = 0.22$.

where Λ_D is the submatrix of nonzero singular values of the data matrix $\mathbf{Y}$ (whose columns are the array snapshot vectors). This expression allows us to find the mean square error in $\hat{\theta}$ by finding the perturbation of J'_θ for $\Delta\theta$, when J'_θ is written as

$$J'_\theta(\mathbf{V}_{D'} + \Delta\mathbf{V}_{D'}, \theta) = \text{Re}\,\{\mathbf{a}^H(\mathbf{V}_{D'} + \Delta\mathbf{V}_{D'})(\mathbf{V}_{D'} + \Delta\mathbf{V}_{D'})^H\mathbf{Ba}\} \tag{4.31}$$

Realizing that $J'_\theta(\mathbf{V}_{D'}, \theta_k) = 0$, we have

$$\Delta\theta = J(\mathbf{V}_{D'} + \Delta\mathbf{V}_{D'})/\partial J(\mathbf{V}_{D'} + \Delta\mathbf{V}_{D'}, \theta)/\partial\theta|_{\theta=\theta_k} \tag{4.32}$$

Taking the indicated partial derivative and keeping only first-order and nonzero terms, yields

$$\Delta\theta = \frac{\text{Re}\,\left\{\mathbf{a}^H\mathbf{V}_D\Lambda_D^{-1}\mathbf{V}_D^H\Delta\mathbf{Y}^H\mathbf{V}_{D'}\mathbf{V}_{D'}^H\mathbf{Ba}\right\}}{\|\mathbf{V}_{D'}^H\mathbf{Ba}\|^2} \tag{4.33}$$

This $\Delta\theta$ is identical to that in [11] for MUSIC. We then also extract from [11] that

$$E\left\{\Delta\theta_k^2\right\} = \frac{\mathbf{R}_s^{-1}(k,k)]\sigma_r^2}{2N\|\mathbf{V}_{D'}^H\mathbf{Ba}(\theta_k)\|^2} \tag{4.34}$$

where N is the number of snapshots, σ_n^2 is the independent noise variance at each sensor, and $\mathbf{R}_s^{-1}(k,k)$ is the signal source covariance matrix's k^{th} diagonal element. For incoherent sources, $\mathbf{R}_s^{-1}(k,k) = \sigma_k^{-2}$, the inverse of the power in the kth source.

Because we have shown that $\mathbf{Ba}$ lies totally in the noise subspace, $\|\mathbf{V}_{D'}^H \mathbf{Ba}(\theta_k)\|^2 = \|\mathbf{Ba}(\theta_k)\|^2$. This is straightforwardly evaluated with (4.18) to give

$$\|\mathbf{Ba}(\theta_k)\|^2 = \begin{cases} 2\sum_{i=1}^{\frac{M-1}{2}} i^2 = \frac{1}{2}M(M^2-1), & M \text{ odd} \quad (4.35a) \\ 2\sum_{i=1}^{M/2}\left(\frac{2i-1}{2}\right)^2 = \frac{1}{2} + M(M^2-4)/12, & M \text{ even} \quad (4.35b) \end{cases}$$

Thus, (4.34) may be more simply written

$$E\left\{\Delta\theta_k^2\right\} = \frac{\sigma_n^2/\sigma_k^2}{2N\|\mathbf{Ba}(\theta_k)\|^2} \tag{4.36}$$

We shall see in the next section that this is the identical expression for the noise variance on the bearing estimator in PLL form if N is equated to the reciprocal of the closed-loop bandwidth, the effective averaging time in the loop.

4.3 THE ITERATIVE, NONSUBSPACE, PLL APPROACH

The PLL requires a local copy of the (deterministic) signal, which is either periodic, as in a pure-tone, or nearly so, as in angle-modulated narrowband signals. We assume the latter from this point. Each source signal has the form

$$s_k(i) = |s_k(i)|e^{j\phi_k(i)} \tag{4.37}$$

where $\phi_k(i)$ may include the integrated carrier frequency, giving a ramp of phase. Typically $\phi_k(i)$ is simply the difference in instantaneous phase from that known (the carrier phase ramp). A local copy of the signal is generated: $\hat{s}_k(i) = |\hat{s}(i)|e^{j(\hat{\phi}_k(i)+\pi/2)}$. This is at phase lock 90° out of phase with the incoming signal. To within a scale factor it is the derivative of the signal. The product $s_k(i)\hat{s}_k^*(i)$ has a reel part proportional to $\sin(\hat{\phi}_k - \phi_k)$, and this becomes an error control signal to advance or retard the local phase $\hat{\phi}_k(i)$.

Now recall Eq. (4.8):

$$J_\theta = \operatorname{Re}\left\{\sum_i \mathbf{x}^H(i)\mathbf{A}\mathbf{s}(i)\right\} \tag{4.38}$$

In this case $\mathbf{s}(i)$ is assumed known; however, we do not know its phase $\phi_k(i)$, therefore we must solve both

$$J'_\theta = \frac{\partial J_\theta}{\partial \theta} = \operatorname{Re}\left\{\sum_i \mathbf{x}^H(i)\mathbf{B}\hat{\mathbf{a}}_k\hat{s}_k(i)\right\} = 0 \tag{4.39}$$

and

$$J'_{\phi} = \frac{\partial J_{\theta}}{\partial \phi_k} = \text{Re}\left\{ \sum_i \mathbf{x}^H(i)\hat{\mathbf{a}}_k \frac{\partial \hat{\mathbf{s}}_k(i)}{\partial \phi_k} \right\} = 0 \tag{4.40}$$

where, using the end sensor as reference

$$\mathbf{B} = j \text{ diag } \{-M-1, -M-2, \ldots, 1, 0\} \tag{4.41}$$

and, if $\hat{s}_k(i) = |\hat{s}(i)|e^{j\hat{\phi}_k(i)}$,

$$\frac{\partial \hat{s}_k(i)}{\partial \phi_k} = j\hat{s}_k(i) \tag{4.42}$$

Note that for the PLL system we have switched to the end sensor as reference. In the following expressions involving M, a substitution may be made to accommodate the array center being the reference.

Recalling that $\mathbf{x}(i) = \sum_{i=1}^{N} \mathbf{a}_d s_d(i) + \mathbf{n}(i)$, we investigate the noise-free solution first by assuming $\mathbf{n}(i) = 0$. Eventually the noise effects on the linearized system will be found separately. In addition, if the soures are uncorrelated and stationary with zero mean, the summation of $s_d^*(i)s_k(i)$ for large N is nearly zero.

Under the noise-free and large N assumptions we may focus on terms in (4.39) and (4.40) which become the cross-coupled equations

$$J'_{\theta}(i) = -\sum_{m=0}^{M-1} \hat{\sigma}_k^2(i) m \sin(\alpha_{mk}(i)) \tag{4.43}$$

$$J'_{\phi}(i) = -\sum_{m=0}^{M-1} \hat{\sigma}_k^2 \sin(\alpha_{mk}(i)) \tag{4.44}$$

where

$$\alpha_{mk}(i) = \hat{\theta}_k(i) - \theta_k(i) + m\hat{\theta}_k(i) - m\theta_k(i) \tag{4.45}$$

and

$$\hat{\sigma}_k^2 = |s_k|\,|\hat{s}_k|.$$

Note that if the array center is taken as reference, the sums in (4.43) become $2\sum_{m=1}^{M/2} \frac{(2m-1)}{2}(\cdot)$, m even, or $2\sum_{m=1}^{\frac{M-1}{2}} m$, m odd. The index limits in (4.44) change similarly.

When the phase and bearing estimators are small, $\sin(\alpha_{mk}(i)) \simeq \alpha_{mk}(i)$ and $J'_{\theta}(i)$ can be used as a control signal to adjust θ_k in the $M-1$ generators of $\exp\{jm\hat{\theta}_k(i)\}$ in $\hat{a}_k$, and $J'_{\theta}(i)$ can adjust $\hat{\theta}_k(i)$ in $\hat{s}_k(i)$.

Although these equations contain the unknown $\hat{\sigma}_k^2$, typical phase-lock applications hardlimit the input signal so that $\hat{\sigma}_k^2$ is forced to a known value. Analysis of the effects of the limiter as a function of input SNR are given in [16]. The basic feature of hardlimiting is that weaker signals are suppressed (the FM "capture" effect).

Combinations of PLLs with sensor arrays have been studied before [12, 13]; however, cross-coupling for dual purposes of direction finding and demodulation has not been reported.

Digital PLL System Description

A block diagram of a sensor array incorporating the CDPLL pair is depicted in Figure 4.4. The first loop with the linearly weighted sum feedback[1] tracks the bearing angle of the incoming signal; the second loop with uniformly weighted sum feedback is the temporal phase tracking loop. Assuming a linear equally spaced sensor array, the noisy incoming signal at the ith sensor is modeled via

$$\begin{aligned} x_i(t) &= s_i(t) + n_i(t) \\ &= A \sin[\omega_0 t + \phi(t) + (i-1)\theta] + n_i(t), \quad i = 1, \ldots, M \qquad (4.46) \\ \theta &= \omega_0 \Delta \sin\alpha / c \qquad (4.47) \end{aligned}$$

where M is the number of sensors, θ is the normalized bearing angle of the source, α is the true incident source bearing, Δ is the sensor spacing, c is the velocity of the wave, and $n(\cdot)$ is the additive white Gaussian noise.

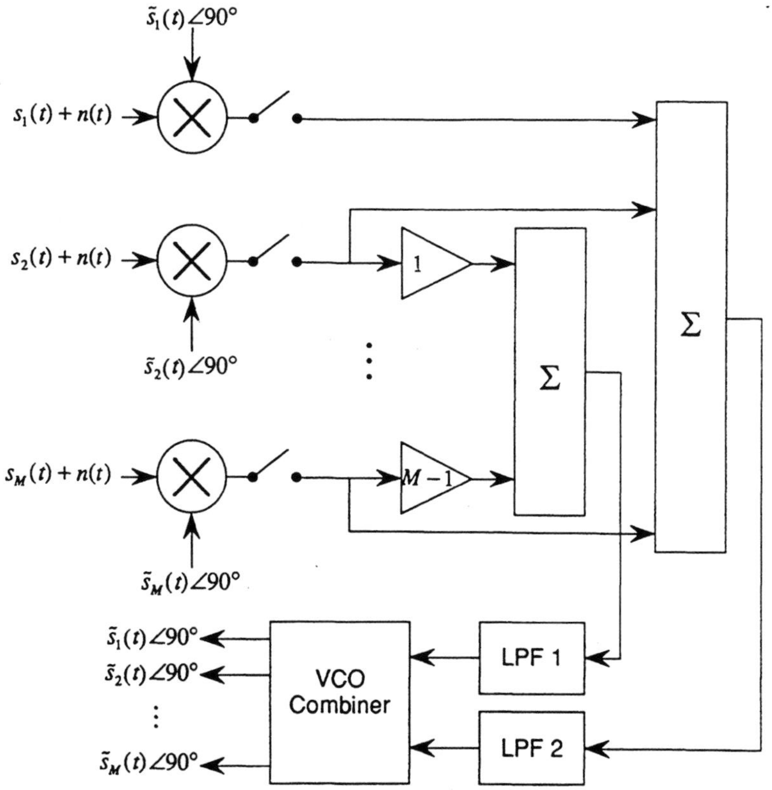

Figure 4.4 Block diagram of array system with CPLLs.

[1] Linearly weighted feedback is the result of the maximum-likelihood solution for bearing when the plane wavefront is present in Gaussian noise.

The zero-crossing digital phase-locked loop (ZC-DPLL) is chosen for implementation due to its simplicity and ease of simulation [14]. The incoming signal at the ith sensor is sampled at time instants $t_i(k)$ determined by the digital clock. The time elapsed between the $(k-1)^{\text{th}}$ and k^{th} samples is denoted by

$$T_i(k) \triangleq t_i(k) - t_i(k-1), \quad k = 1, 2, \ldots, \tag{4.48}$$

while the value of the incoming signal at the ith sensor at $t_i(k)$ is

$$x_i(k) = s_i(k) + n_i(k), \quad k = 0, 1, \ldots \tag{4.49}$$

The signal samples $\{x_i(k)\}, i = 1, \ldots, M$ are multiplied by estimates of their signal components, shifted by 90°. These products are given two separate weightings which form two feedback signals. The feedback signals are filtered and the two filter outputs (see Figure 4.5) $y_1(k)$ and $y_2(k), k = 0, 1, \ldots$ are used to control the next period of the digital clock according to the algorithm

$$\begin{aligned} T_i(k+1) &= T - T_i(k) \\ &= T - (i-1)y_1(k) - y_2(k), \quad k = 0, 1, \ldots \end{aligned} \tag{4.50}$$

where $T = 2\pi/\omega_0$ is the nominal clock period.

We have chosen order one for the first loop, which is used to track the normalizied (fixed or slowly varying) bearing angle θ, and order two for the second loop, which locks onto the time-varying phase $\phi(t)$. The second-order loop can have zero steady-state phase error for a frequency step in the signal.

As can be seen from the figures, the special features of the proposed CDPLL pair which make it differ from conventional phase-locked loops are its two separately weighted sums of phase errors providing two feedback signals and coupling of phase errors between the bearing-angle loop and the temporal phase loop.

With the above assumptions the mathematically equivalent baseband model for the proposed CDPLL pair is shown in Figure 4.5. The first loop has gain G_1 and the second loop filter is

$$F_2(z) = G_2 + \frac{G_3}{1 - z^{-1}} \tag{4.51}$$

The baseband model can be expressed by two coupled, nonlinear, time-dependent stochastic difference equations associated with the phase error process $\psi_1(k)$ and $\psi_2(k)$ where $\psi_1(k) = \theta - \hat{\theta}(k)$ is the normalized bearing error of the first loop and $\psi_2(k) = \phi(k) - \hat{\phi}(k)$ is the temporal phase error of the second loop:

$$\begin{aligned} \psi_1(k+1) = \psi_1(k) &- K_1 \sum_{j=1}^{M-1} j \sin[j\psi_1(k) + \psi_2(k)] \\ &- \frac{K_1}{A} \sum_{j=1}^{M-1} j n_{j+1}(k) + \theta(k+1) - \theta(k) \end{aligned} \tag{4.52}$$

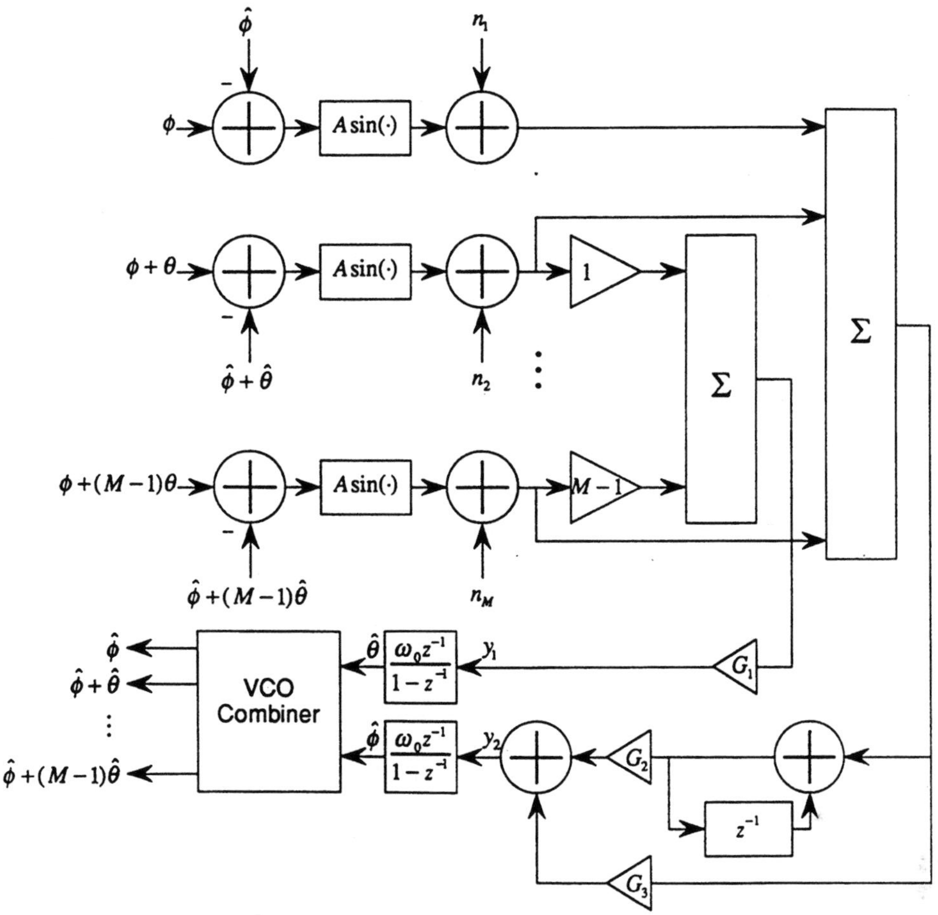

Figure 4.5 Baseband model. Θ is normalized source bearing Φ is temporal phase.

$$\begin{aligned}\psi_2(k+1) = {} & 2\psi_2(k) - \psi_2(k-1) - rK_2\sum_{j=0}^{M-1}\sin[j\psi_1(k) \\ & + \psi_2(k)] + K_2\sum_{j=0}^{M-1}\sin[j\psi_1(k-1)+\psi_2(k-1)] \\ & - \frac{rK_2}{A}\sum_{j=0}^{M-1} n_{j+1}(k) + \frac{K_2}{A}\sum_{j=0}^{M-1} n_{j+1}(k-1) \\ & + \phi(k+1) - 2\phi(k) + \phi(k-1)\end{aligned} \tag{4.53}$$

where $K_1 = \omega_0 A G_1$, $K_2 = \omega_0 A G_2$, and $r = 1 + G_3/G_2$.

If the received signal has a frequency offset $\Delta\omega = \omega_1 - \omega_0$, the phase process to be acquired by the second loop is

$$\phi(t) = (\omega_1 - \omega_0)t + \phi(0) \tag{4.54}$$

By setting $\theta(k) = \theta(k-1)$, which indicates essentially constant source direction, and using Eq. (4.45) in Eqs. (4.43) and (4.44), we have

$$\begin{aligned}\psi_1(k+1) &= \psi_1(k) - K_1 \sum_{j=1}^{M-1} j \sin[j\psi_1(k) + \psi_2(k)] \\ &\quad - \frac{K_1}{A} \sum_{j=1}^{M-1} j n_{j+1}(k)\end{aligned} \tag{4.55}$$

$$\begin{aligned}\psi_2(k+1) &= 2\psi_2(k) - \psi_2(k-1) - rK_2' \sum_{j=0}^{M-1} \sin[j\psi_1(k) \\ &\quad + \psi_2(k)] + K_2' \sum_{j=0}^{M-1} \sin[j\psi_1(k-1) + \psi_2(k-1)] \\ &\quad - \frac{rK_2'}{A} \sum_{j=0}^{M-1} n_{j+1}(k) + \frac{K_2'}{A} \sum_{j=0}^{M-1} n_{j+1}(k-1)\end{aligned} \tag{4.56}$$

where K_1 and r are the same as before and

$$K_2' = K_2 \frac{\omega_1}{\omega_0} = \omega_1 A G_2 \tag{4.57}$$

When the phase errors in both loops are sufficiently small, the model of the CDPLL pair can be linearized, enabling us to employ the z-transform, establish the closed-loop transfer functions, and determine other performance measures. Under the linearizing assumption, Eqs. (4.55) and (4.56) can be written as

$$\begin{aligned}\psi_1(k+1) &= (1 - K_1 M_1)\psi_1(k) - K_1 M_2 \psi_2(k) \\ &\quad - \frac{K_1}{A} \sum_{j=1}^{M-1} j n_{j+1}(k)\end{aligned} \tag{4.58}$$

$$\begin{aligned}\psi_2(k+2) &= (2 - rK_2'M)\psi_2(k+1) + (K_2'M - 1)\psi_2(k) \\ &\quad - K_2'M_2[r\psi_1(k+1) - \psi_1(k)] \\ &\quad - \frac{rK_2'}{A} \sum_{j=0}^{M-1} n_{j+1}(k) + \frac{K_2'}{A} \sum_{j=0}^{M-1} n_{j+1}(k-1)\end{aligned} \tag{4.59}$$

where $M_1 = \sum_{l=1}^{M-1} l^2$ and $M_2 = \sum_{l=1}^{M-1} l$. These quantities are modified somewhat for array center reference and m even or odd. Note that the summations over the noise processes yield one unfiltered driving process in (4.58) and another single (first-order filtered) process driving (4.59).

As mentioned above, the CDPLL pair has unique features in comparison to the conventional DPLLs. In what follows, we consider CDPLL performance first in the absence of noise and later in the presence of noise.

CDPLL Performance Analysis

Operation in the Absence of Noise. It is known from the literature [15] that for a conventional DPLL of any order there is a set of parameters in the stability region which yields optimal convergence. It is interesting to ask if there exists such an optimal set of parameters for CDPLLs. A positive answer is found through the linearized (noise-free) Eqs. (4.58) and (4.59); the optimal parameter set is

$$K_1(k) = \frac{1}{M_1 + p_k M_2}, \; K_2'(k) = \frac{1}{M + M_2/p_k}, \; p_k = \psi_2(k)/\psi_1(k), r = 2 \tag{4.60}$$

Recall from (4.53) and (4.57) that these gains are proportional to G_1 and G_2, respectively. It can be shown from the linear equations that with these parameter values the phase errors equal zero at the next sampling instant no matter what their values at the current step.

Unfortunately, as opposed to the conventional case, this set of parameters depends on time k and is adapted by the phase error process step by step through p_k, and in practice p_k can only be estimated. However, a set of stable parameters can always be found if p_k is fixed at a positive constant. Subsequently, we arrive at stability criteria.

Operation in the Presence of Noise. In the following we derive the error densities for $\psi_i(k)$ at steady state. It is assumed that the sensor noises, signal phase, and source bearing are all mutually uncorrelated. The noise sequences are white and Gaussian with zero mean and variance σ_n^2. Defining $\zeta(k)$ through

$$\psi_2(k) = \zeta(k) - r\zeta(k+1) \tag{4.61}$$

and letting $y_1(k) = \psi_1(k)$, $y_2(k) = \zeta(k)$, and $y_3(k) = \zeta(k+1)$ results in the following set of linear state equations from Eqs. (4.58) and (4.59):

$$\begin{aligned} y_1(k+1) &= (1 - K_1 M_1) y_1(k) - K_1 M_2 y_2(k) + K_1 r M_2 y_3(k) \\ &\quad - \frac{K_1}{A} \sum_{j=1}^{M-1} j n_{j+1}(k) \end{aligned} \tag{4.62}$$

$$y_2(k+1) = y_3(k) \tag{4.63}$$

$$\begin{aligned} y_3(k+1) &= (2 - rK_2' M) y_3(k) + (K_2' M - 1) y_2(k) \\ &\quad + K_2' M_2 y_1(k) + \frac{K_2'}{A} \sum_{j=0}^{M-1} n_{j+1}(k) \end{aligned} \tag{4.64}$$

where M_1 and M_2 are the same parameters as in Eqs. (4.58) and (4.59).

This form of the equations has obviated the filtering of the noise process in driving Eq. (4.59). Further, with obvious substitutions (4.62)–(4.64) give

$$\mathbf{y}(k+1) = \mathbf{F}\mathbf{y}(k) + \mathbf{B}\mathbf{n} \tag{4.65}$$

where $\mathbf{y}$ is 3×1, $\mathbf{F}$ is 3×3, $\mathbf{B}$ is $3 \times M$, and $\mathbf{n}$ is $M \times 1$. The steady-state covariance matrix for $\mathbf{y}$ is the solution for $\mathbf{R}_y$ in

$$\mathbf{R}_y = \mathbf{F}\mathbf{R}_y\mathbf{F}^T + \sigma_n^2\mathbf{B}\mathbf{B}^T \tag{4.66}$$

Since $\psi(k) = y_1(k)$ and $\psi_2 = y_2(k) - ry_3(k)$, and the $y_i(k)$ are linear combinations of the noise, the steady-state joint density of $\psi_1(k)$ and $\psi_2(k)$ must be Guassian with zero means and variances, covariance

$$\sigma_{\psi_1}^2 = \sigma_{y_1}^2 \tag{4.67}$$

$$\sigma_{\psi_2}^2 = (1+r^2)\sigma_{y_2}^2 - 2r\sigma_{y_2y_3} \tag{4.68}$$

$$\sigma_{\psi_1\psi_2} = \sigma_{y_1y_2} - r\sigma_{y_1y_3} \tag{4.69}$$

with $\sigma_{y_iy_j}$ and $\sigma_{y_i}^2$ from R_y in (4.66).

We have calculated $\sigma_{\Psi_1}^2$ and $\sigma_{\Psi_2}^2$ for a range of SNR and various M with K_1 and K_2 giving stable loop conditions derived in the next section. Results are plotted in Figure 4.6.

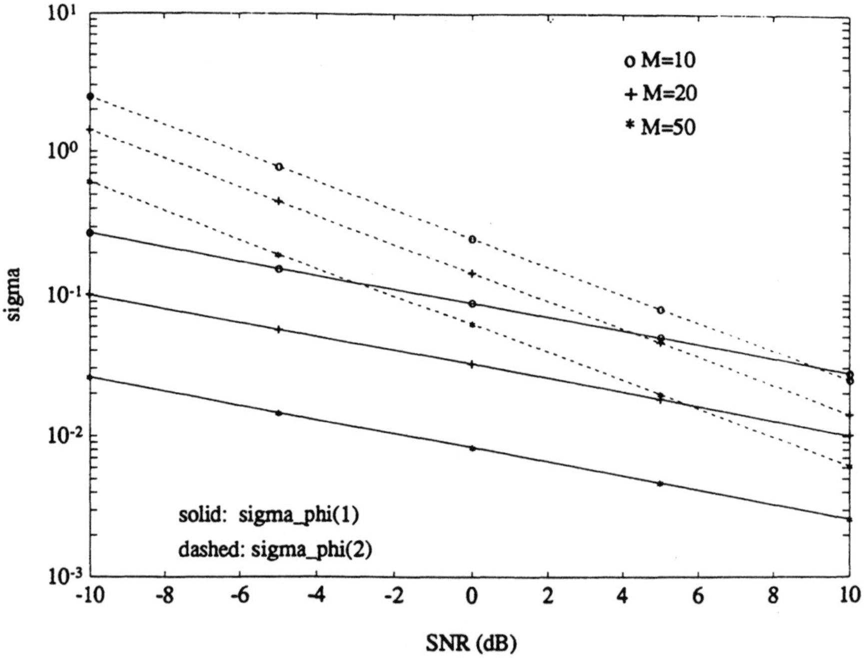

Figure 4.6 Standard deviation δ_{ψ_1} and δ_{ψ_2} versus SNR (from Equations 4.66-4.69 for $r = 1$, $K_1 = (M_1 + M_2)^{-1}$, $K_2 = (M + M_2)^{-1}$, $p = 1$. Values over 1 rad indicate marginal operation.

Coupled Loop Stability, Noise-Free

Applying the z-transform to Eqs. (4.58) and (4.59), we may obtain closed-loop transfer functions, $H_1(z) = \hat{\theta}(z)/\theta(z) = B_1(z)/D(z)$ and $H_2(z) = \hat{\phi}(z)/\phi(z) = B_2(z)/D(z)$. The characteristic equation for the coupled system is

$$\begin{aligned} D(z) = z^3 &+ (rK_2'M + K_1M_1 - 3)z^2 + (rK_1K_2'MM_1 - rK_1K_2'M_2^2 \\ &- 2K_1M_1 - rK_2'M - K_2'M + 3)z + K_1K_2'M_2^2 \\ &- K_1K_2'MM_1 + K_1M_1 + K_2'M - 1 \end{aligned} \tag{4.70}$$

and

$$\begin{aligned} B_1(z) &= K_1M_1z^2 + (K_1K_2'MM_1r - K_1K_2'rM_2^2 - 2K_1M_1)z \\ &\quad + K_1M_1 + K_1K_2'M_2^2 - K_1K_2'M_1M \\ B_2(z) &= K_2'rMz^2 + (K_1K_2'MM_1r - K_1K_2'rM_2^2 - K_2'M - MrK_2')z \\ &\quad + K_2'M + K_1K_2'M_2^2 - K_1K_2'M_1M \end{aligned}$$

To guarantee that $D(z)$ has no zeros outside the unit circle of the z-plane, we use Jury's stability criterion [15]. It is found that this stability region can be described by five inequality constraints:

$$r > 1 \tag{4.71}$$

$$K_1K_2'(MM_1 - M_2^2)(r+1) - 2K_2'M(r+1) - 4K_1M_1 + 8 > 0 \tag{4.72}$$

$$0 < K_1M_1 + k_2'M - K_1K_2'(MM_1 - M_2^2) < 2 \tag{4.73}$$

$$\begin{aligned} &K_1^2K_2'(MM_1 - M_2^2)^2 - K_2'M^2(r-1) - K_1M_2^2(r-1) \\ &\quad + K_1K_2'M(MM_1 - M_2^2)(r-2) - K_1^2M_1(MM_1 - M_2^2) < 0 \end{aligned} \tag{4.74}$$

$$\begin{aligned} &K_1K_2'^2M(MM_1 - M_2^2)(r+2) + 3K_1^2K_2'M_1(MM_1 - M_2^2) \\ &\quad + K_2'M\left[4 - (r+1)K_2'M\right] + 2K_1M_1(2 - K_1M_1) \\ &\quad + K_1K_2'\left[8MM_1 + (r-5)M_2^2\right] - K_1^2K_2'^2(MM_1 - M_2^2)^2 > 0 \end{aligned} \tag{4.75}$$

Once a set of parameters is given, it may be checked for stability using the above constraints. Stability can also be verified by finding the parameter set r, K_1, K_2' which keeps the phase error variance non-negative. For example, with $M = 12, \sigma_n^2/A^2 = 1$, the parameter set $r = 2, K_1 = 1/M_1 = 1/506, K_2' = 1/M = 1/12$, will produce $\sigma_{\psi_1}^2 = -0.0021, \sigma_{\psi_2}^2 = -0.2383$, which are impossible and indicate instability. However, $r = 2, K_1 = (M_1 + M_2)^{-1} = 1/572, K_2' = (M + M_2)^{-1} = 1/78$ will produce $\sigma_{\psi_1}^2 = 0.0023, \sigma_{\psi_2}^2 = 0.0493$.

It may be proved by this method (positive variance) that the parameter set $r = 2, K_1 = (M_1 + pM_2)^{-1}, K_2' = (M + M_2/p)^{-1}$ and $10^{-5} \le p < \infty$ are all in the stability region. Recall from the preceding section that for the optimum parameter set in the noise-free case, p must be chosen as the ratio of ψ_2 over ψ_1 and adapted step-by-step. If we choose p as any positive constant number, the parameter set in

(4.60) is not optimum but stable. We are left to test for linearity violation if σ_ψ^2 is too large.

Gain and Equivalent Bandwidth for an Independent Bearing PLL

In this section we consider the uncoupled bearing loop alone, as if $\Psi_2 = 0$, or signal is known exactly. Although the methodogy would apply to the CDPLL, we are interested in making a comparison to MU and ML eigenstructure algorithms, wherein the signal's temporal phase is not a factor. We conclude with the performance of the uncoupled bearing loop, and we also compare its error variance to that of the CDPLL with equivalent loop gain K_1.

A conventional bandwidth B_e is defined by

$$B_e = \frac{1}{2\pi j} \oint H(z)H(z^{-1})z^{-1}dz \tag{4.76a}$$

for discrete systems and

$$B_e = \frac{1}{2\pi j} \oint H(s)H(-s)ds \tag{4.76b}$$

for analog systems. In the analog system the voltage-controlled oscillator has voltage-to-phase transfer function K_v/s, an integrator. The equivalent in our digital system is $w_0 z^{-1}(1 - z^{-1})^{-1}$. $H(s)$ is the transfer function from the noise input to the phase estimate. Considering the noise sequence as having power $\sigma_n^2/A^2\,\text{rad}^2$ at the input to the linearized loop and assuming the output mean squared phase error σ_ψ^2 as being reduced by the reciprocal of the equivalent bandwidth, we also have

$$\sigma_\psi^2 = (\sigma_n^2/A^2)B_e \tag{4.77}$$

This bandwidth is distinct from the loop bandwidth B_L, which is the bandwidth of the transfer function from θ to $\hat{\theta}$. It can be seen in Figure 4.5 that θ undergoes a gain proportional to sensor number, whereas noise does not. It makes no difference in figuring $H(z)$ whether $n(t)$ is inserted inside the loop as shown in Figure 4.5, or whether we add it to θ outside the loop, first passing it through a gain A^{-1}. If we use $H(z)$ the same as $\hat{\theta}(z)/\theta(z)$, then σ_n^2/A^2 is the input power. In the following we will refer to Figure 4.5, where input noises each have variance σ_n^2.

The noise equivalent bandwidth must be found considering each of the sensor noises as a separate input. Each transfer function has its output phases *mse*; these add to give a total *mse*. The equivalent bandwidth is then

$$B_e = \frac{1}{2\pi j} \sum_{i=1}^{M-1} \oint H_i(z)H_i(-z)z^{-1}dz \tag{4.78}$$

where from Mason's rule applied to each noise-in bearing-out transfer function (see Figure 4.4)

$$H_i(z) = \frac{i(K_1/A)z^{-1}/(1-z^{-1})}{1+\sum_i i^2 K_1 z^{-1}/(1-z^{-1})} \tag{4.79}$$

Insertion of (4.79) into (4.78) gives the bearing-loop-only equivalent bandwidth

$$B_e = \frac{K_1/A^2}{2(1-M_1K_1/2)} \tag{4.80}$$

Similarly, it can be shown by using

$$\hat{\theta}(z)/\theta(z) = \frac{\sum_i i^2 K_1 z^{-1}/(1-z^{-1})}{1+\sum_i i^2 K_1 z^{-1}/(1-z^{-1})} \tag{4.81}$$

that the uncoupled bearing loop bandwidth is $B_{L_1} = \dfrac{K_1 M_1}{2(1-M_1K_1/2)}$

Applying noise power σ_n^2 to B_e in (4.80) gives

$$\sigma_{\psi_1}^2 = (\sigma_n^2/A^2)\frac{K_1}{2(1-M_1K_1/2)} \tag{4.82}$$

and substituting B_{L_1} gives

$$\sigma_{\psi_1}^2 = \frac{\sigma_n^2}{2(A^2/2)M_1} B_{L_1} \tag{4.83}$$

In the form of (4.83), it is clear that if $B_{L_1}^{-1}$ is the averaging time in the loop, then inverse loop bandwidth $B_{L_1}^{-1}$ is effectively equal to N, the effective number of snapshots considered by the loop in estimating $\hat{\theta}$. Equating B_L^{-1} to N and recognizing $A^2/2 = \sigma_k^2$ gives $\sigma_{\psi_1}^2$ identical to $E\{\Delta\theta^2\}$ in (4.36). That is, performance of MU, scorefunction ML, and the DPLL direction-finding loop are identical. However, we note that the PLL requires essentially an angle-modulated carrier or at least a narrowband signal whose envelope is modulated less than 100%.

It is interesting to compare the phase variances $\sigma_{\psi_1}^2$ for the coupled and uncoupled loops. In Figure 4.6 the value $K_1 = (M_1 + M_2)^{-1}$ has been used. Some algebra gives that $K_1 = (M(M^2-1)/2)^{-1}$. Inserting this same value of K_1 into (4.82) results in

$$\sigma_{\psi_1}^2 = \frac{3}{4\,\mathrm{SNR}\left(1-\dfrac{2M-1}{3(M+1)}\right)M(M^2-1)}$$

For large M this approaches

$$\sigma_{\psi_1}^2 = \frac{9/4}{M^3\,\mathrm{SNR}}, \quad M \text{ large}$$

For comparison with CDPLLs choose the SNR = 0 dB point in Figure 4.6, where for $M = 10$, $\sigma_{\psi_1} \simeq 10^{-1}$ or $\sigma_{\psi_1}^2 \simeq 10^{-2}$. However, the DPLL would have $\sigma_{\psi_1}^2 = 1.5 \cdot 10^{-3}$. The implication is that cross-coupling the loops greatly decreases the bandwidths that they would otherwise have.

This phenomenon had been noted in the experimental work in [18]: in order to achieve lock in the CDPLLs used to cancel interfering tones, the design bandwidths of each individual loop had to be opened extremely wide compared to what they would have been for acquisition without cross-coupling.

Computer Simulations

The following experiments were performed by simulating the linear baseband model.

Experiment 4.1. With a randomly chosen initial value set $\{\psi_1(0) = 0.0684, \psi_2(0) = 1.34, \psi_2(1) = 2.08\}$ which is inside the convergence region and a stable set of parameters ($p = 4$), Figure 4.7 shows the phase errors vs. time with noise (SNR $= 0\,\text{dB}$) for $M = 6$ and $M = 12$.

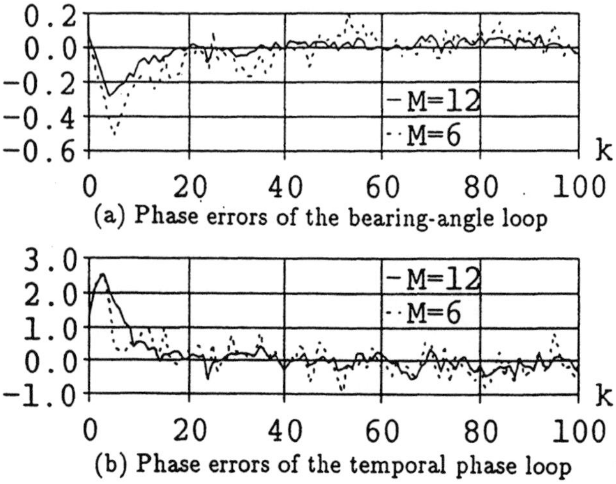

Figure 4.7 Transient phase errors (rad) in the presence of noise.

Experiment 4.2. Let SNR threshold be defined as that input SNR above which the steady-state phase error variance coincides approximately with the variance of the linear analysis. Computer simulations were run to seek this threshold for various numbers of sensors (M). In each simulation the initial phase errors were set to be zero and phase errors over 1,000 sample times were calculated. Figure 4.8 shows the comparison of the simulation results each with the linear approximations for $M = 12$ and $M = 6$. It can be seen from the figures that SNR threshold for $M = 12$ is about -5 dB, and for $M = 6$ is about 0 dB.

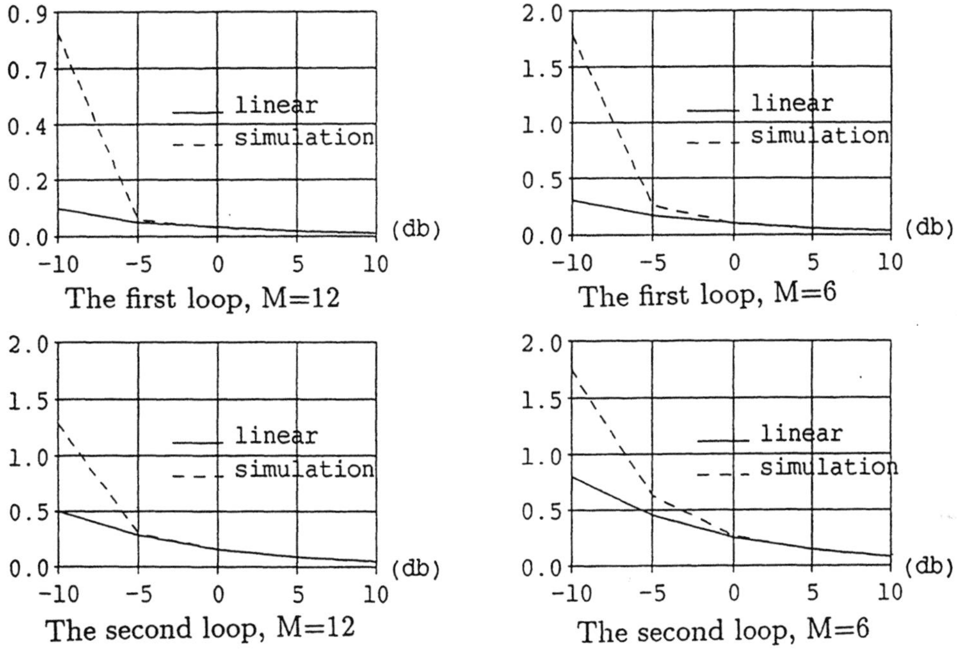

Figure 4.8 The standard deviation of phase errors (rad) of CDPLL versus SNR.

4.4 COUPLED DIGITAL PLLS FOR SEPARATING CO-CHANNEL SIGNALS

When distinct signal sources cannot be distinguished spatially, or when DOA resolvability is marginal, further processing in the time domain will lead to the desired separations under certain conditions. It is clear, for example, that if the signals are spectrally mutually exclusive, then even at the same DOA they may be detected and separately estimated or demodulated. In this section we deal with the most difficult case of simultaneous co-directional *and* co-channel signals. In fact we allow them to have the same carrier, but consider them possibly distinct by virtue of their amplitudes, angle-modulation processes, and carrier phases. Thus, the array pictured in Figure 4.4 may be tracking at bearing θ, but instead of one temporal PLL, we may require two or more, according to the number of signals present at the DOA θ.

Any approach to determining the presence of multiple signals in the same channel should depend on the a priori information about the signals. We consider that they are narrow-band angle-modulated signals. The adjacent and co-channel interference problem has been addressed in the literature for over a decade with various scenarios and solutions posed. One of the earlier works proposed a receiver, derived from maximum a posteriori estimation theory, for suppression of interchannel interference in FM receivers [17]. It was found that this receiver could be efficiently implemented by two, analog cross-coupled phase-locked loops (CPLLs). The analysis was carried out for analog signals and analog phase-locked loops (PLLs).

Bradley and Kirlin studied nonoptimal digital CPLLs for cancelling interfering tones with random narrowband modulation [18]. Polk and Gupta [19] derived a quasi-optimal receiver from Extended Kalman Filter (EKF) theory, which was shown to be realizable by a digital phase locked loop (DPLL). Bradley [20] and Lagunas [21] have separately derived EKF-based estimators which may be realized by cross-coupled digital phase-locked loops (CDPLLs), the digital counterpart to the analog CPLLs of Cassara [17].

One of the commonalities shared by the above is the presence of one or more interferers. Cassara et. al., Bradley, and Lagunas all model the co-channel problem as two independent message processes frequency modulated on spectrally proximate carriers. Each phase-locked loop "locks" onto one of the processes and removes it from the input of the other PLL (see Figure 4.9). However, little effort has been spent in the investigation of the nature of the two (or more) message processes. Cassara et. al. found the CPLL receiver to become unstable if the carrier amplitudes of the sources were equal. Wulich et. al. [22] found that a CPLL receiver could separate two FM sources with equal carrier amplitude if the carrier frequencies were unequal (corresponding to a ramp in phase). In the Lagunas development of the CDPLL estimator, it is assumed that the message processes are orthogonal. Although Bradley did not have this requirement, he arrived at a solution similar to Lagunas. The effect of the signals' distinctness, or orthogonality, on their separability will be investigated. We examine the conditions necessary for separability of such similar signals.

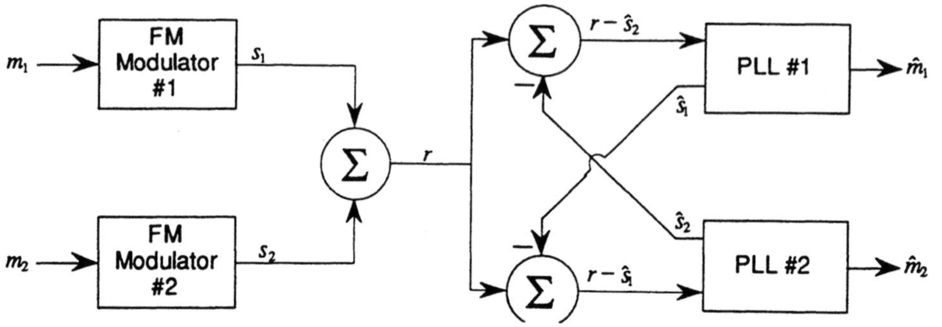

Figure 4.9 Conceptual block diagram of co-channel interference and a coupled PLL-based interference canceller.

Signal Characterization

Treatment of the separability of multiple signals may be decomposed into two related themes: (1) distinctness and (2) observability. This may be illustrated intuitively by looking at examples of a couple of limiting cases. If two scalar signals are identical and the only available observation is, say, their sum, the two signals cannot be distinguished without any other a priori knowledge. In the other extreme, if two signals are orthogonal and each observable, there is no problem in distinguishing the signals (which is a rather uninteresting scenario). However, there is a nearly continuous blending of distinctness and observability between these two extremes in the co-channel interference problem.

As in previous works [19, 20], the problem will be modeled using state-space notation to describe the carrier amplitude and phase processes and their resulting measurement. Because we do not implicate DOA and array parameters in the following, the meanings of some variables such as x, θ, and ϕ have changed from that of previous sections in this chapter, mostly to accommodate conventional Kalman filtering notation.

The Phase Process. Since the problem at hand deals with multiple sources, the state vector will be derived for one source which can then easily be expanded to describe multiple sources. The ith stochastic message process will be modeled by a first-order unit-energy lowpass process

$$\theta_i(k+1) = \phi_i \theta_i(k) + \gamma_i w_i(k) \tag{4.84}$$

where $\phi_i = \exp(-\alpha_i T_s)$ is the state transition constant, $\gamma_i = \sqrt{1-\phi_i^2}(1-\phi_i)$ is the process driving noise power, α_i defines the bandwidth, T_s is the sampling interval, and $w_i(k)$ is the $N : [0, 1]$ process driving noise. An unknown carrier amplitude is also considered and will be modeled by a random constant. Thus, the state vector for one phase-modulated source is

$$\mathbf{x}_i(k) = [\theta_i(k)\mathbf{A}_i(k)]^T \tag{4.85}$$

Multiple sources are modeled by augmenting the state vector

$$\mathbf{x}(k) = \left[\mathbf{x}_1^T(k)\mathbf{x}_2^T(k)\ldots\mathbf{x}_N^T(k)\right]^T \tag{4.86}$$

where N is the number of sources. The multisource state equation is

$$\mathbf{x}(k+1) = \mathbf{\Phi}\mathbf{x}(k) + \mathbf{\Gamma}\mathbf{w}(k) \tag{4.87}$$

where the state transition and process noise gain matrices are defined

$$\mathbf{\Phi} = \begin{bmatrix} \phi_i & 0 \\ 0 & 1 \end{bmatrix} \otimes \underset{N\times N}{\mathbf{I}}, \mathbf{\Gamma} = \begin{bmatrix} \gamma_i & 0 \\ 0 & 0 \end{bmatrix} \otimes \underset{N\times N}{\mathbf{I}} \tag{4.88}$$

where $\otimes$ denotes the Kronecker matrix product. The process driving noise vector is

$$\mathbf{w}(k) = [w_1(k), w_2(k), \ldots, w_{2N}(k)]^T \tag{4.89}$$

where $E[w_i(k)] = 0$ and $E\left[w_i(k)w_j(k)\right] = \delta(i-j)$

State Measurement. The process of phase modulation is inherently nonlinear. For a scalar measurement of N sources, the measurement equation is

$$z(k) = h(\mathbf{x}(k), k) + v(k) \tag{4.90}$$

where, in the case of PM, the observation function is

$$h(\mathbf{x}(k), k) = \sum_{i=1}^{N} A_i(k)\sin(\omega_c k T_s + \theta_i(k)) \tag{4.91}$$

In addition to being nonlinear, $h(\mathbf{x}(k), k)$ is also a function of both state and time. In order to make use of the well-developed state-space theory, the observation function needs to be simplified.

By forming a new state vector $\tilde{\mathbf{x}}(k) = \mathbf{x}(k) - \hat{\mathbf{x}}(k)$, where $\hat{\mathbf{x}}(k)$ is the state estimation, a new observation function may be formed which allows for the separation of the state from the observation function. If we define $\tilde{z}(k)$ to be the measurement estimation error, *for small phase estimation error* it may be shown, in a manner similar to that of Lagunas [21], that

$$\tilde{\mathbf{z}}(k) \approx \mathbf{H}(\hat{\mathbf{x}}(k), k)\tilde{\mathbf{x}}(k) + v(k) \tag{4.92}$$

where

$$\mathbf{H}(\hat{\mathbf{x}}(k), k) = \left[\mathbf{h}_1(\hat{x}_1(k), k)\mathbf{h}_2(\hat{x}_2(k), k)\ldots\mathbf{h}_N(\hat{x}_N(k), k)\right] \tag{4.93}$$

and $h_i(\hat{\mathbf{x}}(k), k)$ is linearized around the current i^{th} state estimate forming

$$\mathbf{h}_i(\hat{x}_i(k), k) = \left[\hat{A}_i(k)\cos(\omega_c kT_s + \hat{\theta}(k))\sin(\omega_c kT_s + \hat{\theta}_i(k))\right] \tag{4.94}$$

This matrix is of the same form, although derived differently, as in the extended Kalman filter where $h(\mathbf{x}(k), k)$ is linearized around the current state prediction

$$\mathbf{h}(\hat{\mathbf{x}}(k|k-1), k) = \left.\frac{\partial h(\mathbf{x}(k), k)}{\partial \mathbf{x}(k)}\right|_{\mathbf{x}(k)=\hat{\mathbf{x}}(k|k-1)} \tag{4.95}$$

Since $\hat{\mathbf{x}}(k)$ is derived from the EKF, it may be shown that the new state $\hat{\mathbf{x}}$ has the same state transition matrix as (4.87)

$$\begin{aligned}
\tilde{\mathbf{x}}(k) &= \mathbf{x}(k) - \hat{\mathbf{x}}(k) \\
&= \mathbf{\Phi}\mathbf{x}(k-1) + \mathbf{\Gamma}\mathbf{w}(k) - \hat{\mathbf{x}}(k|k-1) - \mathbf{K}(k)\left[z(k) - h(\hat{\mathbf{x}}(k|k-1), k)\right] \\
&= \mathbf{\Phi}\left[\mathbf{x}(k-1) - \hat{\mathbf{x}}(k-1)\right] + \mathbf{\Gamma}\mathbf{w}(k) - \mathbf{K}(k)\left[z(k) - h(\hat{\mathbf{x}}(k|k-1), k)\right] \\
&= \mathbf{\Phi}\tilde{\mathbf{x}}(k-1) + \mathbf{\Gamma}\mathbf{w}(k) - \mathbf{K}(k)\left[z(k) - h(\hat{\mathbf{x}}(k|k-1), k)\right]
\end{aligned} \tag{4.96}$$

State Observability

One method of examining the separability of signals is to see if the states are *observable*. If the states are observable, implying that they can be estimated, then the signals they define should be separable.

Observability is a function of the state transition and observaton matrices. The message is modeled by a linear process; however, phase modulation is a nonlinear observation of the state. Thus, the traditional observability conditions [23] may not be directly applied. We first examine the observability conditions by linearizing the observation function. In a previous paper [24] the observability of two states, each described by (4.85) and where $h(\mathbf{x}(k), k)$ was approximated by $\mathbf{H} = [1, 1]$, revealed that the separability of the states was dependent on the relationship of the α's in the state transition matrix. If the α's are equal, the states cannot be separated; however, as their difference increases, so does their observability. By use of (4.95), the effect of phase modulation of observability *of state estimation error* may be studied. By studying the observability of $\tilde{\mathbf{x}}(k)$, some much needed light may be shed on the observability of $\mathbf{x}(k)$.

State observability may be determined by examination of the *observability Gramian*

$$\mathcal{M}(k_0, k_f) = \sum_{k=k_0}^{k_f} \mathbf{\Phi}^T(k, k_0)\mathbf{H}^T(k)\mathbf{H}(k)\mathbf{\Phi}(k, k_0) \tag{4.97}$$

A system is uniformly completely observable if there exists an interval $k_f > k_0$ such that $\mathcal{M}(k_0, k_f)$ is positive definite [23]. Substituting (4.88) and (4.93) into (4.97) forms the observability Gramian for the multisource problem. Algebraic evaluation of (4.97) becomes overly burdensome after a few samples. Therefore, the eigenvalues of $\mathcal{M}(k_0, k_f)$ will be used as a measure of the "positive definiteness" of $\mathcal{M}(k_0, k_f)$. The Gramian is Hermitian and is thus positive semidefinite resulting in all eigenvalues being ≥ 0. A Hermitian matrix is positive definite if and only if its eigenvalues are positive [25]. Therefore, if the eigenvalues (or singular values since the eigenvalues are positive or zero) are positive, then $\mathcal{M}(k_0, k_f)$ is positive definite. The condition of $\mathcal{M}(k_0, k_f)$, where the condition is the ratio of the largest singular value to the smallest singular value, is used to investigate the state observability based on the observability Gramian.

State Observation and Estimation

Using the development presented in the previous section, we concentrate on the state observability and separability of two signals ($N = 2$). The two carrier frequencies are identical but the two phase processes are independent.

Phase Observability. Although both phase and amplitude estimation are desired, phase observability and separability will be discussed first assuming that the carrier amplitude is known. Amplitude estimation will be included subsequently. To study the phase observability independent of the carrier amplitudes, the amplitudes are dropped from the state vector resulting in the phase state transition matrix

$$\mathbf{\Phi}_\theta = \begin{bmatrix} \phi_1 & 0 \\ 0 & \phi_2 \end{bmatrix} \tag{4.98}$$

and the observation matrix

$$\mathbf{H}_\theta(\tilde{\theta}(k), k) = \left[A_1 \cos(\omega_c k T_s + \hat{\theta}_1(k)), A_1 \cos(\omega_c k T_s + \hat{\theta}_2(k))\right] \tag{4.99}$$

where $\tilde{\theta}(k) = \left[\tilde{\theta}_1(k), \tilde{\theta}_2(k)\right]^T$.

The phase observability Gramian is formed by substituting (4.98) and (4.99) into (4.97). Evaluation of $\mathcal{M}(k_0, k_f)$ for two observations reveals that $\mathcal{M}(k_0, k_f)$ is positive definite for all cases except when $\alpha_1 = \alpha_2$ *and* when $\tilde{\theta}_1(k) = \tilde{\theta}_2(k)$. This indicates that if the two signals have identical bandwidth and are in phase, the states are not observable. $\mathcal{M}(k_0, k_f)$ becomes better conditioned with additional measurements, but it remains poorly conditioned. But what if the signals are *nearly* identical? Examination of the eigenvalues of $\mathcal{M}(k_0, k_f)$ reveals that $\mathcal{M}(k_0, k_f)$ is ill conditioned only when the two carriers are nearly in-phase and is well conditioned when the two carriers are in quadrature (see Figure 4.10). Only when the carriers are nearly in-phase does the relationship of the α's affect observability. The inset in Figure 4.10 shows that $\mathcal{M}(k_0, k_f)$ is singular when $\alpha_1 = \alpha_2$ and when the carriers are in-phase. Thus, the state observability is reduced as the carriers become in-phase (or anti-phase).

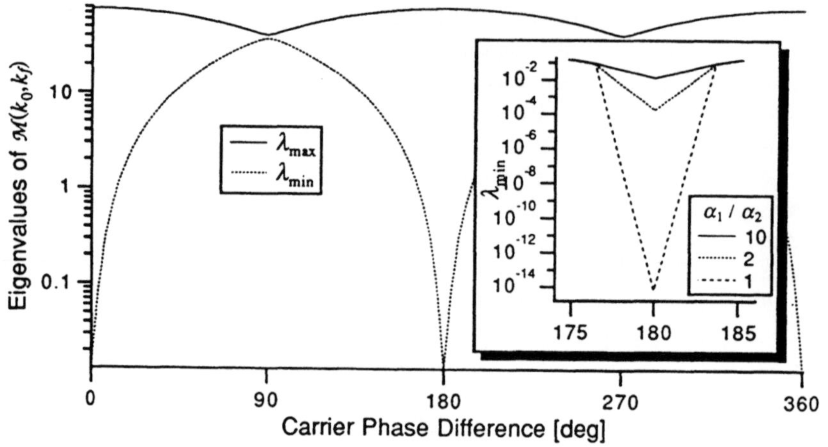

Figure 4.10 Eigenvalues of the observability Gramian as a function of carrier phase difference.

Phase Estimation. To further explore the relationship between observability and separability, the EKF estimator was derived and its ability to separate the two carriers studied. Figure 4.11 shows the mean-square error (MSE) in the phase estimation of two combined PM source as a function of nominal carrier phase difference. Here $\alpha_1 = 10\alpha_2 = f_c/100$, $A_1 = A_2 = 1$, and the MSE is the ensemble average of the estimation error of 100 separate runs. In each run the EKF is started with perfect knowledge of each phase and processes 5 carrier cycles with 8 samples per cycle and generally achieves steady state within one carrier cycle. The MSE of $\hat{\theta}_i$ is normalized by γ_i to account for the different α's. From Figure 4.11 it may be concluded that the error in estimation increases as the state observability decreases.

The effect of state observability on state estimation manifests itself in the Kalman gain matrix, $\mathbf{K}$. The Kalman gain $\mathbf{K}_i$ of phase θ_i is cyclic (see Figure 4.12). The magnitude of $\mathbf{K}_i$ decreases as θ_i becomes less observable, $|\mathbf{K}_i| \rightarrow 0$ as $|\mathbf{H}(\hat{\theta}_i(k|k-1), k)| \rightarrow 0$ from (4.95), resulting in a cyclic $\mathbf{K}_i$. Also the RMS value of $\mathbf{K}_i$ decreases when $|\theta_1(k) - \theta_2(k)|$ is near zero. This is due to the reduced amount of state information available in the measurement and corresponding innovations process causing the EKF to weight the prediction more than the measurement.

In summary, the phase observability and separability is a function of both the carrier phase difference and the relative bandwidths of the phase processes. In the worst case, when the carriers are in phase and the phase processes have the same

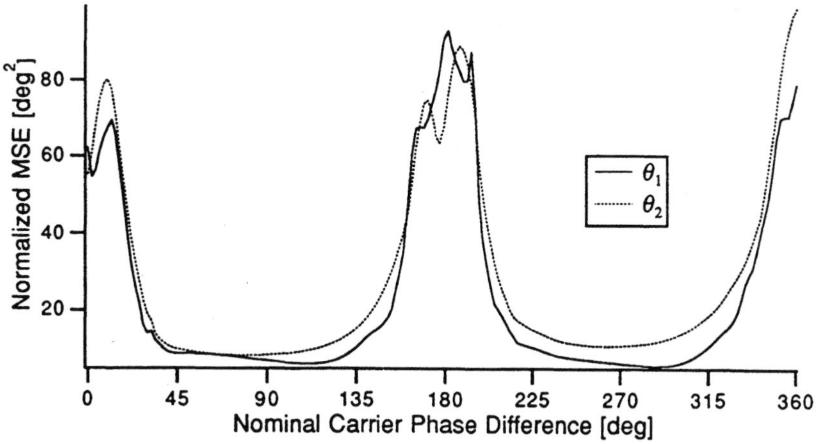

Figure 4.11 MSE of EKF estimates of phase states as a function of carrier phase difference.

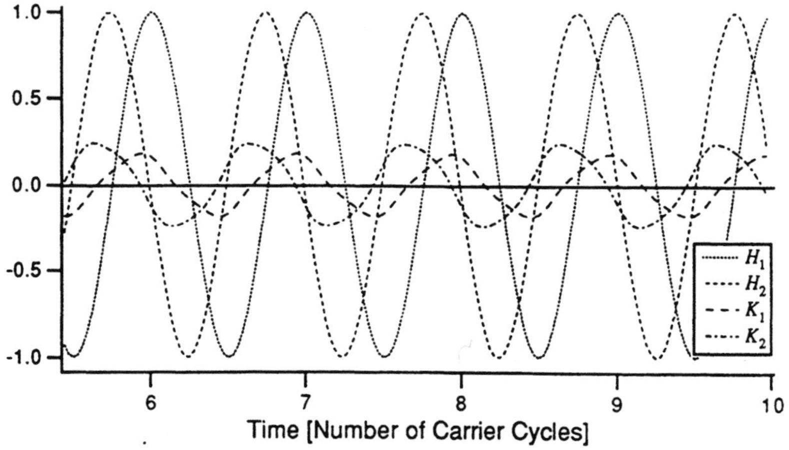

Figure 4.12 Kalman gain and observation matrix for two-source EKF.

bandwidths, the signals are not distinguishable. In the other extreme, when the carriers are in quadrature and the phase processes have the substantially differing bandwidths, the signals are readily separated. Between these two extremes is an almost continuous range of observability and separability. If the phase process bandwidths are equal, then they are separable only if the carrier phases are in near quadrature. Conversely, if the carriers are nearly in phase, then the signals are separable only if the phase process bandwidths are substantially different.

Inclusion of Amplitude Estimation

Inclusion of the carrier amplitudes as states to be estimated adds another dimension to the problem: that of the relative carrier powers. Since we have shown that two in-phase sources cannot be separated, the carriers in the following discussion are in quadrature.

The observability Gramian is formed by substituting (4.88) and (4.93) into (4.97), which is then used to study the dependence of state observability on carrier amplitude ratio (see Figure 4.13). Here it is evident that as one carrier's power dominates the second, the state observability of the lower power signal is diminished.

The effect of differing carrier power levels on signal separability may also be studied by examining how accurately the EKF can estimate the states. In Figure 4.14, the MSE of phase process and carrier amplitude estimates for each carrier are shown as the carrier amplitude ratio is varied. Here the two carriers are in quadrature and it is evident that the EKF is able to track the phase processes and maintain an accurate estimate of the carrier amplitudes over a wide range of relative carrier power levels. It is also evident that as one carrier dominates the other, the EKF is less able to accurately estimate the states of the lesser powered carrier. This agrees

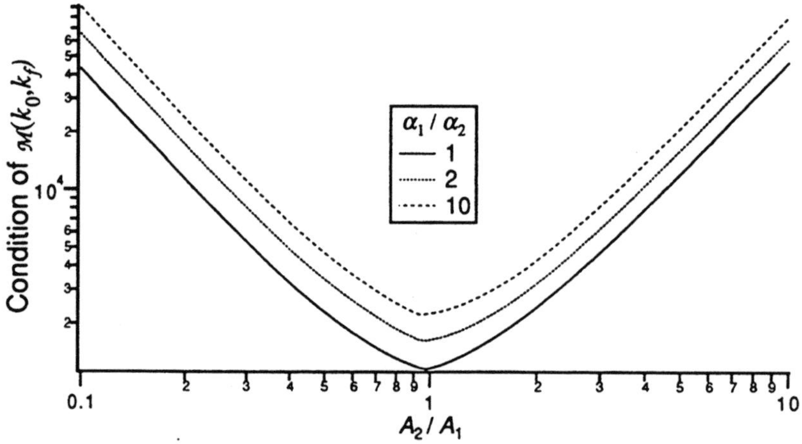

Figure 4.13 Observability as a function of carrier amplitude ratio.

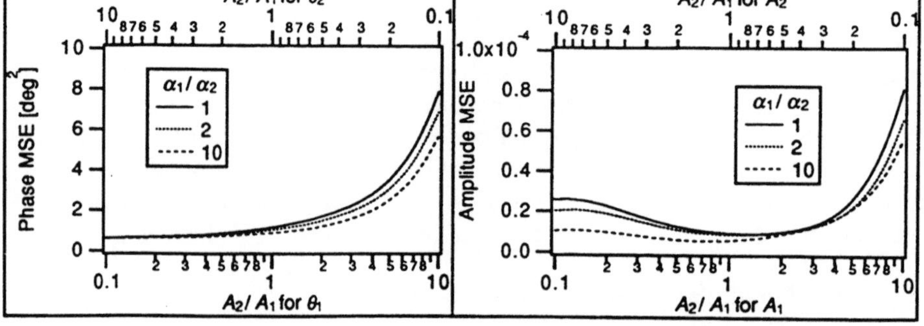

Figure 4.14 Carrier phase and amplitude estimation MSE as a function of carrier amplitudes.

with Figure 4.13. It should also be pointed out that if one carrier is at a substantially lower level than the second, the lower powered signal's contribution to CCI will be substantially reduced.

Summary

The separability of two independent co-channel signals has been studied by examination of (1) state observability and (2) state estimation. It has been shown that if two carriers are in-phase and their state transition matrices are identical, then their modulating processes cannot be observed or estimated. It has also been shown that the carrier amplitudes and phase processes of two PM carriers *in quadrature* can be accurately estimated and separated. Some implications for polarization diversity reception may be drawn from our results. However, the diversity receiver has two input measurements rather than one. Each measurement is like the single one we have considered.

We note that the phase processes (4.84) can be replaced by PN and other deterministic sequences where the state vector is composed of delay states

$$x_i = \begin{bmatrix} 0 \cdots 0 & \underset{d}{\alpha_i} & 0 \cdots 0 \end{bmatrix}^T = \alpha_i 1_d \tag{4.100}$$

where d is the sequence delay. The observation matrix is formed by the time shifted (delayed) sequences. A hypothesis test may then be formed to determine the number of sequences present by examination of the divergence of the various state estimation errors.

The determination of the number of signals present is another problem currently being studied. One method under investigation uses the divergence between the estimation error covariances for two cases: (1) assumes two sources present but only one is actually present, and (2) assumes two sources present when both are actually present. A hypothesis test may be formulated based on this divergence measure.

The acquisition of co-channel signals by CDPLLs also remains to be studied. The nonlinear behavior of the signals and of the DPLLs makes this a very challenging problem. One possible route of investigation may lie in chaos theory of nonlinear dynamics as recently reported by Endo and Chua [26].

APPENDIX 4A ESTIMATION OF THE SOURCE SEQUENCE

In this section the data and the source sequences are considered samples of jointly random processes. With the direction vector matrix **A** assumed known, the source sequence $\mathbf{s}(i)$ can be estimated in terms of the data $\mathbf{x}(i)$.

For white independent Gaussian sources with covariance $\mathbf{P}_s = E\{\mathbf{s}(i)\mathbf{s}(i)^H\}$, diagonal, and with

$$\mathbf{s} = (\mathbf{s}(1)^T \mathbf{s}(2)^T \ldots \mathbf{s}(N)^H)^T \tag{4A.1}$$

the density of **s** is

$$f(\mathbf{s}) = \prod_{i=1}^{N} \frac{1}{\pi \det [\mathbf{P}_s]} e^{-\mathbf{s}^H(i)\mathbf{P}_s^{-1}\mathbf{s}(i)} \tag{4A.2}$$

Because $\mathbf{s}(i)$ is independent of $\mathbf{x}(k), k \neq i$, estimation of $\mathbf{s}(i)$ is a function of $\mathbf{x}(i)$ only. If we let

$$\mathbf{z} = (\mathbf{x}(i)^T \mathbf{s}(i)^T)^T \tag{4A.3}$$

then $\mathbf{z}$ is Gaussian with

$$\mathbf{P}_z = \operatorname{cov}\{\mathbf{z}\} = \begin{pmatrix} \mathbf{P}_x & \mathbf{A}\mathbf{P}_s \\ \mathbf{P}_s\mathbf{A}^H & \mathbf{P}_s \end{pmatrix} \tag{4A.4}$$

$$\mathbf{P}_x = E\{\mathbf{x}(i)\mathbf{x}(i)^H\} = \sigma^2\mathbf{I} + \mathbf{A}\mathbf{P}_s\mathbf{A}^H \tag{4A.5}$$

$$E\{\mathbf{x}(i)\mathbf{s}(i)^H)\} = \mathbf{A}\mathbf{P}_s \tag{4A.6}$$

giving $f(z|A) = N(0, P_z)$. The density of $s(i)$ conditioned on $x(i)$, given A, is also normal [10, p. 300] with

$$f(\mathbf{s}(i)|\mathbf{x}(i), \mathbf{A}) = N(\mathbf{P}_s\mathbf{A}^H\mathbf{P}_x^{-1}\mathbf{x}(i), \mathbf{P}_s - \mathbf{P}_s\mathbf{A}^H\mathbf{P}_x^{-1}\mathbf{A}\mathbf{P}_s) \tag{4A.7}$$

Thus, the estimate of $\mathbf{s}(i)$ is its conditional mean, $\mathbf{P}_s\mathbf{A}^H\mathbf{P}_x^{-1}\mathbf{x}(i)$. This is identical to that derived by use of the orthogonality principle, resulting in Eq. (4.11). This estimate for $\mathbf{s}(i)$ is the least-squares estimate.

APPENDIX 4B

Let the relationship between the direction vector matrix $\mathbf{A}$ and the signal space eigenvectors be given through the $D \times D$ matrix $\mathbf{C}$:

$$\mathbf{V}_D = \mathbf{A}\mathbf{C} \tag{4B.1}$$

Then

$$\begin{aligned} J'_\theta = \hat{\mathbf{a}}^H\mathbf{V}_D\mathbf{V}_D^H\mathbf{B}\hat{\mathbf{a}} &= \hat{\mathbf{a}}^H\mathbf{A}\mathbf{C}\mathbf{C}^H\mathbf{A}^H\mathbf{B}\hat{\mathbf{a}} \\ &= \hat{\mathbf{a}}^H\mathbf{A}\mathbf{G}\mathbf{A}^H\hat{\mathbf{a}}' \end{aligned} \tag{4B.2}$$

where $\hat{\mathbf{a}}' = \mathbf{B}\hat{\mathbf{a}}$, it is easily shown using (4A.1) and $\mathbf{V}_D^H\mathbf{V}_D = \mathbf{I}_D$ that

$$\mathbf{G} = \mathbf{C}\mathbf{C}^H = (\mathbf{A}^H\mathbf{A})^{-1} \tag{4B.3a}$$

$$= \operatorname{adj}(\mathbf{A}^H\mathbf{A})/|\mathbf{A}^H\mathbf{A}| \tag{4B.3b}$$

Thus, the (i, k) element of $\mathbf{G}$ is

$$G_{ik} = (-1)^{i+j} M_{ki}/|\mathbf{A}^H\mathbf{A}|, \tag{4B.3c}$$

where M_{ki} is the determinant of $\mathbf{A}^H\mathbf{A}$ with the ith row and k^{th} column deleted. Rewriting $\hat{\mathbf{a}}^H\mathbf{A}$ in 4B.2 gives

$$J'_\theta = \begin{pmatrix} <\hat{\mathbf{a}}, \mathbf{a}_1> \\ <\hat{\mathbf{a}}, \mathbf{a}_2> \\ \vdots \\ <\hat{\mathbf{a}}, \mathbf{a}_D> \end{pmatrix}^T \mathbf{G} \begin{pmatrix} <\mathbf{a}_1, \hat{\mathbf{a}}'> \\ <\mathbf{a}_2, \hat{\mathbf{a}}'> \\ \vdots \\ <\mathbf{a}_D, \hat{\mathbf{a}}'> \end{pmatrix} \tag{4B.4}$$

Thus, all terms in J'_θ are of the form

$$J'_\theta(i,k) = <\hat{\mathbf{a}}, \mathbf{a}_i> G_{ik} <\mathbf{a}_k, \hat{\mathbf{a}}'> \tag{4B.5}$$

Appendix 4C shows that each inner product in (4B.5) is real and $<\mathbf{a}_k, \hat{\mathbf{a}}_i> = <\mathbf{a}_i, \mathbf{a}_k>$, therefore J'_θ is real.

APPENDIX 4C

Here the closed form inner products for $<\mathbf{a}_i, \mathbf{a}_j>$ and $\mathbf{a}_i, \hat{\mathbf{a}}'$ are derived. From (4.5)

$$<\mathbf{a}_i, \mathbf{a}_l> = \mathbf{a}_i^H \mathbf{a}_l$$

$$= M^{-1} \begin{cases} 2\,\mathrm{Re}\left\{ e^{j(\theta_l-\theta_i)/2} \sum_{m=0}^{\frac{M-2}{2}} e^{mj(\theta_l-\theta_i)} \right\}, M \text{ even} \\ 1 + 2\,\mathrm{Re}\left\{ \sum_{m=1}^{\frac{M-1}{2}} e^{mj(\theta_l-\theta_i)} \right\}, M \text{ odd} \end{cases} \tag{4C.1}$$

$$= M^{-1} \begin{cases} 2\,\mathrm{Re}\left\{ e^{j\theta_{li}/2} \left(\dfrac{1 - e^{j\frac{M}{2}\theta_{li}}}{1 - e^{j\theta_{li}}} \right) \right\}, M \text{ even} \\ 1 + 2\,\mathrm{Re}\left\{ e^{j\theta_{li}} \left(\dfrac{1 - e^{j\frac{M-1}{2}\theta_{li}}}{1 - e^{j\theta_{li}}} \right) \right\}, M \text{ odd} \end{cases} \tag{4C.2}$$

$$= M^{-1} \begin{cases} \dfrac{\sin(M\theta_{li}/2)}{\sin(\theta_{li}/2)}, M \text{ even} \\ \dfrac{2\sin(M\theta_{li}/2)\cos((M-2)\theta_{li}/2)}{\sin(\theta_{li}/2)}, M \text{ odd} \end{cases} \tag{4C.3}$$

where $\theta_{li} = \theta_l - \theta_i$.

Similarly

$$< \mathbf{a}_i, \hat{\mathbf{a}}' > = M^{-1} \begin{cases} \sum_{m=0}^{\frac{M-2}{2}} \frac{2m+1}{2} j \left(e^{\frac{2m+1}{2} j(\hat{\theta}-\theta_i)} - e^{-\frac{2m+1}{2} j(\hat{\theta}-\theta_i)} \right), M \text{ even} \\ \sum_{m=1}^{\frac{M-1}{2}} mj \left(e^{mj(\hat{\theta}-\theta_j)} - e^{-mj(\hat{\theta}-\theta_i)} \right), M \text{ odd} \end{cases} \quad (4C.4)$$

$$= M^{-1} \begin{cases} -\sum_{m=0}^{\frac{M-2}{2}} (2m+1) \sin\left[(2m+1)\delta_i/2\right], M \text{ even} \\ -\sum_{m=1}^{\frac{M-1}{2}} 2m \sin(m\delta_i/2), M \text{ odd} \end{cases} \quad (4C.5)$$

where $\delta_i = \hat{\theta} - \theta_i$.

It can be recognized that the series in (4C.4) are derivatives of series of the forms in (4C.1), thus the derivatives of closed forms (4C.3) give

$$< \mathbf{a}_i, \hat{\mathbf{a}}' > = \frac{(2M)^{-1}}{\sin^2(\delta_i/2)} \begin{cases} M \sin(\delta_i/2) \cos(M\delta_i/2) \\ \quad - \sin(M\delta_1/2) \cos(\delta_i/2); M \text{ even} \\ (M-1) \sin(\delta_i/2) \cos((M-1)\delta_i/2) \\ \quad - \sin((M-1)\delta_i/2) \cos(\delta_i/2), M \text{ odd.} \end{cases} \quad (4C.6)$$

$$= \frac{(2M)^{-1}}{\sin^2(\delta_i/2)} \begin{cases} \frac{M-1}{2} \sin((M+1)\delta_i/2) - \frac{M+1}{2} \\ \sin((M-1)\delta_i/2), M \text{ even}; \\ \frac{M-2}{2} \sin(M\delta_i/2) - \frac{M}{2} \\ \sin((M-2)\delta_i/2), M \text{ odd} \end{cases} \quad (4C.7)$$

APPENDIX 4D OPTIMALITY OF NOISE EIGENVECTOR WEIGHTS

Equation (4.17) shows that $\mathbf{B}\hat{\mathbf{a}} = \partial\hat{\mathbf{a}}/\partial\hat{\theta}$ is the vector to be used in correlation with the vector a. It can be shown that no other function of $\hat{\theta}$ will do better (if we know $a(\theta_k)$) than -$\mathbf{Ba}(\hat{\theta}_k)$, if the sharpness of the null in $|J'_{\theta_k}|$ is the criteria (minimizing error variance[1]). Suppose some other vector function of $\hat{\theta}$, say $z(\hat{\theta})$, is used in (4.19). Then we require

$$-J'_{\theta_k} = \text{Re} \left\{ < \mathbf{z}(\hat{\theta}), \mathbf{a}(\theta_k) > \right\} |_{\hat{\theta}=\theta_k} = 0 \quad (4D.1)$$

and we want $\mathbf{z}(\theta)$ to maximize the magnitude of

$$-\frac{\partial J'_{\theta_k}}{\partial \delta_k} |_{\delta_k=0} = \text{Re} \left\{ < \frac{\partial}{\partial \delta_k} z(\hat{\theta}_k), a(\theta_k) > \right\} |_{\delta_k=0} \quad (4D.2)$$

[1] Maximizing $|\partial J'_{\theta_k}/\partial\hat{\theta}_k|_{\delta_k=0}$ minimizes var$(\hat{\theta}_k)$ according to the Cramér-Rao bound.

where $\hat{\theta}_k = \theta_k + \delta_k$, and $< a, b >= a^H b$. Now note that

$$0 = \frac{\partial}{\partial \theta_k} [< \mathbf{z}(\theta_k + \delta_k), \mathbf{a}(\theta_k) >]_{\delta_k=0}$$

$$= < \mathbf{z}(\theta_k), \mathbf{a}'(\theta_k) > + < \mathbf{z}'(\theta_k), \mathbf{a}(\theta_k) > \quad (4D.3)$$

where $(\cdot)'$ denotes $\partial/\partial\theta_k$. Then using $\partial z(\hat{\theta}_k)/\partial\delta_k = \partial z(\hat{\theta}_k)/\partial\theta_k = z'(\hat{\theta}_k)$ in (4D.3)

$$\text{Re}\left\{< \partial \mathbf{z}(\hat{\theta}_k)/\partial\delta_k, \mathbf{a}(\theta)_k) >\right\} = -\,\text{Re}\left\{< \mathbf{z}(\theta_k), \mathbf{a}'(\theta_k) >\right\} \quad (4D.4)$$

and (4D.2) becomes

$$\frac{\partial J'_{\theta_k}}{\partial \delta_k}\Big|_{\delta_k=0} = \text{Re}\left\{< \mathbf{z}(\theta_k), \mathbf{a}'(\theta_k) >\right\} \quad (4D.5)$$

The magnitude of the RH side of (4D.5) is maximized if the inner product is maximized, and that is accomplished through the Schwarz inequality. Thus (dropping arguments)

$$\left|< \mathbf{z}, \mathbf{a}' >\right|^2 \leq < \mathbf{z}, \mathbf{z} >< \mathbf{a}', \mathbf{a}' > \quad (4D.6)$$

and with equality only if $\mathbf{z}$ is proportional to $\mathbf{a}' = \mathbf{B}\mathbf{a}$. Therefore, the optimum function of $\hat{\theta}$ to use in correlation with $\mathbf{a}(\theta)$ is $\mathbf{B}\mathbf{a}(\hat{\theta}) = \mathbf{B}\hat{\mathbf{a}}$, to within a scale factor.

REFERENCES

1. R. P. SCHMIDT, "Multiple Emitter Location and Signal Parameter Estimation," Proc. RADC Spectral Estimation Workshop, Griffiss AFB, N.Y., 1979.
2. M. WAX, T. SHAN, and T. KAILATH, "Spatiotemporal Spectral Analysis by Eigenstructure Methods," *IEEE Trans. Acoustics, Speech and Signal Processing*, vol. ASSP-32, no. 4, pp. 817–27, Aug. 1984.
3. M. KAVEH, and A. J. BARABELL, "The Statistical Performance of the MUSIC and Minimum-Norm Algorithms in Resolving Plane Waves in Noise," *IEEE Trans. Acoustic, Speech, and Signal Processing*, vol. ASSP-34, no. 2, pp. 331–41, April 1986.
4. R. L. KIRLIN, "Spectral and Sensor Array Analysis Using Maximum Entropy with New Eigenstructure Constraints," "A new maximum entropy spectrum analysis using uncertain eigenstructure constraints," *IEEE Trans. AES*, vol. 28, no. 1, pp. 2–14, Jan. 1992.
5. S. L. MARPLE, Jr. *Digital Spectral Analysis with Application*, Englewood Cliffs, N.J.: Prentice Hall, 1987.
6. N. L. OWSLEY, "Sonar array processing," Chapter 3 of *Array Signal Processing*, S. Haykin, ed., Englewood Cliffs, N.J.: Prentice Hall, 1985.
7. I. ZISKIND, and M. WAX, "Maximum Likelihood Localization of Multiple Sources by Alternating Projection," *IEEE Trans. Acoustics, Speech and Signal Processing*, vol. ASSP-36, no. 10, pp. 1553–60, Oct. 1988.
8. J. J. STIFFLER, *Theory of Synchronous Communication*, Englewood Cliffs, N.J.: Prentice Hall, 1971.

9. R. L. Kirlin, "A Sequential Maximum-likelihood Multiple Source Tracker and the Relation of Phase-locked Loop Theory to Eigenstructure Methods," *Proc. ISCAS*, Portland, pp. 1762–1967, 1989.

10. L. L. Scharf, *Statistical Signal Processing*, Reading, Mass.: Addison-Wesley, 1981.

11. F. Li, H. Lin, and R. J. Vacarro, "Performance Analysis for DOA Estimation Algorithms: Further Unification, Simplification and Observations," *IEEE Trans. Aerospace and Electronic Systems*, vol. AES-29, no. 4, pp. ..., Oct. 1993.

12. C. L. Golliday and R. J. Huff, "Phase-locked Loop Coherent Combiners for Phased Array Sensor Systems," *IEEE Trans. on COM.*, vol. COM-30, pp. 2329–40, Oct. 1982.

13. D. Divsalar and J. H. Yuen, "Carrier Arraying with Coupled Phase-locked Loops for Tracking Improvement," *IEEE Trans. on COM.*, vol. COM-30, pp. 2319–28, Oct. 1982.

14. W. C. Lindsey and C. M. Chie, "A Survey of Digital Phase-locked Loops," *Proc. IEEE*, vol. 69, no. 4, pp. 410–31, April 1981.

15. H. C. Osborne, "Stability Analysis of an Nth Power Digital Phase-locked Loop—part II: Second- and Third-order DPLL's," *IEEE Trans. on COM.*, vol. COM-28, pp. 1355–64, Aug. 1980.

16. W. B. Davenport and W. L. Root, *Random Signals and Noise*, New York: McGraw-Hill, 1958.

17. Y. Bar-ness, F. A. Cassara, H. Schachter, and R. DiFazio, "Cross-coupled Phase-locked Loop with Closed Loop Amplitude Control," *IEEE Trans. on COM., vol. COM-32*, no. 2, pp. 195–99, February 1984.

18. J. Bradley and R. L. Kirlin, "Phase-locked Loop Cancellation of Interfering Tones," to appear in *IEEE Transactions on Signal Processing*.

19. D. R. Polk and S. C. Gupta, "Quasi-optimum Digital Phase-locked Loops," *IEEE Trans. on COM., vol. COM-21*, no. 1, pp. 75–82, January 1973.

20. J. Bradley, "Suppression of Adjacent-channel and Cochannel FM Interference via Extended Kalman Filtering," *Proceedings of ICASSP-92*, vol. IV, pp. 693–96, 1992.

21. M. Lagunas and A. Pages, "Multitone Tracking with Coupled EKFs and High Order Learning," *Proceedings of ICASSP-92*, vol. V, pp. 153–56, 1992. Also personal correspondence.

22. D. Wulich, E. I. Plotkin, M. N. S. Swamy, and E. Kashi, "Separation of Close Sinusoids by Cross-coupled Phase Locked Loop, in ICASSP-89: 1989 International Conference on Acoustics, Speech and Signal Processing, vol. 4, pp. 2128–31, 1989.

23. A. P. Sage and J. L. Melsa, *Estimation Theory with Applications to Communications and Control*, New York: McGraw-Hill, 1971.

24. B. A. Hedstrom and R. L. Kirlin, "Kalman-derived Coupled Digital PLLs for Separation of Co-channel Signal, in COMCON-91 (Communication and Control Conference), pp. 250–60, 1991.

25. G. W. Stewart, *Introduction to Matrix Computation, Computer Science and Applied Mathematics*, New York; Academic Press, 1973.

26. T. Endo and L. O. Chua, "Chaos from Two-coupled Phase-locked Loops, *IEEE Trans. on Circuits and Systems*, vol. 37, no. 9, pp. 1183–87, September 1990.

5 Closed-Form 2D Angle Estimation with Circular Arrays/Apertures via Phase Mode Excitation and ESPRIT

Cherian P. Mathews and Michael D. Zoltowski

5.1 INTRODUCTION

Estimating the directions-of-arrival (DOAs) of propagating plane waves is a problem of interest in a variety of applications including radar, mobile communications, sonar, and seismology. The widely studied uniform linear array (ULA) can provide estimates of source bearings relative to the array axis. However, a planar array is needed if estimates of source azimuth and elevation are required. The following properties of uniform circular arrays (UCAs) make them attractive in the context of DOA estimation. UCAs provide 360° azimuthal coverage, and also provide information on source elevation angles. In addition, directional patterns synthesized with UCAs can be electronically rotated in the plane of the array without significant change of beam shape. ULAs in contrast provide only 180° coverage, and beams formed with ULAs broaden as the array is steered away from boresight. Phase mode excitation of UCAs, which is essentially Fourier analysis of the array excitation function, was studied by researchers in the early 1960s [1, 2]. This theory led to a powerful pattern synthesis technique for UCAs [3]. Davies [4] also showed how the simple phasing techniques normally associated with ULAs (Butler beamforming matrices) could be used to provide the necessary phasing for pattern rotation with UCAs. These attractive features led to the development of experimental systems that employed phase mode excitation for pattern synthesis with UCAs [4, 5]. These systems, however, employed the beamforming principle to obtain DOA estimates; it is well known that beamforming cannot resolve closely spaced sources (spacing less than the main-lobe width of the array pattern). Two recently developed algorithms [6] for DOA estimation with UCAs are described in this chapter. Both algorithms employ phase mode excitation-based beamforming in conjunction with subspace techniques to obtain high-resolution DOA estimates. The first algorithm, UCA-RB (Real-Beamspace) MUSIC, is a beamspace version of the popular MUSIC algorithm [7], wherein the beamspace manifold is *real-valued*, thereby facilitating real-valued matrix-based computations. The second algorithm is coined UCA-ESPRIT because the steps involved in the algorithm are similar to those of TLS-ESPRIT [8]. We note that the applicability of the ESPRIT principle in conjunction with rotationally invariant arrays (such as UCAs) was studied in [9]. It was shown that such techniques cannot provide unique DOA estimates when more than one source is present. Although this is true in element space, the phase mode excitation-based transformation to beamspace induces a beamspace manifold whose structure can be exploited to develop the ESPRIT-like algorithm, UCA-ESPRIT.

The UCA-RB-MUSIC algorithm offers numerous advantages over element space MUSIC. These advantages include the ability to compute subspace estimates via real-valued eigenvalue decompositions (EVDs), improved estimator performance in correlated source scenarios due to an inherent Forward/Backward (FB) average, and the ability to employ Root-MUSIC to obtain azimuth estimates of sources at a given elevation. The algorithm requires a 2D spectral search to obtain DOA estimates: the computational complexity, however, is lower than for element space MUSIC, as samples of the 2D beamspace MUSIC spectrum corresponding to a given elevation can be obtained via an FFT. Further, element space MUSIC requires a complex-valued EVD for computing subspace estimates, and ULA techniques such as Root-MUSIC cannot be employed. Averaging similar to FB averaging can be performed in element space with UCAs, but only when the number, N, of array elements is even. It was shown in [10] that beamspace MUSIC estimators can never outperform the corresponding element space MUSIC estimators. While this is generally true, UCA-RB-MUSIC *can outperform* element space MUSIC in correlated source scenarios when N is odd. This is due to the decorrelating effect of the FB average inherent in UCA-RB-MUSIC, but not available in element space for odd N. Previous work on the application of ULA techniques with UCAs include that of Tewfik and Hong [11], and Friedlander and Weiss [12]. Comparisons between their approaches and UCA-RB-MUSIC are made in section 5.3.

UCA ESPRIT is a novel algorithm that represents a significant advance in the area of 2D angle estimation. It is a *closed-form* algorithm that provides *automatically paired* azimuth and elevation estimates for each source. The term *closed-form* indicates that the algorithm dispenses with the search/optimization procedures that are characteristic of one class of 2D angle estimation algorithms. The MUSIC algorithm, for example, requires a search for peaks in a 2D spectrum to obtain DOA estimates. Maximum likelihood approaches [13] require computationally expensive multidimensional search procedures to obtain the optimal estimates. The other class of 2D angle estimation algorithms, ESPRIT-based algorithms [8, 14], require arrays that contain subarrays possessing displacement invariances in two dimensions. These algorithms do not require search procedures, and provide closed-form estimates of source direction cosines with respect to each displacement axis. However, they require a pairing procedure (usually ad hoc) to properly associate the independently obtained direction cosine estimates. The ESPRIT-based algorithm for arrays with regular geometries described in [15] provides automatically paired angle estimates; a multidimensional search is however required for optimality. Unlike the existing 2D angle estimation algorithms, UCA-ESPRIT provides automatically paired source azimuth and elevation estimates via the eigenvalues of a matrix (that is derived from the least squares solution to an overdetermined system of equations). The eigenvalues have the form $\mu_i = \sin\theta_i e^{j\phi_i}$, where θ_i and ϕ_i are respectively the elevation and azimuth angles of the ith source. Note also that $\mu_i = u_i + jv_i$, where u_i and v_i are respectively the direction cosines with respect to the x and y axes. UCA-ESPRIT is clearly superior to the existing 2D angle estimation algorithms in terms of computational complexity. Another factor that reduces the computational burden of UCA-ESPRIT is that the algorithm can be implemented with just real-valued EVDs. However, simulations reveal that the UCA-RB-MUSIC estimators have lower variance than the UCA-ESPRIT estimators. For improved estimator performance, the azimuth and elevation estimates from UCA-ESPRIT can be used as starting points for localized Newton searches of the two-dimensional UCA-RB-MUSIC spectrum. Phase mode excitation of filled circular arrays is briefly discussed. A method for employing the UCA-ESPRIT invariance principle to develop a closed-form algorithm for 2D angle estimation with filled circular arrays is described.

The effects of mutual coupling on the structure of the UCA element space manifold are discussed. It is shown that the general structure of the original beamspace manifold (in the absence of mutual coupling) is retained even when mutual coupling effects apply. The only difference is the introduction of gain and phase factors in the original beamspace manifold. Consequently, minor modifications enable UCA-RB-MUSIC and UCA-ESPRIT to cope with mutual coupling effects. Accounting for mutual coupling effects is much simpler with UCAs than with other array configurations. The effects of using directional antenna elements in the UCA are investigated. Both UCA-RB-MUSIC and UCA-ESPRIT are applicable if the individual element patterns are omnidirectional in azimuth. UCA-ESPRIT cannot be employed if this condition on element patterns is not met. However, a beamspace algorithm possessing many of the features of UCA-RB-MUSIC is still applicable.

Asymptotic expressions for the variances/covariances of the MUSIC estimator for the 1D angle estimation problem are available in [16]. The MUSIC spectrum in the 2D angle estimation problem is a function of two variables (azimuth and elevation); asymptotic results on the performance of MUSIC in the 2D scenario are derived herein; the derivation is along the lines of the work in [16]. Asymptotic expressions for the variances/covariances of the UCA-RB-MUSIC estimators are also obtained. Results on the asymptotic performance of the UCA-ESPRIT algorithm are presented [17]. The analysis employs techniques used in [18] to study the performance of the ESPRIT algorithm for 1D angle estimation. All the performance analysis results are verified by computer simulations. The asymptotic behavior of the direction cosine estimators corresponding to element space MUSIC, UCA-RB-MUSIC, and the Cramér-Rao bound (CRB) is investigated in some detail for the one- and two-source scenarios. Closed-form expressions for the element space MUSIC direction cosine estimator variances are obtained for these scenarios. The performance study reveals that both the element space MUSIC estimator variances and the CRB are constants (independent of the DOA) in the single-source case. The UCA thus favors sources from all directions equally. In the two-source case, the DOA dependence of the element space MUSIC estimator variances and the CRB is only through the distance between the source locations, and the orientation of the line joining the source locations in direction cosine space. In other words, the direction cosine estimator variances depend on the source DOAs only through the position of one source relative to the other. The above properties are due to the circular symmetry of the UCA; arbitrary array geometries do not in general possess such properties. Although closed-form expressions for the UCA-RB-MUSIC estimator variances cannot be obtained, their behavior is shown to closely follow that of element space MUSIC and the CRB.

The notational conventions employed in the development are as follows: boldface lowercase letters are used to denote column vectors, and boldface uppercase letters denote matrices. An asterisk is used to denote the complex conjugate operation, for example, $\mathbf{A}^*$. The transpose and Hermitian transpose operations are respectively denoted by superscripts T and H, for example, $\mathbf{G}^T, \mathbf{G}^H$. Hats are used to denote estimated values of quantities, for example, $\hat{\mathbf{S}}$. The same symbol is used to denote similar quantities in element space and beamspace; the element space quantities are distinguished by underbars, for example, $\mathbf{S}$ and $\underline{\mathbf{S}}$ respectively span the beamspace and element space signal subspaces.

The outline of this chapter is as follows: section 5.2 presents the relevant background on phase mode excitation of circular arrays. The UCA-RB-MUSIC and UCA-ESPRIT algorithms are next described in section 5.3. Section 5.4 addresses the performance analysis of the algorithms. Results of computer simulations of UCA-RB-MUSIC and UCA-ESPRIT are presented in section 5.5. Section 5.6 concludes the chapter by summarizing the important results.

5.2 PHASE MODE EXCITATION FOR UNIFORM CIRCULAR ARRAYS

The UCA geometry is depicted in Figure 5.1. The antenna elements, assumed to be identical and omnidirectional, are uniformly distributed over the circumference of a circle of radius r in the xy plane. A spherical coordinate system is used to represent the arrival directions of the incoming plane waves. The origin of the coordinate system is located at the center of the array. Source elevation angles $\theta \in [0, \pi/2]$ are measured down from the z axis, and azimuth angles $\phi \in [0, 2\pi]$ are measured counterclockwise from the x axis.

Element n of the array is displaced by an angle $\gamma_n = 2\pi n/N$ from the x axis. The position vector at this location is $\vec{p}_n = (r\cos\gamma_n, r\sin\gamma_n, 0)$. Consider a

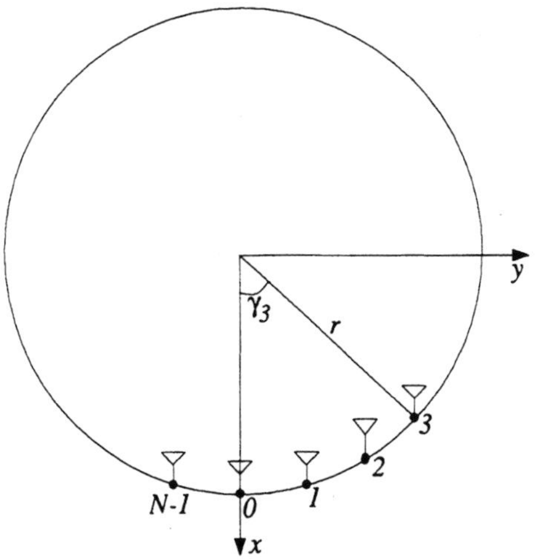

Figure 5.1 Uniform circular array geometry.

narrowband plane wave with wavenumber $k_0 = 2\pi/\lambda$ propagating in the direction $-\hat{r}$ with elevation and azimuth θ and ϕ, respectively. The unit vector $\hat{r}$ has cartesian coordinates $\hat{r} = (u, v, \cos\theta)$: $u = \sin\theta\cos\phi$, and $v = \sin\theta\sin\phi$ are respectively the direction cosines with respect to the x and y axes. The vector $\boldsymbol{\theta} = (\zeta, \phi)$ where $\zeta = k_0 r \sin\theta$, or the vector $\boldsymbol{\beta} = (u, v)$ will be used to represent the DOA of a signal. At any given time instant, the signal complex envelopes received at the origin and at element n differ in phase by $\psi_n = e^{jk_0\hat{r}\cdot\bar{p}_n} = e^{jk_0 r(u\cos\gamma_n + v\sin\gamma_n)} = e^{j\zeta\cos(\phi-\gamma_n)}$. The element space UCA manifold vector thus has the following representations:

$$\underline{\mathbf{a}}(\boldsymbol{\theta}) = \begin{bmatrix} e^{j\zeta\cos(\phi-\gamma_0)} \\ e^{j\zeta\cos(\phi-\gamma_1)} \\ \vdots \\ e^{j\zeta\cos(\phi-\gamma_{N-1})} \end{bmatrix}, \quad \underline{\mathbf{a}}(\boldsymbol{\beta}) = \begin{bmatrix} e^{jk_0 r(u\cos\gamma_0 + v\sin\gamma_0)} \\ e^{jk_0 r(u\cos\gamma_1 + v\sin\gamma_1)} \\ \vdots \\ e^{jk_0 r(u\cos\gamma_{N-1} + v\sin\gamma_{N-1})} \end{bmatrix} \tag{5.1}$$

The representation in terms of $\boldsymbol{\theta}$ will be used in the development of the algorithms. The representation in terms of $\boldsymbol{\beta}$ is more convenient for performance analysis of the algorithms.

Phase Mode Excitation: Continuous Circular Aperture. Consider first the case of a continuous circular aperture. Any excitation function is periodic with period 2π and can therefore be represented in terms of a Fourier series. The arbitrary excitation function $w(\gamma)$ thus has the representation $w(\gamma) = \sum_{m=-8}^{\infty} c_{me^{jm\gamma}}$, where the mth phase mode $w_m(\gamma) = e^{jm\gamma}$ is just a spatial harmonic of the array excitation, and c_m is the corresponding Fourier series coefficient. The normalized far-field pattern resulting from exciting the aperture with the mth phase mode is $f_m^c(\boldsymbol{\theta}) = \frac{1}{2\pi}\int_0^{2\pi} w_m(\gamma)e^{j\zeta\cos(\phi-\gamma)}d\gamma$, where the superscript c signifies a continuous aperture. Substituting for $w_m(\gamma)$, the far-field pattern can be expressed as [3]

$$f_m^c(\boldsymbol{\theta}) = j^m J_m(\zeta)e^{jm\phi} \tag{5.2}$$

where $J_m(\zeta)$ is the Bessel function of the first kind of order m. The far-field pattern has the same azimuthal variation $e^{jm\phi}$ as the excitation function itself. This property allows attractive directional patterns to be synthesized using phase mode excitation [3]. The amplitude (and elevation dependence) of the far-field pattern is through the Bessel function $J_m(\zeta)$. A consequence of this is that only a limited number of modes can be excited by a given circular aperture. Let M denote the highest order mode that can be excited by the aperture at a reasonable strength. A rule of thumb for determining M is [3]

$$M \approx k_0 r. \tag{5.3}$$

This is justified as follows: The visible region $\theta \in [0, \pi/2]$ translates into $\zeta = k_0 r \sin\theta \in [0, k_0 r]$. M is chosen as above because the mode amplitude $J_m(\zeta)$ is small when the Bessel function order m exceeds its argument ζ. For mode orders $|m| \geq M$, $f_m^c(\theta)$ is small over the entire visible region: The beamformer for such a mode m thus severely attenuates sources from all directions.

To illustrate this property, consider a circular aperture of radius $r = \lambda$. Equation 5.3 suggests that the maximum mode order is $M = 6$ (the closest integer to 2π). The Bessel functions of order 0 through 7 are plotted in Figure 5.2. The figure reveals that $J_7(\zeta)$ is indeed small over the entire visible region $\zeta \in [0, 2\pi]$. Thus, only phase modes of order $m \in [-6, 6]$ can be excited at a reasonable strength by the aperture.

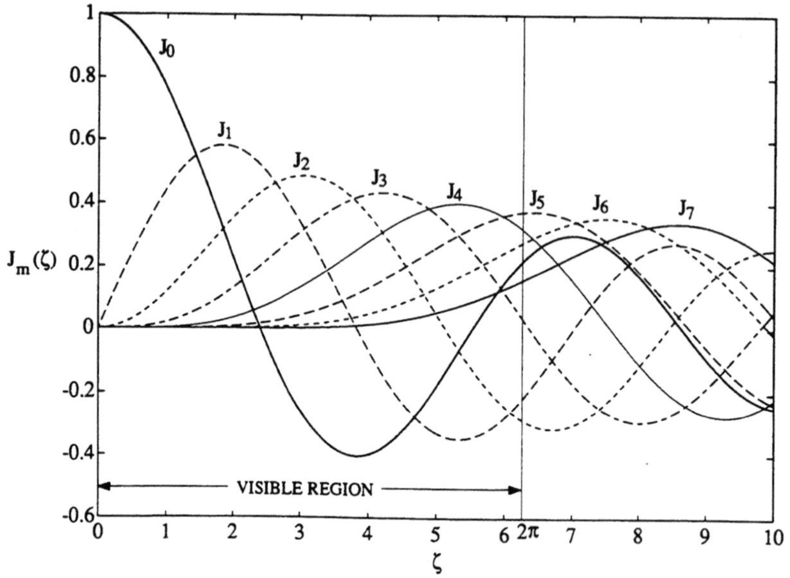

Figure 5.2 Bessel functions.

Phase Mode Excitation: Uniform Cirular Array. We now consider phase mode excitation of an N element UCA. The normalized beamforming weight vector that excites the array with phase mode m, $|m| \leq M$, is

$$\mathbf{w}_m^H = \frac{1}{N}\left[e^{jm\gamma_0}, e^{jm\gamma_1}, \cdots, e^{jm\gamma_{N-1}}\right] = \frac{1}{N}\left[1, e^{j2\pi m/N}, \cdots, e^{j2\pi m(N-1)/N}\right] \tag{5.4}$$

The resulting array pattern $f_m^s(\theta)$, where the superscript s denotes the sampled aperture, is

$$f_m^s(\theta) = \mathbf{w}_m^H \underline{\mathbf{a}}(\theta) = \frac{1}{N} \sum_{n=0}^{N-1} e^{jm\gamma_n} e^{j\zeta \cos(\phi - \gamma_n)} \tag{5.5}$$

For mode orders $|m| < N$, the array pattern can be expressed as follows [3]:

$$f_m^s(\theta) = j^m J_m(\zeta) e^{jm\phi} + \sum_{q=1}^{\infty} \left(j^g J_g(\zeta) e^{-jg\phi} + j^h J_h(\zeta) e^{jh\phi} \right) \tag{5.6}$$

where $g = Nq - m$, and $h = Nq + m$. The first term in this equation, the principal term, is identical to the far-field pattern of Eq. (5.2) corresponding to the continuous circular aperture. The remaining terms arise due to sampling of the continuous aperture, and are called residual or aliasing terms. Examination of Eq. (5.6) reveals that the condition $N > 2|m|$ must be satisfied for the principal term to be the dominant one. The highest mode excited has order M, and we therefore need

$$N > 2M \tag{5.7}$$

array elements. This condition is identical to the Nyquist sampling criterion, as M defines the maximum spatial frequency component in the array excitation. With $M = k_0 r$ as in Eq. (5.3), it is clear that (5.7) requires the circumferential spacing between adjacent array elements to be less than 0.5λ. Note that an interelement spacing of 0.5λ is sufficient to avoid grating lobes with ULAs.

The following discussion shows how the contribution of the residual terms to the pattern of Eq. (5.6) can be made as small as desired by choosing N to be sufficiently large. The residual term that contributes the most arises from the $q = 1$ index in Eq. (5.6), and its amplitude follows the Bessel function of order $N - |m|$. The residual contribution is clearly maximum for mode M, and the amplitude of this residual term follows $J_{N-M}(\zeta)$. Now, $J_{N-M}(\zeta)$ is monotone increasing over the visible region $[0, k_0 r]$ by virtue of the choice of M and N in Eqs. (5.3) and (5.7). $J_{N-M}(k_0 r)$ is therefore an upper bound on the maximum contribution of any residual term. This upper bound can be made as small as desired by making the number, N, of array elements sufficiently large. We return to the example of the UCA of radius $r = \lambda$ with $M = 6$ to illustrate the selection of N. Equation (5.7) requires that the array have $N > 12$ elements. The upper bound on the maximum residual contribution, $J_{N-M}(k_0 r)$, is tabulated in Table 5.1 for various values of N. The table indicates that the residual contribution is "small enough" to be ignored for $N > 15$ elements (circumferential spacing $< 0.42\lambda$). Thus, having $N > 15$ elements ensures that the UCA far-field patterns corresponding to phase mode excitation with mode orders $|m| \leq M = 6$ are virtually identical to the far-field patterns of Eq. (5.2) corresponding to a continuous aperture.

The following section introduces the phase mode excitation-based beamformers employed by the UCA-RB-MUSIC and UCA-ESPRIT algorithms. The development assumes that M and N have been selected according to the design guidelines just established. The principal term thus characterizes the UCA far-field pattern for mode orders $|m| \leq M$: We have

TABLE 5.1 Maximum residual contribution as a function of N for a UCA with $r = \lambda$ and $M = 6$

N	13	14	15	16	17	18	19
$J_{N-M}(k_0 r)$	0.158	0.073	0.029	0.010	0.003	8.8e-4	2.3e-4

$$f_m^s(\theta) \approx j^m J_m(\zeta)e^{jm\phi} = j^{|m|} J_{|m|}(\zeta)e^{jm\phi},\ |m| \leq M \tag{5.8}$$

The final equality above follows from the property $J_{-m}(\zeta) = (-1)^m J_m(\zeta)$ of Bessel functions.

Phase Mode Excitation-Based UCA Beamformers

Three phase mode excitation-based beamformers that synthesize beamspace manifolds of dimension $M' = 2M + 1$ are developed in this section. The beamforming matrices are denoted $\mathbf{F}_\mathbf{e}^H, \mathbf{F}_\mathbf{r}^H$, and $\mathbf{F}_\mathbf{u}^H$; the corresponding beamspace manifolds are $\mathbf{a_e}(\theta), \mathbf{a_r}(\theta)$, and $\mathbf{a_u}(\theta)$, respectively. The subscipts $\mathbf{e}, \mathbf{r}$, and $\mathbf{u}$ respectively stand for "even", "real-valued" and "UCA-ESPRIT" for reasons that will soon be apparent. All three beamformers are orthogonal (have orthonormal rows), and satisfy $\mathbf{F}^H\mathbf{F}=\mathbf{I}$. The orthogonal matrix

$$\mathbf{V} = \sqrt{N}\left[\mathbf{W}_{-M} \vdots \cdots \vdots \mathbf{W}_0 \vdots \cdots \vdots \mathbf{W}_M\right] \tag{5.9}$$

is the building block for the three beamformers. It is evident from Eq. (5.4) that the rows of the $M' \times N$ matrix $\mathbf{V}^H$ are phase mode excitation beamforming weight vectors. These rows are in fact an M' dimensional subset of the N IDFT (Inverse Discrete Fourier Transform) weight vectors.

The beamforming matrix $\mathbf{F}_\mathbf{e}^H$ is defined as follows:

$$\begin{aligned} \mathbf{F}_\mathbf{e}^H &= \mathbf{C_e}\mathbf{V}^H, \text{ where} \\ \mathbf{C_e} &= \text{ diag } \{j^{-M}, \ldots, j^{-1}, j^0, j^{-1}, \ldots, j^{-M}\} \end{aligned} \tag{5.10}$$

The diagonal, unitary matrix $\mathbf{C_e}$ serves to eliminate factors of the form $j^{|m|}$ in Eq. (5.8). The resulting beamspace manifold vectors thus have the form

$$\mathbf{a_e}(\theta) = \mathbf{F}_\mathbf{e}^H \underline{\mathbf{a}}(\theta) \approx \sqrt{N}\mathbf{J}(\zeta)\mathbf{v}(\phi), \text{ where} \tag{5.11}$$

$$\mathbf{J}(\zeta) = \text{diag } \{J_M(\zeta), \ldots, J_1(\zeta), J_0(\zeta), J_1(\zeta), \ldots, J_M(\zeta)\}, \text{ and} \tag{5.12}$$

$$\mathbf{v}(\phi) = \left[e^{-jM\phi}, \ldots, e^{-j\phi}, e^{j0}, e^{j\phi}, \ldots, e^{jM\phi}\right]^T \tag{5.13}$$

The azimuthal variation of $\mathbf{a_e}(\theta)$ is through the vector $\mathbf{v}(\phi)$ that is similar in form to the ULA manifold vector. The elevation dependence takes the form of a symmetric amplitude taper through the matrix $\mathbf{J}(\zeta)$ of Bessel functions. The diagonal elements of $\mathbf{J}(\zeta)$ are even about the center element; thence the subscript $\mathbf{e}$. The beamspace manifold $\mathbf{a_e}(\theta)$ is centro-Hermitian and satisfies $\tilde{\mathbf{I}}\mathbf{a_e}(\theta) = \mathbf{a}_\mathbf{e}^*(\theta)$, where $\tilde{\mathbf{I}}$ is the reverse permutation matrix with ones on the antidiagonal and zeros elsewhere.

Employing the beamformer $\mathbf{F}_{\mathbf{e}}^H$ thus allows a Forward-Backward (FB) average to be performed in beamspace.

The centro-Hermitian nature of $\mathbf{a}_{\mathbf{e}}(\boldsymbol{\theta})$ motivates the development of the beamformer $\mathbf{F}_{\mathbf{r}}^H$ that synthesizes a real-valued beamspace manifold. We have

$$\mathbf{F}_{\mathbf{r}}^H = \mathbf{W}^H \mathbf{F}_{\mathbf{e}}^H = \mathbf{W}^H \mathbf{C}_{\mathbf{e}} \mathbf{V}^H, \text{ and} \tag{5.14}$$

$$\mathbf{a}_{\mathbf{r}}(\boldsymbol{\theta}) = \mathbf{F}_{\mathbf{r}}^H \underline{\mathbf{a}}(\boldsymbol{\theta}) = \mathbf{W}^H \mathbf{a}_{\mathbf{e}}(\boldsymbol{\theta}) = \sqrt{N}\mathbf{W}^H \mathbf{J}(\zeta)\mathbf{v}(\phi) \tag{5.15}$$

It is evident that the beamspace manifold $\mathbf{a}_{\mathbf{r}}(\boldsymbol{\theta})$ is real-valued if the matrix $\mathbf{W}^H$ has centro-Hermitian rows. Orthogonality of $\mathbf{F}_{\mathbf{r}}^H$ is maintained by restricting $\mathbf{W}$ to be unitary. An appropriate sparse matrix $\mathbf{W}^H$ and the corresponding beamspace manifold are given below. The symbol $\mathbf{O}_M$ denotes a zero vector of length M;

$$\mathbf{W}^H = \frac{1}{\sqrt{2}} \begin{bmatrix} \mathbf{I}_M & \mathbf{0}_M & \tilde{\mathbf{I}}_M \\ \mathbf{0}_M^T & \sqrt{2} & \mathbf{0}_M^T \\ j\tilde{\mathbf{I}}_M & \mathbf{0}_M & -j\tilde{\mathbf{I}}_M \end{bmatrix}; \tag{5.16}$$

$$\mathbf{a}_{\mathbf{r}}(\boldsymbol{\theta}) = \sqrt{2N}[J_M(\zeta)\cos(M\phi), \ldots, J_1(\zeta)\cos(\phi), \frac{1}{\sqrt{2}}J_0(\zeta), \\ J_1(\zeta)\sin(\phi), \ldots, J_M(\zeta)\sin(M\phi)]^T \tag{5.17}$$

Another choice of $\mathbf{W}$ and the corresponding real-valued beamspace manifold are as follows:

$$\mathbf{W} = \frac{1}{\sqrt{M'}}[\mathbf{v}(\alpha_{-M}) \vdots \cdots \vdots \mathbf{v}(\alpha_0) \vdots \cdots \vdots \mathbf{v}(\alpha_M)], \text{ and} \tag{5.18}$$

$$\mathbf{a}_{\mathbf{r}}(\boldsymbol{\theta}) = [f(\zeta, \phi - \alpha_{-M}), \ldots, f(\zeta, \phi - \alpha_{-1}), f(\zeta, \phi), f(\zeta, \phi - \alpha_1), \ldots, \\ f(\zeta, \phi - \alpha_M)]^T \tag{5.19}$$

where $f(\zeta, \phi) = \sqrt{\frac{N}{M'}}[J_0(\zeta) + 2\sum_{m=1}^{M} J_M(\zeta)\cos(m\phi)]$, and $\alpha_i = 2\pi i/M', i \in [-M, M]$. With this choice of $\mathbf{W}$, the beamformer $\mathbf{F}_{\mathbf{r}}^H$ synthesizes the basic beam pattern $f(\zeta, \phi)$ which is just the sum of the components of $\mathbf{a}_{\mathbf{e}}(\boldsymbol{\theta})$. Multiple beams are obtained by rotating this basic pattern in azimuth by the angles α_i. Having a real-valued beamspace manifold is advantageous for computational reasons. It allows subspace estimates to be obtained via real-valued EVDs as opposed to complex-valued EVDs, and thus provides computational savings. The UCA-RB-MUSIC algorithm developed in the following section exploits this property.

The last beamformer $\mathbf{F}_{\mathbf{u}}^H$ synthesizes the beamspace manifold $\mathbf{a}_{\mathbf{u}}(\boldsymbol{\theta})$ whose special structure is exploited in the development of the UCA-ESPRIT algorithm. The beamformer is defined by

$$\mathbf{F}_{\mathbf{u}}^H = \mathbf{C}_{\mathbf{u}} \mathbf{V}^H, \text{ where} \tag{5.20}$$

$$\mathbf{C}_{\mathbf{u}} = \text{diag}\{j^M, \ldots, j^1, j^0, j^{-1}, \ldots, j^{-M}\}$$

The diagonal, unitary matrix $\mathbf{C}_{\mathbf{u}}$ serves to eliminate factors of the form j^m in Eq. (5.8). The resulting beamspace manifold is

$$\mathbf{a_u}(\theta) = \mathbf{F_u}^H \underline{a}(\theta) \approx \sqrt{N}\mathbf{J}^-(\zeta)\mathbf{v}(\phi), \text{ where} \tag{5.21}$$

$$\mathbf{J}^-(\zeta) = \text{diag}\,\{J_{-M}(\zeta), \ldots, J_{-1}(\zeta), J_0(\zeta), J_1(\zeta), \ldots, J_M(\zeta)\} \tag{5.22}$$

and $\mathbf{v}(\phi)$ is defined in Eq. (5.13). Examination of Eqs. (5.14) and (5.20) shows that a unitary transformation relates the beamformers $\mathbf{F_u}^H$ and $\mathbf{F_r}^H$: We have $\mathbf{F_u}^H = \mathbf{C_o W F_r}^H$, where

$$\mathbf{C_o} = \mathbf{C_u C_e}^H = \text{diag}\,\{(-1)^M, \ldots, (-1)^1, 1, 1, \ldots, 1\} \tag{5.23}$$

The same transformation thus relates the corresponding beamspace manifold vectors: We have

$$\mathbf{a_u}(\theta) = \mathbf{C_o W a_r}(\theta) \tag{5.24}$$

This property enables computation of the beamspace signal subspace matrix for UCA-ESPRIT via a real-valued EVD.

5.3 DEVELOPMENT OF UCA-RB-MUSIC AND UCA-ESPRIT

The UCA-RB-MUSIC algorithm is described in this section. The advantages offered by UCA-RB-MUSIC over element space MUSIC are also discussed. The UCA-ESPRIT algorithm is developed later in the section. The changes in the UCA manifold due to mutual coupling between array elements is described next. Simple modifications that enable UCA-RB-MUSIC and UCA-ESPRIT to account for mutual coupling effects are presented. Finally, we investigate the use of directional antenna elements with UCAs. Both UCA-RB-MUSIC and UCA-ESPRIT are applicable if the element patterns are omnidirectional in azimuth (only elevation dependent). Although UCA-ESPRIT cannot be employed when the elements have arbitrary patterns, a phase mode excitation-based beamspace MUSIC algorithm is still applicable.

The UCA-RB-MUSIC Algorithm

UCA-RB (Real-Beamspace) MUSIC employs the beamformer $\mathbf{F_r}^H$ that synthesizes the real-valued beamspace manifold $\mathbf{a_r}(\theta) = \sqrt{N}\mathbf{W}^H\mathbf{J}(\zeta)\mathbf{v}(\phi)$ of Eq. (5.15). Depending on the choice of $\mathbf{W}, \mathbf{a_r}(\theta)$ has the form of Eq. (5.17) or Eq. (5.19). Since $\mathbf{a_r}(\theta)$ is real-valued, signal eigenvector estimates can be obtained via a real-valued EVD as described below. This reduction in computation (element space MUSIC requires a complex-valued EVD) is one of the advantages of UCA-RB-MUSIC.

Let $\underline{\mathbf{A}} = \left[\underline{\mathbf{a}}(\theta_1), \ldots, \underline{\mathbf{a}}(\theta_d)\right]$ be the $N \times d$ DOA matrix, assuming that d signals impinge on the array. Using the standard data model, the element space data vector can be represented as follows: $\mathbf{x}(t) = \underline{\mathbf{A}}\mathbf{s}(t) + \mathbf{n}(t)$, where $\mathbf{s}(t)$ is the vector of signal complex envelopes, and $\mathbf{n}(t)$ is the noise vector. The signals and the noises are assumed to be stationary, zero mean, uncorrelated random processes. The incident signals are assumed to be noncoherent, and the source covariance matrix $\mathbf{P} = E\left[\mathbf{s}(t)\mathbf{s}^H(t)\right]$ is thus positive definite. The noise process $\mathbf{n}(t)$ is assumed to be complex Gaussian distributed and spatially white with covariance matrix $\sigma^2\mathbf{I}$. The element space covariance matrix $\underline{\mathbf{R}} = E\left[\mathbf{x}(t)\mathbf{x}^H(t)\right]$ thus has the form

$$\underline{\mathbf{R}} = \underline{\mathbf{A}}\mathbf{P}\underline{\mathbf{A}}^H + \sigma^2\mathbf{I} \tag{5.25}$$

Employing the beamformer $\mathbf{F}_\mathbf{r}^H$ yields the beamspace data vector $\mathbf{y}(t) = \mathbf{F}_\mathbf{r}^H\mathbf{x}(t) = \mathbf{A}_\mathbf{r}\mathbf{s}(t) + \mathbf{F}_\mathbf{r}^H\mathbf{n}(t)$, where $\mathbf{A}_\mathbf{r} = \mathbf{F}_\mathbf{r}^H\underline{\mathbf{A}}$ is the real-valued beamspace DOA matrix. The corresponding beamspace covariance matrix is denoted $\mathbf{R}_\mathbf{y}$. Expressions for $\mathbf{R}_\mathbf{y}$ and the matrix $\mathbf{R}$ =Re $\{\mathbf{R}_\mathbf{y}\}$ are given below:

$$\mathbf{R}_\mathbf{y} = \mathbf{F}_\mathbf{r}^H\underline{\mathbf{R}}\mathbf{F}_\mathbf{r} = \mathbf{A}_\mathbf{r}\mathbf{P}\mathbf{A}_\mathbf{r}^T + \sigma^2\mathbf{I} \tag{5.26}$$

$$\mathbf{R} = \mathrm{Re}\{\mathbf{R}_\mathbf{y}\} = \mathbf{A}_\mathbf{r}\mathbf{P}_R\mathbf{A}_\mathbf{r}^T + \sigma^2\mathbf{I} \tag{5.27}$$

where $\mathbf{P}_R = \mathrm{Re}\{\mathbf{P}\}$. It is clear that the *real-valued* EVD of $\mathbf{R}$ yields bases for the beamspace signal and noise subspaces. Let $\lambda_1 > \cdots > \lambda_d > \lambda_{d+1} = \cdots = \lambda_{M'} = \sigma^2$ be the ordered eigenvalues of $\mathbf{R}$, and let $\mathbf{s}_1, \ldots, \mathbf{s}_d, \mathbf{g}_{d+1}, \ldots, \mathbf{g}_{M'}$ be the corresponding orthonormal eigenvectors. The real-valued matrices $\mathbf{S}$ and $\mathbf{G}$ defined below respectively span beamspace signal and noise subspaces.

$$\mathbf{S} = [\mathbf{s}_1, \ldots, \mathbf{s}_d] \tag{5.28}$$

$$\mathbf{G} = [\mathbf{g}_{d+1}, \ldots, \mathbf{g}_{M'}] \tag{5.29}$$

The UCA-RB-MUSIC spectrum

$$S_b(\theta) = \frac{1}{\mathbf{a}_\mathbf{r}^T(\theta)\mathbf{G}\mathbf{G}^T\mathbf{a}_\mathbf{r}(\theta} \propto \frac{1}{\mathbf{v}^H(\phi)\mathbf{J}(\zeta)\left[\mathbf{W}\mathbf{G}\mathbf{G}^T\mathbf{W}^H\right]\mathbf{J}(\zeta)\mathbf{v}(\phi)} \tag{5.30}$$

has peaks at $\theta = \theta_i$ corresponding to the signal arrival directions. DOA estimates are therefore obtained by searching for d peaks in the two-dimensional UCA-RB-MUSIC spectrum. The major computations required by UCA-RB-MUSIC are a real-valued EVD of the matrix $\mathbf{R}$, and a two-dimensional search for peaks in the spectrum $S_b(\theta) = S_b(\zeta, \phi)$. The elevation dependence of the spectrum is through the parameter $\zeta = k_o r \sin\theta$, where θ is the elevation angle.

Advantages of UCA-RB-MUSIC. The UCA-RB-MUSIC algorithm offers many attractive features that are not available in element space. All these features are available in beamspace because the azimuthal dependence of the beamspace manifold $\mathbf{a}_\mathbf{r}(\theta)$ of Eq. (5.15) is through the vector $\mathbf{v}(\phi)$ whose form is similar to the ULA manifold vector. One advantage mentioned already is the computational savings due to the requirement of only a real-valued EVD. Other advantages include improved estimator performance due to FB averaging, applicability of Root-MUSIC, and the ability to perform coarse searches of the UCA-RB-MUSIC spectrum via an FFT. These features are discussed in further detail below.

Forward/Backward Averaging in Beamspace. As mentioned in section 5.2, the beamformer $\mathbf{F}_\mathbf{e}^H$ synthesizes the centro-Hermitian beamspace manifold $\mathbf{a}_\mathbf{e}(\theta)$. FB averaging is thus possible in beamspace: The corresponding Forward/Backard (FB) averaged covariance matrix is $\mathbf{R}_{fb} = (\mathbf{R}_\mathbf{e} + \tilde{\mathbf{I}}\mathbf{R}_\mathbf{e}^*\tilde{\mathbf{I}})/2$, where $\mathbf{R}_\mathbf{e} = \mathbf{F}_\mathbf{e}^H\underline{\mathbf{R}}\mathbf{F}_\mathbf{e}$ is the beamspace covariance matrix under the beamformer $\mathbf{F}_\mathbf{e}^H$. The property $\tilde{\mathbf{I}}\mathbf{W} = \mathbf{W}^*$

is employed in the manipulations below that show that the matrix $\mathbf{R} = \text{Re}\{\mathbf{R}_y\}$ employed by UCA-RB-MUSIC is derived from the FB averaged covariance matrix $\mathbf{R}_{fb}$.

$$\mathbf{R} = \text{Re}\{\mathbf{W}^H\mathbf{R}_e\mathbf{W}\} = \frac{1}{2}(\mathbf{W}^H\mathbf{R}_e\mathbf{W} + \mathbf{W}^T\mathbf{R}_e^*\mathbf{W}^*) \tag{5.31}$$

$$= \frac{1}{2}\mathbf{W}^H(\mathbf{R}_e + \tilde{\mathbf{I}}\mathbf{R}_e^*\tilde{\mathbf{I}})\mathbf{W} = \mathbf{W}^H\mathbf{R}_{fb}\mathbf{W} \tag{5.32}$$

UCA-RB-MUSIC is thus endowed with benefits associated with FB averaging such as improved estimator performance in correlated source scenarios. FB type averaging can be performed in element space with a UCA when the number of array elements N is even; this property does not appear to have been exploited by researchers prior to this work. FB type averaging is possible for even N because the permuted version $\mathbf{J}\underline{\mathbf{a}}(\Theta)$ of the element space manifold vector, with

$$\mathbf{J} = \begin{bmatrix} \mathbf{I}_{N/2} & \mathbf{0} \\ \mathbf{0} & \tilde{\mathbf{I}}_{N/2} \end{bmatrix}$$

is centro-Hermitian. However, such averaging is not possible in element space when N is odd. UCA-RB-MUSIC can thus outperform element space MUSIC when N is odd. The theoretical performance curves in section 5.5 substantiate this claim.

Spectral Search via FFT. UCA-RB-MUSIC requires a search for peaks in the 2D spectrum of Eq. (5.30) to obtain source azimuth and elevation estimates. This search is expedited by the fact that the computationally efficient FFT can be employed to evaluate the spectrum at each candidate elevation angle. Let $V(\phi;\zeta) = \mathbf{v}^H(\phi)\mathbf{J}(\zeta)\mathbf{W}\mathbf{G}\mathbf{G}^T\mathbf{W}^H\mathbf{J}(\zeta)\mathbf{v}(\phi)$ denote the UCA-RB-MUSIC null spectrum at the elevation specified by $\zeta = k_0 r \sin\theta$. With $\mathbf{Q}_\zeta = \mathbf{J}(\zeta)\mathbf{W}\mathbf{G}\mathbf{G}^T\mathbf{W}^H\mathbf{J}(\zeta)$, the null spectrum can be written in the form

$$V(\phi;\zeta) = \sum_{l=-(M'-1)}^{M'-1} a_\zeta(l)e^{jl\phi}, \text{ where } a_\zeta(l) = \sum_{i,j:j-=l} \mathbf{Q}_\zeta(i,j) \tag{5.33}$$

The matrix $\mathbf{Q}_\zeta$ is Hermitian such that $a_\zeta(-l) = a_\zeta^*(l)$. $V(\phi;\zeta)$ can thus be written in terms of the Discrete Time Fourier Transform of the M' point sequence $a'_\zeta = \{a_\zeta(0), 2a_\zeta(-1), \ldots, 2a_\zeta(-M'+1)\}$: We have $V(\phi;\zeta) = \text{Re}\{A'_\zeta(\phi)\}$, where $A'_\zeta(\phi) = \sum_{l=0}^{M'-1} a'_\zeta(l)e^{-jl\phi}$. The UCA-RB-MUSIC null spectrum $V(\phi;\zeta)$ at the elevation specified by ζ can thus be evaluated at L equispaced azimuth angles $\phi_l = 2\pi l/L, l = 0, 1, \ldots L-1$ via an L-point FFT of the sequence a'_ζ appropriately zero padded. In contrast, the search for peaks in the element space MUSIC spectrum cannot be expedited via an FFT.

Application of Root-MUSIC. The Root-MUSIC algorithm [19] originally developed for use in conjunction with ULAs hinges on the Vandermonde structure of the ULA manifold. Root-MUSIC cannot be employed in element space with UCAs as the UCA manifold vectors $\underline{\mathbf{a}}(\Theta)$ of Eq. (5.1) are not Vandermonde. However, the azimuthal dependence of the UCA-RB-MUSIC null spectrum is through the vector $\mathbf{v}(\Theta)$ of Eq. (5.13) that is Vandermonde except for a multiplicative scale factor.

Root-MUSIC can thus be employed in beamspace to obtain azimuth angles of sources at a given elevation. The Root-MUSIC formulation follows on setting $z = e^{j\phi}$ in Eq. (5.33) and equating the null spectrum $V(\phi; \zeta)$ to zero. The polynomial equation

$$a_\zeta(M'-1)z^{2M'-2} + a_\zeta(M'-2)z^{2M'-3} + \cdots + a_\zeta(-M'+1) = 0$$

results. Roots z_i of this equation which are close to the unit circle yield the azimuth estimates $\phi_i = \arg(z_i)$ of sources at the elevation ζ. UCA-RB-MUSIC thus benefits from the concomitant advantages of Root-MUSIC such as a lower failure rate for closely spaced sources at a given elevation.

Mapping onto ULA Type Manifold. Several researchers [3, 20] have considered the case where all incident sources are confined to a given elevation angle, say ζ_0, and the problem of interest is to estimate the source azimuth angles. It is clear from Eq. (5.11) that the beamformer $\mathbf{F}_{\text{ULA}}^H(\zeta_0) = (1/\sqrt{N})\mathbf{J}^{-1}(\zeta_0)\mathbf{F}_\mathbf{e}^H$ maps the UCA manifold $\underline{\mathbf{a}}(\zeta_0, \phi)$ corresponding to the elevation ζ_0 onto the manifold $\mathbf{v}(\phi)$ of Eq. (5.13). We have

$$\mathbf{F}_{\text{ULA}}^H(\zeta_0)\underline{\mathbf{a}}(\zeta_0, \phi) = \mathbf{v}(\phi) \tag{5.34}$$

and the beamspace manifold corresponding to the elevation ζ_0 is similar to the ULA manifold. Using the beamformer $\mathbf{F}_{\text{ULA}}^H(\zeta_0)$ in such a scenario thus permits *Spatial Smoothing* [21] to be employed in beamspace to combat the rank-reducing effect caused by source coherency. Further, sinc-type azimuthal patterns can be synthesized as with a ULA [22, 23], and the Beamspace Root-MUSIC algorithm [24] can thus be employed. This algorithm allows for parallel sectorwise azimuthal searches for sources via rooting of reduced-order polynomials.

SUMMARY OF UCA-RB-MUSIC

1. Form the array sample covariance matrix $\hat{\underline{\mathbf{R}}} = \frac{1}{K}\sum_{t=1}^{K}\mathbf{x}(t)\mathbf{x}^H(t)$ by averaging over the K data snapshots. Also form the sample beamspace covariance matrix $\hat{\mathbf{R}}_\mathbf{y} = \mathbf{F}_\mathbf{r}^H\hat{\underline{\mathbf{R}}}\mathbf{F}_\mathbf{r}$.
2. Perform the real-valued EVD of the matrix $\hat{\mathbf{R}} = \text{Re}\,\{\hat{\mathbf{R}}_\mathbf{y}\}$, and apply an appropriate detection technique to get an estimate $\hat{d}$ of the number of sources. Let the ordered eigenvalues of $\hat{\mathbf{R}}$ be $\hat{\lambda}_1 \geq \cdots \geq \hat{\lambda}_{M'}$, and the corresponding orthonormal eigenvectors be $\hat{\mathbf{s}}_1, \ldots, \hat{\mathbf{s}}_{\hat{d}}, \hat{\mathbf{g}}_{\hat{d}+1}, \ldots, \hat{\mathbf{g}}_{M'}$. Form the matrices $\hat{\mathbf{S}} = \left[\hat{\mathbf{s}}_1, \ldots, \hat{\mathbf{s}}_{\hat{d}}\right]$ and $\hat{\mathbf{G}} = \left[\hat{\mathbf{g}}_{\hat{d}+1}, \ldots, \hat{\mathbf{g}}_{M'}\right]$ that respectively span the estimated signal and noise subspaces.
3. Search for $\hat{d}$ peaks in the two-dimensional UCA-RB-MUSIC spectrum

$$\hat{S}_b(\boldsymbol{\theta}) = \frac{1}{\mathbf{v}^H(\phi)\mathbf{J}(\zeta)\left[\mathbf{W}\hat{\mathbf{G}}\hat{\mathbf{G}}^T\mathbf{W}^H\right]\mathbf{J}(\zeta)\mathbf{v}(\phi)}$$

The peak locations $\boldsymbol{\theta}_i = (k_0 r \sin\hat{\theta}_i, \hat{\phi}_i)$, $i = 1, \ldots, \hat{d}$ give the DOA estimates. As described earlier, use of the FFT facilitates this 2D spectral search.

4. If a good estimate of source elevation angle is available, Root-MUSIC can be employed to obtain the azimuth angle estimate as described earlier. Root-MUSIC can resolve sources at a given elevation and closely spaced in azimuth even if the UCA-RB-MUSIC spectrum reveals only a single peak in the vicinity.

Previous Work on the Application of ULA Techniques with UCAs. As mentioned in section 5.1, phase mode excitation-based beamformers have been employed to synthesize attractive directional patterns with UCAs, and to obtain DOA estimates via the beamforming principle. Our initial work [22, 23] focused on phase mode excitation-based sinc-type pattern synthesis with UCAs (patterns similar to the cophasal ULA beam patterns). Root MUSIC was then employed to obtain azimuth angle estimates of sources at a given elevation. Other work on the application of ULA techniques with UCAs include that of Tewfik and Hong [11], and Friedlander and Weiss [12].

To compare UCA-RB-MUSIC with the work reported in [11], we recall that the rows of the matrix $\mathbf{V}^H$ (of Eq. (5.9)) that defines the beamforming matrix $\mathbf{F}_e^H$ are inverse DFT weight vectors. Row $m \in [-M, M]$ of $\mathbf{V}^H$ excites the array with phase mode m, and a total of $M' = 2M + 1 < N$ modes are excited. Thus, only M' of the N possible phase modes are excited. The reason for choosing $M' < N$ was to make the contributions of residual terms to the UCA far-field pattern negligible, thus leading to a beamspace manifold whose azimuthal dependence is through the Vandermonde (except for a multiplicative scale factor) vector $\mathbf{v}(\phi)$. A *full* $N \times N$ inverse DFT beamformer was employed in [11] to make the transformation from element space to beamspace. Some of the beams thus have significant contributions from residual terms, and this detracts from the desired Vandermonde structure. The approach proposed in [11] was to employ Root-MUSIC to obtain source azimuth estimates at each elevation angle under consideration. The imperfect Vandermonde structure, however, introduces errors in the estimates. The problem of elevation angle estimation was not addressed in [11].

Friedlander [12] proposed the interpolated array scheme that employs mapping matrices to map the manifold vectors for an arbitrary array onto Vandermonde ULA-type steering vectors. The azimuthal field of view corresponding to each candidate elevation angle is divided into sectors, for each of which a different matrix is designed. The interpolating matrix for a given sector is computed as the least squares solution of an overdetermined system of equations corresponding to the desired mapping. The link between the present work and the interpolated array technique is provided by Eq. (5.34). It reveals that $\mathbf{F}_{\mathrm{ULA}}^H(\zeta_0)$ is the desired mapping matrix that maps the element space UCA manifold $\underline{\mathbf{a}}(\zeta_0, \phi)$ corresponding to the elevation ζ_0 onto the ULA-type manifold vector $\mathbf{v}(\phi)$. Phase mode analysis thus provides closed-form expressions for the mapping matrix for each elevation angle, and the mapping is valid for the entire 360° of azimuth.

The UCA-ESPRIT Algorithm

The UCA-ESPRIT algorithm represents a significant advance in the area of 2D arrival angle estimation. It is a *closed-form* algorithm that provides *automatically paired* source azimuth and elevation angle estimates. In contrast, the algorithms for 2D arrival angle estimation to date have required expensive spectral searches [7], iterative solutions to multidimensional optimization problems [13, 15], or a pairing procedure for associating independently obtained direction cosine estimates [14]. The UCA-ESPRIT algorithm is fundamentally different from ESPRIT in that it is not based on the displacement invariance array structure required by ESPRIT [8]. The development of UCA-ESPRIT hinges rather on a recursive relationship between Bessel functions. The steps in the algorithm, however, are similar to those of TLS-ESPRIT [8]. In the 1D angle estimation scenario, TLS-ESPRIT provides DOA estimates via the eigenvalues of a matrix. UCA-ESPRIT provides closed-form DOA estimates via matrix eigenvalues in the 2D angle estimation scenario: The eigenvalues have the form $\mu_i = \sin\theta_i e^{j\phi_i}$, and thus yield automatically paired source azimuth and elevation angle estimates. Since $\theta_i \in [0, \pi/2]$, the eigenvalues satisfy $|\mu_i| \leq 1$, and lie within or on the unit circle. It is clear that $|\mu_i| = \sin\theta_i$, and $\arg(\mu_i) = \phi_i$ respectively specify the elevation and azimuth angles of the ith source without ambiguity. Note also that $\mu_i = u_i + jv_i$, where $u_i = \sin\theta_i \cos\phi_i$, and $v_i = \sin\theta : \sin\phi_i$ are respectively the direction cosines with respect to the x and y axes. The u and v estimates are preferable to the θ and ϕ estimates when a source is close to boresight (elevation angle θ close to 0°). This is because azimuth is not very meaningful in such case; for example, all azimuth angles are equivalent when $\theta = 0°$. Another similarity between UCA-ESPRIT and ESPRIT is the approximate halving (with respect to the size of the beamspace manifold) in the maximum number of resolvable sources. UCA-ESPRIT can resolve a maximum of $d_{\max} = M - 1$ sources, roughly half the number resolvable with UCA-RB-MUSIC.

The beamformer $\mathbf{F}_\mathbf{u}^H$ of Eq. (5.20) forms the basis for the development of UCA-ESPRIT. The structure of the corresponding beamspace manifold $\mathbf{a_u}(\theta)$ of Eq. (5.21) is crucial to the development of the algorithm. We have

$$\mathbf{a_u}(\theta) = \mathbf{F}_\mathbf{u}^H \underline{\mathbf{a}}(\theta) = \sqrt{N} \begin{bmatrix} J_{-M}(\zeta)e^{-jM\phi} \\ \vdots \\ J_{-1}(\zeta)e^{-j\phi} \\ J_0(\zeta) \\ J_1(\zeta)e^{j\phi} \\ \vdots \\ J_M(\zeta)e^{jM\phi} \end{bmatrix} \tag{5.35}$$

Consider extracting three subvectors of size $M_e = M' - 2$ from the beamspace manifold as follows: $\mathbf{a_i} = \boldsymbol{\Delta}_i \mathbf{a_u}(\theta), i = -1, 0, 1$, where the $M_e \times M'$ selection matrices $\boldsymbol{\Delta}_{-1}$, $\boldsymbol{\Delta}_0$, and $\boldsymbol{\Delta}_1$ pick out the first, middle, and last M_e elements from $\mathbf{a_u}(\theta)$. The property $J_{-m}(\zeta) = (-1)^m J_m(\zeta)$ of Bessel functions leads to the following relationship:

$$\mathbf{a}_1 = \mathbf{D}\tilde{\mathbf{I}}\mathbf{a}_{-1}^*, \text{ where} \tag{5.36}$$

$$\mathbf{D} = \text{diag}\,\{(-1)^{M-2}, \ldots, (-1)^1, (-1)^0, (-1)^1, \ldots, (-1)^M\}.$$

The phases (excluding the signs of the values of the Bessel functions) of the vectors $\mathbf{a}_0$, $e^{j\phi}\mathbf{a}_{-1}$, and $e^{-j\phi}\mathbf{a}_1$ are the same. The recursive relationship $J_{m-1}(\zeta) + J_{m+1}(\zeta) =$

$(2m/\zeta)J_m(\zeta)$ can now be applied to match the magnitude components of the three vectors. This leads to the critical relationship

$$\begin{aligned}\mathbf{\Gamma a}_0 &= \mu \mathbf{a}_{-1} + \mu^* \mathbf{a}_1 \qquad (5.37)\\ &= \mu \mathbf{a}_{-1} + \mu^* \mathbf{D\tilde{I}a}_{-1}^*, \text{ where}\\ \mathbf{\Gamma} &= \frac{\lambda}{\pi r} \operatorname{diag}\{-(M-1), \ldots, -1, 0, 1, \ldots, M-1\}, \text{ and}\\ \mu &= \sin\theta e^{j\phi}\end{aligned}$$

The partitions of the beamspace DOA matrix $\mathbf{A_u} = [\mathbf{a_u}(\theta_1) \vdots \cdots \vdots \mathbf{a_u}(\theta_d)]$ also satisfy the above property. Defining $\mathbf{A}_i = \mathbf{\Delta}_i \mathbf{A}_u, i = -1, 0$, we obtain

$$\begin{aligned}\mathbf{\Gamma A}_0 &= \mathbf{A}_{-1}\mathbf{\Phi} + \mathbf{D\tilde{I}A}_{-1}^*\mathbf{\Phi}^*, \text{ where} \qquad (5.38)\\ \mathbf{\Phi} &= \operatorname{diag}\{\mu_1, \ldots, \mu_d\} = \operatorname{diag}\{\sin\theta_1 e^{j\phi_1}, \ldots, \sin\theta_d e^{j\phi_d}\}\end{aligned}$$

The beamspace signal subspace matrix $\mathbf{S_u}$ that spans $\mathcal{R}\{\mathbf{A_u}\}$ can be obtained via a complex-valued EVD of the beamspace covariance matrix $\mathbf{R_u} = \mathbf{F_u}^H \underline{\mathbf{R}} \mathbf{F_u}$. However, the relationship of Eq. (5.24) allows $\mathbf{S_u}$ to be expressed in terms of the signal subspace matrix $\mathbf{S}$ of Eq. (5.28) that was obtained via a real-valued EVD. We have $\mathbf{A_u} = \mathbf{C_0 W A_r} = \mathbf{C_0 W S T}^{-1}$, where $\mathbf{T}$ is a $d \times d$ *real-valued* nonsingular matrix. Thus

$$\mathbf{A_u} = \mathbf{S_u T}^{-1}, \text{ and } \mathbf{S_u} = \mathbf{C_0 W S} \qquad (5.39)$$

The relationship (5.38) can now be expressed in terms of the partitions $\mathbf{S}_i = \mathbf{\Delta}_i \mathbf{S_u}, i = -1, 0$ of the signal subspace matrix $\mathbf{S_u}$. Substituting $\mathbf{A}_i = \mathbf{S}_i \mathbf{T}^{-1}$ in Eq. (5.38), and using the fact that $\mathbf{T}$ is real-valued leads to the following relationship:

$$\begin{aligned}\mathbf{\Gamma S}_0 &= \mathbf{S}_{-1}\mathbf{\Psi} + \mathbf{D\tilde{I}S}_{-1}^*\mathbf{\Psi}^*, \text{ where} \qquad (5.40)\\ \mathbf{\Psi} &= \mathbf{T}^{-1}\mathbf{\Phi T}\end{aligned}$$

Writing in block matrix form yields the following system of equations:

$$\begin{aligned}\mathbf{E}\underline{\mathbf{\Psi}} &= \mathbf{\Gamma S}_0, \text{ where} \qquad (5.41)\\ \mathbf{E} &= \left[\mathbf{S}_{-1} \vdots \mathbf{D\tilde{I}S}_{-1}^*\right], \text{ and}\\ \underline{\mathbf{\Psi}} &= \begin{bmatrix}\mathbf{\Psi}\\ \mathbf{\Psi}^*\end{bmatrix}\end{aligned}$$

This system of equations is overdetermined when $M_e > 2d$, that is, $d < M$, and has a unique solution $\underline{\mathbf{\Psi}}$ or equivalently, $\mathbf{\Psi}$. From (5.40) we have $\mathbf{\Phi} = \mathbf{T\Psi T}^{-1}$, and the eigenvalues of $\mathbf{\Psi}$ are thus $\mu_i = \sin\theta_i e^{j\phi_i}, i = 1, \ldots, d$. The eigenvalues of $\mathbf{\Psi}$ thus yield automatically paired source azimuth and elevation angles: We have $\theta_i = \sin^{-1}(|\mu_i|$ and $\phi_i = \arg(\mu_i)$. We point out that the eigenvalues μ_i can be obtained via a real-valued EVD in place of the complex-valued EVD of $\mathbf{\Psi} = \mathbf{\Psi}_R + j\mathbf{\Psi}_I$. We have $\mathbf{\Psi\Psi}^* = \mathbf{\Psi}_R^2 + \mathbf{\Psi}_I^2 = \mathbf{T}^{-1}\mathbf{\Phi\Phi}^*\mathbf{T}$, a real-valued matrix.

The real-valued EVD of the matrix $\Psi_R^2 + \Psi_I^2$ thus yields the matrix $\mathbf{T}$ of eigenvectors, and $\mathbf{\Phi}$ is computed according to $\mathbf{\Phi} = \mathbf{T\Psi T}^{-1}$. UCA-ESPRIT cannot be employed when $d \geq M$, and the system of Eq. (5.41) is underdetermined. This is because the system possesses an infinity of solutions having the block conjugate structure of $\underline{\Psi}$, as shown in Appendix 5A.2. The maximum number of sources that UCA-ESPRIT can resolve is thus $d_{\max} = M - 1$, where M is the maximum mode excited.

Under noisy conditions, the matrices $\hat{\mathbf{E}}$ and $\hat{\mathbf{S}}_0$ are formed using signal subspace estimates. The matrix $\underline{\Psi}$ is then obtained as the least squares (LS) solution to the overdetermined system

$$\hat{\mathbf{E}}\underline{\Psi} = \mathbf{\Gamma}\hat{\mathbf{S}}_0 \tag{5.42}$$

Appendix 5A.1 shows that $\hat{\underline{\Psi}}$ has block conjugate structure, as in the noise-free case. The eigenvalues of the upper block $\hat{\Psi}$ yield the source DOA estimates as described earlier. Appendix 5A.1 also shows that the block conjugate structure leads to the following simplification in computing the LS solution. It allows the LS solution to be obtained by solving the system of $2d$ real equations below rather than solving a system of $2d$ complex equations as would otherwise be required.

$$\begin{bmatrix} (\mathbf{B}+\mathbf{C})_R & (\mathbf{C}-\mathbf{B})_I \\ (\mathbf{B}+\mathbf{C})_I & (\mathbf{B}-\mathbf{C})_R \end{bmatrix} \begin{bmatrix} \hat{\Psi}_R \\ \hat{\Psi}_I \end{bmatrix} = \begin{bmatrix} \mathbf{Q}_R \\ \mathbf{Q}_I \end{bmatrix}, \quad \text{where} \tag{5.43}$$

$$\mathbf{B} = \hat{\mathbf{S}}_{-1}^H \hat{\mathbf{S}}_{-1}, \ \mathbf{C} = \hat{\mathbf{S}}_{-1}^H \mathbf{D}\tilde{\mathbf{I}}\hat{\mathbf{S}}_{-1}^*, \ \text{and } \mathbf{Q} = \hat{\mathbf{S}}_{-1}^H \mathbf{\Gamma}\hat{\mathbf{S}}_0 \tag{5.44}$$

The subscripts R and I in the above equation denote the real and imaginary parts, respectively.

UCA-ESPRIT is clearly superior to existing 2D angle estimation algorithms with respect to computational complexity. The significant computations required by UCA-ESPRIT include a real-valued EVD of the $M' \times M'$ matrix $\hat{\mathbf{R}}$, solution of the system (5.43) of $2d$ real equations, and a $d \times d$ EVD of the complex-valued matrix $\hat{\Psi}$ (or a real-valued EVD of $\hat{\Psi}_R^2 + \hat{\Psi}_I^2$). Spectral searches, iterative optimization techniques, and the need to pair independently obtained direction cosine estimates are dispensed with. The simulations in section 5.5 however show that the UCA-RB-MUSIC estimates have lower variances than the UCA-ESPRIT estimates. The UCA-ESPRIT estimates serve as good starting points for iterative Newton searches for peaks in the UCA-RB-MUSIC spectrum. The performance of UCA-RB-MUSIC can thus be realized at the additional cost of a Newton iteration if required.

SUMMARY OF UCA-ESPRIT

1. Obtain the real-valued matrix $\hat{\mathbf{S}}$ via steps 1 and 2 in the Summary of the UCA-RB-MUSIC algorithm of section 5.3. Compute $\hat{\mathbf{S}}_{\mathbf{u}} = \mathbf{C}_o\mathbf{W}\hat{\mathbf{S}}$, where $\mathbf{C}_o$ is defined in Eq. (5.23), and $\mathbf{W}$ is specified by either Eq. (5.16) or Eq. (5.18). Form the submatrices $\hat{\mathbf{S}}_i = \mathbf{\Delta}_i\hat{\mathbf{S}}_{\mathbf{u}}, i = -1, 0$, and construct the matrix $\hat{\mathbf{E}} = [\hat{\mathbf{S}}_{-1} \vdots \mathbf{D}\tilde{\mathbf{I}}\hat{\mathbf{S}}_{-1}^*]$.
2. Obtain the least squares solution $\hat{\Psi} = \hat{\Psi}_{R+j}\hat{\Psi}_I$ by solving the real-valued system of equations in (5.43).

3. Compute the eigenvalues $\mu_i, i = 1, \ldots, d$ of $\hat{\mathbf{\Psi}}$. The eigenvalues $\hat{\mu}_i$ are the diagonal entries of the matrix $\hat{\mathbf{T}}\hat{\mathbf{\Psi}}\hat{\mathbf{T}}^{-1}$, where $\hat{\mathbf{T}}^{-1}$ is the real-valued matrix whose columns are the eigenvectors of $\hat{\mathbf{\Psi}}_R^2 + \hat{\mathbf{\Psi}}_I^2$. The estimates of the elevation and azimuth angles of the ith source are $\hat{\theta}_i = \sin^{-1}(|\hat{\mu}_i|)$ and $\hat{\phi}_i = \arg(\hat{\mu}_i)$, respectively. If direction cosine estimates are desired, we have $\hat{u}_i = \text{Re}\{\hat{\mu}_i\}$, and $\hat{v}_i = \text{Im}\{\hat{\mu}_i\}$.
4. DOA estimates of lower variance can be obtained by using the UCA-ESPRIT estimates from step 3 as starting points for a Newton search for nearby maxima in the two-dimensional UCA-RB-MUSIC spectrum $\hat{S}_b(\zeta, \phi) = 1/\mathbf{v}^H(\phi)\mathbf{J}(\zeta)\mathbf{W}\hat{\mathbf{G}}\hat{\mathbf{G}}^T\mathbf{W}^H\mathbf{J}(\zeta)\mathbf{v}(\phi)$.

Adaptation of UCA-ESPRIT for Filled Circular Arrays. We now describe how the UCA-ESPRIT principle can be employed to develop a closed-form 2D angle estimation algorithm for filled circular apertures/arrays. Several existing phased array radar systems (e.g., the SPY-1A or SPY-1B radars in the AEGIS series [25], and the arrays comprising the PAVE-PAWS surveillance network [26]) have circular apertures, with antenna element locations specified by a hexagonal sampling lattice. Filled circular arrays also have potential application as base station antennas in mobile communication systems. The effort to adapt the UCA-ESPRIT algorithm for filled circular arrays is thus relevant and significant. Expressions for the far-field patterns resulting from phase mode excitation (with a radial amplitude taper) of a continuous circular aperture are derived below. These expressions provide an accurate description of the far-field pattern of the sampled aperture (as was the case with a UCA), provided the interelement spacings are sufficiently small. The far-field phase mode amplitudes turn out to be specified by Bessel functions of the first kind even for filled circular apertures. The UCA-ESPRIT principle can thus be employed even with filled circular arrays/apertures.

The aperture excitation function corresponding to phase mode m (for integer m) is

$$w_m(\rho, \gamma) = j^{-m}\left(\frac{\rho}{r}\right)^m e^{jm\gamma}, \quad \rho \in [0, r], \gamma \in [0, 2\pi] \tag{5.45}$$

The term $e^{jm\gamma}$ excites the mth phase mode, $(\rho/r)^m$ is a mode-dependent radial amplitude taper, and the phase factor j^{-m} ensures (as with the UCA) that the far-field mode amplitude is real-valued. The resulting far-field pattern is

$$\begin{aligned} f_m(\zeta, \phi) = f_m(k_o r \sin\theta, \phi) &= \frac{1}{2\pi}\int_0^R \int_0^{2\pi} w_m(\rho, \gamma) e^{jk_o\rho\sin\theta\cos(\phi-\gamma)} \rho d\rho d\gamma \\ &= \int_0^R (\rho/r)^m J_m(k_o\rho\sin\theta) e^{jm\phi} \rho d\rho \\ &= \left(\frac{r^2}{\zeta}\right) J_{m+1}(\zeta) e^{jm\phi} \end{aligned} \tag{5.46}$$

The relationship $\int x^m J_m(x) = x^m J_{m+1}(x)$ is employed to obtain the final expression above. The far-field pattern above is similar to that resulting from excitation of a

UCA with phase mode m. The differences are the increase in Bessel function order by one, and the ζ dependence in the denominator. As with the UCA, a relationship between the far-field patterns corresponding to three consecutive phase modes can be developed. The critical relationship (which results from matching the mode amplitudes using the recursive relationship between Bessel funtions) is

$$\frac{\lambda}{\pi r}(m+1) f_m(\zeta, \phi) = \mu f_{m-1}(\zeta, \phi) + \mu^* f_{m+1}(\zeta, \phi) \tag{5.47}$$

where $\mu = \sin\theta e^{j\phi}$. The development of the closed-form 2D angle estimation algorithm for filled circular apertures now parallels the development of UCA-ESPRIT.

Accounting for Mutual Coupling Effects

Mutual coupling effects can be quite significant with UCAs of omnidirectional elements [3]. In the presence of mutual coupling, $\underline{\mathbf{a}}(\theta)$ of Eq. (5.1) is no longer an accurate representation of the UCA manifold. The UCA manifold after incorporating mutual coupling effects is denoted $\underline{\mathbf{a}}_m(\theta)$. We have $\underline{\mathbf{a}}_m(\theta) = \mathbf{Y}\underline{\mathbf{a}}(\theta)$, where $\mathbf{Y}$ is the mutual coupling matrix [27]. At first glance it appears that all the phase mode excitation developments are inapplicable due to the presence of $\mathbf{Y}$. However, it is well known [3, 34] that exciting a UCA with phase mode m synthesizes the same phase mode in the far-field pattern even when mutual coupling effects apply. A change in the corresponding mode amplitude in the far-field pattern is the only effect of mutual coupling. As a consequence of this property, UCA-RB-MUSIC and UCA-ESPRIT are easily adapted to account for mutual coupling effects. Accounting for mutual coupling with a UCA turns out to be much simpler than with other array geometries (see [28]).

Before proceeding to discuss the modifications required to adapt the algorithms to cope with mutual coupling, we provide a proof of the above mentioned property. The matrix $\mathbf{V}^H$ defined in Eq. (5.9) excites the UCA with the appropriate phase modes, and the relationship to be proved is

$$\mathbf{V}^H\underline{\mathbf{a}}_m(\theta) = \mathbf{V}^H\mathbf{Y}\underline{\mathbf{a}}(\theta) = \mathbf{\Lambda}_y\mathbf{V}^H\underline{\mathbf{a}}(\theta) \tag{5.48}$$

where $\mathbf{\Lambda}_y$ is a diagonal matrix whose entries represent the change in far-field mode amplitude due to mutual coupling. The mutual coupling matrix $\mathbf{Y}$ is circulant due to circular symmetry of the UCA. Let $\mathbf{y}^T$ be the first row of $\mathbf{Y}$: we have $\mathbf{y}^T = \{y_0, y_1, y_2, y_1\}$ for a four-element UCA. It is well known [29] that the DFT (Discrete Fourier Transform) matrix diagonalizes any circulant matrix; the IDFT (Inverse DFT) matrix thus gives the left eigenvectors of any circulant matrix. As mentioned in section 5.2, the M' rows of the matrix $\mathbf{V}^H$ are a subset of the N IDFT weight vectors. We therefore have

$$\mathbf{V}^H\mathbf{Y} = \mathbf{\Lambda}_y\mathbf{V}^H, \text{ where} \tag{5.49}$$

$$\mathbf{\Lambda}_y = \text{diag}\,\{\lambda_{-M}, \ldots, \lambda_0, \ldots, \lambda_M\} \tag{5.50}$$

is the diagonal matrix whose entries are the appropriate subset of eigenvalues of $\mathbf{Y}$. These equations prove the assertion in (5.48). We also point out that the eigenvalue of $\mathbf{Y}$ associated with the ith column of the DFT matrix (or the ith row of the IDFT matrix) is just the ith element in the DFT of the sequence $\mathbf{y}^T$.

From Eq. (5.48) it is evident that the beamformers

$$\mathbf{F}_{\mathbf{em}}^H = \Lambda_y^{-1}\mathbf{F}_{\mathbf{e}}^H, \mathbf{F}_{\mathbf{rm}}^H = \mathbf{W}^H\mathbf{F}_{\mathbf{em}}^H, \text{ and } \mathbf{F}_{\mathbf{um}}^H = \Lambda_y^{-1}\mathbf{F}_{\mathbf{u}}^H \tag{5.51}$$

respectively synthesize the beamspace manifolds $\mathbf{a_e}(\theta)$, $\mathbf{a_r}(\theta)$, and $\mathbf{a_u}(\theta)$ of section 5.2. It is clear that both UCA-RB-MUSIC and UCA-ESPRIT can be applied in conjunction with the above beamformers. However, there are slight differences in the implementation of the algorithms as $\mathbf{F}_{\mathbf{rm}}^H$ is not an orthogonal beamformer. With σ^2 being the power of the spatially white element space noise, the beamspace noise covariance matrix is $\mathbf{R_n} = \sigma^2\mathbf{F}_{\mathbf{rm}}^H\mathbf{F_{rm}} = \sigma^2\mathbf{W}^H|\Lambda_y|^{-2}\mathbf{W}$, a real-valued, Toeplitz matrix. Since the beamspace noise is nonwhite, a generalized eigenvalue decomposition (GEVD) is required to obtain signal eigenvector estimates. The steps involved in adapting UCA-RB-MUSIC and UCA-ESPRIT to cope with mutual coupling effects are summarized below. The mutual coupling matrix $\mathbf{Y}$ can be obtained experimentally or via theoretical analysis.

ALGORITHM SUMMARY—INCORPORATION OF MUTUAL COUPLING EFFECTS

1. Let λ_i denote the ith bin of the DFT of the vector $\mathbf{y}^T$ that specifies the first row of the mutual coupling matrix $\mathbf{Y}$. Using the fact that $\lambda_{-i} = \lambda_{N-i}$, form the matrix $\Lambda_y = \text{diag}\,\{\lambda_{-M}, \ldots, \lambda_0, \ldots, \lambda_M\}$.
2. Form the sample beamspace covariance matrix $\hat{\mathbf{R}}_y = \mathbf{F}_{\mathbf{rm}}^H\hat{\mathbf{R}}\mathbf{F}_{\mathbf{rm}}^H$, where $\mathbf{F}_{\mathbf{rm}}^H = \mathbf{W}^H\Lambda_y^{-1}\mathbf{F}_{\mathbf{e}}^H$. The beamformer $\mathbf{F}_{\mathbf{e}}^H$ is defined in Eq. (5.10).
3. Perform the real-valued GEVD of $\hat{\mathbf{R}} = \text{Re}\,\{\hat{\mathbf{R}}_y\}$ in the metric of $\mathbf{W}^H|\Lambda|^{-2}\mathbf{W}$. Obtain an estimate $\hat{d}$ of the number of sources, and form the matrices $\hat{\mathbf{S}} = [\hat{s}_1, \ldots, \hat{s}_{\hat{d}}]$ and $\hat{\mathbf{G}} = [\hat{\mathbf{g}}_{\hat{d}+1}, \cdots, \hat{\mathbf{g}}_{M'}]$ by grouping the $\hat{d}$ "largest" and $M' - \hat{d}$ "smallest" generalized eigenvectors.
4. Employ the matrices $\hat{\mathbf{S}}$ and $\hat{\mathbf{G}}$ from the previous step in the algorithm summaries corresponding to UCA-RB-MUSIC and UCA-ESPRIT.

The UCA of Directional Elements

The developments to this point assumed that the UCA consisted of omnidirectional elements. We now consider the case where the UCA employs directional elements disposed such that circular symmetry is retained. Rahim and Davies [30] obtained expressions for the far-field patterns of UCAs of directional elements under phase mode excitation: They showed that the far-field pattern still has the same azimuthal variation $e^{jm\phi}$ as the excitation function. However, the amplitude of the phase mode in the far-field pattern is a sum of Bessel functions rather than just $J_m(\zeta)$. One of the advantages of using directional elements is that attractive azimuthal directional patterns can be synthesized over wide frequency ranges (over an octave). Consider azimuthal pattern synthesis in the array plane ($\theta = 90°$): With omnidirectional elements, the mode amplitude $J_m(2\pi r/\lambda)$ undergoes rapid variations and passes through nulls as λ is varied. This is not suitable for pattern synthesis, as nonzero mode amplitudes are required. With directional elements, the mode amplitude is a sum of Bessel functions, and is observed to be stable over wide frequency ranges. Another advantage of using directional elements is that mutual coupling effects (specifically diametrical coupling across the array) are mitigated in arrays of small radius.

Let $g(\theta, \phi)$ represent the directional response of an individual antenna element. The UCA of directional elements is characterized by the element space manifold.

$$\underline{\mathbf{a}}_d(\theta) = \mathbf{G}_\Theta \underline{\mathbf{a}}(\theta), \text{ where} \tag{5.52}$$
$$\mathbf{G}_\Theta = \operatorname{diag}\{g(\theta, \phi - \gamma_0), \ldots, g(\theta, \phi - \gamma_{N-1})\}$$

and $\gamma_i = 2\pi i/N$ is the angular position of the ith element. First consider the case where the element pattern is only elevation dependent (omnidirectional in azimuth). With the element pattern denoted $g(\theta)$, the corresponding UCA manifold is $\underline{\mathbf{a}}_d(\theta) = g(\theta)\underline{\mathbf{a}}(\theta)$, a scalar multiple of the omnidirectional UCA manifold. The element space data vector thus has the representation $\mathbf{x}(t) = \underline{\mathbf{A}}\mathbf{s}'(t) + \mathbf{n}(t)$, where $\mathbf{s}'(t) = \mathbf{G}_\Theta \mathbf{s}(t)$, and $\mathbf{G}_\Theta = \operatorname{diag}\{g(\theta_1), \ldots, g(\theta_d)\}$. It is evident that UCA-RB-MUSIC and UCA-ESPRIT are both applicable in this scenario—the only change is that the source covariance matrix $\mathbf{P}$ is replaced by $\mathbf{P}_d = \mathbf{G}_\Theta \mathbf{P} \mathbf{G}_\Theta^H$.

Now consider the general case where the element pattern is a function of both azimuth and elevation. Let $g(\phi; \theta) = \sum_{k=-p}^{p} c_\theta(k) e^{jk\phi}$ be the Fourier series expansion for the azimuthal variation of the element pattern at the elevation θ. The far-field pattern (ignoring residual terms) resulting from excitation of the UCA of directional elements with phase mode m is [30]

$$f_m^s(\theta) = \mathbf{w}_m^H \underline{\mathbf{a}}_d(\theta) \approx A_m(\theta) e^{jm\phi}, \text{ where} \tag{5.53}$$
$$A_m(\theta) = \sum_{k=-p}^{p} c_\theta(k) j^{m-k} J_{m-k}(k_0 r \sin\theta)$$

This equation is similar to Eq. (5.8) for the omnidirectional element case. The only difference is that the mode amplitude $A_m(\theta)$ involves a sum of Bessel functions, and is not just $J_m(\zeta)$. The mode amplitudes are easily shown to satisfy $A_{-m}(\theta) = A_m(\theta)$. An element pattern suggested in [30] is $g(\theta, \phi) = 1 + \sin\theta\cos\phi$: The mode amplitudes corresponding to this pattern are $A_m(\theta) = j^m \left[J_m(\zeta) - j\sin\theta J_m'(\zeta)\right]$, where $\zeta = k_0 r \sin\theta$. From Eq. (5.53), it is evident that the beamformer $\mathbf{V}^H$ of Eq. (5.9) synthesizes the beamspace manifold

$$\mathbf{a}_d(\theta) = \mathbf{V}^H \underline{\mathbf{a}}_d(\theta) = \mathbf{J}_d(\theta)\mathbf{v}(\phi), \text{ where} \tag{5.54}$$
$$\mathbf{J}_d(\theta) = \operatorname{diag}\{A_M(\theta), \ldots, A_1(\theta), A_0(\theta), A_1(\theta), \ldots, A_M(\theta)\}$$

The azimuthal dependence of the beamspace manifold $\mathbf{a}_d(\theta)$ is through the vector $\mathbf{v}(\phi)$, as was the case with UCA-RB-MUSIC. The beamspace MUSIC algorithm (employing the beamformer $\mathbf{V}^H$) for the UCA of directional elements thus possesses most of the features of UCA-RB-MUSIC. The features that are lost are the ability to perform FB averaging, and to compute signal eigenvectors via a real-valued EVD. This is because the beamspace manifold $\mathbf{a}_d(\theta)$ is not centro-Hermitian. UCA-

ESPRIT cannot be employed when the elements have directional patterns that are functions of both azimuth and elevation. This is because the components of $J_d(\theta)$ are sums of Bessel functions, and the recursive Bessel function relationship cannot be employed to match the magnitude components of the subvectors of $\mathbf{a}_d(\boldsymbol{\theta})$.

5.4 PERFORMANCE ANALYSIS

The statistical performance of the element space MUSIC, UCA-RB-MUSIC, and UCA-ESPRIT algorithms is investigated in this section. Asymptotic expressions for the variances/covariances of the element space MUSIC estimators for 2D angle estimation are presented in this section. Such performance analysis results are available for the 1D angle estimation case [16]. However, the present work appears to be the first time such results have been made available for the case of 2D angle estimation. Next the section considers the performance of the UCA-RB-MUSIC algorithm. With UCA-RB-MUSIC, signal subspace estimates are obtained via real-valued EVDs, and the analysis differs from that of element space MUSIC in this respect. The final results, however, are similar in form to those for element space MUSIC. The statistical performance of UCA-ESPRIT is then investigated. The analysis is similar to that of the ESPRIT algorithm for 1D angle estimation [18]. The next part presents results on the Cramér-Rao bound (CRB) for the 2D angle estimation problem. Finally, the performance of the algorithms for the one- and two-source cases is investigated in some detail. It has been shown [10] that beamspace MUSIC estimators cannot perform better than the corresponding element space MUSIC estimators. However, FB averaging is possible in beamspace and not possible in element space when the number of array elements N is odd. UCA-RB-MUSIC can thus *outperform* element space MUSIC when N is odd. The theoretical performance curves presented demonstrate this property.

Before beginning the analysis, we restate some of the assumptions made, and introduce some notation. The number of incident signals d is assumed to be known. The signals $\mathbf{s}(t)$ and noises $\mathbf{n}(t)$ are assumed to be stationary, zero mean, uncorrelated random processes. The noise process $\mathbf{n}(t)$ is assumed to be complex Gaussian and spatially white with covariance matrix $\sigma^2\mathbf{I}$. The signals are assumed to be noncoherent, and the source covariance matrix $\mathbf{P}$ in thus positive definite. The number of snapshots of array data is K. The dimension of the element space UCA manifold is N, and the dimension of the real-valued beamspace manifold is M'. UCA-RB-MUSIC works with subspace estimates obtained via an EVD of the real matrix $\mathbf{R}$ of Eq. (5.27). The eigenvalues of $\mathbf{R}$ in descending order are $\{\lambda_i\}_{i=1}^{M'}$. The real, orthonormal matrices $\mathbf{S}$ and $\mathbf{G}$ that respectively span the beamspace signal and noise subspaces are defined in Eqs. (5.28) and (5.29). The same symbol is used to denote similar quantities in element space and beamspace; the element space quantities are distinguished by underbars, for example, $\mathbf{S}$ and $\underline{\mathbf{S}}$, respectively, span the beamspace and element space signal subspaces. Hats are used to denote estimated values of quantities, for example, $\hat{\mathbf{G}}$. In this section, subscripts are used to denote partial derivatives, for example, $\mathbf{b}_\zeta$ and $\mathbf{b}_{\zeta\phi}$, respectively, represent the first partial derivative of $\mathbf{b}$ with respect to ζ, and the mixed partial derivative with respect to ζ and ϕ.

Performance of MUSIC for 2D Angle Estimation

Theorem 5.1 gives asymptotic (large number of snapshots K) expressions for the variances and covariances of the element space MUSIC estimator for 2D angle estimation. The results of the theorem hold for arbitrary array configurations. The following lemma gives asymptotic expressions for the errors in the element space MUSIC arrival angle estimates. The proof of the lemma is based on a first-order Taylor series expansion of the MUSIC null spectrum about the true parameter values, and is similar to the proof of Lemma 5.3 presented below. The only difference is that the manifold vectors and subspace matrices are complex-valued in element space, whereas they are real-valued in beamspace.

Lemma 5.1. *The asymtotic expression for the element space MUSIC estimation error vector,* $\underline{\mathbf{e}}_i = [(\hat{\underline{\zeta}}_i - \zeta_i), (\hat{\underline{\phi}}_i - \phi_i)]^T$, *for source* i *is*

$$\underline{\mathbf{e}}_i = \left\{\underline{\mathbf{E}}^{-1}\underline{\mathbf{p}}\right\}_{\Theta=\Theta_i}, \quad where \tag{5.55}$$

$$\underline{\mathbf{E}} = \begin{bmatrix} \underline{\mathbf{a}}_\zeta^H \underline{\mathbf{G}}\underline{\mathbf{G}}^H \underline{\mathbf{a}}_\zeta & \mathrm{Re}\{\underline{\mathbf{a}}_\phi^H \underline{\mathbf{G}}\underline{\mathbf{G}}^H \underline{\mathbf{a}}_\zeta\} \\ \mathrm{Re}\{\underline{\mathbf{a}}_\phi^H \underline{\mathbf{G}}\underline{\mathbf{G}}^H \underline{\mathbf{a}}_\zeta\} & \underline{\mathbf{a}}_\phi^H \underline{\mathbf{G}}\underline{\mathbf{G}}^H \underline{\mathbf{a}}_\phi \end{bmatrix} = \begin{bmatrix} \underline{a} & \underline{c} \\ \underline{c} & \underline{b} \end{bmatrix}$$

is a symmetric, positive definite matrix with determinant $\underline{\mathbf{\Delta}}$. *The vector*

$$\underline{\mathbf{p}} = \begin{bmatrix} -\mathrm{Re}\{\underline{\mathbf{a}}^H \hat{\underline{\mathbf{G}}}\hat{\underline{\mathbf{G}}}^H \underline{\mathbf{a}}_\zeta\} \\ -\mathrm{Re}\{\underline{\mathbf{a}}^H \hat{\underline{\mathbf{G}}}\hat{\underline{\mathbf{G}}}^H \underline{\mathbf{a}}_\phi\} \end{bmatrix} = \begin{bmatrix} \underline{e} \\ \underline{f} \end{bmatrix}$$

is a random vector.

Although the MUSIC estimation errors are in terms of the matrix $\hat{\underline{\mathbf{G}}}$ that spans the estimated noise subspace, knowledge of the statistics of the signal space eigenvectors is sufficient to obtain expressions for the variances of the DOA estimators. The following lemma gives the well-known result [16] on the statistics of the signal space eigenvectors of the element space sample covariance matrix $\hat{\underline{\mathbf{R}}}$ that is complex Wishart distributed with K degrees of freedom.

Lemma 5.2. *The element space signal eigenvector estimation errors,* $(\hat{\underline{\mathbf{s}}}_i - \underline{\mathbf{s}}_i)$, *are asymptotically jointly Gaussian distributed with zero means. The error covariance matrices are given by*

$$E\left[(\hat{\underline{\mathbf{s}}}_i - \underline{\mathbf{s}}_i)(\hat{\underline{\mathbf{s}}}_j - \underline{\mathbf{s}}_j)^H\right] = \frac{\underline{\lambda}_i}{K}\left[\sum_{\substack{r=1 \\ r\neq i}}^{d} \frac{\underline{\lambda}_r}{(\underline{\lambda}_r - \underline{\lambda}_i)^2}\underline{\mathbf{s}}_r\underline{\mathbf{s}}_r^H + \sum_{r=d+1}^{N} \frac{\sigma^2}{(\underline{\lambda}_i - \sigma^2)^2}\underline{\mathbf{g}}_r\underline{\mathbf{g}}_r^H\right]\delta_{ij} \tag{5.56}$$

The following theorem gives expressions for the variances and covariance of the element space MUSIC arrival angle estimators $\hat{\underline{\zeta}}_i$ and $\hat{\underline{\phi}}_i$ corresponding to the

i^{th} source. The proof of theorem employs Lemmas 5.1 and 5.2, and is similar to the proof of Theorem 5.2 to follow.

Theorem 5.1. *The element space MUSIC estimation error vector* $\underline{\mathbf{e}}_{\rm i} = [(\hat{\underline{\zeta}}_{\rm i} - \zeta_{\rm i}), (\hat{\underline{\phi}}_{\rm i} - \phi_{\rm i})]^{\rm T}$ *for the* $\rm i^{th}$ *source is asymptotically zero mean with covariance matrix*

$$\mathrm{Cov}(\underline{\mathbf{e}}_i) = \begin{bmatrix} \mathrm{Var}(\hat{\underline{\zeta}}_i) & \mathrm{Cov}(\hat{\underline{\zeta}}_i, \hat{\underline{\phi}}_i) \\ \mathrm{Cov}(\hat{\underline{\zeta}}_i, \hat{\underline{\phi}}_i) & \mathrm{Var}(\hat{\underline{\phi}}_i) \end{bmatrix} = \frac{\sigma^2 \underline{\rho}}{2K\underline{\Delta}} \begin{bmatrix} \underline{b} & \underline{c} \\ \underline{c} & \underline{a} \end{bmatrix}_{\theta=\theta_i} \tag{5.57}$$

where $\underline{\mathrm{a}}$, $\underline{\mathrm{b}}$, $\underline{\mathrm{c}}$, *and* $\underline{\Delta}$ *are as defined in Lemma 5.1. Two expressions for the factor* $\underline{\rho}$ *follow. The latter expression is useful for analytical studies of performance.*

$$\begin{aligned} \underline{\rho}(\theta_i) &= \sum_{r=1}^{d} \frac{\underline{\lambda}_r}{(\underline{\lambda}_r - \sigma^2)^2} |\underline{\mathbf{a}}^H(\theta_i)\underline{\mathbf{s}}_r|^2 \\ &= \left[\mathbf{P}^{-1}\right]_{ii} + \sigma^2 \left[\mathbf{P}^{-1} \left(\underline{\mathbf{A}}^H \underline{\mathbf{A}}\right)^{-1} \mathbf{P}^{-1}\right]_{ii} \end{aligned} \tag{5.58}$$

Performance Analysis of UCA-RB-MUSIC

Theorem 5.2 gives asymptotic (large number of snapshots K) expressions for the variances and covariances of the UCA-RB-MUSIC estimators. To avoid double subscripts, the symbols $\mathbf{b}(\theta) = \mathbf{a_r}(\theta)$ and $\mathbf{B} = \mathbf{A_r}$ are used to respectively represent the beamspace manifold vector and the beamspace DOA matrix. The following lemma gives asymptotic expressions for the errors in the UCA-RB-MUSIC arrival angle estimates.

Lemma 5.3. *The asymptotic expression for the UCA-RB-MUSIC estimation error vector,* $\mathbf{e}_{\rm i} = [(\hat{\zeta}_{\rm i} - \zeta_{\rm i}), (\hat{\phi}_{\rm i} - \phi_{\rm i})]^{\rm T}$, *for source* i *is*

$$\mathbf{e}_i = \{\mathbf{E}^{-1}\mathbf{p}\}_{\theta=\theta_i}, \textit{ where} \tag{5.59}$$

$$\mathbf{E} = \begin{bmatrix} \mathbf{b}_\zeta^T \mathbf{G}\mathbf{G}^T \mathbf{b}_\zeta & \mathbf{b}_\phi^T \mathbf{G}\mathbf{G}^T \mathbf{b}_\zeta \\ \mathbf{b}_\phi^T \mathbf{G}\mathbf{G}^T \mathbf{b}_\zeta & \mathbf{b}_\phi^T \mathbf{G}\mathbf{G}^T \mathbf{b}_\phi \end{bmatrix} = \begin{bmatrix} a & c \\ c & b \end{bmatrix}$$

is a symmetric, positive definite matrix with determinant Δ. *The vector*

$$\mathbf{p} = \begin{bmatrix} -\mathbf{b}^T \hat{\mathbf{G}}\hat{\mathbf{G}}^T \mathbf{b}_\zeta \\ -\mathbf{b}^T \hat{\mathbf{G}}\hat{\mathbf{G}}^T \mathbf{b}_\phi \end{bmatrix} = \begin{bmatrix} e \\ f \end{bmatrix}$$

is a random vector.

Proof. The UCA-RB-MUSIC null spectrum is $\mathbf{V}(\theta) = \mathbf{b}^T(\theta)\hat{\mathbf{G}}\hat{\mathbf{G}}^T\mathbf{b}(\theta)$. The null spectrum has a local minimum at $\hat{\theta}_i = (\hat{\zeta}_i, \hat{\phi}_i)$ and we thus have $\mathbf{V}_\zeta(\hat{\theta}_i) = 0$, and $V_\phi(\hat{\theta}_i) = 0$. Now, $\hat{\theta}_i$ is a consistent estimator of θ_i, and a first-order Taylor series expansion yields the following:

$$0 = V_\zeta(\hat{\theta}_i) \approx V_\zeta(\theta_i) + V_{\zeta\zeta}(\theta_i)(\hat{\zeta}_i - \zeta_i) + V_{\zeta_\phi}(\theta_i)(\hat{\phi}_i - \phi_i)$$

$$0 = V_\phi(\hat{\theta}_i) \approx V_\phi(\theta_i) + V_{\phi\zeta}(\theta_i)(\hat{\zeta}_i - \zeta_i) + V_{\phi\phi}(\theta_i)(\hat{\phi}_i - \phi_i)$$

Putting these equations into matrix form, we obtain

$$\begin{bmatrix} V_{\zeta\zeta}(\theta_i) & V_{\zeta\phi}(\theta_i) \\ V_{\phi\zeta}(\theta_i) & V_{\phi\phi}(\theta_i) \end{bmatrix} \begin{bmatrix} (\hat{\zeta}_i - \zeta_i) \\ (\hat{\phi}_i - \phi_i) \end{bmatrix} = - \begin{bmatrix} V_\zeta(\theta_i) \\ V_\phi(\theta_i) \end{bmatrix} \tag{5.60}$$

The expansions for the derivatives occurring in this equation are as given below. Only terms which result in contributions of order $O(1/N)$ in Eq. (5.60) are retained.

$$\begin{aligned}
V_\zeta(\theta_i) &= 2\mathbf{b}^T(\theta_i)\hat{\mathbf{G}}\hat{\mathbf{G}}^T\mathbf{b}_\zeta(\theta_i) \\
V_\phi(\theta_i) &= 2\mathbf{b}^T(\theta_i)\hat{\mathbf{G}}\hat{\mathbf{G}}^T\mathbf{b}_\phi(\theta_i) \\
V_{\zeta\zeta}(\theta_i) &= 2\mathbf{b}_\zeta^T(\theta_i)\hat{\mathbf{G}}\hat{\mathbf{G}}^T\mathbf{b}_\zeta(\theta_i) + 2\mathbf{b}^T(\theta_i)\hat{\mathbf{G}}\hat{\mathbf{G}}^T\mathbf{b}_{\zeta\zeta}(\theta_i) \\
&\approx 2\mathbf{b}_\zeta^T(\theta_i)\mathbf{G}\mathbf{G}^T\mathbf{b}_\zeta(\theta_i) \\
V_{\phi\phi}(\theta_i) &= 2\mathbf{b}_\phi^T(\theta_i)\hat{\mathbf{G}}\hat{\mathbf{G}}^T\mathbf{b}_\phi(\theta_i) + 2\mathbf{b}^T(\theta_i)\hat{\mathbf{G}}\hat{\mathbf{G}}^T\mathbf{b}_{\phi\phi}(\theta_i) \\
&\approx 2\mathbf{b}_\phi^T(\theta_i)\mathbf{G}\mathbf{G}^T\mathbf{b}_\phi(\theta_i) \\
V_{\zeta\phi}(\theta_i) &= 2\mathbf{b}_\phi^T(\theta_i)\hat{\mathbf{G}}\hat{\mathbf{G}}^T\mathbf{b}_\zeta(\theta_i) + 2\mathbf{b}^T(\theta_i)\hat{\mathbf{G}}\hat{\mathbf{G}}^T\mathbf{b}_{\zeta\phi}(\theta_i) = V_{\phi\zeta}(\theta_i) \\
&\approx 2\mathbf{b}_\phi^T(\theta_i)\mathbf{G}\mathbf{G}^T\mathbf{b}_\zeta(\theta_i)
\end{aligned}$$

Substituting these expressions back into Eq. (5.60) and dropping the common factor of two leads to the desired result $\mathbf{E}\mathbf{e}_i = \mathbf{p}$, where $\mathbf{E}$, $\mathbf{e}_i$, and $\mathbf{p}$ are as defined in the lemma. Positive definiteness and hence nonsingularity of $\mathbf{E}$ follow from the Cauchy-Schwarz inequality. ■

UCA-RB-MUSIC works with subspace estimates obtained from the real matrix $\hat{\mathbf{R}} = \text{Re}\,\{\hat{\mathbf{R}}_y\}$ that is derived from an FB averaged covariance matrix. The statistics of the signal space eigenvectors of $\hat{\mathbf{R}}$ are required for the analysis of UCA-RB-MUSIC. The following lemma drawn from [31] gives these statistics.

Lemma 5.4. *The real beamspace signal eigenvector estimation errors,* $(\hat{\mathbf{s}}_i - \mathbf{s}_i)$, *are asymptotically (large* K*) zero mean with covariance matrices given by*

$$E\left[(\hat{\mathbf{s}}_i - \mathbf{s}_i)(\hat{\mathbf{s}}_j - \mathbf{s}_j)^T\right] = \frac{1}{K}\left[\sum_{\substack{r=1\\r\neq i}}^{d}\sum_{\substack{s=1\\s\neq j}}^{d}\frac{\Gamma_{rsji}}{(\lambda_i - \lambda_r)(\lambda_j - \lambda_s)}\mathbf{s}_r\mathbf{s}_s^T + \delta_{ij}\sum_{r=d+1}^{M'}\frac{\lambda_i\sigma^2}{2(\lambda_i - \sigma^2)^2}\mathbf{g}_r\mathbf{g}_r^T\right] \tag{5.61}$$

where

$$\Gamma_{rsji} = \frac{1}{2}\left\{\lambda_i\lambda_s\delta_{ij}\delta_{rs} + \lambda_i\lambda_j\delta_{is}\delta_{jr} + \mathbf{w}_r^T(\mathbf{s}_s\mathbf{s}_j^T + \mathbf{s}_j\mathbf{s}_s^T)\mathbf{w}_i\right\}, \text{ and } \mathbf{w}_i = Im\left\{\mathbf{R}_\mathbf{y}\right\}\mathbf{s}_i$$

The following theorem gives expressions for the variances and covariance of the UCA-RB-MUSIC arrival angle estimators $\hat{\zeta}_i$ and $\hat{\phi}_i$ corresponding to the ith source. The results are similar in form to those of Theorem 5.1 corresponding to the element space case. However, due to the inherent FB average, the results depend only on the real part $\mathbf{P}_R$ of the source covariance matrix $\mathbf{P}$. The decorrelating effect of the FB average allows UCA-RB-MUSIC to outperform element space MUSIC in correlated source scenarios when N is odd.

Theorem 5.2. *The UCA-RB-MUSIC estimation error vector* $\mathbf{e}_i = [(\hat{\zeta}_i - \zeta_i), (\hat{\phi}_i - \phi_i)]^T$ *for the* i^{th} *source is asymptotically zero mean with covariance matrix*

$$\text{Cov}(\mathbf{e}_i) = \begin{bmatrix} \text{Var}(\hat{\zeta}_i) & \text{Cov}(\hat{\zeta}_i, \hat{\phi}_i) \\ \text{Cov}(\hat{\zeta}_i, \hat{\phi}_i) & \text{Var}(\hat{\phi}_i) \end{bmatrix} = \frac{\sigma^2\rho}{2K\Delta}\begin{bmatrix} b & c \\ c & a \end{bmatrix}_{\theta=\theta_i} \tag{5.62}$$

where a, b, c, *and* Δ *are as defined in Lemma 5.3. Two expressions for the factor* ρ *follow. In the latter expression, which is useful for analytical studies of performance,* $\mathbf{P}_R = \text{Re}\{\mathbf{P}\}$.

$$\begin{aligned} \rho(\theta_i) &= \sum_{r=1}^{d} \frac{\lambda_r}{(\lambda_r - \sigma^2)^2}|\mathbf{b}^T(\theta_i)\mathbf{s}_r|^2 \\ &= \left[\mathbf{P}_R^{-1}\right]_{ii} + \sigma^2\left[\mathbf{P}_R^{-1}\left(\mathbf{B}^T\mathbf{B}\right)^{-1}\mathbf{P}_R^{-1}\right]_{ii} \end{aligned}$$

Proof. The MUSIC estimation error vector as given by Lemma 5.3 is

$$\mathbf{e}_i = \begin{bmatrix} \hat{\zeta}_i - \zeta_i \\ \hat{\phi}_i - \phi_i \end{bmatrix} = \left\{\mathbf{E}^{-1}\mathbf{p}\right\}_{\theta=\theta_i} = \frac{1}{\Delta}\begin{bmatrix} b & -c \\ -c & a \end{bmatrix}\begin{bmatrix} e \\ f \end{bmatrix}$$

We proceed to derive the expression for the variance of the estimator $\hat{\zeta}_i$. The remaining results can be obtained in similar fashion. The above equation yields

$$\hat{\zeta}_i - \zeta_i = \frac{b'}{\Delta}, \text{ where } b' = be - cf \tag{5.63}$$

As shown in [16], we have $\mathbf{b}^T(\theta_i)\hat{\mathbf{G}}\hat{\mathbf{G}}^T \approx -\mathbf{b}^T(\theta_i)\mathbf{S}\hat{\mathbf{S}}^T\mathbf{G}\mathbf{G}^T$. This result leads to the following expressions for the random quantities e and f in terms of the estimated signal space eigenvectors, whose statistics are available. The dependence of the expressions on θ_i is dropped for conciseness.

$$e = -\mathbf{b}^T\hat{\mathbf{G}}\hat{\mathbf{G}}^T\mathbf{b}_\zeta \approx \mathbf{b}^T\mathbf{S}\hat{\mathbf{S}}^T\mathbf{G}\mathbf{G}^T\mathbf{b}_\zeta = \mathbf{b}_\zeta^T\mathbf{G}\mathbf{G}^T\hat{\mathbf{S}}\mathbf{S}^T\mathbf{b}$$

$$f = -\mathbf{b}^T\hat{\mathbf{G}}\hat{\mathbf{G}}^T\mathbf{b}_\phi \approx \mathbf{b}^T\mathbf{S}\hat{\mathbf{S}}^T\mathbf{G}\mathbf{G}^T\mathbf{b}_\phi = \mathbf{b}_\phi^T\mathbf{G}\mathbf{G}^T\hat{\mathbf{S}}\mathbf{S}^T\mathbf{b}$$

Substituting in Eq. (5.63) and using the definitions in Lemma 5.3 leads to the following expression for the term b':

$$b' = \mathbf{q}^T \mathbf{z} \tag{5.64}$$

where $\mathbf{q} = (b\mathbf{b}_\zeta - c\mathbf{b}_\phi)$ is a deterministic quantity, and $\mathbf{z} = \mathbf{G}\mathbf{G}^T \hat{\mathbf{S}}\mathbf{S}^T \mathbf{b}$ is a random vector. The vector $\mathbf{z}$ will shortly be shown to have the following statistics:

$$E(\mathbf{z}) = 0, \text{ and} \tag{5.65}$$

$$\text{Cov}(\mathbf{z}) = E\left(\mathbf{z}\mathbf{z}^T\right) = \frac{\sigma^2 \rho}{2K} \mathbf{G}\mathbf{G}^T$$

where $\rho(\theta_i)$ is defined in the theorem statement. Equation (5.64) now yields $E(b') = 0$, and Var $(b') = \frac{\sigma^2 \rho}{2K} \mathbf{q}^T \mathbf{G}\mathbf{G}^T \mathbf{q} = \frac{\sigma^2 \rho b \Delta}{2K}$. The final equality results because $\mathbf{q}^T \mathbf{G}\mathbf{G}^T \mathbf{q} = b\Delta$, a relationship which is easily verified. Employing these results in Eq. (5.63) completes the proof: We obtain $E(\hat{\zeta}_i - \zeta_i) = 0$, and Var $(\hat{\zeta}_i) = \frac{\sigma^2}{2K} \left\{ \frac{\rho b}{\Delta} \right\}_{\theta = \theta_i}$. The proof for the expression of $\rho(\theta_i)$ in terms of $\mathbf{P}_R = \text{Re}\{\mathbf{P}\}$ follows a similar proof in [16].

It now remains to verify the expressions for the statistics of the vector $\mathbf{z}$. We have

$$\mathbf{z} = \mathbf{G}\mathbf{G}^T \hat{\mathbf{S}}\mathbf{S}^T \mathbf{b} = \sum_{k=1}^{d} \mathbf{G}\mathbf{G}^T \hat{\mathbf{s}}_k (\mathbf{s}_k^T \mathbf{b}) = \sum_{k=1}^{d} (\mathbf{b}^T \mathbf{s}_k) \mathbf{G}\mathbf{G}^T (\hat{\mathbf{s}}_k - \mathbf{s}_k) \tag{5.66}$$

We have $E(\hat{\mathbf{s}}_k - \mathbf{s}_k) = 0$ from Lemma 5.4, and thus $E(\mathbf{z}) = 0$ as claimed. Now

$$\text{Cov}(\mathbf{z}) = E\left[\mathbf{z}\mathbf{z}^T\right] = \sum_{k=1}^{d} \sum_{l=1}^{d} (\mathbf{b}^T \mathbf{s}_k)(\mathbf{b}^T \mathbf{s}_l) \mathbf{G}\mathbf{G}^T E\left[(\hat{\mathbf{s}}_k - \mathbf{s}_k)(\hat{\mathbf{s}}_l - \mathbf{s}_l)^T\right] \mathbf{G}\mathbf{G}^T$$

Using the result of Lemma 5.4 on the signal eigenvector statistics, we obtain

$$\begin{aligned}
\text{Cov}(\mathbf{z}) &= \frac{1}{K} \sum_{k=1}^{d} \sum_{l=1}^{d} (\mathbf{b}^T \mathbf{s}_k)(\mathbf{b}^T \mathbf{s}_l) \frac{\lambda_k \sigma^2 \delta_{kl}}{2(\lambda_k - \sigma^2)^2} \sum_{r=d+1}^{N'} \mathbf{g}_r \mathbf{g}_r^T \\
&= \frac{\sigma^2}{2K} \left(\sum_{k=1}^{d} \frac{\lambda_k}{(\lambda_k - \sigma^2)^2} |\mathbf{b}^T(\theta_i)\mathbf{s}_k|^2 \right) \mathbf{G}\mathbf{G}^T \\
&= \frac{\sigma^2 \rho(\theta_i)}{2K} \mathbf{G}\mathbf{G}^T
\end{aligned}$$

as claimed in Eq. (5.65). ■

Performance Analysis of UCA-ESPRIT

The performance analysis of UCA-ESPRIT is drawn from [17]. The development employs techniques similar to those used in [18] to analyze the performance of the ESPRIT algorithm for 1D angle estimation. As described in section 5.3, the eigenvalues of the matrix $\hat{\boldsymbol{\Psi}}$ have the form $\hat{\mu}_i = \sin\hat{\theta}_i e^{j\hat{\phi}_i} = \hat{u}_i + j\hat{v}_i$ and provide automatically paired source DOA estimates. Theorem 5.3 gives asymptotic expressions for the variances and covariance of the UCA-ESPRIT direction cosine estimators $\hat{u}_i$ and $\hat{v}_i$. The theorem also gives approximate expressions (accurate at moderate to high SNRs) for the variances of the estimators $\hat{\zeta}_i' = \sin\hat{\theta}_i$ and $\hat{\phi}_i$. The asymptotic variance expressions in Theorem 5.3 involve the eigenvectors of $\mathbf{\Psi}$. Equation (5.40) gives the spectral decomposition of $\mathbf{\Psi}$: We have $\mathbf{\Psi} = \mathbf{T}^{-1}\mathbf{\Phi}\mathbf{T}$, where $\mathbf{T}$ is *real-valued.* The left and right eigenvectors of $\mathbf{\Psi}$ are thus real valued, and are denoted $\mathbf{q}_i^T$ and $\mathbf{x}_i$, respectively. The UCA-ESPRIT algorithm incorporates beamspace signal subspace computation via a real-valued EVD: We have $\mathbf{S}_u = \mathbf{C}_o\mathbf{W}\mathbf{S}$, where $\mathbf{S}_u$ spans the UCA-ESPRIT signal subspace, and $\mathbf{S}$ is obtained via the EVD of the real-valued matrix $\mathbf{R}$ of Eq. (5.27). The analysis of UCA-ESPRIT thus requires the statistics of the signal eigenvector estimates $\hat{\mathbf{s}}_i$ that form the columns of $\hat{\mathbf{S}}$. Lemma 5.4 gives these statistics. A superscript e is used to denote the error in an estimate in the following developments, for example, $\mathbf{s}_i^e = \hat{s}_i - s_i$ is the error in the ith signal eigenvector estimate. The superscript $+$ is used to denote the Moore-Penrose pseudo-inverse, for example, $\mathbf{E}^+ = (\mathbf{E}^H\mathbf{E})^{-1}\mathbf{E}^H$ is the pseudo-inverse of $\mathbf{E}$.

Theorem 5.3. *The UCA-ESPRIT direction cosine estimators $\hat{u}_i$ and $\hat{v}_i$ are asymptotically unbiased. Asymptotic (large K) expressions for the variances and covariance of these estimators are given below:*

$$\begin{aligned}\text{Var}(\hat{u}_i) &= \boldsymbol{\alpha}_{iR}^T\mathbf{H}_i\boldsymbol{\alpha}_{iR} \\ \text{Var}(\hat{v}_i) &= \boldsymbol{\alpha}_{iI}^T\mathbf{H}_i\boldsymbol{\alpha}_{iI} \\ \text{Cov}(\hat{u}_i, \hat{v}_i) &= \boldsymbol{\alpha}_{iR}^T\mathbf{H}_i\boldsymbol{\alpha}_{iI}\end{aligned} \tag{5.67}$$

The matrices $\mathbf{H}_i$ and the vectors $\boldsymbol{\alpha}_i = \boldsymbol{\alpha}_{iR} + j\boldsymbol{\alpha}_{iI}$ are defined as follows:

$$\mathbf{H}_i = \sum_{j=1}^{d}\sum_{k=1}^{d}\mathbf{x}_{ij}\mathbf{x}_{ik}\,\text{Cov}(\mathbf{s}_j^e, \mathbf{s}_k^e) \tag{5.68}$$

$$\boldsymbol{\alpha}_i^T = \mathbf{q}_i^T\mathbf{E}^{+t}\left[\mathbf{\Gamma}\mathbf{\Delta}_0\mathbf{C}_o\mathbf{W} - \mu_i\mathbf{\Delta}_{-1}\mathbf{C}_o\mathbf{W} - \mu_i^*\mathbf{D}\tilde{\mathbf{I}}\mathbf{\Delta}_{-1}\mathbf{C}_o\mathbf{W}^*\right]$$

where $\mathbf{E}^{+t} = [\mathbf{I}_d \vdots \mathbf{O}_{d\times d}]\mathbf{E}^+$. The matrix $\mathbf{E} = [\mathbf{S}_{-1} \vdots \mathbf{D}\tilde{\mathbf{I}}\mathbf{S}_{-1}^*]$ is formed using the true signal space eigenvectors. The following approximate expressions for the asymptotic variances of the estimators $\hat{\zeta}_i' = \sin\hat{\theta}_i$ and $\hat{\phi}_i$ are accurate at moderate to high SNRs.

$$\text{Var}(\hat{\zeta}_i') \approx \text{Var}(\hat{u}_i)\cos^2\phi_i + \text{Var}(\hat{v}_i)\sin^2\phi_i + \text{Cov}(\hat{u}_i, \hat{v}_i)\sin 2\phi_i \tag{5.69}$$

$$\text{Var}(\hat{\phi}_i) \approx \frac{1}{(\zeta_i')^2}\left[\text{Var}(\hat{v}_i)\cos^2\phi_i + \text{Var}(\hat{u}_i)\sin^2\phi_i - \text{Cov}(\hat{u}_i, \hat{v}_i)\sin 2\phi_i\right] \tag{5.70}$$

Proof. To a first-order approximation, the UCA-ESPRIT eigenvalue error due to errors in subspace estimates is

$$\mu_i^e = \mathbf{q}_i^T\mathbf{\Psi}^e\mathbf{x}_i \tag{5.71}$$

The least squares solution to the overdetermined system $\hat{\mathbf{E}}\underline{\mathbf{\Psi}} = \mathbf{\Gamma}\hat{\mathbf{S}}_0$ of Eq. (5.42) yields the estimate $\hat{\underline{\mathbf{\Psi}}} = [\hat{\mathbf{\Psi}}^T \vdots \hat{\mathbf{\Psi}}^H]^T$. This system can be rewritten as $(\mathbf{E}+\mathbf{E}^e)(\underline{\mathbf{\Psi}} +$

$\underline{\boldsymbol{\Psi}}^e) = \boldsymbol{\Gamma}(\mathbf{S}_0 + \mathbf{S}_0^e)$. Using the fact that $E\underline{\boldsymbol{\Psi}} = \boldsymbol{\Gamma}\mathbf{S}_0$, and retaining only first-order terms, we obtain $\mathbf{E}\underline{\boldsymbol{\Psi}}^e = \boldsymbol{\Gamma}\mathbf{S}_0^e - \mathbf{E}^e\underline{\boldsymbol{\Psi}}$. Employing the least squares solution yields the equation $\underline{\boldsymbol{\Psi}}^e = \mathbf{E}^+(\boldsymbol{\Gamma}\mathbf{S}_0^e - \mathbf{E}^e\underline{\boldsymbol{\Psi}})$. This solution is adequate in that it yields expressions for DOA estimator variance that are accurate to $o(K^{-1})$ [18]. Now $\boldsymbol{\Psi}^e = [\mathbf{I}_d \vdots \mathbf{O}_{d\times d}]\underline{\boldsymbol{\Psi}}^e$ and thus

$$\boldsymbol{\Psi}^e = \mathbf{E}^{+t}\left[\boldsymbol{\Gamma}\mathbf{S}_0^e - \mathbf{E}^e\underline{\boldsymbol{\Psi}}\right] \tag{5.72}$$

where $\mathbf{E}^{+t} = [\mathbf{I}_d \vdots \mathbf{O}_{d\times d}]\mathbf{E}^+$. We have $\mathbf{E}^e = [\mathbf{S}_{-1}^e \vdots \mathbf{D}\tilde{\mathbf{I}}\mathbf{S}_{-1}^{e^*}]$, where $\mathbf{S}_i^e = \boldsymbol{\Delta}_i\mathbf{C}_o\mathbf{W}\mathbf{S}^e$, $i = -1, 0$. Substituting in Eq. (5.72) leads to the following expression:

$$\boldsymbol{\Psi}^e = \mathbf{E}^{+t}\left[\boldsymbol{\Gamma}\boldsymbol{\Delta}_0\mathbf{C}_o\mathbf{W}\mathbf{S}^e - \boldsymbol{\Delta}_{-1}\mathbf{C}_o\mathbf{W}\mathbf{S}^e\boldsymbol{\Psi} - \mathbf{D}\tilde{\mathbf{I}}\boldsymbol{\Delta}_{-1}\mathbf{C}_o\mathbf{W}^*\mathbf{S}^e\boldsymbol{\Psi}^*\right] \tag{5.73}$$

Substituting (5.73) in (5.71) and using the fact that $\mathbf{x}_i$ is a *real-valued* eigenvector of $\boldsymbol{\Psi}$, we obtain the equation

$$\mu_i^e = u_i^e + jv_i^e = \boldsymbol{\alpha}_i^T\mathbf{S}^e\mathbf{x}_i \tag{5.74}$$

where $\boldsymbol{\alpha}_i^T = \boldsymbol{\alpha}_{iR}^T + j\boldsymbol{\alpha}_{iI}^T = \mathbf{q}_i^T\mathbf{E}^{+t}[\boldsymbol{\Gamma}\boldsymbol{\Delta}_0\mathbf{C}_o\mathbf{W} - \mu_i\boldsymbol{\Delta}_{-1}\mathbf{C}_o\mathbf{W} - \mu_i^*\mathbf{D}\tilde{\mathbf{I}}\boldsymbol{\Delta}_{-1}\mathbf{C}_o\mathbf{W}^*]$ is a complex-valued vector. Both $\mathbf{S}^e = [\mathbf{s}_1^e \vdots \cdots \vdots \mathbf{s}_d^e]$ and $\mathbf{x}_i$ are real-valued. From Eq. (5.74), we have

$$u_i^e = \boldsymbol{\alpha}_{iR}^T\mathbf{S}^e\mathbf{x}_i, \text{ and } v_i^e = \boldsymbol{\alpha}_{iI}^T\mathbf{S}^e\mathbf{x}_i \tag{5.75}$$

It is now evident from Lemma 5.4 that $\mathrm{E}(u_i^e) = \mathrm{E}(v_i^e) = 0$. The UCA-ESPRIT direction cosine estimates are thus asymptotically unbiased as claimed. Equations (5.67) and (5.68) that define the variances of the direction cosine estimators follow on straightforward application of the results of Lemma 5.4 in Eq. (5.75).

It now remains to verify Eqs. (5.69) and (5.70) that respectively give asymptotic expressions for the variances of the estimators $\hat{\zeta}_i'$ and $\hat{\phi}_i$. These expressions follow from the geometry depicted in Figure 5.3. The subscript i that denotes the ith source is dropped for notational expedience. Let the UCA-ESPRIT eigenvalue error be $\mu^e = |\mu^e|e^{j\beta}$. The errors in the direction cosine estimates are thus $u^e = |\mu^e|\cos\beta$, and $v^e = |\mu^e|\sin\beta$. From Figure 5.3 we see that $l_1 = |\mu^e|\cos(\beta - \phi) \approx \zeta'^e$. We thus obtain $\zeta'^e \approx u^e\cos\phi + v^e\sin\phi$. This leads to the expression in (5.69) as $\mathrm{E}(\zeta'^e) = 0$, and $\mathrm{Var}(\hat{\zeta}') = \mathrm{Var}(\zeta'^e)$. Figure 5.3 also shows that $l_2 = |\mu^e|\sin(\beta - \phi) \approx \zeta'\phi^e$. Thus $\phi^e \approx \frac{1}{\zeta'}[v^e\cos\phi - u^e\sin\phi]$ and the expression in (5.70) follows as $\mathrm{E}(\phi^e) = 0$, and $\mathrm{Var}(\hat{\phi}) = \mathrm{Var}(\phi^e)$. ■

The approximations made in obtaining (5.69) and (5.70) are accurate provided the eigenvalue error μ^e is small; the expressions are thus accurate for moderate to high SNRs. Note that $\mathrm{Var}(\hat{\phi}) \propto 1/(\sin^2\theta)$: The variance of the azimuth estimator thus increases as the elevation angle θ decreases. This is an intuitive result; we know that azimuth is not a good descriptor of source DOA when θ is small (in fact all azimuth angles are equivalent when $\theta = 0°$).

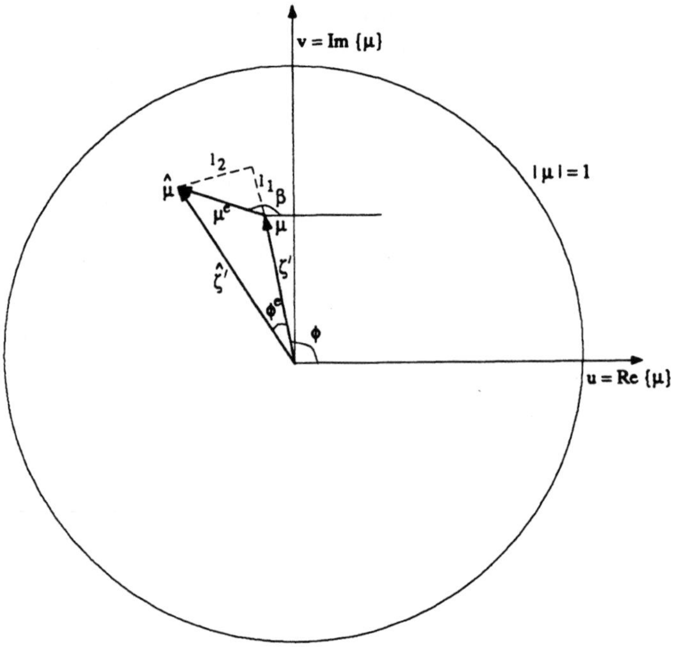

Figure 5.3 UCA-ESPRIT eigenvalue error.

The Cramér-Rao Bound

It is instructive to compare the performance of the element space MUSIC, UCA-RB-MUSIC, and UCA-ESPRIT estimators with the ultimate performance dictated by the Cramér-Rao bound (CRB). The following lemma gives the CRB on the covariance matrix of unbiased estimators of the parameter vector $\boldsymbol{\Theta} = [\zeta_1, \ldots, \zeta_d, \phi_1, \ldots, \phi_d]^T$. The CRB expression below is based on a random signal model, and is known as the unconditional, or stochastic, CRB. The result is a generalization of a similar result in [32] for the 1D angle estimation problem. The symbol $\odot$ is used to denote the Hadamard or element-wise matrix product. The reader is referred to Chapter 2 for further discussions on the CRB for 2D arrays.

Lemma 5.5. *The stochastic CRB for any unbiased estimator of* $\boldsymbol{\Theta}$ *is*

$$\mathrm{CRB}(\boldsymbol{\Theta}) = \frac{\sigma^2}{2K}\left[Re\left\{\mathbf{H} \odot \mathbf{P}_+^T\right\}\right]^{-1}, \ where \tag{5.76}$$

$$\mathbf{P}_+ = \begin{bmatrix} \mathbf{P}' & \mathbf{P}' \\ \mathbf{P}' & \mathbf{P}' \end{bmatrix} \ with\ \mathbf{P}' = \mathbf{P}\underline{\mathbf{A}}^H\underline{\mathbf{R}}^{-1}\underline{\mathbf{A}}\mathbf{P}$$

$$\mathbf{H} = \mathbf{D}^H\left[I - \underline{\mathbf{A}}\left(\underline{\mathbf{A}}^H\underline{\mathbf{A}}\right)^{-1}\underline{\mathbf{A}}^H\right]\mathbf{D},\ and$$

$$\mathbf{D} = \left[\underline{\mathbf{a}}_\zeta(\theta_1), \ldots, \underline{\mathbf{a}}_\zeta(\theta_d), \underline{\mathbf{a}}_\phi(\theta_1), \ldots, \underline{\mathbf{a}}_\phi(\theta_d)\right]$$

We point out that the expression (valid when K is large) for the conditional, or deterministic, CRB can be obtained by making the substitution $\mathbf{P}' = \mathbf{P}$ in the lemma. Due to the simpler expression for $\mathbf{P}_+$, the deterministic CRB is employed in the theoretical performance study presented next. The study is meaningful as the deterministic CRB is a tighter bound than the stochastic CRB. The stochastic CRB is however used as the benchmark for comparison in the simulations of section 5.5. This is because our developments have assumed a random signal model.

Study of Theoretical Performance for the One- and Two-Source Cases

This section investigates the theoretical performance of the element space MUSIC, UCA-RB-MUSIC, and UCA-ESPRIT estimators for the one- and two-source cases. The behavior of the deterministic CRB is also studied. The study focuses on the behavior of the direction cosine estimators $\hat{u}$ and $\hat{v}$ rather than the behavior of $\hat{\zeta}$ and $\hat{\phi}$. This is because the variance of the azimuth estimator $\hat{\phi}$ increases as θ decreases. Further, all values of ϕ are equivalent when the elevation $\theta = 0°$. There is no such ambiguity in the direction cosine space; we have $u = v = 0$. The vector of direction cosines, $\boldsymbol{\beta} = (u, v)$ is thus used to represent the source DOAs. The results of Theorems 5.1 and 5.2 are easily modified to give the variances of the direction cosine estimators. All that is required is to replace the subscripts ζ and ϕ denoting partial derivatives by the subscripts u and v, respectively. The same substitution in Lemma 5.5 gives the CRB expressions for the direction cosine estimators.

The main results of this theoretical performance study are as follows: (a) The deterministic CRB is independent of the source DOA in the single source scenario. For the two-source case, the CRB depends only on the distance τ between the two-source locations in the uv plane, and their relative orientation as specified by the angle ν of the line joining these locations. (b) Closed-form expressions for the element space MUSIC estimator variances are obtained for both the one- and two-source cases. The element space MUSIC estimator variances exhibit the same behavior as the CRB for these cases. (c) It is shown via a study of theoretical performance curves that the DOA dependence of the UCA-RB-MUSIC estimator variances closely follows the behavior of the CRB and element space MUSIC. The performance curves also demonstrate that UCA-RB-MUSIC can outperform element space MUSIC in correlated source scenarios when N is odd. (d) The UCA-ESPRIT estimator variances for the single-source case are independent of azimuth at low elevation angles ($\sin\theta < 0.6$). At higher elevation angles, the estimator variances become azimuth dependent.

The Deterministic CRB. Consider the case of a single source of power $p = \mathrm{E}\,|s(n)|^2$ incident on the UCA from the direction $\beta = (u, v)$. Appendix 5B.2 shows that the deterministic CRB for unbiased estimators of u and v is

$$\mathrm{CRB}(\hat{u}) = \mathrm{CRB}(\hat{v}) = \frac{1}{KN(k_0 r)^2(p/\sigma^2)} \tag{5.77}$$

where p/σ^2 is the signal-to-noise ratio. The CRBs for u and v are identical, and are independent of the arrival angle. The UCA thus favors all arrival angles equally. Not all array configurations have this desirable property; the rectangular array, for example, does not.

Let $\boldsymbol{\beta}_1$ and $\boldsymbol{\beta}_2$ specify the source DOAs for the two-source case. Let the difference vector $\boldsymbol{\beta}_d = \boldsymbol{\beta}_2 - \boldsymbol{\beta}_1$ have the representation $\boldsymbol{\beta}_d = \tau e^{j\nu}$ in polar coordinates. τ is the distance between the two-sources in the uv plane, and ν is the angle of the line joining the two sources. Appendix 5B.2 shows that the dependence of the CRBs on the source DOAs is only through the vector $\boldsymbol{\beta}_d$, or equivalently, through the parameters τ and ν. Thus, the CRBs do not depend on the absolute positions of the sources but only on their positions relative to each other.

Performance of Element Space MUSIC. The variances of the element space MUSIC estimators for the single-source case are shown in Appendix 5B.1 to be

$$\text{Var}(\hat{u}) = \text{Var}(\hat{v}) = \frac{1 + 1/N}{KN(k_0 r)^2 (p/\sigma^2)} \tag{5.78}$$

The u and v estimator variances are equal and independent of the source DOA. Comparison with Eq. (5.77) shows that the element space MUSIC estimators are asymptotically efficient in the single-source scenario.

Theorem 5.1 gives expressions for the element space MUSIC estimator variances. We have $\text{Var}(\hat{u}_i) = (\sigma^2 \underline{\rho}\underline{b}/2K\underline{\Delta})|_{\boldsymbol{\beta}=\boldsymbol{\beta}_i}$, and $\text{Var}(\hat{v}_i) = (\sigma^2 \underline{\rho}\underline{a}/2K\underline{\Delta})|_{\boldsymbol{\beta}=\boldsymbol{\beta}_i}$. Simplified expressions for the parameters $\underline{a}$[1], $\underline{b}$, and $\underline{c}$ can be obtained for the two-source case; Appendix 5B.1 outlines the derivation of the following results:

$$\underline{a}(\boldsymbol{\beta}_1) = \underline{a}(\boldsymbol{\beta}_2) \approx (k_0 r)^2 \left[\frac{N}{2} - q\,\{N J_1(k_0 r \tau)\cos\nu\}^2\right] \tag{5.79}$$

$$\underline{b}(\boldsymbol{\beta}_1) = \underline{b}(\boldsymbol{\beta}_2) \approx (k_0 r)^2 \left[\frac{N}{2} - q\,\{N J_1(k_0 r \tau)\sin\nu\}^2\right]$$

$$\underline{c}(\boldsymbol{\beta}_1) = \underline{c}(\boldsymbol{\beta}_2) \approx -(k_0 r)^2 q\,\{N J_1(k_0 r \tau)\}^2 \sin\nu\cos\nu$$

$$\underline{\rho}(\boldsymbol{\beta}) \approx f(\tau, \mathbf{P})$$

In the above equations, $q = \frac{1}{N[1 - J_0^2(k_0 r \tau)]}$. The final equation above signifies that the quantity $\underline{\rho}$ is a function of τ and the source covariance matrix $\mathbf{P}$. Equation (5.79) shows that $\underline{a}$, $\underline{b}$, $\underline{c}$, and $\underline{\rho}$ depend on the source DOAs only through the parameters τ and ν. Thus, the element space MUSIC estimator variances depend only on τ and ν: this behavior is similar to that of the CRBs. Equation (5.79) also shows that the parameters $\underline{a}$, $\underline{b}$, and $\underline{c}$ are the same for the two-sources. The parameter $\underline{\rho}$ is also the same for both sources, provided they are equipowered. Thus, with equipowered sources, the u and v estimator variances are the same for both sources.

It can be verified that the variances of the u estimates are highest when the angle $\nu = 0$, that is, when the sources have different u coordinates but the same v coordinate. Similarly, the variances of the v estimates are highest when $\nu = \pi/2$. Now, $\underline{c} = 0$ when $\nu = 0$ or $\pi/2$ and hence $\underline{\boldsymbol{\Delta}} = \underline{a}\underline{b}$. The estimator

[1] The scalar $\underline{a}$ is distinct from the UCA manifold vector $\underline{\mathbf{a}}$.

variances corresponding to these orientations are thus given by $\mathrm{Var}(\hat{u}) = \frac{\sigma^2 \underline{\rho}}{2K\underline{a}}$, and $\mathrm{Var}(\hat{v}) = \frac{\sigma^2 \underline{\rho}}{2K\underline{b}}$. The minimum values that $\underline{a}$ and $\underline{b}$ take are identical and equal to $(k_0 r)^2[\frac{N}{2} - qN^2 J_1^2(k_0 r \tau)]$. The worst-case estimator variances for any two-source scenario are thus

$$\max(\mathrm{Var}\,\hat{u}_i) = \max(\mathrm{Var}\,\hat{v}_i) = \frac{\sigma^2 \underline{\rho}(\boldsymbol{\beta}_i)}{2K(k_0 r)^2 \left[\dfrac{N}{2} - qN^2 J_1^2(k_0 r \tau)\right]} \tag{5.80}$$

The expression above depends only on the distance τ, and the signal powers and correlations as specified by the source covariance matrix $\mathbf{P}$.

Figure 5.4 depicts the theoretical performance curves (dashed lines) of the element space MUSIC estimators in a two-source scenario. The scenario is identical to that of Simulation Example 1 in section 5.5: the location of the first source is kept fixed, and the angle ν is changed by shifting the position of the second source. The distance between the sources is maintained at $\tau = 0.25$, corresponding to a spacing of about two-thirds of the main-lobe width of the cophasal beampattern. The graphs depict the estimator performance as a function of the angle ν between the sources. As expected, the highest u and v estimator variances occur at $\nu = 0$, and $\nu = \pi/2$, respectively. The estimator variances are seen to be identical for the two sources. This is in accord with expectations as the sources are equipowered (SNR = 7 dB). The ultimate performance dictated by the CRB is also sketched (dotted lines) in the figure. Note that the element space MUSIC, and CRB performance curves are *independent* of the location of the first source.

Performance of UCA-RB-MUSIC. Theorem 5.2 gives expressions for the UCA-RB-MUSIC estimator variances. We have $\mathrm{Var}(\hat{u}_i) = (\sigma^2 \rho b/2K\Delta)|_{\boldsymbol{\beta}=\boldsymbol{\beta}_i}$, and $\mathrm{Var}(\hat{v}_i) = (\sigma^2 \rho a/2K\Delta)|_{\boldsymbol{\beta}=\boldsymbol{\beta}_i}$. The beamspace manifold $\mathbf{a}_{\mathbf{r}}(\boldsymbol{\theta})$ of Eq. (5.15) (denoted $\mathbf{b}(\boldsymbol{\theta})$ in the performance analysis section) has a complex structure. Thus, unlike with element space MUSIC, simple expressions for the parameters a, b, and c cannot be obtained. The beamformer $\mathbf{F}_{\mathbf{r}}^H$ that makes the transformation to beamspace is orthogonal and one might thus expect the behavior of the UCA-RB-MUSIC estimates to be similar to that of the element space MUSIC estimates. Examination of theoretical performance curves reveals that this is indeed true. Figure 5.4 depicts the theoretical performance of the UCA-RB-MUSIC estimators (solid lines) for the same two-source scenario. The graphs shown are a superposition of the performance curves corresponding to four different locations of the first source. These locations are $\beta = (u, v) = (0, 0)$, $(0.15, 0.15)$, $(-0.3, 0.3)$, and $(0.45, -0.45)$. The curves are almost identical, confirming that the dependence of the UCA-RB-MUSIC estimator variances on the source DOAs is for the most part through the parameters τ and ν.

The graphs in Figure 5.4 also show that UCA-RB-MUSIC outperforms element space MUSIC in the source scenario under consideration. This is due to the decorrelating effect of the FB average inherent in UCA-RB-MUSIC. FB averaging is not possible in element space when N is odd, and UCA-RB-MUSIC thus outperforms element space MUSIC.

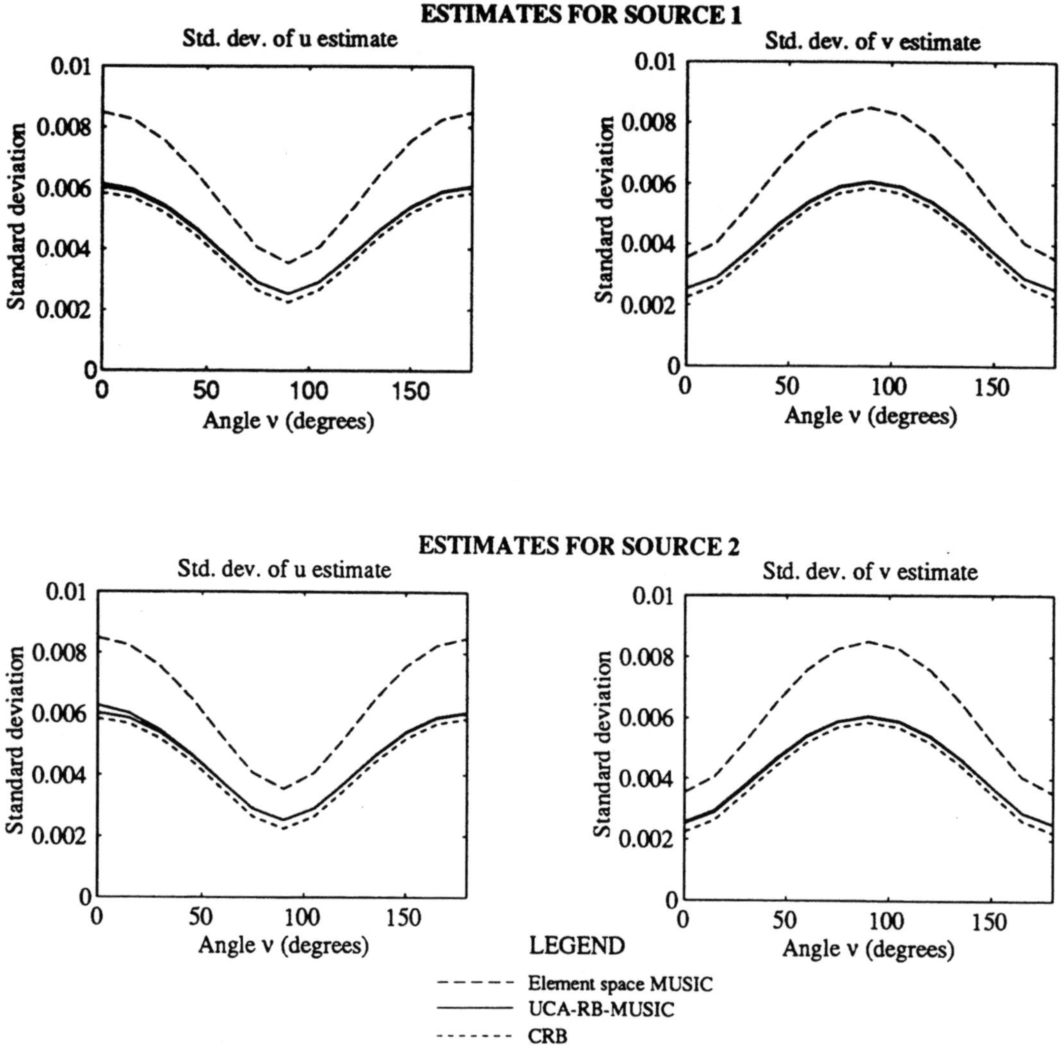

Figure 5.4 Theoretical performance of element space MUSIC and UCA-RB-MUSIC as a function of ν.

Performance of UCA-ESPRIT. Theorem 5.3 gives expressions for the variances of the UCA-ESPRIT direction cosine estimates $\hat{u}_i$ and $\hat{v}_i$. These expressions are quite complicated and cannot be simplified even for the single-source case. Theoretical performance curves are used to investigate the performance of UCA-ESPRIT for the single-source case with an SNR of 5 dB. The standard deviations of the direction cosine estimates as a function of source azimuth angle are plotted in Figure 5.5 for several different elevation angles ($\zeta' = \sin\theta = 0.5$, 0.7, 0.9, and 1.0). The deterministic CRB (which is independent of source location) in also plotted. The theoretical performance curves show that the estimator performance virtually meets the CRB for $\zeta' < 0.5$. The estimator performance degrades as ζ' increases from 0.5

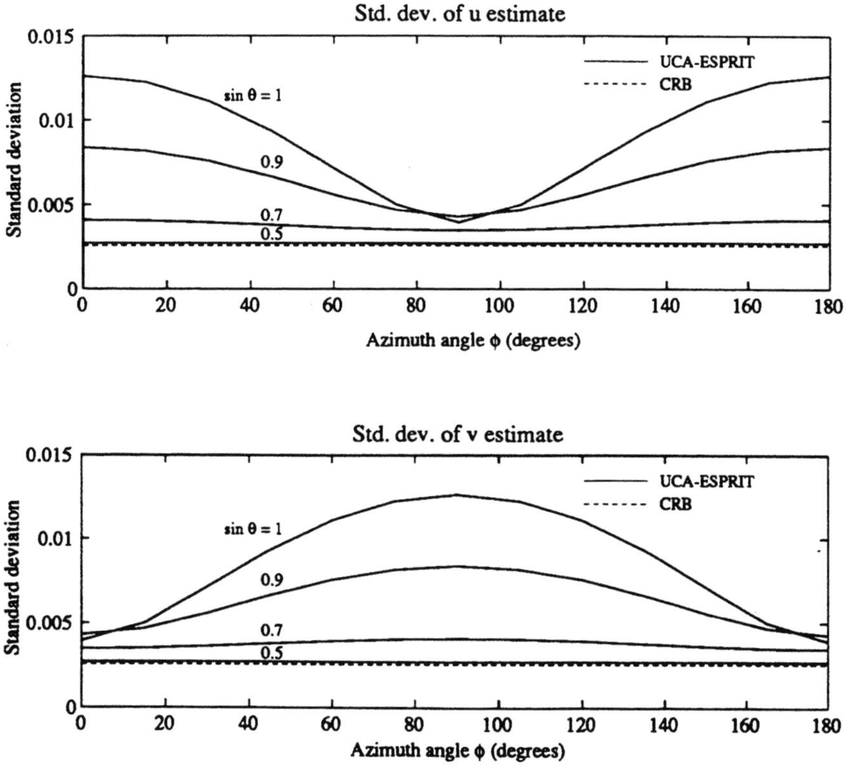

Figure 5.5 Theoretical performance of UCA-ESPRIT for the single-source case.

to 1.0. The performance also becomes azimuth dependent at these higher elevation angles. The worst performance of the u estimate is at an azimuth of 0° (where v performs best), and the worst performance of the v estimate is at an azimuth of 90° (where u performs best). The behavior of Cov (u, v) as given by Eq. (5.67) is responsible for the shape of the graphs. The correlation coefficient between the u and v estimates is observed to be close to zero for $\zeta' < 0.5$. The magnitude of the correlation coefficient increases with increasing ζ', and also becomes azimuth dependent; the correlation coefficient has smallest magnitude at azimuths of 0° and 90°. The u and v estimates are projections of the eigenvalue μ on the real and imaginary axes, respectively. This projection together with the behavior of the correlation coefficient is responsible for the curvature of the graphs at higher elevations. Although the variances of the u and v estimates are azimuth dependent, one would expect the variances of the ζ' and ϕ estimates to be independent of azimuth. Plots of the expressions of Eqs. (5.69) and (5.70) show that the ζ' and ϕ estimator variances are independent of azimuth, but are elevation dependent.

For comparison between UCA-RB-MUSIC and UCA-ESPRIT, we note that the UCA-RB-MUSIC u and v estimator variances are virtually DOA independent for the single-source case. Further, these variances are very close to the CRB in the above single-source scenario. UCA-RB-MUSIC thus performs better than UCA-ESPRIT, with the difference in performance being more pronounced at higher ζ'. The UCA-ESPRIT estimates can be used as starting points for Newton searches for peaks in the UCA-RB-MUSIC spectrum if estimates of better quality are required.

5.5 RESULTS OF COMPUTER SIMULATIONS

This section documents the results of three computer simulations that explore the performance (estimator standard deviation) of UCA-RB-MUSIC and UCA-ESPRIT in a two-source scenario. Simulation examples 1, 2, and 3 respectively investigate the performance of the algorithms as a function of the angle ν between the sources, the common source SNR, and the phase of the correlation coefficient between the signals. The computer simulations show that the experimental results closely match the theoretical performance predictions, thus validating the performance analysis results of section 5.4. The array and source descriptions for the simulations are as follows: The radius of the UCA is $r = \lambda$, and the maximum mode excited is $M = 6$ (this example was considered in section 5.2). The number of array elements is chosen to be $N = 19$; Table 5.1 shows that the maximum residual contribution is negligible with these parameters. The source separation is $\tau = 0.25$ in all the simulations. This separation is about two-thirds of the main-lobe width of the cophasal UCA beam pattern that closely follows the Bessel function J_0 [33]. The correlation between the signals is fairly high (magnitude of correlation coefficient = 0.8) in all the simulations. A moderate number of snapshots ($K = 64$) is employed. The theoretical asymptotic performance expressions are expected to be quite accurate with 64 snapshots. The simulations assume perfect detection of the number of sources. The SNRs quoted in the simulations are per source per array element. All the simulations feature equipowered sources; this implies identical values of the CRB for the two sources (also, the UCA-RB-MUSIC estimator variances for the first source will be very similar to those for the second source).

Simulation Example 1

This simulation investigates the estimator performance as a function of the angle ν of the line joining the two sources in the uv plane. The location of the first source is fixed at $\boldsymbol{\beta}_1 = (u_1, v_1) = (0.3, 0.6)$, corresponding to an elevation $\theta_1 = 42.1°$ and an azimuth $\phi_1 = 63.4°$. The angle ν is varied from 0° to 180° by rotating the second source about the first in the uv plane (the distance τ is set at 0.25). Both sources have SNRs of 7 dB, and the correlation coefficient between the sources is $0.8e^{j\pi/4}$. The results of the simulation are plotted in Figure 5.6. The graphs show that UCA-RB-MUSIC performs better than UCA-ESPRIT. Further, the UCA-RB-MUSIC performance is fairly close to the CRB. Note that the performance curves for UCA-RB-MUSIC and the CRB are *independent* of the location of the first source. In contrast, the performance of the UCA-ESPRIT estimates depends on the absolute position of each source. This is attested by the fact that the variances of the UCA-ESPRIT estimates are higher for the second source than for the first.

In the remaining two simulations, the second source is located at $\boldsymbol{\beta}_2 = (u_2, v_2) = (0.175, 0.8165)$, corresponding to an elevation $\theta_2 = 56.62°$ and an

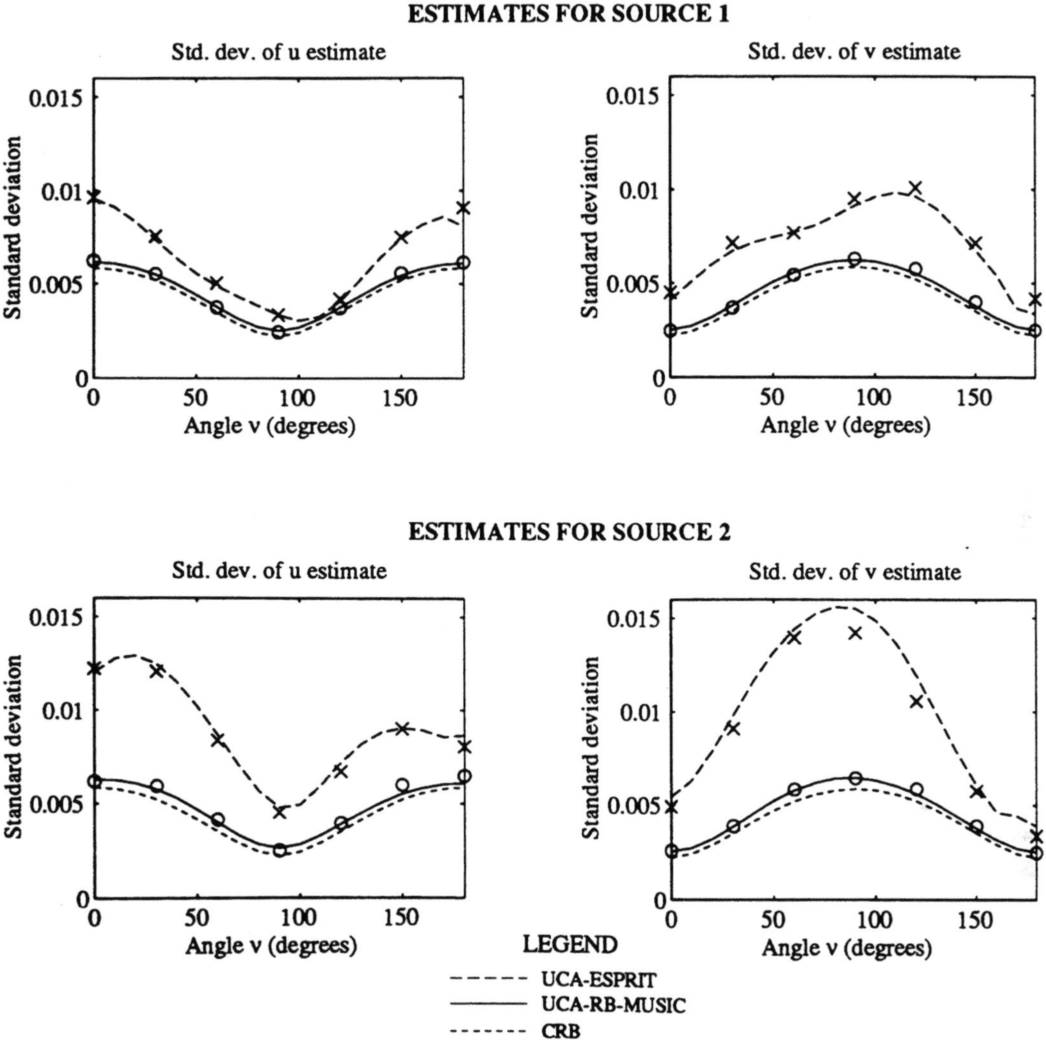

Figure 5.6 Performance of UCA-RB-MUSIC and UCA-ESPRIT as a function of ν.

azimuth $\phi_2 = 77.9°$. The location of the first source is left unchanged. These source locations correspond to a separation $\tau = 0.25$, and an orientation $\nu = 120°$. Examination of Figure 5.6 shows that for $\nu = 120°$, the variance of the UCA-ESPRIT u estimate is higher for the second source than for the first. However, the variance of the UCA-ESPRIT v estimates are approximately the same for the two sources. This behavior is evident in the UCA-ESPRIT performance curves of simulation examples 2 and 3. We point out again that the CRB curves are identical for the two sources (the UCA-RB-MUSIC performance curves for the two sources are virtually identical as well).

Simulation Example 2

This simulation examines the performance of the DOA estimators as a function of the common source SNR. The source locations are as specified in the previous paragraph, and the correlation coefficient between the sources is $0.8e^{j\pi/4}$. Figure 5.7 depicts the results of the simulations. The graphs show that the performance of UCA-RB-MUSIC is fairly close to the CRB even at an SNR of 0 dB. UCA-RB-MUSIC performs better than UCA-ESPRIT—the performance difference, however, is less significant at higher SNRs. The experimental results for UCA-ESPRIT corresponding to 0 dB SNR deviate a little from the theoretical predictions. This is probably because the first-order approximation employed in the analysis is not accurate enough at this low SNR.

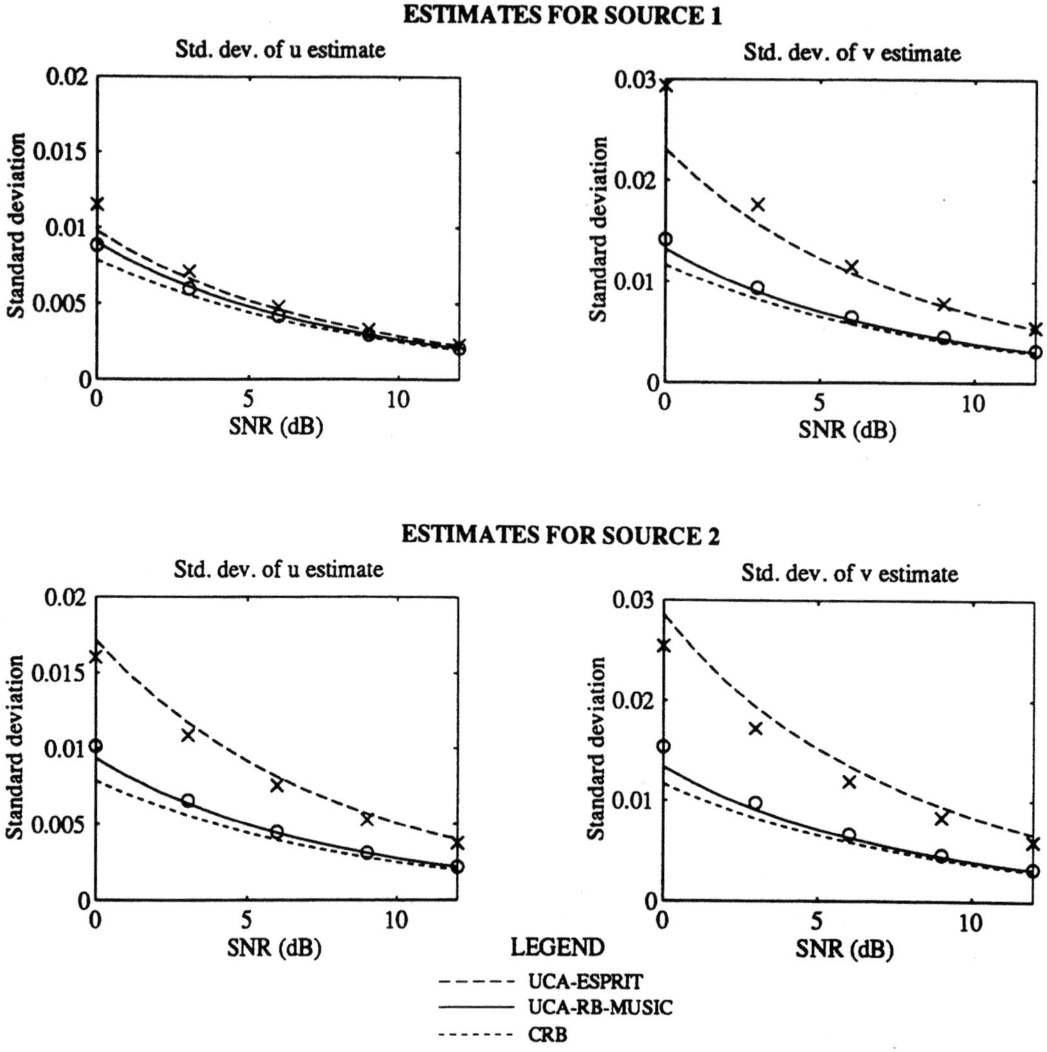

Figure 5.7 Performance of UCA-RB-MUSIC and UCA-ESPRIT as a function of the common source SNR.

Another observation from Figure 5.7 is that the UCA-ESPRIT estimator variances for the second source are a little lower than the theoretical predictions. This difference is more pronounced in Figure 5.9 corresponding to simulation example 3. This behavior can be explained as follows: The far-field pattern corresponding to phase modes 0 and 2 have low gains at the location of the second source. We have $J_0(\zeta_2) = -0.09$ and $J_2(\zeta_2) = -0.04$, where $\zeta_2 = 2\pi \sin\theta_2$. The output powers associated with the second source are thus small in two of the 13 available beams. In contrast, the gain of these two beams is fairly large at the location of the first source: We have $J_0(\zeta_1) = -0.37$ and $J_2(\zeta_1) = 0.31$. This explains the fact that the theoretical analysis predicts higher UCA-ESPRIT estimator variances for the second source than for the first. The nonasymptotic behavior of the FB average is responsibile for the fact that the experimental estimator performance for the second source (with $K = 64$ snapshots) is a little better than the theoretical (asymptotic) predictions.

Figure 5.8 depicts the UCA-ESPRIT eigenvalues $\hat{\mu}_i$ (marked by 'x's). The figure was formed by superimposing the results of 200 runs at an SNR of 10 dB. The true source locations are at the intersections of the dotted radial lines (azimuth angles) and the dotted circles (elevation angles).

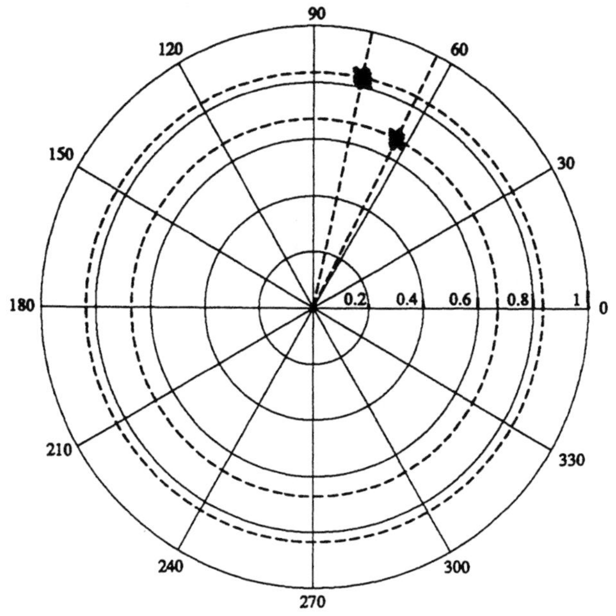

Figure 5.8 Plot of UCA-ESPRIT eigenvalues.

Simulation Example 3

This simulation investigates the performance of the algorithms as a function of the phase of the correlation coefficient between the signals. The source locations are the same as in the previous example. The common source SNR is 7 dB, and the magnitude of the correlation coefficient between the source is 0.8. Figure 5.9 depicts the performance of the DOA estimators as the phase of the correlation coefficient is varied from 0° to 180°. The behavior of the estimators (best performance at a correlation phase of 90°) is due to the FB average inherent in UCA-RB-MUSIC and UCA-ESPRIT. As a consequence of the FB average, the performance of the algorithms depends only on the real part, $\mathbf{P}_R$,

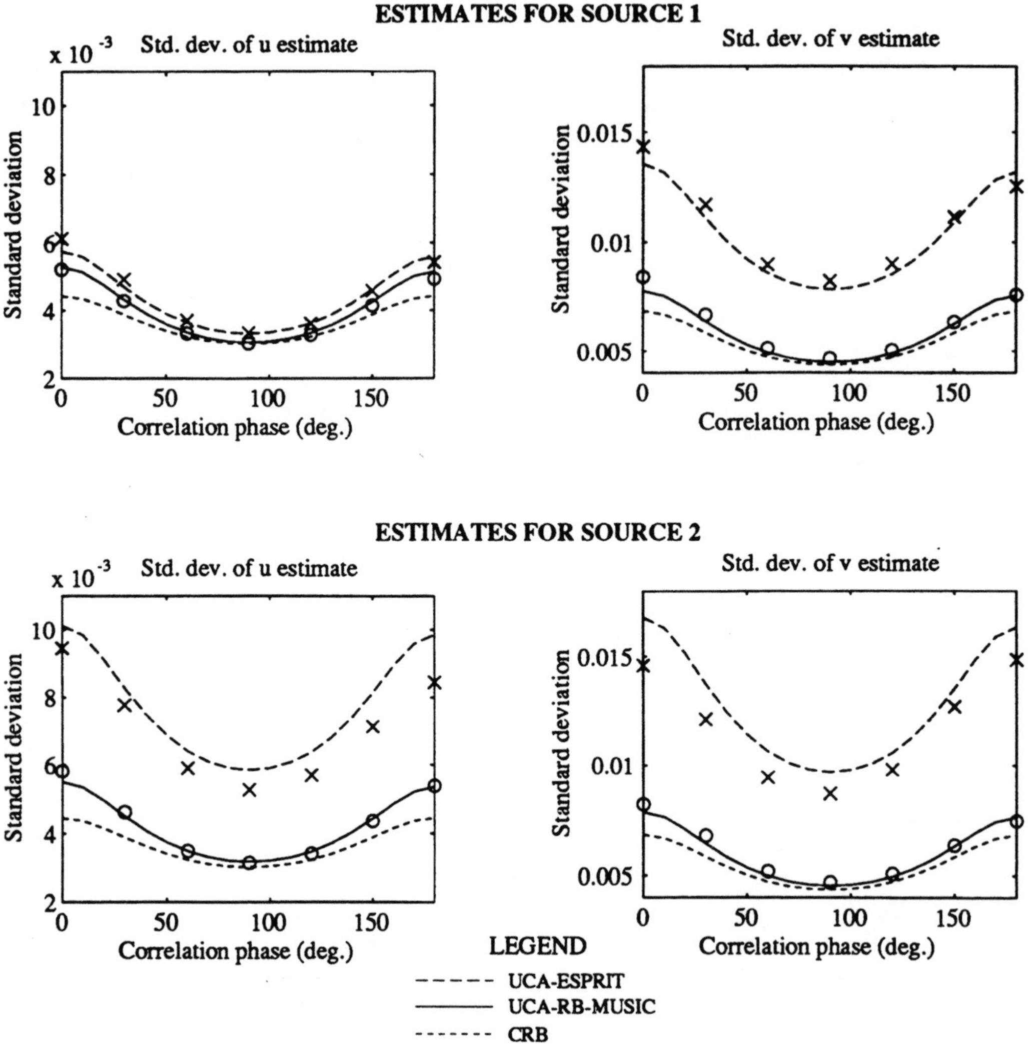

Figure 5.9 Performance of UCA-RB-MUSIC and UCA-ESPRIT as a function of the correlation coefficient phase.

of the source covariance matrix $\mathbf{P}$. The off-diagonal elements of $\mathbf{P}_R$ are zero when the correlation phase is 90°, and the algorithms see the sources as effectively uncorrelated. The FB average in fact enables UCA-RB-MUSIC and UCA-ESPRIT to resolve two coherent sources, provided the phase of the correlation coefficient is neither 0° nor 180°.

5.6 SUMMARY

Two subspace-based algorithms, UCA-RB-MUSIC and UCA-ESPRIT, for 2D angle estimation with UCAs have been developed. Both algorithms employ phase mode excitation-based beamformers, and operate in beamspace. UCA-RB-MUSIC is a beamspace version of MUSIC that offers the following advantages over element space MUSIC: Uniform linear array (ULA) techniques such as FB averaging, Root-MUSIC, and spatial smoothing can be employed in beamspace with the UCA. This is because the structure of the induced beamspace manifold is similar to the ULA manifold. The decorrelating effect of the inherent FB average allows UCA-RB-MUSIC to *outperform* elment space MUSIC in correlated source scenarios when the number, N, of array elements is odd (FB-type averaging is not possible in element space when N is odd). Finally, UCA-RB-MUSIC is more computationally efficient than element space MUSIC. This is because a real-valued EVD provides signal subspace estimates, and the FFT can be employed to facilitate the search for peaks in the beamspace MUSIC spectrum.

UCA-ESPRIT is a *closed-form* algorithm that provides *automatically paired* azimuth and elevation angle estimates for each source. It is the only available closed-form algorithm for 2D angle estimation, and thus represents a significant advance in the area. The eigenvalues of the matrix $\boldsymbol{\Psi}$ (derived from the least squares solution to an overdetermined system of equations) have the form $\mu_i = \sin\theta_i e^{j\phi_i}$, and thus provide the properly associated azimuth and elevation angle estimates. UCA-ESPRIT does not require expensive search procedures, and is thus superior to existing 2D angle estimation algorithms with respect to computational complexity. Another factor that reduces the computational load is that the implementation of UCA-ESPRIT calls for only real-valued EVDs. A method for employing the UCA-ESPRIT invariance principle to develop a closed-form algorithm for 2D angle estimation with filled circular arrays has been described.

The effects of mutual coupling on the structure of the UCA element space manifold have been discussed. It is shown that the general structure of the original beamspace manifold is retained even when mutual coupling effects apply—the only difference is the introduction of gain and phase factors in the beamspace manifold. Consequently, minor modifications enable UCA-RB-MUSIC and UCA-ESPRIT to cope with mututal coupling effects. The properties of the UCA of directional elements have also been studied. Both UCA-RB-MUSIC and UCA-ESPRIT are applicable if the individual element patterns are omnidirectional in azimuth. UCA-ESPRIT cannot be employed if this condition on element patterns is not met. However, a beamspace algorithm possessing many of the features of UCA-RB-MUSIC is still applicable.

The statistical performance of the element space MUSIC, UCA-RB-MUSIC, and UCA-ESPRIT algorithms for 2D angle estimation have been analyzed; asymptotic (large number of snapshots) expressions for the estimator variances/covariances have been obtained. The analysis of element space MUSIC for 2D angle estimation holds for arbitrary array configurations. Prior to this work, the performance of MUSIC had been examined for only the 1D angle estimation problem. Closed-form expressions have been obtained for the element space MUSIC estimator variances in the one- and two-source scenarios. An investigation of the theoretical behavior of the element space MUSIC direction cosine estimator variances and the CRB has provided useful insights. In the single-source case, both the element space MUSIC estimator variances and the CRB are constants (independent of the DOA). In the two-source case, they depend only on the *relative* positions of the sources in the direction cosine space (distance between the source locations, and the orientation of the line joining the sources). The behavior of the UCA-RB-MUSIC estimator variances closely follows that of element space MUSIC and the CRB.

Finally, results of computer simulations that demonstrate the efficacy of the UCA-RB-MUSIC and UCA-ESPRIT algorithms have been presented. The results are in close agreement with the theoretical performance predictions. The simulations thus validate the asymptotic estimator variance/covariance expressions that have been derived.

APPENDIX 5A PROPERTIES OF UCA-ESPRIT

5A.1 BLOCK CONJUGATE STRUCTURE OF THE LS SOLUTION

Consider first the noise-free case, where the true signal subspace matrices are available. Assume that the least squares solution to Eq. (5.41) is $\underline{\boldsymbol{\Psi}} = [\boldsymbol{\Psi}_1^T \vdots \boldsymbol{\Psi}_2^T]^T$. The LS solution is obtained by solving the system $\mathbf{E}^H\mathbf{E}\underline{\boldsymbol{\Psi}} = \mathbf{E}^H\boldsymbol{\Gamma}\mathbf{S}_0$. Substituting for $\mathbf{E}$ from (5.41), this system can be expanded as follows:

$$\begin{bmatrix} \mathbf{S}_{-1}^H \\ \mathbf{S}_{-1}^T\mathbf{D}\tilde{\mathbf{I}} \end{bmatrix} \begin{bmatrix} \mathbf{S}_{-1} \vdots \mathbf{D}\tilde{\mathbf{I}}\mathbf{S}_{-1}^* \end{bmatrix} \begin{bmatrix} \boldsymbol{\Psi}_1 \\ \boldsymbol{\Psi}_2 \end{bmatrix} = \begin{bmatrix} \mathbf{S}_{-1}^H \\ \mathbf{S}_{-1}^T\mathbf{D}\tilde{\mathbf{I}} \end{bmatrix} \boldsymbol{\Gamma}\mathbf{S}_0 \tag{5A.1}$$

Equating the upper and lower blocks of the above equation, we obtain

$$\mathbf{S}_{-1}^H\left[\mathbf{S}_{-1}\boldsymbol{\Psi}_1 + \mathbf{D}\tilde{\mathbf{I}}\mathbf{S}_{-1}^*\boldsymbol{\Psi}_2\right] = \mathbf{S}_1^H\boldsymbol{\Gamma}\mathbf{S}_0, \text{ and} \tag{5A.2}$$

$$\mathbf{S}_{-1}^T\left[\mathbf{D}\tilde{\mathbf{I}}\mathbf{S}_{-1}\boldsymbol{\Psi}_1 + \mathbf{S}_{-1}^*\boldsymbol{\Psi}_2\right] = \mathbf{S}_{-1}^T\mathbf{D}\tilde{\mathbf{I}}\boldsymbol{\Gamma}\mathbf{S}_0 = \mathbf{S}_{-1}^T\boldsymbol{\Gamma}\mathbf{S}_0^*. \tag{5A.3}$$

The property $\mathbf{D}\tilde{\mathbf{I}}\boldsymbol{\Gamma}\mathbf{S}_0 = \boldsymbol{\Gamma}\mathbf{S}_0^*$ that was used in the final equality above can be established by multiplying Eq. (5.40) by $\mathbf{D}\tilde{\mathbf{I}}$. Now, the right-hand sides of Eqs. (5A.2) and (5A.3) are conjugates, and the left-hand sides are therefore conjugates as well. We thus have $\boldsymbol{\Psi}_2 = \boldsymbol{\Psi}_1^*$, and the LS solution $\underline{\boldsymbol{\Psi}}$ has block conjugate structure as expected.

All that is required for the proof to carry over to the case where signal subspace estimates are employed is to show that $\mathbf{D}\tilde{\mathbf{I}}\boldsymbol{\Gamma}\hat{\mathbf{S}}_0 = \boldsymbol{\Gamma}\hat{\mathbf{S}}_0^*$. This relationship was shown to hold in the noise-free case: We have $\mathbf{D}\tilde{\mathbf{I}}_{M_e}\boldsymbol{\Gamma}\mathbf{S}_0 = \boldsymbol{\Gamma}\mathbf{S}_0^*$. The subscript M_e denotes the dimension of the reverse permutation matrix. Substituting $\mathbf{S}_0 = \boldsymbol{\Delta}_0\mathbf{C}_o\mathbf{W}\mathbf{S}$ and using the property $\tilde{\mathbf{I}}_{M'}\mathbf{W} = \mathbf{W}^*$ and the fact that $\mathbf{S}$ is real-valued establishes the following: $\mathbf{D}\tilde{\mathbf{I}}_{M_e}\boldsymbol{\Gamma}\boldsymbol{\Delta}_0\mathbf{C}_o = \boldsymbol{\Gamma}\boldsymbol{\Delta}_0\mathbf{C}_o\tilde{\mathbf{I}}_{M'}$. This property is used in the following sequence of manipulations which complete the proof for the case where signal subspace estimates are employed.

$$\begin{aligned} \mathbf{D}\tilde{\mathbf{I}}_{M_e}\boldsymbol{\Gamma}\hat{\mathbf{S}}_0 &= \mathbf{D}\tilde{\mathbf{I}}_{M_e}\boldsymbol{\Gamma}\boldsymbol{\Delta}_0\mathbf{C}_o\mathbf{W}\hat{\mathbf{S}} = \boldsymbol{\Gamma}\boldsymbol{\Delta}_0\mathbf{C}_o\tilde{\mathbf{I}}_{M'}\mathbf{W}\hat{\mathbf{S}} \\ &= \boldsymbol{\Gamma}\boldsymbol{\Delta}_0\mathbf{C}_o\mathbf{W}^*\hat{\mathbf{S}} = \boldsymbol{\Gamma}\hat{\mathbf{S}}_0^* \end{aligned}$$

We now have $\hat{\Psi}_1 = \hat{\Psi}_2^* = \hat{\Psi}$, and Eq. (5A.2) (written in terms of estimated quantities) thus uniquely specifies the LS solution. The LS solution $\hat{\Psi}$ is obtained by solving the following $d \times d$ complex-valued system of equations:

$$\mathbf{B}\hat{\Psi} + \mathbf{C}\hat{\Psi}^* = \mathbf{Q}, \text{ where}$$

$$\mathbf{B} = \hat{\mathbf{S}}_{-1}^H \hat{\mathbf{S}}_{-1}, \mathbf{C} = \hat{\mathbf{S}}_{-1}^H \mathbf{D}\tilde{\mathbf{I}}\hat{\mathbf{S}}_{-1}^*, \text{ and } \mathbf{Q} = \hat{\mathbf{S}}_{-1}^H \Gamma \hat{\mathbf{S}}_0$$

Writing in terms of the real and imaginary parts shows that $\hat{\Psi}$ can be obtained by solving the $2d \times 2d$ real-valued system of Eq. (5.43).

5A.2 FAILURE OF UCA-ESPRIT WHEN $d \geq M$

The system $\hat{\mathbf{E}}\underline{\hat{\Psi}} = \Gamma\hat{\mathbf{S}}_0$ is underdetermined, and has an infinity of solutions $\underline{\hat{\Psi}}$ when the number of sources $d \geq M$. Following the procedure employed in the previous proof, the minimum norm solution $\underline{\hat{\Psi}}_{\text{min}} = \hat{\mathbf{E}}^H(\hat{\mathbf{E}}\hat{\mathbf{E}}^H)^{-1}\Gamma\hat{\mathbf{S}}_0$ can be shown to have block conjugate structure. We now proceed to show that $\eta(\hat{\mathbf{E}}) = \eta(\hat{\mathbf{E}}^H\hat{\mathbf{E}})$ is spanned by block conjugate vectors. Linear combinations of these vectors can be added to $\underline{\hat{\Psi}}_{\text{min}}$ without destroying the block conjugate structure. There is thus no unique block conjugate solution to Eq. (5.42) when $d \geq M$.

The $2d \times 2d$ matrix $\mathbf{E}' = \hat{\mathbf{E}}^H\hat{\mathbf{E}}$ can be written in block form as follows:

$$\mathbf{E}' = \begin{bmatrix} \mathbf{B} & \mathbf{C} \\ \mathbf{C}^* & \mathbf{B}^* \end{bmatrix}$$

where $\mathbf{B}$ and $\mathbf{C}$ are as defined in Eq. (5.44). Let $\mathbf{z} = [\mathbf{z}_1^T \vdots \mathbf{z}_2^T]^T$ be a vector in $\eta(\mathbf{E}')$. Setting $\mathbf{E}'\mathbf{z} = 0$ leads to the two equations below:

$$\mathbf{B}\mathbf{z}_1 + \mathbf{C}\mathbf{z}_2 = 0, \text{ and} \tag{5A.4}$$

$$\mathbf{C}^*\mathbf{z}_1 + \mathbf{B}^*\mathbf{z}_2 = 0 \tag{5A.5}$$

Forming (5A.4) + (5A.5)* and (5.A4)* + (5A.5) leads to the system

$$\begin{bmatrix} \mathbf{B} & \mathbf{C} \\ \mathbf{C}^* & \mathbf{B}^* \end{bmatrix} \begin{bmatrix} \mathbf{z}_1 + \mathbf{z}_2^* \\ \mathbf{z}_2 + \mathbf{z}_1^* \end{bmatrix} = 0$$

This system shows that there is a block conjugate vector in $\eta(\mathbf{E}')$ corresponding to the vector $\mathbf{z} \in \eta(\mathbf{E}')$. Thus $\eta(\mathbf{E}') = \eta(\hat{\mathbf{E}})$ can be spanned by block conjugate vectors, and the proof is complete.

APPENDIX 5B STUDY OF THEORETICAL ESTIMATOR PERFORMANCE

The following easily proved relationships are employed in obtaining expressions for the variances of the element space MUSIC estimators. For brevity, the dependence of the manifold vectors on the DOA is not always shown, that is, $\underline{\mathbf{a}}$ is used in place of $\underline{\mathbf{a}}(\boldsymbol{\beta})$.

$$\underline{\mathbf{a}}_u = \mathbf{D}_c\underline{\mathbf{a}}, \text{ and } \underline{\mathbf{a}}_v = \mathbf{D}_s\underline{\mathbf{a}}, \text{ where} \tag{5B.1}$$

$$\mathbf{D}_c = (jk_0r)\ \text{diag}\ \{\cos\gamma_0, \ldots, \cos\gamma_{N-1}\}, \text{ and}$$

$$\mathbf{D}_s = (jk_0r)\ \text{diag}\ \{\sin\gamma_0, \ldots, \sin\gamma_{N-1}\}$$

In the definitions above, $\gamma_i = 2\pi i/N$ is the angular location of the ith antenna element. The following relationships can be established using the above equations:

$$\underline{\mathbf{a}}^H\underline{\mathbf{a}}_u = \underline{\mathbf{a}}^H\underline{\mathbf{a}}_v = 0 \tag{5B.2}$$

$$\underline{\mathbf{a}}_u^H\underline{\mathbf{a}}_u = \underline{\mathbf{a}}_v^H\underline{\mathbf{a}}_v = \frac{N(k_0r)^2}{2}$$

Consider the two-source locations $\boldsymbol{\beta}_1 = (u_1, v_1)$ and $\boldsymbol{\beta}_2 = (u_2, v_2)$. Let $\boldsymbol{\beta}_d = (u_d, v_d)$, where $u_d = u_2 - u_1$ and $v_d = v_2 - v_1$. The representation of $\boldsymbol{\beta}_d$ in polar coordinates is (τ, ν), where $\tau = \sqrt{u_d^2 + v_d^2}$ and $\nu = \tan^{-1}(v_d/u_d)$. The relationship below states that the UCA cophasal beam pattern approximately follows the Bessel function J_0. The relationship is accurate for our purposes because the residual contributions have been made negligible by appropriate choice of N. We have

$$\underline{\mathbf{a}}^H(\boldsymbol{\beta}_1)\underline{\mathbf{a}}(\boldsymbol{\beta}_2) \approx NJ_0(k_0r\tau) \tag{5B.3}$$

and the array gain depends only on the distance τ between the locations $\boldsymbol{\beta}_1$ and $\boldsymbol{\beta}_2$. The final relationships to be established are

$$\underline{\mathbf{a}}^H(\boldsymbol{\beta}_1)\underline{\mathbf{a}}_u(\boldsymbol{\beta}_2) \approx -k_0rNJ_1(k_0r\tau)\cos\nu = -\underline{\mathbf{a}}^H(\boldsymbol{\beta}_2)\underline{\mathbf{a}}_u(\boldsymbol{\beta}_1), \text{ and} \tag{5B.4}$$

$$\underline{\mathbf{a}}^H(\boldsymbol{\beta}_1)\underline{\mathbf{a}}_v(\boldsymbol{\beta}_2) \approx -k_0rNJ_1(k_0r\tau)\sin\nu = -\underline{\mathbf{a}}^H(\boldsymbol{\beta}_2)\underline{\mathbf{a}}_v(\boldsymbol{\beta}_1) \tag{5B.5}$$

An outline of the proof of Eq. (5B.4) is provided below. Eq. (5B.5) can be proved in similar fashion. Using the results in (5B.6), we can write $\underline{\mathbf{a}}^H(\boldsymbol{\beta}_1)\underline{\mathbf{a}}_u(\boldsymbol{\beta}_2) = \sum_{n=1}^{N} jk_0r\cos\gamma_n e^{jk_0r(u_d\cos\gamma_n + v_d\sin\gamma_n)}$. The summation of the exponential terms alone is just the cophasal beampattern of Eq. (5B.3). The partial derivative $\frac{\partial}{\partial u_d} NJ_0\left(k_0r\sqrt{u_d^2 + v_d^2}\right)$ thus evaluates the entire summation. Simplifying and using the property $J_0' = -J_1$ yields the desired result.

5B.1 ELEMENT SPACE MUSIC ESTIMATOR VARIANCES

Expressions for the element space MUSIC estimator variances and covariance are given in Theorem 5.1. The variance expressions are in terms of the quantities $\underline{a} = \underline{\mathbf{a}}_u^H\underline{\mathbf{G}}\underline{\mathbf{G}}^H\underline{\mathbf{a}}_u$, $\underline{b} = \underline{\mathbf{a}}_v^H\underline{\mathbf{G}}\underline{\mathbf{G}}^H\underline{\mathbf{a}}_v$, and $\underline{c} = \text{Re}\{\underline{\mathbf{a}}_v^H\underline{\mathbf{G}}\underline{\mathbf{G}}^H\underline{\mathbf{a}}_u\}$. The projection matrix onto the noise subspace can be expressed in terms of the DOA matrix as follows: $\underline{\mathbf{G}}\underline{\mathbf{G}}^H = I - \underline{\mathbf{A}}(\underline{\mathbf{A}}^H\underline{\mathbf{A}})^{-1}\underline{\mathbf{A}}^H$.

Single-Source Case

For the single-source case, we have $\underline{\mathbf{A}} = \underline{\mathbf{a}}$. Using Eq. (5B.2), we obtain $\underline{a} = \underline{b} = (k_0r)^2N/2$ and $\underline{c} = 0$. Substituting the source power p in place of $\mathbf{P}$ in Eq. (5.58), we find that $\sigma^2 \underline{\rho} = (1 + 1/N)/(p/\sigma^2)$. Substituting these results in Theorem 5.1 yields

$$\text{Var}\,(\hat{u}) = \text{Var}\,(\hat{v}) = \frac{1 + 1/N}{KN(k_0 r)^2(p/\sigma^2)}, \text{ and}$$

$$\text{Cov}\,(\hat{u}, \hat{v}) = 0$$

Two-Source Case

Let $\mathbf{Q} = (\underline{\mathbf{A}}^H \underline{\mathbf{A}})^{-1}$. Using the result of Eq. (5B.3), $\mathbf{Q}$ can be expressed as follows:

$$\mathbf{Q} = \begin{bmatrix} q & qJ_0(k_0 r\tau) \\ qJ_0(k_0 r\tau) & q \end{bmatrix}, \text{ where} \tag{5B.6}$$

$$q = \frac{1}{N\left[1 - J_0^2(k_0 r\tau)\right]}$$

The simplified expression for $\underline{a}$ in Eq. (5.79) is derived below. Expressions for $\underline{b}$ and $\underline{c}$ can be obtained in similar fashion. We have

$$\begin{aligned} \underline{a}(\beta_1) &= \underline{\mathbf{a}}_u^H(\beta_1)\left[\mathbf{I} - \underline{\mathbf{A}}\mathbf{Q}\underline{\mathbf{A}}^H\right]\underline{\mathbf{a}}_u(\beta_1) \\ &= N(k_0 r)^2/2 - \left(\underline{\mathbf{a}}_u^H(\beta_1)\underline{\mathbf{A}}\right)\mathbf{Q}\left(\underline{\mathbf{A}}^H\underline{\mathbf{a}}_u(\beta_1)\right) \end{aligned} \tag{5B.7}$$

Using Eqs. (5B.2) and (5B.4), we obtain $\underline{\mathbf{A}}^H\underline{\mathbf{a}}_u(\beta_1) \approx [0 \vdots k_0 r N J_1(k_0 r\tau)\cos\nu]^T$. Substituting in Eq. (5B.7) yields the desired expression

$$\underline{a}(\beta_1) \approx (k_0 r)^2\left[\frac{N}{2} - q\,\{NJ_1(k_0 r\tau)\cos\nu\}^2\right] \tag{5B.8}$$

Finally, Eq. (5.58) shows that the dependence of $\underline{\rho}$ on the source DOAs is only through the matrix $\mathbf{Q} = (\underline{\mathbf{A}}^H\underline{\mathbf{A}})^{-1}$. It is evident from Eq. (5B.6) that $\mathbf{Q}$ and thus $\underline{\rho}$ depends only on the distance τ.

5B.2 THE DETERMINISTIC CRB

From Lemma 5.5 it is clear that the dependence of the deterministic CRB on the source DOAs is through the matrix $\mathbf{H}$. We proceed to show that the CRBs are independent of the DOA in the single-source case. We also show that the CRBs depend on the source DOAs only through the parameters τ and ν in the two-source case.

Single-Source Case

Substituting $\underline{\mathbf{a}}$ in place of $\underline{\mathbf{A}}$ in Eq. (5.76) and using the results of Eq. (5B.2) yields $\mathbf{H} = \frac{N(k_0 r)^2}{2}\mathbf{I}$. Further, all the elements of the matrix $\mathbf{P}_+$ are identical and equal to the source power p. Substituting these results in Eq. (5.76) yields the CRB covariance matrix

$$\mathrm{CRB}\,(\hat{u}, \hat{v}) = \frac{1}{KN(k_0 r)^2 (p/\sigma^2)} \mathbf{I}$$

Two-Source Case

We have $\mathbf{H} = \mathbf{D}^H\mathbf{D} - \mathbf{D}^H\underline{\mathbf{A}}\mathbf{Q}\underline{\mathbf{A}}^H\mathbf{D}$. The entries of the second matrix in this expression have forms similar to the second term in Eq. (5B.7), and thus depend only on τ and ν. The entries of the first matrix $\mathbf{D}^H\mathbf{D}$ have the form $\underline{\mathbf{a}}_x^H(\beta_i)\underline{\mathbf{a}}_y(\beta_j)$ where x and y could be either u or v, and $i, j \in [1, 2]$. Expressions for these terms can be obtained by taking partial derivatives of the cophasal beampattern with respect to u_d and v_d as in the proof of Eq. (5B.4). These terms can also be shown to depend only on τ and ν. Thus, the matrix $\mathbf{H}$ depends on the source DOAs only through τ and ν, and so does the CRB covariance matrix.

REFERENCES

1. J. D. Tillman, C. E. Hickman, and H. P. Neff, "The Theory of a Single Ring Circular Array," *Trans. Amer. Inst. Electr. Engrs.*, vol. 80, pt. 1, p. 110, 1961.
2. I. D. Longstaff, P.E.K. Chow, and D.E.N. Davies "Directional Properties of Circular Arrays," *Proc. IEE,* vol. 114, June 1967.
3. D.E.N. Davies, *The Handbook of Antenna Design,* vol. 2, chapter 12, A. W. Rudge, K. Milne, A. D. Olver, and P. Knight, (eds.), London Peter Peregrinus, 1983.
4. D.E.N. Davies "A Transformation Between the Phasing Techniques Required for Linear and Circular Aerial Arrays," *Proc. IEE,* vol. 112, pp. 2041–45, November 1965.
5. J.R.F. Guy and D.E.N. Davies, "UHF Circular Array Incorporating Open-loop Null Steering for Communications," *Proc. IEE,* vol. 130, pts. F and H, pp. 67–77, February 1983.
6. C. P. Mathews and M. D. Zoltowski, "Eigenstructure Techniques for 2D Angle Estimation with Uniform Circular Arrays," *IEEE Trans. on Signal Processing*, vol. 42, pp. 2395–2407, September 1994.
7. R. O. Schmidt, "Multiple Emitter Location and Signal Parameter Estimation," *IEEE Trans. Antennas Prop.*, vol. 34, pp. 276–80, March 1986.
8. R. Roy and T. Kailath, "ESPRIT-Estimation of Signal Parameters via Rotational Invariance Techniques," *IEEE Trans. Acoust., Speech, Signal Processing,* vol. 37, pp. 984–95, July 1989.
9. A. Swindlehurst, "DOA Identifiability for Rotationally Invariant Arrays," *IEEE Trans. Acoust., Speech, Signal Processing,* vol. 40, pp. 1825–28, July 1992.
10. P. Stoica and A. Nehorai, "Comparative Performance of Element-space and Beam-space MUSIC Estimators," *Circuits, Syst. and Signal Processing*, vol. 10, pp. 285–92, 1991.
11. A. H. Tewfik and W. Hong, "On the Application of Uniform Linear Array Bearing Estimation Techniques to Uniform Circular Arrays," *IEEE Trans. on Signal Processing,* vol. 40, pp. 1008–11, April 1992.
12. B. Friedlander and A. J. Weiss, "Direction Finding Using Spatial Smoothing with Interpolated Arrays," *IEEE Trans. Aerosp. Electron. Syst.,* vol. 28, pp. 574–87, April 1992.

13. M. P. CLARK and L. L. SCHARF, "A Maximum Likelihood Estimation Technique for Spatial-Temporal Modal Analysis," *In Proc. 25th Annu. Asilomar Conf. Signals, Syst., Comput.*, vol. 1, pp. 257–61, 1991.

14. M. D. ZOLTOWSKI and D. STAVRINIDES, "Sensor Array Signal Processing via a Procrustes Rotations Based Eigenanalysis of the ESPRIT Data Pencil," *IEEE Trans. Acoust., Speech, Signal Processing,* vol. 37, pp. 832–61, June 1989.

15. A. L. SWINDLEHURST and T. KAILATH, "Azimuth/Elevation Direction Finding Using Regular Array Geometries," *IEEE Trans. Aerosp. Electron. Syst.,* vol. 29, pp. 145–56, January 1993.

16. P. STOICA and A. NEHORAI, "MUSIC, Maximum Likelihood and Cramer-Rao Bound," *IEEE Trans. Acoust., Speech, Signal Processing,* vol. 37, pp. 720–41, May 1989.

17. C. P. MATHEWS and M. D. ZOLTOWSKI, "Performance Analysis of the UCA-ESPRIT algorithm for Circular Ring Arrays," *IEEE Trans. on Signal Processing,* vol. 42, pp. 2535–39, September 1994.

18. B. D. RAO and K.V.S. HARI, "Performance Analysis of ESPRIT and TAM in Determining the Direction of Arrival of Plane Waves in Noise," *IEEE Trans. Acoust., Speech, Signal Processing,* vol. 37, pp. 1990–95, December 1989.

19. A. J. BARABELL, "Improving the Resolution Performance of Eigenstructure-based Direction-finding Algorithms," *Proc. IEEE Int. Conf. Acoust., Speech, Signal Processing,* pp. 336–39, 1983.

20. E. DORON and M. DORON, "Coherent Wideband Array Processing," *Proc. IEEE Int. Conf. Acoust., Speech, Signal Processing,* vol. 2 pp. 497–500, 1992.

21. S. U. PILLAI and B. H. KNOW, "Forward/Backward Spatial Smoothing Techniques for Coherent Signal Identification," *IEEE Trans. Acoust., Speech, Signal Processing,* vol. 37, pp. 8–15, January 1989.

22. M. D. ZOLTOWSKI and C. P. MATHEWS, "Direction Finding with Uniform Circular Arrays via Phase Mode Excitation and Beamspace Root-MUSIC," *Proc. IEEE Int. Conf. Acoust., Speech, Signal Processing,* vol. 5, pp. 245–48, 1992.

23. C. P. MATHEWS and M. D. ZOLTOWSKI, "Direction Finding with Circular Arrays via Phase Mode, Excitation and Root-MUSIC," *Proc. IEEE AP-S Int. Symposium,* vol. 2, pp. 1019–22, 1992.

24. M. D. ZOLTOWSKI, G. M. KAUTZ, and S. D. SILVERSTEIN, "Beamspace Root-MUSIC," *IEEE Trans. on Signal Processing,* vol. 41, pp. 344–64, January 1993.

25. J. SENSI, JR., *The AEGIS System,* chapter 3, E. Brookner (ed.), Artech House, Boston, Mass.: 1988.

26. E. BROOKNER, "Phased Array Radars," *Scientific American,* pp. 94-102, February 1985.

27. C. ROLLER and W. WASYLKIWSKYJ, "Effects of Mutual Coupling on Super-resolution DF in Linear Arrays," *Proc. IEEE Int. Conf. Acoust., Speech, Signal Processing*, vol. 5, pp. 257–60, 1992.

28. B. FRIEDLANDER and A. J. WEISS, "Direction Finding in the Presence of Mutual Coupling," *IEEE Trans. Antennas Propagat.*, vol. 39, pp. 273–84, March 1991.

29. P. J. DAVIES, *Circulant Matrices,* New York, Wiley, 1979.

30. T. RAHIM and D.E.N. DAVIES, "Effect of Directional Elements on the Directional Response of Circular Antenna Arrays," *Proc. IEE,* vol. 129, pt. H, pp. 180–22, February 1982.

31. M. D. ZOLTOWSKI and G. M. KAUTZ, "Performance Analysis of Eigenstructure Based DOA Estimators Employing Conjugate Centro-symmetric Beamformers," *Proc. 6th SSAP Workshop on Statistical Signal and Array Processing,* pp. 384–87, October 1992.

32. P. STOICA and A. NEHORAI, "Performance Study of Conditional and Unconditional Direction-of-Arrival Estimation," *IEEE Trans. Acoust., Speech, Signal Processing,* vol. 38, pp. 1783–95, October 1990.

33. R. E. COLLIN and F. J. ZUCKER (ed.), *Antenna Theory,* vol. 1, chapter 5, McGraw-Hill, New York, 1969.

34. M. WAX and J. SHEINVALD, "Direction Finding of Coherent Signals via Spatial Smoothing for Uniform Circular Arrays," *IEEE Trans. Antennas Prop.*, vol. 42, pp. 613–20, May 1994.

6 Generalized Correlation Decomposition Applied to Array Processing in Unknown Noise Environments

K. M. Wong, Q. Wu, and P. Stoica

6.1 INTRODUCTION

Array processing, or more accurately, *sensor array signal processing*, is the processing of the output signals of an array of sensors located at different points in space in a wavefield. The purpose of array processing is to extract useful information from the received signals such as the number and location of the signal sources, the propagation velocity of the waves, as well as the spectral properties of the signals. Array processing techniques have been employed in various areas in which very different wave phenomena occur, for example:

In *seismic exploration*, arrays of seismometers are used to collect geophysical data to unravel the physical characteristics of a part of the interior of the earth. In *passive* (listening-only) *sonar*, arrays of hydrophones are used to collect data generated by underwater sound sources so that a directional map of the background sound power can be determined. In *radar*, a receiving array of antenna elements is used to listen to the return caused by the reflections from targets illuminated by the electromagnetic waves transmitted by the transmitting antenna. In *radio astronomy*, antenna groups are used to construct maps of radiating sources emphasizing ambiguity, resolution, and dynamic range of the images. Common to all these applications, there are, in general, two essential purposes in array processing:

(i) To determine the number of signal sources (decision),
(ii) To estimate the locations of these sources (estimation).

Since the contribution of this chapter is on a limited aspect of array processing, our discussion will center only on these two issues with more emphasis on the latter. Excellent references [1–7] are available for other specific applications of array processing. Furthermore, we only consider the case of one-dimensional arrays in this chapter although the discussion can be readily extended to arrays of higher dimensions.

The traditional way of tackling the problems in array processing is based on procedures employing the fast Fourier transform (FFT) [8–13]. This approach is computationally efficient and produces reasonable results for a large class of signal environments. It is also robust in the sense that it does not depend on a signal model. In spite of these advantages, there are several inherent performance limitations of the FFT methods. The most important limitation is that of spatial resolution, that is, the ability to distinguish the spatial response of two or more signal sources. The spatial resolution of an FFT array processor is roughly inversely proportional to the size of the array. A second limitation is due to the implicit weighting of the sensor outputs (array shading) that occurs when processing with the FFT. Array shading manifests itself as "leakage" in the spatial domain, that is, energy in the main lobe of a spatial response "spills" into the sidelobes, obscuring and distorting other spatial responses that are present. In fact, weak signal spatial responses can be masked by higher sidelobes from stronger spatial responses. Careful application of tapered sensor apertures can reduce the sidelobe leakage, but always at the expense of reduced resolution.

In an attempt to overcome the inherent limitations of the FFT approach, many alternative array processing procedures, generally known as *high-resolution methods*, have been proposed in the past several years. Both problems of decision and estimation have been addressed and have usually been treated separately [14–21] [22–45]. These methods generally start with a signal and noise model, and, based on the covariance matrix of the received data, the data space is partitioned into the *signal subspace* and the *noise subspace*. Various methods are then applied to estimate the parameters concerning the number and the directions of arrival (DOA) of the signal sources in these subspaces.

In establishing these high-resolution array processing methods, there are several assumptions made on the signal and noise model. Different methods may have additional assumptions, but the ones most commonly made are:

1.1. The signal and noise processes are stationary over the observation interval and are uncorrelated with each other.

1.2. The number of sensors in the array is greater than the number of signal sources.

1.3. The sensors are properly calibrated at the time of the measurement.

1.4. The background noise is isotropic (spatially white), or, in the case of spatially nonwhite noise, the covariance matrix of the noise is known.

Analyses and simulation tests on some of these high-resolution methods have been carried out [46–55] and it has been shown that under the appropriate assumptions, these methods have superior performance compared to the FFT approach in the cases involving multiple and closely spaced signal sources. However, when the environment departs from its ideal assumptions, the performance of these methods generally deteriorates.

The four assumptions generally made in high-resolution array processing methods have different practical implications. The first assumption can usually be satisfied, at least to a close approximation, in practice. The second assumption does not represent a harsh practical restriction either. However, Assumptions 1.3 and 1.4 are not easily satisfied in practice.

Assumption 1.3 is not easy to satisfy because, in practice, calibrating an antenna array can be time consuming and expensive. In addition to the problem of initial calibration, there is the problem of maintaining proper calibration of the sensors. The factors contributing to the gradual changes in the response of the sensors and their associated circuits are many and varied: Changes in the environment around the array may affect the response (e.g., coupling) of the sensors, changes in the location of the sensors may introduce perturbation of the sensor positions, aging of the components, moisture, and, temperature variation all contribute to changes in the response of the sensors. It is impossible in many practical situations to maintain array calibration to the accuracy required for the proper operation of these high-resolution methods. Analyses and simulation tests [56–58] show that improperly calibrated sensor arrays may lead to unacceptable estimation results. Self-calibration algorithms [59–63] for high-resolution array processing have been proposed to alleviate the problems introduced by imprecise array calibration. These methods generally use the received signals to fine-tune the array calibration and are usually based on the simultaneous estimation of the DOA of the signal sources and of the unknown array parameters. However, estimation of both signal and array parameters is (a) computationally complex, (b) hampered by potential identifiable problems.

Assumption 1.4, on the other hand, is often invalid in practice. Background noise, or *ambient noise*, refers to the noise that remains after all easily identifiable signal sources are eliminated. The sources of ambient noise can be both natural and man-made, with different sources exhibiting different directional and spectral characteristics. In a passive sonar system for instance [64, 65], the natural sources of noise include seismic disturbances, agitation of the sea surfaces by wind, thermal activity of the water molecules, as well as biological sources such as snapping shrimps, dolphins, and other fishes and ocean mammals. The principal man-made component of ambient noise, on the other hand, is the sound generated by distant shipping. Hence, the ambient noise characteristics for such cases are highly dependent on geographic location, the transmission medium, season of the year, and weather, and are, in general, unknown. Thus, Assumption 1.4 is often an over-simplified model which may not represent the true spatial characteristics of the noise. To carry out high-resolution array processing in an unknown, spatially correlated noise environment is an intriguing and challenging research problem because the peaks in the nonwhite spatial spectrum of the ambient noise could be mis-identified as signal sources leading to serious degradation in performance. It is on this problem that our attention is focused in this chapter. In the past few years, researchers have considered methods capable of high-resolution array processing in unknown correlated ambient noise.

Paulraj and Kailath [66] described a technique which utilized two measurements of the array covariance assuming that the unknown noise field had remained spatially invariant. However, in many applications such as sonar, spatial invariance is often violated in a gross sense by geographic location, and in finer sense by the proximity and state of the surface and bottom boundaries. Le Cadre [67] and Tewfik [68] developed different methods based on the parametric ARMA modeling of the noise field. Thus, the methods are applicable to only the cases in which such a model is valid. The MAP algorithm developed by Reilly, Wong, et al. [69–71] is based on Bayesian analysis. A modified version of the criterion was later developed by Wax [72] based on the principle of minimum description length (MDL). Both algorithms are robust and applicable to arbitrary array geometry and to coherent or noncoherent signal sources and are not restricted by parametric noise modelling. However, they are not asymptotically consistent and the result obtained may be biased [71, 73].

In this chapter, a new approach is proposed for array processing in an unknown correlated noise environment. The signal and noise model used is based on the assumption that the data are received by two arrays well separated so that their noise outputs are uncorrelated. A new concept called the *generalized correlation decomposition* (GCD) of the cross-correlation matrix between the two arrays is introduced. The GCD includes many commonly used decomposition techniques. Of particular interest is the *canonical correlation decomposition* (CCD). The analysis of the genealized correlation leads to various interesting geometric and asymptotic properties of the structure of the data space. From these properties. algorithms are developed for the decision of the number as well as for the estimation of the DOA of the signal sources in unknown spatially correlated noise.

Outlining this chapter, we begin in section 6.2 with a review of the signal and noise model for array processing in spatially white noise. Our emphasis is on the estimation of the DOA of signal sources, under the assumption of spatially white noise, employing the MUSIC algorithm and the maximum likelihood principle. In section 6.3, we turn our attention to the signal and noise model when the ambient noise is unknown and correlated. In particular, we establish the model by having two array sensors sufficiently separated so that their noise outputs are uncorrelated. Based on this signal and noise model, we introduce the concept of the GCD of the cross-correlation matrix between the two arrays in section 6.4. The geometric and asymptotic properties of the data space structure obtained by GCD are examined. In array processing, the logical order should be to decide on the number of signals first and then estimate their DOA. However, in this chapter, we treat the procedures in reverse order. This is because the development of the algorithm for the decision of the number of signal necessitates the knowledge of the algorithm for the estimation of the DOA signals. Thus, in section 6.5, we develop the algorithms for estimation of DOA of the signal sources based on the geometric and asymptotic properties of GCD derived in the previous section assuming that the number of signals has been correctly determined. The consistency of the criteria and estimates of these algorithms are examined. Section 6.6 presents analyses of the performance of these algorithms and, based on these analyses, the optimum weighting matrices as well as the optimality of CCD are derived in section 6.7. Computer simulation examples confirm the validity of the results and show that the new methods are effective. In section 6.8, we return to the problem of determining the number of signals in unknown correlated noise. Different approaches are proposed and their performance examined through examples of computer simulations under various circumstances. Section 6.9 summarizes and concludes the chapter.

Notation and Glossary

The following is a list of the notational conventions for the vectors and matrices used in this chapter.

Vectors and matrices in a real space $\mathcal{R}$ or a complex space $\mathcal{C}$ are usually denoted by boldface, lower case and by boldface, upper case, respectively. In addition, for a complex matrix of dimension $M \times N$, we have:

$$
\begin{aligned}
\mathbf{A}^T &= \text{the transpose of } \mathbf{A} \\
\mathbf{A}^* &= \text{the complex conjugate of } \mathbf{A} \\
\mathbf{A}^H &= \text{the conjugate transpose of } \mathbf{A} \\
\mathbf{A}^{-1} &= \text{the inverse of } \mathbf{A} \in \mathcal{C}^{M \times M} \\
\mathbf{A}^+ &= \text{generalized inverse of } \mathbf{A} \\
\|\mathbf{A}\| &= \text{Frobenius norm of } \mathbf{A} \\
\text{Re}\,[\mathbf{A}] &= \text{the real part of } \mathbf{A} \\
\text{Im}\,[\mathbf{A}] &= \text{the imaginary part of } \mathbf{A} \\
\text{tr}\,[\mathbf{A}] &= \text{the trace of } \mathbf{A} \in \mathcal{C}^{M \times M} \\
\det[\mathbf{A}] &= \text{the determinant of } \mathbf{A} \in \mathcal{C}^{M \times M} \\
[\mathbf{A}]_{mn} &= \text{the } mn\text{th element of } \mathbf{A} \\
(\mathbf{A})_{.j} &= \text{the } j\text{th column vector of } \mathbf{A} \\
(\mathbf{A})_{i.} &= \text{the } i\text{th row vector of } \mathbf{A} \\
\text{vec}\,(\mathbf{A}) &= [a_{11} \cdots a_{M1} a_{12} \cdots a_{M2} \cdots a_{1N} \cdots a_{MN}]^T \\
\mathbf{A} \otimes \mathbf{B} &= \text{the Kronecker product of } \mathbf{A} \text{ and } \mathbf{B} \\
&= \begin{bmatrix} a_{11}\mathbf{B} & \cdots & a_{1N}\mathbf{B} \\ \cdot & \cdots & \cdot \\ a_{M1}\mathbf{B} & \cdots & a_{MN}\mathbf{B} \end{bmatrix} \\
\mathbf{A} \odot \mathbf{B} &= \text{the Hadamard product of two matrices of same dimension} \\
&= \begin{bmatrix} a_{11}b_{11} & \cdots & a_{1N}b_{1N} \\ \cdot & \cdots & \cdot \\ a_{M1}b_{M1} & \cdots & a_{MN}b_{MN} \end{bmatrix}.
\end{aligned}
$$

6.2 DOA ESTIMATION IN SPATIALLY WHITE NOISE

Signal Model

We first review some methods of estimating the DOA of signals in a spatially white noise environment. The signal model we consider here consists of an M-dimensional complex data vector $\mathbf{x}(n) \in \mathcal{C}^M$ which represents the data received by an array of M sensors at the nth snapshot. The data vector is composed of plane-wave incident narrowband signals each of angular frequency ω_0 from K distinct sources embedded in Gaussian noise. The received data vector at the nth snapshot can be written as

$$\mathbf{x}(n) = \mathbf{D}(\boldsymbol{\theta})\mathbf{s}(n)\mathbf{v}(n), \quad n = 1, \cdots, N \tag{6.1}$$

where N is the total number of snapshots observed. $\mathbf{s}(n) \in \mathcal{C}^K$ is the unknown K-dimensional complex vector whose elements represent the phasors of the signals at the nth snapshot. $\boldsymbol{\theta} \in \mathcal{R}^K$ is the K-dimensional real vector the elements of which represent the DOA of the K plane-wave signals such that

$$\boldsymbol{\theta} = [\theta_1 \cdots \theta_K]^T \tag{6.2}$$

where ^{T} denotes the transpose of a vector or matrix. The vector $\boldsymbol{\theta}$ in Eq. (6.2) is the vector to be estimated. The matrix $\mathbf{D}(\boldsymbol{\theta}) \in \mathcal{C}^{M \times K}$ has the following structure:

$$\mathbf{D}(\boldsymbol{\theta}) = [\mathbf{d}(\theta_1) \cdots \mathbf{d}(\theta_K)] \triangleq [\mathbf{d}_1 \cdots \mathbf{d}_K] \tag{6.3}$$

where $\mathbf{d}_k \triangleq \mathbf{d}(\theta_k) \in \mathcal{C}^M$, $k = 1, \cdots, K$, is an M-dimensional complex vector usually called the *directional vector* or sometimes also the *steering vector* of the signals. The exact form of the steering vector depends on the array configuration. Most of the discussions in this chapter do not require a specific form of $\mathbf{d}(\theta_k)$, that is, they are valid for any general array geometry. However, the uniform linear array, apart from being most commonly used, may also offer advantageous implementation efficiency of some algorithm. A uniform linear array is shown in Figure 6.1 where the array elements are uniformly spaced. For a velocity of propagation c, the distance between two sensors in a uniform linear array must be $L_0 \le \pi c/\omega_0$ and the corresponding steering matrix is given by

$$\mathbf{d}(\theta_k) = \begin{bmatrix} 1 & e^{j\omega_0 L_0 \sin\theta_k/c} & \cdots & e^{j(M-1)\omega_0 L_0 \sin\theta_k/c} \end{bmatrix}^T \tag{6.4}$$

where $\theta_k \in (-\frac{\pi}{2}, \frac{\pi}{2})$ for all k. The vectors $\mathbf{d}(\theta_k), k = 1, \cdots, K$ corresponding to K different values of θ_k are assumed to be linearly independent. This implies that $M > K$, and

$$\text{rank}\ (\mathbf{D}) = K \tag{6.5}$$

The noise vector sequence $\{\mathbf{v}(n)\}$ is assumed to be zero mean circular Gaussian such that

$$E\left[\mathbf{v}(n_1)\mathbf{v}^H(n_2)\right] = \sigma_v^2 \mathbf{I}\delta_{n_1 n_2} \tag{6.6a}$$

and

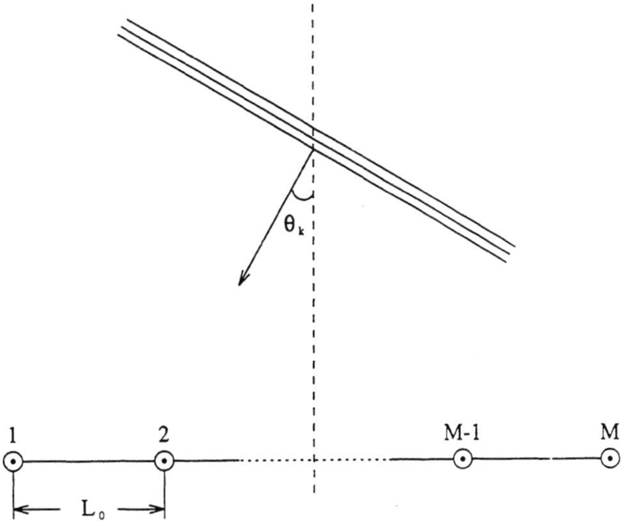

Figure 6.1 An incident signal and a linear uniform sensor array.

$$E\left[\mathbf{v}(n_1)\mathbf{v}^T(n_2)\right] = \mathbf{O} \tag{6.6b}$$

where ^{H} denotes the conjugate transpose operation

$$\delta_{n_1 n_2} = \begin{cases} 0 & n_1 \neq N_2 \\ 1 & n_1 = N_2 \end{cases}$$

and σ_v^2 is the noise power. Furthermore, $\mathbf{v}(n_1)$ and $\mathbf{s}(n_2)$ are assumed to be uncorrelated for all n_1 and n_2. Thus, if we form the correlation matrix of the received data $\mathbf{x}(n)$, using the above assumptions on the noise vector, we have

$$\mathbf{\Sigma}_x \triangleq E\left[\mathbf{x}\mathbf{x}^H\right] = \mathbf{D}\mathbf{\Sigma}_s\mathbf{D}^H + \sigma_v^2\mathbf{I} \tag{6.7}$$

where

$$\mathbf{\Sigma}_s \triangleq E\left[\mathbf{s}\mathbf{s}^H\right] \tag{6.8}$$

and is positive definite.

In practice, the true value of $\mathbf{\Sigma}_x$ is seldom known. Therefore, an estimate $\hat{\mathbf{\Sigma}}_x(N)$ is obtained by averaging the outer products of the data over the total number of snapshots N such that

$$\hat{\mathbf{\Sigma}}_x(N) = \frac{1}{N}\sum_{n=1}^{N} \mathbf{x}(n)\mathbf{x}^H(n) \tag{6.9}$$

where $\hat{}$ denotes the estimated value of a quantity. In the ensuing material, the symbol $\hat{}$ appearing over a variable indicates the estimate of the variable based on $\hat{\mathbf{\Sigma}}_x(N)$. It can be shown [18, 21] that under certain conditions the estimate $\hat{\mathbf{\Sigma}}_x(N)$ converges to $\mathbf{\Sigma}$ with probability 1 and at a rate proportional to $(\log\log N/N)^{1/2}$, that is,

$$\lim_{N\to\infty} \hat{\mathbf{\Sigma}}_x(N) \underset{a.s.}{=} \mathbf{\Sigma}_x \tag{6.10}$$

where "$\underset{a.s.}{=}$" indicates that the equality holds with probability 1, or almost surely. In particular, if the signal $\mathbf{s}$ is Gaussian, then under some regular conditions, the convergence of $\hat{\mathbf{\Sigma}}_x$ to $\mathbf{\Sigma}_x$ is at the rate of $O(N^{-1/2})$ in probability [54, 82]. In the following text, the explicit dependence of $\hat{\mathbf{\Sigma}}_x$ on N will be omitted to maintain conciseness of the expression.

In the past decade or two, various high-resolution methods of estimating the DOA of the signal sources based on the above model have been developed. Most of these methods are based on the partition of the space of the observed data into the signal and the noise subspaces assuming that the number of the signals K is known. Different techniques are then employed to locate the DOA of the signals using these estimated subspaces. Two methods have withstood the test of time and have remained popular among researchers in high-resolution array processing.

The MUSIC Algorithm

The *Multiple Signal Classification* (MUSIC) algorithm [31] of DOA estimation in high-resolution array processing assumes, in addition to Assumptions 1.1–1.4, that the K signal sources are incoherent with each other, and that the number K has been correctly determined.

We begin by examining the eigenvalues λ_m and the corresponding normalized eigenvectors $\boldsymbol{\upsilon}_m, m = 1, \cdots, M$ of $\mathbf{\Sigma}_x$. Since $\mathbf{\Sigma}_x$ is Hermitian and positive definite [95], its eigenvalues are real and positive, and the eigenvectors are orthonormal. Furthermore, $\mathbf{\Sigma}_x$ can be decomposed such that [74, 75, 95]

$$\mathbf{\Sigma}_x = \sum_{m=1}^{M} \lambda_m \boldsymbol{\upsilon}_m \boldsymbol{\upsilon}_m^H \tag{6.11}$$

where $\lambda_1 > \lambda_2 > \cdots > \lambda_K > \lambda_{K+1} = \cdots = \lambda_M = \sigma_v^2$, and

$$\boldsymbol{\upsilon}_m^H \boldsymbol{\upsilon}_n = 0 \quad \text{for } m \neq n \tag{6.12}$$

From the definition of eigenvalues and eigenvectors, we can write

$$(\mathbf{\Sigma}_x - \sigma_v^2 \mathbf{I})\boldsymbol{\upsilon}_m = \mathbf{O} \quad m = K+1, \cdots, M \tag{6.13}$$

Using Eq. (6.7), Eq. (6.13) can be rewritten as

$$\mathbf{D}\mathbf{\Sigma}_s\mathbf{D}^H \boldsymbol{\upsilon}_m = \mathbf{O} \quad m = K+1, \cdots, M \tag{6.14}$$

Since the signals are incoherent, $\mathbf{\Sigma}_s$ is full rank. Also, since the signal sources are from K distinct directions, $\mathbf{D}$ is of full column rank. Thus, we can conclude that

$$\mathbf{D}^H \boldsymbol{\upsilon}_m = \mathbf{O} \quad m = K+1, \cdots, M \tag{6.15}$$

Equation (6.15) tells us that each of the K directional vectors $\mathbf{d}_k$ is orthogonal to the eigenvectors $\{\boldsymbol{\upsilon}_m\}, m = K+1, \cdots, M$, of the matrix $\mathbf{\Sigma}_x$. But from Eq. (6.12), we see that the eigenvectors $\{\upsilon_k\}, k = 1, \cdots, K$, span the orthogonal complement of the space spanned by $\{\upsilon_m, m = K+1, \cdots, M\}$. Hence we can deduce that

$$\text{span}\,\{\mathbf{d}_1, \cdots, \mathbf{d}_K\} = \text{span}\,\{\upsilon_1, \cdots, \upsilon_K\} \tag{6.16}$$

Therefore, we can partition the eigenvectors of $\mathbf{\Sigma}_x$ into two groups such that

$$\mathbf{V}_s \triangleq [\upsilon_1 \upsilon_2 \cdots \upsilon_K] \tag{6.17a}$$

and

$$\mathbf{V}_\nu \triangleq [\upsilon_{K+1} \upsilon_{K+2} \cdots \upsilon_M] \tag{6.17b}$$

From Eqs. (6.15) and (6.16), we see that the eigenvectors of $\mathbf{\Sigma}_x$ span two disjoint subspaces. The one spanned by the eigenvectors corresponding to the K largest eigenvalues of $\mathbf{\Sigma}_x$ is called the *signal subspace* bacause it is also spanned by the directional vectors of the signals. The one spanned by the eigenvectors corresponding to the $(M-K)$ smallest eigenvalues of $\mathbf{\Sigma}_x$ is called the *noise subspace* because of its association with the noise power σ_ν^2. The principle of the MUSIC algorithm is to determine the DOA of the signals by searching for the directional vectors $\mathbf{d}_k, k = 1, \cdots, K$, that are orthogonal to the noise subspace.

As mentioned before, in practice we can only estimate $\mathbf{\Sigma}_x$ by $\hat{\mathbf{\Sigma}}_x$ given by Eq. (6.9). Since $\hat{\mathbf{\Sigma}}_x$ is still Hermitian and positive definite, its eigenvalues, denoted by $\hat{\lambda}_m, m = 1, \cdots, M$, are still positive, and the associated eigenvectors, denoted by $\{\hat{\upsilon}_m\}, m = 1, \cdots, M$, are still mutually orthonormal. However, the smallest $(M-K)$ eigenvalues $\hat{\lambda}_{K+1}, \cdots, \hat{\lambda}_M$ will not be equal any more, and the estimated noise subspace spanned by $\{\hat{\upsilon}_m\}, m = K+1, \cdots, M$, will no longer be orthogonal to the true signal subspace spanned by $\{\mathbf{d}_k\}, k = 1, \cdots, K$. The best that we can do is to search for the directional vectors which are closest to being orthogonal to the noise subspace. Accordingly, we form an *orthogonal projector* [75, 76] such that

$$\hat{\mathbf{P}}_\nu = \hat{\mathbf{V}}_\nu(\hat{\mathbf{V}}_\nu^H \hat{\mathbf{V}}_\nu)^{-1}\hat{\mathbf{V}}_\nu^H = \hat{\mathbf{V}}_\nu \hat{\mathbf{V}}_\nu^H \tag{6.18}$$

where $\hat{\mathbf{V}}_\nu = \left[\hat{\upsilon}_{K+1}, \cdots, \hat{\upsilon}_M\right]$. The second equality in Eq. (6.18) is due to the orthonormality of the eigenvectors. We note that

$$\hat{\mathbf{P}}_\nu^H = \hat{\mathbf{P}}_\nu \tag{6.19a}$$

and

$$\hat{\mathbf{P}}_\nu^H \hat{\mathbf{P}}_\nu = \hat{\mathbf{P}}_\nu \tag{6.19b}$$

We can now use $\hat{\mathbf{P}}_\nu$ to project the general directional vector $\mathbf{d}(\theta)$ onto the estimated noise subspace spanned by the column vectors of $\hat{\mathbf{V}}_\nu$. The values of $\hat{\theta}_k, k = 1, \cdots, K$, which give the K highest peaks of the quantity $1/\|\hat{\mathbf{P}}_\nu \mathbf{d}(\theta)\|^2$ are the estimated DOA of the K signal sources. In other words, the estimates of DOA for the MUSIC algorithm can be written as

$$\hat{\theta}_k = \arg\min_{\theta_k}\{\|\hat{\mathbf{P}}_\nu \mathbf{d}(\theta_k)\|^2\} = \arg\min_{\theta_k}\{\mathbf{d}^H(\theta_k)\hat{\mathbf{P}}_\nu \mathbf{d}(\theta_k)\} \tag{6.20}$$

where $\arg\min_{\theta_k}(\cdot)$ denotes the minimization of the bracketed quantity with respect to the argument θ_k.

Suggestions have been made to weight the projection $\hat{\mathbf{P}}_\nu \mathbf{d}(\theta_k)$ in different ways resulting in various criteria of *weighted MUSIC* algorithms [27, 34]. The MUSIC method remains one of the most popular methods of DOA estimation in high-resolution array processing because of its relative simplicity in computation as well as its relative robustness and efficiency [46–51, 53–55].

The Maximum Likelihood Estimation (MLE)

In the theory of parameter estimation we deal with the problem of extracting information on the properties of stochastic processes given a set of observed data samples. DOA estimation of signals in high-resolution array processing is a problem of this kind. *Maximum likelihood* (ML) [77] is by far the most general and most powerful method of estimation. It can be applied, at least in principle, to all problems of estimation as long as a joint conditional probability density function (PDF) of the observed data, or of the sufficient statistics, can be formulated. The idea of MLE is relatively simple: Different populations generate different data samples and a given set of data samples is more likely to have come from some population than from others.

Let $p(\mathbf{y}|\boldsymbol{\alpha})$ denote the conditional joint PDF of the random vector $\mathbf{y}$, and $\boldsymbol{\alpha}$ is a parameter vector. The maximum likelihood principle states that the estimated value of the parameter vector given the observed sample of $\mathbf{y}$ is that for which $p(\mathbf{y}|\boldsymbol{\alpha})$ is at maximum. The likelihood function, denoted by $\Lambda(\boldsymbol{\alpha})$, is the conditional joint PDF $p(\mathbf{y}|\boldsymbol{\alpha})$ viewed as a function of the parameter vector $\boldsymbol{\alpha}$, that is

$$\Lambda(\boldsymbol{\alpha}) = p(\mathbf{y}|\boldsymbol{\alpha}) \tag{6.21}$$

Often, it is more convenient to work with the logarithm of the likelihood function. The log-likelihood function is given by

$$\ell(\boldsymbol{\alpha}) = \ln[\Lambda(\boldsymbol{\alpha})] = \ln[p(\mathbf{y}|\boldsymbol{\alpha})] \tag{6.22}$$

Since the logarithm function is a monotonic mapping meaning that whenever $\Lambda(\boldsymbol{\alpha})$ increases, $\ell(\boldsymbol{\alpha})$ also increases, then the parameter vector for which $\Lambda(\boldsymbol{\alpha})$ is at maximum is exactly the same parameter vector for which $\ell(\boldsymbol{\alpha})$ is at maximum.

The application of MLE to DOA estimation of signal sources in spatially white noise has been studied by various researchers [32, 33, 37, 42, 43]. If the general random vector $\mathbf{y}$ in Eq. (6.21) is taken to be the observed data $\mathbf{x}$ in our signal model of Eq. (6.1), a likelihood function can be formulated from the conditional joint PDF of the data. Two different criteria of estimation can then be obtained by the ML principle depending on the signal model. The details of the derivation of these two criteria have been discussed in Chapter 1. Here we recapitulate the development in the notation of this chapter.

Deterministic Signal Model. In this model, $\mathbf{s}(n)$ in Eq. (6.1) is a deterministic but unknown quantity. As such, it is a parameter of the PDF of the observed data, and must be estimated. Under this model, the parameters are $\boldsymbol{\theta}, \mathbf{s}$, and σ_v^2, and because of the assumption that the noise is Gaussian and spatially white, the joint conditional PDF of the observed data over N snapshots, and hence the log-likelihood function, can be obtained. By maximizing the log-likelihood function with respect to the parameters and ignoring the irrelevant terms, the ML estimates of DOA in the case of deterministic signals can be written as

$$\hat{\boldsymbol{\theta}} = \arg\max_{\boldsymbol{\theta}} \operatorname{tr}\{\mathbf{D}(\boldsymbol{\theta})\left[\mathbf{D}^H(\boldsymbol{\theta})\mathbf{D}(\boldsymbol{\theta})\right]^{-1}\mathbf{D}^H(\boldsymbol{\theta})\hat{\boldsymbol{\Sigma}}_x\} = \arg\max_{\boldsymbol{\theta}} \operatorname{tr}\{\mathbf{P}_s(\boldsymbol{\theta})\hat{\boldsymbol{\Sigma}}_x\} \tag{6.23}$$

where

$$\mathbf{P}_s(\boldsymbol{\theta}) \triangleq \mathbf{D}(\boldsymbol{\theta})\left[\mathbf{D}^H(\boldsymbol{\theta})\mathbf{D}(\boldsymbol{\theta})\right]^{-1}\mathbf{D}^H(\boldsymbol{\theta}) \tag{6.24}$$

and $\operatorname{tr}(\cdot)$ denotes the matrix trace operation. We note that $\mathbf{P}_s^H(\boldsymbol{\theta}) = \mathbf{P}_s(\boldsymbol{\theta})$ and is in fact an *orthogonal projector* onto the signal subspace spanned by the K column vectors of $\mathbf{D}(\boldsymbol{\theta})$. Thus, the MLE in Eq. (6.23) can be interpreted as the location of the optimal K-dimensional parameter vector $\boldsymbol{\theta}$ which maximizes the norm of the projection of the data onto the signal subspace.

The problem of Eq. (6.23) for a general $\mathbf{D}(\boldsymbol{\theta})$ is a complex nonlinear optimization problem and does not yield a closed-form solution. Iterative methods of solving such a multidimensional optimization problem are usually considered. However, if the array is a uniform linear array, the procedure can be reduced to an iterative linear least-squares problem. Appendix 6C outlines the solution Eq. (6.23) for this particulare case. A noniterative large-sample realization of the MLE for deterministic signals is also available [40]. It should also be noted that for uncorrelated signals, the MUSIC algorithm is a large-sample realization of the MLE algorithm for arbitrary array geometries and thus is a computationally convenient substitute in such cases.

Stochastic Signal Model. In this model, it is assumed that there exists some statistical description of the signals. The signal model adopted here is that $\mathbf{s}(n), n = 1, \cdots, N$, are independent identically distributed (IID) zero-mean Gaussian random vectors with unknown covariance matrix $\boldsymbol{\Sigma}_s$. Thus, a log-likelihood function can be obtained by formulating the joint conditional PDF of the data over N snapshots. The parameters in this case are $\boldsymbol{\theta}$, $\boldsymbol{\Sigma}_s$ and σ_v^2. Again, by maximizing the log-likelihood function with respect to $\boldsymbol{\Sigma}_s$ and σ_v^2 for a fixed $\boldsymbol{\theta}$, and substituting the resulting $\hat{\boldsymbol{\Sigma}}_s$ and $\hat{\sigma}_v^2$ as a function of $\boldsymbol{\theta}$ and $\hat{\boldsymbol{\Sigma}}_x$ back into the log-likelihood function, the ML estimates of DOA in the case of stochastic signals can be written as [33, 53]

$$\hat{\boldsymbol{\theta}} = \arg\max_{\boldsymbol{\theta}}\{-\ln\det[\mathbf{P}_s(\boldsymbol{\theta})\hat{\boldsymbol{\Sigma}}_x\mathbf{P}_s(\boldsymbol{\theta}) + \{\mathbf{I} - \mathbf{P}_s(\boldsymbol{\theta})\}\hat{\sigma}_v^2]\} \tag{6.25}$$

where $\mathbf{P}_s(\boldsymbol{\theta})$ is given by Eq. (6.24) and

$$\hat{\sigma}_v^2 = \operatorname{tr}\{[\mathbf{I} - \mathbf{P}_s(\boldsymbol{\theta})]\hat{\boldsymbol{\Sigma}}_x\}/(M-K) \tag{6.26}$$

The ML estimation method for both the deterministic and the stochastic models poses a difficult nonlinear optimization problem. Neither Eq. (6.23) nor Eq. (6.25) readily yields a closed-form solution and in general, numerical solutions have to be employed. The application of the *expectation-maximization* (EM) algotrithm to this problem seems to be particularly attractive [43] because of its association with maximum-likelihood problems, its monotonicity in the log-likelihood function, and the manner in which it decouples the computation for a possible parallel processor implementation. A detailed description of how the EM algorithm can be formulated to solve the ML estimation of DOA of signals for both the deterministic and the stochastic models is found in [43].

There are other methods for DOA estimation based on the ML principle being applied on the conditional density functions of sufficient statistics. Sharman and Durrani [34] examine the inner product of the directional vector and the estimated noise eigenvectors, that is

$$\boldsymbol{\varepsilon} = \left[\langle \mathbf{d}(\theta), \hat{\mathbf{\upsilon}}_{K+1}\rangle \cdots \langle \mathbf{d}(\theta), \hat{\mathbf{\upsilon}}_M\rangle\right]^T \tag{6.27}$$

By maximizing the likelihood function of $\boldsymbol{\varepsilon}$, an "improved MUSIC" algorithm [34] is arrived at. Stoica and Sharman [40, 42] extended this principle by considering the linear combinations of the sample noise eigenvectors $\{\mathbf{D}^H(\theta)\hat{\mathbf{\upsilon}}_m\}$, $m = K+1, \cdots, M$ or, equivalently, by considering the linear combinations of sample signal eigenvectors $\{\mathbf{B}^H\hat{\mathbf{\upsilon}}_k\}$, $k = 1, \cdots, K$, where $\mathbf{B}$ consists of column vectors spanning the noise subspace. By showing that these sets of vectors are asymptotically circular Gaussian with zero mean, two likelihood functions can be obtained. Letting N tend to infinity for asymptotic considerations and maximizing the two likelihood functions leads to identical results and hence to the same DOA estimation method which is called MODE (method of direction estimation). The MODE criterion can be written as

$$\hat{\mathbf{B}} = \arg\min_{\mathbf{B}} \operatorname{tr}\left[\hat{\mathbf{V}}_s^H\mathbf{B}(\mathbf{B}^H\mathbf{B})^{-1}\mathbf{B}^H\hat{\mathbf{V}}_s\hat{\mathbf{\Lambda}}_s^{-1}\mathring{\mathbf{\Lambda}}_s^2\right] \tag{6.28}$$

where $\hat{\mathbf{V}}_s$ is the sample signal eigenvector matrix defined in Eq. (6.17a), $\hat{\mathbf{\Lambda}}_s = \operatorname{diag}(\hat{\lambda}_1, \cdots, \hat{\lambda}_K)$ consists of the sample eigenvalues of the data covariance matrix, and

$$\mathring{\mathbf{\Lambda}}_s \triangleq \hat{\mathbf{\Lambda}}_s - \hat{\sigma}_v^2\mathbf{I} \tag{6.29}$$

with the estimated noise power given by

$$\hat{\sigma}_v^2 = \frac{1}{M-K}\sum_{m=K+1}^{M}\hat{\lambda}_m \tag{6.30}$$

Since $\mathbf{B}(\mathbf{B}^H\mathbf{B})^{-1}\mathbf{B}^H = \mathbf{I} - \mathbf{D}(\mathbf{D}^H\mathbf{D})^{-1}\mathbf{D}^H$, the estimate of $\mathbf{B}$ will yield an estimate of $\mathbf{D}$ which in turn yields the DOA estimates. It can be shown [40,42] that in the case of a uniform linear array, using an idea parallel to "the re-parameterization" outlined in Appendix 6C, and applying the MODE algorithm, a noninterative algorithm results and the computation can be significantly reduced. Furthermore, it can be shown that MODE is a large-sample realization of the MLE for the stochastic signals.

The MLE, in its various forms, shows superior performance in the estimation of DOA over other methods and has remained perhaps the most powerful technique in high-resolution array processing.

6.3 ARRAY PROCESSING IN UNKNOWN NOISE

The high-resolution of DOA estimation methods described in the previous section are effective under the assumption of spatially white noise. Unfortunately, their performance may significantly degrad in situations where the noise is unknown and spatially correlated. It has been shown [79] that a large bias or spurious estimates may result if the methods discussed in the previous sections are applied to an unknown spatially correlated noise environment. In this section, we establish a signal model for array processing in an unknown noise environment. From this model, analysis of the joint correlation matrix can be performed, new concepts on the signal and noise subspaces can be established, and new methods of array processing in unknown noise can be developed.

The signal and noise model used in this chapter is based on the assumption that the data are received by two arrays having M_1 and M_2 sensors, respectively. Some possible configurations of the two sensor arrays are shown in Figure 6.2. The two data vectors from the arrays are composed of plane-wave incident narrowband signals. We denote the relative DOA of the K signals with respect to the two arrays by $\boldsymbol{\theta}_1$ and $\boldsymbol{\theta}_2$ respectively such that

$$\boldsymbol{\theta}_i = [\theta_{i1}\,\theta_{i2}\cdots\theta_{iK}]^T \tag{6.31}$$

where $\theta_{ik}, i = 1, 2, k = 1, \cdots, K$, is the DOA of the kth signal with reference to the position of the ith array. We assume that $K < \min\{M_1, M_2\}$. Since the estimation of either $\boldsymbol{\theta}_1$ or $\boldsymbol{\theta}_2$ is sufficient to determine the DOA of the signals, our attention is focused on the development of methods to estimate one of the two vectors. However, in the case when both arrays are calibrated, then it may be advantageous to make use of the estimates of both quantities to arrive at an improved estimate.

At the nth snapshot, the two output vectors from the array can be written as

$$\begin{aligned}\mathbf{x}_1(n) &= \mathbf{D}_1(\boldsymbol{\theta}_1)\mathbf{s}_1(n) + \boldsymbol{\nu}_1(n)\\ \mathbf{x}_2(n) &= \mathbf{D}_2(\boldsymbol{\theta}_2)\mathbf{s}_2(n) + \boldsymbol{\nu}_2(n) \quad n = 1, \cdots, N\end{aligned} \tag{6.32}$$

In Eq. (6.32), $\mathbf{x}_1(n)$ and $\mathbf{x}_2(n)$ are data vectors of dimensions M_1 and M_2 respectively. N is the total number of snapshots and $\mathbf{D}_i(\boldsymbol{\theta}_i)$ is an $M_i \times K$ unambiguous directional matrix of the signals with respect to the geometry of the ith array meaning that while all the K column vectors of $\mathbf{D}_i(\boldsymbol{\theta}_i)$ are linearly independent, no other value of $\boldsymbol{\theta}_i$ will result in the column vectors of $\mathbf{D}_i(\boldsymbol{\theta}_i)$ spanning the same K-dimensional subspace. $\mathbf{s}_i(n), i = 1, 2$ are the $K \times 1$ signal vectors received by two arrays and modeled to be complex zero-mean jointly Gaussian vectors. If the traveling of the signals from Array 1 to Array 2 involves no distortion or attenuation, then $\mathbf{s}_2$ is merely a delayed version of $\mathbf{s}_1$. In this case, the narrowband signals appearing in the two arrays may be modeled identically and the actual delay between the two arrays can be absorbed in the directional matrices. The noise vectors $\boldsymbol{\nu}_1(n)$ and $\boldsymbol{\nu}_2(n)$ are of dimensions M_1 and M_2 respectively and are assumed to be stationary, zero-mean, Gaussian with joint covariance given by

$$E\left\{\begin{bmatrix}\boldsymbol{\nu}_1\\ \boldsymbol{\nu}_2\end{bmatrix}\begin{bmatrix}\boldsymbol{\nu}_1^H & \boldsymbol{\nu}_2^H\end{bmatrix}\right\} = \begin{bmatrix}\boldsymbol{\Sigma}_{1\nu} & \mathbf{O}\\ \mathbf{O} & \boldsymbol{\Sigma}_{2\nu}\end{bmatrix} \tag{6.33}$$

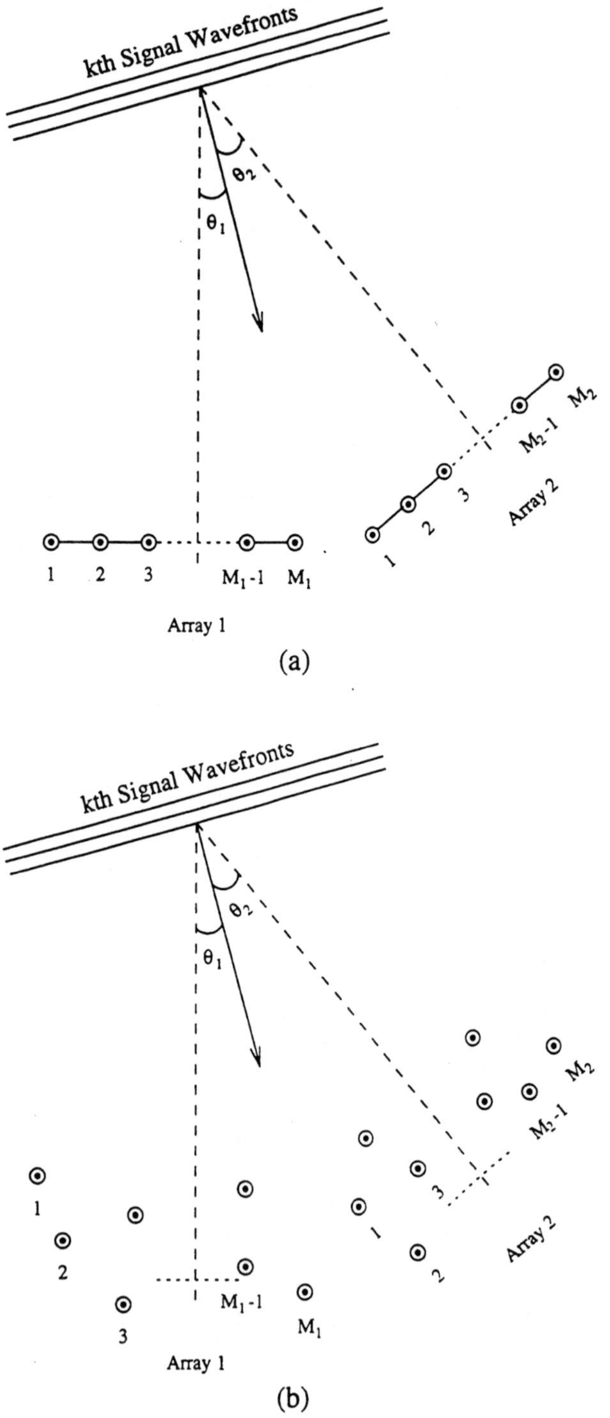

Figure 6.2 Configurations of the two arrays for the estimation of DOA in unknown noise (a) uniform linear arrays (b) two general arrays.

where Σ_{1v} and Σ_{2v} are unknown covariance matrices of the noise in the two arrays. In other words, the noise field is assumed to be correlated only over a limited distance, and the two arrays are sufficiently separated so that the noise outputs of them are uncorrelated. This assumption of spatially limited noise correlation is realistic and has been verified by real data [80, 81]. In practice, the distance for significant spatial correlation of noise is usually in the order of a few wavelengths.

The joint correlation matrix of the received data in Eq. (6.32) can therefore be written as

$$\Sigma \triangleq E\left\{\begin{bmatrix}\mathbf{x}_1\\ \mathbf{x}_2\end{bmatrix}\begin{bmatrix}\mathbf{x}_1^H & \mathbf{x}_2^H\end{bmatrix}\right\} = \begin{bmatrix}\Sigma_{11} & \Sigma_{12}\\ \Sigma_{21} & \Sigma_{22}\end{bmatrix} \tag{6.34}$$

where the submatrices in Eq. (6.34) are given by

$$\Sigma_{ii} = \mathbf{D}_i(\theta_i)\Sigma_{s_i}\mathbf{D}_i^H(\theta_i) + \Sigma_{iv}, \quad i = 1, 2 \tag{6.35a}$$

$$\Sigma_{12} = \Sigma_{21}^H = \mathbf{D}_1(\theta_i)\Sigma_{s_{12}}\mathbf{D}_2^H(\theta_2) \tag{6.35b}$$

with Σ_{s_i} and $\Sigma_{s_{12}}$ being the auto- and cross-correlation matrices of the signals received by the two arrays, that is, $\Sigma_{s_i} = E\{\mathbf{s}_i\mathbf{s}_i^H\}$ and $\Sigma_{s_{12}} = E\{\mathbf{s}_1\mathbf{s}_2^H\}$, both of which are assumed to be of full rank. Here, we have allowed for the general case that the traveling of the signals from Array 1 to Array 2 may involve some distortion of the signal such that $\mathbf{s}_2$ may no longer be merely a delayed version of $\mathbf{s}_1$. As mentioned in section 6.2, in practice, we have to estimate the correlation matrix Σ by an averaging process such that

$$\hat{\Sigma} = \frac{1}{N}\sum_{n=1}^{N}\begin{bmatrix}\mathbf{x}_1(n)\\ \mathbf{x}_2(n)\end{bmatrix}\begin{bmatrix}\mathbf{x}_1^H(n) & \mathbf{x}_2^H(n)\end{bmatrix} = \begin{bmatrix}\hat{\Sigma}_{11} & \hat{\Sigma}_{12}\\ \hat{\Sigma}_{21} & \hat{\Sigma}_{22}\end{bmatrix} \tag{6.36}$$

That the estimate $\hat{\Sigma}$ converges to Σ has been discussed in the previous section already. Of course, the rank of the estimated submatrix $\hat{\Sigma}_{12}$ will no longer be K but will be $M_1(\leq M_2)$ due to the uncertainty in estimation.

Since the output noise vectors of the two arays are assumed to be uncorrelated to each other, the submatrix $\Sigma_{12}(= \Sigma_{21}^H)$ in Eq. (6.34) contains only the correlation of the signal components in the outputs of the two arrays. Thus, estimating the rank of Σ_{12} using its estimate $\hat{\Sigma}_{12}$ in Eq. (6.36) enables us to determine the number of signals arriving at the array. Furthermore, one can conceive that the signal subspace which is spanned by the columns of $\mathbf{D}_i(\theta_i)$ can also be estimated from $\hat{\Sigma}_{12}$. From the estimated signal subspace, the DOA of the signals can then be estimated using methods parallel to those discussed in section 6.2 for spatially white noise environment. To facilitate the development of the algorithms for the determination of the number of signals and the estimation of their DOA, it is necessary to introduce the analysis of the decomposition of Σ and $\hat{\Sigma}$.

6.4 GENERALIZED CORRELATION ANALYSIS

In this section, we first develop a general set of decomposition techniques for estimating the signal subspace and its complement form $\hat{\mathbf{\Sigma}}_{12}$. Such a decomposition set, called *Generalized Correlation Decomposition*(GCD), includes the commonly used decomposition techniques in statistical signal processing. We also develop some properties of GCD and examine how GCD can be applied to array processing.

Suppose we have two positive definite Hermitian matrices $\mathbf{\Pi}_1$ and $\mathbf{\Pi}_2$ of dimensions $M_1 \times M_1$ and $M_2 \times M_2$ respectively. We note that the positive definite square roots $\mathbf{\Pi}_1^{1/2}$ and $\mathbf{\Pi}_2^{1/2}$ are uniquely defined [74,75]. Then, referring to the joint covariance matrix $\mathbf{\Sigma}$ in Eq. (6.34), and assuming $M_1 \leq M_2$ without loss of generality, we can form the $M_1 \times M_2$ matrix product $\mathbf{\Pi}_1^{-1/2}\mathbf{\Sigma}_{12}\mathbf{\Pi}_2^{-1/2}$ on which a singular value decomposition (SVD) [75] can be performed such that

$$\mathbf{\Pi}_1^{-1/2}\mathbf{\Sigma}_{12}\mathbf{\Pi}_2^{-1/2} = \mathbf{U}_1\mathbf{\Gamma}_0\mathbf{U}_2^H \tag{6.37}$$

where $\mathbf{U}_1 \in \mathcal{C}^{M_1 \times M_1}$ and $\mathbf{U}_2 \in \mathcal{C}^{M_2 \times M_2}$ are unitary matrices and $\mathbf{\Gamma}_0$ is an $M_1 \times M_2$ matrix given by

$$\mathbf{\Gamma}_0 = \left[\begin{array}{c|c} \mathbf{\Gamma} & \mathbf{O} \\ \hline \mathbf{O} & \mathbf{O} \end{array}\right] \tag{6.38}$$

Since there are K signals, from Eq. (6.35b), $\mathbf{\Sigma}_{12}$ is of rank K. Therefore, $\mathbf{\Gamma}$ in Eq. (6.38) is given by

$$\mathbf{\Gamma} = \text{diag}\,(\gamma_1, \cdots, \gamma_K) \tag{6.39}$$

where $\gamma_k, k = 1, \cdots, K$, are real and positive such that $\gamma_1 \geq \cdots \geq \gamma_K > 0$. Eq. (6.37) can be rewritten as

$$\mathbf{\Pi}_1^{-1}\mathbf{\Sigma}_{12}\mathbf{\Pi}_2^{-1} = \mathbf{\Pi}_1^{-1/2}\mathbf{U}_1\mathbf{\Gamma}_0\mathbf{U}_2^H\mathbf{\Pi}_2^{-1/2} \tag{6.40}$$

Accordingly, we have the following definition

Definition 6.1. *From Eq. (6.40), the generalized correlation vector matrices of $\mathbf{\Sigma}_{12}$ and $\mathbf{\Sigma}_{21}$ are respectively defined as*

$$\mathbf{L}_1 = \mathbf{\Pi}_1^{-1/2}\mathbf{U}_1, \qquad \mathbf{L}_2 = \mathbf{\Pi}_2^{-1/2}\mathbf{U}_2 \tag{6.41a}$$

and the nonzero elements γ_k of $\mathbf{\Gamma}_0$ are called the generalized correlation coefficients of $\mathbf{\Sigma}_{12}$ and $\mathbf{\Sigma}_{21}$. In addition, the reciprocal generalized correlation vector matrices of $\mathbf{\Sigma}_{12}$ and $\mathbf{\Sigma}_{21}$ are respectively defined as

$$\mathbf{R}_1 = \mathbf{\Pi}_1^{1/2}\mathbf{U}_1, \qquad \mathbf{R}_2 = \mathbf{\Pi}_2^{1/2}\mathbf{U}_2 \tag{6.41b}$$

Definition 6.2 [85]. *An $M \times M$ matrix $\mathbf{H}$ is said to be Hermitian in the metric of a positive definite Hermitian matrix $\mathbf{\Pi}$ if $\mathbf{\Pi H}$ is Hermitian, that is, $\mathbf{\Pi H} = \mathbf{H}^H\mathbf{\Pi}^H$. $\mathbf{H}$ is also said to be generalized Hermitian.*

We now show that $\mathbf{L}_i$ and $\mathbf{R}_i$ are actually the eigenvector matrices of some matrices which are Hermitian in the metric of $\boldsymbol{\Pi}_i$. Let $\bar{i}$ denote the complement of the index i, that is if $i = 1$, then $\bar{i} = 2$ and *vice versa*.

Theorem 6.1. *Define the matrices*

$$\mathbf{S}_i = (\boldsymbol{\Pi}_i^{-1}\boldsymbol{\Sigma}_{i\bar{i}}\boldsymbol{\Pi}_{\bar{i}}^{-1}\boldsymbol{\Sigma}_{\bar{i}i}), \quad i = 1, 2, \tag{6.42}$$

Then, $\mathbf{S}_i$ *and* $\mathbf{S}_i^H$ *are Hermitian in the metrics of* $\boldsymbol{\Pi}_i$ *and* $\boldsymbol{\Pi}_i^{-1}, i = 1, 2$, *respectively. Furthermore,* $\mathbf{S}_i$ *and* $\mathbf{S}_i^H$ *both have the same eigenvalues being equal to* $\{\gamma_1^2, \cdots, \gamma_K^2, 0, \cdots, 0\}$ *and have eigenvectors being the columns of* $\mathbf{L}_i$ *and* $\mathbf{R}_i$, *respectively.*

Proof. That $\mathbf{S}_i$ and $\mathbf{S}_i^H$ are Hermitian in the metric of $\boldsymbol{\Pi}_i$ can easily be shown directly from Definition 6.2 and using the facts that $\boldsymbol{\Pi}_i^{-1} = (\boldsymbol{\Pi}_i^{-1})^H$ and $\boldsymbol{\Sigma}_{12}^H = \boldsymbol{\Sigma}_{21}$.

To find the eigenvectors of $\mathbf{S}_1$, we see that from Eq. (6.37), we can write

$$\boldsymbol{\Gamma}_0 = \mathbf{U}_1^H\boldsymbol{\Pi}_1^{-1/2}\boldsymbol{\Sigma}_{12}\boldsymbol{\Pi}_2^{-1/2}\mathbf{U}_2 \tag{6.43a}$$

and

$$\boldsymbol{\Gamma}_0^H = \mathbf{U}_2^H\boldsymbol{\Pi}_2^{-1/2}\boldsymbol{\Sigma}_{21}\boldsymbol{\Pi}_1^{-1/2}\mathbf{U}_1 \tag{6.43b}$$

Multiplying the two equations, we can write

$$\boldsymbol{\Gamma}_0\boldsymbol{\Gamma}_0^H = (\mathbf{U}_1^H\boldsymbol{\Pi}_1^{-1/2}\boldsymbol{\Sigma}_{12}\boldsymbol{\Pi}_2^{-1/2}\mathbf{U}_2)(\mathbf{U}_2^H\boldsymbol{\Pi}_2^{-1/2}\boldsymbol{\Sigma}_{21}\boldsymbol{\Pi}_1^{-1/2}\mathbf{U}_1) \tag{6.44a}$$

$$= \mathbf{L}_1^{-1}(\boldsymbol{\Pi}_1^{-1}\boldsymbol{\Sigma}_{12}\boldsymbol{\Pi}_2^{-1}\boldsymbol{\Sigma}_{21})\mathbf{L}_1 = \mathbf{L}_1^{-1}\mathbf{S}_1\mathbf{L}_1 \tag{6.44b}$$

But $\boldsymbol{\Gamma}_0\boldsymbol{\Gamma}_0^H = \text{diag}(\gamma_1^2, \cdots, \gamma_K^2, 0, \cdots, 0)$, and is an $M_1 \times M_1$ diagonal matrix, thus, Eq. (6.44b) represents a similarity transformation of $\mathbf{S}_1$, and the eigenvalues of $\mathbf{S}_1$ are $\{\gamma_1^2, \cdots, \gamma_K^2, 0, \cdots, 0\}$ with the corresponding eigenvectors being the columns of $\mathbf{L}_1$. Finally, from Eqs. (6.41a) and (6.41b), we see that

$$\mathbf{R}_1^H = \mathbf{L}_1^{-1} \tag{6.45}$$

Then substituting Eq. (6.45) into Eq. (6.44b) and taking the conjugate transpose, we obtain

$$\mathbf{R}_1^{-1}\mathbf{S}_1^H\mathbf{R}_1 = \text{diag}\,(\gamma_1^2, \cdots, \gamma_K^2, 0, \cdots, 0) \tag{6.46}$$

Hence, the eigenvalues of $\mathbf{S}_1^H$ are also $\{\gamma_1^2, \cdots, \gamma_K^2, 0, \cdots, 0\}$ with the columns of $\mathbf{R}_1$ being the corresponding eigenvectors. Similarly, we can prove the result for $i = 2$ by starting with the product $\boldsymbol{\Gamma}_0^H\boldsymbol{\Gamma}_0$ in Eq. (6.44a). ■

Remark 6.1. Theorem 6.1 is equally valid if GCD is performed on the estimated correlation matrix $\hat{\boldsymbol{\Sigma}}$. *The resulting matrix* $\hat{\mathbf{S}}_i$, *however, will be of rank* M_1 *since* $\hat{\boldsymbol{\Sigma}}_{i\bar{i}}$ *will be of rank* M_1 *instead of* K. *The nonzero eigenvalues of* $\hat{\mathbf{S}}_i$ *and* $\hat{\mathbf{S}}_i^H$ *will then be* $\{\hat{\gamma}_1^2, \cdots, \hat{\gamma}_{M_1}^2\}$ *and the eigenvectors will be the columns of* $\hat{\mathbf{L}}_i$ *and* $\hat{\mathbf{R}}_i$ *respectively.*

It is obvious that different choices of $\boldsymbol{\Pi}_i, i = 1, 2$, in the GCD of Eq. (6.37) lead to different decompositions.

Example 6.1

If $\mathbf{\Pi}_i = \mathbf{I}_{M_i}$, where $\mathbf{I}_{M_i}, i = 1, 2$, is an $M_i \times M_i$ identity matrix, then from Eqs. (6.41a) and (6.41b), we have

$$\mathbf{L}_i = \mathbf{R}_i, \quad i = 1, 2$$

$\mathbf{L}_i$ and $\mathbf{L}_{\bar{i}}$ are respectively the left and right singular vector matrices [75] of $\mathbf{\Sigma}_{i\bar{i}}$.

Example 6.2

Consider the special case when we have only one array with M_1 sensors, that is, we only have the first array available. Since there is only one array, $\mathbf{x}_1 \equiv \mathbf{x}_2$ and $\mathbf{\Sigma}_{12} = \mathbf{\Sigma}_{21} = \mathbf{\Sigma}_{11}$. Note that in this special case, the rank of $\mathbf{\Sigma}_{12}$ is M_1 instead of K due to the correlation of noise within the array and $\mathbf{\Gamma}_0$ is a square diagonal matrix of dimension $M_1 \times M_1$. Now, if we choose $\mathbf{\Pi}_1 = \mathbf{I}$ and $\mathbf{\Pi}_2 = \mathbf{\Sigma}_{11}$, then the matrix $\mathbf{S}_1$ in Theorem 6.1 is reduced to $\mathbf{S}_1 = \mathbf{\Sigma}_{11}$ and Eq. (6.44) is reduced to

$$\mathbf{\Gamma}_0\mathbf{\Gamma}_0^H = (\mathbf{U}_1^H\mathbf{\Pi}_1^{-1/2}\mathbf{\Sigma}_{12}\mathbf{\Pi}_2^{-1/2}\mathbf{U}_2)(\mathbf{U}_2^H\mathbf{\Pi}_2^{-1/2}\mathbf{\Sigma}_{21}\mathbf{\Pi}_1^{-1/2}\mathbf{U}_1) \tag{6.47a}$$

$$= \mathbf{L}_1^{-1}\mathbf{\Sigma}_{11}\mathbf{L}_1 \tag{6.47b}$$

We observe that $\mathbf{\Gamma}_0\mathbf{\Gamma}_0^H = \mathrm{diag}(\gamma_1^2, \cdots, \gamma_{M_1}^2)$, and Eq. (6.47b) is an eigen-decomposition of $\mathbf{\Sigma}_{11}$ with $\{\gamma_m^2\}, m = 1, \cdots, M_1$, being its eigenvalues and $\mathbf{L}_1$ being its eigenvector matrix. We note that in this case, $\mathbf{R}_1 = \mathbf{L}_1$.

Example 6.3

Again consider the special case in Example 6.2 in which $\mathbf{\Sigma}_{12} = \mathbf{\Sigma}_{21} = \mathbf{\Sigma}_{11}$. If we choose $\mathbf{\Pi}_1 = \mathbf{\Sigma}_{1\nu}$ and $\mathbf{\Pi}_2 = \mathbf{\Sigma}_{11}$, then Eq. (6.47a) becomes

$$\mathbf{\Gamma}_0\mathbf{\Gamma}_0^H = \mathbf{\Gamma}_0^2 = \mathbf{U}_1^H\mathbf{\Sigma}_{1\nu}^{-1/2}\mathbf{\Sigma}_{11}\mathbf{\Sigma}_{1\nu}^{-1/2}\mathbf{U}_1 \tag{6.48}$$

Pre-multiplying both sides of Eq. (6.48) by $\mathbf{L}_1 = \mathbf{\Sigma}_{1\nu}^{-1/2}\mathbf{U}_1$, we obtain

$$\mathbf{L}_1\mathbf{\Gamma}_0^2 = \mathbf{\Sigma}_{1\nu}^{-1}\mathbf{\Sigma}_{11}\mathbf{L}_1 \tag{6.49}$$

Thus, by the choices of $\mathbf{\Pi}_1$ and $\mathbf{\Pi}_2$ in this special example, the generalized correlation decomposition of $\mathbf{\Sigma}$ is reduced to the generalized eigen-decomposition of the matrix pencil $(\mathbf{\Sigma}_{11}, \mathbf{\Sigma}_{1\nu})$ with $\{\gamma_m^2\}, m = 1, \cdots, M_1$, being the generalized eigenvalues and the columns of $\mathbf{L}_1$ being the generalized eigenvectors.

Example 6.4

Returning to our model when there are two arrays available, if we choose $\mathbf{\Pi}_i = \mathbf{\Sigma}_{ii}, i = 1, 2$, where $\mathbf{\Sigma}_{ii}$ are the sub-matrices along the major diagonal of $\mathbf{\Sigma}$ in Eq. (6.34), the resulting decomposition is called the *Canonical Correlation Decomposition (CCD)* [77, 82, 83].

Canonical Correlation Decomposition (CCD)

Specifically, CCD has been developed and studied in the area of multivariate statistics and time series analysis [77, 82–85]. Since CCD plays an important and special role in this study as will be evidenced in the ensuing sections, we use $\tilde{\mathbf{L}}_i$, $\tilde{\mathbf{R}}_i$ and $\tilde{\mathbf{\Gamma}}$

to denote the matrices $\mathbf{L}_i$, $\mathbf{R}_i$ and $\mathbf{\Gamma}$, respectively, when CCD is employed, that is, $\mathbf{\Pi}_i = \mathbf{\Sigma}_{ii}, i = 1, 2$. Thus, in our notation, Eqs. (6.37), (6.41a), and (6.41b) are, in the case of CCD, written as

$$\mathbf{\Sigma}_{11}^{-1/2}\mathbf{\Sigma}_{12}\mathbf{\Sigma}_{22}^{-1/2} = \tilde{\mathbf{U}}_1\tilde{\mathbf{\Gamma}}_0\tilde{\mathbf{U}}_2^H \tag{6.50}$$

and

$$\tilde{\mathbf{L}}_i = \mathbf{\Sigma}_{ii}^{-1/2}\tilde{\mathbf{U}}_i, \quad \tilde{\mathbf{R}}_i = \mathbf{\Sigma}_{ii}^{1/2}\tilde{\mathbf{U}}_i, \quad i = 1, 2 \tag{6.51}$$

where the nonzero elements $\{\tilde{\gamma}_1, \cdots, \tilde{\gamma}_K\}$ of $\tilde{\mathbf{\Gamma}}_0$ are called the *canonical correlation coefficients*, $\tilde{\mathbf{L}}_i$ and $\tilde{\mathbf{R}}_i, i = 1, 2$, are respectively called the *canonical vector matrices* and the *reciprocal canonical vector matrices* corresponding to the data $\mathbf{x}_i$.

CCD attempts to characterize the correlation structure between two sets of variables $\mathbf{x}_1$ and $\mathbf{x}_2$ by replacing them with two new sets. If we consider the linear combinations $\boldsymbol{\alpha}_{1k}^H\mathbf{x}_1$ and $\boldsymbol{\alpha}_{2k}^H\mathbf{x}_2, k = 1, \cdots, K$, of the two sets of variables, where $\boldsymbol{\alpha}_{1k}$ and $\boldsymbol{\alpha}_{2k}$ are M_1-tuples and M_2-tuples respectively, the correlation coefficient between any two corresponding sets $\boldsymbol{\alpha}_{1k}^H\mathbf{x}_1$ and $\boldsymbol{\alpha}_{2k}^H\mathbf{x}_2$ is defined as

$$\begin{aligned} r_k &\triangleq E\left[\boldsymbol{\alpha}_{1k}^H\mathbf{x}_1\mathbf{x}_2^H\boldsymbol{\alpha}_{2k}\right]/\{E\left[\boldsymbol{\alpha}_{1k}^H\mathbf{x}_1\mathbf{x}_1^H\boldsymbol{\alpha}_{1k}\right]E\left[\boldsymbol{\alpha}_{2k}^H\mathbf{x}_2\mathbf{x}_2^H\boldsymbol{\alpha}_{2k}\right]\}^{1/2} \\ &= \boldsymbol{\alpha}_{1k}^H\mathbf{\Sigma}_{12}\boldsymbol{\alpha}_{2k}/\{\boldsymbol{\alpha}_{1k}^H\mathbf{\Sigma}_{11}\boldsymbol{\alpha}_{1k}\boldsymbol{\alpha}_{2k}^H\mathbf{\Sigma}_{22}\boldsymbol{\alpha}_{2k}\}^{1/2}, \quad k = 1, \cdots, K \end{aligned} \tag{6.52}$$

Property 6.1. *The canonical correlation coefficiants, $\{\tilde{\gamma}_k\}$, $k = 1, \cdots, K$, are the maximum values attained by r_k with $\boldsymbol{\alpha}_{ik}^H\mathbf{\Sigma}_{ii}\boldsymbol{\alpha}_{il} = 0, k \neq l$. These maximum correlations can be achieved by having $\boldsymbol{\alpha}_{ik} = \tilde{\mathbf{l}}_{ik}, i = 1, 2$, where $\tilde{\mathbf{l}}_{ik}$ is the kth column of $\tilde{\mathbf{L}}_i$.*

This property can be shown by applying the Schwarz inequality such that from Eq. (6.52)

$$\begin{aligned} r_k^2 &= (\boldsymbol{\alpha}_{1k}^H\mathbf{\Sigma}_{12}\boldsymbol{\alpha}_{2k})^2/(\boldsymbol{\alpha}_{1k}^H\mathbf{\Sigma}_{11}\boldsymbol{\alpha}_{1k}\boldsymbol{\alpha}_{2k}^H\mathbf{\Sigma}_{22}\boldsymbol{\alpha}_{2k}) \\ &= (\boldsymbol{\alpha}_{1k}^H\mathbf{\Sigma}_{12}\mathbf{\Sigma}_{22}^{-1/2}\mathbf{\Sigma}_{22}^{1/2}\boldsymbol{\alpha}_{2k})^2/(\boldsymbol{\alpha}_{1k}^H\mathbf{\Sigma}_{11}\boldsymbol{\alpha}_{1k}\boldsymbol{\alpha}_{2k}^H\mathbf{\Sigma}_{22}\boldsymbol{\alpha}_{2k}) \\ &\leq (\boldsymbol{\alpha}_{1k}^H\mathbf{\Sigma}_{12}\mathbf{\Sigma}_{22}^{-1}\mathbf{\Sigma}_{21}\boldsymbol{\alpha}_{1k}\boldsymbol{\alpha}_{2k}^H\mathbf{\Sigma}_{22}\boldsymbol{\alpha}_{2k})/(\boldsymbol{\alpha}_{1k}^H\mathbf{\Sigma}_{11}\boldsymbol{\alpha}_{1k}\boldsymbol{\alpha}_{2k}^H\mathbf{\Sigma}_{22}\boldsymbol{\alpha}_{2k}) \\ &= \boldsymbol{\alpha}_{1k}^H\mathbf{\Sigma}_{12}\mathbf{\Sigma}_{22}^{-1}\mathbf{\Sigma}_{21}\boldsymbol{\alpha}_{1k}/\boldsymbol{\alpha}_{1k}^H\mathbf{\Sigma}_{11}\boldsymbol{\alpha}_{1k} \end{aligned} \tag{6.53}$$

Equality in Eq. (6.53) is attained if and only if

$$\boldsymbol{\alpha}_{2k} = c\mathbf{\Sigma}_{22}^{-1}\mathbf{\Sigma}_{21}\boldsymbol{\alpha}_{1k} \tag{6.54}$$

Now, applying Theorem 6.1 to CCD, we have $\tilde{\mathbf{l}}_{ik}$ being an eigenvector of $\tilde{\mathbf{S}}_i$, that is,

$$(\mathbf{\Sigma}_{11}^{-1}\mathbf{\Sigma}_{12}\mathbf{\Sigma}_{22}^{-1}\mathbf{\Sigma}_{21})\tilde{\mathbf{l}}_{1k} = \tilde{\gamma}_k^2\tilde{\mathbf{l}}_{1k} \tag{6.55a}$$

and

$$(\mathbf{\Sigma}_{22}^{-1}\mathbf{\Sigma}_{21}\mathbf{\Sigma}_{11}^{-1}\mathbf{\Sigma}_{12})\tilde{\mathbf{l}}_{2k} = \tilde{\gamma}_k^2\tilde{\mathbf{l}}_{2k} \tag{6.55b}$$

Equations (6.55a) and (6.55b) are satisfied if we choose $\tilde{\mathbf{l}}_{2k} \propto \mathbf{\Sigma}_{22}^{-1}\mathbf{\Sigma}_{21}\tilde{\mathbf{l}}_{1k}$ showing that $\tilde{\mathbf{l}}_{1k}$ and $\tilde{\mathbf{l}}_{2k}$ satisfy Eq. (6.54). For the choice $\boldsymbol{\alpha}_{ik} = \tilde{\mathbf{l}}_{ik}$, the maximum value of r_k^2

in Eq. (6.53) is given by

$$\begin{aligned} r_{k\max}^2 &= \tilde{\mathbf{l}}_{1k}^H \boldsymbol{\Sigma}_{12}\boldsymbol{\Sigma}_{22}^{-1}\boldsymbol{\Sigma}_{21}\tilde{\mathbf{l}}_{1k}/\tilde{\mathbf{l}}_{1k}^H\boldsymbol{\Sigma}_{11}\tilde{\mathbf{l}}_{1k} \\ &= \tilde{\mathbf{l}}_{1k}^H(\tilde{\gamma}_k^2\boldsymbol{\Sigma}_{11}\tilde{\mathbf{l}}_{1k})/\tilde{\mathbf{l}}_{1k}^H\boldsymbol{\Sigma}_{11}\tilde{\mathbf{l}}_{1k} = \tilde{\gamma}_k^2 \end{aligned} \tag{6.56}$$

where Eq. (6.55a) has been employed. For the choice $\boldsymbol{\alpha}_{ik} = \tilde{\mathbf{l}}_{ik}$, that the condition $\boldsymbol{\alpha}_{ik}^H\boldsymbol{\Sigma}_{ii}\boldsymbol{\alpha}_{il} = 0$ for $k \neq l$ is also satisfied can easily be verified as shown in Property 6.2 below. Hence the property follows. We note that since $\tilde{\gamma}_k$ are all positive and real and is the maximum correlation between two variables, then

$$0 \leq \tilde{\gamma}_k \leq 1, \quad k = 1, \cdots, K \tag{6.57}$$

Property 6.2. *The linear combinations* $\tilde{\mathbf{l}}_{ik}^H\mathbf{x}_i$ *and* $\tilde{\mathbf{l}}_{jl}^H\mathbf{x}_j$ *are uncorrelated for* $k \neq l, i, j = 1, 2$.

This property can be shown quite simply by forming a matrix $\tilde{\mathbf{L}}$ such that

$$\tilde{\mathbf{L}}^H = \begin{bmatrix} \tilde{\mathbf{L}}_1 & \mathbf{O} \\ \mathbf{O} & \tilde{\mathbf{L}}_2 \end{bmatrix}$$

From the definition of $\tilde{\mathbf{L}}_i$, $i = 1, 2$, in Eq. (6.51), using the unitary properties of $\tilde{\mathbf{U}}_i$ and Eq. (6.50), we obtain

$$\tilde{\mathbf{L}}^H\boldsymbol{\Sigma}\tilde{\mathbf{L}} = \begin{bmatrix} \tilde{\mathbf{L}}_1^H & \mathbf{O} \\ \mathbf{O} & \tilde{\mathbf{L}}_2^H \end{bmatrix} \begin{bmatrix} \boldsymbol{\Sigma}_{11} & \boldsymbol{\Sigma}_{12} \\ \boldsymbol{\Sigma}_{21} & \boldsymbol{\Sigma}_{22} \end{bmatrix} \begin{bmatrix} \tilde{\mathbf{L}}_1 & \mathbf{O} \\ \mathbf{O} & \tilde{\mathbf{L}}_2 \end{bmatrix} = \begin{bmatrix} \mathbf{I}_{M_1} & \tilde{\boldsymbol{\Gamma}}_0^H \\ \tilde{\boldsymbol{\Gamma}}_0 & \mathbf{I}_{M_2} \end{bmatrix} \tag{6.58}$$

where $\mathbf{I}_{M_1}$ and $\mathbf{I}_{M_2}$ are identity matrices of dimensions $M_1 \times M_1$ and $M_2 \times M_2$ respectively, and $\tilde{\boldsymbol{\Gamma}}_0$ is an $M_1 \times M_2$ matrix similar to Eq. (6.38). Hence, the property follows.

Fast Computation of CCD. In anticipation of the importance of CCD in the ensuing sections, we now present a fast algorithm for the computation of CCD.

Define

$$\mathbf{X}_i = [\mathbf{x}_i(1)\mathbf{x}_i(2)\cdots\mathbf{x}_i(N)] \tag{6.59}$$

Then, $\hat{\boldsymbol{\Sigma}}_{ij}$ can be represented as

$$\hat{\boldsymbol{\Sigma}}_{ij} = \frac{1}{N}\mathbf{X}_i\mathbf{X}_j^H \quad i, j = 1, 2 \tag{6.60}$$

If the "QR" decomposition of $\mathbf{X}_i^H$ is written as

$$\frac{1}{N^{1/2}}\mathbf{X}_i^H = \hat{\mathbf{K}}_i\hat{\boldsymbol{\Upsilon}}_i \quad i = 1, 2 \tag{6.61}$$

where $\hat{\boldsymbol{\Upsilon}}_i$ is an $M_i \times M_i$ upper triangular matrix, then it follows

$$\hat{\boldsymbol{\Sigma}}_{ii} = \hat{\boldsymbol{\Upsilon}}_i^H\hat{\boldsymbol{\Upsilon}}_i, \quad \text{and} \quad \hat{\boldsymbol{\Sigma}}_{i\bar{i}} = \hat{\boldsymbol{\Upsilon}}_i^H\hat{\mathbf{K}}_i^H\hat{\mathbf{K}}_{\bar{i}}\hat{\boldsymbol{\Upsilon}}_{\bar{i}} \tag{6.62}$$

On the other hand, we have shown in Theorem 6.1 and Remark 6.1 that $\hat{\hat{\mathbf{L}}}_i$ consists of the eigenvectors of the matrix

$$\hat{\mathbf{S}}_i = \hat{\boldsymbol{\Sigma}}_{ii}^{-1}\hat{\boldsymbol{\Sigma}}_{i\bar{i}}\hat{\boldsymbol{\Sigma}}_{\bar{i}\bar{i}}^{-1}\hat{\boldsymbol{\Sigma}}_{\bar{i}i} \tag{6.63}$$

Substituting Eq. (6.62) into Eq. (6.63), we see

$$\hat{\boldsymbol{\Sigma}}_{ii}^{-1}\hat{\boldsymbol{\Sigma}}_{i\bar{i}}\hat{\boldsymbol{\Sigma}}_{\bar{i}\bar{i}}^{-1}\hat{\boldsymbol{\Sigma}}_{\bar{i}i} = \hat{\boldsymbol{\Upsilon}}_i^{-1}\hat{\mathbf{K}}_i^H\hat{\mathbf{K}}_{\bar{i}}\hat{\mathbf{K}}_{\bar{i}}^H\hat{\mathbf{K}}_i\hat{\boldsymbol{\Upsilon}}_i \tag{6.64}$$

Represent the matrix $\hat{\mathbf{K}}_i^H\hat{\mathbf{K}}_{\bar{i}}$ by its singular value decomposition:

$$\hat{\mathbf{K}}_i^H\hat{\mathbf{K}}_{\bar{i}} = \hat{\hat{\mathbf{V}}}_i\begin{bmatrix}\hat{\hat{\boldsymbol{\Gamma}}} & \mathbf{O}\end{bmatrix}\hat{\hat{\mathbf{V}}}_{\bar{i}}^H \tag{6.65}$$

where $\hat{\hat{\boldsymbol{\Gamma}}} = \text{diag}[\hat{\hat{\gamma}}_1, \cdots, \hat{\hat{\gamma}}_{M_1}]$ (note it has been assumed $M_1 \leq M_2$). Substituting Eq. (6.65) into Eq. (6.64) and comparing with Eq. (6.44b), we can identify that the columns of $\hat{\boldsymbol{\Upsilon}}_i^{-1}\hat{\hat{\mathbf{V}}}_i$ are the eigenvectors of the matrix in Eq. (6.63). Hence, we obtain

$$\hat{\hat{\mathbf{L}}}_i = \hat{\boldsymbol{\Upsilon}}_i^{-1}\hat{\hat{\mathbf{V}}}_i, \quad i = 1, 2 \tag{6.66}$$

$$\hat{\hat{\mathbf{R}}}_i = \hat{\hat{\mathbf{V}}}_i^H\hat{\boldsymbol{\Upsilon}}_i, \quad i = 1, 2 \tag{6.67}$$

Equations (6.66) and (6.67) present the computational formulae for $\hat{\hat{\mathbf{L}}}_i$ and $\hat{\hat{\mathbf{R}}}_i$. The main computations involved in this algorithm are two QR decompositions of Eq. (6.61) and one SVD of Eq. (6.65). The inverse of the upper triangular matrix $\hat{\boldsymbol{\Upsilon}}_i$ can be obtained rather efficiently in general [106]. However, in many applications, we may only need $\hat{\hat{\mathbf{R}}}_i$ and $\hat{\hat{\boldsymbol{\Gamma}}}$. $\hat{\hat{\mathbf{L}}}_i$ may not be needed at all. In such cases, the inverse of $\hat{\boldsymbol{\Upsilon}}_i$ can even be avoided. Furthermore, Eq. (6.61) operates directly on the data matrix without having to form the covariance matrix. This means that the computation load is reduced significantly compared to solving the CCD by Eqs. (6.50) and (6.51). As well, the algorithm will also be better conditioned due to the avoidance of forming the covariance matrix.

Geometric Properties of GCD

We now return to the more general decomposition of GCD. We examine the geometric interpretation of the coluumns of $\mathbf{L}_i$ and $\mathbf{R}_i$ in terms of the signal subspaces and their complements. Let us partition $\mathbf{L}_i$ and $\mathbf{R}_i$, $i = 1, 2$, such that

$$\mathbf{L}_i = \left[\mathbf{L}_{is}|\mathbf{L}_{i\nu}\right] = \left[\boldsymbol{\Pi}_i^{-1/2}\mathbf{U}_{is}|\boldsymbol{\Pi}_i^{-1/2}\mathbf{U}_{i\nu}\right] \tag{6.68a}$$

$$\mathbf{R}_i = \left[\mathbf{R}_{is}|\mathbf{R}_{i\nu}\right] = \left[\boldsymbol{\Pi}_i^{1/2}\mathbf{U}_{is}|\boldsymbol{\Pi}_i^{1/2}\mathbf{U}_{i\nu}\right] \tag{6.68b}$$

where $\mathbf{L}_{is}$ and $\mathbf{L}_{i\nu}$ are the first K columns and the last $(M_i - K)$ columns of $\mathbf{L}_i$ respectively, $\mathbf{R}_{is}$ and $\mathbf{R}_{i\nu}$ are the first K columns and the last $(M_i - K)$ columns of $\mathbf{R}_i$ respectively, and $\mathbf{U}_i$ is similarly partitioned into $\mathbf{U}_{is}$ and $\mathbf{U}_{i\nu}$.

Theorem 6.2. *For* Π_1 *and* Π_2 *being positive definite Hermitian matrices of dimensions* $M_1 \times M_1$ *and* $M_2 \times M_2$ *respectively, and for rank* $(\Sigma_{12}) = K$, *we have the following relations:*

$$span\{\mathbf{R}_{is}\} = span\{\mathbf{D}_i(\theta_i)\} \quad i = 1, 2 \tag{6.69}$$

$$span\{\mathbf{L}_{i\nu}\} = \overline{span}\{\mathbf{D}_i(\theta_i)\} \quad i = 1, 2 \tag{6.70}$$

where $\overline{\text{span}}\{\mathbf{D}_i\}$ *denotes the orthogonal complement of* $\text{span}\{\mathbf{D}_i\}$.

Proof. From Eq. (6.37), we have

$$\Sigma_{i\bar{i}}\Pi_{\bar{i}}^{-1/2} = \Pi_i^{1/2}\mathbf{U}_i\Gamma_0\mathbf{U}_{\bar{i}}^H = \Pi_i^{1/2}\mathbf{U}_{is}\Gamma\mathbf{U}_{\bar{i}s}^H = \mathbf{R}_{is}\Gamma\mathbf{U}_{\bar{i}s}^H \tag{6.71}$$

where we have used Eq. (6.38). Now, using Eq. (6.35b) in Eq. (6.71), we can write

$$\mathbf{D}_i(\theta_i)\Sigma_{s_{i\bar{i}}}\mathbf{D}_{\bar{i}}^H(\theta_{\bar{i}})\Pi_{\bar{i}}^{-1/2} = \mathbf{R}_{is}\Gamma\mathbf{U}_{\bar{i}s}^H \tag{6.72}$$

Since $\mathbf{D}_{\bar{i}}^H(\theta_{\bar{i}})$ is of rank K, then $\mathbf{D}_{\bar{i}}^H(\theta_{\bar{i}})\Pi_{\bar{i}}^{-1/2}$ is a $K \times M_{\bar{i}}$ matrix having a pseudo-inverse which is an $M_{\bar{i}} \times K$ matrix of rank K. If we post-multiply both sides of Eq. (6.72) by this pseudo-inverse, we obtain

$$\mathbf{D}_i(\theta_i)\Sigma_{s_{i\bar{i}}} = \mathbf{R}_{is}\Gamma\mathbf{U}_{\bar{i}s}^H(\mathbf{D}_{\bar{i}}(\theta_{\bar{i}})\Pi_{\bar{i}}^{-1/2})^+ \tag{6.73}$$

where $+$ denotes the pseudo-inverse operation. We note that $\mathbf{D}_i(\theta_i)$ and $\mathbf{R}_{is}$ each contains K linearly independent vectors as columns. Since the left side of Eq. (6.73) is of rank K, it implies that $\Gamma\mathbf{U}_{\bar{i}s}^H(\mathbf{D}_{\bar{i}}(\theta_{\bar{i}})\Pi_{\bar{i}}^{-1/2})^+$ is also of rank K. Therefore, the left side and right side of Eq. (6.73) represent a nonsingular linear transformation of the K vectors contained in $\mathbf{D}_i(\theta_i)$ and $\mathbf{R}_{is}$ respectively within the same subspace. Hence, the vectors of $\mathbf{D}_i(\theta_i)$ and those of $\mathbf{R}_{is}$ must span the same subspace and Eq. (6.69) follows.

Now, since $\mathbf{U}_1$ and $\mathbf{U}_2$ are unitary matrices, then using the definitions in Eqs. (6.68),

$$\mathbf{R}_{is}^H\mathbf{L}_{i\nu} = \mathbf{U}_{is}^H(\Pi_i^{1/2})^H\Pi_i^{-1/2}\mathbf{U}_{i\nu} = \mathbf{O} \tag{6.74}$$

and Eq. (6.70) follows. ■

Theorem 6.2 has very important implications. It tells us that by performing a GCD on the correlation matrix Σ, we can partition the data space into the signal subspace and its orthogonal complement which are respectively spanned by the first K columns of $\mathbf{R}_i$ and the last $M_i - K$ columns of $\mathbf{L}_i$.

More relations between the submatrices of $\mathbf{L}_i$ and $\mathbf{R}_i$, defined in Eqs. (6.68), are given in the following theorem.

Theorem 6.3. *For* $\mathbf{L}_{is}, \mathbf{L}_{i\nu}, \mathbf{R}_{is}$ *and* $\mathbf{R}_{i\nu}$ *defined in Eqs. (6.68), we have the following identities:*

$$\mathbf{L}_{i\nu}^H\mathbf{R}_{is} = \mathbf{R}_{i\nu}^H\mathbf{L}_{is} = \mathbf{O}, \quad i = 1, 2 \tag{6.75}$$

$$\mathbf{L}_{i\nu}^H\mathbf{R}_{i\nu} = \mathbf{I}_{M_i-K}, \quad i = 1, 2 \tag{6.76a}$$

$$\mathbf{L}_{is}^H\mathbf{R}_{is} = \mathbf{I}_K, \quad i = 1, 2 \tag{6.76b}$$

that is, the columns of $\mathbf{L}_i$ *and* $\mathbf{R}_i$ *form reciprocal sets. Furthermore,*

$$\mathbf{L}_{is}\mathbf{R}_{is}^H + \mathbf{L}_{i\nu}\mathbf{R}_{i\nu}^H = \mathbf{I}_{M_i}, \quad i = 1, 2 \tag{6.77}$$

Proof. These identities follow directly from the definitions in Eqs. (6.68). ■

We note that if a GCD is performed on the estimated correlation matrix $\hat{\boldsymbol{\Sigma}}$, the orthogonal properties of the columns of the resulting matrices $\hat{\mathbf{L}}_i$ and $\hat{\mathbf{R}}_i$ shown in Theorem 6.3 still maintain. The identification of the signal subspace and its orthogonal complement implied by Theorem 6.2 together with the orthogonal properties of the columns of $\mathbf{L}_i$ and $\mathbf{R}_i$ indicated by Theorem 6.3 enable us to utilize some of the concepts in high-resolution DOA estimation. We now introduce the concept of eigen-subspaces and eigenprojectors [85, 86] for a generalized Hermitian matrix.

Let $\mathbf{H}$ be an $M \times M$ matrix which is Hermitian in the metric of a positive definite Hermitian matrix $\boldsymbol{\Pi}$ as defined in Definition 6.2. Let $\{\lambda_m : m = 1, \cdots, M\}$ be the set of eigenvalueas of $\mathbf{H}$. The eigen-subspaces $\mathcal{E}_m$ associated with λ_m is defined as

$$\mathcal{E}_m = span\{\boldsymbol{\upsilon}_{mj} : j = 1, \cdots, \mu(\lambda_m); \mathbf{H}\boldsymbol{\upsilon}_{mj} = \lambda_m\boldsymbol{\upsilon}_{mj}\} \tag{6.78}$$

where $\{\boldsymbol{\upsilon}_{m1}, \cdots, \boldsymbol{\upsilon}_{m\mu(\lambda_m)}\}$ are the eigenvectors associated with the eigenvalue λ_m which is of multiplicity $\mu(\lambda_m)$. We note that if λ_m and λ_l are two distinct eigenvalues of $\mathbf{H}$, then $\mathcal{E}_m$ and $\mathcal{E}_l$ are orthogonal subspaces in the metric of $\boldsymbol{\Pi}$, that is, if $\mathbf{z}_m \in \mathcal{E}_m$ and $\mathbf{z}_l \in \mathcal{E}_l$, then $\mathbf{z}_m^H\boldsymbol{\Pi}\mathbf{z}_l = 0$, for $l \neq m$. The entire M-dimensional space C^M is the direct sum of all the eigen-subspaces. The eigenprojector of $\mathbf{H}$ associated with λ_m, denoted by $\mathbf{P}_{\lambda_m}$, is defined as the projection operator onto $\mathcal{E}_m$ in the metric of $\boldsymbol{\Pi}$. For any set of vectors $\{\boldsymbol{\upsilon}_{mj} : j = 1, \cdots, \mu(\lambda_m)\} \in \mathcal{E}_m$ such that

$$\boldsymbol{\upsilon}_{mj}^H\boldsymbol{\Pi}\boldsymbol{\upsilon}_{mk} = \delta_{jk} \tag{6.79}$$

where δ_{jk} denotes the Kronecker delta, then

$$\mathbf{P}_{\lambda_m} = \sum_{j=1}^{\mu(\lambda_m)} \boldsymbol{\upsilon}_{mj}\boldsymbol{\upsilon}_{mj}^H\boldsymbol{\Pi} \tag{6.80}$$

Thus, $\mathbf{P}_{\lambda_m}$ is Hermitian in the metric $\boldsymbol{\Pi}$. If Ω is a subset of $\{\lambda_m : m = 1, \cdots, M\}$, then the total eigenprojection for $\mathbf{H}$ associated with the eigenvalues in Ω is defined as

$$\mathbf{P}_\Omega = \sum_{\lambda_m \in \Omega} \mathbf{P}_{\lambda_m} \tag{6.81}$$

Since $\mathbf{P}_\Omega$ is linear and bounded, it is also continuous [87]. It is straigthforward to verify, using Eq. (6.79), that $\mathbf{P}_\Omega$ is *idempotent* ($\mathbf{P}_\Omega\mathbf{P}_\Omega = \mathbf{P}_\Omega$) and is therefore a projector. However, $\mathbf{P}_\Omega$ is not *self-adjoint* ($\mathbf{P}_\Omega \neq \mathbf{P}_\Omega^H$) and therefore, the projection is not orthogonal [75, 76].

We now apply the concept of eigen-subspaces and eigen-projection to the matrices $\mathbf{S}_i$ and $\mathbf{S}_i^H$ as defined in Eq. (6.42). From Theorem 6.1, we know that $\mathbf{S}_i$ and $\mathbf{S}_i^H$ are Hermitian in the metrics of $\boldsymbol{\Pi}_i$ and $\boldsymbol{\Pi}_i^{-1}$ respectively, both have the same

M_i eigenvalues $\{\gamma_1^2, \cdots, \gamma_K^2, 0, \cdots, 0\}$ and respectively have eigenvectors being the columns of $\mathbf{L}_i$ and $\mathbf{R}_i$. Therefore, to construct an eigenprojector $\mathbf{P}_{i\nu}$ associated with the $M_i - K$ zero eigenvalues of $\mathbf{S}_i$ and projecting onto the subspace spanned by $\{\mathbf{l}_{im}\}, m = K+1, \cdots, M_i$, we have

$$\mathbf{P}_{i\nu} = \sum_{m=K+1}^{M_i} \mathbf{l}_{im}\mathbf{l}_{im}^H \mathbf{\Pi}_i = \mathbf{L}_{i\nu}\mathbf{L}_{i\nu}^H \mathbf{\Pi}_i = \mathbf{L}_{i\nu}\mathbf{R}_{i\nu}^H, \quad i = 1, 2 \tag{6.82}$$

Similarly, to construct an eigenprojectior $\mathbf{P}_{is}$ associated with the K nonzero eigenvalues $\{\gamma_1^2, \cdots, \gamma_K^2\}$ of $\mathbf{S}_i$, and projecting onto the subspace spanned by $\{\mathbf{l}_{ik}\}, k = 1, \cdots, K$, we have

$$\mathbf{P}_{is} = \sum_{k=1}^{K} \mathbf{l}_{ik}\mathbf{l}_{ik}^H \mathbf{\Pi}_i = \mathbf{L}_{is}\mathbf{L}_{is}^H \mathbf{\Pi}_i = \mathbf{L}_{is}\mathbf{R}_{is}^H, \quad i = 1, 2 \tag{6.83}$$

It is easy to verify that $\mathbf{P}_{i\nu}$ and $\mathbf{P}_{is}$ are Hermitian in the metric $\mathbf{\Pi}_i$. We also observe that the eigenprojectors $\mathbf{P}_{i\nu}$ and $\mathbf{P}_{is}, i = 1, 2$, are composed of two parts, the left factors, $\mathbf{L}_{i\nu}$ and $\mathbf{L}_{is}$, and the right factors, $\mathbf{R}_{i\nu}$ and $\mathbf{R}_{is}$.

For $\mathbf{S}_i^H$, the eigenprojector associated with the $M_i - K$ zero eigenvalues and projecting onto the subspace spanned by $\{\mathbf{r}_{im}\}, m = K+1, \cdots, M_i$, is given by

$$\begin{aligned} \sum_{m=K+1}^{M_i} \mathbf{r}_{im}\mathbf{r}_{im}^H \mathbf{\Pi}_i^{-1} &= \mathbf{R}_{i\nu}\mathbf{R}_{i\nu}^H \mathbf{\Pi}_i^{-1} \\ &= \mathbf{R}_{i\nu}\mathbf{L}_{i\nu}^H = \mathbf{P}_{i\nu}^H \end{aligned} \tag{6.84}$$

where Eqs. (6.68) have been used. Similarly, the eigenprojector of $\mathbf{S}_i^H$ associated with the K nonzero eigenvalues and projecting onto the subspace spanned by $\{\mathbf{r}_{ik}\}, k = 1, \cdots, K$, is

$$\begin{aligned} \sum_{k=1}^{K} \mathbf{r}_{ik}\mathbf{r}_{ik}^H \mathbf{\Pi}_i^{-1} &= \mathbf{R}_{is}\mathbf{R}_{is}^H \mathbf{\Pi}_i^{-1} \\ &= \mathbf{R}_{is}\mathbf{L}_{is}^H = \mathbf{P}_{is}^H \end{aligned} \tag{6.85}$$

Now, from Theorem 6.2, we see that the signal subspace and its orthogonal complement are respectively spanned by the columns of $\mathbf{R}_{is}$ and $\mathbf{L}_{i\nu}$. For convenience, we designate the subspace $\overline{\text{span}}\{\mathbf{D}_i(\boldsymbol{\theta}_i)\}$ the *"noise" subspace.* Hence from Eqs. (6.82) and (6.85), we can deduce that $\mathbf{P}_{is}^H$ *and* $\mathbf{P}_{i\nu}$ *project onto the signal subspace and noise subspace respectively*. The subspace spanned by $\{\mathbf{l}_{ik}\}, k = 1, \cdots, K$, is not necessarily the signal subspace spanned by $\mathbf{D}_i(\boldsymbol{\theta}_i)$, nor is the subspace spanned by $\{\mathbf{r}_{im}\}, m = K+1, \cdots, M_i$, necessarily the noise subspace. Thus, we conclude that the projectors $\mathbf{P}_{is}$ and $\mathbf{P}_{i\nu}^H$ do not respectively project onto the signal subspace and the noise subspace. We designate the subspaces onto which $\mathbf{P}_{is}$ and $\mathbf{P}_{i\nu}^H$ project the *supplementary signal subspace* and the *supplementary noise subspace* respectively. From Theorem 6.3, we see that $\mathbf{P}_{i\nu}$ and $\mathbf{P}_{is}$ are related by

$$\mathbf{P}_{i\nu} = \mathbf{L}_{i\nu}\mathbf{R}_{i\nu}^H = \mathbf{I}_{M_i} - \mathbf{L}_{is}\mathbf{R}_{is}^H = \mathbf{I}_{M_i} - \mathbf{P}_{is}, \quad i = 1, 2 \tag{6.86}$$

We also observe that

$$\mathbf{P}_{i\nu}\mathbf{P}_{i\nu} = \mathbf{P}_{i\nu}, \qquad \mathbf{P}_{i\nu}\mathbf{P}_{i\nu}^H \neq \mathbf{P}_{i\nu} \tag{6.87a}$$

$$\mathbf{P}_{is}\mathbf{P}_{is} = \mathbf{P}_{is}, \qquad \mathbf{P}_{is}\mathbf{P}_{is}^H \neq \mathbf{P}_{is} \tag{6.87b}$$

We note that if we use Eqs. (6.82) and (6.83) to form the projectors $\hat{\mathbf{P}}_{i\nu}$ and $\hat{\mathbf{P}}_{is}$ employing the estimated matrices $\hat{\mathbf{L}}_{i\nu}, \hat{\mathbf{L}}_{is}, \hat{\mathbf{R}}_{i\nu}$ and $\hat{\mathbf{R}}_{is}$, then Eqs. (6.86) and (6.87) will still be valid, and $\hat{\mathbf{P}}_{is}^H$ and $\hat{\mathbf{P}}_{i\nu}$ will project onto the estimated signal subspace and the estimated noise subspace respectively.

As we have seen, the high-resolution array processing methods described in section 6.2 for estimating the DOA of signals in white noise utilize the framework of subspace fitting [44, 45]. These methods exploit the orthogonality between the signal and noise subspaces such that the estimated noise subspace is fitted to the known signal subspace, or vice versa, by minimizing certain criteria.

In the case of the high-resolution array processing in unknown noise, once the signal and noise subspaces are identified, the same basic framework of subspace fitting can be employed. Here, the properties of the reciprocal sets in Theorem 6.3 can be exploited such that $\hat{\mathbf{L}}_{i\nu}$ which is estimated from the data is fitted to $\mathbf{R}_{is}$ which is a function of θ_i, or alternatively, $\hat{\mathbf{R}}_{is}$ is fitted to $\mathbf{L}_{i\nu}$. The accuracy of using subspace fitting to obtain $\hat{\theta}_i$ depends on the accuracy of the estimate $\hat{\mathbf{L}}_{i\nu}$ (or $\hat{\mathbf{R}}_{is}$), or, equivalently, on the accuracy of the eigenprojector $\hat{\mathbf{P}}_{i\nu}$ (or $\hat{\mathbf{P}}_{is}$). Therefore, in order to develop the consistent DOA estimations, we examine the asymptotic properties of these estimated subspaces obtained from GCD.

Asymptotic Properties of the Eigensubspaces

We now examine the asymptotic properties of the estimated eigensubspaces onto which $\hat{\mathbf{P}}_{i\nu}$ and $\hat{\mathbf{P}}_{is}$ project. We have mentioned in section 6.2 that $\hat{\Sigma}$ converges to Σ, at the rate of $O(N^{-1/2})$ in probability. This statistic uncertainty introduces random errors on the estimates of the eigensubspaces spanned by the columns of the matrices $\hat{\mathbf{L}}_{i\nu}, \hat{\mathbf{L}}_{is}, \hat{\mathbf{R}}_{i\nu}$, and $\hat{\mathbf{R}}_{is}$, which are, in turn, estimated by substituting Σ_{ij} in the GCD by $\hat{\Sigma}_{ij}$. We shall examine the convergence and perturbations of these estimated subspaces as well as the asymptotic distributions of the perturbations.

We first propose the following theorem which describes the convergence properties of the subspaces and the eigenprojectors.

Theorem 6.4. *Let Π_i and $\Pi_{\bar{i}}$ be positive definite matrices. Correspondingly we may have $\hat{\Pi}_i$ and $\hat{\Pi}_{\bar{i}}$ being their respective estimates converging asymptotically to the true values at the rate of $O(N^{-1/2})$ in probability. If $\hat{\Sigma}_{ij}$ also converges to Σ_{ij}, $i, j = 1, 2$ at the same rate and in the same sense, then $\hat{\mathbf{L}}_{is}, \hat{\mathbf{R}}_{is}, \hat{\mathbf{P}}_{is}$, and $\hat{\mathbf{P}}_{i\nu}$ all converge to $\mathbf{L}_{is}, \mathbf{R}_{is}, \mathbf{P}_{is}$, and $\mathbf{P}_{i\nu}$ respectively at the rate of $O(N^{-1/2})$ in probability.*

Proof. We denote $(\hat{\Pi}_i - \Pi_i)$ and $(\hat{\Sigma}_{i\bar{i}} - \Sigma_{i\bar{i}})$ by $\Delta\Pi_i$ and $\Delta\Sigma_{i\bar{i}}$ respectively. From the definition of the matrix $\mathbf{S}_i$ in Eq. (6.42), we have

$$\begin{aligned}\Delta\mathbf{S}_i = \hat{\mathbf{S}}_i - \mathbf{S}_i = &-\Pi_i^{-1}(\Delta\Pi_i)\Pi_i^{-1}\Sigma_{i\bar{i}}\Pi_{\bar{i}}^{-1}\Sigma_{\bar{i}i} + \Pi_i^{-1}(\Delta\Sigma_{i\bar{i}})\Pi_{\bar{i}}^{-1}\Sigma_{\bar{i}i} \\ &-\Pi_i^{-1}\Sigma_{i\bar{i}}\Pi_{\bar{i}}^{-1}(\Delta\Pi_{\bar{i}})\Pi_{\bar{i}}^{-1}\Sigma_{\bar{i}i} + \Pi_i^{-1}\Sigma_{i\bar{i}}\Pi_{\bar{i}}^{-1}\Delta\Sigma_{\bar{i}i} + \mathbf{O}(N^{-1})\end{aligned} \tag{6.88}$$

where $\hat{\mathbf{S}}_i$ is obtained from Eq. (6.42) using the estimated quantities $\hat{\mathbf{\Pi}}_i$, $\hat{\mathbf{\Pi}}_{\bar{i}}$, and $\hat{\Sigma}_{i\bar{i}}$.

Let us denote the Frobenius norm of a matrix by $\|\cdot\|$. Since $\|\hat{\mathbf{\Pi}}_i\| = \|\mathbf{\Pi}_i\| + O(N^{-1/2})$ *prob.*, and $\|\hat{\Sigma}_{i\bar{i}}\| = \|\Sigma_{i\bar{i}}\| + O(N^{-1/2})$ *prob.*, then $\|\hat{\mathbf{S}}_i - \mathbf{S}_i\| = \|\Delta\mathbf{S}_i\| = O(N^{-1/2})$ *prob.*, where "*prob.*" denotes that the equation holds in probability. Since $\hat{\mathbf{L}}_{is}$ and $\hat{\mathbf{R}}_{is}$ consist of eigenvectors corresponding to the first K distinct eigenvalues of $\hat{\mathbf{S}}_i$ and $\hat{\mathbf{S}}_i^H$ respectively, then [46, 82], the convergence of $\hat{\mathbf{L}}_{is}$ and $\hat{\mathbf{R}}_{is}$ are all at the rate of $O(N^{-1/2})$ in probability.

Furthermore, we can expand a complex eigenprojector in a Taylor series [88, chapter 2] such that

$$\hat{\mathbf{P}}_{is} = \mathbf{P}_{is} + \sum_{m=K+1}^{M_i} [\mathbf{l}_{im}\mathbf{l}_{im}^H\mathbf{\Pi}_i\Delta\mathbf{S}_i(\mathbf{S}_i - \gamma_m^2\mathbf{I})^+ + (\mathbf{S}_i - \gamma_m^2\mathbf{I})^+\Delta\mathbf{S}_i\mathbf{l}_{im}\mathbf{l}_{im}^H\mathbf{\Pi}_i] + \mathbf{O}(N^{-1})\,prob. \tag{6.89}$$

where $\mathbf{l}_{im}$ is the mth column of $\mathbf{L}_i$ that is associated with the eigenvalue γ_m^2 of $\mathbf{S}_i$, $\mathbf{O}(N^{-1})$ denotes a matrix the norm of which is $O(N^{-1})$*prob.*, and $(\cdot)^+$ is a generalized inverse of a matrix. We note that this generalized inverse is defined in the metric of $\mathbf{\Pi}_i$, and for a matrix $\mathbf{A}$, $(\mathbf{AA}^+)$ is Hermitian in the corresponding metric. Therefore, we see that $\hat{\mathbf{P}}_{is}$ converges at the same rate as that of $\Delta\mathbf{S}_i$ which is $O(N^{-1/2})$. Since $\hat{\mathbf{P}}_{iv} = \mathbf{I} - \hat{\mathbf{P}}_{is}$, the convergence rate of $\hat{\mathbf{P}}_{iv}$ is also $O(N^{-1/2})$ in probability. ■

The above theorem shows that $\hat{\mathbf{L}}_{is}$ and $\hat{\mathbf{R}}_{is}$ converges to $\mathbf{L}_{is}$ and $\mathbf{R}_{is}$. However, it is worth noting that $\hat{\mathbf{L}}_{iv}$ and $\hat{\mathbf{R}}_{iv}$, in general, do not converge to definite matrices since the columns of $\mathbf{L}_{iv}$ and $\mathbf{R}_{iv}$ are eigenvectors corresponding to the zero eigenvalues of $\mathbf{S}_i$ and $\mathbf{S}_i^H$ and are therefore not uniquely determined. However, as shown by the above theorem, the eigenprojector $\hat{\mathbf{P}}_{iv} = \hat{\mathbf{L}}_{iv}\hat{\mathbf{R}}_{iv}^H$ converges, therefore, the columns of $\hat{\mathbf{L}}_{iv}$ and $\hat{\mathbf{R}}_{iv}$ do converge to the definite subspaces.

Now, we turn our attention to the perturbation of these subspaces. To measure, for instance, the perturbation of $\hat{\mathbf{L}}_{iv}$, we can calculate the projection error from the subspace $\overline{\text{span}}\{\mathbf{D}_i\}$ to the subspace spanned by $\hat{\mathbf{L}}_{iv}$. Let $\mathbf{Y}_{iv} \subseteq \overline{\text{span}}\{\mathbf{D}_i\}$, that is, the columns of $\mathbf{Y}_{iv}$ are vectors in $\overline{\text{span}}\{\mathbf{D}_i\}$. Since $\mathbf{P}_{iv}$ project onto the noise subspace, such an error can be represented as

$$\mathbf{Y}_{iv} - \hat{\mathbf{P}}_{iv}\mathbf{Y}_{iv} = \hat{\mathbf{P}}_{is}\mathbf{Y}_{iv} \tag{6.90}$$

Here we have used the relation $\hat{\mathbf{P}}_{is} = \mathbf{I} - \hat{\mathbf{P}}_{iv}$ in Eq. (6.86). This error will be zero when $\hat{\mathbf{P}}_{iv} = \mathbf{P}_{iv}$. Similarly, the perturbation of $\hat{\mathbf{R}}_{is}$ can be represented by

$$\mathbf{Y}_{is} - \hat{\mathbf{P}}_{is}^H\mathbf{Y}_{is} = \hat{\mathbf{P}}_{iv}^H\mathbf{Y}_{is} \tag{6.91}$$

where $\mathbf{Y}_{is} \subseteq \text{span}(\mathbf{D}_i)$.

The following theorem and its corollary show the order of magnitude of these projection errors.

Theorem 6.5. *For the ith array having a data model given by Eq. (6.32), let* $\mathbf{Y}_{i\nu} \subseteq \overline{\text{span}}(\mathbf{D}_i)$ *and* $\mathbf{\Pi}_i$ *and* $\mathbf{\Pi}_{\bar{i}}$ *be positive definite matrices which may be stochastic matrices with convergence rate of* $O(N^{-1/2})$ *in probability, then* $\hat{\mathbf{P}}_{is}\mathbf{Y}_{i\nu}$ *can be written as*

$$\hat{\mathbf{P}}_{is}\mathbf{Y}_{i\nu} = \mathbf{L}_{is}\mathbf{\Gamma}^{-1}\mathbf{L}_{\bar{i}s}^{H}\Delta\mathbf{\Sigma}_{\bar{i}i}\mathbf{Y}_{i\nu} + \mathbf{O}(N^{-1})\,prob. \quad i = 1, 2 \tag{6.92}$$

where $\Delta\mathbf{\Sigma}_{\bar{i}i} \equiv \hat{\mathbf{\Sigma}}_{\bar{i}i} - \mathbf{\Sigma}_{\bar{i}i}$.

Proof. We have seen that the eigenvalues of $\mathbf{S}_i$ satisfy that $\gamma_1^2 > \cdots > \gamma_K^2 > \gamma_{K+1}^2 = \cdots = \gamma_{M_i}^2 = 0$. It is noted that $\gamma_m^2 = 0$, for $m \in \{K+1, \cdots, M_i\}$, then $(\mathbf{S}_i - \gamma_m^2\mathbf{I})^{+}$ in Eq. (6.89) can be constructed [75, 85, 86] as

$$(\mathbf{S}_i - \gamma_m^2\mathbf{I})^{+} = \mathbf{L}_{is}\mathbf{\Gamma}^{-2}\mathbf{R}_{is}^{H}, \quad \gamma_m^2 \in \{\gamma_{K+1}^2 \cdots \gamma_{M_i}^2\} \tag{6.93}$$

Using Eqs. (6.89) and (6.92), we can write

$$\begin{aligned}\hat{\mathbf{P}}_{is}\mathbf{Y}_{i\nu} = \mathbf{P}_{is}\mathbf{Y}_{i\nu} + \sum_{m=K+1}^{M_i} \big[&\mathbf{l}_{im}\mathbf{l}_{im}^{H}\mathbf{\Pi}_i\Delta\mathbf{S}_i\mathbf{L}_{is}\mathbf{\Gamma}^{-2}\mathbf{R}_{is}^{H}\mathbf{Y}_{i\nu} \\ &+ \mathbf{L}_{is}\mathbf{\Gamma}^{-2}\mathbf{R}_{is}^{H}\Delta\mathbf{S}_i\mathbf{l}_{im}\mathbf{l}_{im}^{H}\mathbf{\Pi}_i\mathbf{Y}_{i\nu}\big] + \mathbf{O}(N^{-1})\,prob.\end{aligned} \tag{6.94}$$

Here since $\mathbf{Y}_{i\nu} \subseteq \overline{\text{span}}(\mathbf{D}_i)$,

$$\mathbf{L}_{is}\mathbf{\Gamma}^{-2}\mathbf{R}_{is}^{H}\mathbf{Y}_{i\nu} = \mathbf{O} \tag{6.95}$$

and then the first term under the summation sign in Eq. (6.94) vanishes. Furthermore, we have

$$\mathbf{\Sigma}_{\bar{i}i}\mathbf{l}_{im} = \mathbf{O} \quad \text{for} \quad m = K+1, \cdots, M_i \tag{6.96}$$

Subsituting Eqs. (6.89) and (6.96) into Eq. (6.94) and using that $\mathbf{P}_{is}\mathbf{Y}_{i\nu} = \mathbf{O}$ yields

$$\hat{\mathbf{P}}_{is}\mathbf{Y}_{i\nu} = \sum_{m=K+1}^{M_i} \mathbf{L}_{is}\mathbf{\Gamma}^{-2}\mathbf{R}_{is}^{H}\mathbf{\Pi}_i^{-1}\mathbf{\Sigma}_{i\bar{i}}\mathbf{\Pi}_{\bar{i}}^{-1}\Delta\mathbf{\Sigma}_{\bar{i}i}\mathbf{l}_{im}\mathbf{l}_{im}^{H}\mathbf{\Pi}_i\mathbf{Y}_{i\nu} + \mathbf{O}(N^{-1})\,prob. \tag{6.97}$$

Now, from Eqs. (6.38), (6.40), and (6.41a), we have

$$\mathbf{\Pi}_i^{-1}\mathbf{\Sigma}_{i\bar{i}}\mathbf{\Pi}_{\bar{i}}^{-1} = \mathbf{L}_{is}\mathbf{\Gamma}\mathbf{L}_{\bar{i}s}^{H} \tag{6.98}$$

Substituting Eq. (6.98) into Eq. (6.97), we obtain

$$\hat{\mathbf{P}}_{is}\mathbf{Y}_{i\nu} = \mathbf{L}_{is}\mathbf{\Gamma}^{-2}\mathbf{R}_{is}^{H}\mathbf{L}_{is}\mathbf{\Gamma}\mathbf{L}_{\bar{i}s}^{H}\Delta\mathbf{\Sigma}_{\bar{i}i}\left(\sum_{m=K+1}^{M_i}\mathbf{l}_{im}\mathbf{l}_{im}^{H}\mathbf{\Pi}_i\right)\mathbf{Y}_{i\nu} + \mathbf{O}(N^{-1})\,prob. \tag{6.99}$$

and Eq. (6.92) follows by noting $\mathbf{R}_{is}^{H}\mathbf{L}_{is} = \mathbf{I}$ and

$$\left(\sum_{m=K+1}^{M_i}\mathbf{l}_{im}\mathbf{l}_{im}^{H}\mathbf{\Pi}_i\right)\mathbf{Y}_{i\nu} = \mathbf{P}_{i\nu}\mathbf{Y}_{i\nu} = \mathbf{Y}_{i\nu} \tag{6.100}$$

The above proof is, of course, also valid if $\mathbf{\Pi}_i$ and $\mathbf{\Pi}_{\bar{i}}$ are constant matrices. ■

In the special case when only the first array is available, then $\mathbf{x}_1 = \mathbf{x}_2$, and

$\boldsymbol{\Sigma}_{12} = \boldsymbol{\Sigma}_{21} = \boldsymbol{\Sigma}_{11}$. $\boldsymbol{\Gamma}_0$ given by Eq. (6.38) is a full-ranked $M_1 \times M_1$ matrix such that $\boldsymbol{\Gamma}_0 = \text{diag}(\gamma_1, \cdots, \gamma_{M_1})$. If we choose $\boldsymbol{\Pi}_1 = \mathbf{I}$ and $\boldsymbol{\Pi}_2 = \boldsymbol{\Sigma}_{11}$, then from Example 6.2, $\{\gamma_m^2\}, m = 1, \cdots, M_1$, are the eigenvalues and the columns of $\mathbf{L}_1$ are the eigenvectors of the matrix $\boldsymbol{\Sigma}_{11}$. If, in addition, the sensor noise is spatially white, then $\gamma_{K+1}^2 = \cdots = \gamma_{M_i}^2 = \sigma_\nu^2$. For this special case, the approximation of the projection $\hat{\mathbf{P}}_{1s}\mathbf{Y}_{1\nu}$ can be described as follows:

Corollary 6.5a. *For the special case when* $\mathbf{x}_1 = \mathbf{x}_2, \boldsymbol{\Sigma}_{12} = \boldsymbol{\Sigma}_{21} = \boldsymbol{\Sigma}_{11}, \boldsymbol{\Pi}_1 = \mathbf{I}$, *and* $\boldsymbol{\Pi}_2 = \boldsymbol{\Sigma}_{11}$, *then for* $\mathbf{Y}_{1\nu} \subseteq \overline{\text{span}}(\mathbf{D}_1)$, *we have*

$$\hat{\mathbf{P}}_{1s}\mathbf{Y}_{1\nu} = \mathbf{L}_{1s}\mathring{\boldsymbol{\Lambda}}_s^{-1}\boldsymbol{\Gamma}\mathbf{L}_{1s}^H\Delta\boldsymbol{\Sigma}_{11}\mathbf{Y}_{1\nu} + \mathbf{O}(N^{-1})\,prob. \tag{6.101}$$

where $\boldsymbol{\Gamma} = diag(\gamma_1, \cdots, \gamma_K)$ *and* $\mathring{\boldsymbol{\Lambda}}_s = diag\ (\gamma_1^2 - \sigma_\nu^2, \cdots, \gamma_{M_i}^2 - \sigma_\nu^2)$.

Proof. Substituting $\Delta\boldsymbol{\Sigma}_{i\bar{i}} = \Delta\boldsymbol{\Sigma}_{ii}$ into Eq. (6.97) and noting that for $\gamma_{K+1}^2 = \cdots = \gamma_{M_i}^2 = \sigma_\nu^2$, the Moore-Penrose generalized inverse of Eq. (6.93) becomes

$$(\mathbf{S}_i - \gamma_m^2\mathbf{I})^+ = \sum_{k=1}^{K} \frac{1}{\gamma_k^2 - \sigma_\nu^2}\mathbf{l}_k\mathbf{l}_k^H\boldsymbol{\Pi}_i = \mathbf{L}_{is}\mathring{\boldsymbol{\Lambda}}_s^{-1}\mathbf{R}_{is}^H \tag{6.102}$$

Using Eq. (6.102) and Eq. (6.89), we obtain a parallel of Eq. (6.94) and following the same development from then onward we arrive at the result. ■

In the same parallel, we can examine the projection error from the subspace span($\mathbf{D}_i$). Since $\mathbf{P}_{is}^H$ projects onto the signal subspace, such an error can be represented as Eq. (6.91) and a corollary of Theorem 6.5 can be obtained.

Corollary 6.5b. *For the* i*th array, let* $\mathbf{Y}_{is} \subseteq \text{span}(\mathbf{D}_i)$. *Then*

$$\hat{\mathbf{P}}_{i\nu}^H\mathbf{Y}_{is} = -\mathbf{R}_{i\nu}\mathbf{L}_{i\nu}^H\hat{\mathbf{P}}_{is}^H\mathbf{Y}_{is} + \mathbf{O}(N^{-1})\,prob. \tag{6.103}$$

Proof. From Eqs. (6.76a) and (6.86), we have $\hat{\mathbf{R}}_{i\nu}^H\hat{\mathbf{L}}_{i\nu} = \mathbf{I}$ and $\hat{\mathbf{P}}_{i\nu} = \hat{\mathbf{L}}_{i\nu}\hat{\mathbf{R}}_{i\nu}^H$, thus we can write

$$\mathbf{Y}_{is}^H\hat{\mathbf{P}}_{i\nu} = \mathbf{Y}_{is}^H\hat{\mathbf{L}}_{i\nu}\hat{\mathbf{R}}_{i\nu}^H = \mathbf{Y}_{is}^H\hat{\mathbf{L}}_{i\nu}\hat{\mathbf{R}}_{i\nu}^H\hat{\mathbf{L}}_{i\nu}\hat{\mathbf{R}}_{i\nu}^H \tag{6.104}$$

But from Theorem 6.4 we know

$$\hat{\mathbf{L}}_{i\nu}\hat{\mathbf{R}}_{i\nu}^H = \mathbf{L}_{i\nu}\mathbf{R}_{i\nu}^H + \mathbf{O}(N^{-1/2})\,prob. \tag{6.105}$$

and

$$\mathbf{Y}_{is}^H\hat{\mathbf{L}}_{i\nu}\hat{\mathbf{R}}_{i\nu}^H = \mathbf{O}(N^{-1/2})\,prob. \tag{6.106}$$

Substituting Eqs. (6.105) and (6.106) into (6.104), we have

$$\begin{aligned}\mathbf{Y}_{is}^H\hat{\mathbf{L}}_{i\nu}\hat{\mathbf{R}}_{i\nu}^H &= \mathbf{Y}_{is}^H\hat{\mathbf{L}}_{i\nu}\hat{\mathbf{R}}_{i\nu}^H\mathbf{L}_{i\nu}\mathbf{R}_{i\nu}^H + \mathbf{O}(N^{-1})\,prob.\\ &= \mathbf{Y}_{is}^H(\mathbf{I} - \hat{\mathbf{L}}_{is}\hat{\mathbf{R}}_{is}^H)\mathbf{L}_{i\nu}\mathbf{R}_{i\nu}^H + \mathbf{O}(N^{-1})\,prob.\\ &= -\mathbf{Y}_{is}^H\hat{\mathbf{P}}_{is}\mathbf{L}_{i\nu}\mathbf{R}_{i\nu}^H + \mathbf{O}(N^{-1})\,prob.\end{aligned} \tag{6.107}$$

which is the conjugate-transposed version of Eq. (6.103). ■

Theorem 6.5 and its corollaries respectively express the projection error from the subspace $\overline{\text{span}}(\mathbf{D}_i)$ to the estimated subspace spanned by $\hat{\mathbf{L}}_{i\nu}$ and from the signal subspace span($\mathbf{D}_i$) to the estimated signal subspace spanned by $\hat{\mathbf{R}}_{is}$. The respective projection errors $\hat{\mathbf{P}}_{is}\mathbf{Y}_{i\nu}$ and $\hat{\mathbf{P}}_{i\nu}^H\mathbf{Y}_{is}$ are of great importance in the development of a high-resolution array processing method of estimating the DOA of signal sources as will be evidenced in the next section. Let us now examine the asymptotic distribution of these errors are described by the following theorem and its corollaries. To represent the random elements of the matrices $\hat{\mathbf{P}}_{is}\mathbf{Y}_{i\nu}$ and $\hat{\mathbf{P}}_{i\nu}^H\mathbf{Y}_{is}$, we employ the vec operation and the Kronecker product of matrices (see Appendix 6A).

Theorem 6.6. *For the ith array,* $i = 1, 2$, *if* $\mathbf{Y}_{i\nu} \subseteq \overline{\text{span}}(\mathbf{D}_i)$, *then the random vectors* $\text{vec}(\hat{\mathbf{P}}_{is}\mathbf{Y}_{i\nu})$ *are asymptotically complex Gaussian with zero mean and covariance matrices*

$$E\left[\text{vec}\,(\hat{\mathbf{P}}_{is}\mathbf{Y}_{i\nu})\,\text{vec}^H(\hat{\mathbf{P}}_{is}\mathbf{Y}_{i\nu})\right] = \frac{1}{N}\left[\mathbf{Y}_{i\nu}^H\boldsymbol{\Sigma}_{ii}\mathbf{Y}_{i\nu}\right]^T \otimes \left[\mathbf{L}_{is}\boldsymbol{\Gamma}^{-1}\mathbf{L}_{\bar{i}s}^H\boldsymbol{\Sigma}_{\bar{i}\bar{i}}\mathbf{L}_{\bar{i}s}\boldsymbol{\Gamma}^{-1}\mathbf{L}_{is}^H\right] \tag{6.108a}$$

and

$$E\left[\text{vec}\,(\hat{\mathbf{P}}_{is}\mathbf{Y}_{i\nu})\,\text{vec}^T(\hat{\mathbf{P}}_{is}\mathbf{Y}_{i\nu})\right] = \mathbf{O} \tag{6.108b}$$

Futhermore, asymptotically we have

$$E\left[\text{vec}\,(\hat{\mathbf{P}}_{is}\mathbf{Y}_{i\nu})\,\text{vec}^H(\hat{\mathbf{P}}_{\bar{i}s}\mathbf{Y}_{\bar{i}\nu})\right] = \mathbf{O} \tag{6.108c}$$

Proof. We need the well-established formula (see Appendix 6A) that for properly dimensioned matrices **A**, **B**, **C**

$$\text{vec}\,(\mathbf{ABC}) = (\mathbf{C}^T \otimes \mathbf{A})\text{vec}\,(\mathbf{B}) \tag{6.109}$$

Using Eqs. (6.91) and (6.109), for $i = 1, 2$, we can write

$$\text{vec}\,(\hat{\mathbf{P}}_{is}\mathbf{Y}_{i\nu}) = \{\mathbf{Y}_{i\nu}^T \otimes (\mathbf{L}_{is}\boldsymbol{\Gamma}^{-1}\mathbf{L}_{\bar{i}s}^H)\}\text{vec}\,(\Delta\boldsymbol{\Sigma}_{\bar{i}i}) + \mathbf{O}(N^{-1})\,prob. \tag{6.110}$$

It is well-known that vec($\Delta\boldsymbol{\Sigma}_{\bar{i}i}$) is asymptotically jointly Gaussian with zero mean [82, 83], thus vec($\hat{\mathbf{P}}_{is}\mathbf{Y}_{i\nu}$) is also asymptotically jointly Gaussian with zero mean. The asymptotic covariance matrix of vec($\hat{\mathbf{P}}_{is}\mathbf{Y}_{i\nu}$), from Eq. (6.110), is then

$$E\left[\text{vec}\,(\hat{\mathbf{P}}_{is}\mathbf{Y}_{i\nu})\,\text{vec}^H(\hat{\mathbf{P}}_{is}\mathbf{Y}_{i\nu})\right] = \{\mathbf{Y}_{i\nu}^T \otimes (\mathbf{L}_{is}\boldsymbol{\Gamma}^{-1}\mathbf{L}_{\bar{i}s}^H)\}E\left[\text{vec}\,(\Delta\boldsymbol{\Sigma}_{\bar{i}i})\,\text{vec}^H(\Delta\boldsymbol{\Sigma}_{\bar{i}i})\right]\{\mathbf{Y}_{i\nu}^* \otimes (\mathbf{L}_{\bar{i}s}\boldsymbol{\Gamma}^{-1}\mathbf{L}_{is}^H)\} \tag{6.111}$$

where the superscript * denotes the conjugate. Also, from Appendix 6B, Eq. (6B.13), we can write

$$E\left[\text{vec}\,(\Delta\boldsymbol{\Sigma}_{\bar{i}i})\,\text{vec}^H(\Delta\boldsymbol{\Sigma}_{\bar{i}i})\right] = \frac{1}{N}\boldsymbol{\Sigma}_{ii}^T \otimes \boldsymbol{\Sigma}_{\bar{i}\bar{i}} \tag{6.112}$$

Substituting Eq. (6.112) into Eq. (6.111) and employing the following formula (Appendix 6A)

$$(\mathbf{A} \otimes \mathbf{B})(\mathbf{C} \otimes \mathbf{D}) = \mathbf{AC} \otimes \mathbf{BD} \tag{6.113}$$

Eq. (6.108a) follows.

Using Eq. (6.110), we obtain

$$\begin{aligned} &E\left[\text{vec}\,(\hat{\mathbf{P}}_{is}\mathbf{Y}_{iv})\,\text{vec}^T(\hat{\mathbf{P}}_{is}\mathbf{Y}_{iv})\right] \\ &= \{\mathbf{Y}_{\bar{i}v}^T \otimes (\mathbf{L}_{is}\boldsymbol{\Gamma}^{-1}\mathbf{L}_{\bar{i}s}^H)\}E\left[\text{vec}\,(\Delta\boldsymbol{\Sigma}_{\bar{i}i})\,\text{vec}^T(\Delta\boldsymbol{\Sigma}_{\bar{i}i})\right]\{\mathbf{Y}_{iv} \otimes (\mathbf{L}_{\bar{i}s}\boldsymbol{\Gamma}^{-1}\mathbf{L}_{is}^H)^*\} \end{aligned} \tag{6.114}$$

Again, from Appendix 6B, Eq. (6B.15), we obtain

$$E\left[\text{vec}\,(\Delta\boldsymbol{\Sigma}_{\bar{i}i})\,\text{vec}^T(\Delta\boldsymbol{\Sigma}_{\bar{i}i})\right] = \frac{1}{N}\left\{\sum_{l=1}^{M_i}\sum_{m=1}^{M_{\bar{i}}}\mathbf{E}_{lm} \otimes \mathbf{E}_{lm}^T\right\}(\boldsymbol{\Sigma}_{\bar{i}i} \otimes \boldsymbol{\Sigma}_{\bar{i}i}^T) \tag{6.115}$$

where $\mathbf{E}_{lm}$ is an $M_{\bar{i}} \times M_i$ matrix having the (lm)th element being unity and zero elsewhere. Substituting Eq. (6.115) into Eq. (6.114) and remembering that $\boldsymbol{\Sigma}_{\bar{i}i}\mathbf{Y}_{iv} = \mathbf{O}$, Eq. (6.108b) follows.

Using Eq. (6.110) again, we obtain

$$\begin{aligned} &E\left[\text{vec}\,(\hat{\mathbf{P}}_{is}\mathbf{Y}_{iv})\,\text{vec}^H(\hat{\mathbf{P}}_{\bar{i}s}\mathbf{Y}_{\bar{i}v})\right] \\ &= \{\mathbf{Y}_{\bar{i}v}^T \otimes (\mathbf{L}_{is}\boldsymbol{\Gamma}^{-1}\mathbf{L}_{\bar{i}s}^H)\}E\left[\text{vec}\,(\Delta\boldsymbol{\Sigma}_{\bar{i}i})\,\text{vec}^H(\Delta\boldsymbol{\Sigma}_{i\bar{i}})\right]\{\mathbf{Y}_{\bar{i}v}^* \otimes (\mathbf{L}_{is}\boldsymbol{\Gamma}^{-1}\mathbf{L}_{\bar{i}s}^H)\} \end{aligned} \tag{6.116}$$

Again, from Appendix 6B, Eq. (6B.14), we have

$$E\left[\text{vec}\,(\Delta\boldsymbol{\Sigma}_{\bar{i}i})\,\text{vec}^H(\Delta\boldsymbol{\Sigma}_{i\bar{i}})\right] = \frac{1}{N}\boldsymbol{\Sigma}_{\bar{i}i}^T \otimes \boldsymbol{\Sigma}_{\bar{i}i} \tag{6.117}$$

Substituting Eq. (6.117) into Eq. (6.116) and remembering that $\boldsymbol{\Sigma}_{i\bar{i}}\mathbf{Y}_{\bar{i}v} = \mathbf{Y}_{iv}^H\boldsymbol{\Sigma}_{i\bar{i}} = \mathbf{O}$, Eq. (6.108c) follows. ■

Note that if canonical correlation decomposition is employed, then from Eq. (6.51) we have

$$\tilde{\mathbf{L}}_{is}^H\boldsymbol{\Sigma}_{ii}\tilde{\mathbf{L}}_{is} = \tilde{\mathbf{U}}_{is}^H\tilde{\mathbf{U}}_{is} = \mathbf{I}, \quad i = 1, 2 \tag{6.118}$$

and Eq. (6.108a) further reduces to

$$E\left[\text{vec}\,(\hat{\tilde{\mathbf{P}}}_{is}\mathbf{Y}_{iv})\,\text{vec}^H(\hat{\tilde{\mathbf{P}}}_{is}\mathbf{Y}_{iv})\right] = \frac{1}{N}\left[\mathbf{Y}_{iv}^H\boldsymbol{\Sigma}_{ii}\mathbf{Y}_{iv}\right]^T \otimes \left[\tilde{\mathbf{L}}_{is}\tilde{\boldsymbol{\Gamma}}^{-2}\tilde{\mathbf{L}}_{is}^H\right] \tag{6.119}$$

Again, in the case when only the first array is available and the ambient noise is spatially white, we have the following:

Corollary 6.6a. *For the special case when* $\mathbf{x}_1 = \mathbf{x}_2$, $\boldsymbol{\Sigma}_{12} = \boldsymbol{\Sigma}_{21} = \boldsymbol{\Sigma}_{11}$, $\boldsymbol{\Pi}_1 = \mathbf{I}$, *and* $\boldsymbol{\Pi}_2 = \boldsymbol{\Sigma}_{11}$, *then for* $\mathbf{Y}_{1v} \subseteq \overline{\text{span}}\{\mathbf{D}_1\}$, *the random vector* vec $(\hat{\mathbf{P}}_{1s}\mathbf{Y}_{1v})$ *is asymptotically complex Gaussian with zero mean and covariance matrices*

$$E\left[\text{vec}\,(\hat{\mathbf{P}}_{1s}\mathbf{Y}_{1v})\,\text{vec}^H(\mathbf{P}_{1s}\mathbf{Y}_{1v})\right] = \frac{1}{N}\left[\mathbf{Y}_{1v}^H\boldsymbol{\Sigma}_{11}\mathbf{Y}_{1v}\right]^T \otimes \left[\mathbf{L}_{1s}\mathring{\boldsymbol{\Lambda}}_s^{-2}\boldsymbol{\Gamma}^2\mathbf{L}_{1s}^H\right] \tag{6.120}$$

and

$$E\left[\text{vec}\,(\hat{\mathbf{P}}_{1s}\mathbf{Y}_{1\nu})\,\text{vec}^T(\hat{\mathbf{P}}_{1s}\mathbf{Y}_{1\nu})\right] = \mathbf{O} \tag{6.121}$$

Proof. We note that since we have only one array $\mathbf{\Sigma}_{11} = \mathbf{\Sigma}_{22}$, from Eq. (6.76b), we have

$$\begin{aligned} \mathbf{R}_{2s}^H\mathbf{L}_{2s} &= (\mathbf{\Pi}_2\mathbf{L}_{2s})^H\mathbf{L}_{2s} \\ &= \mathbf{L}_{2s}^H\mathbf{\Sigma}_{22}\mathbf{L}_{2s} = \mathbf{I} \end{aligned} \tag{6.122}$$

Also, in this case, Eq. (6.112) becomes

$$E\left[\text{vec}\,(\Delta\mathbf{\Sigma}_{ii})\,\text{vec}\,(\Delta\mathbf{\Sigma}_{\bar{i}\bar{i}})\right] = \frac{1}{N}\mathbf{\Sigma}_{ii}^T \otimes \mathbf{\Sigma}_{ii} \tag{6.123}$$

Using Corollary 6.5a and following the same steps in the proof of Theorem 6.6, the results follow. ■

In particular, if $\mathbf{Y}_{1\nu} = \mathbf{L}_{1\nu}$ since $\mathbf{\Pi}_1 = \mathbf{I}$, and $\mathbf{R}_{1\nu}^H\mathbf{L}_{1\nu} = \mathbf{I}$ then Eq. (6.120) becomes

$$\begin{aligned} E\left[\text{vec}\,(\hat{\mathbf{P}}_{1s}\mathbf{L}_{1\nu})\,\text{vec}^H(\hat{\mathbf{P}}_{1s}\mathbf{L}_{1\nu})\right] &= \frac{1}{N}\left[\mathbf{L}_{i\nu}^H(\mathbf{D}\mathbf{\Sigma}_s\mathbf{D}^H + \sigma_\nu^2\mathbf{I})\mathbf{L}_{1\nu}\right]^T \otimes \left[\mathbf{L}_{1s}\mathring{\mathbf{\Lambda}}_s^{-2}\mathbf{\Gamma}^2\mathbf{L}_{1s}^H\right] \\ &= \frac{\sigma_\nu^2}{N}\left[\mathbf{I} \otimes \mathbf{L}_{1s}\mathring{\mathbf{\Lambda}}_s^{-2}\mathbf{\Gamma}^2\mathbf{L}_{1s}^H\right] \end{aligned} \tag{6.124}$$

where we have used the fact that since the ambient noise is white, $\mathbf{\Sigma}_{11} = \mathbf{\Sigma}_x$ which is given by Eq. (6.7).

Corollary 6.6b. *For the* i*th array,* i $= 1, 2$, *let* $\mathbf{Y}_{\text{is}} \subseteq \text{span}(\mathbf{D}_{\text{i}})$. *Then* $\text{vec}(\mathbf{Y}_{\text{is}}^{\text{H}}\hat{\mathbf{P}}_{\text{i}\nu})$ *is asymptotically Gaussian distributed with zero mean and covariance matrices*

$$\begin{aligned} &E\left[\text{vec}\,(\mathbf{Y}_{is}^H\hat{\mathbf{P}}_{i\nu})\,\text{vec}^H(\mathbf{Y}_{is}^H\hat{\mathbf{P}}_{i\nu})\right] \\ &\quad = \frac{1}{N}\left[\mathbf{P}_{i\nu}^H\mathbf{\Sigma}_{ii}\mathbf{P}_{i\nu}\right]^T \otimes \left[\mathbf{Y}_{is}^H\mathbf{L}_{is}\mathbf{\Gamma}^{-1}\mathbf{L}_{\bar{i}s}^H\mathbf{\Sigma}_{\bar{i}\bar{i}}\mathbf{L}_{\bar{i}s}\mathbf{\Gamma}^{-1}\mathbf{L}_{is}^H\mathbf{Y}_{is}\right] \end{aligned} \tag{6.125}$$

Furthermore, asymptotically we have

$$E\left[\text{vec}\,(\mathbf{Y}_{is}^H\hat{\mathbf{P}}_{i\nu})\,\text{vec}^T(\mathbf{Y}_{is}^H\hat{\mathbf{P}}_{i\nu})\right] = \mathbf{O} \tag{6.126a}$$

$$E\left[\text{vec}\,(\mathbf{Y}_{is}^H\hat{\mathbf{P}}_{i\nu})\,\text{vec}^H(\mathbf{Y}_{\bar{i}s}^H\hat{\mathbf{P}}_{\bar{i}\nu})\right] = \mathbf{O} \tag{6.126b}$$

Proof. From Eq. (6.82), we can write $\hat{\mathbf{P}}_{i\nu} = \hat{\mathbf{L}}_{i\nu}\hat{\mathbf{R}}_{i\nu}^H$. From Eq. (6.107) together with Eq. (6A.7), we can write

$$\text{vec}\,(\mathbf{Y}_{is}^H\hat{\mathbf{P}}_{i\nu}) = -(\mathbf{R}_{i\nu}^* \otimes \mathbf{Y}_{is}^H)\text{vec}\,(\hat{\mathbf{P}}_{is}\mathbf{L}_{i\nu}) + \mathbf{O}(N^{-1})\,prob. \tag{6.127}$$

Hence, the asymptotic distribution of $\text{vec}(\mathbf{Y}_{is}^H\hat{\mathbf{P}}_{i\nu})$ is of the same type as that of $\text{vec}(\hat{\mathbf{P}}_{is}\mathbf{L}_{i\nu})$ which, as shown in Theorem 6.6, is asymptotically complex Gaussian with zero mean. The covariance matrix of $\text{vec}(\mathbf{Y}_{is}^H\hat{\mathbf{P}}_{i\nu})$ can be directly obtained from Eq. (6.127) after ignoring terms involving $\mathbf{O}(N^{-1})$ so that

$$
\begin{aligned}
&E\left[\operatorname{vec}(\mathbf{Y}_{is}^H\hat{\mathbf{P}}_{iv})\operatorname{vec}^H(\mathbf{Y}_{is}^H\hat{\mathbf{P}}_{iv})\right] \\
&\qquad = (\mathbf{R}_{iv}^* \otimes \mathbf{Y}_{is}^H)E\left[\operatorname{vec}(\hat{\mathbf{P}}_{is}\mathbf{L}_{iv})\operatorname{vec}^H(\hat{\mathbf{P}}_{is}\mathbf{L}_{iv})\right](\mathbf{R}_{iv}^T \otimes \mathbf{Y}_{is})
\end{aligned} \tag{6.128}
$$

Using Eq. (6.108a) to obtain the expectation term in Eq. (6.128), we have

$$
\begin{aligned}
&E\left[\operatorname{vec}(\mathbf{Y}_{is}^H\hat{\mathbf{P}}_{iv})\operatorname{vec}^H(\mathbf{Y}_{is}^H\hat{\mathbf{P}}_{iv})\right] \\
&\quad = \frac{1}{N}(\mathbf{R}_{iv}^* \otimes \mathbf{Y}_{is}^H)\left[(\mathbf{L}_{iv}^H\boldsymbol{\Sigma}_{ii}\mathbf{L}_{iv})^T \otimes (\mathbf{L}_{is}\boldsymbol{\Gamma}^{-1}\mathbf{L}_{\bar{i}s}^H\boldsymbol{\Sigma}_{\bar{i}\bar{i}}\mathbf{L}_{\bar{i}s}\boldsymbol{\Gamma}^{-1}\mathbf{L}_{is}^H)\right](\mathbf{R}_{iv}^T \otimes \mathbf{Y}_{is}) \\
&\quad = \frac{1}{N}(\mathbf{R}_{iv}\mathbf{L}_{iv}^H\boldsymbol{\Sigma}_{ii}\mathbf{L}_{iv}\mathbf{R}_{iv}^H)^T \otimes (\mathbf{Y}_{is}^H\mathbf{L}_{is}\boldsymbol{\Gamma}^{-1}\mathbf{L}_{\bar{i}s}\boldsymbol{\Sigma}_{\bar{i}\bar{i}}\mathbf{L}_{\bar{i}s}\boldsymbol{\Gamma}^{-1}\mathbf{L}_{is}^H\mathbf{Y}_{is})
\end{aligned} \tag{6.129}
$$

and the result of Eq. (6.125) follows by noting that $\mathbf{P}_{iv} = \mathbf{L}_{iv}\mathbf{R}_{iv}^H$.

By following exactly the same approach as above, we arrive at Eqs. (6.126a) and (6.126b) using the results in Eqs. (6.108b) and (6.108c). ■

The material presented in this section is fundamental to the ensuing sections and accordingly will be used extensively.

6.5 DOA ESTIMATION IN UNKNOWN CORRELATED NOISE

In this section, we examine the DOA estimation in unknown correlated noise environments. The material developed in the previous section will be utilized to develop consistent methods of DOA estimation. The development of the algorithms is based on the signal model established in section 6.3. Also, we assume that the number of signals, K, has been correctly determined. The ideas of these algorithms were first proposed in [107–110]. The following presents a detailed development of the algorithms.

The UN-MUSIC Algorithm

The MUSIC algorithm forms, as described in Section 6.2, the estimated covariance matrix of the data collected by an array of M sensors and then formulates a projection of the direction manifold onto the estimated noise subspace by utilizing the eigenvectors corresponding to the $M - K$ smallest eigenvalues of the estimated covariance matrix. The MUSIC spatial spectrum is then defined as the reciprocal of the square-magnitude of the projection of the steering vector onto the estimated noise subspace for different values of θ.

We can employ the concept of the MUSIC algorithm to arrive at an algorithm known as the *Unknown Noise-Multiple Signal Classification* (UN-MUSIC) for the estimation of DOA in unknown correlated noise environments. Based on the signal model established in Section 6.3, from Eq. (6.90b), we see that the projection of the steering vector onto the estimated noise subspace can be viewed as the perturbation on the steering vector due to the estimated signal subspace. Now, using the ith sensor array, the projector onto the estimated signal subspace, as shown in Eq. (6.85) is given by $\hat{\mathbf{P}}_{is}^H, i = 1, 2$, thus the perturbation of the steering vector due to the estimated signal subspace is given by

$$\mathbf{d}_i(\theta) - \hat{\mathbf{P}}_{is}^H \mathbf{d}_i(\theta) = (\mathbf{I} - \hat{\mathbf{P}}_{is}^H)\mathbf{d}_i(\theta) = \hat{\mathbf{P}}_{i\nu}^H \mathbf{d}_i(\theta) \tag{6.130}$$

Accordingly, by minimizing this error, we can locate the DOA of the signals. Thus, the spatial spectrum of the UN-MUSIC algorithm for each of the two arrays can be defined as

$$S_i(\theta) = 1/\left[\mathbf{d}_i^H(\theta)\hat{\mathbf{P}}_{i\nu}\hat{\mathbf{P}}_{i\nu}^H \mathbf{d}_i(\theta)\right] = 1/\left[\mathbf{d}_i^H(\theta)\hat{\mathbf{L}}_{i\nu}\hat{\mathbf{R}}_{i\nu}^H\hat{\mathbf{R}}_{i\nu}\hat{\mathbf{L}}_{i\nu}^H \mathbf{d}_i(\theta)\right], \quad i = 1, 2 \tag{6.131a}$$

The K highest peaks of $S_i(\theta)$ at $\theta = \hat{\theta}_{ik}, k = 1, \cdots, K$ will show the estimated DOA of the signals with respect to the ith array. Around each of the K highest peaks, we can re-write the UN-MUSIC algorithm as

$$\hat{\theta}_{ik} = \arg\min_{\theta_{ik}}\{\mathbf{d}_i^H(\theta_{ik})\hat{\mathbf{P}}_{i\nu}\hat{\mathbf{P}}_{i\nu}^H \mathbf{d}_i(\theta_{ik})\} \quad k = 1, \cdots, K; i = 1, 2 \tag{6.131b}$$

We can generalize the criterion in Eq. (6.131b) by weighting the components of the projection error such that

$$\begin{aligned} \hat{\theta}_{ik} &= \arg\min_{\theta_{ik}}\{\mathbf{d}_i^H(\theta_{ik})\hat{\mathbf{P}}_{i\nu}\mathbf{W}_{i\nu}\hat{\mathbf{P}}_{i\nu}^H \mathbf{d}_i(\theta_{ik})\} \\ &\triangleq \arg\min_{\theta_{ik}} h(\theta_{ik}, \hat{\mathbf{P}}_{i\nu}, \mathbf{W}_{i\nu}), \quad i = 1, 2, \quad k = 1, \cdots, K \end{aligned} \tag{6.131c}$$

where $\mathbf{W}_{i\nu}$ is a weighting matrix. Eq. (6.131c) is the UN-MUSIC algorithm with weighting matrix $\mathbf{W}_{i\nu}$.

Equation (6.131b) bears strong resemblance to Eq. (6.20) which is the MUSIC algorithm. It should be noted, however, that unlike $\hat{\mathbf{P}}_\nu$ in Eq. (6.20), the projector $\hat{\mathbf{P}}_{i\nu}^H$, as shown in Section 6.4, is not orthogonal, neither does it project directly onto the estimated noise subspace.

Equation (6.131b) yields two spatial spectra for the UN-MUSIC algorithm, one corresponding to each of the arrays. In the case that the two arrays are both calibrated, with the same orientation, we can say that $\theta_i = \theta_{\bar{i}} = \theta$. To utilize the information contained in both $\hat{\mathbf{P}}_{i\nu}, i = 1, 2$, simultaneously, we propose the θ estimate of UN-MUSIC to be simply obtained by

$$\hat{\theta}_k = \frac{1}{2}\sum_{i=1}^{2}\hat{\theta}_{ik}, \quad k = 1, \cdots, K \tag{6.132}$$

This is because of the fact that $\{\hat{\theta}_{ik}\}, i = 1, 2$, are asymptotically independent of each other (see Corollary 6.6b).

The UN-CLE Algorithm

In Section 6.2, we have introduced how the principle of MLE can be applied to the estimation of DOA of signals in spatially white noise. We also mentioned that the ML principle of DOA estimation can also be employed if the conditional PDF of the sufficient statistics can be obtained. We follow this methodology in this section and derive algorithms for the estimation of DOA in unknown correlated noise environments. Again using the same signal model as in Section 6.3, we examine the product $\hat{\mathbf{R}}_{is}^H\mathbf{L}_{iv}$. We note, from Eqs. (6.37)–(6.41), that $\mathbf{L}_{iv}$ is an implicit matrix function of θ_i and therefore contains all the information on the DOA of the signals. Thus if the asymptotical distribution of the random matrix $\hat{\mathbf{R}}_{is}^H\mathbf{L}_{iv}$ can be derived, then an estimate, $\hat{\theta}_i$, can be obtained by locating the maximum of the likelihood function so obtained. To obtain the asymptotic distribution for $\hat{\mathbf{R}}_{is}^H\mathbf{L}_{iv}$, we note that from Theorem 6.3, for the ith array, $\hat{\mathbf{L}}_{is}^H\hat{\mathbf{R}}_{is} = \mathbf{I}_K$, and from Theorem 6.4,

$$\hat{\mathbf{R}}_{is} = \mathbf{R}_{is} + \mathbf{O}(N^{-1/2})\,prob. \tag{6.133}$$

and

$$\hat{\mathbf{L}}_{is}\hat{\mathbf{R}}_{is}^H\mathbf{L}_{iv} = \mathbf{O}(N^{-1/2})\,prob. \tag{6.134}$$

Thus, with the term $\mathbf{O}(N^{-1})$ omitted, we have

$$\begin{aligned}
\text{vec}\,\hat{\mathbf{R}}_{is}^H\mathbf{L}_{iv} &= \text{vec}\,(\hat{\mathbf{R}}_{is}^H\hat{\mathbf{L}}_{is}\hat{\mathbf{R}}_{is}^H\mathbf{L}_{iv}) && (6.135\text{a})\\
&\simeq \text{vec}\,(\mathbf{R}_{is}^H\hat{\mathbf{L}}_{is}\hat{\mathbf{R}}_{is}^H\mathbf{L}_{iv}) && (6.135\text{b})\\
&= (\mathbf{I}_K \otimes \mathbf{R}_{is}^H)\text{vec}\,(\hat{\mathbf{L}}_{is}\hat{\mathbf{R}}_{is}^H\mathbf{L}_{iv}) = (\mathbf{I}_K \otimes \mathbf{R}_{is}^H)\text{vec}\,(\hat{\mathbf{P}}_{is}\mathbf{L}_{iv})\,prob. && (6.135\text{c})
\end{aligned}$$

where the approximation is made by replacing first $\hat{\mathbf{R}}_{is}^H$ in the right-hand side of Eq. (6.135a) by $\mathbf{R}_{is}$ without affecting the asymptotic analysis and the identity of Eq. (6A.6) in Appendix 6A has been employed in obtaining Eq. (6.135c). But from Theorem 6.5, we have established that for $\mathbf{Y}_{iv} \subseteq \overline{\text{span}}(\mathbf{D}_i)$, vec$(\hat{\mathbf{P}}_{is}\mathbf{Y}_{iv})$ is asymptotically complex Gaussian with zero mean, and since $\overline{\text{span}}(\mathbf{L}_{iv}) = \overline{\text{span}}(\mathbf{D}_i)$. Hence from Eq. (6.135) we conclude that vec$(\hat{\mathbf{R}}_{is}^H\mathbf{L}_{iv})$ is asymptotically complex Gaussian with zero mean. With the use of Eq. (6.135) and the result in Theorem 6.5, its asymptotic covariance matrix can be obtained as

$$E\left[\text{vec}\,(\hat{\mathbf{R}}_{is}^H\mathbf{L}_{iv})\text{vec}^H(\hat{\mathbf{R}}_{is}^H\mathbf{L}_{iv})\right] = \frac{1}{N}(\mathbf{L}_{iv}^H\boldsymbol{\Sigma}_{ii}\mathbf{L}_{iv})^T \otimes (\boldsymbol{\Gamma}^{-1}\mathbf{L}_{\bar{i}s}^H\boldsymbol{\Sigma}_{\bar{i}\bar{i}}\mathbf{L}_{\bar{i}s}\boldsymbol{\Gamma}^{-1}) \quad i = 1, 2 \tag{6.136}$$

Thus, the log-likelihood function of vec$(\hat{\mathbf{R}}_{is}^H\mathbf{L}_{iv})$ becomes

$$\begin{aligned}
&\ell(\hat{\mathbf{R}}_{is}^H\mathbf{L}_{iv}|\theta_i, \boldsymbol{\Sigma}_{ii}, \boldsymbol{\Sigma}_{\bar{i}\bar{i}}, \boldsymbol{\Sigma}_{i\bar{i}} && (6.137)\\
&\propto -\log\det\{(\mathbf{L}_{iv}^H\boldsymbol{\Sigma}_{ii}\mathbf{L}_{iv})^T \otimes (\boldsymbol{\Gamma}^{-1}\mathbf{L}_{\bar{i}s}^H\boldsymbol{\Sigma}_{\bar{i}\bar{i}}\mathbf{L}_{\bar{i}s}\boldsymbol{\Gamma}^{-1})\}\\
&- N\text{tr}\,\{\text{vec}^H(\hat{\mathbf{R}}_{is}^H\mathbf{L}_{iv})\left[(\mathbf{L}_{iv}^H\boldsymbol{\Sigma}_{ii}\mathbf{L}_{iv})^T \otimes (\boldsymbol{\Gamma}^{-1}\mathbf{L}_{\bar{i}s}^H\boldsymbol{\Sigma}_{\bar{i}\bar{i}}\mathbf{L}_{\bar{i}s}\boldsymbol{\Gamma}^{-1})\right]^{-1}\text{vec}\,(\hat{\mathbf{R}}_{is}^H\mathbf{L}_{iv})\}\\
&\simeq -N\text{tr}\,\{\text{vec}^H(\hat{\mathbf{R}}_{is}^H\mathbf{L}_{iv})\left[(\mathbf{L}_{iv}^T\boldsymbol{\Sigma}_{ii}^T\mathbf{L}_{iv}^*)^{-1} \otimes (\boldsymbol{\Gamma}^{-1}\mathbf{L}_{\bar{i}s}^H\boldsymbol{\Sigma}_{\bar{i}\bar{i}}\mathbf{L}_{\bar{i}s}\boldsymbol{\Gamma}^{-1})^{-1}\right]\text{vec}\,(\hat{\mathbf{R}}_{is}^H\mathbf{L}_{iv})\}
\end{aligned}$$

where the relationship in Eq. (6A.13) of Appendix 6A has been used and the first term log(det{·}) is omitted since for the large N, the second term is dominant. From Eq. (6A.12) in Appendix 6A, we have

$$\text{vec}\,(\hat{\mathbf{R}}_{is}^H\mathbf{L}_{iv}) = (\mathbf{L}_{iv}^T \otimes \hat{\mathbf{R}}_{is}^H)\text{vec}\,(\mathbf{I}) \tag{6.138}$$

Furthermore, using Eq. (6A.3) and the following relationship (Eq. 6A.17, Appendix 6A),

$$\mathrm{tr}\,\{\mathrm{vec}\,(\mathbf{I})\,\mathrm{vec}^H(\mathbf{A})\} = \mathrm{tr}\,\mathbf{A}$$

where $\mathbf{A}$ is Hermitian, Eq. (6.137) can be rewritten as

$$\begin{aligned} &\ell(\hat{\mathbf{R}}_{is}^H \mathbf{L}_{iv} | \boldsymbol{\theta}_i, \boldsymbol{\Sigma}_{ii}, \boldsymbol{\Sigma}_{\bar{i}\bar{i}}, \boldsymbol{\Sigma}_{i\bar{i}} \qquad (6.139)\\ &\propto -\mathrm{tr}\,\{\mathrm{vec}^H(\mathbf{I})\left[\mathbf{L}_{iv}(\mathbf{L}_{iv}^H \boldsymbol{\Sigma}_{ii} \mathbf{L}_{iv})^{-1}\mathbf{L}_{iv}^H\right]^* \otimes \left[\hat{\mathbf{R}}_{is}(\boldsymbol{\Gamma}^{-1}\mathbf{L}_{\bar{i}s}^H \boldsymbol{\Sigma}_{\bar{i}\bar{i}} \mathbf{L}_{\bar{i}s}\boldsymbol{\Gamma}^{-1})^{-1}\hat{\mathbf{R}}_{is}^H\right] \mathrm{vec}\,(\mathbf{I})\}\\ &= -\mathrm{tr}\,\{\mathbf{L}_{iv}(\mathbf{L}_{iv}^H \boldsymbol{\Sigma}_{ii}\mathbf{L}_{iv})^{-1}\mathbf{L}_{iv}^H \hat{\mathbf{R}}_{is}(\boldsymbol{\Gamma}^{-1}\mathbf{L}_{\bar{i}s}^H \boldsymbol{\Sigma}_{\bar{i}\bar{i}}\mathbf{L}_{\bar{i}s}\boldsymbol{\Gamma}^{-1})^{-1}\hat{\mathbf{R}}_{is}^H\} \end{aligned}$$

Therefore, for a chosen method of GCD, the MLE of the DOA based on the statistics of $\hat{\mathbf{R}}_{is}^H \mathbf{L}_{iv}$ in unknown noise is given by

$$\begin{aligned} \hat{\boldsymbol{\theta}}_i &= \arg\max_{\boldsymbol{\theta}} \ell(\hat{\mathbf{R}}_{is}^H \mathbf{L}_{iv} | \boldsymbol{\theta}, \boldsymbol{\Sigma}_{ii}, \boldsymbol{\Sigma}_{\bar{i}\bar{i}}, \boldsymbol{\Sigma}_{i\bar{i}})\\ &= \arg\min_{\boldsymbol{\theta}}\ \mathrm{tr}\,\{\mathbf{L}_{iv}(\mathbf{L}_{iv}^H \boldsymbol{\Sigma}_{ii}\mathbf{L}_{iv})^{-1}\mathbf{L}_{iv}^H \hat{\mathbf{R}}_{is}(\boldsymbol{\Gamma}^{-1}\mathbf{L}_{\bar{i}s}^H \boldsymbol{\Sigma}_{\bar{i}\bar{i}}\mathbf{L}_{\bar{i}s}\boldsymbol{\Gamma}^{-1})^{-1}\hat{\mathbf{R}}_{is}^H\} \quad i=1,2 \end{aligned} \qquad (6.140)$$

By using the fact that $\mathbf{R}_{iv}^H \mathbf{L}_{iv} = \mathbf{I}$, and that $\mathbf{L}_{iv}\mathbf{R}_{iv}^H = \mathbf{P}_{iv}$ as shown in Eqs. (6.76a) and (6.84) respectively, we can rewrite the criterion of Eq. (6.140) as

$$\begin{aligned} \hat{\boldsymbol{\theta}}_i &= \arg\max_{\boldsymbol{\theta}} \ell(\hat{\mathbf{R}}_{is}^H \mathbf{L}_{iv} | \boldsymbol{\theta}, \boldsymbol{\Sigma}_{ii}, \boldsymbol{\Sigma}_{\bar{i}\bar{i}}, \boldsymbol{\Sigma}_{i\bar{i}}) \qquad (6.141)\\ &= \arg\min_{\boldsymbol{\theta}}\ \mathrm{tr}\,\{\mathbf{L}_{iv}\mathbf{R}_{iv}^H\mathbf{L}_{iv}(\mathbf{L}_{iv}^H \boldsymbol{\Sigma}_{ii}\mathbf{L}_{iv})^{-1}\mathbf{L}_{iv}^H \mathbf{R}_{iv}\mathbf{L}_{iv}^H \hat{\mathbf{R}}_{is}(\boldsymbol{\Gamma}^{-1}\mathbf{L}_{\bar{i}s}^H \boldsymbol{\Sigma}_{\bar{i}\bar{i}}\mathbf{L}_{\bar{i}s}\boldsymbol{\Gamma}^{-1})^{-1}\hat{\mathbf{R}}_{is}^H\}\\ &= \arg\min_{\boldsymbol{\theta}}\ \mathrm{tr}\,\{\mathbf{P}_{iv}\mathbf{L}_{iv}(\mathbf{L}_{iv}^H \boldsymbol{\Sigma}_{ii}\mathbf{L}_{iv})^{-1}\mathbf{L}_{iv}^H \mathbf{P}_{iv}^H \hat{\mathbf{R}}_{is}(\boldsymbol{\Gamma}^{-1}\mathbf{L}_{\bar{i}s}^H \boldsymbol{\Sigma}_{\bar{i}\bar{i}}\mathbf{L}_{\bar{i}s}\boldsymbol{\Gamma}^{-1})^{-1}\hat{\mathbf{R}}_{is}^H\} \end{aligned}$$

Since $\mathbf{P}_{iv}^H \hat{\mathbf{R}}_{is} = \left[\mathbf{I} - \mathbf{P}_{is}^H\right]\hat{\mathbf{R}}_{is}$, Eq. (6.141) shows that the criterion is a measure of the projection error from the estimated signal subspace to the true signal subspace with the weighting matrices $\mathbf{L}_{iv}(\mathbf{L}_{iv}^H \boldsymbol{\Sigma}_{ii}\mathbf{L}_{iv})^{-1}\mathbf{L}_{iv}^H$ and $(\boldsymbol{\Gamma}^{-1}\mathbf{L}_{\bar{i}s}^H \boldsymbol{\Sigma}_{\bar{i}\bar{i}}\mathbf{L}_{\bar{i}s}\boldsymbol{\Gamma}^{-1})^{-1}$. Consequently, from the viewpoint of model fitting, using two weighting matrices $\mathbf{W}_{iv}$ and $\mathbf{W}_{is}$, we may rewrite the criterion in a more general form such that

$$\begin{aligned} \hat{\boldsymbol{\theta}}_i &= \arg\min_{\boldsymbol{\theta}_i} f(\hat{\mathbf{R}}_{is}, \mathbf{W}_{iv}, \mathbf{W}_{is})\\ &= \arg\min_{\boldsymbol{\theta}_i}\ \mathrm{tr}\,\{\mathbf{P}_{iv}\mathbf{W}_{iv}\mathbf{P}_{iv}^H\hat{\mathbf{R}}_{is}\mathbf{W}_{is}\hat{\mathbf{R}}_{is}\}, \quad i = 1, 2 \end{aligned} \qquad (6.142)$$

Alternatively, we can use $\left[\mathbf{I} - \hat{\mathbf{P}}_{is}^H\right]\mathbf{R}_{is}$, that is the projection error from $\mathbf{R}_{is}$ to the subspace spanned by $\hat{\mathbf{R}}_{is}$, under the weighting $\mathbf{W}'_{iv}$ and $\mathbf{W}'_{is}$ and arrive at another general criterion such that

$$\begin{aligned} \hat{\boldsymbol{\theta}}_i &= \arg\min_{\boldsymbol{\theta}_i} g(\hat{\mathbf{P}}_{iv}, \mathbf{W}'_{iv}, \mathbf{W}'_{is})\\ &= \arg\min_{\boldsymbol{\theta}_i}\ \mathrm{tr}\,\{\hat{\mathbf{P}}_{iv}\mathbf{W}'_{iv}\hat{\mathbf{P}}_{iv}^H\mathbf{R}_{is}\mathbf{W}'_{is}\mathbf{R}_{is}\}, \quad i = 1, 2 \end{aligned} \qquad (6.143)$$

Both criteria of Eqs. (6.142) and (6.143) are algorithms under the general designation of *Unknown Noise—Correlation, and Location Estimation* (UN-CLE).

Thus, we can see that Eq. (6.141) is a special case of UN-CLE in which particular weighting matrices $\bar{\mathbf{W}}_{i\nu}$ and $\bar{\mathbf{W}}_{is}$ are employed in Eq. (6.142) where

$$\bar{\mathbf{W}}_{i\nu} = \mathbf{L}_{i\nu}(\mathbf{L}_{i\nu}^H \mathbf{\Sigma}_{ii} \mathbf{L}_{i\nu})^{-1} \mathbf{L}_{i\nu}^H \tag{6.144}$$

and

$$\bar{\mathbf{W}}_{is} = (\mathbf{\Gamma}^{-1} \mathbf{L}_{\bar{i}s}^H \mathbf{\Sigma}_{\bar{i}\bar{i}} \mathbf{L}_{\bar{i}s} \mathbf{\Gamma}^{-1})^{-1} \tag{6.145}$$

It should be noted that generally the weighting matrices $\mathbf{W}_{i\nu}, \mathbf{W}_{is}, \mathbf{W}'_{i\nu}$, and $\mathbf{W}'_{is}$ in Eqs. (6.131c), (6.142), and (6.143) are functions of $\mathbf{L}_{i\nu}, \mathbf{L}_{is}, \mathbf{\Sigma}_{ii}$ and $\mathbf{\Gamma}$, which, in turn, implies that these weighting matrices are functions of θ_i. However, as shown in Section 6.6, the asymptotic performances of corresponding criteria will not be affected if these matrices are treated as constant matrices.

For the case that $\theta_i = \theta_{\bar{i}} = \theta$, since in Theorem 6.5 it is known that $\hat{\mathbf{R}}_{is}^H \mathbf{L}_{i\nu}$ is asymptotically uncorrelated with $\hat{\mathbf{R}}_{\bar{i}s}^H \mathbf{L}_{\bar{i}\nu}$, then we can commbine the UN-CLE criteria for this case so that

$$\hat{\theta}_i = \arg\min_{\theta_i} \sum_{i=1}^{2} f(\hat{\mathbf{R}}_{is}, \mathbf{W}_{i\nu}, \mathbf{W}_{is}) \tag{6.146}$$

Neither criterion $f(\mathbf{R}_{is}, \mathbf{W}_{i\nu}, \mathbf{W}_{is})$ nor $g(\hat{\mathbf{P}}_{i\nu}, \mathbf{W}'_{i\nu}, \mathbf{W}'_{is})$ in its present form is convenient to use since neither $\mathbf{P}_{i\nu}$ in Eq. (6.142) nor $\mathbf{R}_{is}$ in Eq. (6.143) is an explicit function of θ_i and the direct minimization with respect to θ_i is not easy. In Appendix 6C, we show that for the special case when $\mathbf{W}_{i\nu} = \bar{\mathbf{W}}_{i\nu}$ and $\mathbf{W}_{is} = \bar{\mathbf{W}}_{is}$, the UN-CLE criterion can be reduced to a convenient form for direct application, and it is shown that for a general array

$$f(\hat{\mathbf{R}}_{is}, \bar{\mathbf{W}}_{i\nu}, \bar{\mathbf{W}}_{is}) = \operatorname{tr}\left\{\left[\mathbf{I} - \mathbf{\Sigma}_{ii}^{-1/2} \mathbf{D}_i (\mathbf{D}_i^H \mathbf{\Sigma}_{ii}^{-1} \mathbf{D}_i)^{-1} \mathbf{D}_i^H \mathbf{\Sigma}_{ii}^{-1/2}\right] \mathbf{\Sigma}_{ii}^{-1/2} \hat{\mathbf{R}}_{is} \bar{\mathbf{W}}_{is} \hat{\mathbf{R}}_{is}^H \mathbf{\Sigma}_{ii}^{-1/2}\right\} \tag{6.147}$$

or, equivalently

$$f(\hat{\mathbf{R}}_{is}, \bar{\mathbf{W}}_{i\nu}, \bar{\mathbf{W}}_{is}) = \operatorname{tr}\{\mathbf{B}_{i\nu}(\mathbf{B}_{i\nu}^H \mathbf{\Sigma}_{ii} \mathbf{B}_{i\nu})^{-1} \mathbf{B}_{i\nu}^H \hat{\mathbf{R}}_{is} \bar{\mathbf{W}}_{is} \hat{\mathbf{R}}_{is}^H\} \tag{6.148}$$

where $\mathbf{B}_{i\nu}$ is a matrix of $M_i \times (M_i - K)$ and $\mathbf{B}_{i\nu} \perp \mathbf{D}_i$. The functional form of $\mathbf{B}_{i\nu}$ for a uniform linear array is very simple and is given by Eq. (6C.3). Both Eqs. (6.147) and (6.148) can be optimized directly with respect to θ_i since both $\mathbf{D}_i$ and $\mathbf{B}_{i\nu}$ are explicit functions of θ_i.

Asymptotic Properties of the UN-MUSIC and UN-CLE Criteria

From Eq. (6.131c) as well as Eqs. (6.142) and (6.143), we see that the general expressions of the UN-MUSIC and UN-CLE criteria contain weghting matrices. In the case of the UN-MUSIC algorithm, the weighting matrix $\mathbf{W}_{i\nu}$ puts weights on the projection of the directional vector $\mathbf{d}_i(\theta_{ik})$ onto the estimated supplementary noise subspace spanned by the columns of $\hat{\mathbf{R}}_{i\nu}$. In the case of the UN-CLE algorithm, using Eq. (6.142), the weighting matrix $\mathbf{W}_{is}$ puts weights on the columns of $\hat{\mathbf{R}}_{is}$

in the estimated signal subspace and then $\mathbf{W}_{i\nu}$ puts weights on the projection of these weighted columns onto the true supplementary noise subspace spanned by the columns of $\mathbf{R}_{i\nu}$. The use of Eq. (6.143) of the UN-CLE algorithm can be similarly interpreted with the exception that the roles of the estimated and the true subspaces be interchanged. It in intuitively obvious that the optimum utilization of these weighting matrices is to emphasize or de-emphasize the information from these vectors in the respective sub-spaces. However, in order not to cause any loss of information in the sub-spaces, it is clear that

$$\text{rank}\,(\mathbf{P}_{i\nu}\mathbf{W}_{i\nu}\mathbf{P}_{i\nu}^H) = \text{rank}\,\mathbf{P}_{i\nu} \tag{6.149a}$$

and

$$\text{rank}\,(\mathbf{R}_{is}\mathbf{W}_{is}\mathbf{R}_{is}^H) = \text{rank}\,\mathbf{R}_{is} \tag{6.149b}$$

It is easy to show that both weighting matrices $\bar{\mathbf{W}}_{i\nu}$ and $\bar{\mathbf{W}}_{is}$ in Eqs. (6.144) and (6.145) satisfy these conditions in Eqs. (6.149).

Now, let us examine some asymptotic properties of the criteria given by Eqs. (6.131c), (6.142), and (6.143). Since as discussed in Sections 6.2 and 6.3, the estimated covariance $\hat{\Sigma}$ converges to its true value Σ with probability one, it is expected that the various criteria of the UN-MUSIC and UN-CLE algorithms will converge as the number of snapshots increases. This indeed the case. We first examine the convergence of the UN-MUSIC criterion $h(\theta_{ik}, \hat{\mathbf{P}}_{i\nu}, \mathbf{W}_{i\nu})$ given by Eq. (6.131c).

Since $\lim\limits_{N\to\infty} \hat{\Sigma}_{i\bar{i}} \underset{a.s.}{=} \Sigma_{i\bar{i}}$, using the same argument as in Theorem 6.4, we see that

$$\lim_{N\to\infty} \hat{\mathbf{P}}_{i\nu} \underset{a.s.}{=} \mathbf{P}_{i\nu} \tag{6.150}$$

$$\lim_{N\to\infty} \hat{\mathbf{R}}_{is} \underset{a.s.}{=} \mathbf{R}_{is} \tag{6.151}$$

Since $h(\theta_{ik}, \hat{\mathbf{P}}_{i\nu}, \mathbf{W}_{i\nu})$ is measurable in $\hat{\mathbf{P}}_{i\nu}$ and continuous in θ_{ik}, then

$$\lim_{N\to\infty} h(\theta_{ik}, \hat{\mathbf{P}}_{i\nu}, \mathbf{W}_{i\nu}) \underset{a.s.}{=} h(\theta_{ik}, \mathbf{P}_{i\nu}, \mathbf{W}_{i\nu}) \tag{6.152}$$

Furthermore, as $N \to \infty$, with probability 1, $h(\theta_{ik}, \hat{\mathbf{P}}_{i\nu}, \mathbf{W}_{i\nu})$ converges uniformly to $h(\theta_{ik}, \mathbf{P}_{i\nu}, \mathbf{W}_{i\nu})$, that is

$$\lim_{N\to\infty} \sup |h(\theta_{ik}, \hat{\mathbf{P}}_{i\nu}, \mathbf{W}_{i\nu}) - h(\theta_{ik}, \mathbf{P}_{i\nu}, \mathbf{W}_{i\nu})| \underset{a.s.}{=} 0 \tag{6.153}$$

To show this, we re-write Eq. (6.153) such that

$$\begin{aligned}
&\lim_{N\to\infty} \sup |\ \mathbf{d}_i^H \hat{\mathbf{P}}_{i\nu}\mathbf{W}_{i\nu}\hat{\mathbf{P}}_{i\nu}\mathbf{d}_i - \mathbf{d}_i^H \mathbf{P}_{i\nu}\mathbf{W}_{i\nu}\mathbf{P}_{i\nu}\mathbf{d}_i\ | \\
&= \lim_{N\to\infty} \sup |\ \mathbf{d}_i^H (\hat{\mathbf{P}}_{i\nu}\mathbf{W}_{i\nu}\hat{\mathbf{P}}_{i\nu} - \mathbf{P}_{i\nu}\mathbf{W}_{i\nu}\mathbf{P}_{i\nu})\mathbf{d}_i\ | \\
&\le \lim_{N\to\infty} \sup [|\ \mathbf{d}_i^H (\hat{\mathbf{P}}_{i\nu} - \mathbf{P}_{i\nu})\mathbf{W}_{i\nu}\hat{\mathbf{P}}_{i\nu}^H\mathbf{d}_i\ | \\
&+ |\ \mathbf{d}_i^H \mathbf{P}_{i\nu}\mathbf{W}_{i\nu}(\hat{\mathbf{P}}_{i\nu}^H - \mathbf{P}_{i\nu})\mathbf{d}_i\ |]
\end{aligned} \tag{6.154}$$

Since

$$\lim_{N\to\infty} \hat{\mathbf{P}}_{i\nu} = \mathbf{P}_{i\nu} \tag{6.155}$$

then Eq. (6.153) follows. Therefore, we conclude that

$$\lim_{N\to\infty} \arg\min_{\theta_{ik}} h(\theta_{ik}, \hat{\mathbf{P}}_{i\nu}, \mathbf{W}_{i\nu}) \underset{a.s.}{=} \arg\min_{\theta_{ik}} h(\theta_{ik}, \mathbf{P}_{i\nu}, \mathbf{W}_{i\nu}) \tag{6.156}$$

Equation (6.156) tells us that as $N \to \infty$, the point $\hat{\theta}$ which minimizes the criterion evaluated using quantities obtained from the sample data covariance matrix $\hat{\Sigma}$, converges to the point, which will minimize the criterion évaluated using quantities obtained from the true data covariance matrix Σ.

Using similar arguments, we can show that both criteria of the UN-CLE algorithm in Eqs. (6.142) and (6.143) converge uniformly and

$$\lim_{N\to\infty} \arg\min_{\theta} f(\hat{\mathbf{R}}_{is}, \mathbf{W}_{i\nu}, \mathbf{W}_{is}) \underset{a.s.}{=} \arg\min_{\theta} f(\mathbf{R}_{is}, \mathbf{W}_{i\nu}, \mathbf{W}_{is}) \tag{6.157a}$$

and

$$\lim_{N\to\infty} \arg\min_{\theta} g(\hat{\mathbf{P}}_{i\nu}, \mathbf{W}'_{i\nu}, \mathbf{W}'_{is}) \underset{a.s.}{=} \arg\min_{\theta} g(\mathbf{P}_{i\nu}, \mathbf{W}'_{i\nu}, \mathbf{W}'_{is}) \tag{6.157b}$$

where the elements of $\boldsymbol{\theta}$ belong to the set Ω.

Now, let us examine the consistency of the estimates obtained by the UN-MUSIC and UN-CLE algorithms.

Theorem 6.7. *For an array which gives an unambiguous directional matrix of the signals, the true DOA* $\theta_{ik}, k = 1, \cdots, K$, *are the only values minimizing the criterion of the UN-MUSIC algorithm* $h(\theta_{ik}, \hat{\mathbf{P}}_{i\nu}, \mathbf{W}_{i\nu})$ *as* $N \to \infty$, *where* $\mathbf{W}_{i\nu}$ *satisfies Eq. (6.149a).*

Proof. The result of this theorem follows directly from the uniform convergence of the UN-MUSIC criterion in Eq. (6.157) and the unambiguity of the directional vectors. Noting that the criterion h of the UN-MUSIC algorithm is always non-negative, when the true values of θ_{ik} are substituted into the directional vector, because of the unambiguity, these are the only vectors which are orthogonal to the true noise sub-space onto which, as $N \to \infty$, $\hat{\mathbf{P}}_{i\nu}$ projects. Thus, $\mathbf{d}(\theta_{ik}), k = 1, \cdots, K$ are the only vectors which minimize h. ■

Theorem 6.8. *For an array which yields an unambiguous directional matrix of the signals, the true DOA* θ_i *uniquely minimizes the criterion* $f(\hat{\mathbf{R}}_{is}, \mathbf{W}_{i\nu}, \mathbf{W}_{is})$ *of the UN-CLE algorithm as* $N \to \infty$ *where* $\mathbf{W}_{i\nu}$ *and* $\mathbf{W}_{is}$ *satisfy Eqs. (6.149).*

Proof. The proof of θ_i uniquely minimizing f and g follows from essentially the same argument as in Theorem 6.7 that both criteria converge uniformly and that the unambiguity of the direction vectors leads to unique estimations of the signal and noise subspaces. The result follows from the orthogonality of these subspaces when $N \to \infty$. ■

Relations of UN-MUSIC and UN-CLE to Algorithms in Spatially White Noise

In this section, we have developed new algorithms for the estimation of DOA in unknown noise utilizing the concepts and properties of GCD derived in previous sections. These algorithms are obtained using principles similar to the methods of MUSIC and MODE for white noise environments. In Section 6.4, we have seen that GCD includes the special case of correlation analysis in white or known noise environment, thus it is reasonable to expect that the MUSIC and MODE algorithms can be obtained form UN-MUSIC and UN-CLE as special cases. This is indeed true as shown by the following: For the algorithms of MUSIC and MODE, the ambient noise is assumed to be white and the data are obtained using a single array. Thus, as mentioned in Examples 6.2 and 6.3, we have $\mathbf{x}_1(n) = \mathbf{x}_2(n)$, and $\boldsymbol{\Sigma}_{12} = \boldsymbol{\Sigma}_{21} = \boldsymbol{\Sigma}_{11} = \boldsymbol{\Sigma}_x$ where $\boldsymbol{\Sigma}_x$ is defined in Eq. (6.7). Now, as shown in Example 6.2, if we choose $\boldsymbol{\Pi}_1 = \mathbf{I}$ and $\boldsymbol{\Pi}_2 = \hat{\boldsymbol{\Sigma}}_{11}$, then $\hat{\mathbf{R}}_1 = \hat{\mathbf{L}}_1$ and has columns which are the eigenvectors of $\hat{\boldsymbol{\Sigma}}_{11}$. Hence, the columns of $\hat{\mathbf{L}}_{1s} = \hat{\mathbf{R}}_{1s}$ form an orthonormal basis for the signal subspace. Likewise, the columns of $\hat{\mathbf{L}}_{1\nu}$ also form an orthonormal basis for the noise subspace. Thus, the eigenprojectors onto the estimated noise and signal subspaces as defined in Eqs. (6.82) and (6.83) respectively are, in this special case, reduced to

$$\hat{\mathbf{P}}_{1\nu} = \hat{\mathbf{L}}_{1\nu}\hat{\mathbf{L}}_{1\nu}^H \tag{6.158a}$$

$$\hat{\mathbf{P}}_{1s}^H = \hat{\mathbf{L}}_{1s}\hat{\mathbf{L}}_{1s}^H \tag{6.158b}$$

We note that these projectors are self-adjoint and are therefore orthogonal projectors. Using Eq. (6.158a) in the UN-MUSIC algorithm of Eq. (6.131b), we obtain, for the case of spatially white noise

$$\hat{\theta}_{ik} = \arg\min_{\theta_k}\{\mathbf{d}_i^H(\theta_k)\hat{\mathbf{L}}_{i\nu}\hat{\mathbf{L}}_{i\nu}^H\mathbf{d}_i(\theta_k)\}, \quad k = 1, \cdots, K, \tag{6.159}$$

which is exactly the same as the MUSIC algorithm given by Eq. (6.20).

We now turn our attention to the comparison between the UN-CLE and the MODE algorithms. Comparing the results in Eq. (6.108a) of Theorem 6.6 and Eq. (6.120) of Corollary 6.6a, we see that in the case of white ambient noise, the factor $(\boldsymbol{\Gamma}^{-1}\mathbf{L}_{is}^H\boldsymbol{\Sigma}_{\bar{i}\bar{i}}\mathbf{L}_{\bar{i}s}\boldsymbol{\Gamma}^{-1})$ is replaced by $\mathring{\boldsymbol{\Lambda}}_s^{-2}\boldsymbol{\Gamma}^2$. Thus, following exactly the same development of the UN-CLE algorithm from Eq. (6.136) through Eq. (6.146), and using only Array 1, we obtain the corresponding weighting matrices for the case of white ambient noise such that

$$\bar{\mathbf{W}}_{1\nu} = \mathbf{L}_{1\nu}(\mathbf{L}_{1\nu}^H\boldsymbol{\Sigma}_{11}\mathbf{L}_{1\nu})^{-1}\mathbf{L}_{1\nu}^H \tag{6.160}$$

and

$$\bar{\mathbf{W}}_{1s} = \boldsymbol{\Gamma}^{-1}\mathring{\boldsymbol{\Lambda}}_s^2\boldsymbol{\Gamma}^{-1} \tag{6.161}$$

Substituting these two weighting matrices into the criterion of UN-CLE in Eq. (6.142), and using the transformation $\mathbf{L}_{1\nu} = \mathbf{B}_{1\nu}\mathbf{T}_{1\nu}$ as indicated by Eq. (6C.10) in Appendix 6C, we obtain

$$\hat{\theta} = \arg\min_{\theta} f(\hat{\mathbf{R}}_{1s}, \bar{\mathbf{W}}_{1\nu}, \bar{\mathbf{W}}_{1s})$$
$$= \arg\min_{\theta} \operatorname{tr}\{\mathbf{B}_{1\nu}(\mathbf{B}_{1\nu}^H \boldsymbol{\Sigma}_{11}\mathbf{B}_{1\nu})^{-1}\mathbf{B}_{1\nu}\hat{\mathbf{L}}_{1s}\boldsymbol{\Gamma}^{-1}\mathring{\boldsymbol{\Lambda}}_s^2\boldsymbol{\Gamma}^{-1}\hat{\mathbf{L}}_{1s}^H\} \tag{6.162}$$

where we have used the fact that $\hat{\mathbf{R}}_{1s} = \hat{\mathbf{L}}_{1s}$. Now, for the case of white ambient noise, using only one array, $\boldsymbol{\Sigma}_{11}$ is the covariance matrix of the data received by the sensor array and can be expressed as in Eq. (6.7). Since $\mathbf{B}_{1\nu}$ consists of column vectors which form a basis in the noise subspace and are therefore othogonal to the signal part of $\boldsymbol{\Sigma}_{11}$, Eq. (6.162) can be reduced to

$$\hat{\theta} = \arg\min_{\theta} \operatorname{tr}\{\mathbf{B}_{1\nu}(\mathbf{B}_{1\nu}^H \mathbf{B}_{1\nu})^{-1}\mathbf{B}_{1\nu}\hat{\mathbf{L}}_{1s}\boldsymbol{\Gamma}^{-1}\mathring{\boldsymbol{\Lambda}}_s^2\boldsymbol{\Gamma}^{-1}\hat{\mathbf{L}}_{1s}^H\} \tag{6.163}$$

Noting that $\boldsymbol{\Gamma} = \operatorname{diag}(\gamma_1. \cdots, \gamma_K)$ with $\{\gamma_k^2\}$ being the signal eigenvalues and $\hat{\mathbf{L}}_{1s}$ consists of columns which are the estimated signal eigenvectors, it can be seen that Eq. (6.163) is exactly the same as the criterion for MODE given by Eq. (6.34).

6.6 PERFORMANCE ANALYSIS OF UN-MUSIC AND UN-CLE

We now examine the asymptotic performance of the two algorithms UN-MUSIC and UN-CLE. Our attention is focused on the asymptoic means and asymptotic covariances of the two algorithms. For the analysis of an asymptotic mean, we are only concerned with terms greater than or equal to $O(N^{-1/2})$, and for an asymptotic covariance, only with terms greater than or equal to $O(N^{-1})$.

Bias and Variance of the UN-MUSIC Estimates

The UN-MUSIC estimate of the DOA of the kth signal using the ith array in unknown noise environment is, as presented in the last section, given by

$$\hat{\theta}_{ik} = \arg\min_{\theta_{ik}} h(\theta_{ik}) \tag{6.164}$$

where

$$h(\theta_{ik}) = \mathbf{d}_i^H(\theta_{ik})\hat{\mathbf{P}}_{i\nu}\mathbf{W}_{i\nu}\hat{\mathbf{P}}_{i\nu}^H\mathbf{d}_i(\theta_{ik}) \tag{6.165}$$

Let $\Delta\theta_{ik}$ be the error involved in estimating the DOA of the kth signal using the ith array, that is

$$\Delta\theta_{ik} = \hat{\theta}_{ik} - \theta_{ik} \tag{6.166}$$

Theorem 6.9. *The UN-MUSIC estimates, $\hat{\theta}_{ik}$, of the DOA in an unknown noise environment are asymptotically Gaussian distributed, with zero bias and a covariance matrix given by*

$$E[\Delta\theta_{ik}\Delta\theta_{il}] = \frac{2}{N\ddot{h}(\theta_{ik})\ddot{h}(\theta_{il})}\operatorname{Re}\Big\{\big[\dot{\mathbf{d}}_i^H(\theta_{ik})\mathbf{P}_{i\nu}\mathbf{W}_{i\nu}\mathbf{P}_{i\nu}^H\boldsymbol{\Sigma}_{ii}\mathbf{P}_{i\nu}\mathbf{W}_{i\nu}\mathbf{P}_{i\nu}^H\dot{\mathbf{d}}_i(\theta_{il})\big]$$
$$\big[\mathbf{d}_i^H(\theta_{ik})\mathbf{L}_{is}\boldsymbol{\Gamma}^{-1}\mathbf{L}_{\bar{i}s}^H\boldsymbol{\Sigma}_{\bar{i}\bar{i}}\mathbf{L}_{\bar{i}s}\boldsymbol{\Gamma}^{-1}\mathbf{L}_{is}^H\mathbf{d}_i(\theta_{il})\big]\Big\} \tag{6.167}$$
$$i = 1, 2 \quad k, l \in [1, \cdots, K]$$

where

$$\ddot{h}(\theta_{ik}) = 2\left[\dot{\mathbf{d}}_i^H(\theta_{ik})\mathbf{P}_{iv}\mathbf{W}_{iv}\mathbf{P}_{iv}^H\dot{\mathbf{d}}_i(\theta_{ik})\right] \tag{6.168}$$

and $\dot{\mathbf{d}}_i$ *denotes the first derivative of* $\mathbf{d}_i$ *respect to* θ_{ik}.

Proof. For the asymptotical analysis of estimation errors, we can expand the first-order derivative of $h(\theta_{ik})$ in a Taylor series up to the first order term $\Delta\theta_{ik}$ and equate it to zero since the UN-MUSIC algorithm seeks the minimum of h so that

$$\dot{h}(\hat{\theta}_{ik}) = \dot{h}(\theta_{ik}) + \ddot{h}(\bar{\theta}_{ik})\Delta\theta_{ik} = 0 \tag{6.169}$$

where $\dot{h}(\theta_{ik})$ and $\ddot{h}(\theta_{ik})$ are respectively the first-order and second-order derivative of h with respect to θ_{ik}, and $\bar{\theta}_{ik}$ is a point between $\hat{\theta}_{ik}$ and θ_{ik}, Therefore, we can write

$$\Delta\theta_{ik} = -\dot{h}(\theta_{ik})/\ddot{h}(\bar{\theta}_{ik}) \tag{6.170}$$

We first evaluate $\dot{h}(\theta_{ik})$ by differentiating Eq. (6.165) with respect to θ_{ik} so that

$$\begin{aligned}\dot{h}(\theta_{ik}) &= 2\text{Re}\left[\mathbf{d}_i^H(\theta_{ik})\hat{\mathbf{P}}_{iv}\mathbf{W}_{iv}\hat{\mathbf{P}}_{iv}^H\dot{\mathbf{d}}_i(\theta_{ik})\right] + \mathbf{d}_i^H(\theta_{ik})\hat{\mathbf{P}}_{iv}\frac{\partial \mathbf{W}_{iv}}{\partial\theta_{ik}}\hat{\mathbf{P}}_{iv}^H\mathbf{d}_i(\theta_{ik}) \\ &= 2\text{Re}\left[\mathbf{d}_i^H(\theta_{ik})\hat{\mathbf{P}}_{iv}\mathbf{W}_{iv}\hat{\mathbf{P}}_{iv}^H\dot{\mathbf{d}}_i(\theta_{ik})\right] + o(N^{-1/2})\,prob.\end{aligned} \tag{6.171}$$

where $o(N^{-1/2})$ is a sequence converging to zero faster than $N^{-1/2}$. In the second step of Eq. (6.171), we have used the fact that $\hat{\mathbf{P}}_{iv} = \mathbf{P}_{iv} + \mathbf{O}(N^{-1/2})\,prob.$ (see Theorem 6.4), which implies that $\mathbf{d}_i^H(\theta_{ik})\hat{\mathbf{P}}_{iv} = \mathbf{O}(N^{-1/2})\,prob.$ Thus, we can replace $\hat{\mathbf{P}}_{iv}$ by $\mathbf{P}_{iv}$ in the approximation error $o(N^{-1/2})$ which denotes a sequence that converges to zero faster than $N^{-1/2}$.

Now, we evaluate $\ddot{h}(\bar{\theta}_{ik})$ by differentiating Eq. (6.165) twice with respect to θ_{ik}, and substituting $\bar{\theta}_{ik}$ for θ_{ik} to obtain (a step made possible by the fact that $\bar{\theta}_{ik} \to \theta_{ik}$ due to the consistency of the UN-MUSIC estimates)

$$\begin{aligned}\ddot{h}(\bar{\theta}_{ik}) &= 2\text{Re}\left[\dot{\mathbf{d}}_i^H(\bar{\theta}_{ik})\hat{\mathbf{P}}_{iv}\mathbf{W}_{iv}\hat{\mathbf{P}}_{iv}^H\dot{\mathbf{d}}_i(\bar{\theta}_{ik}) + \mathbf{d}_i^H(\bar{\theta}_{ik})\hat{\mathbf{P}}_{iv}\mathbf{W}_{iv}\hat{\mathbf{P}}_{iv}^H\ddot{\mathbf{d}}_i(\bar{\theta}_{ik})\right] \\ &\quad + 2\text{Re}\left[\mathbf{d}_i^H(\theta_{ik})\hat{\mathbf{P}}_{iv}\frac{\partial \mathbf{W}_{iv}}{\partial\theta_{ik}}\hat{\mathbf{P}}_{iv}^H\dot{\mathbf{d}}_i(\theta_{ik})\right] \\ &\quad + \mathbf{d}_i^H(\theta_{ik})\hat{\mathbf{P}}_{iv}\frac{\partial^2 \mathbf{W}_{iv}}{\partial\theta_{ik}^2}\hat{\mathbf{P}}_{iv}^H\mathbf{d}_i(\theta_{ik})\end{aligned} \tag{6.172a}$$

$$= 2\left[\dot{\mathbf{d}}_i^H(\theta_{ik})\mathbf{P}_{iv}\mathbf{W}_{iv}\mathbf{P}_{iv}^H\dot{\mathbf{d}}_i(\theta_{ik})\right] + O(\Delta\theta_{ik}) + O(N^{-1/2})\,prob. \tag{6.172b}$$

Equation (6.172b) results from the facts that replacing $\bar{\theta}_{ik}$ by θ_{ik} in $\dot{\mathbf{d}}$ yields the residual term $\mathbf{O}(\Delta\theta_{ik})$ and that replacing $\hat{\mathbf{P}}_{iv}$ by $\mathbf{P}_{iv}$ and ignoring the term involving $\mathbf{d}_i^H(\theta_{ik})\hat{\mathbf{P}}_{iv}$ gives rise to the residual term $\mathbf{O}(N^{-1/2})$. The two residual terms in Eq. (6.172b) are small compared to the first term, which is noted to be real. Thus, substituting Eqs. (6.171) and (6.172b) into Eq. (6.170), we have

$$\Delta\theta_{ik} = -\frac{1}{\ddot{h}(\theta_{ik})}2\text{Re}\left[\mathbf{d}_i^H(\theta_{ik})\hat{\mathbf{P}}_{iv}\mathbf{W}_{iv}\mathbf{P}_{iv}^H\dot{\mathbf{d}}_i(\theta_{ik})\right] + o(N^{-1/2})\,prob. \tag{6.173}$$

where $\ddot{\bar{h}}(\theta_{ik})$ is defined in Eq. (6.168). We note that $\ddot{\bar{h}}(\theta_{ik})$ is positive. This is because UN-MUSIC is asymptotically consistent, implying that $\bar{h}(\theta_{ik}) \triangleq \lim_{N\to\infty} h(\theta_{ik})$ is a unique local minimum, and $\ddot{\bar{h}}(\theta_{ik})$ is continuous in θ_{ik}.

Now, from Corollary 6.5b, replacing $\mathbf{Y}_{is}$ by $\mathbf{d}_i(\theta_{ik})$ we have

$$\mathbf{d}_i^H(\theta_{ik})\hat{\mathbf{P}}_{iv} = -\mathbf{d}_i^H(\theta_{ik})\hat{\mathbf{P}}_{is}\mathbf{L}_{iv}\mathbf{R}_{iv}^H + \mathbf{O}(N^{-1})\,prob. \tag{6.174}$$

From Theorem 6.6, we have seen that $\hat{\mathbf{P}}_{is}\mathbf{Y}_{iv}$ is asymptotically Gaussian distributed with zero mean for $\mathbf{Y}_{iv} \subseteq \overline{\text{span}}\{\mathbf{D}_i(\theta_i)\}$. Hence, we deduce that $\mathbf{d}_i^H(\theta_{ik})\hat{\mathbf{P}}_{iv}$ is also asymptotically Gaussian distributed with zero mean. Thus, from Eq. (6.173), we can conclude that $\Delta\theta_{ik}$ is asymptotically Gaussian distributed with zero mean, and therefore $\hat{\theta}_{ik}$ is unbiased and is asymptotically Gaussian distributed.

We now examine the covariance of the estimates $\hat{\theta}_{ik}$. Let use define the quantity in the square brackets of Eq. (6.173) such that

$$\begin{aligned} u_{ik} &\triangleq \mathbf{d}_i^H(\theta_{ik})\hat{\mathbf{P}}_{iv}\mathbf{W}_{iv}\mathbf{P}_{iv}^H\dot{\mathbf{d}}_i(\theta_{ik}) \\ &= (\dot{\mathbf{d}}_i^H(\theta_{ik})\mathbf{P}_{iv}\mathbf{W}_{iv})^*\text{vec}\,(\mathbf{d}_i^H(\theta_{ik})\hat{\mathbf{P}}_{iv}) \end{aligned} \tag{6.175}$$

where * denotes conjugate of a complex quantity. Substituting Eq. (6.175) in Eq. (6.173), and ignoring the terms $o(N^{-1/2})$, we obtain the covariance of the errors of two separate DOA estimates θ_{ik} and θ_{il} using the ith array such that

$$E\left[\Delta\theta_{ik}\Delta\theta_{il}\right] = \frac{2}{\ddot{\bar{h}}(\theta_{ik})\ddot{\bar{h}}(\theta_{il})}E\left[\text{Re}\left\{u_{ik}u_{il}^* + u_{ik}u_{il}\right\}\right] \tag{6.176}$$

where we have employed the identity that $\text{Re}\,(u_{ik})\text{Re}\,(u_{il}) \equiv \frac{1}{2}\text{Re}\,(u_{ik}u_{il}^* + u_{ik}u_{il})$. Now, using the definition of u_{ik} in Eq. (6.175), we can evaluate the quantities in Eq. (6.176) as

$$\begin{aligned} &E\left[u_{ik}u_{il}^*\right] \\ &= (\dot{\mathbf{d}}_i^H(\theta_{ik})\mathbf{P}_{iv}\mathbf{W}_{iv})^*E\left[\text{vec}\,(\mathbf{d}_i^H(\theta_{ik})\hat{\mathbf{P}}_{iv})\text{vec}^H(\mathbf{d}_i^H(\theta_{il})\hat{\mathbf{P}}_{iv})\right](\mathbf{W}_{iv}^H\mathbf{P}_{iv}^H\dot{\mathbf{d}}_i(\theta_{il}))^* \end{aligned} \tag{6.177}$$

Applying the results of Eq. (6.125) in Corollary 6.6b to the expectation term in Eq. (6.177), we have

$$\begin{aligned} E\left[u_{ik}u_{il}^*\right] = \frac{1}{N}\Big\{ &\left[\dot{\mathbf{d}}_i^H(\theta_{ik})\mathbf{P}_{iv}\mathbf{W}_{iv}\mathbf{P}_{iv}^H\boldsymbol{\Sigma}_{ii}\mathbf{P}_{iv}\mathbf{W}_{iv}\mathbf{P}_{iv}^H\dot{\mathbf{d}}_i(\theta_{il})\right]^* \\ &\left[\mathbf{d}_i^H(\theta_{ik})(\mathbf{L}_{is}\boldsymbol{\Gamma}^{-1}\mathbf{L}_{\bar{i}s}^H\boldsymbol{\Sigma}_{\bar{i}\bar{i}}\mathbf{L}_{\bar{i}s}\boldsymbol{\Gamma}^{-1}\mathbf{L}_{is}^H)\mathbf{d}_i(\theta_{il})\right]\Big\} \end{aligned} \tag{6.178}$$

Also, using Eq. (6.126) in Corollary 6.6b, we obtain

$$E\left[u_{ik}\,u_{il}\right] = 0 \tag{6.179}$$

and substituting Eqs. (6.178) and (6.179) into Eq. (6.176), the result of Eq. (6.167) follows. ■

The above theorem gives us the asymptotic performance of the UN-MUSIC algorithm. It should be noted that the analysis of the asymptotic performance is valid for any general weighting matrix $\mathbf{W}_{i\nu}$ and for any generalized correlation decomposition, that is, for any general choice of $\mathbf{\Pi}_i$. Furthermore, through Eq. (6.171) and Eqs. (6.172) we have seen that $\Delta\theta_{ik}$, given in Eq. (6.173), is asymptotically invariant no matter if $\mathbf{W}_{i\nu}$ is treated as constant or not. Tha same conclusion can be said about the performance of UN-MUSIC.

Bias and Variance of the UN-CLE Estimates

In the previous section, two criteria $f_i \triangleq f(\hat{\mathbf{R}}_{is}, \mathbf{W}_{i\nu}, \mathbf{W}_{is})$ and $g_i \triangleq g(\hat{\mathbf{P}}_{i\nu}, \mathbf{W}'_{i\nu}, \mathbf{W}'_{is})$ were presented as general UN-CLE algorithms for the estimation of DOA in unknown correlated noise. These criteria, given respectively by Eqs. (6.142) and (6.143), were developed based on two different measures of the projection errors between the estimated and true signal subspaces. Since the estimated signal subspace converges to the true signal subspace, and $\hat{\mathbf{R}}_{is}^H\hat{\mathbf{L}}_{i\nu} = \mathbf{O}$, it is reasonable to expect that the two criteria have the same performance. This is indeed the case, as shown by the following theorem:

Theorem 6.10. *Employing the same weighting matrices, that is,* $\mathbf{W}_{i\nu} = \mathbf{W}'_{i\nu}$ *and* $\mathbf{W}_{is} = \mathbf{W}'_{is}$, *the two criteria* $f_i \triangleq f(\hat{\mathbf{R}}_{is}, \mathbf{W}_{i\nu}, \mathbf{W}_{is})$ *and* $g_i \triangleq g(\hat{\mathbf{P}}_{i\nu}, \mathbf{W}_{i\nu}, \mathbf{W}_{is})$ *have equivalent asymptotic performance.*

Proof. For the ith array, the UN-CLE algorithms seek for the minimum of g_i so that

$$\dot{\mathbf{g}}_i(\hat{\theta}_i) \triangleq \nabla_{\theta_i} g \,|_{\theta_i=\hat{\theta}_i} = \mathbf{O} \tag{6.180a}$$

where

$$\nabla_{\theta_i} \triangleq [\partial/\partial\theta_{i1} \cdots \partial/\partial\theta_{iK}]^T \tag{6.180b}$$

Representing $\dot{\mathbf{g}}_i(\hat{\theta}_i)$ around θ_i by the mean value theorem and noting $\dot{\mathbf{g}}_i(\hat{\theta}_i) = \mathbf{O}$, we have

$$\Delta\theta_i = \hat{\theta}_i - \theta_i = -\left[\ddot{\mathbf{g}}_i(\bar{\theta}_i)\right]^{-1} \dot{\mathbf{g}}_i(\theta_i) \tag{6.181}$$

where $\bar{\theta}_i$ is a point on the line joining $\hat{\theta}_i$ and θ_i. Let us define a matrix $\ddot{\bar{\mathbf{g}}}_i(\theta_i)$ whose (jk)th element is given by

$$\begin{aligned}(\ddot{\bar{\mathbf{g}}}_i)_{jk} &\triangleq \lim_{N\to\infty}\left[\frac{\partial^2}{\partial\theta_{ij}\partial\theta_{ik}} g(\hat{\mathbf{P}}_{i\nu}, \mathbf{W}_{i\nu}, \mathbf{W}_{is})\right] \\ &= \frac{\partial^2}{\partial\theta_{ij}\partial\theta_{ik}} g(\mathbf{P}_{i\nu}, \mathbf{W}_{i\nu}, \mathbf{W}_{is}) = \frac{\partial^2}{\partial\theta_{ij}\partial\theta_{ik}}\bar{g}\end{aligned} \tag{6.182}$$

Note that $\text{span}\{\mathbf{R}_{is}\} = \text{span}\{\mathbf{D}_i(\theta_i)\}$ so that

$$\mathbf{R}_{is} = \mathbf{D}_i\mathbf{T}_{is} \tag{6.183}$$

where $\mathbf{T}_{is}$ is a $K \times K$ non-singular matrix and we have omitted the obvious notation of the dependence of $\mathbf{D}_i$ on θ_i. Thus, using Eq. (6.183) in the definition in Eq. (6.143), we can write

$$g_i = \text{tr}\left[\hat{\mathbf{P}}_{iv}\mathbf{W}_{iv}\hat{\mathbf{P}}_{iv}^H\mathbf{D}_i\mathbf{T}_{is}\mathbf{W}_{is}\mathbf{T}_{is}^H\mathbf{D}_i^H\right] \tag{6.184}$$

From Eq. (6.184), the kth elements of the vector $\dot{\mathbf{g}}_i(\boldsymbol{\theta}_i)$ is given by

$$\begin{aligned}\dot{g}_{ik} = \text{tr}\Big[&(\hat{\mathbf{P}}_{iv}\mathbf{W}_{iv}\hat{\mathbf{P}}_{iv}^H)\{\dot{\mathbf{D}}_{ik}\mathbf{T}_{is}\mathbf{W}_{is}\mathbf{T}_{is}^H\mathbf{D}_i^H + \mathbf{D}_i\mathbf{T}_{is}\mathbf{W}_{is}\mathbf{T}_{is}^H\dot{\mathbf{D}}_{ik}^H \\ &+ \mathbf{D}_i\frac{\partial(\mathbf{T}_{is}\mathbf{W}_{is}\mathbf{T}_{is}^H)}{\partial\theta_{ik}}\mathbf{D}_i^H\} + (\hat{\mathbf{P}}_{iv}\frac{\partial\mathbf{W}_{iv}}{\partial\theta_{ik}}\hat{\mathbf{P}}_{iv}^H)(\mathbf{D}_i\mathbf{T}_{is}\mathbf{W}_{is}\mathbf{T}_{is}^H\mathbf{D}_i^H)\Big]\end{aligned} \tag{6.185}$$

where $\dot{\mathbf{D}}_{ik}$ represents the derivative of the matrix $\mathbf{D}_i$ with respect to θ_{ik}. Using the fact that $\hat{\mathbf{P}}_{iv}^H\mathbf{D}_i = \mathbf{O}(N^{-1/2})$, we can write

$$\dot{g}_{ik} \triangleq \frac{\partial g_i}{\partial\theta_{ik}} = 2\text{tr}\left\{\text{Re}\left[\mathbf{W}_{iv}\mathbf{P}_{iv}^H\dot{\mathbf{D}}_{ik}\mathbf{T}_{is}\mathbf{W}_{is}\mathbf{R}_{is}^H\hat{\mathbf{P}}_{iv}\right]\right\} + O(N^{-1/2})\,prob. \tag{6.186}$$

Differentiating Eq. (6.185) with respect to θ_{il} and noting that $\lim_{N\to\infty}\hat{\mathbf{P}}_{iv}^H\mathbf{D}_i = \mathbf{P}_{iv}^H\mathbf{D}_i = \mathbf{O}$, we obtain the (kl)th element of the matrix $\ddot{\bar{\mathbf{g}}}_i(\boldsymbol{\theta}_i)$ as

$$\left[\ddot{\bar{\mathbf{g}}}_i\right]_{kl} = 2\text{tr}\left\{\text{Re}\left[(\mathbf{P}_{iv}\mathbf{W}_{iv}\mathbf{P}_{iv}^H)\dot{\mathbf{D}}_{il}\mathbf{T}_{is}\mathbf{W}_{is}\mathbf{T}_{is}^H\dot{\mathbf{D}}_{ik}^H\right]\right\} \tag{6.187}$$

We note that $\ddot{\bar{\mathbf{g}}}_i$ is positive definite. This is because of the continuity of the Hessian matrix of $\bar{g}_i$ with respect to $\boldsymbol{\theta}_i$ and the consistency of UN-CLE implying that a unique minimum occurs at $\boldsymbol{\theta}_i$. Thus, we can see that by replacing $\ddot{\mathbf{g}}_i$ by $\ddot{\bar{\mathbf{g}}}_i$, we have introduced terms of $\mathbf{O}(N^{-1/2})$. Furthermore, since $\bar{\boldsymbol{\theta}}_i$ is a point between $\hat{\boldsymbol{\theta}}_i$ and $\boldsymbol{\theta}_i$, then by replacing $\bar{\boldsymbol{\theta}}_i$ by $\boldsymbol{\theta}_i$, we have introduced terms of $\mathbf{O}(\Delta\boldsymbol{\theta}_i)$. Thus

$$\ddot{\mathbf{g}}_i(\bar{\boldsymbol{\theta}}_i) = \ddot{\bar{\mathbf{g}}}_i(\boldsymbol{\theta}_i) + \mathbf{O}(\Delta\boldsymbol{\theta}_i) + \mathbf{O}(N^{-1/2})\,prob. \tag{6.188}$$

Ignoring terms $\mathbf{O}(\Delta\boldsymbol{\theta}_i)$ and $\mathbf{O}(N^{-1/2})$ in Eq. (6.188) and substituting Eqs. (6.186) and (6.188) into Eq. (6.181), we obtain

$$\Delta\boldsymbol{\theta}_i = -\left[\ddot{\bar{\mathbf{g}}}_i(\boldsymbol{\theta}_i)\right]^{-1}\dot{\mathbf{g}}_i(\boldsymbol{\theta}_i) + \mathbf{o}(N^{-1/2})\,prob. \tag{6.189}$$

Using the same reasoning that for the ith array the UN-CLE algorithm seeks for the minimum of f, we have

$$\Delta\boldsymbol{\theta}_i = \hat{\boldsymbol{\theta}}_i - \boldsymbol{\theta}_i = -\left[\ddot{\mathbf{f}}_i(\bar{\boldsymbol{\theta}}_i)\right]^{-1}\dot{\mathbf{f}}_i(\boldsymbol{\theta}_i) \tag{6.190}$$

In a similar way, we define $\bar{f}_i \triangleq \lim_{N\to\infty} f(\hat{\mathbf{R}}_{is}, \mathbf{W}_{iv}, \mathbf{W}_{is}) = f(\mathbf{R}_{is}, \mathbf{W}_{iv}, \mathbf{W}_{is})$, and $\ddot{\bar{\mathbf{f}}}_i(\boldsymbol{\theta}_i)$ such that its (kl)th element is

$$(\ddot{\bar{\mathbf{f}}}_i)_{kl} = \frac{\partial^2}{\partial\theta_{ik}\partial\theta_{il}} f(\mathbf{R}_{is}, \mathbf{W}_{iv}, \mathbf{W}_{is}) \tag{6.191}$$

Also, since $\text{span}\{\mathbf{L}_{iv}\} = \overline{\text{span}}(\mathbf{D}_i)$, we can find a basis $\mathbf{B}_{iv}$ of $\overline{\text{span}}(\mathbf{D}_i)$ so that

$$\mathbf{L}_{iv} = \mathbf{B}_{iv}\mathbf{T}_{iv} \tag{6.192}$$

where $\mathbf{T}_{iv}$ is an $(M-K)\times(M-K)$ nonsingular matrix. Therefore, using the same steps as those by which we arrive at $\dot{\mathbf{g}}_i$ and $\ddot{\bar{\mathbf{g}}}_i$, we obtain the kth element of $\dot{\mathbf{f}}_i$ to be

$$\dot{f}_{ik} = 2\,\mathrm{tr}\left\{\mathrm{Re}\left[\mathbf{W}_{iv}\mathbf{R}_{iv}\mathbf{T}_{iv}^H\dot{\mathbf{B}}_{ivk}^H\mathbf{D}_i\mathbf{T}_{is}\mathbf{W}_{is}\hat{\mathbf{R}}_{is}^H\mathbf{P}_{iv}\right]\right\} + \mathbf{o}(N^{-1/2})\,prob. \tag{6.193}$$

and the (kl)th element of $\ddot{\mathbf{f}}_i$ to be

$$\left[\ddot{\mathbf{f}}_i\right]_{kl} = 2\,\mathrm{tr}\left\{\mathrm{Re}\left[(\dot{\mathbf{B}}_{ivk}\mathbf{T}_{iv}\mathbf{R}_{iv}^H\mathbf{W}_{iv}\mathbf{R}_{iv}\mathbf{T}_{iv}^H\dot{\mathbf{B}}_{ivl}^H)\mathbf{R}_{is}\mathbf{W}_{is}\mathbf{R}_{is}^H\right]\right\} \tag{6.194}$$

where $\dot{\mathbf{B}}_{ivk} \triangleq \dfrac{\partial}{\partial\theta_{ik}}\mathbf{B}_{iv}$. Again, ignoring terms of $\mathbf{O}(\Delta\boldsymbol{\theta}_i)$ and $\mathbf{O}(N^{-1/2})$ when $\ddot{\mathbf{f}}_i(\bar{\boldsymbol{\theta}}_i)$ is replaced by $\ddot{\mathbf{f}}_i(\boldsymbol{\theta}_i)$, Eq. (6.190) becomes

$$\Delta\boldsymbol{\theta}_i = -\left[\ddot{\mathbf{f}}_i(\boldsymbol{\theta}_i)\right]^{-1}\dot{\mathbf{f}}_i(\boldsymbol{\theta}_i) + \mathbf{o}(N^{-1/2})\,prob. \tag{6.195}$$

To show the equivalence of Eqs. (6.189) and (6.195), it remains to be shown that $\dot{\mathbf{g}}_i(\boldsymbol{\theta}_i) = \dot{\mathbf{f}}_i(\boldsymbol{\theta}_i) + \mathbf{o}(N^{-1/2})\,prob.$ and that $\ddot{\mathbf{g}}_i(\boldsymbol{\theta}_i) = \ddot{\mathbf{f}}_i(\boldsymbol{\theta}_i)$.

Differentiating the identity $\mathbf{B}_{iv}^H\mathbf{D}_i = \mathbf{O}$, with respect to θ_{ik}, we have

$$\dot{\mathbf{B}}_{ivk}^H\mathbf{D}_i = -\mathbf{B}_{iv}^H\dot{\mathbf{D}}_{ik} \tag{6.196}$$

and post-multiplying by $\mathbf{T}_{is}$ on both sides, we see

$$\dot{\mathbf{B}}_{ivk}^H\mathbf{R}_{is} = -\mathbf{B}_{iv}^H\dot{\mathbf{D}}_{ik}\mathbf{T}_{is} \tag{6.197}$$

Also, since $\dot{\mathbf{P}}_{is}^H\hat{\mathbf{R}}_{is} = \hat{\mathbf{R}}_{is}$, and $\hat{\mathbf{P}}_{iv}^H = \mathbf{I} - \hat{\mathbf{P}}_{is}^H$, we can write

$$\mathbf{P}_{iv}^H\hat{\mathbf{R}}_{is} = \mathbf{P}_{iv}^H\hat{\mathbf{P}}_{is}^H\mathbf{R}_{is} + \mathbf{o}(N^{-1/2}) = -\mathbf{P}_{iv}^H\hat{\mathbf{P}}_{iv}^H\mathbf{R}_{is} + \mathbf{o}(N^{-1/2})\,prob. \tag{6.198}$$

Since $\mathbf{P}_{iv}^H = \hat{\mathbf{P}}_{iv}^H + \mathbf{O}(N^{-1/2})\,prob.$ and $\hat{\mathbf{P}}_{iv}^H\hat{\mathbf{P}}_{iv}^H = \hat{\mathbf{P}}_{iv}^H$, we obtain

$$\mathbf{P}_{iv}^H\hat{\mathbf{R}}_{is} = -\hat{\mathbf{P}}_{iv}^H\mathbf{R}_{i\pm}^H\mathbf{o}(N^{-1/2})\,prob. \tag{6.199}$$

Substituting Eqs. (6.196) and (6.199) into (6.193) and using $\mathbf{P}_{iv} = \mathbf{L}_{iv}\mathbf{R}_{iv}^H = \mathbf{B}_{iv}\mathbf{T}_{iv}\mathbf{R}_{iv}^H$, we have

$$\begin{aligned}\dot{f}_{ik} &= 2\,\mathrm{tr}\left\{\mathrm{Re}\left[\mathbf{W}_{iv}\mathbf{R}_{iv}\mathbf{T}_{iv}^H\mathbf{B}_{iv}^H\dot{\mathbf{D}}_{ik}\mathbf{T}_{is}\mathbf{W}_{is}\mathbf{R}_{is}^H\hat{\mathbf{P}}_{iv}\right]\right\} + \mathbf{o}(N^{-1/2})\,prob.\\ &= \dot{g}_{ik} + \mathbf{o}(N^{-1/2})\,prob.\end{aligned} \tag{6.200}$$

Using $\mathbf{P}_{iv} = \mathbf{B}_{iv}\mathbf{T}_{iv}\mathbf{R}_{iv}^H$ and $\mathrm{tr}(\mathbf{AB}) = \mathrm{tr}(\mathbf{BA})$ in Eq. (6.187) as well as using Eq. (6.197), we have

$$\begin{aligned}\left[\ddot{\mathbf{g}}_i\right]_{kl} &= 2\,\mathrm{tr}\left\{\mathrm{Re}\left[\mathbf{T}_{is}^H\dot{\mathbf{D}}_{ik}^H\mathbf{B}_{iv}\mathbf{T}_{iv}\mathbf{R}_{iv}^H\mathbf{W}_{iv}\mathbf{R}_{iv}\mathbf{T}_{iv}^H\mathbf{B}_{iv}^H\dot{\mathbf{D}}_{il}\mathbf{T}_{is}\mathbf{W}_{is}\right]\right\}\\ &= 2\,\mathrm{tr}\left\{\mathrm{Re}\left[\mathbf{R}_{is}^H\dot{\mathbf{B}}_{ivk}\mathbf{T}_{iv}\mathbf{R}_{iv}^H\mathbf{W}_{iv}\mathbf{R}_{iv}\mathbf{T}_{iv}^H\dot{\mathbf{B}}_{ivl}^H\mathbf{R}_{is}\mathbf{W}_{is}\right]\right\}\\ &= \left[\ddot{\mathbf{f}}_i\right]_{kl} \quad \blacksquare\end{aligned} \tag{6.201}$$

The above theorem shows that both UN-CLE criteria f and g are asymptotically equivalent when the same weighting matrices $\mathbf{W}_{is}$ and $\mathbf{W}_{iv}$ are employed. Thus, we can examine the asymptotic performance of the UN-CLE algorithm through the analysis of either criterion. Also, as in the case for UN-MUSIC, Eqs. (6.185), (6.186), and (6.187) reveal that whether the weighting matrices $\mathbf{W}_{is}$ and $\mathbf{W}_{iv}$ are treated as constants or as functions of θ_i, the asymptotic performance of the UN-CLE criteria is not affected.

We now turn our attention to the evaluation of the asymptotic performance of the UN-CLE algorithm. This is stated in the following theorem:

Theorem 6.11. *For the ith array, i = 1, 2, the estimation errors $\Delta\theta_i = \hat{\theta}_i - \theta_i$, using the UN-CLE algorithm, are asymptotically Gaussian distributed with zero mean and covariance matrices*

$$\begin{aligned}\mathrm{cov}(\Delta\theta_i) &\triangleq E\left[(\hat{\theta}_i - \theta_i)(\hat{\theta}_i - \theta_i)^T\right]\\ &= \frac{1}{2N}\left[\mathrm{Re}\,\{(\mathbf{D}_i'^H \mathbf{P}_{iv}\mathbf{W}_{iv}\mathbf{P}_{iv}^H \mathbf{D}_i') \odot (\mathbf{T}_{is}\mathbf{W}_{is}\mathbf{T}_{is}^H)^T\}\right]^{-1}\\ &\quad \mathrm{Re}\,\{(\mathbf{D}_i'^H \mathbf{P}_{iv}\mathbf{W}_{iv}\mathbf{P}_{iv}^H \boldsymbol{\Sigma}_{ii}\mathbf{P}_{iv}\mathbf{W}_{iv}\mathbf{P}_{iv}^H \mathbf{D}_i')\\ &\quad \odot (\mathbf{T}_{is}\mathbf{W}_{is}\boldsymbol{\Gamma}^{-1}\mathbf{L}_{\bar{i}s}^H \boldsymbol{\Sigma}_{\bar{i}\bar{i}}\mathbf{L}_{\bar{i}s}\boldsymbol{\Gamma}^{-1}\mathbf{W}_{is}\mathbf{T}_{is}^H)^T\}\\ &\quad \left[\mathrm{Re}\,\{(\mathbf{D}_i'^H \mathbf{P}_{iv}\mathbf{W}_{iv}\mathbf{P}_{iv}^H \mathbf{D}_i') \odot (\mathbf{T}_{is}\mathbf{W}_{is}\mathbf{T}_{is}^H)^T\}\right]^{-1}\end{aligned} \tag{6.202}$$

where $\mathbf{T}_{is}$ *is a* K × K *matrix such that* $\mathbf{R}_{is} = \mathbf{D}_i\mathbf{T}_{is}$, $\mathbf{D}_i' = \sum_{k=1}^{K} \dot{\mathbf{D}}_{ik}$, *and* $\odot$ *denotes the Hadamard product of two matrices of same dimensions such that* $\mathbf{A} \odot \mathbf{B} \triangleq \left[a_{ij} \cdot b_{ij}\right]$.

Proof. From Eq. (6.187), we note that all the columns of $\dot{\mathbf{D}}_{il}$ are zero except the lth one; in addition, all the rows of $\dot{\mathbf{D}}_{ik}^H$ are zero except the kth one. Thus, to obtain the element $\left[\ddot{\mathbf{g}}_i\right]_{kl}$, only the ($lk$)th element of $(\mathbf{T}_{is}\mathbf{W}_{is}\mathbf{T}_{is})$ is used. Using the definition of Hadamard product and the fact that tr(**AB**) = tr(**BA**), we can write

$$\ddot{\mathbf{g}}_i = 2\,\mathrm{Re}\left[(\mathbf{D}_i'^H \mathbf{P}_{iv}\mathbf{W}_{iv}\mathbf{P}_{iv}^H \mathbf{D}_i') \odot (\mathbf{T}_{is}\mathbf{W}_{is}\mathbf{T}_{is}^H)^T\right] \tag{6.203}$$

where

$$\mathbf{D}_i' = \sum_{k=1}^{K} \dot{\mathbf{D}}_{ik} \tag{6.204}$$

Now, substituting Eqs. (6.203) and (6.186) into Eq. (6.189), we obtain

$$\Delta\theta_i = -\{\mathrm{Re}\,(\mathbf{D}_i'^H \mathbf{P}_{iv}\mathbf{W}_{iv}\mathbf{P}_{iv}^H \mathbf{D}_i') \odot (\mathbf{T}_i\mathbf{W}_{is}\mathbf{T}_i^H)^T\}^{-1}\,\mathrm{Re}\,(\beta_i) + \mathbf{o}(N^{-1/2})\,\mathit{prob}. \tag{6.205}$$

where

$$\boldsymbol{\beta}_i = [\beta_{i1}, \cdots, \beta_{iK}]^T,$$

with

$$\beta_{ik} = \mathrm{tr}\,\{(\mathbf{T}_i\mathbf{W}_{is})\mathbf{R}_{is}^H \hat{\mathbf{P}}_{iv}\mathbf{W}_{iv}\mathbf{P}_{iv}^H \dot{\mathbf{D}}_{ik}\}, \quad k = 1, \cdots, K \tag{6.206}$$

But from Corollary 6.5b, $\mathbf{R}_{is}^H \hat{\mathbf{P}}_{iv}$ is asymptotically Gaussian distributed with zero mean. Hence, we can conclude that $\boldsymbol{\beta}_i$ is asymptotically Gaussian distributed with zero mean.

The covariance of the error of the estimates is given by

$$
\begin{aligned}
& E\left[(\hat{\boldsymbol{\theta}}_i - \boldsymbol{\theta}_i)(\hat{\boldsymbol{\theta}}_i - \boldsymbol{\theta}_i)^T\right] \\
&= \left[\mathrm{Re}\,\{(\mathbf{D}_i'^H \mathbf{P}_{iv} \mathbf{W}_{iv} \mathbf{P}_{iv}^H \mathbf{D}_i') \odot (\mathbf{T}_{is} \mathbf{W}_{is} \mathbf{T}_{is}^H)^T\}\right]^{-1} E\left[\mathrm{Re}\,(\boldsymbol{\beta}_i)\mathrm{Re}\,(\boldsymbol{\beta}_i)^T\right] \\
& \left[\mathrm{Re}\,\{(\mathbf{D}_i'^H \hat{\mathbf{P}}_{iv} \mathbf{W}_{iv} \hat{\mathbf{P}}_{iv}^H \mathbf{D}_i') \odot (\mathbf{T}_{is} \mathbf{W}_{is} \mathbf{T}_{is}^H)^T\}\right]^{-1}
\end{aligned}
\tag{6.207}
$$

To evaluate $E\left[\mathrm{Re}\,(\boldsymbol{\beta}_i)\mathrm{Re}\,(\boldsymbol{\beta}_i)^T\right]$, we see from Eq. (6.206) that the kth element of $\boldsymbol{\beta}_i$ is given by

$$
\begin{aligned}
\beta_{ik} &= (\mathbf{T}_{is}\mathbf{W}_{is})_{k.} \mathbf{R}_{is}^H \hat{\mathbf{P}}_{iv} \mathbf{W}_{iv} \mathbf{P}_{iv}^H (\dot{\mathbf{D}}_{ik})_{.k} \\
&= (\mathbf{T}_{is}\mathbf{W}_{is})_{k.} \mathbf{R}_{is}^H \hat{\mathbf{P}}_{iv} \mathbf{W}_{iv} \mathbf{P}_{iv}^H \dot{\mathbf{d}}_{ik}
\end{aligned}
\tag{6.208}
$$

where $(\mathbf{A})_{k.}$ is a row vector denoting the kth row of $\mathbf{A}$ and $(\mathbf{A})_{.k}$ denotes the kth column vector of $\mathbf{A}$. From the identity in Eq. (6A.7), we can rewrite Eq. (6.208) as

$$
\beta_{ik} = \{(\mathbf{W}_{iv}\mathbf{P}_{iv}^H \dot{\mathbf{d}}_{ik})^T \otimes (\mathbf{T}_{is}\mathbf{W}_{is})_{k.}\}\mathrm{vec}\,(\mathbf{R}_{is}^H \hat{\mathbf{P}}_{iv})
\tag{6.209}
$$

Now, let the (kl)th element of the matrix $E\left[\mathrm{Re}\,(\boldsymbol{\beta}_i)\mathrm{Re}\,(\boldsymbol{\beta}_i)^T\right]$ be $E\,[\mathrm{Re}\,(\beta_{ik})\mathrm{Re}\,(\beta_{il})]$, $k, l = 1, \cdots, K$. We have the identity

$$
E\,[\mathrm{Re}\,(\beta_{ik})\mathrm{Re}\,(\beta_{il})] = \frac{1}{2} E\left[\mathrm{Re}\,(\beta_{ik}\beta_{il}^*) + \mathrm{Re}\,(\beta_{ik}\beta_{il})\right]
\tag{6.210}
$$

Using Eq. (6.209) and (6.210), we can write

$$
\begin{aligned}
& E\left[\beta_{ik}\beta_{il}^*\right] \\
&= E\left[\{(\mathbf{T}_{is}\mathbf{W}_{is})_{k.} \mathbf{R}_{is}^H \hat{\mathbf{P}}_{iv} \mathbf{W}_{iv} \mathbf{P}_{iv}^H \dot{\mathbf{d}}_{ik}\}\{(\mathbf{T}_{is}\mathbf{W}_{is})_{l.} \mathbf{R}_{is}^H \hat{\mathbf{P}}_{iv} \mathbf{W}_{iv} \mathbf{P}_{iv}^H \dot{\mathbf{d}}_{il}\}^H\right] \\
&= \left[\{\mathbf{W}_{iv}\mathbf{P}_{iv}^H \dot{\mathbf{d}}_{ik}\}^T \otimes (\mathbf{T}_{is}\mathbf{W}_{is})_{k.}\right] E\left[\mathrm{vec}\,(\mathbf{R}_{is}^H \hat{\mathbf{P}}_{iv})\mathrm{vec}^H(\mathbf{R}_{is}^H \hat{\mathbf{P}}_{iv})\right] \\
& \left[\{\mathbf{W}_{iv}\mathbf{P}_{iv}^H \dot{\mathbf{d}}_{il}\}^* \otimes (\mathbf{W}_{is}\mathbf{T}_{is}^H)_{.l}\right]
\end{aligned}
\tag{6.211}
$$

The expectation term in Eq. (6.211) can be obtained by using Eq. (6.125) of Corollary 6.6b, and remembering that $\mathbf{R}_{is}^H \mathbf{L}_{is} = \mathbf{I}$ from Theorem 6.3, we can write

$$
\begin{aligned}
& E\left[\beta_{ik}\beta_{il}^*\right] \\
&= \frac{1}{N}\left[\{\mathbf{W}_{iv}\mathbf{P}_{iv}^H \dot{\mathbf{d}}_{ik}\}^T \otimes (\mathbf{T}_{is}\mathbf{W}_{is})_{k.}\right]\left[(\mathbf{P}_{iv}^H \boldsymbol{\Sigma}_{ii} \mathbf{P}_{iv})^* \otimes (\boldsymbol{\Gamma}^{-1}\mathbf{L}_{\bar{i}s}^H \boldsymbol{\Sigma}_{\bar{i}\bar{i}} \mathbf{L}_{\bar{i}s}\boldsymbol{\Gamma}^{-1})\right] \\
& \left[\{\mathbf{W}_{iv}\mathbf{P}_{iv}^H \dot{\mathbf{d}}_{il}\}^* \otimes (\mathbf{W}_{is}\mathbf{T}_{is}^H)_{.l}\right] \\
&= \frac{1}{N}\{\dot{\mathbf{d}}_{ik}^H \mathbf{P}_{iv} \mathbf{W}_{iv} \mathbf{P}_{iv}^H \boldsymbol{\Sigma}_{ii} \mathbf{P}_{iv} \mathbf{W}_{iv} \mathbf{P}_{iv}^H \dot{\mathbf{d}}_{il}\}^* \\
& (\mathbf{T}_{is})_{k.} \mathbf{W}_{is} \boldsymbol{\Gamma}^{-1} \mathbf{L}_{\bar{i}s}^H \boldsymbol{\Sigma}_{\bar{i}\bar{i}} \mathbf{L}_{\bar{i}s} \boldsymbol{\Gamma}^{-1} \mathbf{W}_{is} (\mathbf{T}_{is}^H)_{.l}
\end{aligned}
\tag{6.212}
$$

Also, using Eq. (6.126) of Corollary 6.6b, we can see that $E\,[\beta_{ik}\beta_{il}] = 0$. Thus, in matrix form, we have

$$E\left[\mathrm{Re}\,(\boldsymbol{\beta}_i)\,\mathrm{Re}\,(\boldsymbol{\beta}_i)^T\right]$$
$$= \frac{1}{2N}\mathrm{Re}\,[(\mathbf{D}_i'^H \mathbf{P}_{i\nu}\mathbf{W}_{i\nu}\mathbf{P}_{i\nu}^H \boldsymbol{\Sigma}_{ii}\mathbf{P}_{i\nu}\mathbf{W}_{i\nu}\mathbf{P}_{i\nu}^H \mathbf{D}_i') \tag{6.213}$$
$$\odot\,(\mathbf{T}_{is}\mathbf{W}_{is}\boldsymbol{\Gamma}^{-1}\mathbf{L}_{\bar{i}s}^H \boldsymbol{\Sigma}_{\bar{i}\bar{i}}\mathbf{L}_{\bar{i}s}\boldsymbol{\Gamma}^{-1}\mathbf{W}_{is}\mathbf{T}_{is}^H)^T]$$

Substituting Eq. (6.213) into Eq. (6.207), the result of Eq. (6.202) follows. ∎

The above theorem informs us that for any weighting matrices $\mathbf{W}_{is}$ and $\mathbf{W}_{i\nu}$ satisfying Eqs. (6.149) and for any choice of GCD, that is, any choice of $\boldsymbol{\Pi}_i$, the UN-CLE estimates are unbiased and their covariance matrices can be expressed in a closed form.

6.7 OPTIMUM UN-MUSIC AND OPTIMUM UN-CLE

In the previous section, the perfomance analyses for both the UN-MUSIC and UN-CLE estimates for any general weighting matrix and any choice of general correlation decomposition have been presented. The natural questions that arise are: (a) What weighting matrices should we use in order to obtain best performance from the algorithms? (b) Is there an optimum choice from the members of the set of GCD by which the algorithms yield best performance? These are the questions we examine in this section.

Optimum Weighting Matrices

We note that there is one weighting matrix $\mathbf{W}_{i\nu}$ of dimension $(M_i \times M_i)$ in the UN-MUSIC criterion of Eq. (6.131c) and that there are two weighting matrices, $\mathbf{W}_{i\nu}$ of dimension $(M_i \times M_i)$, and $\mathbf{W}_{is}$ of dimension $K \times K$, in the UN-CLE criteria of Eqs. (6.142) and (6.143). Since these are weighting matrices, as mentioned in Section 6.5, in order not to cause any loss of information in the sub-spaces, $\mathbf{W}_{i\nu}$ must have eigenvectors corresponding to its non-zero eigenvalues spanning the supplementary noise sub-space, and likewise, $\mathbf{W}_{is}$ must have eigenvectors corrsponding to its non-zero eigenvalues spanning the signal subspace. We now examine the choice of these matrices.

Theorem 6.12. *For the* i*th array,* i $= 1, 2$, *and a particular choice of GCD, the mean-square errors,* $\mathrm{E}\left[(\Delta\theta_{ik})^2\right]$, $k = 1, \cdots, K$ *of the DOA estimates obtained by employing the UN-MUSIC algorithm are minimized by applying the optimum weighting matrix* $\bar{\mathbf{W}}_{i\nu}$ *in the algorithm, where*

$$\bar{\mathbf{W}}_{i\nu} = \mathbf{L}_{i\nu}(\mathbf{L}_{i\nu}^H \boldsymbol{\Sigma}_{ii}\mathbf{L}_{i\nu})^{-1}\mathbf{L}_{i\nu}^H \tag{6.214}$$

The corresponding minimum mean-square errors $\mathrm{E}\left[(\Delta\theta_{ik})^2\right]$ *are given by*

$$E\left[(\Delta\theta_{ik})^2\right] = \frac{1}{2N}\left\{\frac{\mathbf{d}_i^H(\theta_{ik})\mathbf{L}_{is}\boldsymbol{\Gamma}^{-1}(\mathbf{L}_{\bar{i}s}^H \boldsymbol{\Sigma}_{\bar{i}\bar{i}}\mathbf{L}_{\bar{i}s})\boldsymbol{\Gamma}^{-1}\mathbf{L}_{is}^H \mathbf{d}_i(\theta_{ik})}{\dot{\mathbf{d}}_i^H(\theta_{ik})\mathbf{L}_{i\nu}(\mathbf{L}_{i\nu}^H \boldsymbol{\Sigma}_{ii}\mathbf{L}_{i\nu})^{-1}\mathbf{L}_{i\nu}^H \dot{\mathbf{d}}_i(\theta_{ik})}\right\} \tag{6.215}$$

Proof. We utilize the results obtained in Theorem 6.9. For $k = l$, Eq. (6.167) yields the value of $E\left[(\Delta\theta_{ik})^2\right]$, and in that case, we note that the two terms within the braces are always positive, and the last term $\left[\dot{\mathbf{d}}_i^H(\theta_{ik})\mathbf{L}_{is}\mathbf{\Gamma}^{-1}\mathbf{L}_{\bar{is}}^H\mathbf{\Sigma}_{\bar{ii}}\mathbf{L}_{\bar{is}}\mathbf{\Gamma}^{-1}\mathbf{L}_{is}^H\dot{\mathbf{d}}_i(\theta_{ik})\right]$ is independent of $\mathbf{W}_{iv}$. Thus, we only have to minimize the fraction formed by the first term in the braces of Eq. (6.167) and the denominator. The quantity to be minimized with respect to $\mathbf{W}_{iv}$ can be written as

$$\begin{aligned}&[\dot{\mathbf{d}}_i^H(\theta_{ik})\mathbf{L}_{iv}\mathbf{R}_{iv}^H\mathbf{W}_{iv}\mathbf{R}_{iv}\mathbf{L}_{iv}^H\mathbf{\Sigma}_{ii}\mathbf{L}_{iv}\mathbf{R}_{iv}^H\mathbf{W}_{iv}\mathbf{R}_{iv}\mathbf{L}_{iv}^H\dot{\mathbf{d}}_i(\theta_{ik})]\\ &\quad /[\dot{\mathbf{d}}_i^H(\theta_{ik})\mathbf{L}_{iv}\mathbf{R}_{iv}^H\mathbf{W}_{iv}\mathbf{R}_{iv}\mathbf{L}_{iv}^H\dot{\mathbf{d}}_i(\theta_{ik})]^2 \\ &\quad = \langle\mathbf{F}_i\dot{\mathbf{d}}_i(\theta_{ik}),\mathbf{F}_i\dot{\mathbf{d}}_i(\theta_{ik})\rangle/\|\langle\mathbf{G}_i\dot{\mathbf{d}}_i(\theta_{ik}),\mathbf{F}_i\dot{\mathbf{d}}_i(\theta_{ik})\rangle\|^2\end{aligned} \tag{6.216}$$

where $\mathbf{F}_i \triangleq (\mathbf{L}_{iv}^H\mathbf{\Sigma}_{ii}\mathbf{L}_{iv})^{1/2}\mathbf{R}_{iv}^H\mathbf{W}_{iv}\mathbf{R}_{iv}\mathbf{L}_{iv}^H$ and $\mathbf{G}_i \triangleq (\mathbf{L}_{iv}^H\mathbf{\Sigma}_{ii}\mathbf{L}_{iv})^{-1/2}\mathbf{L}_{iv}^H$ and the inner product of two vectors is defined as $\langle\mathbf{u},\boldsymbol{\upsilon}\rangle = \mathbf{u}^H\boldsymbol{\upsilon}$. To find a lower bound for Eq. (6.216), we employ the Schwarz inequality to the denominator of Eq. (6.216) such that, for any weighting matrix $\mathbf{W}_{iv}$

$$\|\langle\mathbf{G}_i\dot{\mathbf{d}}_i(\theta_{ik}),\mathbf{F}_i\dot{\mathbf{d}}_i(\theta_{ik})\rangle\|^2 \le \langle\mathbf{G}_i\dot{\mathbf{d}}_i(\theta_{ik}),\mathbf{G}_i\dot{\mathbf{d}}_i(\theta_{ik})\rangle\langle\mathbf{F}_i\dot{\mathbf{d}}_i(\theta_{ik}),\mathbf{F}_i\dot{\mathbf{d}}_i(\theta_{ik})\rangle \tag{6.217}$$

and equality holds iff $\mathbf{F}_i\dot{\mathbf{d}}_i(\theta_{ik}) = \alpha\mathbf{G}_i\dot{\mathbf{d}}_i(\theta_{ik})$ where α is a scalar. Substituting Eq. (6.217) into Eq. (6.216), we have

$$\begin{aligned}\frac{\langle\mathbf{F}_i\dot{\mathbf{d}}_i(\theta_{ik}),\mathbf{F}_i\dot{\mathbf{d}}_i(\theta_{ik})\rangle}{\|\langle\mathbf{G}_i\dot{\mathbf{d}}_i(\theta_{ik}),\mathbf{F}_i\dot{\mathbf{d}}_i(\theta_{ik})\rangle\|^2} &\ge \frac{1}{\langle\mathbf{G}_i\dot{\mathbf{d}}_i(\theta_{ik}),\mathbf{G}_i\dot{\mathbf{d}}_i(\theta_{ik})\rangle}\\ &= \frac{1}{\dot{\mathbf{d}}_i^H(\theta_{ik})\mathbf{L}_{iv}(\mathbf{L}_{iv}^H\mathbf{\Sigma}_{ii}\mathbf{L}_{iv})^{-1}\mathbf{L}_{iv}^H\dot{\mathbf{d}}_i(\theta_{ik})}\end{aligned} \tag{6.218}$$

To achieve this lower bound, we use the condition for equality in Eq. (6.217). A simple choice is $\mathbf{F}_i = \mathbf{G}_i$ and we obtain the condition for an optimum weigthing matrix $\bar{\mathbf{W}}_{iv}$ such that

$$\mathbf{R}_{iv}^H\bar{\mathbf{W}}_{iv}\mathbf{R}_{iv} = (\mathbf{L}_{iv}^H\mathbf{\Sigma}_{ii}\mathbf{L}_{iv})^{-1} \tag{6.219}$$

Premultiplying and postmultiplying Eq. (6.219) by $\mathbf{L}_{iv}$ and $\mathbf{L}_{iv}^H$ respectively and using the relation $\mathbf{L}_{iv}^H\mathbf{R}_{iv} = \mathbf{I}$, we have

$$\bar{\mathbf{W}}_{iv} = \mathbf{L}_{iv}(\mathbf{L}_{iv}^H\mathbf{\Sigma}_{ii}\mathbf{L}_{iv})^{-1}\mathbf{L}_{iv}^H \tag{6.220}$$

By substituting this optimum weighting matrix into Eq. (6.216), we obtain

$$\begin{aligned}&\frac{\dot{\mathbf{d}}_i^H(\theta_{ik})\mathbf{L}_{iv}\mathbf{R}_{iv}^H\mathbf{L}_{iv}(\mathbf{L}_{iv}^H\mathbf{\Sigma}_{ii}\mathbf{L}_{iv})^{-1}\mathbf{L}_{iv}^H\mathbf{R}_{iv}(\mathbf{L}_{iv}^H\mathbf{\Sigma}_{ii}\mathbf{L}_{iv})^{-1}\mathbf{R}_{iv}^H\mathbf{L}_{iv}(\mathbf{L}_{iv}^H\mathbf{\Sigma}_{ii}\mathbf{L}_{iv})^{-1}\mathbf{L}_{iv}^H\mathbf{R}_{iv}\mathbf{L}_{iv}^H\dot{\mathbf{d}}_i(\theta_{ik})}{\left[\dot{\mathbf{d}}_i^H(\theta_{ik})\mathbf{L}_{iv}\mathbf{R}_{iv}^H\mathbf{L}_{iv}(\mathbf{L}_{iv}^H\mathbf{\Sigma}_{ii}\mathbf{L}_{iv})^{-1}\mathbf{L}_{iv}^H\mathbf{R}_{iv}\mathbf{L}_{iv}^H\dot{\mathbf{d}}_i(\theta_{ik})\right]^2}\\ &\quad = 1/\left[\dot{\mathbf{d}}_i^H(\theta_{ik})\mathbf{L}_{iv}(\mathbf{L}_{iv}^H\mathbf{\Sigma}_{ii}\mathbf{L}_{iv})^{-1}\mathbf{L}_{iv}^H\dot{\mathbf{d}}_i(\theta_{ik})\right]\end{aligned} \tag{6.221}$$

which is indeed the lower bound given by the right side of Eq. (6.218), Thus, we conclude that $\bar{\mathbf{W}}_{iv}$ given by Eq. (6.220) is an optimum weighting matrix which yields the minimum variance of the estimate of the kth DOA, $k = 1, \cdots, K$.

Substituting the minimum value of Eq. (6.221) into Eq. (6.167) for $k = l$, the result of Eq. (6.215) follows. ■

The above theorem tells us that for a particular choice of GCD, if we employ the UN-MUSIC algorithm to estimate the DOA in unknown correlated noise, by applying the optimum weighting matrix $\bar{\mathbf{W}}_{i\nu}$ given by Eq. (6.214), we arrive at minimum mean-square errors of the estimates.

We now turn our attention to the choice of weighting matrices in the employment of the UN-CLE algorithms.

Theorem 6.13. *For the ith array, i = 1, 2, and a particular choice of GCD, the covariance matrix of the errors in the estination DOA employing the UN-CLE algorithm is minimized by applying the optimum weighting matrices*

$$\bar{\mathbf{W}}_{i\nu} = \mathbf{L}_{i\nu}(\mathbf{L}_{i\nu}^H \mathbf{\Sigma}_{ii} \mathbf{L}_{i\nu})^{-1} \mathbf{L}_{i\nu}^H \tag{6.222a}$$

$$\bar{\mathbf{W}}_{is} = \mathbf{\Gamma}(\mathbf{L}_{\bar{i}s}^H \mathbf{\Sigma}_{\bar{i}\bar{i}} \mathbf{L}_{\bar{i}s})^{-1} \mathbf{\Gamma} \tag{6.222b}$$

i.e.,

$$E\left[(\hat{\theta}_i - \theta_i)(\hat{\theta}_i - \theta_i)^T\right]_{\bar{\mathbf{W}}_{i\nu}, \bar{\mathbf{W}}_{is}} \le E\left[(\hat{\theta}_i - \theta_i)(\hat{\theta}_i - \theta_i)^T\right]_{\mathbf{W}_{i\nu}, \mathbf{W}_{is}} \tag{6.223}$$

and

$$E\left[(\hat{\theta}_i - \theta_i)(\hat{\theta}_i - \theta_i)^T\right]_{\bar{\mathbf{W}}_{i\nu}, \bar{\mathbf{W}}_{is}} = \frac{1}{2N}\left[\mathrm{Re}\,\{(\mathbf{D}_i'^H \mathbf{P}_{i\nu} \bar{\mathbf{W}}_{i\nu} \mathbf{P}_{i\nu}^H \mathbf{D}_i') \odot (\mathbf{T}_{is} \bar{\mathbf{W}}_{is} \mathbf{T}_{is}^H)^T\}\right]^{-1} \tag{6.224}$$

where, for matrices **A** *and* **B**, **A** $\le$ **B** *means that* (**B** − **A**) *is positive semi-definite.*

Proof. As mentioned in Theorem 6.10 of the previous section, the two UN-CLE criteria f_i and g_i are asymptotically equivalent in performance if the same weighting matrices $\mathbf{W}_{i\nu}$ and $\mathbf{W}_{is}$ are employed. We use the performance analysis based on g_i in Eq. (6.202) of Theorem 6.11 to prove the present theorem.

We start by constructing two positive semi-definite matrices $\mathbf{\Phi}_i$ and $\mathbf{\Psi}_i$ such that

$$\begin{aligned}
\mathbf{\Phi}_i &= \begin{bmatrix} \mathbf{\Phi}_{i11} & \mathbf{\Phi}_{i12} \\ \mathbf{\Phi}_{i21} & \mathbf{\Phi}_{i22} \end{bmatrix} \\
&\triangleq \begin{bmatrix} \mathbf{D}_i'^H \mathbf{P}_{i\nu} \\ \mathbf{D}_i'^H \mathbf{P}_{i\nu} \mathbf{W}_{i\nu} \mathbf{P}_{i\nu}^H \mathbf{\Sigma}_{ii} \mathbf{P}_{i\nu} \end{bmatrix} \mathbf{L}_{i\nu}(\mathbf{L}_{i\nu}^H \mathbf{\Sigma}_{ii} \mathbf{L}_{i\nu})^{-1} \mathbf{L}_{i\nu}^H \begin{bmatrix} \mathbf{D}_i'^H \mathbf{P}_{i\nu} \\ \mathbf{D}_i'^H \mathbf{P}_{i\nu} \mathbf{W}_{i\nu} \mathbf{P}_{i\nu}^H \mathbf{\Sigma}_{ii} \mathbf{P}_{i\nu} \end{bmatrix}^H \\
&= \begin{bmatrix} \mathbf{D}_i'^H \mathbf{P}_{i\nu} \mathbf{L}_{i\nu}(\mathbf{L}_{i\nu}^H \mathbf{\Sigma}_{ii} \mathbf{L}_{i\nu})^{-1} \mathbf{L}_{i\nu}^H \mathbf{P}_{i\nu}^H \mathbf{D}_i' & \mathbf{D}_i'^H \mathbf{P}_{i\nu} \mathbf{W}_{i\nu} \mathbf{P}_{i\nu}^H \mathbf{D}_i' \\ \mathbf{D}_i'^H \mathbf{P}_{i\nu} \mathbf{W}_{i\nu} \mathbf{P}_{i\nu}^H \mathbf{D}_i' & \mathbf{D}_i'^H \mathbf{P}_{i\nu} \mathbf{W}_{i\nu} (\mathbf{P}_{i\nu}^H \mathbf{\Sigma}_{ii} \mathbf{P}_{i\nu}) \mathbf{W}_{i\nu} \mathbf{P}_{i\nu}^H \mathbf{D}_i' \end{bmatrix} \\
&\ge \mathbf{O}
\end{aligned} \tag{6.225}$$

where $\mathbf{D}_i'$ is defined in Eq. (6.204), and we have used the facts that $\mathbf{L}_{i\nu}^H \mathbf{R}_{i\nu} = \mathbf{I}$, that $\mathbf{P}_{i\nu} = \mathbf{L}_{i\nu} \mathbf{R}_{i\nu}^H$, and that $\mathbf{P}_{i\nu}^H \mathbf{P}_{i\nu}^H \mathbf{D}_i' = \mathbf{P}_{i\nu}^H \mathbf{D}_i'$, and

$$
\begin{aligned}
\boldsymbol{\Psi}_i &= \begin{bmatrix} \boldsymbol{\Psi}_{i11} & \boldsymbol{\Psi}_{i12} \\ \boldsymbol{\Psi}_{i21} & \boldsymbol{\Psi}_{i22} \end{bmatrix} \\
&\triangleq \begin{bmatrix} \{\mathbf{T}_{is}(\boldsymbol{\Gamma}^{-1}\mathbf{L}_{\bar{i}s}^H\boldsymbol{\Sigma}_{\bar{i}\bar{i}}\mathbf{L}_{\bar{i}s}\boldsymbol{\Gamma}^{-1})^{-1}\mathbf{T}_{is}^H\}^T & (\mathbf{T}_{is}\mathbf{W}_{is}\mathbf{T}_{is}^H)^T \\ (\mathbf{T}_{is}\mathbf{W}_{is}\mathbf{T}_{is}^H)^T & \{\mathbf{T}_{is}\mathbf{W}_{is}\boldsymbol{\Gamma}^{-1}\mathbf{L}_{\bar{i}s}^H\boldsymbol{\Sigma}_{\bar{i}\bar{i}}\mathbf{L}_{\bar{i}s}\boldsymbol{\Gamma}^{-1}\mathbf{W}_{is}\mathbf{T}_{is}^H\}^T \end{bmatrix} \\
&\geq \mathbf{O}
\end{aligned} \tag{6.226}
$$

From the property of positive semidefinite matrices (Eq. 6A.19 in Appendix 6A), the Hadamard product of $\boldsymbol{\Phi}_i$ and $\boldsymbol{\Psi}_i$ is also positive semidefinite, that is

$$
\boldsymbol{\Phi}_i \odot \boldsymbol{\Psi}_i = \begin{bmatrix} \boldsymbol{\Phi}_{i11} \odot \boldsymbol{\Psi}_{i11} & \boldsymbol{\Phi}_{i12} \odot \boldsymbol{\Psi}_{i12} \\ \boldsymbol{\Phi}_{i21} \odot \boldsymbol{\Psi}_{i21} & \boldsymbol{\Phi}_{i22} \odot \boldsymbol{\Psi}_{i22} \end{bmatrix} \geq \mathbf{O} \tag{6.227}
$$

Furthermore, applying the property of the submatrices of a positive semidefinite matrix (Eq. 6A.22), to the matrix in Eq. (6.227), we see that

$$
\begin{aligned}
&\{\mathrm{Re}\,(\boldsymbol{\Phi}_{i11} \odot \boldsymbol{\Psi}_{i11})\}^{-1} \\
&= \left\{\mathrm{Re}\left[(\mathbf{D}_i'^H\mathbf{P}_{iv}\mathbf{L}_{iv}(\mathbf{L}_{iv}^H\boldsymbol{\Sigma}_{ii}\mathbf{L}_{iv})^{-1}\mathbf{L}_{iv}^H\mathbf{P}_{iv}^H\mathbf{D}_i') \odot \{\mathbf{T}_{is}(\boldsymbol{\Gamma}^{-1}\mathbf{L}_{\bar{i}s}^H\boldsymbol{\Sigma}_{\bar{i}\bar{i}}\mathbf{L}_{\bar{i}s}\boldsymbol{\Gamma}^{-1})^{-1}\mathbf{T}_{is}\}^T\right]\right\}^{-1} \\
&\leq \{\mathrm{Re}\,(\boldsymbol{\Phi}_{i12} \odot \boldsymbol{\Psi}_{i12})\}^{-1}\mathrm{Re}\,(\boldsymbol{\Phi}_{i22} \odot \boldsymbol{\Psi}_{i22})\{\mathrm{Re}\,(\boldsymbol{\Phi}_{i12} \odot \boldsymbol{\Psi}_{i12})\}^{-1} \\
&= \{\mathrm{Re}\left[(\mathbf{D}_i'^H\mathbf{P}_{iv}\mathbf{W}_{iv}\mathbf{P}_{iv}^H\mathbf{D}_i') \odot (\mathbf{T}_{is}\mathbf{W}_{is}\mathbf{T}_{is}^H)^T\right]\}^{-1} \\
&\quad \mathrm{Re}\left[(\mathbf{D}_i'^H\mathbf{P}_{iv}\mathbf{W}_{iv}(\mathbf{P}_{iv}^H\boldsymbol{\Sigma}_{ii}\mathbf{P}_{iv})\mathbf{W}_{iv}\mathbf{P}_{iv}^H\mathbf{D}_i') \odot (\mathbf{T}_{is}\mathbf{W}_{is}\boldsymbol{\Gamma}^{-1}\mathbf{L}_{\bar{i}s}^H\boldsymbol{\Sigma}_{\bar{i}\bar{i}}\mathbf{L}_{\bar{i}s}\boldsymbol{\Gamma}^{-1}\mathbf{W}_{is}\mathbf{T}_{is}^H)^T\right] \\
&\quad \{\mathrm{Re}\left[(\mathbf{D}_i'^H\mathbf{P}_{iv}\mathbf{W}_{iv}\mathbf{P}_{iv}^H\mathbf{D}_i') \odot (\mathbf{T}_{is}\mathbf{W}_{is}\mathbf{T}_{is}^H)^T\right]\}^{-1}
\end{aligned} \tag{6.228}
$$

Comparing Eq. (6.202) to the right side of Eq. (6.228), we see that they are identical except for the scalar $1/2N$. Thus, the right side of Eq. (6.228) is proportional to the covariance matrix of the UN-CLE algorithm in which the general weighting matrices $\mathbf{W}_{iv}$ and $\mathbf{W}_{is}$ are employed. On the other hand, if we let

$$
\mathbf{W}_{iv} = \bar{\mathbf{W}}_{iv} = \mathbf{L}_{iv}(\mathbf{L}_{iv}^H)\boldsymbol{\Sigma}_{ii}\mathbf{L}_{iv})^{-1}\mathbf{L}_{iv}^H
$$

and

$$
\mathbf{W}_{is} = \bar{\mathbf{W}}_{is} = (\boldsymbol{\Gamma}^{-1}\mathbf{L}_{\bar{i}s}^H\boldsymbol{\Sigma}_{\bar{i}\bar{i}}\mathbf{L}_{\bar{i}s}\boldsymbol{\Gamma}^{-1})^{-1} = \boldsymbol{\Gamma}(\mathbf{L}_{\bar{i}s}^H\boldsymbol{\Sigma}_{\bar{i}\bar{i}}\mathbf{L}_{\bar{i}s})^{-1}\boldsymbol{\Gamma}
$$

and noting that

$$
\begin{aligned}
\bar{\mathbf{W}}_{iv}\mathbf{P}_{iv}^H\boldsymbol{\Sigma}_{ii}\mathbf{P}_{iv}\bar{\mathbf{W}}_{iv} &= \mathbf{L}_{iv}(\mathbf{L}_{iv}^H\boldsymbol{\Sigma}_{ii}\mathbf{L}_{iv})^{-1}\mathbf{L}_{iv}^H\mathbf{R}_{iv}\mathbf{L}_{iv}^H\boldsymbol{\Sigma}_{ii}\mathbf{L}_{iv}\mathbf{R}_{iv}^H\mathbf{L}_{iv}(\mathbf{L}_{iv}^H\boldsymbol{\Sigma}_{ii}\mathbf{L}_{iv})^{-1}\mathbf{L}_{iv}^H \\
&= \mathbf{L}_{iv}(\mathbf{L}_{iv}^H\boldsymbol{\Sigma}_{ii}\mathbf{L}_{iv})^{-1}\mathbf{L}_{iv}^H = \bar{\mathbf{W}}_{iv}
\end{aligned} \tag{6.229}
$$

then, the right side of Eq. (6.228) is reduced to the left side and the equality sign holds. Hence, we can conclude that the covariance matrix of the DOA estimate is minimized if we choose the weighting matrices to be those given by Eqs. (6.222). ■

The above theorem tells us that if we employ the UN-CLE algorithm for the estimation of the DOA in an unknown correlated noise environment, for a particular choice of GCD, minimum mean-square errors of the estimates could be arrived at if the optimum weighting matrices given by Eqs. (6.222) are applied. We note that these optimum weighting matrices coincide with those in Eqs. (6.144) and (6.145) obtained by deriving the UN-CLE algorithm from the principle of maximum likelihood estimation. We also note that the optimum weighting matrix $\bar{\mathbf{W}}_{i\nu}$ for the UN-MUSIC algorithm is identical to one of the optimum weighting matrices $\bar{\mathbf{W}}_{i\nu}$ for the UN-CLE algorithm.

In practice, we do not know the true values of $\mathbf{\Sigma}_{ij}, i, j = 1, 2$, and therefore, may not be able to use the exact values of the optimum weighting matrices. However, without affecting the asymptotic performance, we can employ the consistent estimates of the optimum weighting matrices, $\hat{\bar{\mathbf{W}}}_{i\nu}$ and $\hat{\bar{\mathbf{W}}}_{is}$ such that

$$\hat{\bar{\mathbf{W}}}_{i\nu} = \hat{\mathbf{L}}_{i\nu}(\hat{\mathbf{L}}_{i\nu}^H \hat{\mathbf{\Sigma}}_{ii} \hat{\mathbf{L}}_{i\nu})^{-1} \hat{\mathbf{L}}_{i\nu}^H \tag{6.230a}$$

$$\hat{\bar{\mathbf{W}}}_{is} = \hat{\mathbf{\Gamma}}(\hat{\mathbf{L}}_{\bar{i}s}^H \hat{\mathbf{\Sigma}}_{\bar{i}\bar{i}} \hat{\mathbf{L}}_{\bar{i}s})^{-1} \hat{\mathbf{\Gamma}} \tag{6.230b}$$

That $\hat{\bar{\mathbf{W}}}_{i\nu}$ and $\hat{\bar{\mathbf{W}}}_{is}$ are consistent estimates of $\bar{\mathbf{W}}_{i\nu}$ and $\bar{\mathbf{W}}_{is}$ is shown in the following theorem.

Theorem 6.14. *The optimum weighting matrices $\bar{\mathbf{W}}_{i\nu}$ and $\bar{\mathbf{W}}_{is}$ for the UN-MUSIC and the UN-CLE algorithms can be estimated consistently by those given by Eqs. (6.230a) and (6.230b) respectively such that*

$$\hat{\bar{\mathbf{W}}}_{i\nu} = \bar{\mathbf{W}}_{i\nu} + \mathbf{O}(N^{-1/2}) \tag{6.231a}$$

and

$$\hat{\bar{\mathbf{W}}}_{is} = \bar{\mathbf{W}}_{is} + \mathbf{O}(N^{-1/2}) \tag{6.231b}$$

Proof. To prove the consistency of $\hat{\bar{\mathbf{W}}}_{i\nu}$ is relatively involved, since $\hat{\mathbf{L}}_{i\nu}$ does not converge to a unique $\mathbf{L}_{i\nu}$. Consider the factor $\mathbf{L}_{i\nu}^H \mathbf{\Sigma}_{ii} \mathbf{L}_{i\nu}$ in the expression of the optimum weighting matrix $\bar{\mathbf{W}}_{i\nu}$ in Eq. (6.222a). Postmultiplying by $(\mathbf{R}_{i\nu}^H \tilde{\mathbf{L}}_{i\nu} \tilde{\mathbf{L}}_{i\nu}^H \mathbf{R}_{i\nu})$, we have

$$\mathbf{L}_{i\nu}^H \mathbf{\Sigma}_{ii} \mathbf{L}_{i\nu} \mathbf{R}_{i\nu}^H \tilde{\mathbf{L}}_{i\nu} \tilde{\mathbf{L}}_{i\nu}^H \mathbf{R}_{i\nu} = \mathbf{L}_{i\nu}^H \mathbf{\Sigma}_{ii} \tilde{\mathbf{L}}_{i\nu} \tilde{\mathbf{L}}_{i\nu}^H \mathbf{R}_{i\nu} = \mathbf{L}_{i\nu}^H \tilde{\mathbf{R}}_{i\nu} \tilde{\mathbf{L}}_{i\nu}^H \mathbf{R}_{i\nu} \tag{6.232}$$

where we have applied the facts that $\mathbf{L}_{i\nu} \mathbf{R}_{i\nu}^H \tilde{\mathbf{L}}_{i\nu} = \tilde{\mathbf{L}}_{i\nu}$ and $\mathbf{\Sigma}_{ii} \tilde{\mathbf{L}}_{i\nu} = \tilde{\mathbf{R}}_{i\nu}$ with $\tilde{\mathbf{L}}_{i\nu}$ and $\tilde{\mathbf{R}}_{i\nu}$ being the matrices $\mathbf{L}_{i\nu}$ and $\mathbf{R}_{i\nu}$ obtained when CCD is employed. Again, since $\tilde{\mathbf{L}}_{i\nu} \tilde{\mathbf{R}}_{i\nu}^H \mathbf{L}_{i\nu} = \mathbf{L}_{i\nu}$, therefore Eq. (6.232) becomes

$$(\mathbf{L}_{i\nu}^H \mathbf{\Sigma}_{ii} \mathbf{L}_{i\nu}) \mathbf{R}_{i\nu}^H \tilde{\mathbf{L}}_{i\nu} \tilde{\mathbf{L}}_{i\nu}^H \mathbf{R}_{i\nu} = \mathbf{L}_{i\nu}^H \mathbf{R}_{i\nu} = \mathbf{I} \tag{6.233}$$

Hence we conclude that $\mathbf{R}_{i\nu}^H \tilde{\mathbf{L}}_{i\nu} \tilde{\mathbf{L}}_{i\nu}^H \mathbf{R}_{i\nu}$ is the inverse of $\mathbf{L}_{i\nu}^H \mathbf{\Sigma}_{ii} \mathbf{L}_{i\nu}$, and hence the expression of $\bar{\mathbf{W}}_{i\nu}$ in Eq. (6.222a) becomes

$$\bar{\mathbf{W}}_{i\nu} = \mathbf{L}_{i\nu} \mathbf{R}_{i\nu}^H \tilde{\mathbf{L}}_{i\nu} \tilde{\mathbf{L}}_{i\nu}^H \mathbf{R}_{i\nu} \mathbf{L}_{i\nu}^H \tag{6.234}$$

Now

$$\hat{\mathbf{L}}_{i\nu} \hat{\mathbf{R}}_{i\nu}^H = \mathbf{L}_{i\nu} \mathbf{R}_{i\nu}^H + \mathbf{O}(N^{-1/2})\, prob. \tag{6.235}$$

and

$$\hat{\tilde{\mathbf{L}}}_{iv}\hat{\tilde{\mathbf{L}}}_{iv}^H = \tilde{\mathbf{L}}_{iv}\tilde{\mathbf{L}}_{iv}^H + \mathbf{O}(N^{-1/2})\,prob. \tag{6.236}$$

Thus, replacing all the quantities of the right side of Eq. (6.234) by their respective estimates and using Eqs. (6.235) and (6.236), we obtain

$$\hat{\mathbf{L}}_{iv}\hat{\mathbf{R}}_{iv}^H\hat{\tilde{\mathbf{L}}}_{iv}\hat{\tilde{\mathbf{L}}}_{iv}^H\hat{\mathbf{R}}_{iv}\hat{\mathbf{L}}_{iv}^H = \mathbf{L}_{iv}\mathbf{R}_{iv}^H\tilde{\mathbf{L}}_{iv}\tilde{\mathbf{L}}_{iv}^H\mathbf{R}_{iv}\mathbf{L}_{iv}^H + \mathbf{O}(N^{-1/2})$$
$$= \bar{\mathbf{W}}_{iv} + \mathbf{O}(N^{-1/2}) \tag{6.237}$$

Now, replacing all quantities on the left side of Eq. (6.233) with their respective estimates, we have

$$(\hat{\mathbf{L}}_{iv}^H\hat{\mathbf{\Sigma}}_{ii}\hat{\mathbf{L}}_{iv})\hat{\mathbf{R}}_{iv}^H\hat{\tilde{\mathbf{L}}}_{iv}\hat{\tilde{\mathbf{L}}}_{iv}^H\hat{\mathbf{R}}_{iv}$$

$$= \hat{\mathbf{L}}_{iv}^H(\mathbf{\Sigma}_{ii} + \mathbf{O}(N^{-1/2}))(\mathbf{L}_{iv}\mathbf{R}_{iv}^H + \mathbf{O}(N^{-1/2}))(\tilde{\mathbf{L}}_{iv}\tilde{\mathbf{L}}_{iv}^H + \mathbf{O}(N^{-1/2}))\hat{\mathbf{R}}_{iv}$$

$$= \hat{\mathbf{L}}_{iv}^H(\tilde{\mathbf{R}}_{iv}^H\tilde{\mathbf{L}}_{iv}^H + \mathbf{O}(N^{-1/2}))\hat{\mathbf{R}}_{iv} \tag{6.238}$$

$$= \hat{\mathbf{L}}_{iv}^H(\mathbf{I} - \tilde{\mathbf{R}}_{is}^H\tilde{\mathbf{L}}_{is}^H + \mathbf{O}(N^{-1/2}))\hat{\mathbf{R}}_{iv}$$

$$= \mathbf{I} + \mathbf{O}(N^{-1/2})\,prob.$$

Therefore we can conclude that

$$\hat{\mathbf{R}}_{iv}^H\hat{\tilde{\mathbf{L}}}_{iv}\hat{\tilde{\mathbf{L}}}_{iv}^H\hat{\mathbf{R}}_{iv} = (\hat{\mathbf{L}}_{iv}^H\hat{\mathbf{\Sigma}}_{ii}\hat{\mathbf{L}}_{iv})^{-1} + \mathbf{O}(N^{-1/2})\,prob. \tag{6.239}$$

Applying Eq. (6.239) of Eq. (6.237), we see that a consistent estimate of the optimum weighting matrix $\bar{\mathbf{W}}_{iv}$ is

$$\hat{\bar{\mathbf{W}}}_{iv} = \hat{\mathbf{L}}_{iv}(\hat{\mathbf{L}}_{iv}^H\hat{\mathbf{\Sigma}}_{ii}\hat{\mathbf{L}}_{iv})^{-1}\hat{\mathbf{L}}_{iv}^H$$
$$= \bar{\mathbf{W}}_{iv} + \mathbf{O}(N^{-1/2})\,prob. \tag{6.240}$$

From the consistency of $\hat{\mathbf{\Gamma}}$ and $\hat{\mathbf{L}}_{is}$ in Theorem 6.4 as well as that of $\hat{\mathbf{\Sigma}}_{ii}$ we can see that a consistent estimate of the optimum weighting matrix $\bar{\mathbf{W}}_{is}$ is given by

$$\hat{\bar{\mathbf{W}}}_{is} = \hat{\mathbf{\Gamma}}(\hat{\mathbf{L}}_{is}^H\hat{\mathbf{\Sigma}}_{\bar{i}\bar{i}}\hat{\mathbf{L}}_{\bar{i}s})^{-1}\hat{\mathbf{\Gamma}}$$
$$= \bar{\mathbf{W}}_{is} + \mathbf{O}(N^{-1/2})\,prob. \tag{6.241}$$

∎

In view of the consistency of the estimated optimum weighting matrices $\hat{\bar{\mathbf{W}}}_{iv}$ and $\hat{\bar{\mathbf{W}}}_{is}$, we can conclude that, in practice, instead of using the criteria given by Eqs. (6.131c), (6.142), and (6.143), we may modify the algorithms so that the criterion for optimally weighted UN-MUSIC is

$$\hat{\theta}_{ik} = \arg\min_{\theta_{ik}}\{\mathbf{d}_i^H(\theta_{ik})\hat{\mathbf{P}}_{iv}\hat{\bar{\mathbf{W}}}_{iv}\hat{\mathbf{P}}_{iv}^H\mathbf{d}_i(\theta_{ik})\} \tag{6.242}$$

and the criteria for optimally weighted UN-CLE are

$$\hat{\theta}_i = \arg\min_{\theta_i}\text{tr}\,\{\mathbf{P}_{iv}\hat{\bar{\mathbf{W}}}_{iv}\mathbf{P}_{iv}^H\hat{\mathbf{R}}_{is}\hat{\bar{\mathbf{W}}}_{is}\hat{\mathbf{R}}_{iv}^H\} \tag{6.243a}$$

or

$$\hat{\theta}_i = \arg\min_{\theta_i} \operatorname{tr}\{\hat{\mathbf{P}}_{iv}\hat{\hat{\mathbf{W}}}_{iv}\hat{\mathbf{P}}_{iv}^H\hat{\mathbf{R}}_{is}\hat{\hat{\mathbf{W}}}_{is}\hat{\mathbf{R}}_{is}^H\} \tag{6.243b}$$

These modified criteria will have asymptotically the same performance as those in which the true values of the optimum weighting matrices are used.

Optimality of Canonical Correlation Decomposition

The above derivations of the optimum weighting matrices in the UN-MUSIC and the UN-CLE algorithms are applicable for any chosen generalized correlation decomposition, that is, any choice of the positive definite matrics $\mathbf{\Pi}_i$. A valid question that remains to be answered is: "Under optimum weighting, what is the best choice of decomposition among all kinds of GCD?" From section 6.4, we have seen that the canonical correlation decomposition maximizes the correlations between the linear combinations of the two sets of outputs $\mathbf{x}_1$ and $\mathbf{x}_2$. It can be surmised, therefore, that the application of CCD may yield the optimum performance in both the UN-MUSIC and the UN-CLE algorithms. This is indeed the case as indicated by the theorems that follow. Using Property 6.2 of the CCD, the corresponding optimum weighting matrices in both UN-MUSIC and UN-CLE when CCD is employed are simplified to

$$\tilde{\tilde{\mathbf{W}}}_{iv} = \tilde{\mathbf{L}}_{iv}(\tilde{\mathbf{L}}_{iv}^H\mathbf{\Sigma}_{ii}\tilde{\mathbf{L}}_{iv})^{-1}\tilde{\mathbf{L}}_{iv}^H = \tilde{\mathbf{L}}_{iv}\tilde{\mathbf{L}}_{iv}^H \tag{6.244}$$

and

$$\tilde{\tilde{\mathbf{W}}}_{is} = \tilde{\mathbf{\Gamma}}(\tilde{\mathbf{L}}_{\bar{i}s}^H\mathbf{\Sigma}_{\bar{i}\bar{i}}\tilde{\mathbf{L}}_{\bar{i}s})^{-1}\tilde{\mathbf{\Gamma}} = \tilde{\mathbf{\Gamma}}^2 \tag{6.245}$$

Now, let us examine the optimality of applying CCD in the UN-MUSIC algorithm.

Theorem 6.15. *Under optimum weighting, the UN-MUSIC estimates of the DOA in an unknown correlated noise environment employing CCD yield the minimum mean-square errors, that is,*

$$E\left[(\Delta\tilde{\theta}_{ik})^2\right] \le E\left[(\Delta\theta_{ik})^2\right] \quad i = 1, 2, \quad k = 1, \cdots, K \tag{6.246}$$

where $\Delta\tilde{\theta}_{ik}$ and $\Delta\theta_{ik}$ are the estimation errors obtained by using the UN-MUSIC algorithm employing CCD and employing any other kind of GCD respectively. The corresponding mean-square error for the estimates by UN-MUSIC using CCD is given by

$$E\left[(\Delta\tilde{\theta}_{ik})^2\right] = \frac{1}{2N}\left\{\frac{\mathbf{d}_i^H(\theta_{ik})\tilde{\mathbf{L}}_{\bar{i}s}\tilde{\mathbf{\Gamma}}^{-2}\tilde{\mathbf{L}}_{\bar{i}s}^H\mathbf{d}_i(\theta_{ik})}{\dot{\mathbf{d}}_i^H(\theta_{ik})\tilde{\mathbf{L}}_{iv}\tilde{\mathbf{L}}_{iv}^H\dot{\mathbf{d}}_i(\theta_{ik})}\right\} \tag{6.247}$$

Proof. For any choice of GCD, the mean-square error of the estimates given by the optimally weighted UN-MUSIC algorithm has been shown (Eq. 6.215) to be

$$E\left[(\Delta\theta_{ik})^2\right] = \frac{1}{2N}\left\{\frac{\mathbf{d}_i^H(\theta_{ik})\mathbf{L}_{is}\mathbf{\Gamma}^{-1}(\mathbf{L}_{\bar{i}s}^H\mathbf{\Sigma}_{\bar{i}\bar{i}}\mathbf{L}_{\bar{i}s})\mathbf{\Gamma}^{-1}\mathbf{L}_{is}^H\mathbf{d}_i(\theta_{ik})}{\dot{\mathbf{d}}_i^H(\theta_{ik})\mathbf{L}_{iv}(\mathbf{L}_{iv}^H\mathbf{\Sigma}_{ii}\mathbf{L}_{iv})^{-1}\mathbf{L}_{iv}^H\dot{\mathbf{d}}_i(\theta_{ik})}\right\}$$
$$k = 1, \cdots, K, \quad i = 1, 2 \tag{6.248}$$

In particular, if CCD is employed in the UN-MUSIC algorithm, applying Eq. (6.58) to Eq. (6.248), the mean-suqare error of the estimate under optimum weigthing is

$$E\left[(\Delta\tilde{\theta}_{ik})^2\right] = \frac{1}{2N}\left\{\frac{\mathbf{d}_i^H(\theta_{ik})\tilde{\mathbf{L}}_{is}\tilde{\mathbf{\Gamma}}^{-2}\tilde{\mathbf{L}}_{is}^H\mathbf{d}_i(\theta_{ik})}{\dot{\mathbf{d}}_i^H(\theta_{ik})\tilde{\mathbf{L}}_{iv}\tilde{\mathbf{L}}_{iv}^H\dot{\mathbf{d}}_i(\theta_{ik})}\right\} \quad k = 1, \cdots, K, \quad i = 1, 2 \tag{6.249}$$

Now, we show that $E\left[(\Delta\tilde{\theta}_{ik})^2\right] \le E\left[(\Delta\theta_{ik})^2\right]$ for $k = 1, \cdots, K$, $i = 1, 2$. Since both $\mathbf{L}_{iv}$ and $\tilde{\mathbf{L}}_{iv}$ are in the same subspace as $\overline{\text{span}}\{\mathbf{D}_i(\theta_i)\}$, the two denominators in Eqs. (6.248) and (6.249) have the same value. This can be easily seen by choosing a common basis $\mathbf{B}_{iv}$ in $\overline{\text{span}}\{\mathbf{D}_i(\theta_i)\}$ and expressing both $\mathbf{L}_{iv}$ and $\tilde{\mathbf{L}}_{iv}$ in terms of this basis so that

$$\mathbf{L}_{iv} = \mathbf{B}_{iv}\mathbf{T}_{iv}, \quad \tilde{\mathbf{L}}_{iv} = \mathbf{B}_{iv}\tilde{\mathbf{T}}_{iv} \tag{6.250}$$

where $\mathbf{T}_{iv}$ and $\tilde{\mathbf{T}}_{iv}$ are both nonsingular $(M_i - K) \times (M_i - K)$ matrices. Applying the representations of Eqs. (6.250) to the denominators of Eqs. (6.248) and (6.249) respectively, we have

$$\begin{aligned}\dot{\mathbf{d}}_i^H(\theta_{ik})\mathbf{L}_{iv}(\mathbf{L}_{iv}^H\mathbf{\Sigma}_{ii}\mathbf{L}_{iv})^{-1}\mathbf{L}_{iv}^H\dot{\mathbf{d}}_i(\theta_{ik}) &= \dot{\mathbf{d}}_i^H(\theta_{ik})\tilde{\mathbf{L}}_{iv}\tilde{\mathbf{L}}_{iv}^H\dot{\mathbf{d}}_i(\theta_{ik}) \\ &= \dot{\mathbf{d}}_i^H(\theta_{ik})\mathbf{B}_{iv}(\mathbf{B}_{iv}^H\mathbf{\Sigma}_{ii}\mathbf{B}_{iv})^{-1}\mathbf{B}_{iv}^H\dot{\mathbf{d}}_i(\theta_{ik})\end{aligned} \tag{6.251}$$

Thus, we only need to show that the numerator in Eq. (6.248) is larger than or equal to the numerator in Eq. (6.249). To do so, we have to find an expression of $\tilde{\mathbf{\Gamma}}^{-1}$ in terms of $\mathbf{\Gamma}^{-1}$. From the definitions of GCD and CCD, we have

$$\mathbf{\Pi}_i^{-1/2}\mathbf{\Sigma}_{i\bar{i}}\mathbf{\Pi}_{\bar{i}}^{-1/2} = \mathbf{U}_{is}\mathbf{\Gamma}\mathbf{U}_{\bar{i}s}^H$$
$$\mathbf{\Sigma}_{ii}^{-1/2}\mathbf{\Sigma}_{i\bar{i}}\mathbf{\Sigma}_{\bar{i}\bar{i}}^{-1/2} = \tilde{\mathbf{U}}_{is}\tilde{\mathbf{\Gamma}}\tilde{\mathbf{U}}_{\bar{i}s}^H$$

from which we can write

$$\mathbf{\Sigma}_{i\bar{i}} = \mathbf{\Sigma}_{ii}^{1/2}\tilde{\mathbf{U}}_{is}\tilde{\mathbf{\Gamma}}\tilde{\mathbf{U}}_{\bar{i}s}^H\mathbf{\Sigma}_{\bar{i}\bar{i}}^{1/2} = \mathbf{\Pi}_i^{1/2}\mathbf{U}_{is}\mathbf{\Gamma}\mathbf{U}_{\bar{i}s}^H\mathbf{\Pi}_{\bar{i}}^{1/2} \tag{6.252}$$

But by definitions, $\tilde{\mathbf{R}}_{is} = \mathbf{\Sigma}_{ii}^{1/2}\tilde{\mathbf{U}}_{is}$ and $\mathbf{R}_{is} = \mathbf{\Pi}_i^{1/2}\mathbf{U}_{is}$, therefore we have the equality:

$$\tilde{\mathbf{R}}_{is}\tilde{\mathbf{\Gamma}}\tilde{\mathbf{R}}_{\bar{i}s}^H = \mathbf{R}_{is}\mathbf{\Gamma}\mathbf{R}_{\bar{i}s}^H \tag{6.253}$$

Premultiplying and postmultiplying both sides of Eq. (6.253) by $\mathbf{L}_{is}^H$ and $\mathbf{L}_{\bar{i}s}$ respectively and noting that $\mathbf{L}_{is}^H\mathbf{R}_{is} = \mathbf{I}$, and $\mathbf{L}_{is}^H\tilde{\mathbf{R}}_{is}$ are invertible for $i = 1, 2$, we obtain

$$\tilde{\mathbf{\Gamma}} = (\mathbf{L}_{is}^H\tilde{\mathbf{R}}_{is})^{-1}\mathbf{\Gamma}(\tilde{\mathbf{R}}_{\bar{i}s}^H\mathbf{L}_{\bar{i}s})^{-1} \tag{6.254}$$

But as $\tilde{\mathbf{\Gamma}}$ and $\mathbf{\Gamma}$ are diagonal and hence Hemitian, we can write

$$\tilde{\boldsymbol{\Gamma}}^{-1} = (\tilde{\mathbf{R}}_{\bar{i}s}^H \mathbf{L}_{\bar{i}s})\boldsymbol{\Gamma}^{-1}(\mathbf{L}_{\bar{i}s}^H \tilde{\mathbf{R}}_{is}) = (\tilde{\mathbf{R}}_{\bar{i}s}^H \mathbf{L}_{is})\boldsymbol{\Gamma}^{-1}(\mathbf{L}_{\bar{i}s}^H \tilde{\mathbf{R}}_{\bar{i}s}) \tag{6.255}$$

Now let

$$\mathbf{z}_{ik} = \mathbf{L}_{\bar{i}s}\boldsymbol{\Gamma}^{-1}\mathbf{L}_{is}^H \mathbf{d}_i(\theta_{ik}) \tag{6.256}$$

Then the numerator of Eq. (6.248) becomes

$$N_r(\theta_{ik}) = \mathbf{z}_{ik}^H \boldsymbol{\Sigma}_{\bar{i}\bar{i}} \mathbf{z}_{ik} \tag{6.257}$$

and, using Eq. (6.255) as well as the fact that $\tilde{\mathbf{L}}_{is}\tilde{\mathbf{R}}_{is}^H \mathbf{L}_{is} = \tilde{\mathbf{P}}_{is}\mathbf{L}_{is} = \mathbf{L}_{is}$, the numerator of Eq. (6.249) becomes

$$\begin{aligned}
\tilde{N}_r(\theta_{ik}) &= \mathbf{d}_i^H(\theta_{ik})\tilde{\mathbf{L}}_{is}\tilde{\mathbf{R}}_{is}^H \mathbf{L}_{is}\boldsymbol{\Gamma}^{-1}\mathbf{L}_{\bar{i}s}^H \tilde{\mathbf{R}}_{\bar{i}s}\tilde{\mathbf{R}}_{\bar{i}s}^H \mathbf{L}_{\bar{i}s}\boldsymbol{\Gamma}^{-1}\mathbf{L}_{is}^H \tilde{\mathbf{R}}_{is}\tilde{\mathbf{L}}_{is}^H \mathbf{d}_i(\theta_{ik}) \\
&= \mathbf{d}_i^H(\theta_{ik})\mathbf{L}_{is}\boldsymbol{\Gamma}^{-1}\mathbf{L}_{\bar{i}s}^H \tilde{\mathbf{R}}_{\bar{i}s}\tilde{\mathbf{R}}_{\bar{i}s}^H \mathbf{L}_{\bar{i}s}\boldsymbol{\Gamma}^{-1}\mathbf{L}_{is}^H \mathbf{d}_i(\theta_{ik}) \\
&= \mathbf{z}_{ik}^H \tilde{\mathbf{R}}_{\bar{i}s}(\tilde{\mathbf{L}}_{\bar{i}s}^H \boldsymbol{\Sigma}_{\bar{i}\bar{i}}\tilde{\mathbf{L}}_{\bar{i}s})\tilde{\mathbf{R}}_{\bar{i}s}^H \mathbf{z}_{ik}
\end{aligned} \tag{6.258}$$

where the middle term $(\tilde{\mathbf{L}}_{\bar{i}s}^H \boldsymbol{\Sigma}_{\bar{i}\bar{i}}\tilde{\mathbf{L}}_{\bar{i}s}) = \mathbf{I}$ has been inserted in the last step without affecting the value of the numerator. Recalling the result of Eq. (6.77), we can write the difference between the numerator of Eq. (6.248) and that of Eq. (6.249) as

$$\begin{aligned}
N_r(\theta_{ik}) - \tilde{N}_r(\theta_{ik}) &= \mathbf{z}_{ik}^H \boldsymbol{\Sigma}_{\bar{i}\bar{i}}\mathbf{z}_{ik} - \mathbf{z}_{ik}^H \tilde{\mathbf{R}}_{\bar{i}s}\tilde{\mathbf{L}}_{\bar{i}s}^H \boldsymbol{\Sigma}_{\bar{i}\bar{i}}\tilde{\mathbf{L}}_{\bar{i}s}\tilde{\mathbf{R}}_{\bar{i}s}^H \mathbf{z}_{ik} \\
&= \mathbf{z}_{ik}^H(\tilde{\mathbf{R}}_{\bar{i}v}\tilde{\mathbf{L}}_{\bar{i}v}^H + \tilde{\mathbf{R}}_{\bar{i}s}\tilde{\mathbf{L}}_{\bar{i}s}^H)\boldsymbol{\Sigma}_{\bar{i}\bar{i}}(\tilde{\mathbf{L}}_{\bar{i}v}\tilde{\mathbf{R}}_{\bar{i}v}^H + \tilde{\mathbf{L}}_{\bar{i}s}\tilde{\mathbf{R}}_{\bar{i}s}^H)\mathbf{z}_{ik} \\
&\quad - \mathbf{z}_{ik}^H \tilde{\mathbf{R}}_{\bar{i}s}\tilde{\mathbf{L}}_{\bar{i}s}^H \boldsymbol{\Sigma}_{\bar{i}\bar{i}}\tilde{\mathbf{L}}_{\bar{i}s}\tilde{\mathbf{R}}_{\bar{i}s}^H \mathbf{z}_{ik} \\
&= \mathbf{z}_{ik}^H \tilde{\mathbf{R}}_{\bar{i}v}\tilde{\mathbf{R}}_{\bar{i}v}^H \mathbf{z}_{ik} \\
&= \|\tilde{\mathbf{R}}_{\bar{i}v}^H \mathbf{z}_{ik}\|^2 \\
&\geq 0
\end{aligned} \tag{6.259}$$

where the fact that $\tilde{\mathbf{L}}_{is}^H \boldsymbol{\Sigma}_{ii}\tilde{\mathbf{L}}_{iv} = \tilde{\mathbf{L}}_{is}^H \tilde{\mathbf{R}}_{iv} = \mathbf{O}$ has been employed. The result in Eq. (6.259) leads directly to the conclusion of the theorem. ■

Now, let us turn our attention to examine the optimality of applying CCD in the UN-CLE algorithm. Equation (6.224) shows the resulting covariance matrix of the UN-CLE estimates when optimum weighting matrices $\bar{\mathbf{W}}_{iv}$ and $\bar{\mathbf{W}}_{is}$ are applied. If in addition, CCD is employed in the UN-CLE algorithm under optimum weighting, then substituting Eqs. (6.244) and (6.245) into Eq. (6.224) and noticing Eq. (6.251), we have

$$\operatorname{cov}(\Delta\tilde{\theta}) = \frac{1}{2N}\left[\operatorname{Re}\{(\mathbf{D}_i'^H \mathbf{B}_{iv}(\mathbf{B}_{iv}^H \boldsymbol{\Sigma}_{ii}\mathbf{B}_{iv})^{-1}\mathbf{B}_{iv}^H \mathbf{D}_i') \odot (\tilde{\mathbf{T}}_{is}\tilde{\boldsymbol{\Gamma}}^2\tilde{\mathbf{T}}_{is}^H)^T\}\right]^{-1} \tag{6.260}$$

where $\tilde{\mathbf{T}}_{is}$ is a $K \times K$ nonsingular matrix such that $\tilde{\mathbf{R}}_{is} = \mathbf{D}_i\tilde{\mathbf{T}}_{is}$. The simplicity of Eq. (6.260) is due to the fact that when CCD is employed $\tilde{\mathbf{L}}_{is}^H \boldsymbol{\Sigma}_{ii}\tilde{\mathbf{L}}_{is} = \mathbf{I}$ for $i = 1, 2$. We now proceed on to show that the covariance matrix expressed in Eq. (6.260) is the minimum among all possible choices of GCD. To do this we need the following lemmas:

Lemma 6.16a. *For* Σ_{ii}, i = 1, 2, *being positive definite, we have*

$$\Sigma_{ii}^{-1} \geq \mathbf{L}_{is}(\mathbf{L}_{is}^H \Sigma_{ii} \mathbf{L}_{is})^{-1}\mathbf{L}_{is}^H \quad \text{for} \quad i = 1, 2 \tag{6.261}$$

Proof. Let $\mathbf{F}_i = \Sigma_{ii}^{1/2}\mathbf{L}_{is}(\mathbf{L}_{is}^H \Sigma_{ii}\mathbf{L}_{is})^{-1}\mathbf{L}_{is}^H$ and $\mathbf{G}_i = \Sigma_{ii}^{-1/2}$ and let ζ_i be any M_i-dimensional vector. Then, by the Schwarz's inequality

$$\| \langle \mathbf{F}_i\zeta_i, \mathbf{G}_i\zeta_i \rangle \|^2 \leq \langle \mathbf{F}_i\zeta_i, \mathbf{F}_i\zeta_i \rangle \langle \mathbf{G}_i\zeta_i, \mathbf{G}_i\zeta_i \rangle$$

that is

$$\| \zeta_i^H \mathbf{L}_{is}(\mathbf{L}_{is}^H \Sigma_{ii}\mathbf{L}_{is})^{-1}\mathbf{L}_{is}^H\zeta_i \|^2 \leq (\zeta_i^H \mathbf{L}_{is}(\mathbf{L}_{is}^H\Sigma_{ii}\mathbf{L}_{is})^{-1}\mathbf{L}_{is}^H\zeta_i)(\zeta_i^H\Sigma_{ii}^{-1}\zeta_i)$$

or

$$\zeta_i^H \mathbf{L}_{is}(\mathbf{L}_{is}^H\Sigma_{ii}\mathbf{L}_{is})^{-1}\mathbf{L}_{is}^H\zeta_i \leq \zeta_i^H\Sigma_{ii}^{-1}\zeta_i$$

and Eq. (6.261) follows. ■

Lemma 6.16b. *For* $\mathbf{T}_{is}$ *and* $\tilde{\mathbf{T}}_{is}$ *being* $K \times K$ *nonsingular matrices such that* $\mathbf{R}_{is} = \mathbf{D}_i\mathbf{T}_{is}$ *and* $\tilde{\mathbf{R}}_{is} = \mathbf{D}_i\tilde{\mathbf{T}}_{is}$ *we have*

$$\tilde{\mathbf{T}}_{is}\Gamma^2\tilde{\mathbf{T}}_{is}^H \geq \mathbf{T}_{is}\Gamma(\mathbf{L}_{\bar{i}s}^H\Sigma_{\bar{i}\bar{i}}\mathbf{L}_{\bar{i}s})^{-1}\Gamma\mathbf{T}_{is}^H \tag{6.262}$$

Proof. Recall Eq. (6.253):

$$\tilde{\mathbf{R}}_{is}\tilde{\Gamma}\tilde{\mathbf{R}}_{\bar{i}s}^H = \mathbf{R}_{is}\Gamma\mathbf{R}_{\bar{i}s}^H \tag{6.263}$$

Postmultiplying $\mathbf{L}_{\bar{i}s}$ both sides of Eq. (6.263) we obtain

$$\tilde{\mathbf{R}}_{is}\tilde{\Gamma} = \mathbf{R}_{is}\Gamma\mathbf{R}_{\bar{i}s}^H\tilde{\mathbf{L}}_{\bar{i}s} \tag{6.264}$$

On the other hand, we have

$$\mathbf{R}_{is} = \mathbf{D}_i\mathbf{T}_{is} \tag{6.265a}$$

and

$$\tilde{\mathbf{R}}_{is} = \mathbf{D}_i\tilde{\mathbf{T}}_{is} \tag{6.265b}$$

Substituting Eqs. (6.265) into Eq. (6.264), we see

$$\mathbf{D}_i\tilde{\mathbf{T}}_{is}\tilde{\Gamma} = \mathbf{D}_i\mathbf{T}_{is}\Gamma\mathbf{R}_{\bar{i}s}^H\tilde{\mathbf{L}}_{\bar{i}s} \tag{6.266}$$

Since $\mathbf{D}_i$ is full columns rank, Eq. (6.266) implies

$$\tilde{\mathbf{T}}_{is}\tilde{\Gamma} = \mathbf{T}_{is}\Gamma\mathbf{R}_{\bar{i}s}^H\tilde{\mathbf{L}}_{\bar{i}s} \tag{6.267}$$

Then, multiplying both sides of Eq. (6.267) by their respective conjugate transposes, we have

$$\tilde{\mathbf{T}}_{is}\tilde{\Gamma}^2\tilde{\mathbf{T}}_{is}^H = \mathbf{T}_{is}\Gamma\mathbf{R}_{\bar{i}s}^H\tilde{\mathbf{L}}_{\bar{i}s}\tilde{\mathbf{L}}_{\bar{i}s}^H\mathbf{R}_{\bar{i}s}\Gamma\mathbf{T}_{is}^H \tag{6.268a}$$

Using $\tilde{\mathbf{L}}_{\bar{i}s}\tilde{\mathbf{L}}_{\bar{i}s}\mathbf{R}_{\bar{i}s} = \Sigma_{\bar{i}\bar{i}}^{-1}\Sigma_{\bar{i}\bar{i}}\tilde{\mathbf{L}}_{\bar{i}s}\tilde{\mathbf{L}}_{\bar{i}s}^H\mathbf{R}_{\bar{i}s} = \Sigma_{\bar{i}\bar{i}}^{-1}\tilde{\mathbf{R}}_{\bar{i}s}\tilde{\mathbf{L}}_{\bar{i}s}^H\mathbf{R}_{\bar{i}s} = \Sigma_{\bar{i}\bar{i}}^{-1}\tilde{\mathbf{P}}_{\bar{i}s}^H\mathbf{R}_{\bar{i}s} = \Sigma_{\bar{i}\bar{i}}^{-1}\mathbf{R}_{\bar{i}s}$, we can rewrite Eq. (6.268a) as

$$\tilde{\mathbf{T}}_{is}\tilde{\Gamma}^2\tilde{\mathbf{T}}_{is}^H = \mathbf{T}_{is}\Gamma\mathbf{R}_{\bar{i}s}^H\Sigma_{\bar{i}\bar{i}}^{-1}\mathbf{R}_{\bar{i}s}\Gamma\mathbf{T}_{is}^H \tag{6.268b}$$

Replacing $\Sigma_{\bar{i}\bar{i}}^{-1}$ in Eq. (6.268b) by the inequality of Eq. (6.261) in Lemma 6.16a, we arrive at the inequality in Eq. (6.262) where we have used the fact that $\mathbf{L}_{\bar{i}s}^H \mathbf{R}_{\bar{i}s} = \mathbf{I}$. ■

Theorem 6.16. *Under optimum weighting, the UN-CLE estimates of the DOA in an uknown correlated noise environment employing CCD yield the minimum covariance, that is,*

$$\text{cov}(\Delta\tilde{\theta}_i) \le \text{cov}(\Delta\theta_i) \tag{6.269}$$

where $\text{cov}(\Delta\tilde{\theta}_i)$ *and* $\text{cov}(\Delta\theta_i)$ *are the covariance matrices of the estimation errors obtained by the UN-CLE algorithm employing CCD and employing any othe kind of GCD respectively.*

Proof. Following directly from the result given in Eq. (6.251), we can write

$$\mathbf{D}_i'^H \mathbf{L}_{iv}(\mathbf{L}_{iv}^H \Sigma_{ii} \mathbf{L}_{iv})^{-1} \mathbf{L}_{iv}^H \mathbf{D}_i' = \mathbf{D}_i'^H \mathbf{B}_{iv}(\mathbf{B}_{iv}^H \Sigma_{ii} \mathbf{B}_{iv})^{-1} \mathbf{B}_{iv}^H \mathbf{D}_i' \tag{6.270}$$

where $\mathbf{D}_i'$ has been defined in Theorem 6.11.

Using Eqs. (6.222) in Eq. (6.224), we can write

$$\begin{aligned}
&\text{cov}^{-1}(\Delta\theta_i) \\
&= 2N \,\text{Re}\left\{(\mathbf{D}_i'^H \mathbf{P}_{iv} \mathbf{L}_{iv}(\mathbf{L}_{iv}^H \Sigma_{ii} \mathbf{L}_{iv})^{-1} \mathbf{L}_{iv} \mathbf{P}_{iv}^H \mathbf{D}_i') \odot \left[\mathbf{T}_{is} \Gamma (\mathbf{L}_{\bar{i}s}^H \Sigma_{\bar{i}\bar{i}} \mathbf{L}_{\bar{i}s})^{-1} \Gamma \mathbf{T}_{is}\right]^T\right\} \\
&= 2N \,\text{Re}\left\{(\mathbf{D}_i' \mathbf{B}_{iv}(\mathbf{B}_{iv}^H \Sigma_{ii} \mathbf{B}_{iv})^{-1} \mathbf{B}_{iv}^H \mathbf{D}_i') \odot \left[\mathbf{T}_{is} \Gamma (\mathbf{L}_{\bar{i}s}^H \Sigma_{\bar{i}\bar{i}} \mathbf{L}_{\bar{i}s})^{-1} \Gamma \mathbf{T}_{is}\right]^T\right\}
\end{aligned} \tag{6.271}$$

where we have used the fact that $\mathbf{P}_{iv}\mathbf{L}_{iv} = \mathbf{L}_{iv}$ as well as the results of Eq. (6.270). Now, combining Eqs. (6.260) and (6.271), and using Lemma (6.16b), we can write

$$\begin{aligned}
&\text{cov}^{-1}(\Delta\tilde{\theta}_i) - \text{cov}^{-1}(\Delta\theta_i) \\
&= 2N\text{Re}\Big\{(\mathbf{D}_i'^H \mathbf{B}_{iv}(\mathbf{B}_{iv}^H \Sigma_{ii} \mathbf{B}_{iv})^{-1} \mathbf{B}_{iv}^H \mathbf{D}_i') \\
&\quad \odot \left[\tilde{\mathbf{T}}_{is} \tilde{\Gamma}^2 \tilde{\mathbf{T}}_{is}^H - \mathbf{T}_{is} \Gamma (\mathbf{L}_{\bar{i}s}^H \Sigma_{\bar{i}\bar{i}} \mathbf{L}_{\bar{i}s})^{-1} \Gamma \mathbf{T}_{is}^H\right]^T \Big\} \\
&\ge \mathbf{0}
\end{aligned} \tag{6.272}$$

and the result of Eq. (6.269) follows directly. ■

Theorems 6.15 and 6.16 have profound implications. Because of the optimal characteristics possessed by the use of CCD as indicated by Properties 6.1 and 6.2, it has been our suspicion that the application of CCD to DOA estimation in an unknown noise environment will yield better results than the use of any other kind of GCD. These two theorems confirm that this is indeed the case. Thus, for optimal DOA estimation, we should employ CCD in our algorithms of UN-MUSIC and UN-CLE. The fast computation method for CCD developed in Section 6.4 would greatly facilitate the use of CCD in practice.

Numerical Experiments

As an illustration of the analysis on the performance of the UN-MUSIC algorithm, we have included a few examples of computer simulation.

Example 6.5

In this example, we employ two separate linear eight-sensor arrays having the same orientations and each having the same background noise which is chosen as an AR model of order 2 with coefficients $\mathbf{a} = [1 \ -1 \ 0.2]$. More specifically, the noise field is generated as

$$\nu_{ij}(n) = \nu_{i(j-1)}(n) - 0.2\nu_{i(j-2)}(n) + w_{ij}(n) \tag{6.273}$$

where $\nu_{ij}(n)$ is the jth element of $\boldsymbol{\nu}_i(n)$ and $w_{ij}(n)$ is the jth element of the spatial white noise vector $\mathbf{w}_i(n)$. There are two independent Gaussian signals with equal power arriving at $-5°$ and $5°$ to the normals of the arrays. We use the UN-MUSIC algorithm to estimate the DOA of the signals and the RMS of the combined error of the two DOA estimates are evaluated over 100 trials in each case with $N = 100$. Figure 6.3 shows the RMS errors of the estimates against the SNR. The two cases shown here are as follows:

1. The UN-MUSIC estimate with optimum weighting employing CCD (i.e., $\boldsymbol{\Pi}_i = \hat{\boldsymbol{\Sigma}}_{ii},\ i = 1, 2$)

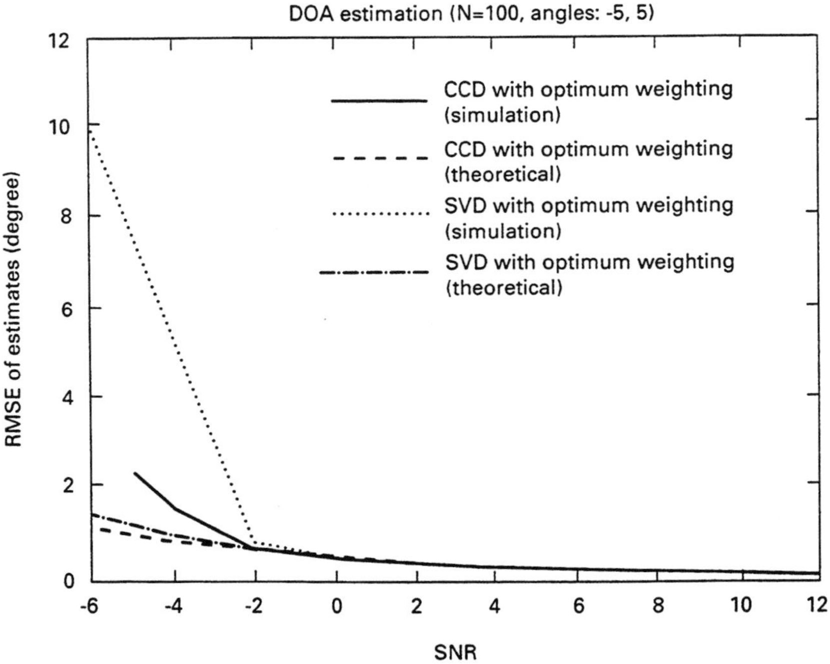

Figure 6.3 Performance of the UN-MUSIC algorithm using two separate arrays having similar background noise.

2. The UN-MUSIC estimate with optimum weighting employing SVD (i.e., $\mathbf{\Pi}_i = \mathbf{I},\ i = 1, 2$)

The theoretical evaluations of the performances of the two cases are also plotted for reference. As can be observed, the simulation results closely agree with the theoretical evaluation when the SNR is above −2 dB. However, for SNR below −2 dB, both cases show a threshold effect with the SVD method being more severe in deterioration of performance. In addition, we can observe that at high SNR, the performance of the two cases are very close to each other.

Example 6.6

In this example, we study the effects of optimum weighting. The signal and noise environments here are identical to those in Example 6.5 above. We employ both the CCD and SVD methods with optimum weighting and identity weighting, that is, $\mathbf{W}_{i\upsilon} = \mathbf{I}$. Figure 6.4 shows the comparison of the performance of these cases. The perfomance of the methods of CCD and SVD with optimum weighting are identical to those in Example 6.5 above. However, the methods of CCD and SVD with identity weighting show deteriorations in performance. The performance deterioration is more severe with the SVD method, having the threshold raised by 2 dB. This example shows that the SVD method is more sensitive to weighting variation.

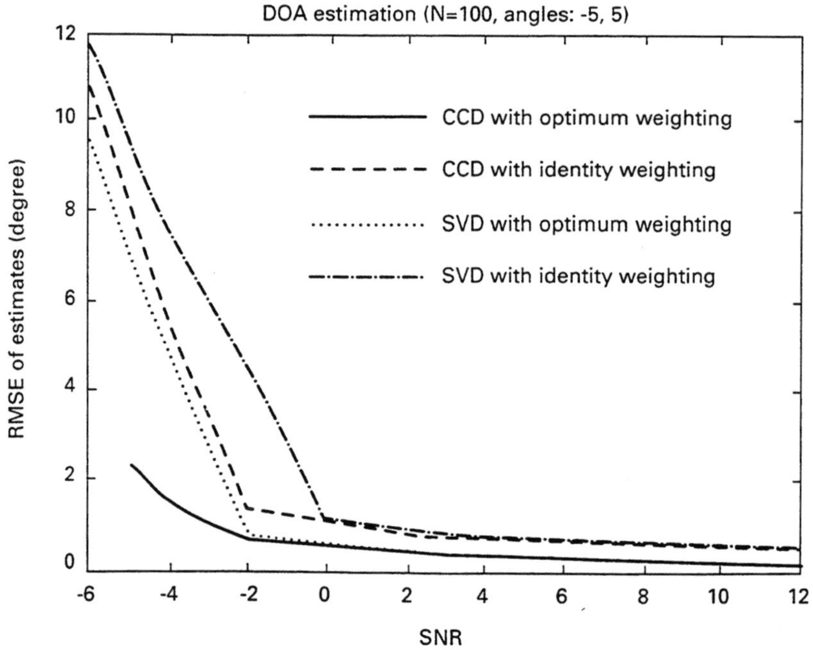

Figure 6.4 Performance of the UN-MUSIC algorithm by simulations using two separate arrays having same background noise.

Example 6.7

In this example we examine the effects of varying the noise conditions in the two arrays. Again we use two separate arrays with the same orientations, and the signal condition is identical to those in the two examples above. However, the background noise associated with these two arrays have different AR models. For Array1, the coefficients are $\mathbf{a}_1 = [1 \ \ -1 \ \ 0.2]$, where as for Array2, they are $\mathbf{a}_2 = [1 \ \ -1.5 \ \ 0.2]$. Again, the RMS of the combined errors of the two DOA estimates are evaluated over 100 trials each having $N = 100$. Figure 6.5 shows the performance of the CCD and SVD methods with both optimum weightings and identity weightings. It can be observed that while the performance of the CCD methods (with both types of weightings) remains essentially unchanged, the performance of the SVD methods (with both types of weightings) have much worse performance compared to the previous examples of identical noise environment for both arrays. The thresholds in the cases of using SVD have risen by about 6 dB in SNR. This example reveals that the SVD methods are very sensitive to the variation in noise environments in the two arrays whereas the CCD methods are very robust, and in the real situations, the noise spatial covariances in the two arrays are rarely identical, which implies that CCD should be employed whenever possible.

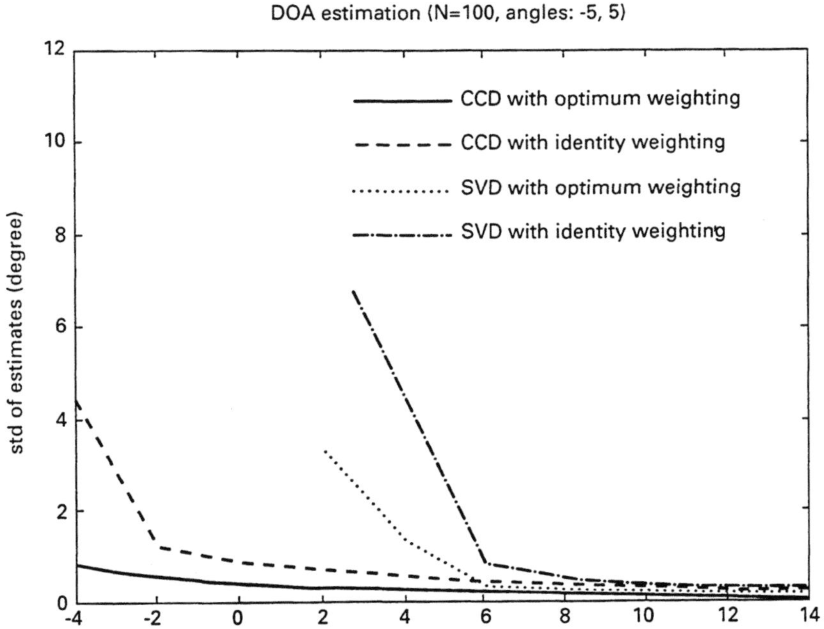

Figure 6.5 Performance of the UN-MUSIC algorithm by simulations using two separate arrays having different background noise.

We now show some computer simulation examples to illustrate the performance of the UN-CLE algorithm.

Example 6.8

In this example, we employ two separate eight-sensor arrays having the same orientations and having the same background noise modeled in the same way as in Example 6.5. The signal model is also the same as that in Example 6.5. We apply the UN-CLE algorithm under optimum weightings to estimate the DOA of the signals and the RMS of the combined errors of the two estimates are evaluated over 100 trials each having $N = 100$. Figure 6.6 shows the RMS errors of the estimates against the SNR. The two cases shown here are:

1. The UN-CLE estimate employing CCD ($\mathbf{\Pi}_{ii} = \mathbf{\Sigma}_{ii}, \ i = 1, 2$).
2. The UN-CLE estimate employing SVD (i.e., $\mathbf{\Pi}_{ii} = \mathbf{I}, \ i = 1, 2$).

The theoretical evaluations of the performance of the two cases are also plotted for reference. As in Example 6.5, the simulation results agree closely with the theoretical evaluation for higher SNR. The threshold effect shows when the SNR is below −2 dB for the SVD case and below −4 dB for the CCD case. The performance of the two methods at higher SNR are very close with the use of CCD marginally better than the use of SVD.

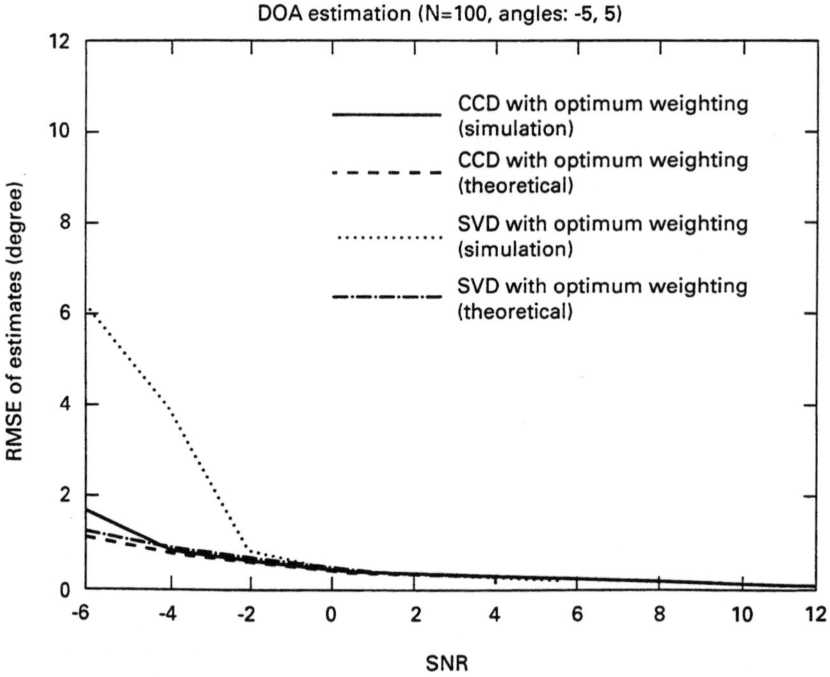

Figure 6.6 Performance of the UN-CLE algorithm using CCD and SVD with optimum weightings.

Example 6.9

In this example, we study the effects of weighting on the UN-CLE algorithm. The signal and noise environments here are identical to those in Example 6.8 above. We apply the CCD and SVD with both optimum and identity weightings. Figure 6.7 shows the comparison of the performances of these cases. The performances of the methods with CCD and SVD having optimum weightings are identical to those in Example 6.8 above. However, the cases in which identity weightings are used, deterioration in performance occurs, while only marginally in the case of CCD, the deterioration is much more apparent in the case of SVD. In fact using CCD with identity weighting outperforms SVD with optimum weighting for all SNR. This example shows that the use of SVD is sensitive to the variation of weighting whereas the use of CCD is more robust.

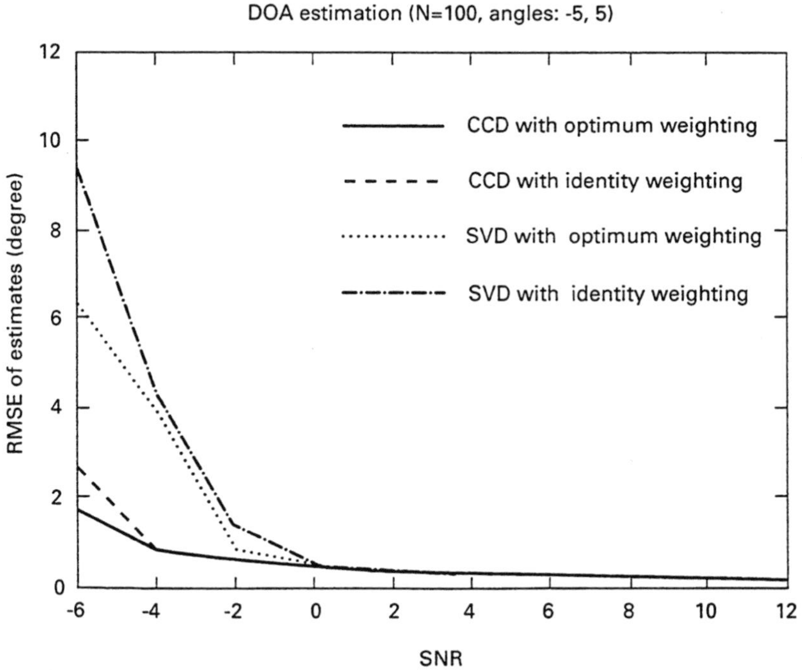

Figure 6.7 Performance of the UN-CLE algorithm by simulations using two separate arrays having similar background noise.

Example 6.10

In this example, we examine the effects of varying the noise conditions in the two arrays on the use of the UN-CLE algorithm. Using the same arrays and signals as in the two examples above, we alter the background noise associated with the two arrays so that they have different autoregressive (AR) models. For Array1, the coefficients are $\mathbf{a}_1 = [1 \ \ -1 \ \ 0.2]$, whereas for Array2, they are $\mathbf{a}_1 = [1 \ \ -1.5 \ \ 0.2]$. Figure 6.8 shows the RMS errors of the combined errors of the two DOA estimates against SNR. It can be observed that while the performance of the UN-CLE using CCD (with both types of weighting) remains essentially unchanged, the performance using SVD (with

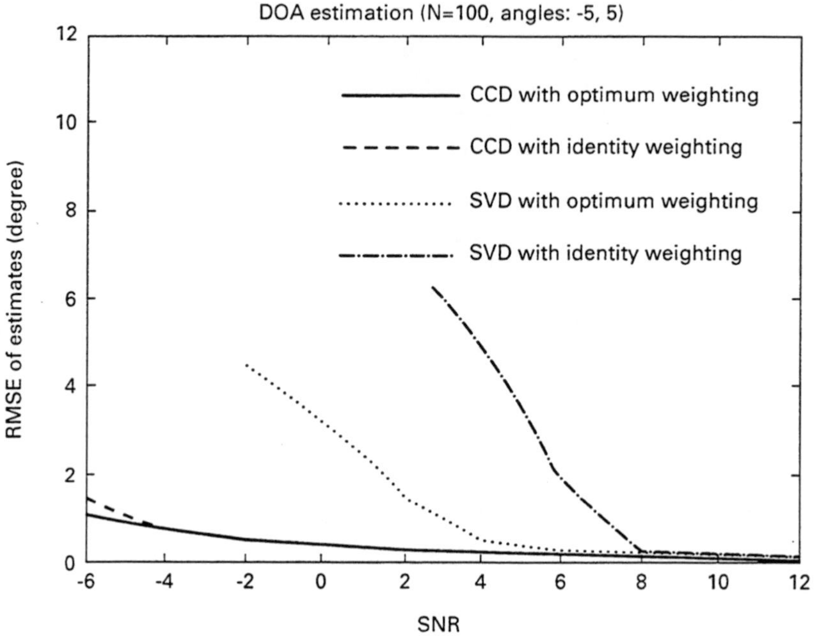

Figure 6.8 Performance of the UN-CLE algorithm by simulations using two separate arrays having different background noise.

both types of weighting) is very poor compared to the previous examples of identical noise environment for both arrays. Again, the example confirms that the use of SVD is very sensitive to the variation in noise environments in the two arrays whereas the use of CCD is robust.

Example 6.11

In this example, we apply the UN-CLE algorithm to a correlated signal environment. The two arrays used here are the same as those in Example 6.8 and they have identical background noise. However, the two signals arriving at $-5°$ and $5°$ to the normals of the arrays are correlated with a correlation coefficient $r = 0.9$. The RMS of the combined errors of the two DOA estimates against SNR are plotted in Figure 6.9. It can be observed that the algorithm with CCD having optimum weighting has performance similar to the cases when the signals are independent, and the algorithm with CCD having identity weighting has higher threshold effect and slightly worse performance for high SNR. However, in the cases when SVD is employed, the performance is very poor. With optimum weighting in both cases, the use of SVD shows poorer performance than that of the UN-CLE employing CCD. With identity weighting, the performance of the UN-CLE using SVD is unacceptable even under very high SNR. This example shows the significance for using the CCD and the optimum weighting matrices for the UN-CLE in the case of correlated signals.

We now present two computer simulation examples to illustrate the performance of the UN-MUSIC and UN-CLE algorithms in unknown correlated noise environments compared to that of MUSIC, MLE for deterministic signals, and MODE. The

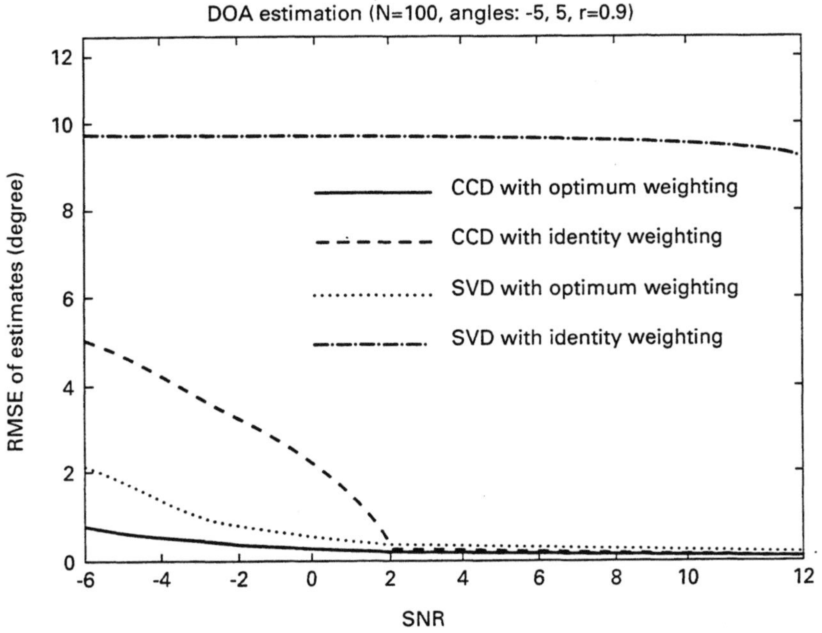

Figure 6.9 Performance of the UN-CLE algorithm by simulations using two separate arrays having similar background noise with two correlated signals. $\gamma = 0.9$.

arrangement of the arrays and the noise conditions are the same as those in Example 6.5. We employ CCD and optimum weightings for both UN-MUSIC and UN-CLE in the following examples. In the cases of MUSIC, MLE, and MODE, we employ a single array with 16 sensors.

Example 6.12

Figure 6.10 shows the RMS errors of the various methods of estimating the DOA of two uncorrelated Gaussian signals in the colored noise case as described in Example 6.5. It can be observed that the thresholds of errors for MLE, MUSIC, and MODE occur at relatively high SNR, being at 6db, 4db, and 3db respectively. On the other hand, UN-MUSIC and UN-CLE do not show any sign of error threshold for much lower SNR.

Example 6.13

Figure 6.11 shows the performance of the various methods of estimating the DOA of two correlated Gaussian signals, the correlation coefficient of which is equal to 0.9. Since MUSIC and UN-MUSIC are not suitable for correlated signals, we only employ the MLE, the MODE, and the UN-CLE methods. Again, the UN-CLE method shows superior performance to either of the other two methods which exhibit error thresholds at relatively high SNR (2db for MODE and 4db for MLE).

The DOA estimation methods in unknown noise using GCD are based on the assumption that the two arrays are separated sufficiently so that their output noises are uncorrelated. However, in some circumstances, there may be only one array

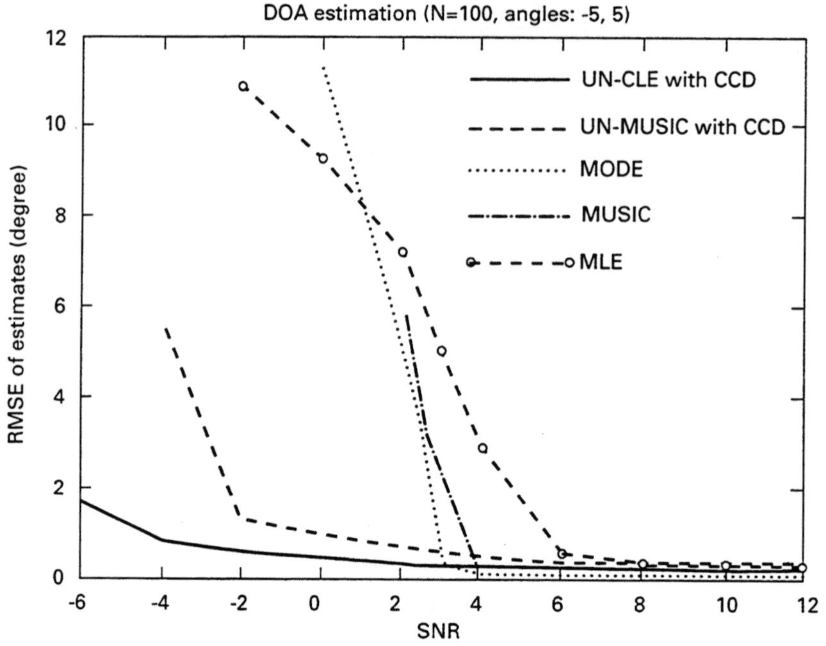

Figure 6.10 Performance of various methods in unknown correlated noise (two independent signals).

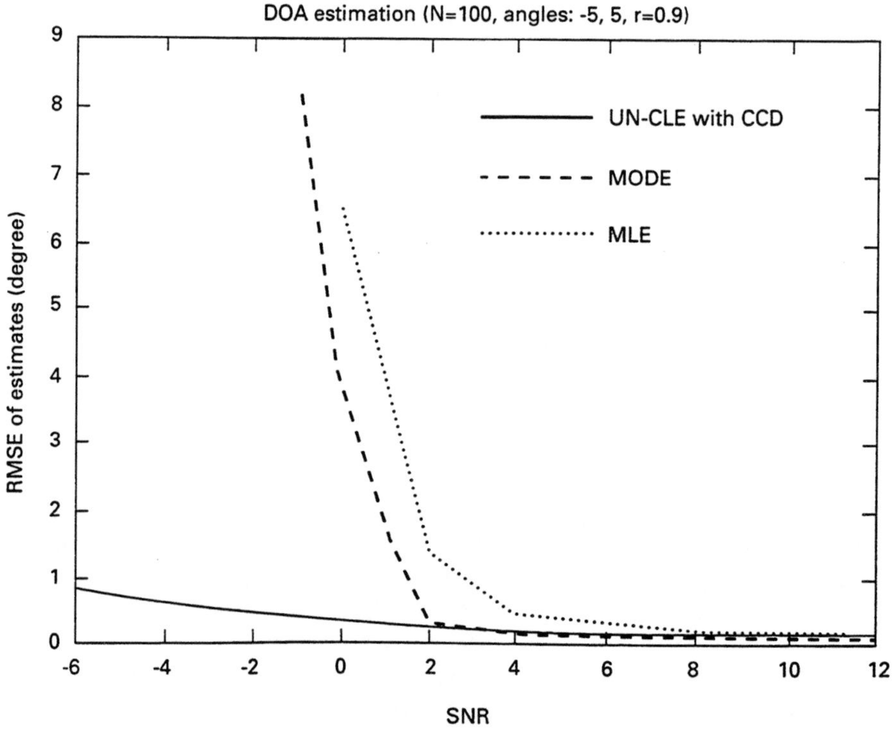

Figure 6.11 Performance of various methods for two correlated signals in unknown correlated noise.

available. Here, we demonstrate that the two methods developed in the last section are applicable even when a single array is used as long as the length of the noise correlation is smaller than the length of the array.

Example 6.14

The composite array used is shown in Figure 6.12a in which there are two subarrays each having 7 equally spaced sensors aligned along a straight line. The separation, d_0, between the sensors is equal to half the wavelength of the signals. The two subarrays are separated by a distance of $3d_0$. In other words, the composite array can be considered as a single linear array having 16 equally spaced sensors from which the two central sensors are removed. Again, the two uncorrelated Gaussian signals are arriving at $-5°$ and $5°$ to the normal of the composite array. The colored noise covariance matrix is chosen to be $\Sigma_v = [\sigma_{lm}]$ where $\sigma_{lm} = 0.9^{|l-m|}\exp\{j\pi(l-m)\}$. The number of snapshots in this case is $N = 40$ and the RMS errors of estimation by the various methods are evaluated over 100 trials. Figure 6.12b shows the comparison between the various methods. Again, while MLE, MUSIC, and MODE

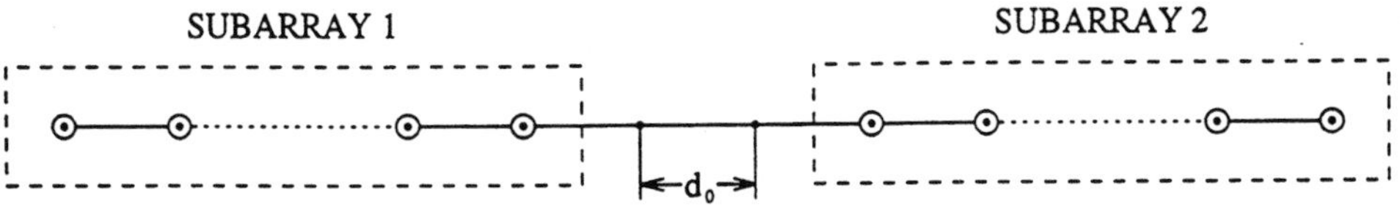

Figure 6.12a A composite linear array.

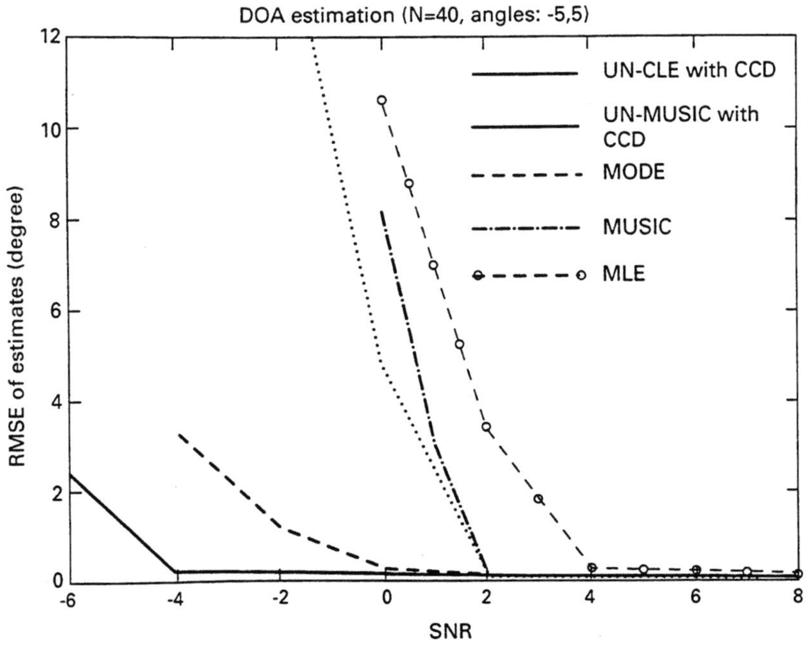

Figure 6.12b Performance of the various methods in unknown correlated noise using one single composite array.

all show thresholds of errors at relatively high SNR, the two new methods of UN-MUSIC, and UN-CLE maintain good perfomance even at low SNR.

6.8 DETERMINATION OF THE NUMBER OF SIGNALS IN UNKNOWN NOISE ENVIRONMENTS

Properties of the UN-CLE Criterion

We mentioned in section 6.1 that there are two essential purposes in array processing, viz.

(i) To determine the number of signal sources.
(ii) To estimate the locations of these sources.

In the previous sections, we have focused our attention on the problem of estimating the DOA of signals in unknown noise environment assuming that the number of signals, K, is known. This, as mentioned in section 6.1, of course is in the wrong logic order. In practice, we always have to determine the number of signals before their locations can be estimated. The reason we treat the two problems in reverse order is because one of the methods we develop in this section to determine the number of signals in an unknown noise environment utilizes the criterion $f(\hat{\mathbf{R}}_{is}, \bar{\mathbf{W}}_{iv}, \bar{\mathbf{W}}_{is})$ of the UN-CLE algorithm and necessitates the knowledge on which the estimation algorithm has been developed.

Let us recall from section 6.3 that the correlation submatrices of the outputs of the two arrays are given by

$$\boldsymbol{\Sigma}_{ii} = \mathbf{D}_i(\boldsymbol{\theta}_i)\boldsymbol{\Sigma}_s\mathbf{D}_i^H(\boldsymbol{\theta}_i) + \boldsymbol{\Sigma}_{iv} \tag{6.274a}$$

$$\boldsymbol{\Sigma}_{i\bar{i}} = \boldsymbol{\Sigma}_{\bar{i}i}^H = \mathbf{D}_i(\boldsymbol{\theta}_i)\boldsymbol{\Sigma}_s\mathbf{D}_{\bar{i}}^H(\boldsymbol{\theta}_{\bar{i}}) \tag{6.274b}$$

where $\boldsymbol{\Sigma}_s$ is the correlation matrix of the signals. Thus far, we have assumed that $\boldsymbol{\Sigma}_s$ is of full rank. However, due to the correlation between signals, this may not be the case. Now, let us assume that the rank of $\boldsymbol{\Sigma}_s$ is K' where $K' \leq K$ with K being the true number of signals.

We also recall that in Section 6.5, the UN-CLE algorithm has been shown to be asymptotic consistent, that is

$$\lim_{N\to\infty} \arg\min_{\boldsymbol{\theta}} f(\hat{\mathbf{R}}_{is}, \mathbf{W}_{iv}, \mathbf{W}_{is}) \underset{a.s.}{=} \arg\min_{\boldsymbol{\theta}} f(\mathbf{R}_{is}, \mathbf{W}_{iv}, \mathbf{W}_{is}) \tag{6.275}$$

Furthermore, in Theorem 6.8, we have shown that the true DOA $\boldsymbol{\theta}_i$ uniquely minimizes the criterion $f(\mathbf{R}_{is}, \bar{\mathbf{W}}_{iv}, \bar{\mathbf{W}}_{is})$ as $N \to \infty$. Also, from Eqs. (6.190) and (6.193) we can see that the local convergence rate of $\hat{\boldsymbol{\theta}}_i$ to $\boldsymbol{\theta}_i$ is $O(N^{-1/2})$. Although the asymptotic consistency of $f(\hat{\mathbf{R}}_{is}, \mathbf{W}_{iv}, \mathbf{W}_{is})$ together with the uniqueness of $\boldsymbol{\theta}_i$ minimizing the criterion and the convergence rate of $O(N^{-1/2})$ have been derived assuming that $\text{rank}(\boldsymbol{\Sigma}_{12}) = K$, the same conclusions can be reached if $K' < K$. In that case we start with the expression in Eq. (6.148) and follows similar arguments (see also [45]).

The accuracy of the determination of the number of signals depends on the behavior of an established criterion under different hypothesized signal number k. This behavior, if favorable, will be used to construct a detector for the determination of the number of signals. Let us examine the asymptotic behavior of $f(\hat{\mathbf{R}}_{is}, \bar{\mathbf{W}}_{iv}, \bar{\mathbf{W}}_{is})$, rewritten as $f(\hat{\mathbf{R}}_{is}, \theta_i)$ and often simply short-handed to f_i, knowing that we are employing optimum weighting matrices $\bar{\mathbf{W}}_{iv}$ and $\bar{\mathbf{W}}_{is}$ given by Eqs. (6.222). To do that we recall that we are employing arrays which are unambiguous. This can be equivalently stated as

$$\mathbf{R}_{is} = \mathbf{D}_i(\hat{\theta}_i)\mathbf{T}_{is} \Rightarrow \hat{\theta}_i = \theta_i \tag{6.276}$$

Here since the rank of $\mathbf{R}_{is}$ is K', and $\mathbf{D}_i(\hat{\theta}_i)$ is of rank K, then $\mathbf{T}_{is}$ is a $K \times K'$ full-column rank. It has been shown [93, 94] that in order to be able to determine θ_i uniquely, $K < (M_i - 1 + K')/2$. Throughout this section, we assume that the number of sensors in our array is large enough to satisfy this requirement. We have the following theorem:

Theorem 6.17. *For an array capable of determining the DOA uniquely, if* k *and* K *are respectively the hypothesized and the true number of signals, then, asymptotically, for* $K' \leq k \leq K$

$$\lim_{N\to\infty} \min_{\theta_i} f(\hat{\mathbf{R}}_{is}, \theta_i) \geq 0 \tag{6.277}$$

Equality in Eq. (6.277) holds only if $k = K$*, in which case the convergence rate is*

$$\min_{\theta_i} f(\hat{\mathbf{R}}_{is}, \theta_i) = O(N^{-1})\,prob. \tag{6.278}$$

Proof. Since $\text{rank}(\mathbf{R}_{is}) = \text{rank}(\Sigma_{i\bar{i}}) = K'$ and from Theorem 6.2, $\text{span}\{\mathbf{R}_{is}\} \subseteq \text{span}\{\mathbf{D}_i(\theta_i)\}$, we can thus write uniquely

$$\mathbf{R}_{is} = \mathbf{D}_i(\theta_i)\mathbf{T}_{is} \tag{6.279}$$

where $\mathbf{T}_{is}$ is a $K \times K'$ nonsingular linear transformation. Furthermore, from the consistency of the UN-CLE criterion as shown in Section 6.5, $\lim_{N\to\infty} \mathbf{P}_{iv}^H \hat{\mathbf{R}}_{is} = \mathbf{O}$, implying that $\lim_{N\to\infty} \min_{\theta} f(\hat{\mathbf{R}}_{is}, \theta_i) = \mathbf{O}$. Thus, $f(\hat{\mathbf{R}}_{is}, \theta_i)$ is an asymptotic minimum if the parameters θ_i assume the true values. Now if $k < K$, then the estimated DOA vector $\hat{\theta}_i$ is k-dimensional, which is less than the dimension of the true DOA θ_i. The matrix $\hat{\mathbf{R}}_{is}$ which now has k columns will no longer be in the same subspace as $\mathbf{D}_i(\theta_i)$ and the uniqueness condition of Eq. (6.276) will be violated. Thus $f(\hat{\mathbf{R}}_{is}, \hat{\theta}_i)$ will not achieve the asymptotic minimum.

To prove Eq. (6.278) for $k = K$, we express $f(\hat{\mathbf{R}}_{is}, \hat{\theta}_i)$ in a Taylor series such that

$$f(\hat{\mathbf{R}}_{is}, \hat{\theta}_i) \simeq f(\hat{\mathbf{R}}_{is}, \theta_i) + \dot{\mathbf{f}}^T(\hat{\mathbf{R}}_{is}, \theta_i)\Delta\theta_i + \frac{1}{2}(\Delta\theta_i)^T \ddot{\mathbf{f}}(\hat{\mathbf{R}}_{is}, \theta_i)\Delta\theta \tag{6.280}$$

where the third- and higher-order terms of $\Delta\theta_i$ have been ignored. But $\hat{\theta}_i$ is an extremal point for f_i, therefore we can also express $\dot{\mathbf{f}}(\hat{\mathbf{R}}_{is}, \hat{\theta}_i)$ in a Taylor series, again, ignoring the second- and and higer-order terms, yielding

$$\dot{\mathbf{f}}(\hat{\mathbf{R}}_{is}, \hat{\theta}_i) \simeq \dot{\mathbf{f}}(\hat{\mathbf{R}}_{is}, \theta_i) + \ddot{\mathbf{f}}(\hat{\mathbf{R}}_{is}, \theta_i)\Delta\theta = \mathbf{O} \tag{6.281}$$

which gives

$$\Delta\theta \simeq -\ddot{\mathbf{f}}^{-1}(\hat{\mathbf{R}}_{is}, \theta_i)\dot{\mathbf{f}}(\hat{\mathbf{R}}_{is}, \theta_i) \tag{6.282}$$

Substituting Eq. (6.282) in (6.280), we obtain

$$\begin{aligned} f(\hat{\mathbf{R}}_{is}, \hat{\theta}_i) &\simeq f(\hat{\mathbf{R}}_{is}, \theta_i) - \dot{\mathbf{f}}^T(\hat{\mathbf{R}}_{is}, \theta_i)\ddot{\mathbf{f}}^{-1}(\hat{\mathbf{R}}_{is}, \theta_i)\dot{\mathbf{f}}(\hat{\mathbf{R}}_{is}, \theta_i) \\ &\quad + \frac{1}{2}\dot{\mathbf{f}}^T(\hat{\mathbf{R}}_{is}, \theta_i)\ddot{\mathbf{f}}^{-1}(\hat{\mathbf{R}}_{is}, \theta_i)\dot{\mathbf{f}}(\hat{\mathbf{R}}_{is}, \theta_i) \qquad (6.283) \\ &= f(\hat{\mathbf{R}}_{is}, \theta_i) - \frac{1}{2}\dot{\mathbf{f}}^T(\hat{\mathbf{R}}_{is}, \theta_i)\ddot{\mathbf{f}}^{-1}(\hat{\mathbf{R}}_{is}, \theta_i)\dot{\mathbf{f}}(\hat{\mathbf{R}}_{is}, \theta_i) \end{aligned}$$

From Theorem 6.8 $\dot{\mathbf{f}}(\hat{\mathbf{R}}_{is}, \theta_i) = \mathbf{O}(N^{-1/2)}$ *prob.* Also, since $\mathbf{P}_{iv}^H\hat{\mathbf{R}}_{is} = \mathbf{O}(N^{-1/2})$ *prob.*, which from the definition in Eq. (6.142), implies that $f(\hat{\mathbf{R}}_{is}, \theta_i) = O(N^{-1})$*prob.* The result of Eq. (6.278) immediately follows when these two facts are used in Eq. (6.283). ■

Form the above theorem, we can see that if the parameter K is estimated correctly, then for a large N, the value of $f(\hat{\mathbf{R}}_{is}, \hat{\theta}_i)$ is bounded by $O(N^{-1})$*prob.* We now examine the distribution of $f(\hat{\mathbf{R}}_{is}, \hat{\theta}_i)$. In particular, it has been shown in Section 6.7 that the use of CCD leads to optimal results in estimating the DOA. This is true even if $K' < K$ since the derivations of Lemmas 6.16a and 6.16b together with Theorem 6.16 are equally valid when the signals are coherent. Thus, we will confine our consideration on the UN-CLE criterion when CCD is employed. Accordingly, the UN-CLE criterion in Eq. (6.148), using estimated quantities, and when $k = K$ can be written as

$$f(\hat{\hat{\mathbf{R}}}_{is}, \hat{\theta}_i) = \mathrm{tr}\,\{\mathbf{B}_{iv}(\hat{\theta}_i)\left[\mathbf{B}_{iv}^H(\hat{\theta}_i)\hat{\boldsymbol{\Sigma}}_{ii}\mathbf{B}(\hat{\theta}_i)\right]^{-1}\mathbf{B}_{iv}^H(\hat{\theta}_i)\hat{\hat{\mathbf{R}}}_{is}\hat{\hat{\boldsymbol{\Gamma}}}^2\hat{\hat{\mathbf{R}}}_{is}^H \tag{6.284}$$

We have the following theorem.

Theorem 6.18. *For* K *being the number of signals,* K′ *the rank of the matrix* $\boldsymbol{\Sigma}_s$ *as defined in Eq. (6.274a), and* M_i *the number of sensors in the* i*th array, the quantity* $2Nf(\hat{\hat{\mathbf{R}}}_{is}, \hat{\theta}_i)$ *is of a* χ^2 *distribution with* $2K'(M_i - K) - K$ *degrees of freedom for large* N, *where* $f(\hat{\hat{\mathbf{R}}}_{is}, \hat{\theta}_i)$ *is obtained by Eq. (6.284).*

Proof. Define

$$\hat{\mathbf{z}}_i \triangleq \mathrm{vec}\,\{\tilde{\boldsymbol{\Gamma}}\hat{\hat{\mathbf{R}}}_{is}^H\mathbf{B}_{iv}(\mathbf{B}_{iv}^H\boldsymbol{\Sigma}_{ii}\mathbf{B}_{iv})^{-1/2}\} \tag{6.285a}$$

$$= \left[[(\mathbf{B}_{iv}^H\boldsymbol{\Sigma}_{ii}\mathbf{B}_{iv})^*]^{-1/2} \otimes \tilde{\boldsymbol{\Gamma}}\right]\mathrm{vec}\,(\hat{\hat{\mathbf{R}}}_{is}^H\mathbf{B}_{iv}) \tag{6.285b}$$

where Eq. (6A.7) in Appendix 6A has been used, and

$$\mathbf{h}_{ik} \triangleq 2\,\text{vec}\left[\left[(\mathbf{B}_{iv}^H \boldsymbol{\Sigma}_{ii} \mathbf{B}_{iv})^{-1/2} \frac{\partial \mathbf{B}_{iv}^H}{\partial \theta_k} \tilde{\mathbf{R}}_{is} \tilde{\boldsymbol{\Gamma}}\right]^H\right] \tag{6.286}$$

Let us also form the matrix

$$\mathbf{H}_i \triangleq [\mathbf{h}_{i1} \mathbf{h}_{i2} \cdots \mathbf{h}_{iK}]^H \tag{6.287}$$

Now, from Eq. (6.193), an approximation of the kth element in the gradient vector $\dot{\mathbf{f}}_i$, with the appropriate weigthing matrices inserted and also Eq. (6.183), is given by

$$\begin{aligned} \dot{f}_{ik} &\triangleq \frac{\partial f(\hat{\tilde{\mathbf{R}}}_{is}, \theta_i)}{\partial \theta_{ik}} \\ &\simeq 2\,\text{tr}\left\{\text{Re}\left[\mathbf{B}_{iv}(\mathbf{B}_{iv}^H \boldsymbol{\Sigma}_{ii} \mathbf{B}_{iv})^{-1} \dot{\mathbf{B}}_{ivk}^H \mathbf{R}_{is} \tilde{\boldsymbol{\Gamma}}^2 \hat{\tilde{\mathbf{R}}}_{is}^H\right]\right\} \end{aligned} \tag{6.288}$$

where $\dot{\mathbf{B}}_{ivk} \triangleq \dfrac{\partial \mathbf{B}_{iv}}{\partial \theta_k}$. Equation (6.288) can be re-written as

$$\dot{f}_{ik} \simeq 2\,\text{Re}\left\{\text{tr}\left[((\mathbf{B}_{iv}^H \boldsymbol{\Sigma}_{ii} \mathbf{B}_{iv})^{-1/2} \dot{\mathbf{B}}_{ivk}^H \tilde{\mathbf{R}}_{is} \tilde{\boldsymbol{\Gamma}}) \tilde{\boldsymbol{\Gamma}} \hat{\tilde{\mathbf{R}}}_{is}^H \mathbf{B}_{iv} (\mathbf{B}_{iv}^H \boldsymbol{\Sigma}_{ii} \mathbf{B}_{iv})^{-1/2})\right]\right\} \tag{6.289a}$$

$$= \text{Re}\,(\mathbf{h}_{ik}^H \hat{\mathbf{z}}_i) \tag{6.289b}$$

where the relations $\text{tr}\,(\mathbf{AC}) = \text{tr}\,(\mathbf{CA})$ and $\text{tr}\,(\mathbf{AC}) = \text{vec}^H(\mathbf{A}^H)\,\text{vec}\,(\mathbf{C})$ have been employed. Furthermore, comparing Eqs. (6.284) and (6.285a), we see that

$$f(\hat{\tilde{\mathbf{R}}}_{is}, \hat{\theta}_i) = \hat{\mathbf{z}}_i^H \hat{\mathbf{z}}_i \tag{6.290}$$

Substituting Eqs. (6.289) and (6.290) into Eq. (6.283) with the obvious change of notation showing that CCD has been employed, we can write

$$f(\hat{\tilde{\mathbf{R}}}_{is}, \hat{\theta}_i) = \hat{\mathbf{z}}_i^H \hat{\mathbf{z}}_i - \frac{1}{2}\left\{\text{Re}\,[\mathbf{H}_i \hat{\mathbf{z}}_i]^T (\ddot{\mathbf{f}}_i^{-1}) \text{Re}\,[\mathbf{H}_i \hat{\mathbf{z}}_i]\right\} \tag{6.291}$$

or,

$$\begin{aligned} f(\hat{\tilde{\mathbf{R}}}_{is}, \hat{\theta}_i) &= \left[\text{Re}\,(\hat{\mathbf{z}}_i^H)\ \text{Im}\,(\hat{\mathbf{z}}_i^H)\right] \begin{bmatrix} \text{Re}\,(\hat{\mathbf{z}}_i) \\ \text{Im}\,(\hat{\mathbf{z}}_i) \end{bmatrix} \\ &\quad - \frac{1}{2}\left[\text{Re}\,(\mathbf{H}_i)\text{Re}\,(\hat{\mathbf{z}}_i) - \text{Im}\,(\mathbf{H}_i)\text{Im}\,(\hat{\mathbf{z}}_i)\right]^T \left[\ddot{\mathbf{f}}_i\right]^{-1} \\ &\quad \left[\text{Re}\,(\mathbf{H}_i)\text{Re}\,(\hat{\mathbf{z}}_i) - \text{Im}\,(\mathbf{H}_i)\text{Im}\,(\hat{\mathbf{z}}_i)\right] \\ &= \left[\text{Re}\,(\hat{\mathbf{z}}_i^H)\ \text{Im}\,(\hat{\mathbf{z}}_i^H)\right] \left\{\mathbf{I} - \frac{1}{2}\begin{bmatrix} \text{Re}\,(\mathbf{H}_i^H) \\ -\text{Im}\,(\mathbf{H}_i^H) \end{bmatrix} \left[\ddot{\mathbf{f}}_i\right]^{-1} [\text{Re}\,(\mathbf{H}_i) - \text{Im}\,(\mathbf{H}_i)]\right\} \\ &\quad \begin{bmatrix} \text{Re}\,(\hat{\mathbf{z}}_i) \\ \text{Im}\,(\hat{\mathbf{z}}_i) \end{bmatrix} \end{aligned} \tag{6.292}$$

Also, from Eq. (6.194), with the appropriate weighting matrices inserted, the (kl)th element of the Hessian matrix $\ddot{\mathbf{f}}_i$ is given by

$$\left[\ddot{\mathbf{f}}_i\right]_{kl} \triangleq \frac{\partial^2 \bar{f}_i}{\partial\theta_k \partial\theta_l} = 2\,\mathrm{tr}\left\{\mathrm{Re}\left[\dot{\mathbf{B}}_{ivk}(\mathbf{B}_{iv}^H \boldsymbol{\Sigma}_{ii}\mathbf{B}_{iv})^{-1}\dot{\mathbf{B}}_{ivk}^H \tilde{\mathbf{R}}_{is}\tilde{\boldsymbol{\Gamma}}^2\tilde{\mathbf{R}}_{is}\right]\right\} \tag{6.293}$$

Comparing Eqs. (6.286) and (6.293) we can write

$$\left[\ddot{\mathbf{f}}_i\right]_{kl} = \frac{1}{2}\,\mathrm{Re}\left[\mathbf{h}_{ik}^H \mathbf{h}_{il}\right] \tag{6.294}$$

Or, in other words, the complete Hessian matrix can be written as

$$\ddot{\mathbf{f}}_i = \frac{1}{2}\,\mathrm{Re}\left\{\mathbf{H}_i\mathbf{H}_i^H\right\} = \frac{1}{2}\left[\mathrm{Re}\,(\mathbf{H}_i)\mathrm{Re}\,(\mathbf{H}_i)^T + \mathrm{Im}\,(\mathbf{H}_i)\,\mathrm{Im}\,(\mathbf{H}_i)^T\right] \tag{6.295}$$

Replacing $\ddot{\mathbf{f}}_i$ by $\ddot{\bar{\mathbf{f}}}_i$ in Eq. (6.292) does not affect the asymptotic value, as has been shown in section 6.6. Here we can write

$$f(\hat{\tilde{\mathbf{R}}}_{is}, \hat{\tilde{\theta}}_i) = \left[\mathrm{Re}\,(\hat{\mathbf{z}}_i)^T\ \mathrm{Im}\,(\hat{\mathbf{z}}_i)^T\right]\left[\mathbf{I} - \tilde{\mathbf{Q}}_{is}\right]\begin{bmatrix}\mathrm{Re}\,(\hat{\mathbf{z}}_i)\\ \mathrm{Im}\,(\hat{\mathbf{z}}_i)\end{bmatrix} \tag{6.296}$$

where

$$\tilde{\mathbf{Q}}_{is} = \begin{bmatrix}\mathrm{Re}\,(\mathbf{H}_i^T)\\ -\mathrm{Im}\,(\mathbf{H}_i^T)\end{bmatrix}\left[\mathrm{Re}\,(\mathbf{H}_i)\mathrm{Re}\,(\mathbf{H}_i)^T + \mathrm{Im}\,(\mathbf{H}_i)\,\mathrm{Im}\,(\mathbf{H}_i)^T\right]^{-1} \cdot \left[\mathrm{Re}\,(\mathbf{H}_i) - \mathrm{Im}\,(\mathbf{H}_i)\right] \tag{6.297}$$

From the definition of $\mathbf{H}_i$ in Eq. (6.287) together with Eq. (6.297), it can easily be observed that $\tilde{\mathbf{Q}}_{is}$ is a projector of rank K. Now, from Theorem 6.6, with the obvious change of notation denoting the employment of CCD together with the replacement of $\mathbf{Y}_{iv}$ by $\mathbf{B}_{iv}$, we have $\mathrm{vec}(\hat{\tilde{\mathbf{P}}}_{is}\mathbf{B}_{iv})$ is asymptotically Gaussian with zero mean and covariance matrix

$$E\left[\mathrm{vec}\,(\hat{\tilde{\mathbf{P}}}_{is}\mathbf{B}_{iv})\mathrm{vec}^H(\hat{\tilde{\mathbf{P}}}_{is}\mathbf{B}_{iv})\right] = \frac{1}{N}\left[\mathbf{B}_{iv}^H\boldsymbol{\Sigma}_{ii}\mathbf{B}_{iv}\right]^T \otimes \left[\tilde{\mathbf{L}}_{is}\tilde{\boldsymbol{\Gamma}}^{-2}\tilde{\mathbf{L}}_{is}^H\right] \tag{6.298}$$

Since $\mathrm{vec}(\hat{\tilde{\mathbf{R}}}_{is}^H\mathbf{B}_{iv}) = \mathrm{vec}\,(\hat{\tilde{\mathbf{R}}}_{is}^H\hat{\tilde{\mathbf{L}}}_{is}\hat{\tilde{\mathbf{R}}}_{is}^H\mathbf{B}_{iv}) = \mathrm{vec}\,(\tilde{\mathbf{R}}_{is}^H\mathbf{B}_{iv}) + \mathbf{o}(N^{-1/2})\,prob.$, using Eq. (6A.7) in Appendix 6A, we can write asymptotically

$$\mathrm{vec}\,(\hat{\tilde{\mathbf{R}}}_{is}^H\mathbf{B}_{iv}) = (\mathbf{I}\otimes\tilde{\mathbf{R}}_{is}^H)\mathrm{vec}\,(\hat{\tilde{\mathbf{P}}}_{is}\mathbf{B}_{iv}) + \mathbf{o}(N^{-1/2}) \tag{6.299}$$

Hence, $\mathrm{vec}(\hat{\tilde{\mathbf{R}}}_{is}^H\mathbf{B}_{iv})$ is asymptotically Gaussian distributed with zero mean and covariance matrix

$$E\left[\mathrm{vec}\,(\hat{\tilde{\mathbf{R}}}_{is}^H\mathbf{B}_{iv})\,\mathrm{vec}^H(\hat{\tilde{\mathbf{R}}}_{is}^H\mathbf{B}_{iv})\right] = (\mathbf{I}\otimes\tilde{\mathbf{R}}_{is}^H)E\left[\mathrm{vec}\,(\hat{\tilde{\mathbf{P}}}_{is}^H\mathbf{B}_{iv})\,\mathrm{vec}^H(\hat{\tilde{\mathbf{P}}}_{is}^H\mathbf{B}_{iv})\right] (\mathbf{I}\otimes\tilde{\mathbf{R}}_{is}^H)^H \tag{6.300}$$

Using Eq. (6.298) together with Eq. (6A.4) in Appendix 6A, we obtain

$$\begin{aligned} E&\left[\mathrm{vec}\,(\hat{\tilde{\mathbf{R}}}_{is}^H\mathbf{B}_{iv})\,\mathrm{vec}^H(\hat{\tilde{\mathbf{R}}}_{is}^H\mathbf{B}_{iv})\right] \\ &= \frac{1}{N}\left[\mathbf{B}_{iv}^H\boldsymbol{\Sigma}_{ii}\mathbf{B}_{iv}\right]^T \otimes \left[\tilde{\mathbf{R}}_{is}^H\tilde{\mathbf{L}}_{is}\tilde{\boldsymbol{\Gamma}}^{-2}\tilde{\mathbf{L}}_{is}^H\tilde{\mathbf{R}}_{is}\right] \\ &= \frac{1}{N}\left[\mathbf{B}_{iv}^H\boldsymbol{\Sigma}_{ii}\mathbf{B}_{iv}\right]\otimes\tilde{\boldsymbol{\Gamma}}^{-2} \end{aligned} \tag{6.301}$$

Using Eqs. (6.285b) and (6.301), we obtain

$$E\left[\hat{\mathbf{z}}_i\hat{\mathbf{z}}_i^H\right] = \{(\mathbf{B}_{iv}^H\boldsymbol{\Sigma}_{ii}\mathbf{B}_{iv})^{-1/2}\otimes\tilde{\boldsymbol{\Gamma}}\}E\left[\text{vec}\,(\hat{\tilde{\mathbf{R}}}_{is}^H\mathbf{B}_{iv})\,\text{vec}\,(\hat{\tilde{\mathbf{R}}}_{is}^H\mathbf{B}_{iv})\right]\{(\mathbf{B}_{iv}^H\boldsymbol{\Sigma}_{ii}\mathbf{B}_{iv})^{-1/2}\otimes\tilde{\boldsymbol{\Gamma}}\}^H$$
$$= \frac{1}{N}\{(\mathbf{B}_{iv}^H\boldsymbol{\Sigma}_{ii}\mathbf{B}_{iv})^{-1/2}\otimes\tilde{\boldsymbol{\Gamma}}\}\{(\mathbf{B}_{iv}^H\boldsymbol{\Sigma}_{ii}\mathbf{B}_{iv})\otimes\tilde{\boldsymbol{\Gamma}}^{-2}\}\{(\mathbf{B}_{iv}^H\boldsymbol{\Sigma}_{ii}\mathbf{B}_{iv})^{-1/2}\otimes\tilde{\boldsymbol{\Gamma}}\}^H$$
$$= \frac{1}{N}\mathbf{I}_{M_i-K}\otimes\mathbf{I}_{K'} \tag{6.302}$$

where $\mathbf{I}_m$ denotes an identity matrix of size m and Eq. (6A.4) in Appendix 6A has been used. Therefore, we can conclude that $\left[\text{Re}\,(\hat{\mathbf{z}}_i^T\;\;\text{Im}\,(\hat{\mathbf{z}}_i^T)\right]^T$ is asymptotically Gaussian distributed with zero mean and covariance matrix

$$E\begin{bmatrix}\text{Re}\,(\hat{\mathbf{z}}_i)\\ \text{Im}\,(\hat{\mathbf{z}}_i)\end{bmatrix}\left[\text{Re}\,(\hat{\mathbf{z}}_i^T)\;\;\text{Im}\,(\hat{\mathbf{z}}_i^T)\right] = \frac{1}{2N}\mathbf{I}_{2K'(M_i-K)} \tag{6.303}$$

Using Eq. (6.303) in Eq. (6.296), the result of the theorem follows. ■

The Likelihood Function

The above two theorems present to us some essential properties of the criterion $f(\hat{\tilde{\mathbf{R}}}_{is}, \hat{\theta}_i)$ which will be used to develop an efficient method to determine the number of signals in unknown correlated noise, for both incoherent and coherent signals, that is, $K' < K$. This method requires the knowledge of K'. Thus, we need to develop methods which are capable of determining K', that is, capable of determining the number of signals in correlated noise when the signals are all incoherent. To this end, we first need to examine the function $\ell(\mathbf{X} \mid \Omega_k)$ which denotes the log-likelihood function [77] of the observed data matrix $\mathbf{X}$ having parameters in Ω_k which is a parameter space containing $\boldsymbol{\Sigma}$ and θ_i under the hypothesis $\mathcal{H}_k$ (there are k signals). This log-likelihood function is essential to the development of the methods.

To obtain the log-likelihood function, we start with the joint PDF of the received data. From the signal model presented in Section 6.3, let $\mathbf{x}(1), \cdots, \mathbf{x}(N)$ denote the N independent samples of $\mathbf{x}(n)$ where

$$\mathbf{x}(n) = \begin{bmatrix}\mathbf{x}_1(n)\\ \mathbf{x}_2(n)\end{bmatrix}, \quad n = 1, \cdots, N \tag{6.304}$$

Then the joint PDF of $\mathbf{x}(1), \cdots, \mathbf{x}(N)$ is given by

$$p(\mathbf{X} \mid \boldsymbol{\Sigma}) \triangleq p(\mathbf{x}(1), \cdots, \mathbf{x}(N) \mid \boldsymbol{\Sigma})$$
$$= \{\pi^{(M_1+M_2)}\det(\boldsymbol{\Sigma})\}^{-N}\exp\left\{-\sum_{n=1}^{N}\mathbf{x}^H(n)\boldsymbol{\Sigma}^{-1}\mathbf{x}(n)\right\} \tag{6.305}$$

The corresponding log-likelihood function of the observed data is, to within a constant, given by

$$\ell \propto -N\left\{\ln\det(\boldsymbol{\Sigma}) + \frac{1}{N}\sum_{n=1}^{N}\mathbf{x}^H(n)\boldsymbol{\Sigma}^{-1}\mathbf{x}(n)\right\} \tag{6.306}$$

The form of the log-likelihood function in Eq. (6.306) is not appropriate for application since it has to be maximized. To this end we present the following theorem:

Theorem 6.19. *Tha maximization of ℓ in Eq. (6.306) is given by*

$$\max_{\Sigma} N\left\{\ln\det(\Sigma)+\frac{1}{N}\sum_{n=1}^{N}\mathbf{x}^H(n)\Sigma^{-1}\mathbf{x}(n)\right\} \propto -N\ln\left\{\prod_{m=1}^{k}\left(1-\hat{\hat{\gamma}}_m^2\right)\right\} \tag{6.307}$$

$$\propto -N\ln\left\{\prod_{m=k+1}^{M_1}\left(1-\hat{\hat{\gamma}}_m^2\right)\right\} \tag{6.308}$$

where Σ is given by Eq. (6.34) in which rank (Σ_{12}) is assumed to be k, *and $\hat{\hat{\gamma}}_{\mathrm{m}}$ is the* m*th sample canonical correlation coefficient.*

Proof. Assuming, without loss of generality, $M_1 \leq M_2$, we let

$$\Xi_a = \Sigma_{22} \tag{6.309a}$$

$$\Xi_b = \Sigma_{11}-\Sigma_{12}\Sigma_{22}^{-1}\Sigma_{21} = \Sigma_{11}-\Sigma_{12}\Sigma_{22}^{-1}\Sigma_{12}^H \tag{6.309b}$$

$$\Xi_c = \Sigma_{12}\Sigma_{22}^{-1} \tag{6.309c}$$

We note that there is a one-to-one mapping between $\{\Sigma_{11}, \Sigma_{12}, \Sigma_{22}; \text{rank}\,(\Sigma_{12}) = k\}$ and $\{\Xi_a, \Xi_b, \Xi_c; \text{rank}\,(\Xi_c) = k\}$. From Eqs. (6A.24) and (6A.25) in Appendix 6A, and using Eqs. (6.309), we can rewrite Eq. (6.306) as

$$\ell \propto -N(\ln\det(\Xi_a)+\ln\det(\Xi_b)$$

$$+\frac{1}{N}\sum_{n=1}^{N}\{[\mathbf{x}_2(n)\Xi_a^{-1}\mathbf{x}_2(n)]+[\mathbf{x}_1(n)-\Xi_c\mathbf{x}_2(n)]^H\,\Xi_b^{-1}\,[\mathbf{x}_1(n)-\Xi_c\mathbf{x}_2(n)]\}) \tag{6.310}$$

Minimization of the negative log-likelihood function in Eq. (6.310) with respect to Ξ_a and Ξ_b yields the estimates [96]

$$\hat{\Xi}_a = \frac{1}{N}\sum_{n=1}^{N}\mathbf{x}_2(n)\mathbf{x}_2^H(n) \triangleq \hat{\Sigma}_{22} \tag{6.311}$$

and

$$\hat{\Xi}_b = \frac{1}{N}\sum_{n=1}^{N}[\mathbf{x}_1(n)-\Xi_c\mathbf{x}_2(n)]\,[\mathbf{x}_1(n)-\Xi_c\mathbf{x}_2(n)]^H$$

$$= \hat{\Sigma}_{11}-\hat{\Sigma}_{12}\Xi_c^H-\Xi_c\hat{\Sigma}_{12}^H+\Xi_c\hat{\Sigma}_{22}\Xi_c^H \tag{6.312}$$

provided that the right sides of Eqs. (6.311) and (6.312) are positive definite. Clearly $\hat{\Sigma}_{22}$ in Eq. (6.311) is positive definite for $N \geq M_2$. That the right side of Eq. (6.312) also satisfies the positive definite condition will be verified later after the quantity $\hat{\Xi}_c$ has been obtained. Substituting Eqs. (6.311) and (6.312) into (6.310), we obtain the concentrated log-likelihood function

$$\ell \propto -N\{\ln\det(\hat{\Sigma}_{22}) + \ln\det(\hat{\Xi}_b)\} \propto -N\ln\det(\hat{\Xi}_b) \tag{6.313}$$

To obtain $\hat{\Xi}_c$, we maximize Eq. (6.313), or, equivalently minimize $\ln\det(\hat{\Xi}_b)$ with respect to Ξ_c with the constraint that $\text{rank}(\Xi_c) = k$. Let

$$\Xi_c = \mathbf{Z}_1\mathbf{Z}_2^H \tag{6.314}$$

where $\mathbf{Z}_1$ and $\mathbf{Z}_2$ are $M_1 \times k$ and $M_2 \times k$ matrices both of which are of rank k. By using Eq. (6.314) the constrained minimization of $\ln\det(\hat{\Xi}_b)$ with respect to Ξ_c is tranformed to an unconstrained minimization with respect to $\mathbf{Z}_1$ and $\mathbf{Z}_2$, and then the resulting matrices $\mathbf{Z}_1$ and $\mathbf{Z}_2$ are tested to see if the rank requirements are satisfied. Substituting Eq. (6.314) into Eq. (6.312), we have

$$\begin{aligned}\hat{\Xi}_b = &\left[\mathbf{Z}_1 - \hat{\Sigma}_{12}\mathbf{Z}_2(\mathbf{Z}_2^H\hat{\Sigma}_{22}\mathbf{Z}_2)^{-1}\right]\left[\mathbf{Z}_2^H\hat{\Sigma}_{22}\mathbf{Z}_2\right]\left[\mathbf{Z}_1 - \hat{\Sigma}_{12}\mathbf{Z}_2(\mathbf{Z}_2^H\hat{\Sigma}_{22}\mathbf{Z}_2)^{-1}\right]^H \\ &+\left[\hat{\Sigma}_{11} - \hat{\Sigma}_{12}\mathbf{Z}_2(\mathbf{Z}_2^H\hat{\Sigma}_{22}\mathbf{Z}_2)^{-1}\mathbf{Z}_2^H\hat{\Sigma}_{12}^H\right]\end{aligned} \tag{6.315}$$

Clearly, $\ln\det(\hat{\Xi}_b)$ is minimized with respect to $\mathbf{Z}_1$ at

$$\hat{\mathbf{Z}}_1 = \hat{\Sigma}_{12}\mathbf{Z}_2(\mathbf{Z}_2^H\hat{\Sigma}_{22}\mathbf{Z}_2)^{-1} \tag{6.316}$$

Substituting this value of $\hat{\mathbf{Z}}_1$ into Eq. (6.315), and the resulting $\hat{\Xi}_b$ in Eq. (6.313), we can rewrite Eq. (6.313) as

$$\begin{aligned}\ell &\propto -N\ln\det\{\hat{\Sigma}_{11} - \hat{\Sigma}_{12}\mathbf{Z}_2(\mathbf{Z}_2^H\hat{\Sigma}_{22}\mathbf{Z}_2)^{-1}\mathbf{Z}_2^H\hat{\Sigma}_{12}^H\} \\ &\propto -N\ln\det\{\mathbf{I} - \hat{\Sigma}_{11}^{-1}\hat{\Sigma}_{12}\mathbf{Z}_2(\mathbf{Z}_2^H\hat{\Sigma}_{22}\mathbf{Z}_2)^{-1}\mathbf{Z}_2^H\hat{\Sigma}_{12}^H\}\end{aligned} \tag{6.317}$$

We now define the matrix

$$\hat{\mathbf{J}} = \hat{\Sigma}_{11}^{-1/2}\hat{\Sigma}_{12}\hat{\Sigma}_{22}^{-1/2} \tag{6.318}$$

From Eq. (6.50), it is obvious that

$$\hat{\mathbf{J}}\hat{\mathbf{J}}^H = \hat{\hat{\mathbf{U}}}_1(\hat{\hat{\Gamma}}_0\hat{\hat{\Gamma}}_0^H)\hat{\hat{\mathbf{U}}}_1^H \tag{6.319a}$$

and

$$\hat{\mathbf{J}}^H\hat{\mathbf{J}} = \hat{\hat{\mathbf{U}}}_2(\hat{\hat{\Gamma}}_0^H\hat{\hat{\Gamma}}_0)\hat{\hat{\mathbf{U}}}_2^H \tag{6.319b}$$

Recall that $\hat{\hat{\Gamma}}_0$ is an $M_1 \times M_2$ matrix with nonzero elements being equal to $\hat{\hat{\gamma}}_m$ only at the (mm)th position, thus we can say that $(\hat{\mathbf{J}}\hat{\mathbf{J}}^H)$ has eigenvalues $\{\hat{\hat{\gamma}}_1^2 \geq \hat{\hat{\gamma}}_2^2 \geq \cdots \geq \hat{\hat{\gamma}}_{M_1}^2\}$ such that $1 > \hat{\hat{\gamma}}_m^2 \geq 0$ for $m = 1, \cdots, M_1$ with corresponding eigenvectors being the columns of $\hat{\hat{\mathbf{U}}}_1$. Similarly, the matrix $(\hat{\mathbf{J}}^H\hat{\mathbf{J}})$ has M_2 eigenvalues $\{\hat{\hat{\gamma}}_1^2 \geq \hat{\hat{\gamma}}_2^2 \geq \cdots \geq \hat{\hat{\gamma}}_{M_1}^2 \geq 0 = \cdots = 0\}$, the first M_1 of which are identical to those of $(\hat{\mathbf{J}}\hat{\mathbf{J}}^H)$ while the rest are zero. The corresponding M_2 eigenvectors are the column vectors of $\hat{\hat{\mathbf{U}}}_2$. Also we define an $M_2 \times k$ unitary matrix $\mathbf{V}$, that is, $\mathbf{V}^H\mathbf{V} = \mathbf{I}$, such that

$$\mathbf{V}^H = (\mathbf{Z}_2^H\hat{\Sigma}_{22}\mathbf{Z}_2)^{-1/2}\mathbf{Z}_2^H\hat{\Sigma}_{22}^{1/2} \tag{6.320}$$

Then, using Eq. (6A.27) that $\det(\mathbf{I} - \mathbf{AB}) = \det(\mathbf{I} - \mathbf{BA})$ for any conformable matrices $\mathbf{A}$ and $\mathbf{B}$, we can rewrite Eq. (6.317) as

$$\ell \propto -N \ln \det\{\mathbf{I} - \mathbf{V}^H(\hat{\mathbf{J}}^H \hat{\mathbf{J}})\mathbf{V}\} \tag{6.321}$$

Let $\lambda_1(\mathbf{Z}_2) \geq \lambda_2(\mathbf{Z}_2) \geq \cdots \geq \lambda_k(\mathbf{Z}_2)$ be the eigenvalues of the $k \times k$ matrix $\mathbf{V}^H(\hat{\mathbf{J}}^H \hat{\mathbf{J}})\mathbf{V}$. Then, from the Poincaré separation theorem [97]

$$\lambda_m(\mathbf{Z}_2) \leq \hat{\hat{\gamma}}_m^2, \quad m = 1, \cdots, k \tag{6.322}$$

with equality in (6.322) achieved when

$$\mathbf{V} = \hat{\hat{\mathbf{U}}}_{2s} \tag{6.323}$$

where $\hat{\hat{\mathbf{U}}}_{2s}$ consists of the first K columns of $\hat{\hat{\mathbf{U}}}_2$, that is, the first k eigenvectors of $\hat{\mathbf{J}}^H\hat{\mathbf{J}}$ associated with the eigenvalues $\{\hat{\hat{\gamma}}_1^2, \cdots, \hat{\hat{\gamma}}_k^2\}$. It can easily be seen from Eq. (6.320) that, to satisfy Eq. (6.323), we have

$$\hat{\mathbf{Z}}_2 = \hat{\mathbf{\Sigma}}_{22}^{-1/2}\hat{\hat{\mathbf{U}}}_{2s} = \hat{\hat{\mathbf{L}}}_{2s} \tag{6.324}$$

Since Eq. (6.321) can be rewritten as

$$\ell \propto -N \ln\{\prod_{m=1}^{k} [1 - \lambda_m(\mathbf{Z}_2)]\} \tag{6.325}$$

and since $\hat{\mathbf{Z}}_2 = \hat{\hat{\mathbf{L}}}_{2s}$ maximizes all $\{\lambda_m(\mathbf{Z}_2); m = 1, \cdots, k\}$, then Eq. (6.324) yields the condition for which ℓ is maximized. Combining Eqs. (6.316) and (6.324), we obtain

$$\hat{\mathbf{\Xi}}_c = \hat{\mathbf{\Sigma}}_{12}\hat{\hat{\mathbf{L}}}_{2s}\hat{\hat{\mathbf{L}}}_{2s}^H \tag{6.326}$$

It can easily be seen that both $\hat{\mathbf{Z}}_1 = \hat{\mathbf{\Sigma}}_{12}\hat{\hat{\mathbf{L}}}_{2s}$ and $\hat{\mathbf{Z}}_2 = \hat{\hat{\mathbf{L}}}_{2s}$ are of rank k rendering $\hat{\mathbf{\Xi}}_c$ being of rank k. Furthermore, substituting the ML estimate $\hat{\mathbf{\Xi}}_c$ of Eq. (6.326) into Eq. (6.312) we have

$$\begin{aligned} \hat{\mathbf{\Xi}}_b\Big|_{\hat{\mathbf{\Xi}}_c} &= \hat{\mathbf{\Sigma}}_{11} - \hat{\mathbf{\Sigma}}_{12}\hat{\mathbf{\Sigma}}_{22}^{-1/2}\hat{\hat{\mathbf{U}}}_2\hat{\hat{\mathbf{U}}}_2^H\hat{\mathbf{\Sigma}}_{22}^{-1/2}\hat{\mathbf{\Sigma}}_{12}^H \\ &\geq \hat{\mathbf{\Sigma}}_{11} - \hat{\mathbf{\Sigma}}_{12}\hat{\mathbf{\Sigma}}_{22}^{-1}\hat{\mathbf{\Sigma}}_{12}^H \end{aligned} \tag{6.327}$$

The right-hand quantity of the inequality (6.327) is the Schur complement of $\hat{\mathbf{\Sigma}}$ which is positive definite with probability 1; hence $\hat{\mathbf{\Xi}}_b\Big|_{\hat{\mathbf{\Xi}}_c}$ is also positive definite.

The minimum value of the negative log-likelihood function in Eq. (6.325) is achieved when equality holds in (6.322), that is,

$$\ell_{\max} \propto -N \ln\left\{\prod_{m=1}^{k}\left[1 - \hat{\hat{\gamma}}_m^2\right]\right\} \tag{6.328a}$$

However, $\det(\mathbf{I} - \hat{\mathbf{J}}\hat{\mathbf{J}}^H) = \prod_{m=1}^{M_1} (1 - \hat{\hat{\gamma}}_m^2) = \text{constant}$. Therefore, we can also write that

$$\ell_{\max} \propto N \ln\{ \prod_{m=k+1}^{M_1} \left[1 - \hat{\hat{\gamma}}_m^2 \right] \} \tag{6.328b}$$

and the result of the theorem follows. ■

The results of the above theorem were first derived in [98]. The proof presented here is more rigorous and contains corrections of some errors therein. Also, from the standpoint of computation, since $M_1 \gg k$ in general Eq. (6.328a) appears to be preferred compared to (6.328b).

Information-Theoretic Criteria

To determine the number of signals arriving at the array in a white noise environment, various methods have been proprosed [14–21]. In particular, the information theoretic criteria which were first proposed by Wax and Kailath [17] using the Akaike information criterion (AIC) [14] and the minimum description length (MDL) criterione introduced by Rissanen [15] and Schwartz [16] have attracted the attention of many researchers in the field. Information-theoretic criteria of this kind, applicable only when the signals are all incoherent, consist of two parts: The first is the likelihood function which represents the information gained from the received data. The second is the penalty term which is a function of the number of snapshots and is proportional to the number of free parameters in the probabilistic model, representing the penalty for the uncertainty introduced by the unknown parameters. Thus, the information-theoretic criteria hava a general expression such that

$$C = -\max_{\Omega_k} \ln p(\mathbf{X} \mid \Omega_k) + a(N)N_f \tag{6.329}$$

where $p(.)$ is a family of conditional PDF dependent on the assumed number of signals k, $\mathbf{X}$ is the given set of N observed data, Ω_k is the parameter space containing the ML estimates of the parameters in the model, and N_f is the number of free parameters in Ω_k. $a(N)$ is the coefficient of the penalty function dependent on the number of snapshots N. For the AIC, $a(N) = 1$; for the MDL criterion, $a(N) = \frac{1}{2}\ln N$. A number of modified criteria have also been proposed to improve the statistical performance of the AIC and the MDL methods [19, 21]. All of these are, of course, only applicable under the assumption of spatially white noise.

To apply the concept of information-theoretic criteria to the determination of the number of signals in unknown spatially correlated noise, we first employ the results in Theorem 6.19 so that the maximized likelihood function in our signal can be employed in the general form of Eq. (6.329). We then proceed to obtain a penalty term which depends on the number of free parameters in the probabilistic model. Hence, we need to derive N_f, the number of free parameters.

Since the submatrix $\mathbf{\Sigma}_{12}$ in our model of Eq. (6.34) supplies all information needed for the determination of the number of signals, we can obtain the number of free parameters by examining the number of free parameters in $\mathbf{\Sigma}_{12}$. To facilitate this, we perform a singular value decomposition (SVD) on $\mathbf{\Sigma}_{12}$, assumed to be of rank k, such that

$$\mathbf{\Sigma}_{12} = \bar{\mathbf{U}}_1 \bar{\mathbf{\Gamma}}_0 \bar{\mathbf{U}}_2^H \tag{6.330}$$

where $\bar{\mathbf{U}}_1$ and $\bar{\mathbf{U}}_2$ are, respectively, $(M_1 \times k)$ and $(M_2 \times k)$ unitary matrices. We note that Eq. (6.330) is a special case of Eq. (6.38) when $\mathbf{\Pi}_1 = \mathbf{I}_{M_1}$ and $\mathbf{\Pi}_2 = \mathbf{I}_{M_2}$. Equation (6.330) yields k nonzero singular values together with two unitary matrices which, altogether, have $[2(M_1 + M_2)k + k]$ parameters. The factor 2 is due to the fact that the elements in the unitary matrices are complex. However, not all these parameters are freely adjustable. The normality and the mutual orthogonality of the k vectors in each of $\bar{\mathbf{U}}_1$ and $\bar{\mathbf{U}}_2$ imposes $[2k + 2k(k-1)] = 2k^2$ constraints on the elements of the vectors altogether. Hence, the dimension of the parameter set spanned by the elements of $\bar{\mathbf{\Gamma}}_0$, $\bar{\mathbf{U}}_1$ and $\bar{\mathbf{U}}_2$ in Eq. (6.330) is reduced to

$$2(M_1 + M_2)k + k - 2k^2 = 2k(M_1 + M_2 - k) + k \tag{6.331}$$

Also, the SVD in Eq. (6.330) is not unique. Indeed, all triplets $(\bar{\mathbf{U}}_1\mathbf{\Phi}, \bar{\mathbf{\Gamma}}_0, \bar{\mathbf{U}}_2\mathbf{\Phi})$ where $\mathbf{\Phi}$ is an arbitrary diagonal and unitary $k \times k$ matrix, will yield valid SVD of $\mathbf{\Sigma}_{12}$. In order to eliminate the nonuniqueness induced by the arbitrary $\mathbf{\Phi}$, we can constrain one element in each of the columns of $\bar{\mathbf{U}}_1$ to be of real value. This imposes k additional real-valued equations that must be satisfied by the elements of $\bar{\mathbf{U}}_1$ and we obtain

$$N_f = 2k(M_1 + M_2 - k) \tag{6.332}$$

If we employ the same penalty functions as those used by the AIC and the MDL criterion, then referring to Eq. (6.329), the penalty terms for the new criteria for unknown correlated noise should, respectvely, be

$$P_0 = N_f = 2k(M_1 + M_2 - k) \tag{6.333a}$$

$$P_1 = k(M_1 + M_2 - k) \ln N \tag{6.333b}$$

These penalty terms are obtained based on the various criteria of optimization. P_0 is a result of maximizing the Boltzmann's entropy, whereas P_1 is obtained by minimizing the description length of the code word associated with modeling the data. However, it is known [47, 49, 52] that while the AIC tends to underpenalize, the MDL criterion tends to overpenalize. As a result, the AIC yields a relatively large probability of overestimation (false alarm) of the number of signals even when the SNR is high, whereas the MDL criterion underestimates (misses) the number of signals when the SNR is low. Realizing that the parameter space used in the original development [17] of the AIC and MDL criterion in array processing contains a huge number of nuisance parameters which considerably reduce the estimation accuracy of the unknown relevant parameters, the use of the marginal distribution of sample eigenvalues has been proposed [19] resulting in new information theoretic-criteria with different maximized likelihood functions and penalty functions. These new criteria moderate the overestimation and underestimation suffered by the AIC and the MDL criterion respectively. Correction terms have been incorporated into the penalty functions of Eqs. (6.333a) and (6.333b) such that the moderated penalty functions are, respectively, given by

$$P_2 = \frac{1}{2}(N_f - 2k)\ln N + k + \frac{1}{2}\sum_{i=1}^{M_1-k} \ln \Gamma(i) \qquad (6.333c)$$

and

$$P_3 = \frac{1}{2}(\ln N)N_f + 2\sum_{i=1}^{M_1-k} \ln \Gamma(i) \qquad (6.333d)$$

where $\Gamma(i)$ denotes the gamma function.

Using the results of Eq. (6.308) in Eq. (6.329) together with these penalty functions, a set of four different information-theoretic criteria for the determination of the number of signals in unknown spatially correlated noise can be proposed such that

$$C_i(k) = N \ln \prod_{m=1}^{k}(1 - \hat{\tilde{\gamma}}_m^2) + P_i \quad i = 0, \cdots, 3 \qquad (6.334a)$$

$$= -N \ln \prod_{m=k+1}^{M_1}(1 - \hat{\tilde{\gamma}}_m^2) + P_i \quad i = 0, \cdots, 3 \qquad (6.334b)$$

The above set of criteria called the information-theoretic criteria (ITC) was first proposed in [98] in which an error occurred in the derivation of the number of free parameters N_f the correct value of which is given by Eq. (6.332) above. In practical situations, the difference between using the correct or erroneous values of N_f is usually small. However, the use of the different penalty functions in Eqs. (6.333) can lead to significant differences in the performance of the criteria. Indeed, $C_0(k)$ in which the penalty term used is given by Eq. (6.333a) has resulted in rather inferior performance [98]. The relative performance of these information-theoretic criteria and other criteria will be shown later. At present, it is important to realize that in the original derivation of the AIC and the MDL criterion, the likelihood function was maximized without any constraint. However, the result in Theorem 6.19 in which the likelihood function is maximized is performed under the constraint that rank $(\Sigma_{12}) = k$. Thus, the ITC proposed in Eq. (6.334) are not standard and while they are still applicable, they may result in the deterioration of the performance.

Methods Based on Constant False Alarm Rate

The Neyman-Pearson Criterion of detection [77, 87, 101] constrains the probability of false alarm to be less than or equal to a certain chosen value and designs a test to minimize the probability of missing. This leads to a likelihood ratio test in which the threshold τ is chosen such that the constraint on the probability of false alarm is satisfied. In the following, based on the Neyman-Pearson principle, we are going to develop two test procedures for the determination of the number of signals in unknown correlated noise. The first method, called CCT, is for the determination of incoherent signals only, that is, $K' = K$, whereas the second, called PARADE, is for both correlated and uncorrelated signals, that is, $K' \leq K$.

The CCT Likelihood Ratio Test

We now establish a procedure for the determination of the number of signals assuming that they are incoherent, that is, the determination of the rank of Σ_{12} assuming that $K' = K$. This procedure is based on a likelihood ratio (LR) test in which the LR is defined as

$$\rho_k = \frac{\max_{\Omega_{k-1}} \Lambda(\mathbf{X} \mid \Omega_{k-1})}{\max_{\Omega_K} \Lambda(\mathbf{X} \mid \Omega_k)} \tag{6.335}$$

where $\Lambda(\mathbf{X} \mid \Omega_k)$ is the likelihood function of the observed data matrix $\mathbf{X}$ when the hypothesis $\mathcal{H}_k$ is true. From Theorem 6.19, we have seen that

$$\max_{\Omega_K} \Lambda(\mathbf{X} \mid \Omega_k) = c_0 \left[\prod_{m=k+1}^{M_1} (1 - \hat{\hat{\gamma}}_m^2) \right]^N \tag{6.336}$$

where $\hat{\hat{\gamma}}_m$ is the mth estimated canonical coefficients. Using this result, the LR in Eq. (6.335) can be written as

$$\rho_k = \frac{\max_{\Omega_{k-1}} \Lambda(\mathbf{X} \mid \Omega_{k-1})}{\max_{\Omega_K} \Lambda(\mathbf{X} \mid \Omega_k)} = \left[1 - \hat{\hat{\gamma}}_k^2\right]^N \tag{6.337}$$

The LR expressed in Eq. (6.337) is not convenient to use because the distribution of the quantity $(1 - \hat{\hat{\gamma}}_k^2)$ on the right side is unknown. However, we can make use of the Bartlett's theorem [99] which states that for $k = K'$, the asymptotic distribution of the quantity

$$F_N(k) = -N_0 \sum_{m=k+1}^{M_1} \log(1 - \hat{\hat{\gamma}}_m^2) \tag{6.338}$$

where $N_0 = 2N - (M_1 + M_2 + 1)$, is a χ^2 distribution with $2(M_1 - k)(M_2 - k)$ degrees of freedom. Notice that if $k < K'$, the quantity in Eq. (6.338) contains some $\hat{\hat{\gamma}}_m$ which involve the correlation of the signals in the two arrays, thereby rendering $F_N(k)$ much larger and is no longer χ^2-distributed. Using this result, we can propose a hypothesis test for determining the number of signals under the assumption that all signals are incoherent, that is, the rank of Σ_{12} is identical to the number of signals. For an assumed number of signals k, we propose to use the following test which employs Bartlett's theorem such that

$$F_N(k) = -N_0 \sum_{m=k+1}^{M_1} \ln(1 - \hat{\hat{\gamma}}_m^2) \underset{\mathcal{H}_k}{\overset{\mathcal{H}_{k+}}{\gtrless}} \tau_C(k) \tag{6.339}$$

where $\mathcal{H}_{k+}$ is the hypothesis that there are more than k signals. The distribution of the left side of Eq. (6.339) can be obtained directly from the Bartlett's theorem. The threshold $\tau_C(k)$ should be set to be the allowable probability of false alarm. For a given allowable probability of false alarm, α_C, we can establish the threshold value $\tau_C(k)$

using the tail end of the χ^2 distribution of $F_N(k)$ in Bartlett's theorem such that

$$\alpha_C = P_{F|k} = \int_{\tau_C(k)}^{\infty} \frac{1}{2^{m_C/2}\Gamma(m_C/2)} y^{(m_C-2)/2} e^{y/2} dy, \quad k = 0, 1, \cdots, M_1 - 1 \tag{6.340}$$

where $P_{F|k}$ is the probability of determining that there are more than k signals while $\mathcal{H}_k$ is ture, and $m_C = 2(M_1 - k)(M_2 - k)$. Therefore, for a specified false alarm rate, a set of thresholds $\{\tau_C(k), k = 0, 1, \cdots, M_1 - 1\}$ can be calculated, and a sequential hypothesis testing procedure can be constructed using the incremental assumed values of k starting from $k = 0$ until the likelihood ratio test is satisfied. The above decision procedure is called the *canonical coefficient test* (CCT) [100] which is an effective test procedure in unknown correlated noise when the signals are completely uncorrelated. However, when the signals are coherent, the accuracy of the CCT deteriorates since the procedure only estimates the rank of the matrix Σ_{12}.

PARADE A New Criterion for Determining the Number of Signals

To overcome the shortcoming of the CCT method, we resort to the UN-CLE algorithm which, as we recall, is developed based on the ML principle so that even when the signals are coherent, the method is still applicable.

Thus, we can use the method of CCT to obtain the initial estimate $\hat{K}'$, and we know that the true number of signals is such that $K \geq K'$ because of the possible existence of coherent signals. We can now set the initial value of the assumed number of signals k to $\hat{K}'$ and employ UN-CLE to estimate the DOA of the signals and obtain the k-dimensional vector $\hat{\theta}_i$. This vector represents an estimated parameter of our signal model based on the assumed value k, and using this vector, we can evaluate the quantity $f(\hat{\tilde{\mathbf{R}}}_{is}, \hat{\theta}_i, k)$. To arrive at a decision of the number of signals $\hat{K}$, we now make use of Theorem 6.18 which states that the asymptotic distribution of $2Nf(\hat{\tilde{\mathbf{R}}}_{is}, \hat{\theta}_i, K)$ is a χ^2-distribution having $2K'(M_i - K) - K$ degrees of freedom. We next employ the following hypothesis test:

$$2Nf(\hat{\tilde{\mathbf{R}}}_{is}, \hat{\theta}_i, k) \underset{\mathcal{H}_k}{\overset{\mathcal{H}_{k+}}{\gtrless}} \tau_P(k) \tag{6.341}$$

where $\mathcal{H}_{k+}$ is the hypothesis that there are more than k signals. The threshold $\tau_P(k)$ is, as in CCT, determined by the predetermined allowable probability of false alarm α_P such that

$$\alpha_P = \int_{\tau_P(k)}^{\infty} \frac{1}{2^{m_P/2}\Gamma(m_P/2)} y^{(m_P-2)/2} e^{y/2} dy, \quad k = 0, 1, \cdots, M_1 - 1 \tag{6.342}$$

where α_P is the allowable probability of error when the decision is that there are more than k signals while $\mathcal{H}_k$ is true, and $m_P = 2K'(M_i - k) - k$, the degrees of freedom of the distribution. From Theorem 6.18, we know that since for a large N, $f(\hat{\tilde{\mathbf{R}}}_{is}, \hat{\theta}_i, K) < f(\hat{\tilde{\mathbf{R}}}_{is}, \hat{\theta}_i, k)$, for $k < K$, then the left side of Eq. (6.341) has a higher probability of exceeding the threshold $\tau_P(k)$ when $k < K$ than when $k = K$.

The above procedure for determining the number of signals is called *parametric detection* (PARADE) in unknown noise, and can be summarized as follows:

1. Compute the canonical decomposition of the sampled data, and apply the CCT method to evaluate $\hat{K}'$, the estimate of the rank of Σ_{12}.
2. Set $k = \hat{K}'$ and assume that $\mathcal{H}_k$ is true.
3. Find $\hat{\theta}_i$ with the given k using the UN-CLE method and then calculate $2Nf(\hat{\hat{\mathbf{R}}}_{is}, \hat{\theta}_i, k)$.
4. From the χ^2 distribution with $2K'(M_1 - k) - k$ degrees of freedom, set the threshold $\tau_P(k)$ according to the desired false alarm rate α_P as shown in Eq. (6.342) for the hypothesis test.
5. If $2Nf(\hat{\hat{\mathbf{R}}}_{is}, \hat{\theta}_i, k) \leq \tau_P(k)$, accept $\mathcal{H}_k : \hat{K} = k$, and stop. Otherwise, reject $\mathcal{H}_k$, let $k = k + 1$, and go back to 3.

In establishing the test procedure above, we have set up the thresholds only according to the desired probabilities of false alarm. This in true for both the CCT and the PARADE methods. In CCT, we realize that if $k < K'$, then $F_N(k)$ in Eq. (6.339) will involve some of the canonical correlation coefficients $\hat{\hat{\gamma}}_m$ which are due to the signals rendering the log-likelihood ratio much larger. However, since $\hat{\hat{\gamma}}_m$ is a random variable, there is a finite probability that it may remain very small even though it is due to the signals. That is the case of missing. Similar situations arise in the PARADE hypothesis test in Eq. (6.341) in which the asymptotic expected value of $f(\hat{\hat{\mathbf{R}}}_{is}, \hat{\theta}_i, k)$ is larger than that of $f(\hat{\hat{\mathbf{R}}}_{is}, \hat{\theta}_i, K)$ when $k < K$. However, $f(\hat{\hat{\mathbf{R}}}_{is}, \hat{\theta}_i, k)$ being a random variable, has a finite probability of being less than the threshold, in which case a "miss" occurs. Unfortunately, the general distributions of both $F_N(k)$ and $f(\hat{\hat{\mathbf{R}}}_{is}, \hat{\theta}_i, k)$ for $k < K$ cannot be derived due to the dependence of the quantities on the SNR, the DOA, separation between the signals, as well as the number of signals. Due to this difficulty, we only consider the case of fixed probability of false alarm and ignore the probability of missing which may be negligible under high SNR or large N. For this strategy, the PARADE method is optimum.

Numerical Experiments

Several numerical experiments with different scenarios have been carried out using computer simulations. Since all parametric detectors in white noise [21, 45, 102] need the DOA estimates, and all the DOA estimators based on the spatial white noise model fail in unknown correlated noise (see Section 6.7), these parametric detectors will fail to determine the number of signals effectively in unknown noise environments. Furthermore, nonparametric detectors such as the AIC and MDL methods have been found to be unacceptable in performance in unknown noise environments [98, 100]. It is thus sufficient in our simulations to compare the performance of the workable methods in unknown correlated noise. It has also been shown [98] that C_0 of the ITC does not perform satisfactorily; we therefore only compare the methods of PARADE, CCT, and C_1, C_2, and C_3 of the ITC in these numerical experiments.

The noise used in these experiments is an AR model of order 2 with a coefficient vector $\mathbf{a} = [1 \; -1 \; 0.2]$ for both Array1 and Array2. The two arrays have identical linear uniform structures with 8 sensors each. The adjacent sensor separation is one-half of the wavelength of the narrowband signals. In the simulations, there are two incident signals which are assigned to be identically distributed Gaussian processes. Since highly correlated signals are found in practice in environments of multipath propagation, the signals in our simulations may be assigned to be independent or correlated with each other, being controlled by a correlation coefficient r. The directions of arrival of the two signals are chosen to be at $-5°$ and $5°$, respectively, to the normals of the arrays. We then employ the methods of PARADE, CCT, and ITC described above for the determination of the number of the signals under various SNR.

Example 6.15

In this example, the two signals are uncorrelated and we set the number of trials in the simulation to 100. The average of the performance of the various methods are then plotted under different SNR. For $N = 100$, that is, there are 100 snapshots,

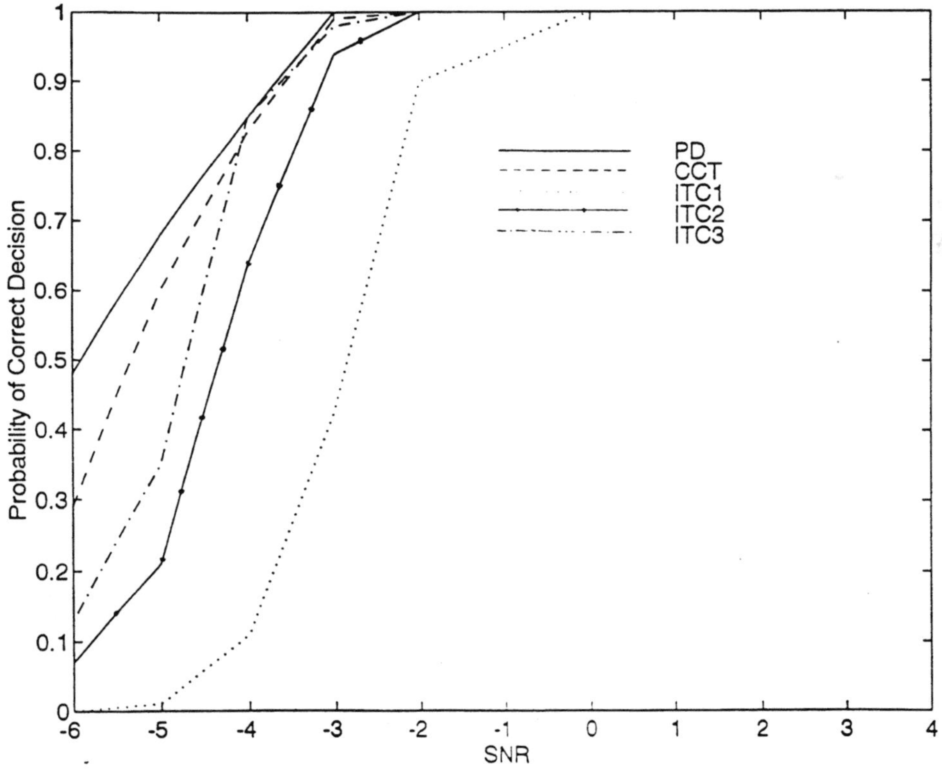

Figure 6.13 Detection performance of uncorrelated signals ($N = 100$, angles: $-5°, 5°$, F.A. rate $= 0.001$).

the detection performances of PARADE and CCT are shown in Figs. 6.13 and 6.14 in which the allowable false alarm probabilities are chosen as $\alpha_C = \alpha_P = 0.001$ and $\alpha_C = \alpha_P = 0.0001$, respectively. Correspondingly, the thresholds τ_C and τ_P are calculated according to Eqs. (6.340) and (6.342). We also show in these figures the performance of the three ITC methods using criteria C_1, C_2, and C_3 (Eq. (6.334)) respectively. As seen in Figs. 6.13 and 6.14, the new parametric detector PARADE denoted by PD and the CCT method have similar performance since the signals are uncorrelated, making the rank of Σ_{12} the same as the number of signals. Figures 6.13 and 6.14 also demonstrate that for a given N, the ITC methods C_2 and C_3 have performance comparable to those of PARADE and CCT and the performance of C_1 is worse than the others by more than 2 dB. We note that if we choose a higher value for the allowable probability of false alarm, the corresponding probability of missing for the PARADE and the CCT methods will decrease and thereby the probability of error for the two methods will decrease correspondingly. This is clearly demonstrated in Figs. 6.13 and 6.14 in which the allowable false alarm rates are 10^{-3} and 10^{-4} respectively. On the other hand, the performance of the ITC methods are independent of the choice of false alarm rate. Thus, there is an additional flexibility for the methods of PARADE and CCT. A further reduction of the allowable false alarm rate (say 10^{-2}) will show even further superiority of the performance of PARADE and CCT over that

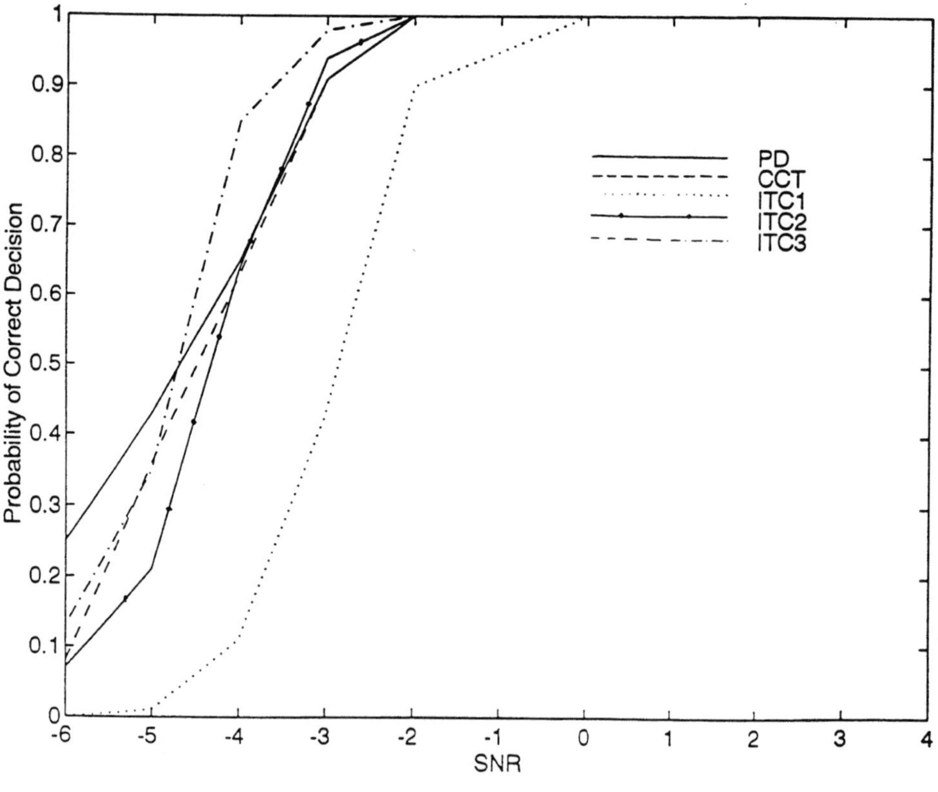

Figure 6.14 Detection performance of uncorrelated signals ($N = 100$, angles: $-5°, 5°$, F.A. rate = 0.0001).

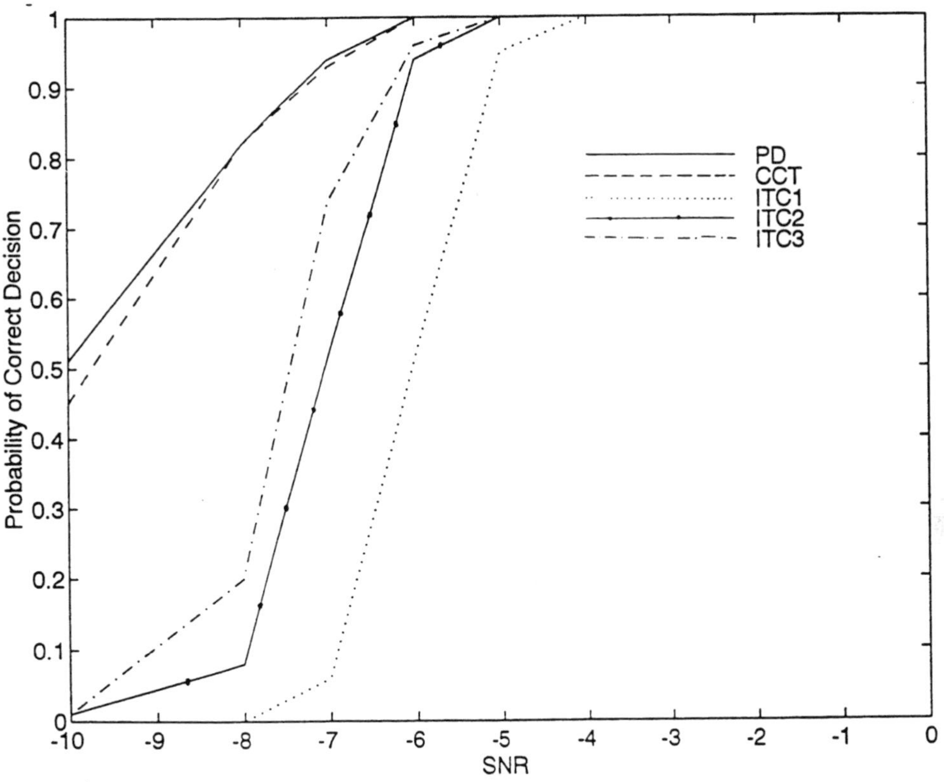

Figure 6.15 Detection performance of uncorrelated signals ($N = 200$, angles: $-5°, 5°$, F.A. rate $= 0.001$).

of the ITC methods. In Fig. 6.15, we show the comparison of the same signal detectors for $N = 200$ and $\alpha_C = \alpha_P = 0.001$. Here, we observe that the performances of PARADE and CCT are still similar as expected, but all the three ITC methods degrade more than 1 dB, indicating that the penalty terms of the ITC methods are not chosen optimally according to the number of snapshots and the allowable probability of false alarm.

Example 6.16

This experiment is carried out under the same scenario as that in Example 6.15, except that the two signals are correlated. The correlation coefficient of the two signals in the first case is chosen to be 0.8. The performances of the PARADE, CCT, and ITC methods are shown in Figs. 6.16 and 6.17 respectively for false alarm rates of 10^{-3} and 10^{-4}. It is seen from Figs. 6.16 and 6.17 that the CCT and the ITC methods degrade considerably in performance compared to the cases when the signals are independent. However, the PARADE method maintains its effectiveness and outperforms CCT and ITC by over 10 dB.

Figure 6.18 shows the case when the correlation between the two signals is unity. The false alarm rates in this case are 10^{-3} and 10^{-4}. Here the CCT and the ITC methods fail completely. However, the PARADE method remains effective.

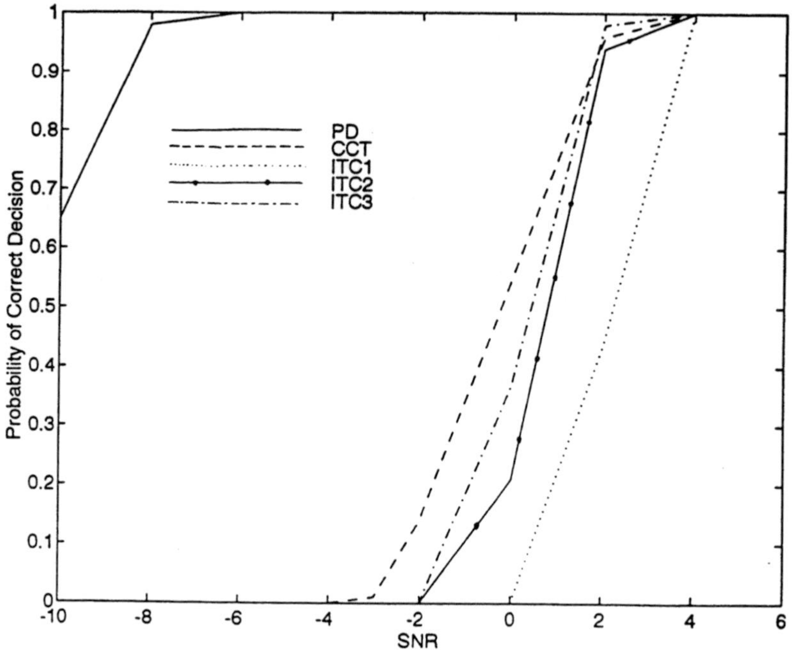

Figure 6.16 Detection performance of correlated signals ($N = 100$, angles: $-5°, 5°$, $\rho = 0.8$, F.A. rate = 0.001).

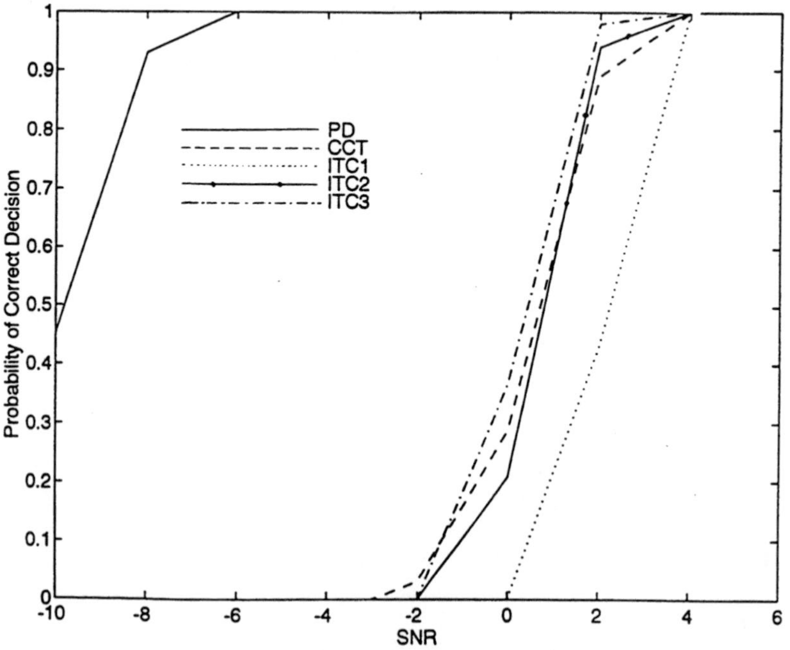

Figure 6.17 Detection performance of correlated signals ($N = 100$, angles: $-5°, 5°$, $\rho = 0.8$, F.A. rate = 0.0001).

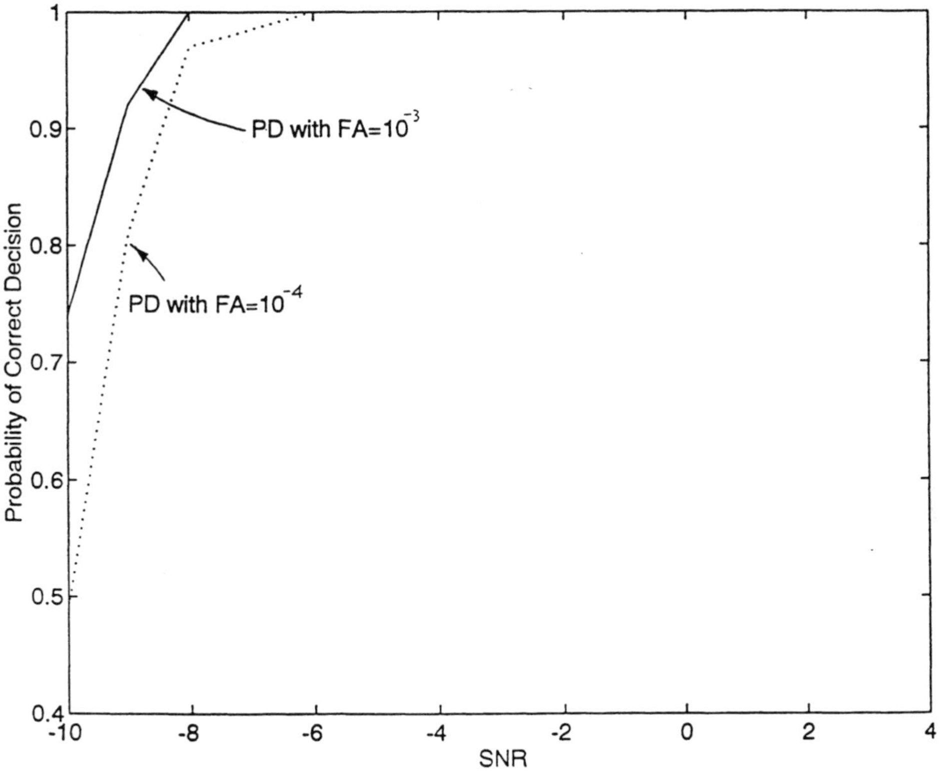

Figure 6.18 Detection performance of correlated signals ($N = 100$, angles: $-5°, 5°$, $\rho = 1$, F.A. rate = 0.001, 0.0001).

Many other simulation examples have also been carried out for equal and unequal power signal, as well as for signals closely spaced signals. Under extreme conditions such as signals having highly different power and/or signals very closely spaced, all the methods of PARADE, CCT, and ITC degrade in performance with the thresholds of the performance occurring at higher SNR for all the methods. However, the performance of the methods relative to each other remain essentially the same in the sense that for uncorrelated signals, the performances of PARADE and CCT are similar for the same allowable probability of false alarm, while for correlated signals, the performance of PARADE is far superior to either CCT or ITC.

6.9 SUMMARY AND CONCLUSIONS

In this chapter, we have examined the problem of high-resolution sensor array processing in an unknown correlated noise environment. The main assumption made is that there are two sets of data collected such that the signal components of the two sets are correlated, whereas the noise components are uncorrelated. Physically this implies that the data are collected by two arrays of sensors which are sufficiently far apart from each other so that their noise components are spatially uncorrelated.

Based on this assumption, we have developed a mathematical technique known as the *generalized correlation decomposition* (GCD) to analyze the correlation matrix of the two sets of data. Through the geometrical properties of GCD, we are able to locate the signal and noise subspaces. An addition, we can obtain operators to project (obliquely) onto these subspaces. The asymptotic properties of these estimated subspaces have also been studied. These properties are then utilized to formulate two algorithms known as the UN-MUSIC and the UN-CLE algorithms so that, with the number of signals being correctly predetermined, the directions of arrival of these signals can be estimated. The criteria used in these algorithms as well as the estimated DOA have been shown to be asymptotically consistent. The performance of these DOA estimation algorithms are then analyzed and, through the analyses, we have established the expressions of the optimum weighting matrices employed in the algorithms. More importantly, through the analyses, we have shown that the use of CCD (a special case of GCD) always yields the optimum estimation results for both algorithms. Numerical experiments have been carried out to confirm the results of these analyses and the superiority of CCD. The practical importance of CCD is further enhanced when it is shown that the computation involved can be substantially reduced by the use of a special algorithm developed in this chapter. Indeed, the UN-CLE algorithm weighted with $\bar{\mathbf{W}}_{iv}$ and $\bar{\mathbf{W}}_{is}$ employing CCD can be shown [111] to be asymptotically statistically efficient, that is, its covariance matrix reaches the Cramér-Rao bound asymptotically. Finally, we have also addressed the problem of the determination of the number of signals arriving at the array. Maximizing the likelihood function for the data model together with some chosen penalty functions, some information-theoretic criteria have been established facilitating the determination of the number of signals. On the other hand, utilizing a likelihood ratio test, the CCT criterion has also been established based on a predetermined probability of false alarm. A parametric method known as PARADE has also been developed utilizing the criterion of the UN-CLE algorithm. It has been shown that all the three methods, ITC, CCT, and PARADE, perform well under unknown correlated noise when the signals are independent. However, the method of PARADE is the only reliable method when the signals are correlated.

This chapter presents a comprehensive study of high-resolution array processing in unknown correlated noise. The main contribution here is the introduction of the concept of GCD which has been shown to be a powerful technique in both the decision of the number of signals and the estimation of the DOA. Although the theory here has been developed under the assumption that there are two sets of data collected in separate locations in space, the same concept can be applied to the situation when the data are collected separately in the time domain as long as the same assumptions are valid. At about the same time as the writing of this chapter, a similar idea [103, 104] of applying *instrumental variables* (IV) [105] to array processing in spatially correlated noise fields has been proposed. A vector of IV is a collection of array outputs at various lag times. The cross-covariance matrix of an array output vector and the vector of IV can be obtained, and a SVD can then be performed on the weighted cross-covariance matrix. Then, the weighted subspace fitting [44, 45] method can be applied to obtain the DOA estimates. The optimum weighting matrices have been derived. This, in some ways, is similar in principle to the application of GCD to sets of data collected in the time domain. However, since the approach taken in [103, 104] to derive an optimum algorithm is different from that proposed in this chapter, the question of which decomposition may be optimum has not been clearly addressed.

Another important aspect of the application of GCD is that as long as one array is calibrated, the methods will work. The other array does not have to be calibrated. Furthermore, as demonstrated in Example 6.14, the two arrays do not have to be far separated from each other physically. They could be parts of the same array divided into subarrays.

With the fast computation method of the CCD developed here, together with the effectiveness of the methods of both detection and estimation in unknown correlated noise fields, we predict that the application of GCD (CCD in particular) to other problems of signal processing will become increasingly important and popular.

APPENDIX 6A SOME RESULTS FROM MATRIX OPERATION

The Vec Operator, Kronecker Products, and Hadamard Products

For a matrix $\mathbf{A} = [a_{ij}]$ of order $m \times n$, the vector valued function vec $\mathbf{A}$ is defined as [47]

$$\text{vec}\,\mathbf{A} = [a_{11} \cdots a_{m1} a_{12} \cdots a_{m2} \cdots a_{1n} \cdots a_{mn}]^T \tag{6A.1}$$

Consider a matrix $\mathbf{A} = [a_{ij}]$ of order $m \times n$ and a matrix $\mathbf{B} = [b_{ij}]$ of order $k \times l$. The Kronecker product of the two matrices, denoted by $\mathbf{A} \otimes \mathbf{B}$ is defined as the $mk \times nl$ matrix [47] such that

$$\mathbf{A} \otimes \mathbf{B} = \begin{bmatrix} a_{11}\mathbf{B} & \cdots & a_{1n}\mathbf{B} \\ \vdots & \cdots & \vdots \\ a_{m1}\mathbf{B} & \cdots & a_{mn}\mathbf{B} \end{bmatrix} \tag{6A.2}$$

The Hadamard product of two matrices of same dimensions is defined as

$$\mathbf{A} \odot \mathbf{B} = [a_{ij} \cdot b_{ij}] \tag{6A.3}$$

The following algebraic results are useful:

1.

$$(\mathbf{A} \otimes \mathbf{B})(\mathbf{C} \otimes \mathbf{D}) = (\mathbf{AC}) \otimes (\mathbf{BD}) \tag{6A.4}$$

where $\mathbf{A}, \mathbf{B}, \mathbf{C}, \mathbf{D}$ are matrices of compatible dimensions.

Proof. The (i, j)th block of the left-hand side of Eq. (6A.4) is obtained by taking the product of the ith row block of $\mathbf{A} \otimes \mathbf{B}$ and the jth column block of $\mathbf{C} \otimes \mathbf{D}$, this is of the following form:

$$[a_{i1}\mathbf{B}\, a_{i2}\mathbf{B} \cdots a_{in}\mathbf{B}] \begin{bmatrix} c_{1j}\mathbf{D} \\ c_{2j}\mathbf{D} \\ \vdots \\ c_{nj}\mathbf{D} \end{bmatrix} = \sum_l a_{il} c_{lj} \mathbf{BD} \tag{6A.5}$$

The (i, j)th block of the right-hand side of Eq. (6A.4) is $g_{ij}\mathbf{BD}$ where g_{ij} is the (i, j)th element of $\mathbf{AC}$. But

$$g_{ij} = \sum_{l} a_{il}c_{lj} \tag{6A.6}$$

Since the (i, j)th blocks are equal, the result follows. ■

2.

$$\text{vec}\,(\mathbf{ABC}) = (\mathbf{C}^T \otimes \mathbf{A})\text{vec}\,\mathbf{B} \tag{6A.7}$$

Proof. We prove Eq. (6A.7) for $\mathbf{A}$, $\mathbf{B}$, and $\mathbf{C}$ each of order $n \times n$. The result, however, is also true for compatible rectangular matrices.

Denoting the jth column vector of a matrix by $(.)_{\cdot j}$, we have

$$\mathbf{A}_{\cdot j} = \mathbf{A}\mathbf{e}_j \tag{6A.8a}$$

and

$$a_{ij} = \mathbf{e}_i^T\mathbf{A}\mathbf{e}_j \tag{6A.8b}$$

where $\mathbf{e}_j$ is an nth order vector with unity as its jth element and zero elsewhere. Thus, we have

$$(\mathbf{AB})_{\cdot k} = (\mathbf{AB})\mathbf{e}_k = \mathbf{A}(\mathbf{B}\mathbf{e}_k) = \mathbf{A}(\mathbf{B})_{\cdot k} \tag{6A.9}$$

From the rule of matrix multiplication, we can also write

$$(\mathbf{AB})_{\cdot k} = \sum_{j} b_{jk}\mathbf{A}_{\cdot j} \tag{6A.10}$$

Hence, we have

$$\begin{aligned}(\mathbf{ABC})_{\cdot k} &= \sum_{j} c_{jk}(\mathbf{AB})_{\cdot j} = \sum_{j}(c_{jk}\mathbf{A})\mathbf{B}_{\cdot j} \\ &= [c_{1k}\mathbf{A} \cdots c_{nk}\mathbf{A}]\begin{bmatrix}\mathbf{B}_{\cdot 1}\\ \mathbf{B}_{\cdot 2}\\ \vdots\\ \mathbf{B}_{\cdot n}\end{bmatrix} \\ &= (\mathbf{C}_{\cdot k}^T \otimes \mathbf{A})\text{vec}\,\mathbf{B}\end{aligned} \tag{6A.11}$$

The result in Eq. (6A.7) thus follows.

In particular, if we let $\mathbf{B} = \mathbf{I}$ in Eq. (6A.7), we have

$$\text{vec}\,(\mathbf{AC}) = (\mathbf{C}^T \otimes \mathbf{A})\text{vec}\,\mathbf{I} \tag{6A.12}$$

■

3. Given $\mathbf{A}$ ($m \times m$ matrix) and $\mathbf{B}$ ($n \times n$ matrix), and that $\mathbf{A}^{-1}$ and $\mathbf{B}^{-1}$ exist, then

$$(\mathbf{A} \otimes \mathbf{B})^{-1} = \mathbf{A}^{-1} \otimes \mathbf{B}^{-1} \tag{6A.13}$$

Proof. Using Eq. (6A.4)

$$(\mathbf{A} \otimes \mathbf{B})(\mathbf{A}^{-1} \otimes \mathbf{B}^{-1}) = \mathbf{AA}^{-1} \otimes \mathbf{BB}^{-1} = \mathbf{I}_m \otimes \mathbf{I}_n = \mathbf{I}_{mn} \qquad (6A.14)$$

and the result of Eq. (6A.13) follows. ■

4.

$$\operatorname{tr} \mathbf{AB} = (\operatorname{vec} \mathbf{A}^T)^T \operatorname{vec} \mathbf{B} \quad \text{for } \mathbf{A} \text{ and } \mathbf{B} \text{ of order } n \times n \qquad (6A.15)$$

Proof

$$\begin{aligned} \operatorname{tr} \mathbf{AB} &= \sum_i \mathbf{e}_i^T \mathbf{ABe}_i \\ &= \sum_i \mathbf{A}_{i\cdot} \mathbf{B}_{\cdot i} \\ &= \sum_i (\mathbf{A}^T)_{\cdot i}^T \mathbf{B}_{\cdot i} \end{aligned} \qquad (6A.16)$$

where $(.)_{i\cdot}$ denotes the ith row of a matrix. Hence

$$\begin{aligned} \operatorname{tr} \mathbf{AB} &= \left[(\mathbf{A}^T)_{\cdot 1}^T (\mathbf{A}^T)_{\cdot 2}^T \cdots (\mathbf{A}^T)_{\cdot n}^T\right] \begin{bmatrix} \mathbf{B}_{\cdot 1} \\ \mathbf{B}_{\cdot 2} \\ \vdots \\ \mathbf{B}_{\cdot n} \end{bmatrix} \\ &= (\operatorname{vec} \mathbf{A}^T)^T \operatorname{vec} \mathbf{B} \end{aligned} \qquad (6A.17)$$

In particular, if $\mathbf{A} = \mathbf{I}$ and $\mathbf{B} = \mathbf{B}^H$, then Eq. (6A.17) can be reduced to

$$\begin{aligned} \operatorname{tr} \mathbf{B} &= (\operatorname{vec} (\mathbf{A}^T))^T \operatorname{vec} \mathbf{B} \\ &= \operatorname{tr} \left[\operatorname{vec} \mathbf{I} (\operatorname{vec} \mathbf{B})^T\right] = \operatorname{tr} \left[\operatorname{vec} \mathbf{I} \operatorname{vec}^H \mathbf{B}\right] \end{aligned} \qquad (6A.18)$$

since the diagonal elements of $\mathbf{B}$ are real. ■

5. Let $\mathbf{A}, \mathbf{B} \in \mathcal{C}^{n\times n}$ such that $\mathbf{A} \geq \mathbf{O}$ and $\mathbf{B} \geq \mathbf{O}$. Then

$$\mathbf{A} \odot \mathbf{B} \geq \mathbf{O} \qquad (6A.19)$$

$$\operatorname{Re} (\mathbf{A} \odot \mathbf{B}) \geq \mathbf{O} \qquad (6A.20)$$

Proof. Since $\mathbf{B} \geq \mathbf{O}$, it can be written as

$$\mathbf{B} = \mathbf{H}^H \mathbf{H}$$

Let $\mathbf{h}_i$ denote the ith column of $\mathbf{H}$; then the ijth element of $\mathbf{A} \odot \mathbf{B}$ is given by

$$(\mathbf{A} \odot \mathbf{B})_{ij} = a_{ij} \mathbf{h}_i^H \mathbf{h}_j$$

Thus, we can write

$$\begin{aligned} \mathbf{A} \odot \mathbf{B} &= \begin{bmatrix} \mathbf{h}_1^H & & \mathbf{O} \\ & \ddots & \\ \mathbf{O} & & \mathbf{h}_n^H \end{bmatrix} \begin{bmatrix} a_{11}\mathbf{I} & \cdots & a_{1n}\mathbf{I} \\ \cdot & \cdots & \cdot \\ a_{n1}\mathbf{I} & \cdots & a_{nn}\mathbf{I} \end{bmatrix} \begin{bmatrix} \mathbf{h}_1 & & \mathbf{O} \\ & \ddots & \\ \mathbf{O} & & \mathbf{h}_n \end{bmatrix} \\ &= \bar{\mathbf{H}}^H (\mathbf{A} \otimes \mathbf{I}) \bar{\mathbf{H}} \end{aligned}$$

where $\bar{\mathbf{H}}$ is the $n^2 \times n$ matrix

$$\bar{\mathbf{H}} = \begin{bmatrix} \mathbf{h}_1 & & \mathbf{O} \\ & \ddots & \\ \mathbf{O} & & \mathbf{h}_n \end{bmatrix}$$

Since $\mathbf{A} \otimes \mathbf{I}$ is positive semidefinite, so is $\mathbf{A} \otimes \mathbf{B}$, which implies that Re $[\mathbf{A} \odot \mathbf{B}] \geq \mathbf{O}$. ■

6. For a positive semidefinite matrix such that

$$\begin{bmatrix} \mathbf{A}_{11} & \mathbf{A}_{12} \\ \mathbf{A}_{21} & \mathbf{A}_{22} \end{bmatrix} \geq \mathbf{O} \tag{6A.21}$$

where $\mathbf{A}_{ij}$ is of dimensions $n \times n$ for $i, j = 1, 2$, then

$$\{\text{Re}\,(\mathbf{A}_{11})\}^{-1} \leq \{\text{Re}\,(\mathbf{A}_{12})\}^{-1}\text{Re}\,(\mathbf{A}_{22})\{\text{Re}\,(\mathbf{A}_{12})\}^{-1} \tag{6A.22}$$

when $(\mathbf{A}_{11}^{-1})$ and $(\mathbf{A}_{12}^{-1})$ exist.

Proof. For Eq. (6A.21) to be true, we have $\mathbf{A}_{ii} = \mathbf{A}_{ii}^H$ and $\mathbf{A}_{12}^H = \mathbf{A}_{21}$. Now we can form another positive semidefinite matrix such that

$$\begin{bmatrix} \mathbf{I} & \mathbf{O} \\ -\mathbf{A}_{21}\mathbf{A}_{11}^{-1} & \mathbf{I} \end{bmatrix} \begin{bmatrix} \mathbf{A}_{11} & \mathbf{A}_{12} \\ \mathbf{A}_{21} & \mathbf{A}_{22} \end{bmatrix} \begin{bmatrix} \mathbf{I} & -\mathbf{A}_{11}^{-1}\mathbf{A}_{12} \\ \mathbf{O} & \mathbf{I} \end{bmatrix} = \begin{bmatrix} \mathbf{A}_{11} & \mathbf{O} \\ \mathbf{O} & \mathbf{A}_{22} - \mathbf{A}_{21}\mathbf{A}_{11}^{-1}\mathbf{A}_{12} \end{bmatrix} \geq \mathbf{O}$$

Thus,

$$\mathbf{A}_{22} - \mathbf{A}_{21}\mathbf{A}_{11}^{-1}\mathbf{A}_{12} \geq \mathbf{O}$$

Therefore,

$$\mathbf{A}_{21}^{-1}\mathbf{A}_{22}\mathbf{A}_{12}^{-1} \geq \mathbf{A}_{11}^{-1}$$

and the result of Eq. (6A.22) follows. ■

7. Consider a matrix

$$\mathbf{A} = \begin{bmatrix} \mathbf{A}_{11} & \mathbf{A}_{12} \\ \mathbf{A}_{21} & \mathbf{A}_{22} \end{bmatrix} \tag{6A.23}$$

where $\mathbf{A}_{11}$ and $\mathbf{A}_{22}$ are square matrices. Assume that $\mathbf{A}_{22}$ and $(\mathbf{A}_{11} - \mathbf{A}_{12}\mathbf{A}_{22}^{-1}\mathbf{A}_{21})$ are nonsingular. Then

$$\mathbf{A}^{-1} = \begin{bmatrix} \mathbf{O} & \mathbf{O} \\ \mathbf{O} & \mathbf{A}_{22}^{-1} \end{bmatrix} + \begin{bmatrix} \mathbf{I} \\ -\mathbf{A}_{22}^{-1}\mathbf{A}_{21} \end{bmatrix} (\mathbf{A}_{11} - \mathbf{A}_{12}\mathbf{A}_{22}^{-1}\mathbf{A}_{21})^{-1} [\mathbf{I} - \mathbf{A}_{12}\mathbf{A}_{22}^{-1}] \tag{6A.24}$$

Proof. Direct multiplication gives

$$\begin{bmatrix} \mathbf{A}_{11} & \mathbf{A}_{12} \\ \mathbf{A}_{21} & \mathbf{A}_{22} \end{bmatrix} \left\{ \begin{bmatrix} \mathbf{O} & \mathbf{O} \\ \mathbf{O} & \mathbf{A}_{22}^{-1} \end{bmatrix} + \begin{bmatrix} \mathbf{I} \\ -\mathbf{A}_{22}^{-1}\mathbf{A}_{21} \end{bmatrix} (\mathbf{A}_{11} - \mathbf{A}_{12}\mathbf{A}_{22}^{-1}\mathbf{A}_{21})^{-1} [\mathbf{I} - \mathbf{A}_{12}\mathbf{A}_{22}^{-1}] \right\}$$

$$= \begin{bmatrix} \mathbf{O} & \mathbf{A}_{12}\mathbf{A}_{22}^{-1} \\ \mathbf{O} & \mathbf{I} \end{bmatrix} + \begin{bmatrix} \mathbf{A}_{11} - \mathbf{A}_{12}\mathbf{A}_{22}^{-1}\mathbf{A}_{21} \\ \mathbf{O} \end{bmatrix} (\mathbf{A}_{11} - \mathbf{A}_{12}\mathbf{A}_{22}^{-1}\mathbf{A}_{21})^{-1} [\mathbf{I} - \mathbf{A}_{12}\mathbf{A}_{22}^{-1}]$$

$$= \begin{bmatrix} \mathbf{O} & \mathbf{A}_{12}\mathbf{A}_{22}^{-1} \\ \mathbf{O} & \mathbf{I} \end{bmatrix} + \begin{bmatrix} \mathbf{I} \\ \mathbf{O} \end{bmatrix} [\mathbf{I} - \mathbf{A}_{12}\mathbf{A}_{22}^{-1}] = \begin{bmatrix} \mathbf{I} & \mathbf{O} \\ \mathbf{O} & \mathbf{I} \end{bmatrix} \quad ■$$

8. For the matrix

$$\mathbf{A} = \begin{bmatrix} \mathbf{A}_{11} & \mathbf{A}_{12} \\ \mathbf{A}_{21} & \mathbf{A}_{22} \end{bmatrix}$$

where $\mathbf{A}_{11}^{-1}$ and $\mathbf{A}_{22}^{-1}$ exist, then

$$\det \mathbf{A} = \det \mathbf{A}_{11} \det (\mathbf{A}_{22} - \mathbf{A}_{21}\mathbf{A}_{11}^{-1}\mathbf{A}_{12}) \tag{6A.25}$$

and

$$\det \mathbf{A} = \det \mathbf{A}_{22} \det (\mathbf{A}_{11} - \mathbf{A}_{12}\mathbf{A}_{22}^{-1}\mathbf{A}_{21}) \tag{6A.26}$$

Proof

$$\det \mathbf{A} = \det \left\{ \mathbf{A} \begin{bmatrix} \mathbf{I} & -\mathbf{A}_{11}^{-1}\mathbf{A}_{12} \\ \mathbf{O} & \mathbf{I} \end{bmatrix} \right\}$$

$$= \det \begin{bmatrix} \mathbf{A}_{11} & \mathbf{O} \\ \mathbf{A}_{21} & \mathbf{A}_{22} - \mathbf{A}_{21}\mathbf{A}_{11}^{-1}\mathbf{A}_{12} \end{bmatrix} = \det \mathbf{A}_{11} \det (\mathbf{A}_{22} - \mathbf{A}_{21}\mathbf{A}_{11}^{-1}\mathbf{A}_{12})$$

Similarly,

$$\det \mathbf{A} = \det \left\{ \mathbf{A} \begin{bmatrix} \mathbf{I} & \mathbf{O} \\ -\mathbf{A}_{22}^{-1}\mathbf{A}_{21} & \mathbf{I} \end{bmatrix} \right\}$$

$$= \det \mathbf{A}_{22} \det (\mathbf{A}_{11} - \mathbf{A}_{12}\mathbf{A}_{22}^{-1}\mathbf{A}_{21}) \quad \blacksquare$$

9. If $\mathbf{A}_{12}$ and $\mathbf{A}_{21}$ are $M_1 \times M_2$ and $M_2 \times M_1$ matrices respectively, then

$$\det (\mathbf{I}_{M_1} - \mathbf{A}_{12}\mathbf{A}_{21}) = \det (\mathbf{I}_{M_2} - \mathbf{A}_{21}\mathbf{A}_{12}) \tag{6A.27}$$

Proof. Let

$$\mathbf{A} = \begin{bmatrix} \mathbf{I}_{M_1} & \mathbf{A}_{12} \\ \mathbf{A}_{21} & \mathbf{I}_{M_2} \end{bmatrix} \tag{6A.28}$$

Then using Eqs. (6A.25) and (6A.26) in Eq. (6A.28), the result follows. ■

Results from Matrix Calculus. The following results of differentiating a matrix with respect to a scalar are useful: For a matrix $\mathbf{A}$ whose elements a_{ij} are functions of a scalar ξ, we have

1.

$$\frac{\partial}{\partial \xi}\mathbf{A} \triangleq \left[\frac{\partial a_{ij}}{\partial \xi} \right] \tag{6A.29}$$

2.

$$\frac{\partial}{\partial \xi}\operatorname{tr}(\mathbf{A}) = \operatorname{tr}\frac{\partial \mathbf{A}}{\partial \xi} \tag{6A.30}$$

3. Since

$$\mathbf{A}^{-1}\mathbf{A} = \mathbf{I}$$

differentiating both sides of the above equation with respect to ξ, we have

$$\left(\frac{\partial}{\partial \xi}\mathbf{A}^{-1}\right)\mathbf{A} + \mathbf{A}\left(\frac{\partial}{\partial \xi}\mathbf{A}\right) = \mathbf{O}$$

Hence

$$\frac{\partial}{\partial \xi}\mathbf{A}^{-1} = -\mathbf{A}^{-1}\left(\frac{\partial}{\partial \xi}\mathbf{A}\right)\mathbf{A}^{-1} \tag{6A.31}$$

4.

$$\frac{\partial}{\partial \xi}(\det \mathbf{A}) = \sum_i \sum_j \frac{\partial}{\partial \alpha_{ij}}(\det \mathbf{A})\frac{\partial \alpha_{ij}}{\partial \xi}$$

Now since the determinant of a matrix can be expanded in terms of its row elements multiplied by the corresponding co-factors, then

$$\frac{\partial}{\partial \alpha_{ij}}(\det \mathbf{A}) = c_{ij} = (\operatorname{adj} \mathbf{A})_{ji}$$

where c_{ij} is the co-factor of α_{ij} and $(.)_{ji}$ denotes the jith element of a matrix. Thus, we have

$$\begin{aligned}\frac{\partial}{\partial \xi}(\det \mathbf{A}) &= (\det \mathbf{A})\sum_i \sum_j (\mathbf{A}^{-1})_{ji}\frac{\partial \alpha_{ij}}{\partial \xi} \\ &= (\det \mathbf{A})\operatorname{tr}\left[\mathbf{A}^{-1}(\frac{\partial}{\partial \xi}\mathbf{A})\right] \end{aligned} \tag{6A.32}$$

APPENDIX 6B THE FIRST AND SECOND MOMENTS OF vec$\Delta\Sigma$

Let $\mathbf{x}(n)$ be an M-dimensional vector representing the received data. We assume that $\mathbf{x}(n)$ is complex and Gaussian distributed with zero mean. Let

$$\hat{\mathbf{C}} = N\hat{\mathbf{\Sigma}} = \sum_{n=1}^{N} \mathbf{x}(n)\mathbf{x}^H(n) \tag{6B.1}$$

Then $\hat{\mathbf{C}}$ has a density function known as the complex Wishart distribution [89] given by

$$p(\hat{\mathbf{C}}) = \frac{1}{\bar{\Gamma}_M(N)(\det \mathbf{\Sigma})^N}\operatorname{etr}(-\mathbf{\Sigma}^{-1}\hat{\mathbf{C}})(\det \hat{\mathbf{C}})^{N-M} \tag{6B.2}$$

where etr(.) denotes exp [tr(.)] and

$$\bar{\Gamma}_M(N) \triangleq \pi^{M(M-1)/2}\prod_{m=1}^{N}\Gamma(N-m+1) \tag{6B.3}$$

$\Gamma(n)$ being the Euler's Gamma function. To evaluate the first and second moments of $\Delta\mathbf{\Sigma}$, we first evaluate the first and second moments of the elements of $\hat{\mathbf{C}}$ for which we have to use the moment generating function of $p(\hat{\mathbf{C}})$. The moment generating function for the real Wishart distribution is well documented [83,84].

To obtain the moment generating function for the complex Wishart distribution, let $\hat{c}_{mn}$ be the mnth element of $\hat{\mathbf{C}}$. We see that we can obtain the first and second moments of the element $\hat{c}_{mn}$ by applying a complex gradient operator $\left(\frac{\partial}{\partial \omega_R} + j\frac{\partial}{\partial \omega_I}\right)$, where ω_R and ω_I are the real and imaginary parts of ω respectively, to a generating function $\mathrm{E}\left[\exp \mathrm{Re}(\omega \hat{c}^*_{mn})\right]$ and evaluating the result at $\omega = 0$, we obtain

$$\begin{aligned}\left(\frac{\partial}{\partial \omega_R} + j\frac{\partial}{\partial \omega_I}\right)\mathrm{E}\left[\exp \mathrm{Re}(\omega \hat{c}^*_{mn})\right]_{\omega=0} &= \mathrm{E}\left[\exp\{\mathrm{Re}(\omega \hat{c}^*_{mn})\}(\hat{c}_{R_{mn}} + j\hat{c}_{I_{mn}})\right]_{\omega=0} \\ &= \mathrm{E}\left[\hat{c}_{mn}\right] \end{aligned} \tag{6B.4}$$

with $\hat{c}_{R_{mn}}$ and $\hat{c}_{I_{mn}}$ being the real and imaginary parts of $\hat{c}_{mn}$ respectively. Applying the conjugate of the gradient operator to Eq. (6B.4) and again evaluating at $\omega = 0$, we have

$$\begin{aligned}&\left(\frac{\partial}{\partial \omega_R} - j\frac{\partial}{\partial \omega_I}\right)\mathrm{E}\left[\exp\{\mathrm{Re}(\omega \hat{c}^*_{mn})\}(\hat{c}_{R_{mn}} + j\hat{c}_{I_{mn}})\right]_{\omega=0} \\ &\qquad = \mathrm{E}\left[(\hat{c}_{R_{mn}} + j\hat{c}_{I_{mn}})(\hat{c}_{R_{mn}} - j\hat{c}_{I_{mn}})\right] = \mathrm{E}\left[\hat{c}_{mn}\hat{c}^*_{mn}\right]\end{aligned} \tag{6B.5}$$

In view of the above discussion, we can deduce that the moment generating function for the complex Wishart distribution is given by

$$\mu(\boldsymbol{\Omega}) = \mathrm{E}\left[\exp(\sum_m \sum_n \omega_{mn}\hat{c}^*_{mn})\right] \tag{6B.6a}$$

$$= \mathrm{E}\left[\mathrm{etr}(\boldsymbol{\Omega}^H\hat{\mathbf{C}})\right] = \{\det(\mathbf{I} - \boldsymbol{\Omega}^H\boldsymbol{\Sigma})\}^{-N} \tag{6B.6b}$$

where ω_{mn} is the mnth complex element of the $M \times M$ Hermitian matrix $\boldsymbol{\Omega}$. Now, define a complex gradient operator

$$\nabla_{\omega_{mn}} = \frac{1}{2}\left[\frac{\partial}{\partial \omega_{R_{mn}}} - j\frac{\partial}{\partial \omega_{I_{mn}}}\right] \tag{6B.7}$$

with $\omega_{R_{mn}}$ and $\omega_{I_{mn}}$ being the real and imaginary parts of ω_{mn}. By applying $\nabla^*_{\omega_{mn}}$ to $\mu(\boldsymbol{\Omega})$ in Eq. (6B.6) and evaluating the result at $\boldsymbol{\Omega} = \mathbf{O}$, we have

$$\begin{aligned}\mathrm{E}\left[\hat{c}_{mn}\right] &= \nabla^*_{\omega_{mn}}\mu(\boldsymbol{\Omega}) \\ &= \nabla^*_{\omega_{mn}}\{\det(\mathbf{I} - \boldsymbol{\Omega}^H\boldsymbol{\Sigma})\}^{-N}\big|_{\Omega=O} \\ &= -N\{\det(\mathbf{I} - \boldsymbol{\Omega}^H\boldsymbol{\Sigma})\}^{-N}\mathrm{tr}\left[(\mathbf{I} - \boldsymbol{\Omega}^H\boldsymbol{\Sigma})^{-1}(-\nabla^*_{\omega_{mn}}\boldsymbol{\Omega}^H\boldsymbol{\Sigma})\right]\big|_{\omega=O} \\ &= N\mathrm{tr}\left[\mathbf{E}_{nm}\boldsymbol{\Sigma}\right] = N\sigma_{mn}\end{aligned} \tag{6B.8}$$

where $\mathbf{E}_{mn}$ is an $M \times M$ matrix with the mnth element being unity and the rest of the elements zero and σ_{mn} is the mnth element of $\boldsymbol{\Sigma}$. To arrive at the final result in Eq. (6B.8), Eqs. (6A.30), (6A.31), and (6A.32) in Appendix A have been employed. Now, applying $\nabla_{\omega_{kl}}$ to Eq. (6B.8) and again evaluating the result at $\boldsymbol{\Omega} = \mathbf{O}$, we obtain

$$\begin{aligned}
\mathrm{E}[\hat{c}_{mn}\hat{c}_{kl}^*] &= \nabla_{\omega_{kl}} \nabla^*_{\omega_{mn}} \{\det(\mathbf{I}-\boldsymbol{\Omega}^H\boldsymbol{\Sigma})\}^{-N}\big|_{\Omega=O} \\
&= N^2\{\det(\mathbf{I}-\boldsymbol{\Omega}^H\boldsymbol{\Sigma})\}^{-N} \\
&\quad \mathrm{tr}\left[(\mathbf{I}-\boldsymbol{\Omega}^H\boldsymbol{\Sigma})^{-1}(-\nabla_{\omega_{kl}}\boldsymbol{\Omega}^H\boldsymbol{\Sigma})\right]\mathrm{tr}\left[(\mathbf{I}-\boldsymbol{\Omega}^H\boldsymbol{\Sigma})^{-1}(-\nabla^*_{\omega_{mn}}\boldsymbol{\Omega}^H\boldsymbol{\Sigma})\right] \\
&\quad + N\{\det(\mathbf{I}-\boldsymbol{\Omega}^H\boldsymbol{\Sigma})\}^{-N} \\
&\quad \mathrm{tr}\left[(\mathbf{I}-\boldsymbol{\Omega}^H\boldsymbol{\Sigma})^{-1}(-\nabla_{\omega_{kl}}\boldsymbol{\Omega}^H\boldsymbol{\Sigma})(\mathbf{I}-\boldsymbol{\Omega}^H\boldsymbol{\Sigma})^{-1}(-\nabla^*_{\omega_{mn}}\boldsymbol{\Omega}^H\boldsymbol{\Sigma})\right]\big|_{\Omega=O} \\
&= N^2\mathrm{tr}\,[\mathbf{E}_{kl}\boldsymbol{\Sigma}]\,\mathrm{tr}\,[\mathbf{E}_{mn}\boldsymbol{\Sigma}] + N\mathrm{tr}\,[\mathbf{E}_{kl}\boldsymbol{\Sigma}\mathbf{E}_{mn}\boldsymbol{\Sigma}] \\
&= N^2\sigma^*_{kl}\sigma_{mn} + N\sigma_{mk}\sigma^*_{nl}
\end{aligned} \tag{6B.9}$$

But $\mathrm{E}\left[\hat{c}_{mn}\hat{c}^*_{kl}\right] = N^2\mathrm{E}\left[\hat{\sigma}_{mn}\hat{\sigma}^*_{kl}\right]$ and $\boldsymbol{\Delta\Sigma} = \hat{\boldsymbol{\Sigma}} - \boldsymbol{\Sigma}$, thus together with Eq. (6B.4), we can obtain the second moment of the elements of $\boldsymbol{\Delta\Sigma}$ such that

$$\begin{aligned}
\mathrm{E}\left[\Delta\sigma_{mn}\,\Delta^*\sigma_{kl}\right] &= \mathrm{E}\left[\hat{\sigma}_{mn}\hat{\sigma}^*_{kl}\right] - \sigma_{mn}\sigma^*_{kl} \\
&= \frac{1}{N}\sigma_{mk}\sigma^*_{nl}
\end{aligned} \tag{6B.10}$$

where $\Delta\sigma_{mn}$ is the mnth element of $\boldsymbol{\Delta\Sigma}$.

Similarly, we have

$$\begin{aligned}
\mathrm{E}\left[c_{mn}c_{kl}\right] &= \Delta^*_{\omega_{kl}}\,\Delta^*_{\omega_{mn}}\,\{\det(\mathbf{I}-\boldsymbol{\Omega}^H\boldsymbol{\Sigma})\}^{-N}\big|_{\Omega=O} \\
&= N^2\sigma_{kl}\sigma_{mn} + N\sigma_{kn}\sigma_{ml}
\end{aligned} \tag{6B.11}$$

and

$$\mathrm{E}\left[\Delta\sigma_{mn}\,\Delta\,\sigma_{kl}\right] = \frac{1}{N}\sigma_{kn}\sigma_{ml} \tag{6B.12}$$

From Eq. (6B.10) and remembering that $\boldsymbol{\Sigma}_{ii}$ is Hermitian, it can easily be verified that

$$\mathrm{E}\left[\mathrm{vec}(\boldsymbol{\Delta\Sigma}_{\bar{i}i})\mathrm{vec}^H(\boldsymbol{\Delta\Sigma}_{\bar{i}i})\right] = \frac{1}{N}\boldsymbol{\Sigma}^T_{ii}\otimes\boldsymbol{\Sigma}_{\bar{i}\bar{i}} \tag{6B.13}$$

and

$$\mathrm{E}\left[\mathrm{vec}(\boldsymbol{\Delta\Sigma}_{\bar{i}i})\mathrm{vec}^H(\boldsymbol{\Delta\Sigma}_{i\bar{i}})\right] = \frac{1}{N}\boldsymbol{\Sigma}^T_{\bar{i}i}\otimes\boldsymbol{\Sigma}_{\bar{i}i} \tag{6B.14}$$

Also, from Eq. (6B.12), it can easily be seen that

$$\mathrm{E}\left[\mathrm{vec}(\boldsymbol{\Delta\Sigma}_{\bar{i}i})\mathrm{vec}^T(\boldsymbol{\Delta\Sigma}_{\bar{i}i})\right] = \frac{1}{N}\left\{\sum_{l=1}^{M_i}\sum_{m=1}^{M_i}(\mathbf{E}_{lm}\otimes\mathbf{E}^T_{lm})(\boldsymbol{\Sigma}_{\bar{i}i}\otimes\boldsymbol{\Sigma}^T_{\bar{i}i})\right\} \tag{6B.15}$$

APPENDIX 6C COMPUTATION OF THE UN-CLE CRITERION

Reparameterization of the Log-Likelihood Function in MLE

In the context of array processing, the idea of reparameterization of the log-likekihood function to facilitate computation in the case when a uniform linear array is used was proposed by Bresler and Macovski [32] as well as by Kumaresan, Scharf, and Shaw [92]. We observe that the criterion of Eq. (6.27) for the ML estimation of the DOA of deterministic signal sources contains the parameters $\{\theta_k\}, k = 1, \cdots, K$. It can also be reparameterized in terms of another set of parameters $\{b_k\}$ defined by

$$b_0 z^K + b_1 z^{K-1} + \cdots + b_K = b_0 \prod_{k=1}^{K} (z - e^{j\omega_0 L_0 \sin\theta_k / c}) \tag{6C.1}$$

with $\theta_k \in \left(-\frac{\pi}{2}, \frac{\pi}{2}\right)$.

To show this, we define the set

$$\mathcal{B} = \{b_k : \sum_{k=0}^{K} b_k z^{K-k} \neq 0 \quad \text{for } |z| \neq 1\} \tag{6C.2}$$

and we note that the mapping in Eq. (6C.1) from distinct values of θ_k to $b_k \in \mathcal{B}, k = 1, \cdots, K$ is one-to-one. Also, we define the $(M - K) \times M$ matrix

$$\mathbf{B}_0^H = \begin{bmatrix} b_K & \cdots & b_1 & b_0 & & & \mathbf{O} \\ & \cdot & \cdots & \cdot & \cdot & & \\ & & \cdot & \cdots & \cdot & \cdot & \\ & & & \cdot & \cdots & \cdot & \cdot \\ \mathbf{O} & & & b_K & \cdots & b_1 & b_0 \end{bmatrix} \tag{6C.3}$$

and note that

$$\mathbf{B}_0^H \mathbf{D} = \mathbf{O} \tag{6C.4}$$

Since Rank($\mathbf{D}$) $= K$ and Rank($\mathbf{B}_0$) $= M - K$, then the columns of $\mathbf{B}_0$ span the null space of $\mathbf{D}$, and thus the orthogonal projector onto the null space of $\mathbf{D}$ can be written as

$$\mathbf{I} - \mathbf{D}(\mathbf{D}^H\mathbf{D})^{-1}\mathbf{D}^H = \mathbf{B}_0(\mathbf{B}_0^H\mathbf{B}_0)^{-1}\mathbf{B}_0^H \tag{6C.5}$$

Inserting Eq. (6C.5) into Eq. (6.27), we have

$$\hat{\mathbf{b}} = \arg\min_{\mathbf{b}\in\mathcal{B}} \text{tr}\{\mathbf{B}_0(\mathbf{B}_0^H\mathbf{B}_0)^{-1}\mathbf{B}_0^H\hat{\Sigma}_x\} \tag{6C.6}$$

where

$$\mathbf{b} = [b_0\, b_1 \cdots b_K]^T \tag{6C.7}$$

Minimization of Eq. (6C.6) with respect to $\mathbf{b} \in \mathcal{B}$ results in the ML estimates of the set $\{b_k, k = 1, \cdots, K\}$, from which the ML estimates of $\{\theta_k\}, k = 1, \cdots, K$ can be obtained by solving the polynomial equation of (6C.1). The constraint $\mathbf{b} \in \mathcal{B}$ can be replaced by the weaker relationship $b_k = b_{K-k}^*$ and the minimization of Eq. (6C.6) can be carried out as follows:

Let $\hat{\mathbf{b}}^{(i)}$ denote the estimate of $\mathbf{b}$ at iteration i. Start with $\hat{B}_0^{(0)}$ such that $\hat{\mathbf{B}}_0^{H(0)}\mathbf{B}_0^{(0)} = \mathbf{I}$. Determine $\hat{\mathbf{b}}^{(i)}$ such that

$$\hat{\mathbf{b}}^{(i)} = \arg\min_{\mathbf{b}\in\mathcal{B}} \text{tr}\left[\mathbf{B}_0^{(i)}(\hat{\mathbf{B}}_0^{H(i-1)}\hat{\mathbf{B}}_0^{(i-1)})^{-1}\mathbf{B}_0^{H(i)}\Sigma_x\right] \quad i = 1, 2, \cdots \tag{6C.8}$$

until $\|\hat{\mathbf{b}}^{(i-1)} - \hat{\mathbf{b}}^{(i)}\| < \varepsilon_0$ where ε_0 is a predetermined constant. In Eq. (6C.8), the matrix $\hat{\mathbf{B}}_0^{(i-1)}$ is of the form given by Eq. (6C.3) with elements $\hat{b}_k^{(i-1)}, k = 1, \cdots, K$. The minimization of Eq. (6C.8), as a result of the structure of $\mathbf{B}_0$, is a set of linear least-squares problems which can be solved efficiently by standard methods [93]. If greater accuracy is desired, a further step can be taken such that using $\hat{\mathbf{b}}^{(i)}$ from Eq. (6C.8) as initial values, we perform a multidimensional search to locate the minimum point of $\mathrm{tr}\left[\mathbf{B}_0(\mathbf{B}_0^H\mathbf{B}_0)^{-1}\mathbf{B}_0^H\hat{\boldsymbol{\Sigma}}_x\right]$.

Implementation of the UN-CLE Algorithm for $W_{iv} = \bar{W}_{iv}$ and $W_{is} = \bar{W}_{is}$

To obtain a more useful form for optimization, consider the UN-CLE criterion shown in Eq. (6.142) with $\mathbf{W}_{iv} = \bar{\mathbf{W}}_{iv}$ and $\mathbf{W}_{is} = \bar{\mathbf{W}}_{is}$. Since $\mathbf{W}_{iv}$ and $\mathbf{W}_{is}$ can be viewed as constant matrices or as functions of $\boldsymbol{\theta}_i$, in the following derivation, $\bar{W}_{iv}$ is considered as a function of $\boldsymbol{\theta}_i$ and $\bar{\mathbf{W}}_{is}$ as a constant matrix. Hence, Eq. (6.142) can be written as

$$\begin{aligned} f(\hat{\mathbf{R}}_{is}, \bar{\mathbf{W}}_{iv}, \bar{\mathbf{W}}_{is}) &= \mathrm{tr}\,\{\mathbf{P}_{iv}\bar{\mathbf{W}}_{iv}\mathbf{P}_{iv}^H\hat{\mathbf{R}}_{is}\bar{\mathbf{W}}_{is}\hat{\mathbf{R}}_{is}^H\} \\ &= \mathrm{tr}\,\{\mathbf{L}_{iv}\mathbf{R}_{iv}^H\mathbf{L}_{iv}(\mathbf{L}_{iv}^H\boldsymbol{\Sigma}_{ii}\mathbf{L}_{iv})^{-1}\mathbf{L}_{iv}^H\mathbf{R}_{iv}\mathbf{L}_{iv}^H\hat{\mathbf{R}}_{is}\bar{\mathbf{W}}_{is}\hat{\mathbf{R}}_{is}^H\} \qquad (6C.9) \\ &= \mathrm{tr}\,\{\mathbf{L}_{iv}(\mathbf{L}_{iv}^H\boldsymbol{\Sigma}_{ii}\mathbf{L}_{iv})^{-1}\mathbf{L}_{iv}^H\hat{\mathbf{R}}_{is}\bar{\mathbf{W}}_{is}\hat{\mathbf{R}}_{is}^H\} \end{aligned}$$

where the last equality is obtained by using $\mathbf{R}_{iv}^H\mathbf{L}_{iv} = \mathbf{I}$.

Now, since $\mathrm{span}\,\{\mathbf{L}_{iv}\} = \overline{\mathrm{span}}\{\mathbf{D}_i(\boldsymbol{\theta}_i)\}$, we can write

$$\mathbf{L}_{iv} = \mathbf{B}_{iv}\mathbf{T}_{iv} \qquad (6C.10)$$

where $\mathbf{B}_{iv}$ is a basis in $\overline{\mathrm{span}}\{\mathbf{D}_i(\boldsymbol{\theta}_i)\}$ and $\mathbf{T}_{iv}$ is a nonsingular square matrix. Substituting Eq. (6C.10) in Eq. (6C.9) and simplifying, we have

$$f(\hat{\mathbf{R}}_{is}, \bar{\mathbf{W}}_{iv}, \bar{\mathbf{W}}_{is}) = \mathrm{tr}\,\{\mathbf{B}_{iv}(\mathbf{B}_{iv}^H\boldsymbol{\Sigma}_{ii}\mathbf{B}_{iv})^{-1}\mathbf{B}_{iv}^H\hat{\mathbf{R}}_{is}\bar{\mathbf{W}}_{is}\hat{\mathbf{R}}_{is}^H\} \qquad (6C.11)$$

As mentioned earlier, for a uniform linear array, $\mathbf{B}_{iv}$, the column vectors of which span the orthogonal complement of $\mathbf{D}_i(\boldsymbol{\theta}_i)$, can be written in the form given by Eq. (6C.3) having the coefficients $\{b_k\}, k = 1, \cdots, K$ satisfying Eq. (6C.1). The minimization of Eq. (6C.11) with respect to $\mathbf{B}_{iv}$ then becomes a least-squares problem.

For a general array, however, we can rewrite Eq. (6C.11) as

$$\begin{aligned} &f(\hat{\mathbf{R}}_{is}, \bar{\mathbf{W}}_{iv}, \bar{\mathbf{W}}_{is}) \\ &= \mathrm{tr}\left\{\boldsymbol{\Sigma}_{ii}^{1/2}\mathbf{B}_{iv}\left[(\boldsymbol{\Sigma}_{ii}^{1/2}\mathbf{B}_{iv})^H(\boldsymbol{\Sigma}_{ii}^{1/2}\mathbf{B}_{iv})\right]^{-1}\left(\boldsymbol{\Sigma}_{ii}^{1/2}\mathbf{B}_{iv}\right)^H\boldsymbol{\Sigma}_{ii}^{-1/2}\hat{\mathbf{R}}_{is}\bar{\mathbf{W}}_{is}\hat{\mathbf{R}}_{is}^H\boldsymbol{\Sigma}_{ii}^{-1/2}\right\} \\ &= \mathrm{tr}\left\{\mathbf{Q}_{iv}\boldsymbol{\Sigma}_{ii}^{-1/2}\hat{\mathbf{R}}_{is}\bar{\mathbf{W}}_{is}\hat{\mathbf{R}}_{is}^H\boldsymbol{\Sigma}_{ii}^{-1/2}\right\} \qquad (6C.12) \end{aligned}$$

where $\mathbf{Q}_{iv}$ is a projector onto the subspace spanned by $\boldsymbol{\Sigma}_{ii}^{1/2}\mathbf{B}_{iv}$ such that

$$\mathbf{Q}_{iv} = (\mathbf{\Sigma}_{ii}^{1/2}\mathbf{B}_{iv})\left[(\mathbf{\Sigma}_{ii}^{1/2}\mathbf{B}_{iv})^H(\mathbf{\Sigma}_{ii}^{1/2}\mathbf{B}_{iv})\right]^{-1}(\mathbf{\Sigma}_{ii}^{1/2}\mathbf{B}_{iv})^H \tag{6C.13}$$

Since $(\mathbf{\Sigma}_{ii}^{1/2}\mathbf{B}_{iv})^H(\mathbf{\Sigma}_{ii}^{-1/2}\mathbf{D}_i(\theta_i)) = \mathbf{O}$ due to the fact that $\text{span}\{\mathbf{B}_{iv}\} = \overline{\text{span}}\{\mathbf{D}_i(\theta_i)\}$, we can conclude that the orthogonal complement of the subspace spanned by $\mathbf{\Sigma}_{ii}^{1/2}\mathbf{B}_{iv}$ is spanned by $\mathbf{\Sigma}_{ii}^{-1/2}\mathbf{D}_i(\theta_i)$. Thus, the projector $\mathbf{Q}_{iv}$ can be expressed as

$$\mathbf{Q}_{iv} = \mathbf{I} - \mathbf{\Sigma}_{ii}^{-1/2}\mathbf{D}_i(\mathbf{D}_i^H\mathbf{\Sigma}_{ii}^{-1}\mathbf{D}_i)^{-1}\mathbf{D}_i^H\mathbf{\Sigma}_{ii}^{-1/2} \tag{6C.14}$$

where the notation for the explicit dependence on θ_i of $\mathbf{D}_i$ has been omitted for convenience. Substituting Eq. (6C.14) into Eq. (6C.12), we have

$$f(\hat{\mathbf{R}}_{is}, \bar{\mathbf{W}}_{iv}, \bar{\mathbf{W}}_{is})1 \\ = \text{tr}\{\left[\mathbf{I} - \mathbf{\Sigma}_{ii}^{-1/2}\mathbf{D}_i(\mathbf{D}_i^H\mathbf{\Sigma}_{ii}^{-1}\mathbf{D}_i)^{-1}\mathbf{D}_i^H\mathbf{\Sigma}_{ii}^{-1/2}\right]\mathbf{\Sigma}_{ii}^{-1/2}\hat{R}_{is}\bar{\mathbf{W}}_{is}\hat{\mathbf{R}}_{is}^H\mathbf{\Sigma}_{ii}^{-1/2}\} \tag{6C.15}$$

and then the criterion in Eq. (6C.9) is equivalent to

$$\hat{\theta}_i = \arg\max_{\theta}\text{tr}\{\mathbf{D}_i(\mathbf{D}_i^H\mathbf{\Sigma}_{ii}^{-1}\mathbf{D}_i)^{-1}\mathbf{D}_i^H\mathbf{\Sigma}_{ii}^{-1}\hat{\mathbf{R}}_{is}\bar{\mathbf{W}}_{is}\hat{\mathbf{R}}_{is}^H\mathbf{\Sigma}_{ii}^{-1}\} \tag{6C.16}$$

In practice, we do not know the true values of $\mathbf{\Sigma}_{ii}$ or $\bar{\mathbf{W}}_{is}$; therefore, we use their consistent estimates $\hat{\mathbf{\Sigma}}_{ii}$ and $\hat{\bar{\mathbf{W}}}_{is}$ in the criteria for both the uniform linear array in Eq. (6C.11) and the general array in Eq. (6C.16) where

$$\hat{\bar{\mathbf{W}}}_{is} = (\hat{\mathbf{\Gamma}}^{-1}\hat{\mathbf{L}}_{\bar{is}}^H\hat{\mathbf{\Sigma}}_{\bar{ii}}\hat{\mathbf{L}}_{\bar{is}}\hat{\mathbf{\Gamma}}^{-1})^{-1} \tag{6C.17}$$

The consistency of $\hat{\bar{\mathbf{W}}}_{is}$ directly follows from Theorem 6.4 and the fact that $\hat{\mathbf{\Sigma}}_{ii} \to \mathbf{\Sigma}_{ii}$. We note that if the CCD is employed, the criterion further simplifies to: For general arrays

$$\hat{\theta}_i = \arg\max_{\theta}\text{tr}\{\mathbf{D}_i(\mathbf{D}_i^H\hat{\mathbf{\Sigma}}_{ii}^{-1}\mathbf{D}_i)^{-1}\mathbf{D}_i^H\hat{\mathbf{L}}_{is}\hat{\hat{\mathbf{\Gamma}}}^2\hat{\hat{\mathbf{L}}}_{is}^H\} \tag{6C.18}$$

For uniform linear arrays

$$\hat{\theta}_i = \arg\min_{\theta}\text{tr}\{\mathbf{B}_{iv}(\mathbf{B}_{iv}^H\hat{\mathbf{\Sigma}}_{ii}\mathbf{B}_{iv})^{-1}\mathbf{B}_{iv}^H\hat{\hat{\mathbf{R}}}_{is}\hat{\hat{\mathbf{\Gamma}}}^2\hat{\hat{\mathbf{R}}}_{is}^H\} \tag{6C.19}$$

where we have used $\hat{\mathbf{\Sigma}}_{ii}^{-1}\hat{\hat{\mathbf{R}}}_{is} = \hat{\hat{\mathbf{L}}}_{is}$ and $\hat{\mathbf{L}}_{\bar{is}}^H\hat{\mathbf{\Sigma}}_{ii}\hat{\mathbf{L}}_{\bar{is}} = \mathbf{I}_k$.

Now, Eq. (6C.16) is expressed as an explicit function of θ_i since $\mathbf{D}_i$ is a function of θ_i and therefore the maximization with respect to θ_i in a K-dimensional space can be directly performed. The optimization of criteria of forms given by Eq. (6C.16) has been extensively studied [37, 42, 45] and efficient computational algorithms have been proposed and can be directly applied here.

In a similar manner, we can attempt to express $g(\hat{\mathbf{P}}_{iv}, \bar{\mathbf{W}}_{iv}, \bar{\mathbf{W}}_{is})$ as an explicit function of $\mathbf{D}_i(\theta_i)$ so that the optimization with respect to θ_i may be carried out directly. However, the resulting expression is very much more cumbersome than Eq. (6C.16). Therefore, we can conclude that for computational efficiency we should use the criterion $f(\hat{\mathbf{R}}_{is}, \bar{\mathbf{W}}_{iv}, \bar{\mathbf{W}}_{is})$, which can be written in the form of Eq. (6C.16). This is especially true when a uniform linear sensor array is employed.

REFERENCES

1. S. Haykin, ed. *Array Processing: Applications to Radar*, Stroudsburg, PA: Dowden, Hutchinsons, and Ross, 1980.
2. S. Haykin, ed. *Array Signal Processing*, Englewood Cliffs, NJ: Prentice Hall, 1985.
3. S. Haykin and A. Steinhardt, eds. *Adaptive Radar Detection and Estimation*, New York: Wiley-Interscience, 1992.
4. A. B. Baggeroer. "Sonar Signal Processing," in *Applications of Digital Signal Processing*, A. V. Oppenheim, ed., Englewood Cliffs, NJ: Prentice Hall, 1978.
5. W. C. Knight, R. C. Pridham, and S. M. Kay. "Digital Signal Processing for Sonar," *Proc. IEEE*, vol. 69, no. 11, pp. 1451–1506, Nov. 1981.
6. R. A. Monzingo and T. W. Miller. *Introduction to Adaptive Arrays*, New York: Wiley-Interscience, 1980.
7. J. E. Hudson. *Adaptive Array Principles*, London: Peter Peregrinus, 1981.
8. J. R. Williams. "Fast Beamforming Algorithms," *J. Acoust. Soc. Amer.*, vol. 44, pp. 1454–55, 1968.
9. P. Rudnick. "Digital Beamforming in the Frequency Domain," *J. Acoust. Soc. Amer.*, vol. 46, pp. 1089–90, 1969.
10. D. E. Dudgeon. "Fundamentals of Digital Array Processing," *Proc. IEEE*, vol. 65, pp. 898–904, June 1977.
11. F. J. Harris. "On the Use of Windows for Harmonic Analysis with the Discrete Fourier Transform," *Proc. IEEE*, vol. 66, pp. 51–83, Jan. 1978.
12. R. G. Pridham and R. A. Mucci. "A Novel Approach to Digital Beamforming," *J. Acoust. Soc. Amer.*, vol. 63, pp. 425–34, Feb. 1978.
13. H. Fan, E. I. El-Masry, and W. K. Jenkins. "Resolution Enhancement of Digital Beamformers," *IEEE Trans. ASSP*, vol. 32, no. 5, pp. 1041–51, Oct. 1984.
14. H. Akaike. "A New Look at Statistical Model Identification," *IEEE Trans. AC*, vol. 19, pp. 716–23, Dec. 1974.
15. J. Rissanen. "Modelling by Shortest Data Description," *Automatica*, vol. 14, pp. 465–71, 1978.
16. G. Schwartz. "Estimating the Dimension of a Model," *Ann. Stat.*, vol. 6, no. 2, pp. 461–64, 1978.
17. M. Wax and T. Kailath. "Detection of Signals by Information Theoretic Criteria," *IEEE Trans. ASSP*, vol. 33, pp. 387–92, April 1985.
18. Y. Yin and P. Krishnaiah. "On Some Nonparametric Methods for detection of the Number of Signals," *IEEE Trans. ASSP*, vol. 35, pp. 1533–38, 1987.
19. K. M. Wong, Q. T. Zhang, J. P. Reilly, and P. C. Yip. "On Information Theoretic Criteria for the Determination of the Number of Signals in High Resolution Array Processing," *IEEE Trans. ASSP*, vol. 38, no. 11, pp. 1959–71, Nov. 1990.
20. W. G. Chen, K. M. Wong, and J. P. Reilly. "Detection of the Number of Signals—a Predictive Eigen-threshold Approach," *IEEE Trans. SP*, vol. 20, no. 5, pp. 1088–98, May 1991.
21. Q. Wu and D. R. Fuhrmann. "A Parametric Method for Determining the Number of Signals in Narrowband Direction-finding," *IEEE Trans, SP*, vol. 39, no. 8, pp. 1848–57, Aug. 1991.

22. J. Capon. "High Resolution Frequency Wavenumber Spectrum Analysis," *Proc. IEEE*, vol. 57, pp. 1408–18, Aug. 1969.

23. V. F. Pisarenko. "The Retrieval of Harmonics from a Covariance Function," *Geophys. J. Roy. Astron. Soc.*, vol. 33, pp. 347–66, 1973.

24. D. H. Johnson and S. R. DeGraff. "Improving the Resolution of Bearing in Passive Sonar Arrays by Eigenvalue Analysis," *IEEE Trans. ASSP*, vol. 30, no. 4, pp. 638–47, Aug. 1982.

25. D. H. Johnson. "The Application of Spectral Estimation Methods to Bearing Estimation Problems," *Proc. IEEE*, vol. 70, pp. 1018–28, Sept. 1982.

26. D. W. Tufts and R. Kumaresan. "Estimation of Frequencies of Multiple Sinusoids: Making Linear Predictions Perform Like Maximum Likelihood," *Proc. IEEE*, vol. 70, no. 9, pp. 975–89, Sept. 1982.

27. R. Kumaresan and D. W. Tufts. "Estimating the Angles of Arrival of Multiple Plane Waves," *IEEE Trans. AES*, vol. 19, pp. 134-39, Jan. 1983.

28. G. Bienvenue and L. Kopp. "Optimality of High Resolution Array Processing Using the Eigensystem Approach," *IEEE Trans. ASSP*, vol. 31, no. 5, pp. 1235–47, Oct. 1983.

29. G. Su and M. Morf. "Signal Subspace Approach for Multiple Wideband Emitter Location," *IEEE Trans. ASSP*, vol. 31, no. 6, pp. 1502–22, Dec. 1983.

30. H. Wang and M. Kaveh. "Coherent Signal-subspace Processing for the Detection and Estimation of Angles of Arrival of Multiple Wideband Sources," *IEEE Trans. ASSP*, vol. 33, no. 4, pp. 823–31, Aug. 1985.

31. R. O. Schmidt. "Multiple Emitter Location and Signal Parameter Estimation," *IEEE Trans. AP*, vol. 34, pp. 276–80, March 1986.

32. Y. Bresler and A. Macovski. "Exact Maximum-likelihood Parameter Estimation of Superimposed Exponential Signals in Noise," *IEEE Trans. ASSP*, vol. 34, no. 5, pp. 1081–89, Oct. 1986.

33. J. F. Böhme. "Estimation of Spectral Parameters of Correlated Signals in Wavefields," *Signal Processing*, vol. 11, pp. 329–37, 1986.

34. K. Sharman and T. S. Durrani. "A Comparative Study of Modern Eigenstructure Methods for Bearing Estimation—A New High Performance Approach," *Proc. 25 IEEE Conf. Dec. Contr.*, pp. 1737–42, Athens, Greece, Dec. 1986.

35. K. M. Buckley and L. J. Griffiths. "Broadband Signal Subspace Spatial Spectrum (BASS-ALE) Estimation," *IEEE Trans. ASSP*, vol. 36, no. 7, pp. 953–64, July 1988.

36. H. Hung and M. Kaveh. "Focusing Matrices for Coherent Signal Subspace Processing," *IEEE Trans. ASSP*, vol. 36, no. 8, pp. 1272–81, Aug. 1988.

37. I. Ziskind and M. Wax. "Maximum Likelihood Localization of Multiple Sources by Alternating Projection," *IEEE Trans. ASSP*, vol. 36, no. 10, pp. 1553–60, Oct. 1988.

38. R. Roy and T. Kailath. "ESPRIT—Estimation of Signal Parameters via Rotational Invariance Techniques," *IEEE Trans. ASSP*, vol. 37, no. 7, pp. 984–95, July 1989.

39. J. Krolik and D. N. Swingler. "Multiple Broadband Source Location Using Steered Covariance Matrices," *IEEE Trans. ASSP*, vol. 37, no. 10, pp. 1481–94, Oct. 1989.

40. P. Stoica and K. C. Sharman. "Novel Eigenanalysis Method for Direction Estimation," *Proc. IEEE, Pt. F*, pp. 19–26, Feb. 1990.

41. J. Krolik and D. N. Swingler. "Focused Wideband Array Processing via Spatial Resampling," *IEEE Trans. ASSP*, vol. 38, no. 2, pp. 356–60, Feb. 1990.

42. P. Stoica and K. C. Sharman. "Maximum Likelihood Methods for Direction-of-Arrival Estimation," *IEEE Trans ASSP*, vol. 38, no. 7, pp. 1132–43, July 1990.

43. M. Miller and D. Fuhrmann. "Maximum Likelihood Narrow-band Direction Finding and the EM Algorithm," *IEEE Trans. ASSP*, vol. 38, pp. 1560–77, Sept. 1990.

44. M. Viberg and B. Ottersten. "Sensor Array Processing Based on Subspace Fitting," *IEEE Trans. SP*, vol. 39, no. 5, pp. 1110–21, May 1991.

45. M. Viberg, B. Ottersten, and T. Kailath. "Detection and Estimation in Sensor Arrays Using Weighted Subspace Fitting," *IEEE Trans. SP*, vol. 39, no. 11, pp. 2436-49, Nov. 1991.

46. M. Kaveh and A. J. Barabell. "The Statistical Performance of the MUSIC and the Minimum-Norm Algorithms in Resolving Plane Waves in Noise," *IEEE Trans. ASSP*, vol. 34, pp. 331–41, April 1986.

47. H. Wang and M. Kaveh. "Performance of Signal Subspace Processing Methods, Pt. I: Narrowband Systems," *IEEE Trans. ASSP*, vol. 34, no. 5, pp. 1201–09, Oct. 1986.

48. H. Wang and M. Kaveh. "Performance of Signal Subspace Processing Methods, Pt. II: Coherent Wideband Systems," *IEEE Trans. ASSP*, vol. 35, no. 11, pp. 1583–91, Nov. 1987.

49. M. Kaveh, H. Wang and H. Hung. "On the Theoretical Performance of a Class of Estimators of the Number of Narrowband Sources," *IEEE Trans. ASSP*, vol. 35, no. 9, pp. 1350–52, Sept. 1987.

50. B. Porat and B. Friedlander. "Analysis of the Relative Efficiency of the MUSIC Algorithm," *IEEE Trans. ASSP*, vol. 36, no. 4, pp. 532–34, April 1988.

51. P. Stoica and A. Nehorai. "MUSIC, Maximum Likelihood, and Cramér-Rao Bound," *IEEE Trans. ASSP*, vol. 37, no. 5, pp. 720–41, May 1989.

52. Q. T. Zhang, K. M. Wong, P. C. Yip, and J. P. Reilly. "Statistical Analysis of the Performance of Information Theoretic Criteria in the Detection of the Number of Signals in Array Processing," *IEEE Trans. ASSP*, vol. 37, no. 10, pp. 1557–67, Oct. 1989.

53. P. Stoica and A. Nehorai. "Performance Study of Conditional and Unconditional Direction of Arrival Estimation," *IEEE Trans. ASSP*, vol. 38, no. 10, pp. 1783–95, Oct. 1990.

54. P. Stoica and A. Nehorai. "MUSIC, Maximum Likelihood, and Cramér-Rao Bound: Further Results and Comparisons," *IEEE Trans. ASSP*, vol. 38, no. 12, pp. 2140–50, Dec. 1990.

55. P. Stoica and T. Söderström. "Statistical Analysis of MUSIC and Subspace Rotation Estimates of Sinusoidal Frequencies," *IEEE Trans. SP*, vol. 39, no. 8, pp. 1836–57, Aug. 1991.

56. P. M. Schultheiss and J. P. Ianniello. "Optimum Range and Bearing Estimation with Randomly Perturbed Arrays," *J. Acoust. Soc. Am.*, vol. 68, pp. 167–73, 1980.

57. K. M. Wong, R. S. Walker, and G. Niezgoda. "Effects of Random Sensor Motion on Bearing Estimation by the MUSIC Algorithm," *IEEE Proc.*, vol. 135, pt. F, no. 3, pp. 233-50, June 1988.

58. B. Friedlander. "A Sensitivity Analysis of the MUSIC Algorithm," *IEEE Trans. ASSP*, vol. 38, no. 10, pp. 1740–51, Oct. 1990.

59. Y. Rockah and P. M. Schultheiss. "Array Shape Calibration Using Sources in Unknown Locations, Pt. I: Far-field Sources," *IEEE Trans. ASSP*, vol. 35, no. 3, pp. 286–99, March 1987.

60. Y. Rockah and P. M. Schultheiss. "Array Shape Calibration Using Sources in Unknown Locations, Pt. II: Near-field Sources and Estimation Implementation," *IEEE Trans. ASSP*, vol. 35, no. 6, pp. 724–35, June 1987.

61. B. Friedlander and A. J. Weiss. "Direction Finding in the Presence of Mutual Coupling," *22nd Asilomar Conf. on Sig., Systems, and Computers*, Oct. 31–Nov. 2, 1987, Pacific Grove, CA.

62. B. Friedlander and A. J. Weiss. "Eigenstructure Methods for Direction Finding with Sensor Gain and Phase Uncertainties," *ICASSP'88*, April 1988.

63. A. J. Weiss, A. S. Willsky, and B. C. Levy. "Eigenstructure Approach for Array Processing with Unknown Intensity Coefficients," *IEEE Trans. ASSP*, vol. 36, no. 10, pp. 1613–17, Oct. 1988.

64. G. M. Wenz. "Acoustic Ambient Noise in the Ocean: Spectra and Sources," *Acoust. Soc. Am.*, vol. 34, no. 12, pp. 1936–56, Dec. 1962.

65. R. J. Urick. *Principles of Underwater Sound*, 3rd ed., New York: McGraw-Hill, 1983.

66. A. Paulraj and T. Kailath. "Eigenstructure Methods for Direction of Arrival Estimation in the Presence of Unknown Noise Fields," *IEEE Trans. ASSP*, vol. 34, no. 1, pp. 13–20, Feb. 1986.

67. J. Le Cadre. "Parametric Methods for Spatial Signal Processing in the Presence of Unknown Colored Noise Fields," *IEEE Trans. ASSP*, vol. 37, pp. 965–83, July 1989.

68. A. H. Tewfik. "Harmonic Retrieval in the Presence of Colored Noise," *Proc. ICASSP'89*, pp. 2069–72, 1989.

69. J. P. Reilly, K. M. Wong, and P. M. Reilly. "Direction of Arrival Estimation in the Presence of Noise with Unknown Arbitrary Covariance Matrices," *Proc. ICASSP'89*, pp. 2609–12, 1989.

70. K. M. Wong, J. P. Reilly, Q. Wu, and S. Qiao. "Estimation of the Direction of Arrival of Signals in Unknown Correlated Noise, Pt. I: The MAP Approach and Its Implementation," *IEEE SP*, vol. 40, no. 8, pp. 2007–17, Aug. 1992.

71. K. M. Wong and J. P. Reilly. "Estimation of the Direction of Arrival of Signals in Unknown Correlated Noise, Pt. II: Asymptotic Behaviour and Performance of the MAP Approach," *IEEE Trans. SP*, vol. 40, no. 8, pp. 2018–28, Aug. 1992.

72. M. Wax. "Detection and Localization of Multiple Sources in Noise with Unknown Covariance," *IEEE Trans. SP*, vol. 40, no. 40, no. 1, pp. 245–49, Jan. 1992.

73. P. Stoica and T. Söderström. "On Array Signal Processing in Spatially Correlated Noise Fields," *IEEE Trans. CAS—II: Analog and Digital Signal Processing*, vol. 39, no. 12, pp. 879–82, Dec. 1992.

74. R. Bellman. *Introduction to Matrix Analysis*, 2nd ed., New York, McGraw-Hill, 1972.

75. P. Lancaster and M. Tismenetsky. *The Theory of Matrices*, 2nd ed., New York: Academic Press, 1985.

76. A. W. Naylor and G. R. Sell. *Linear Operator Theory*, New York: Holt, Reinhart, and Winston, 1971.

77. M. G. Kendall and A. Stuart. *The Advanced Theory of Statistics*, vol. 2, London: Charles Griffin & Co., Ltd., 1961.

78. A. G. Jaffer. "Maximum Likelihood Direction Finding of Stochastic Sources: A Separable Solution," *Proc. ICASSP'88*, pp. 2893–96, New York, 1988.

79. F. Li and R. J. Vaccaro. "Performance Degradation of DOA Estimations Due to Unknown Noise Fields," *IEEE Trans, SP*, vol. 40, no. 3, pp. 686–90, March 1992.

80. W. S. BURDIC. *Underwater Acoustic System Analysis*, Englewood Cliffs, NJ: Prentice Hall, 1984.

81. P. H. THOMAS and S. HAYKIN. "Stochastic Modeling of Radar Clutter," *IEE Proc. Pt. F*, vol. 133, no. 5, Aug. 1986.

82. D. R. BRILLINGER. *Time Series*, San Francisco: Holden-Day, Inc., 1981.

83. R. J. MUIRHEAD. *Aspects of Multivariate Statistical Theory*, New York: John Wiley & Sons, 1982.

84. T. W. ANDERSON. *An Introduction to Multivariate Statistical Analysis*, New York: John Wiley & Sons, 1984.

85. D. TYLER. *Redundancy Analysis and Associated Asymptotic Distribution Theory*, Ph. D. dissertation, Dept. of Statistics, Princeton University, 1979.

86. D. TYLER. "Asymptotic Inference of Eigenvectors," *Annals of Statistics*, vol. 9, no. 4, pp. 725–36, 1981.

87. L. E. FRANKS. *Signal Theory*, Rev. ed. Stroussburg, PA: Dowden & Culver, 1981.

88. T. KATO. *Perturbation Theory for Linear Operators*, Berlin: Springer-Verlag, 1976.

89. N. R. GOODMAN. "Statistical Analysis Based on a Certain Multivariate Complex Gaussian Distribution," *Annals of Mathematical Statistics*, vol. 34, pp. 152–71, 1963.

90. K. S. MILLER. *Complex Stochastic Processes*, Reading, MA: Addison-Wesley, 1974.

91. A. GRAHAM. *Kronecker Products and Matrix Calculus with Applications*, New York: John Wiley & Sons, 1981.

92. J. H. WILKINSON. *The Algebraic Eigenvalue Problem*, Oxford: Oxford University Press, 1965.

93. Y. BRESLER and A. MACOVSKI. "On the Number of Signal Resolvable by a Uniform Linear Array," *IEEE Trans. ASSP*, vol. 34, no. 10, pp. 1361–75, Dec. 1986.

94. M. WAX and I. ZISKIND. "On Unique Localization of Multiple Sources by Passive Sonar Arrays," *IEEE Trans. ASSP*, vol. 37, no. 7, pp. 996–1000, July 1989.

95. F. A. GRAYBILL. *Matrices with Applications in Statistics*, 2nd ed., Belmont, Calif.: Wadsworth, 1983.

96. T. SÖDERSTRÖM and P. STOICA. *System Identification*, New York: Prentice Hall International, 1989.

97. C. R. RAO. *Linear Statistical Inference and its Applications*, 2nd ed., New York: John Wiley & Sons, 1973.

98. Q. T. ZHANG and K. M. WONG. "Information Theoretic Criteria for the Determination of the Number of Signals in Spatially Correlated Noise," *IEEE Trans. SP*, vol. 41, no. 4, pp. 1652–63, April 1993.

99. M. S. BARTLETT. "Further Aspects of the Theory of Multiple Regression," *Proc., Cambridge Phil. Soc.*, vol. 34, p. 33, 1938.

100. W. G. CHEN, J. P. REILLY, and K. M. WONG. "Detection of the Number of Signals in the Presence of Noise with Unknown Banded Covariance Matrices," *Proc. ICASSP'92*, pp. v. 377–80, San Francisco, CA, March. 1992.

101. H. L. VAN TREES. *Detection, Estimation, and Modulation Theory*, Pt. I, New York: John Wiley & Sons, 1968.

102. M. WAX and I. ZISKIND. "Detection of the Number of Coherent Signals by the MDL Principle," *IEEE Trans. ASSP*, vol. 37, no. 8, pp. 1190–96, Aug. 1989.

103. P. Stoica, B. Ottersten, and M. Viberg. "An Instrumental Variable Approach to Array Processing in Spatially Correlated Noise Fields," *Proc. ICASSP'92*, San Francisco, CA, March 1992.

104. M. Viberg, P. Stoica, and B. Ottersten. "Array Processing in Correlated Noise Fields Based on Instrumental Variables and Subspace Fitting," submitted to *IEEE Trans. SP.*

105. T. Söderström and P. Stoica. *Instrumental Variable Methods for System Identification*, Berlin: Springer-Verlag, 1983.

106. G. H. Golub and C. F. Van Loan. *Matrix Computations*, 2nd ed., Baltimore: The Johns Hopkins University Press, 1989.

107. Q. Wu and J. R. Reilly. "Extension of ESPRIT Method to Unknown Noise Environments," *Proc. ICASSP'91*, pp. 3365–68, Toronto, Canada, May 1991.

108. Q. Wu, K. M. Wong, and J. P. Reilly. "Maximum Likelihood Estimation for Array Processing in Unknown Noise Environments," *Proc. ICASSP'92*, pp. v.241–v.244, San Francisco, March 1992.

109. Q. Wu, K. M. Wong, and J. R. Reilly. "Optimal Array Signal Processing in Unknown Noise Environments via Eigendecomposition Approaches," *IEEE International Symposium on Antenna Technology and Applied Electronics*, pp. 721–27, Winnipeg, Canada, August 1992.

110. Q. Wu and K. M. Wong. "Estimation of the Directions of Arrival of Signals in Unknown Correlated Noise: Application of Generalized Correlation Analysis," in J. M. F. Moura and I. M. G. Lourtie, ed., *Acoustic Signal Processing for Ocean Exploration*, Proc. NATO-ASI, July–August 1992, Kluwer Academic Publisher, 1993 Dordrecht, Germany.

111. P. Stoica, K. M. Wong, and Q. Wu. "Maximum Likelihood Bearing Estimation with Partly Calibrated Arrays in Spatially Correlated Noise Fields," submitted to *Sig. Proc.*

112. R. Kumaresan, L. L. Scharf, and A. K. Shaw. "An Algorithm for Pole-Zero Modeling and Spectral Analysis," *IEEE Trans. ASSP*, vol. 34, no. 6, pp. 637–40, June 1986.

7

Detection and Localization of Multiple Signals Using Subarrays Data

Jacob Sheinvald and Mati Wax

7.1 INTRODUCTION

Detection and localization of multiple narrowband sources by a passive sensor array is a fundamental problem in radar, communication, sonar, seismology, and radio-astronomy. The existing super-resolution solutions to this problem (see [12], [8]) require that the whole array be sampled simultaneously and consequently require that the number of receivers equal the number of sensors. This requirement may impede practical implementation of such solutions especially in cases where the number of sensors is large and where the receivers are expensive. To facilitate the implementation of super-resolution techniques in practice, we propose in this chapter a different approach where the array is sampled by parts, that is, different subarrays are sampled sequentially, with the subarrays being arbitrary and not necessarily mutually exclusive. Using this approach, and keeping the size of the subarrays small, the number of receivers needed, and consequently the associated hardware involved, is significantly smaller than in the conventional approach. Thus, for instance, only two receivers may suffice to sample the output of a large array if different pairs of sensors are sequentially switched to their inputs. In fact, the general idea is not new: conventional interferometers are using such a switching technique to sample different "baselines" of the array; see for example [11].

To illuminate the difficulties involved in performing detection and localization of multiple sources from subarrays data, it is instructive to consider the case where pairs of sensors are sampled sequentially, that is, each subarray consists of two sensors only. Notice first that in this case no single subarray can perform detection or localization on its own. Moreover, when only part of the different pairs of sensors are sampled, the existing high-resolution techniques are not applicable since no estimate of the whole array covariance, nor of any large subarray, can be constructed. Even if all the different pairs of sensors are sampled, and an estimate of the whole array covariance matrix $\hat{\mathbf{R}}$ is obtained in an element-by-element manner, with the (i, j)-th element's estimate, $\hat{R}_{ij}$, obtained from the samples of the (i, j)-th sensor pair, eigenvalue-based detection criteria, such as [16], may fail since the eigenvalues of $\hat{\mathbf{R}}$ are not guaranteed to be non-negative (as is the case when the whole array is sampled simultaneously).

The estimation method we use in conjuction with the subarrays sampling scheme is based on first deriving the Maximum Likelihood Estimator (MLE) for the problem, and then, since the computational load involved in its implementation is too heavy, approximating it by a Generalized Least-Squares (GLS) estimator that is computationally much simpler. This new estimator is proved to be both consistent and efficient. It also handles the important case of coherent signals arising, for instance, in specular multipath propagation.

The chapter is organized as follows. In Section 7.2 we formulate the problem and in Section 7.3 we present the MLE. In Section 7.4 we present the Generalized Least-Squares estimator which is the key to the proposed approach. In Sections 7.5 and 7.6 we then apply this estimator to solve the localization and detection problems. In Section 7.7 we present simulation results that demonstrate the estimator's efficiency by comparing it to the Cramér-Rao bound, and in Section 7.8 we present some concluding remarks.

7.2 PROBLEM FORMULATION

Consider q wavefronts impinging from locations $\theta_1, \ldots, \theta_q$ on an array consisting of p sensors. For simplicity, assume that the sensors and the sources are all located on the same plane and that the sources are in the far-field of the array so that $\{\theta_i\}$ represent the Directions-of-Arrival (DOAs).

Assume also that the sources emit narrowband signals (i.e., the source's bandwidth is much smaller than the reciprocal of the time delay across the array) all centered around a common frequency. Let $s_i(t)$ denote the complex envelope of the i-th source signal, and let $\mathbf{x}_c(t) = (x_{c1}(t), x_{c3}(t), \ldots, x_{cp_c}(t))^T$ denote the vector of complex envelopes formed from the signals received by the c-th subarray of sensors, with p_c denoting the number of sensors in the subarray, and T denoting transposition.

In the presence of additive noise, this received vector can be expressed as

$$\mathbf{x}_c(t) = \sum_{k=1}^{q} \mathbf{a}_c(\theta_k)s_k(t) + \mathbf{n}_c(t) \tag{7.1}$$

where $\mathbf{a}_c(\theta)$ is the steering vector of the subarray expressing its complex response to a planar wavefront arriving from direction θ, and $\mathbf{n}_c(t)$ is the complex envelope of the c-th subarray noise. This expression can be written more compactly as

$$\mathbf{x}_c(t) = \mathbf{A}_c(\boldsymbol{\theta})\mathbf{s}(t) + \mathbf{n}_c(t) \tag{7.2}$$

where $\boldsymbol{\theta} \stackrel{\text{def}}{=} (\theta_1 \ldots \theta_q)^T$, and $\mathbf{A}_c(\boldsymbol{\theta}) \stackrel{\text{def}}{=} [\mathbf{a}_c(\theta_1), \ldots, \mathbf{a}_c(\theta_q)]$ is a $p_c \times q$ matrix, and $\mathbf{s}(t) \stackrel{\text{def}}{=} (s_1(t), \ldots, s_q(t))^T$ is a vector formed from the emitted signals. We shall assume that the steering vectors $\{\mathbf{a}_c(\theta)\}$ are known for all c and all $\theta \in \boldsymbol{\Theta}$, where $\boldsymbol{\Theta}$ denotes the field-of-view.

Let the subarrays be sampled sequentially, with $\mathbf{X}_c \stackrel{\text{def}}{=} [\mathbf{x}_c(t_1^c), \ldots, \mathbf{x}_c(t_{m_c}^c)]$ denoting the c-th subarray samples, m_c denoting the number of samples taken from this subarray, and $t_1^c, \ldots, t_{m_c}^c$ denoting the sampling instants.

The matrix $\mathbf{X}_c$ can be expressed as:

$$\mathbf{X}_c = \mathbf{A}_c(\boldsymbol{\theta})\mathbf{S}_c + \mathbf{N}_c \tag{7.3}$$

where $\mathbf{S}_c \stackrel{\text{def}}{=} [\mathbf{s}(t_1^c), \ldots, \mathbf{s}(t_{m_c}^c)]$ and $\mathbf{N}_c \stackrel{\text{def}}{=} [\mathbf{n}_c(t_1^c), \ldots, \mathbf{n}_c(t_{m_c}^c)]$.

Now, the problem can be stated as follows: Given the samples of K different subarrays $\mathbf{X} \stackrel{\text{def}}{=} \{\mathbf{X}_1, \mathbf{X}_2, \ldots, \mathbf{X}_K\}$, estimate the number of sources q and their directions $\boldsymbol{\theta}$.

To solve this problem, we make the following assumptions:

A1. The noise-vector $\mathbf{n}(t)$ of the whole array is a zero-mean complex-Gaussian wide-sense stationary process with a covariance matrix $\sigma^2\mathbf{I}$, where σ^2 is an unknown positive scalar and $\mathbf{I}$ is the identity matrix, and $\{\mathbf{n}(t_i^c)\}$ are statistically independent $\forall c, i$.

A2. The signal-vector $\mathbf{s}(t)$ is a zero-mean complex-Gaussian wide-sense stationary process independent of the noise vector and having an unknown arbitrary covariance matrix $\mathbf{P}$. The signal samples $\{\mathbf{s}(t_i^c)\}$ are statistically independent $\forall c, i$.

Notice that assumption A2 does not exclude the possibility of the signals being coherent.

Based on these assumptions and using (7.2), the covariance matrix of the c-th subarray is given by

$$\mathbf{R}_c(\boldsymbol{\phi}, q) = \mathbf{A}_c(\boldsymbol{\theta})\mathbf{P}\mathbf{A}_c^H(\boldsymbol{\theta}) + \sigma^2\mathbf{I} \tag{7.4}$$

where ϕ represents all the unknown parameters $\phi = \{\boldsymbol{\theta}, \mathbf{P}, \sigma^2\}$, and $()^H$ denotes complex-conjugate transposition. To ensure the uniqueness of the solution, we also need the following assumption:

A3. The number of signals q obeys $q \le q_{\max}$, where $q_{\max}$ is the maximum number of signals for which q and ϕ can be uniquely determined. Notice that since the probability density function (pdf) of the data in the Gaussian case is completely defined by the covariances $\{\mathbf{R}_c\}$, then $q_{\max}$ is the greatest number for which

$$\{\mathbf{R}_c(\phi_1, q_1) = \mathbf{R}_c(\phi_2, q_2); c = 1, \ldots, K\} \text{ if and only if } \{q_1 = q_2 \text{ and } \phi_1 = \phi_2\} \tag{7.5}$$

with the convention that zero-power signals are not counted.

7.3 THE MAXIMUM LIKELIHOOD ESTIMATOR

From assumptions A1 and A2 it follows that the pdf for the samples of the c-th subarray is given by

$$p(\mathbf{X}_c \mid \boldsymbol{\theta}, \mathbf{P}, \sigma^2) = \left(\pi^{-p_c}|\mathbf{R}_c|^{-1} \exp\left[-\operatorname{tr}(\mathbf{R}_c^{-1}\hat{\mathbf{R}}_c)\right]\right)^{m_c} \tag{7.6}$$

where $|\cdot|$ denotes determinant, tr( ) denotes trace $\hat{\mathbf{R}}_c \overset{\text{def}}{=} \mathbf{X}_c\mathbf{X}_c^H/m_c$ is the sample-covariance matrix, and $\mathbf{R}_c$ is the covariance matrix of this subarray. From assumptions A1 and A2 it also follows that the pdf of the whole batch of data $\mathbf{X}$ is the product of the subarray pdfs; that is

$$p(\mathbf{X} \mid \boldsymbol{\theta}, \mathbf{P}, \sigma^2) = \prod_{c=1}^{K} \left\{\pi^{-p_c}|\mathbf{R}_c|^{-1} \exp\left[-\operatorname{tr}(\mathbf{R}_c^{-1}\hat{\mathbf{R}}_c)\right]\right\}^{m_c} \tag{7.7}$$

The log-likelihood function is therefore given by

$$\begin{aligned} L(\boldsymbol{\theta}, \mathbf{P}, \sigma^2) &\overset{\text{def}}{=} \log p(\mathbf{X} \mid \boldsymbol{\theta}, \mathbf{P}, \sigma^2) \\ &= \sum_{c=1}^{K} \{L_c(\boldsymbol{\theta}, \mathbf{P}, \sigma^2)\} \end{aligned} \tag{7.8}$$

with $L_c(\boldsymbol{\theta}, \mathbf{P}, \sigma^2)$ being the log-likelihood of the c-th subarray

$$\begin{aligned} L_c(\boldsymbol{\theta}, \mathbf{P}, \sigma^2) &\overset{\text{def}}{=} \log p(\mathbf{X}_c \mid \boldsymbol{\theta}, \mathbf{P}, \sigma^2) \\ &= -m_c \log|\mathbf{R}_c| - m_c \operatorname{tr}(\mathbf{R}_c^{-1}\hat{\mathbf{R}}_c) - m_c p_c \log \pi \end{aligned} \tag{7.9}$$

Now, the MLE is given by:

$$(\hat{\boldsymbol{\theta}}, \hat{\mathbf{P}}, \hat{\sigma}^2) = \arg \max_{\boldsymbol{\theta}, \mathbf{P}, \sigma^2} \left[L(\boldsymbol{\theta}, \mathbf{P}, \sigma^2)\right] \tag{7.10}$$

Unfortunately, due to the large number of free parameters involved, the computational load involved in this maximization is very heavy. Indeed, there are q^2 free real parameters in the Hermitian matrix $\mathbf{P}$, q free parameters in $\boldsymbol{\theta}$, and one additional parameter σ^2, amounting to a total of $q^2 + q + 1$ free real parameters. Yet, unlike in the MLE estimator for simultaneous sampling [3] [10], we were not able, except for the case of a single source, to reduce the computational load by eliminating $\mathbf{P}$ and σ^2 analytically. For the case of a single source, the exact reduction of (7.8) to a one-dimensional maximization problem is presented in Appendix 7A.

7.4 THE GENERALIZED LEAST-SQUARES (GLS) ESTIMATOR

In this section we derive a Generalized Least-Squares estimator for the signals covariance $\hat{\mathbf{P}}$ and the noise variance $\hat{\sigma}^2$ for any given $\boldsymbol{\theta}$.

Our estimation method can be regarded as a variant of the method of moments. The method of moments is based on expressing some moments of the distribution as a funciton of the unknown parameters and then equating these expressions to the corresponding sample moments. This yields a set of equations whose solution is the desired parameter's estimates. Notice that under the zero-mean Gaussian assumption, it suffices to use second-order moments since all the other moments can be derived from these.

From (7.2) and (7.4) we find that the sample covariance of the subarrays can be expressed as:

$$\hat{\mathbf{R}}_c = \left[\mathbf{A}_c(\boldsymbol{\theta})\mathbf{P}\mathbf{A}_c^H(\boldsymbol{\theta}) + \sigma^2\mathbf{I}\right] + \mathbf{E}_c; \quad c = 1, \ldots, K \tag{7.11}$$

where $\mathbf{E}_c$ is a zero-mean "error-matrix" expressing the difference between the sample-covariance $\hat{\mathbf{R}}_c$ and the true covariance $\mathbf{R}_c$.

Notice that since for a given $\boldsymbol{\theta}$ [and hence given $\mathbf{A}_c(\boldsymbol{\theta})$] the above set of equations is linear in the elements of $\mathbf{P}$ and σ^2, we can derive closed-form weighted-least-squares estimates for these elements. Specifically, we shall first apply a linear transformation that "whitens" the elements of $\mathbf{E}_c$ and then derive the least-squares solutions $\hat{\mathbf{P}}, \hat{\sigma}^2$ for the transformed system, expressed in terms of the matrices $\{\mathbf{A}_c(\boldsymbol{\theta})\}$.

Let $\mathbf{T}_c$ denote the "whitening" transformation. The estimator we propose is given by

$$(\hat{\mathbf{P}}, \hat{\sigma}^2) = \arg\min_{\mathbf{P},\sigma^2}\left\{\sum_c \|\mathbf{T}_c\left[\hat{\mathbf{R}}_c - \mathbf{A}_c(\boldsymbol{\theta})\mathbf{P}\mathbf{A}_c^H(\boldsymbol{\theta}) - \sigma^2\mathbf{I}\right]\mathbf{T}_c^H\|_F^2\right\} \tag{7.12}$$

where $\|\mathbf{A}\|_F = (\operatorname{tr}[\mathbf{A}^H\mathbf{A}])^{1/2}$ denotes the Frobenius norm. This estimator can be regarded as a variant of what in the statistical literature [4], [2] is known as the Generalized Least-Squares (GLS) estimator.

To solve the minimization problem we first multiply (7.11) from the left by $\mathbf{T}_c$ and from the right by $\mathbf{T}_c^H$ and mark the transformed matrices by a tilde. We thus get

$$\tilde{\hat{\mathbf{R}}}_c = \left[\tilde{\mathbf{A}}_c(\boldsymbol{\theta})\mathbf{P}\tilde{\mathbf{A}}_c^H(\boldsymbol{\theta}) + \mathbf{T}_c\sigma^2\mathbf{T}_c^H\right] + \tilde{\mathbf{E}}_c; \quad c = 1, \ldots, K \tag{7.13}$$

where

$$\tilde{\hat{\mathbf{R}}}_c \stackrel{\text{def}}{=} \mathbf{T}_c\hat{\mathbf{R}}_c\mathbf{T}_c^H \tag{7.14}$$

$$\tilde{\mathbf{A}}_c(\boldsymbol{\theta}) \stackrel{\text{def}}{=} \mathbf{T}_c\mathbf{A}_c(\boldsymbol{\theta}) \tag{7.15}$$

$$\tilde{\mathbf{E}}_c \stackrel{\text{def}}{=} \mathbf{T}_c\mathbf{E}_c\mathbf{T}_c^H \tag{7.16}$$

Using Kronecker products [6], we can rearrange (7.13) so that the elements of $\mathbf{P}$ and σ^2 are in vector form. To this end, let $\operatorname{vec}(\mathbf{Z})$ denote the vector of order

$mn \times 1$ formed from the elements of an $m \times n$ matrix $\mathbf{Z}$ by concatenating its columns, that is, if $\mathbf{Z}_{\cdot i}$ denotes the i-th column of $\mathbf{Z}$, then

$$\text{vec}\,(\mathbf{Z}) \overset{\text{def}}{=} \begin{bmatrix} \mathbf{Z}_{\cdot 1} \\ \mathbf{Z}_{\cdot 2} \\ \vdots \\ \mathbf{Z}_{\cdot n} \end{bmatrix}$$

Also, let $\mathbf{Z} \otimes \mathbf{Y}$ denote the Kronecker product of the two $m \times n, r \times s$ matrices $\mathbf{Z}, \mathbf{Y}$ (respectively), defined as the following $mr \times ns$ matrix:

$$\mathbf{Z} \otimes \mathbf{Y} \overset{\text{def}}{=} \begin{bmatrix} Z_{11}\mathbf{Y} & \cdots & Z_{1n}\mathbf{Y} \\ \vdots & \vdots & \vdots \\ Z_{m1}\mathbf{Y} & \cdots & Z_{mn}\mathbf{Y} \end{bmatrix}$$

Using this notation and the following identity [6]

$$\text{vec}\,(\mathbf{XYZ}) = (\mathbf{Z}^T \otimes \mathbf{X})\,\text{vec}\,(\mathbf{Y})$$

we can rewrite (7.13) as:

$$\text{vec}\,(\tilde{\hat{\mathbf{R}}}_c) = \left[\tilde{\mathbf{A}}_c^*(\theta) \otimes \tilde{\mathbf{A}}_c(\theta)\right] \text{vec}\,(\mathbf{P}) + \sigma^2 \text{vec}\,(\mathbf{T}_c\mathbf{T}_c^H) + \text{vec}\,(\tilde{\mathbf{E}}_c); \quad c = 1, \ldots, K \tag{7.17}$$

where $*$ denotes the complex conjugate. This can also be written as:

$$\mathbf{r} = \mathcal{A}(\theta) \begin{bmatrix} \text{vec}\,(\mathbf{P}) \\ \sigma^2 \end{bmatrix} + \mathbf{e} \tag{7.18}$$

where

$$\mathbf{r} \overset{\text{def}}{=} \begin{bmatrix} \text{vec}\,(\tilde{\hat{\mathbf{R}}}_1) \\ \text{vec}\,(\tilde{\hat{\mathbf{R}}}_2) \\ \vdots \\ \text{vec}\,(\tilde{\hat{\mathbf{R}}}_K) \end{bmatrix} \quad \mathcal{A}(\theta) \overset{\text{def}}{=} \begin{bmatrix} \tilde{\mathbf{A}}_1^*(\theta) \otimes \tilde{\mathbf{A}}_1(\theta) & \boldsymbol{\tau}_1 \\ \tilde{\mathbf{A}}_2^*(\theta) \otimes \tilde{\mathbf{A}}_2(\theta) & \boldsymbol{\tau}_2 \\ \vdots & \vdots \\ \tilde{\mathbf{A}}_K^*(\theta) \otimes \tilde{\mathbf{A}}_K(\theta) & \boldsymbol{\tau}_K \end{bmatrix} \quad \mathbf{e} \overset{\text{def}}{=} \begin{bmatrix} \text{vec}\,(\tilde{\mathbf{E}}_1) \\ \text{vec}\,(\tilde{\mathbf{E}}_2) \\ \vdots \\ \text{vec}\,(\tilde{\mathbf{E}}_K) \end{bmatrix}$$

with the $p_c^2 \times 1$ vectors $\{\boldsymbol{\tau}_c\}$ given by

$$\boldsymbol{\tau}_c = \text{vec}\,(\mathbf{T}_c\mathbf{T}_c^H); \quad c = 1, \ldots, K \tag{7.19}$$

Using the vector form (7.18), we can rewrite (7.12) as

$$(\hat{\mathbf{P}}, \hat{\sigma}^2) = \arg\min_{\mathbf{P}, \sigma^2} \left\{ \|\mathbf{r} - \mathcal{A}(\theta) \begin{bmatrix} \text{vec}\,(\mathbf{P}) \\ \sigma^2 \end{bmatrix} \|^2 \right\} \tag{7.20}$$

whose solution is given by the well-known expression

$$\begin{bmatrix} \text{vec}\,(\hat{\mathbf{P}}) \\ \hat{\sigma}^2 \end{bmatrix} = (\mathcal{A}^{\mathbf{H}}(\theta)\mathcal{A}(\theta))^{-1}\mathcal{A}^{\mathbf{H}}(\theta)\mathbf{r} \tag{7.21}$$

As mentioned above, the transformation matrices $\{\mathbf{T}_c\}$ are selected to "whiten" the elements of the error matrices $\{\tilde{\mathbf{E}}_c\}$. To motivate the selection of $\{\mathbf{T}_c\}$, we prove in Appendix 7B that selecting $\mathbf{T}_c = \sqrt[4]{m_c}\mathbf{R}_c^{-1/2}$ whitens the elements of $\tilde{\mathbf{E}}_c$, that is, guarantees that the cross-correlation between any two elements of $\tilde{\mathbf{E}}_c$ is zero, while the variance of any element is 1, namely

$$E\left[\mathbf{e}\mathbf{e}^H\right] = \mathbf{I}$$

Yet, since the true covariances $\{\mathbf{R}_c\}$ are unknown, we use the sample covariances instead, that is

$$\mathbf{T}_c = \sqrt[4]{m_c}\hat{\mathbf{R}}_c^{-1/2}$$

assuming that $\hat{\mathbf{R}}_c$ is full-rank $\forall c$. This choice does not guarantee an exact whitening. However, it can be proved that this choice results in an estimator which is both consistent and efficient; see Appendices 7C and 7E. In addition, this choice simplifies the computation. Indeed, since in this case $\hat{\mathbf{R}}_c = \mathbf{I}_{p_c}$, with $\mathbf{I}_{p_c}$ denoting the $p_c \times p_c$ identity matrix, (7.21) reduces to

$$\begin{bmatrix} \text{vec}\,(\hat{\mathbf{P}}) \\ \hat{\sigma}^2 \end{bmatrix}_{GLS} = (\mathcal{A}^{\mathbf{H}}(\theta)\mathcal{A}(\theta))^{-1}\mathcal{A}^{\mathbf{H}}(\theta)\mathbf{r} \tag{7.22}$$

where

$$\mathbf{r} = \begin{bmatrix} \text{vec}\,(\mathbf{I}_{p_1}) \\ \text{vec}\,(\mathbf{I}_{p_2}) \\ \vdots \\ \text{vec}\,(\mathbf{I}_{p_K}) \end{bmatrix} \tag{7.23}$$

and

$$\mathcal{A}(\theta) = \begin{bmatrix} \tilde{\mathbf{A}}_1^*(\theta) \otimes \tilde{\mathbf{A}}_1(\theta) & \sqrt{m_1}\,\text{vec}\,(\hat{\mathbf{R}}_1^{-1}) \\ \tilde{\mathbf{A}}_2^*(\theta) \otimes \tilde{\mathbf{A}}_2(\theta) & \sqrt{m_2}\,\text{vec}\,(\hat{\mathbf{R}}_2^{-1}) \\ \vdots & \vdots \\ \tilde{\mathbf{A}}_K^*(\theta) \otimes \tilde{\mathbf{A}}_K(\theta) & \sqrt{m_K}\,\text{vec}\,(\hat{\mathbf{R}}_K^{-1}) \end{bmatrix} \tag{7.24}$$

with $\tilde{\mathbf{A}}_c(\theta) = \sqrt[4]{m_c}\hat{\mathbf{R}}_c^{-1/2}\mathbf{A}_c(\theta)$.

Notice that in the above derivation of the GLS estimator, $\hat{\mathbf{P}}$ has not been constrained to be Hermitian, and neither has $\hat{\sigma}^2$ been constrained to be real-valued. This could however be easily accomplished by inserting into (7.18), $\mathbf{P} = \mathbf{P}_R + j\mathbf{P}_I$, where $\mathbf{P}_R$ and $\mathbf{P}_I$ denote the real and imaginary parts of $\mathbf{P}$, and constraining $\mathbf{P}_R$ and $\mathbf{P}_I$ to be symmetric and antisymmetric, respectively, thus reducing the total number of unknown components by half. Then, by equating the real parts and the imaginary parts of both sides of (7.18), we get a set of *real* equations with σ^2 and the elements of the upper triangles of $\mathbf{P}_R, \mathbf{P}_I$ as *real* unknowns. The resulting real LS estimator is more accurate and computationally more effective due to the reduction of the parameter's dimensionality. The detailed derivation is somewhat lengthy and is presented in Appendix 7F.

In case the signals are known to be independent (thus making $\mathbf{P}$ diagonal), the above equations can be easily modified to get the LS estimate for $\begin{bmatrix} \text{diag}(\mathbf{P}) \\ \sigma^2 \end{bmatrix}$, where diag ($\mathbf{P}$) is a $q \times 1$ vector containing the diagonal elements of $\mathbf{P}$. The resulting $\mathcal{A}(\boldsymbol{\theta})$ matrix in this case is the matrix obtained from the general $\mathcal{A}(\boldsymbol{\theta})$ by deleting columns corresponding to zero elements in the vector vec ($\mathbf{P}$).

7.5 ESTIMATION OF THE DIRECTIONS-OF-ARRIVAL

One approach to obtain an estimator for $\boldsymbol{\theta}$ is to substitute $\hat{\mathbf{P}}(\boldsymbol{\theta})$ and $\hat{\sigma}^2(\boldsymbol{\theta})$ from (7.22) into the likelihood function (7.8). This yields the following estimator for $\boldsymbol{\theta}$:

$$\text{GLS-ML:} \quad \hat{\boldsymbol{\theta}} = \arg\max_{\boldsymbol{\theta}} \left\{ L(\boldsymbol{\theta}, \hat{\mathbf{P}}(\boldsymbol{\theta}), \hat{\sigma}^2(\boldsymbol{\theta})) \right\} \tag{7.25}$$

where $L(\boldsymbol{\theta}, \hat{\mathbf{P}}(\boldsymbol{\theta}), \hat{\sigma}^2(\boldsymbol{\theta}))$ is a function of $\boldsymbol{\theta}$ only:

$$\begin{aligned} L(\boldsymbol{\theta}, \hat{\mathbf{P}}(\boldsymbol{\theta}), \hat{\sigma}^2(\boldsymbol{\theta})) = \sum_c \{ & -m_c \log |\mathbf{R}_c(\boldsymbol{\theta}, \hat{\mathbf{P}}(\boldsymbol{\theta}), \hat{\sigma}^2(\boldsymbol{\theta}))| \\ & - m_c \operatorname{tr}(\mathbf{R}_c^{-1}(\boldsymbol{\theta}, \hat{\mathbf{P}}(\boldsymbol{\theta}), \hat{\sigma}^2(\boldsymbol{\theta}))\hat{\mathbf{R}}_c) \\ & - m_c p_c \log \pi \} \end{aligned} \tag{7.26}$$

with

$$\mathbf{R}_c(\boldsymbol{\theta}, \hat{\mathbf{P}}(\boldsymbol{\theta}), \hat{\sigma}^2(\boldsymbol{\theta})) = \mathbf{A}_c(\boldsymbol{\theta})\hat{\mathbf{P}}(\boldsymbol{\theta})\mathbf{A}_c^H(\boldsymbol{\theta}) + \hat{\sigma}^2(\boldsymbol{\theta})\mathbf{I}$$

A simpler alternative is to use the GLS funciton (7.12) to estimate $\boldsymbol{\theta}$, that is

$$\text{GLS:} \quad \hat{\boldsymbol{\theta}} = \arg\min_{\boldsymbol{\theta}} \left\{ L_{GLS}(\boldsymbol{\theta}, \hat{\mathbf{P}}(\boldsymbol{\theta}), \hat{\sigma}^2(\boldsymbol{\theta})) \right\} \tag{7.27}$$

where $L_{GLS}(\boldsymbol{\theta}, \mathbf{P}, \sigma^2)$ is given by

$$L_{GLS}(\boldsymbol{\theta}, \mathbf{P}, \sigma^2) \stackrel{\text{def}}{=} \sum_c \|\mathbf{T}_c \left[\hat{\mathbf{R}}_c - \mathbf{R}_c(\boldsymbol{\theta}, \mathbf{P}, \sigma^2)\right] \mathbf{T}_c^H \|_F^2 \tag{7.28}$$

Using the equivalence of (7.12) and (7.20), this can be written as

$$\text{GLS:} \quad \hat{\boldsymbol{\theta}} = \arg\min_{\boldsymbol{\theta}} \left\{ \|\mathbf{r} - \mathcal{A}(\boldsymbol{\theta}) \begin{bmatrix} \text{vec}(\hat{\mathbf{P}}) \\ \hat{\sigma}^2 \end{bmatrix} \|_F^2 \right\} \tag{7.29}$$

which by substituting the GLS estimates $\hat{\mathbf{P}}$ and $\hat{\sigma}^2$ from (22) can be written as

$$\text{GLS:} \quad \hat{\boldsymbol{\theta}} = \arg\min_{\boldsymbol{\theta}} \left\{ \|\mathbf{r} - \mathcal{A}(\boldsymbol{\theta})(\mathcal{A}^{\mathbf{H}}(\boldsymbol{\theta})\mathcal{A}(\boldsymbol{\theta}))^{-1}\mathcal{A}^{\mathbf{H}}(\boldsymbol{\theta})\mathbf{r}\|^2 \right\} \tag{7.30}$$

or

$$\text{GLS:} \quad \hat{\boldsymbol{\theta}} = \arg\min_{\boldsymbol{\theta}} \left\{ \mathbf{r}^T \mathbf{P}_{\mathcal{A}(\boldsymbol{\theta})}^{\perp} \mathbf{r} \right\} \tag{7.31}$$

where $\mathbf{P}_{\mathcal{A}(\boldsymbol{\theta})}^{\perp}$ is the projection matrix on the subspace orthogonal to the subspace spanned by the columns of $\mathcal{A}(\boldsymbol{\theta})$:

$$\mathbf{P}^{\perp}_{\mathcal{A}(\boldsymbol{\theta})} \stackrel{\text{def}}{=} \mathbf{I} - \mathcal{A}(\boldsymbol{\theta})(\mathcal{A}^{\mathbf{H}}(\boldsymbol{\theta})\mathcal{A}(\boldsymbol{\theta}))^{-1}\mathcal{A}^{\mathbf{H}}(\boldsymbol{\theta})$$

Notice that for $\mathbf{T}_c = \sqrt[4]{m_c}\hat{\mathbf{R}}_c^{-1/2}$ the vector $\mathbf{r}$ in (7.31) is just a binary vector, thus simplifying computations.

As proved in Appendices 7C and 7E, the above estimators are both consistent and efficient (i.e., asymptotically attain the Cramér-Rao lower bound—(CRB).

Notice that the dimensionality of the maximization problem has been reduced to q. The maximization can be efficiently accomplished either by alternatingly searching along the coordinates of $\boldsymbol{\theta}$ in a manner similar to the Alternating Projections (AP) method described in [18], or by any other multidimensional maximization technique.

A necessary condition for the uniqueness of these solutions is that the total number of free parameters is not greater than the total number of equations. If the number of samples equals or exceeds the subarray size, that is, $m_c \geq p_c$, the number of free real components in the Hermitian matrix $\hat{\mathbf{R}}_c$ is p_c^2 while the total number of free real parameters is $q^2 + q + 1$. The above condition is therefore met if

$$\sum_c p_c^2 \geq q^2 + q + 1 \tag{7.32}$$

7.6 DETECTION OF THE NUMBER OF SIGNALS

To solve the detection problem we use the Minimum Description Length (MDL) criterion [13]. As is well known, given data $\mathbf{X}$ and a family of competing parameterized pdfs $\{p(\mathbf{X}|\phi^{(q)}); q = 1, 2, \ldots, q_{\max}\}$, the MDL criterion is given by

$$\hat{q} = \arg \min_{0 \leq q \leq q\,\max} \left[-\log p(\mathbf{X}|\hat{\phi}^{(q)}) + \frac{k_q}{2} \log m \right] \tag{7.33}$$

where $\hat{\phi}^{(q)}$ is the ML estimate, k_q is the number of free parameters in $\hat{\phi}^{(q)}$, and m is the number of samples in $\mathbf{X}$.

To apply this criterion to the problem at hand, we have to resolve two problems. First, we cannot use the ML estimates since they are computationally complex. Second, it is not clear what to use for m since our likelihood is a sum of likelihoods of subarrays, each with m_c samples. Our solutions to both problems are rather natural. Instead of the ML estimates we use the GLS estimates, and for the value of m we use $m = \sum_c m_c$.

Thus, the MDL criterion becomes

$$\hat{q} = \arg \min_{0 \leq q \leq q\,\max} \left[-L(\hat{\boldsymbol{\theta}}, \hat{\mathbf{P}}(\hat{\boldsymbol{\theta}}), \hat{\sigma}^2(\hat{\boldsymbol{\theta}})) + \frac{q + q^2 + 1}{2} \log m \right] \tag{7.34}$$

where $\hat{\mathbf{P}}$, $\hat{\sigma}^2$, and $\hat{\boldsymbol{\theta}}$ are the GLS estimates, and $L(\boldsymbol{\theta}, \hat{\mathbf{P}}(\boldsymbol{\theta}), \hat{\sigma}^2(\boldsymbol{\theta}))$ is given by (7.26).

As proved in Appendix 7D, this criterion yields a consistent estimator of q.

7.7 SIMULATION RESULTS

To demonstrate the performance of the described method, we simulated a 5ttt omnidirectional-element uniform circular array with a diameter of 0.6λ. The 3-dB beamwidth of the array is approximately 26°. Two equipower uncorrelated Gaussian sources were located at 100° and 120°. The subarrays consisted of the five longest 2-element subarrays (baselines). 128 independent samples were taken from each subarray. In a first experiment a set of 100 montecarlo runs was carried out for each SNR, with the DOAs estimated in each run and the RMS DOA error (averaged over the two sources) computed from the whole set. Figure 7.1 compares the results obtained by using the GLS and GLS-ML estimators with the CRB (derived in Appendix 7E). The efficiency of both estimators is evident. It should be noted that in the search for the maximum of the GLS-ML function (7.26), we encountered $\boldsymbol{\theta}$-grid-points that yielded negative-definite $\hat{\mathbf{P}}(\boldsymbol{\theta})$ or $\hat{\sigma}^2(\boldsymbol{\theta})$. We omitted these points from the search grid, so that the actual search was performed on a reduced $\boldsymbol{\theta}$-grid [i.e., only those $\boldsymbol{\theta}$-grid points that yielded non-negative-definite $\hat{\mathbf{P}}(\boldsymbol{\theta})$ or $\hat{\sigma}^2(\boldsymbol{\theta})$]. (Notice that any Gaussian likelihood function is not defined for negative-definite covariance $\mathbf{R}$.)

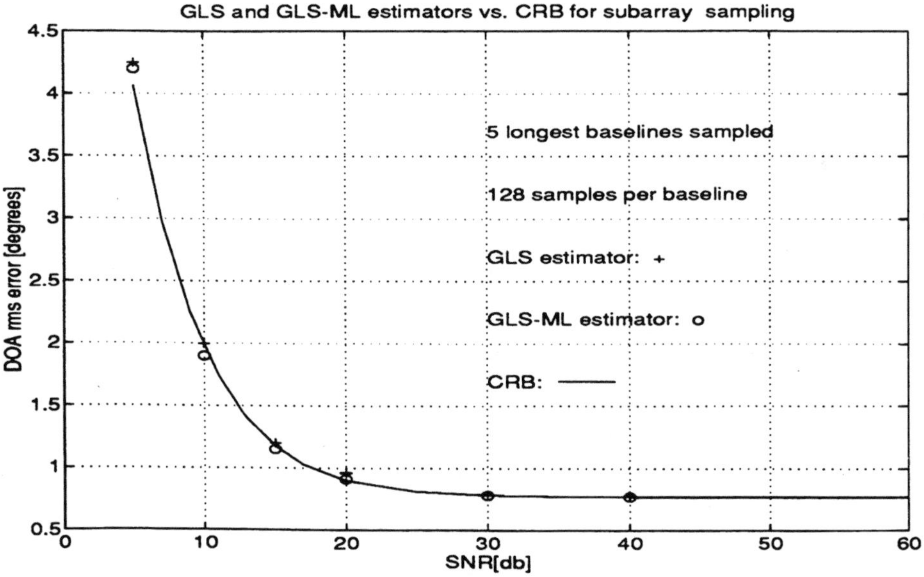

Figure 7.1 The RMS DOA error of subarray sampling as a function of SNR for two equipower uncorrelated sources located at 100° and 120° impinging on a 5-element circular array with diameter 0.6 λ. Only the 5 longest baselines are sampled.

In a second experiment, we sampled all the ten possible 2-element subarrays (baselines), but the number of samples from each subarray was reduced by a factor of two, that is, $m_c = 64$, thus retaining the same overall number of samples used. The results are shown in Figure 7.2. Comparing Figure 7.2 with Figure 7.1, it turns

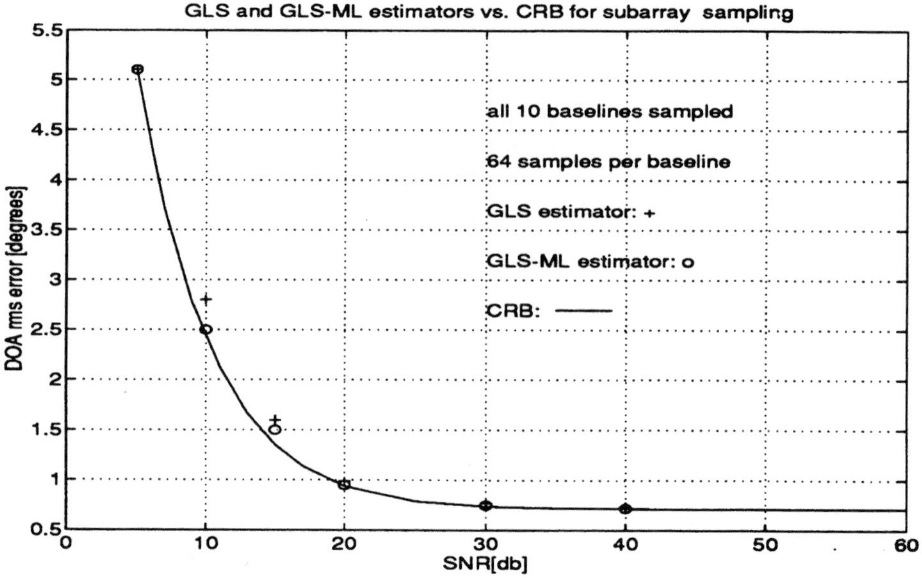

Figure 7.2 The RMS DOA error as a function of SNR for two equipower uncorrelated sources located at 100° and 120° impinging on a 5-element circular array with diameter 0.6 λ. All the 10 baselines are consecutively sampled.

out, not surprisingly, that it is advantageous to sample only the largest baselines rather than spend time on the smaller baselines.

Notice that in Figures 7.1 and 7.2 the DOA errors do not vanish for high SNR values. This phenomenon is seen more clearly in Figures 7.3 and 7.4, where the CRB for subarray sampling is compared with the CRB for simultaneous sampling. The number of snapshots in the simultaneous sampling is $m = 64$, that is, equal to the number of samples taken from each subarray. Notice that at low SNR there is essentially no difference in the DOA errors of the two DF methods.

To demonstrate the advantage of the proposed estimators, we compared them with a "naive" estimator constructed as follows: from the various 2-element samples we create a "sample-covariance" $\hat{\hat{\mathbf{R}}} = \{\hat{\hat{R}}_{ij}\}$ of the whole array in an element-by-element way, with $\hat{\hat{R}}_{ij} = \hat{\hat{R}}^*_{ji}$ created from the samples of the subarray containing elements i and j, and $\hat{\hat{R}}_{ii}$ created from samples of all the subarrays containing element i. We then use the Alternating Projections algorithm [18] in conjuction with $\hat{\hat{\mathbf{R}}}$ to estimate the DOAs. (In fact, any other estimation method, such as MUSIC [14], could have been used intead.) The results are shown in Figure 7.5. Clearly, the sources are not resolvable by this naive estimator.

To demonstrate the performance of the MDL criterion we present the results obtained by applying it to the above two experiments. Figure 7.6 shows the detection performance as a function of the SNR. The probability of correct detection improves as the SNR grows, but not necessarily approaches 1. On the other hand, the

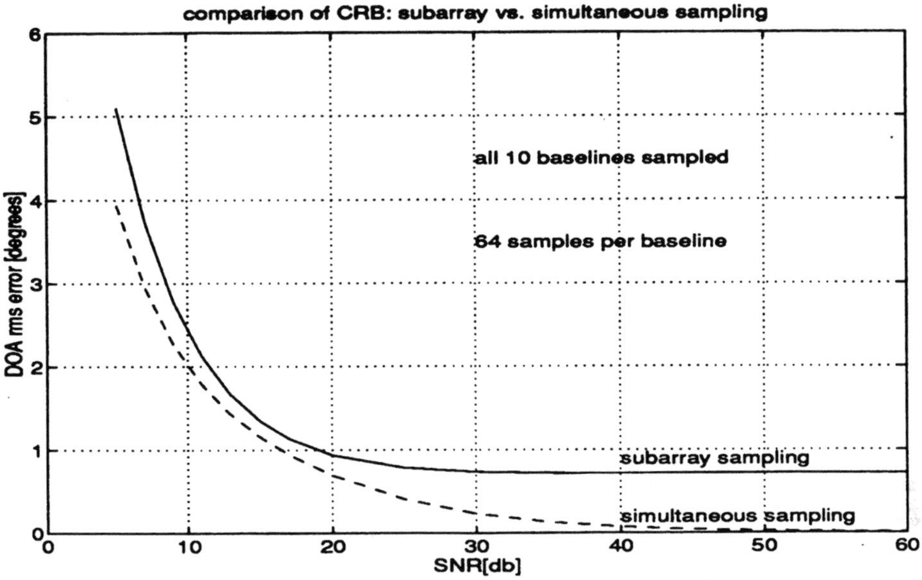

Figure 7.3 The RMS DOA error as a function of SNR for two equipower uncorrelated sources located at 100° and 120° impinging on a 5-element circular array with diameter 0.6 λ. The solid line represents the CRB for subarray sampling of all the 10 baselines. The dashed line represents the CRB for simultaneous sampling of the whole array.

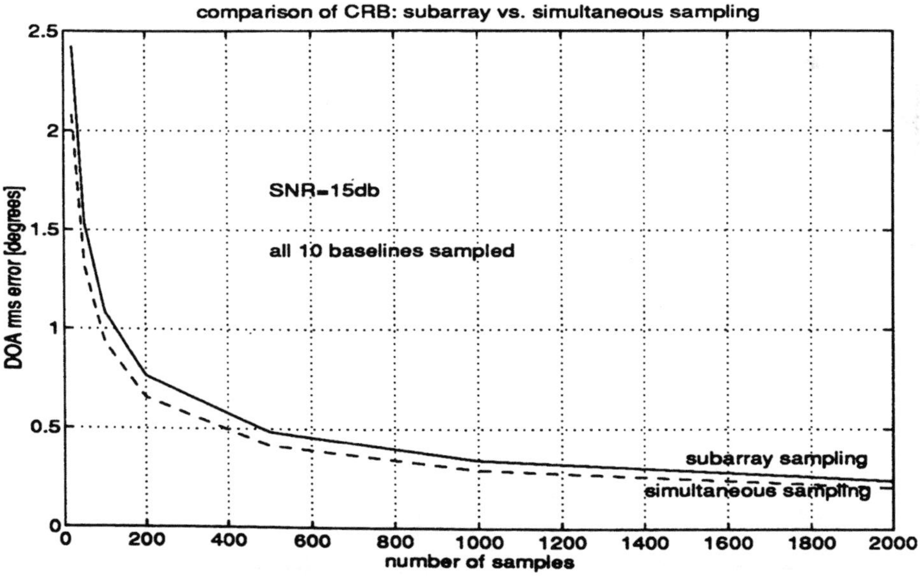

Figure 7.4 The RMS DOA error as a function of the number of samples for two equipower uncorrelated sources located at 100° and 120° impinging on a 5-element circular array with diameter 0.6 λ. The solid line represents the CRB for subarray sampling of all the 10 baselines. The dashed line represents the CRB for simultaneous sampling of the whole array.

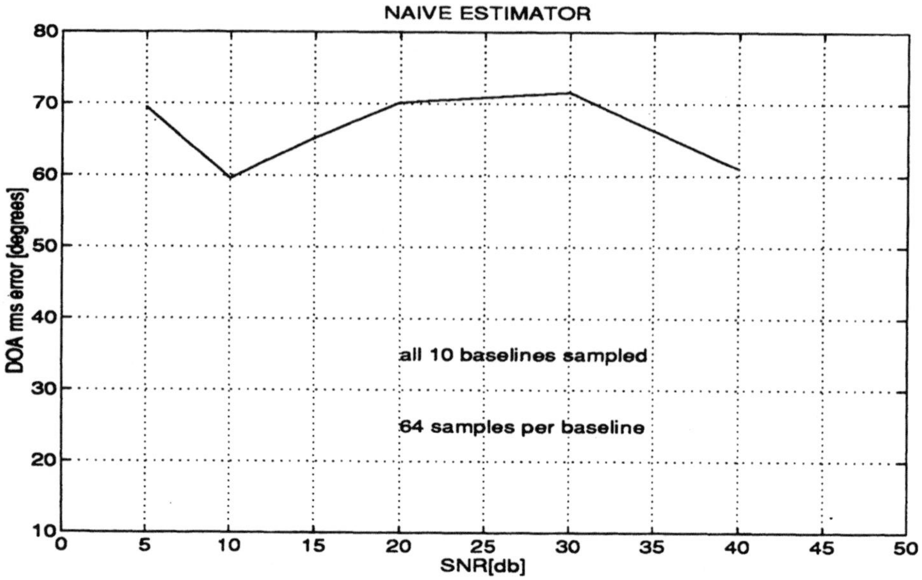

Figure 7.5 The RMS DOA error of the "naive" estimator as a function of SNR for two equipower uncorrelated sources located at 100° and 120° impinging on a 5-element circular array with diameter 0.6 λ. All the 10 baselines are consecutively samples.

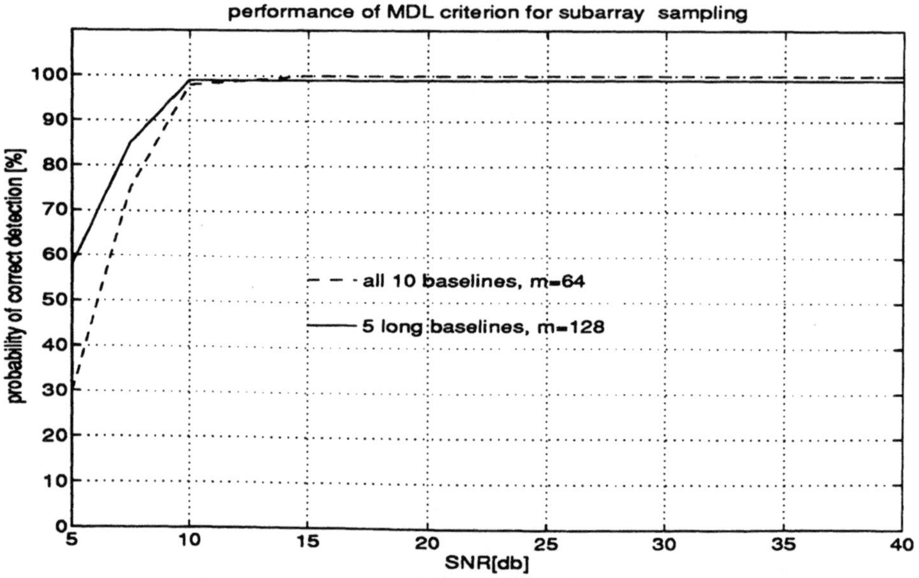

Figure 7.6 The probability of detecting correctly the number of signals by subarray sampling as a function of SNR for two equipower uncorrelated sources located at 100° and 120° impinging on a 5-element circular array with diameter 0.6 λ.

probability of correct detection approached 1 in both experiments when we let the number of samples grow.

7.8 CONCLUDING REMARKS

We have presented a new method for high-resolution direction finding that is based on sampling subarrays in a sequential manner, which significantly reduces the amount of hardware required for the implementation. The DOA estimates derived are efficient and have moderate computational complexity. The detection is based on the MDL criterion and is proved to be consistent.

It should be pointed out that in case the number of signals is less than the number of sensors in each subarray, signal subspace methods such as MUSIC [14] or subspace-fitting [15] may be employed, and the emitter's waveforms may be estimated.

APPENDIX 7A THE MLE FOR A SINGLE SOURCE

In this appendix we derive the exact MLE for the case of a single source and an array consisting of identical elements with the subarrays having an identical number of elements.

The general likelihood function for subarray sampling is given in (7.8) and (7.9) as:

$$L = \sum_{c=1}^{K} L_c; \quad L_c = -m_c \log |\mathbf{R}_c| - m_c \operatorname{tr}(\mathbf{R}_c^{-1}\hat{\mathbf{R}}_c) - m_c p_c \log \pi \tag{7A.1}$$

For a single source we have:

$$\mathbf{R}_c = \beta \mathbf{a}_c \mathbf{a}_c^H + \mathbf{I}\sigma^2 \tag{7A.2}$$

where β denotes the signal power, and $\mathbf{a}_c$ is a shorthand for $\mathbf{a}_c(\theta)$. Using Woodburry's matrix inversion formula [5] we get:

$$\mathbf{R}_c^{-1} = \sigma^{-2}\left[\mathbf{I} - (\beta^{-1}\sigma^2 + \|\mathbf{a}_c\|^2)^{-1}\mathbf{a}_c\mathbf{a}_c^H\right] \tag{7A.3}$$

Note that (7A.2) implies that the vector $\mathbf{a}_c$ is an eigenvector of $\mathbf{R}_c$. Indeed, we have

$$\mathbf{R}_c\mathbf{a}_c = (\sigma^2 + \beta\|\mathbf{a}_c\|^2)\mathbf{a}_c \tag{7A.4}$$

implying that $\lambda_{c1} = (\sigma^2 + \beta\|\mathbf{a}_c\|^2)$ is the corresponding eigenvalue. The rest of the eigenvalues $\{\lambda_{ci}, i \geq 2\}$ are all equal to σ^2. Hence, since the determinant of an Hermitian matrix is the product of its eigenvalues, we have:

$$|\mathbf{R}_c| = \prod_{i=1}^{p_c} \lambda_{ci} = \sigma^{2p_c}(1 + \beta\sigma^{-2}\|\mathbf{a}_c\|^2) \tag{7A.5}$$

Substituting (7A.3), (7A.5) into the likelihood function, we get:

$$L(\theta, \beta, \sigma^2) = \sum_c m_c \left[-p_c \log(\pi\sigma^2) - \log(1 + \beta\sigma^{-2}\|\mathbf{a}_c\|^2) - \gamma_c\sigma^{-2} \right] \quad (7A.6)$$

where

$$\gamma_c \stackrel{\text{def}}{=} \text{tr}\,(\hat{\mathbf{R}}_c) - (\beta^{-1}\sigma^{-2}\|\mathbf{a}_c\|^2)^{-1}\mathbf{a}_c^H\hat{\mathbf{R}}_c\mathbf{a}_c$$

Now, the MLE is given by:

$$\hat{\theta} = \arg\max_{\theta,\beta,\sigma^2} L(\theta, \beta, \sigma^2) \quad (7A.7)$$

To solve this maximization problem we first maximize with respect to β while holding θ and σ^2 fixed, then substitute the maximizing value $\hat{\beta}(\theta, \sigma^2)$ back into the likelihood function, and finally maximize with respect to σ^2, thus obtaining an expression for the likelihood as a function of θ only.

To maximize with respect to β we set

$$\frac{\partial L}{\partial \beta} = 0$$

which by assuming that $\|\mathbf{a}_c(\theta)\|^2$ is identical for all c (as, for instance, is the case when all sensors have the same radiation pattern and all subarrays consist of the same number of elements), yields after some simple algebra

$$\hat{\beta}(\theta, \sigma^2) = \frac{\sum_c m_c \left[\dfrac{\mathbf{a}_c^H(\theta)\hat{\mathbf{R}}_c\mathbf{a}_c(\theta) - \sigma^2\|\mathbf{a}_c(\theta)\|^2}{\|\mathbf{a}_c(\theta)\|^4} \right]}{\sum_c m_c} \quad (7A.8)$$

Substituting this $\hat{\beta}(\theta, \sigma^2)$ back into (7A.6) we get the likelihood $L(\theta, \hat{\beta}(\theta, \sigma^2), \sigma^2)$ as a function of θ and σ^2 only. Then, setting

$$\frac{\partial L(\theta, \hat{\beta}(\theta, \sigma^2), \sigma^2)}{\partial \sigma^2} = 0$$

we get (after some simple algebra)

$$\hat{\sigma}^2 = \frac{\sum_c m_c\, \text{tr}\left[\mathbf{P}^{\perp}_{\mathbf{a}_c(\theta)}\hat{\mathbf{R}}_c\right] / (p_c - 1)}{\sum_c m_c} \quad (7A.9)$$

where $\mathbf{P}^{\perp}_{\mathbf{a}_c(\theta)}$ is the projection matrix on the subspace orthogonal to $\mathbf{a}_c(\theta)$:

$$\mathbf{P}^{\perp}_{\mathbf{a}_c(\theta)} \stackrel{\text{def}}{=} \mathbf{I} - \frac{\mathbf{a}_c(\theta)\mathbf{a}_c^H(\theta)}{\|\mathbf{a}_c(\theta)\|^2}$$

Substituting this σ^2 back into $L(\theta, \hat{\beta}(\theta, \sigma^2), \sigma^2)$ we finally get:

$$\begin{aligned} L(\theta) = & -(\overline{p} - m)\log((\overline{r} - \lambda(\theta)) - m\log(\lambda(\theta)) \\ & + m\log m + (\overline{p} - m)\log(\overline{p} - m) - \overline{p}(1 + \log\pi) \end{aligned} \quad (7A.10)$$

where

$$\bar{p} \stackrel{\text{def}}{=} \sum_c m_c p_c$$

$$m \stackrel{\text{def}}{=} \sum_c m_c$$

$$\bar{r} \stackrel{\text{def}}{=} \sum_c m_c \operatorname{tr}(\hat{\mathbf{R}}_c)$$

$$\lambda(\theta) \stackrel{\text{def}}{=} \sum_c m_c \frac{\mathbf{a}_c^H(\theta)\hat{\mathbf{R}}_c\mathbf{a}_c(\theta)}{\mathbf{a}_c^H(\theta)\mathbf{a}_c(\theta)}$$

Notice that $\lambda(\theta)$ in (7A.10) is the only θ-dependent term. (When searching for the optimal θ, the terms not containing $\lambda(\theta)$ can be omitted. However, they were left in (7A.10) since they are needed when the MDL criterion is applied.) Notice also that $\lambda(\theta)$ can also be represented in terms of $\mathbf{P}_{\mathbf{a}(\theta)}$, the projection matrix on $\mathbf{a}(\theta)$, as:

$$\lambda(\theta) = \sum_c m_c \operatorname{tr}\left[\mathbf{P}_{\mathbf{a}_c(\theta)}\hat{\mathbf{R}}_c\right]$$

Notice that the estimate $\hat{\sigma}^2$ given in (7A.9) can be represented as

$$\hat{\sigma}^2 = \frac{\sum_c m_c \hat{\sigma}_c^2}{\sum_c m_c} \tag{7A.11}$$

where $\hat{\sigma}_c^2$ is the ML estimate of the noise variance based on the c-th subarray data only [3]:

$$\hat{\sigma}_c^2 = \operatorname{tr}\left[\mathbf{P}^{\perp}_{\mathbf{a}_c(\theta)}\hat{\mathbf{R}}_c\right]/(p_c - 1) \tag{7A.12}$$

Thus, $\hat{\sigma}^2$ is a weighted average of the $\{\hat{\sigma}_c^2\}$.

Similarly, the estimate $\hat{\beta}$ given in (7A.8) can be represented as the following weighted average (for any given θ and σ^2):

$$\hat{\beta} = \frac{\sum_c m_c \hat{\beta}_c}{\sum_c m_c} \tag{7A.13}$$

where $\hat{\beta}_c$ is the ML estimate of β based on the c-th subarray data only. Curiously enough, $\hat{\beta}_c$ can also be shown to be the LS estimate obtained from the c-th subarray data. Indeed, from (7.11) we can write

$$\hat{\mathbf{R}}_c = \left[\mathbf{a}_c\mathbf{a}_c^H\right]\beta + \sigma^2\mathbf{I} + \mathbf{E}_c \tag{7A.14}$$

which in vector form is given by

$$\operatorname{vec}(\hat{\mathbf{R}}_c) = \left[\mathbf{a}_c^* \otimes \mathbf{a}_c\right]\beta + \sigma^2 \operatorname{vec}(\mathbf{I}) + \operatorname{vec}(\mathbf{E}_c) \tag{7A.15}$$

Therefore, the LS estimate of β is given by

$$\hat{\beta}_c = \frac{[\mathbf{a}_c^* \otimes \mathbf{a}_c]^H (\text{vec}(\hat{\mathbf{R}}_c) - \sigma^2 \text{vec}(\mathbf{I}))}{\|\mathbf{a}_c^* \otimes \mathbf{a}_c\|^2} \tag{7A.16}$$

Using the following easily verified identities

$$[\mathbf{a}_c^* \otimes \mathbf{a}_c]^H (\text{vec}(\hat{\mathbf{R}}_c) = \mathbf{a}_c^H \hat{\mathbf{R}}_c \mathbf{a}_c$$

$$[\mathbf{a}_c^* \otimes \mathbf{a}_c]^H (\text{vec}(\mathbf{I}) = \|\mathbf{a}_c\|^2$$

$$\|\mathbf{a}_c^* \otimes \mathbf{a}_c\|^2 = \|\mathbf{a}_c\|^4$$

Equation (7A.16) can be rewritten as

$$\hat{\beta}_c = \frac{\mathbf{a}_c^H \hat{\mathbf{R}}_c \mathbf{a}_c - \sigma^2 \|\mathbf{a}_c\|^2}{\|\mathbf{a}_c\|^4}$$

Notice that this LS estimate was derived without whitening the error-vector, vec $(\mathbf{E}_c)$.

APPENDIX 7B DERIVATION OF THE WHITENING MATRIX

In this appendix we prove that $\mathbf{T} = \sqrt[4]{m}\,\mathbf{R}^{-1/2}$ whitens the elements of the transformed error matrix $\tilde{\mathbf{E}}$ in (7.13) (we have omitted the subscript c for notational convenience). As a by-product, we also get expressions for the cross-correlation between the elements of the error matrix $\mathbf{E}$. This is an extension for the complex case of similar results derived, for example, in [1], for the real case.

Let $\mathbf{E} = \hat{\mathbf{R}} - \mathbf{R}$ denote the error matrix, and let R_{ij} and E_{ij} denote the elements of $\mathbf{R}$ and $\mathbf{E}$ (respectively). The elements of $\mathbf{E}$ are obviously zero-mean since:

$$E(\mathbf{E}) = E(\hat{\mathbf{R}}) - \mathbf{R} = \mathbf{R} - \mathbf{R} = 0$$

The cross-correlation between any two elements of $\mathbf{E}$ is given by:

$$E(E_{ij} E_{kl}^*) = E(\hat{R}_{ij} \hat{R}_{kl}^*) - R_{ij} R_{kl}^* \tag{7B.1}$$

To compute the term $E(\hat{R}_{ij} \hat{R}_{kl}^*)$ we use the following formula [9], valid for any complex-Gaussian variables u, v, x, y:

$$E(uv^*xy^*) = E(uv^*)E(xy^*) + E(v^*x)E(uy^*)$$

We get (assuming for simplicity that the sampling instants are the integers $t = 1, \ldots, m$):

$$E(\hat{R}_{ij} \hat{R}_{kl}^*) = E\left[\left(\frac{1}{m}\sum_{t_1=1}^{m} x_i(t_1) x_j^*(t_1)\right)\left(\frac{1}{m}\sum_{t_2=1}^{m} x_k(t_2) x_l^*(t_2)\right)^*\right]$$

$$= \frac{1}{m^2} E\left[\sum_{t=1}^{m} \left(x_i(t) x_j^*(t) x_l(t) x_k^*(t)\right)\right]$$

$$+\frac{1}{m^2}\sum_{t_1\neq t_2} E\left(x_i(t_1)x_j^*(t_1)\right)E^*\left(x_k(t_2)x_l^*(t_2)\right)$$

$$=\frac{1}{m^2}\sum_{t=1}^{m}\left[E\left(x_i(t)x_j^*(t)\right)E\left(x_l(t)x_k^*(t)\right)+E\left(x_l(t)x_j^*(t)\right)E\left(x_i(t)x_k^*(t)\right)\right]$$

$$+\frac{1}{m^2}\sum_{t_1\neq t_2} E\left(x_i(t_1)x_j^*(t_1)\right)E^*\left(x_k(t_2)x_l^*(t_2)\right)$$

$$= R_{ij}R_{kl}^* + R_{ik}R_{jl}^*/m$$

Substituting these results back into (7B.1) we finally get:

$$E(E_{ij}E_{kl}^*) = R_{ik}R_{jl}^*/m \tag{7B.2}$$

Now, for $\mathbf{T}=\sqrt[4]{m}\mathbf{R}^{-1/2}$, we get

$$\tilde{\mathbf{R}} = \mathbf{T}\mathbf{R}\mathbf{T}^H = \sqrt{m}\mathbf{I}$$

and hence, from (7B.2), the elements of the transformed error matrix $\tilde{\mathbf{E}}$ obey

$$E(\tilde{E}_{ij}\tilde{E}_{kl}^*) = \delta_{ik}\delta_{jl} \tag{7B.3}$$

where δ_{ik} denotes the Kronecker delta defined by

$$\delta_{ik} = \begin{cases} 1 & \text{if } i = k \\ 0 & \text{otherwise} \end{cases}$$

Thus, the elements of the transformed error matrix $\tilde{\mathbf{E}}$ are zero-mean, uncorrelated, and all have an identical variance 1.

Notice that $\tilde{\mathbf{E}}$ is Hermitian since both $\hat{\mathbf{R}}$ and $\mathbf{R}$ are Hermitian. Therefore, $\tilde{E}_{ii}$ is real, $\tilde{E}_{ij} = \tilde{E}_{ji}^*$, and the real part and the imaginary part of any nondiagonal element fo $\tilde{\mathbf{E}}$ have equal variances $\frac{1}{2}$.

APPENDIX 7C CONSISTENCY OF THE PARAMETER ESTIMATES

In this appendix we prove that the estimators are consistent even if the underlying distribution is not necessarily Gaussian. To this end, we rewrite the GLS-ML estimator[1] (7.12), (7.25), (7.26) as:

$$(\hat{\mathbf{P}}(\boldsymbol{\theta}), \hat{\sigma}^2(\boldsymbol{\theta})) = \arg\min_{\mathbf{P},\sigma^2}\left[\sum_c \delta_c\left(\mathbf{R}_c(\boldsymbol{\theta},\mathbf{P},\sigma^2), \hat{\mathbf{R}}_c\right)\right] \tag{7C.1}$$

$$\hat{\boldsymbol{\theta}} = \arg\min_{\boldsymbol{\theta}}\left[\sum_c \frac{m_c}{m}\eta_c\left(\mathbf{R}_c(\boldsymbol{\theta},\hat{\mathbf{P}}(\boldsymbol{\theta}),\hat{\sigma}^2(\boldsymbol{\theta})), \hat{\mathbf{R}}_c\right)\right] \tag{7C.2}$$

[1] The proof for the GLS estimator (7.12), (7.27), (7.28) can be established along similar lines.

where $m \stackrel{\text{def}}{=} \sum_c m_c$, and where $\{\delta_c(\mathbf{Y},\mathbf{Z})\}$ and $\{\eta_c(\mathbf{Y},\mathbf{Z}\}$ are two different distance measures measuring the distance between any two positive definite Hermitian matrices $\mathbf{Y},\mathbf{Z} \in \mathbf{C}^{p_c \times p_c}$:

$$\delta_c(\mathbf{Y},\mathbf{Z}) \stackrel{\text{def}}{=} \|\mathbf{T}_c(\mathbf{Y}-\mathbf{Z})\mathbf{T}_c^H\|_F^2$$

$$\eta_c(\mathbf{Y},\mathbf{Z}) \stackrel{\text{def}}{=} -\log|\mathbf{Y}^{-1}\mathbf{Z}| + \operatorname{tr}(\mathbf{Y}^{-1}\mathbf{Z}) - p_c$$

Lemma 7C.1. *Assuming that $\{\mathbf{T}_c\}$ are positive definite, the above distance measures obey:*

$$\delta_c(\mathbf{Y},\mathbf{Z}) \geq 0 \qquad \forall c, \mathbf{Y}, \mathbf{Z} \tag{7C.3}$$

$$\eta_c(\mathbf{Y},\mathbf{Z}) \geq 0 \qquad \forall c, \mathbf{Y}, \mathbf{Z} \tag{7C.4}$$

with equality if and only if $\mathbf{Y} = \mathbf{Z}$.

Proof. The proof of (7C.3) is obvious. To prove (7C.4) we first rewrite $\eta_c(\mathbf{Y},\mathbf{Z})$ as:

$$\eta_c(\mathbf{Y},\mathbf{Z}) = -\log|\mathbf{U}^H\mathbf{Y}^{-1}\mathbf{U}| + \operatorname{tr}(\mathbf{U}^H\mathbf{Y}^{-1}\mathbf{U}) - p_c$$

where $\mathbf{U}$ is the Hermitian square root of $\mathbf{Z}$, that is, $\mathbf{Z} = \mathbf{U}\mathbf{U}^H$. We then denote the eigenvalues of the positive definite Hermitian matrix $\mathbf{U}^H\mathbf{Y}^{-1}\mathbf{U}$ by $\{\lambda_i\}$ and their geometric mean by $\tilde{\lambda} \stackrel{\text{def}}{=} (\prod_i \lambda_i)^{\frac{1}{p_c}}$, and their arithmetic mean by $\bar{\lambda} \stackrel{\text{def}}{=} \frac{1}{p_c}\sum_i \lambda_i$. Using the facts that the determinant of a matrix is the product of its eigenvalues and that the trace of a matrix is the sum of its eigenvalues, we have:

$$\frac{1}{p_c}\eta_c(\mathbf{Y},\mathbf{Z}) = \log\left(\frac{\bar{\lambda}}{\tilde{\lambda}}\right) + \bar{\lambda} - \log\bar{\lambda} - 1$$

Now, the lemma readily follows from the fact that the geometric mean of any sequence of numbers is not greater than its arithmetic mean, and these means are equal if and only if all the numbers are identical, and the fact that the function

$$f(z) = z - \log z - 1$$

is convex with a unique minimum of zero at $z = 1$. ■

To establish the consistency, let $\boldsymbol{\theta}_0, \mathbf{P}_0, \sigma_0^2$ denote the true values of the parameters. Asymptotically as $m_c \to \infty; \forall c$ we clearly have

$$\hat{\mathbf{R}}_c \to \mathbf{R}_c(\boldsymbol{\theta}_0, \mathbf{P}_0, \sigma_0^2); \quad \forall c$$

Now, by Lemma 7C.1,

$$\sum_c \delta_c\left(\mathbf{R}_c(\hat{\boldsymbol{\theta}}, \hat{\mathbf{P}}, \hat{\sigma}^2), \hat{\mathbf{R}}_c\right) \geq 0$$

$$\sum_c \frac{m_c}{m}\eta_c\left(\mathbf{R}_c(\hat{\boldsymbol{\theta}}, \hat{\mathbf{P}}(\hat{\boldsymbol{\theta}}), \hat{\sigma}^2(\hat{\boldsymbol{\theta}})), \hat{\mathbf{R}}_c\right) \geq 0$$

and since both distance measures are continuous, the minima are asymptotically achieved at $\hat{\theta} = \theta_0, \hat{\mathbf{P}} = \mathbf{P}_0, \hat{\sigma}^2 = \sigma_0^2$ since then

$$\sum_c \delta_c \left(\mathbf{R}_c(\hat{\theta}, \hat{\mathbf{P}}, \hat{\sigma}^2), \hat{\mathbf{R}}_c \right) = 0$$

$$\sum_c \frac{m_c}{m} \eta_c \left(\mathbf{R}_c(\hat{\theta}, \hat{\mathbf{P}}(\hat{\theta}), \hat{\sigma}^2(\hat{\theta})), \hat{\mathbf{R}}_c \right) = 0$$

Thus, asymptotically the minimum of (7C.1) and (7C.2) is at the true values, implying that the estimates are consistent.

APPENDIX 7D CONSISTENCY OF THE DETECTION CRITERION

In this appendix we prove that our MDL-based detection criterion for estimating the number of signals is consistent even if the underlying distribution is not necessarily Gaussian. To this end we rewrite the MDL criterion (7.34) as:

$$\hat{q} = \arg \min_{0 \le q \le q_{\max}} [\text{MDL}\,(q)] \tag{7D.1}$$

where

$$\text{MDL}\,(q) \overset{\text{def}}{=} m \left[\sum_c \frac{m_c}{m} \eta_c(\mathbf{R}_c(\hat{\phi}^{(q)}), \hat{\mathbf{R}}_c)) + \frac{k_q}{2} \frac{\log m}{m} \right] \tag{7D.2}$$

where $\phi^{(q)} \overset{\text{def}}{=} \{\theta, \mathbf{P}, \sigma^2\}$ represent all the parameters, with the superscript$^{(q)}$ noting that the size of ϕ is q-dependent, $\eta(\cdot, \cdot)$ is the distance measure defined in (7C.2), $q_{\max}$ is the maximal number of signals for which q and $\phi^{(q)}$ can be uniquely determined, and $m \overset{\text{def}}{=} \sum_c m_c$.

Let $q_o \le q_{\max}$ denote the true number of signals and let $\phi^{(q_o)}$ be their true parameters. Assume that asymptotically

$$\frac{m_c}{m} = \mu_c > 0; \quad c = 1, \ldots, K$$

where $\{\mu_c\}$ are some constants.

Now, to establish consistency we must prove that asymptotically

$$\text{MDL}\,(q) - \text{MDL}\,(q_o) > 0 \quad \text{a.s. for} \quad q \ne q_o, \quad 0 \le q \le q_{\max} \tag{7D.3}$$

To this end we first consider the case $0 \le q < q_o$:
From (7D.2) we get

$$\frac{1}{m}\{\text{MDL}\,(q) - \text{MDL}\,(q_o)\} = \left[\sum_c \mu_c \left[\eta_c(\mathbf{R}_c(\hat{\phi}^{(q)}), \hat{\mathbf{R}}_c)) - \eta_c(\mathbf{R}_c(\hat{\phi}^{(q_o)}), \hat{\mathbf{R}}_c)) \right] \right] + \left[\frac{k_q - k_{q_o}}{2} \frac{\log m}{m} \right] \tag{7D.4}$$

where we asymptotically clearly have

$$\hat{\mathbf{R}}_c = \mathbf{R}_c(\hat{\phi}^{(q_o)}); \quad c = 1, \ldots, K$$

Now, $\mathbf{R}_c(\hat{\phi}^{(q)}) \neq \mathbf{R}_c(\hat{\phi}^{(q_o)})$, for at least some c since otherwise the uniqueness requirement (7.5) is violated. For this c the distance

$$\eta_c(\mathbf{R}_c(\hat{\phi}^{(q)}), \mathbf{R}_c(\hat{\phi}^{(q_o)})) > 0$$

is some positive constant. Therefore, the first term in (7D.4) is a positive constant, while the second term vanishes asymptotically, thus establishing (7D.3) for $q < q_o$.

We now consider the case $q_o \leq q \leq q_{\max}$:
From (7D.2) we get

$$\{\mathrm{MDL}\,(q) - \mathrm{MDL}\,(q_o)\} = \left[\sum_c m_c \left[\eta_c(\mathbf{R}_c(\hat{\phi}^{(q)}), \hat{\mathbf{R}}_c)) - \eta_c(\mathbf{R}_c(\hat{\phi}^{(q_o)}), \hat{\mathbf{R}}_c))\right]\right] + \left[\frac{k_q - k_{q_o}}{2} \log m\right] \tag{7D.5}$$

To evaluate the asymptotic value of the first term we follow [17] and use the following result proved in [7]:

Lemma 7D.1. *Suppose* $\{\mathrm{e_i}, \mathrm{i} \geq 1\}$ *is a stationary real* φ*-mixing sequence with* $\mathrm{E(e_i)} = 0, \mathrm{E(|e_i|^2)} < \infty$, *and* φ *is decreasing with* $\sum_{\mathrm{j}=1}^{\infty} \varphi^{1/2}(\mathrm{j}) < \infty$. *Then*

$$\limsup_{m\to\infty} \frac{\sum_{i=1}^{m} e_i}{(2m\delta^2 \log\log m\delta^2)^{1/2}} = 1 \qquad a.s.$$

where $\delta^2 \overset{\mathrm{def}}{=} \mathrm{E(e_1^2)} + 2\sum_{\mathrm{i}=1}^{\infty} \mathrm{E(e_1 e_{1+i})} \neq 0$.

Specifically we apply this lemma to the elements of the error matrices $m_c(\hat{\mathbf{R}}_c - \mathbf{R}_c(\hat{\phi}^{(q)}))$ and $m_c(\hat{\mathbf{R}}_c - \mathbf{R}_c(\hat{\phi}^{(q_o)}))$ and get

$$\hat{\mathbf{R}}_c - \mathbf{R}_c(\hat{\phi}^{(q)}) = O\left(\sqrt{\frac{\log\log m_c}{m_c}}\right) \quad \mathbf{a.s.} \quad c = 1, \ldots, K$$

$$\hat{\mathbf{R}}_c - \mathbf{R}_c(\hat{\phi}^{(q_o)}) = O\left(\sqrt{\frac{\log\log m_c}{m_c}}\right) \quad \mathbf{a.s.} \quad c = 1, \ldots, K \tag{7D.6}$$

Now, expanding $\eta_c(\mathbf{R}_c, \hat{\mathbf{R}}_c)$ as a Taylor power series of the elements of the difference between its arguments $\mathbf{R}_c - \hat{\mathbf{R}}_c$ using (7D.6), and the fact that

$$\eta_c(\hat{\mathbf{R}}_c, \hat{\mathbf{R}}_c) = 0; \quad c = 1, \ldots, K$$

and also (see [6])

$$\frac{\partial \eta_c(\mathbf{R}_c, \hat{\mathbf{R}}_c)}{\partial \mathbf{R}_c}\Big|_{\mathbf{R}_c = \hat{\mathbf{R}}_c} = 0; \quad c = 1, \ldots, K$$

we get

$$m_c \eta_c(\mathbf{R}_c(\hat{\phi}^{(q_o)}), \hat{\mathbf{R}}_c) = O(\log\log m_c) \quad \textbf{a.s.} \quad c = 1, \ldots, K$$

$$m_c \eta_c(\mathbf{R}_c(\hat{\phi}^{(q)}), \hat{\mathbf{R}}_c) = O(\log\log m_c) \quad \textbf{a.s.} \quad c = 1, \ldots, K$$

This implies that the first term on the right hand of (7D.5) is asymptotically negligible when compared to the second term, which is positive since $k_q > k_{q_o}$, thus establishing (7D.3) for $q > q_o$.

The so-called "penalty term" used in the MDL criterion is $\frac{k_q}{2} \log m$. Notice, however, that in fact we proved the consistency of the MDL criterion for any penalty term h_q obeying

$$\lim_{m\to\infty} h_q / \log\log m \to \infty \quad \text{and} \quad \lim_{m\to\infty} h_q/m \to 0$$

APPENDIX 7E THE EFFICIENCY OF THE ESTIMATOR AND THE CRB FOR SUBARRAY SAMPLING

In this appendix we investigate the asymptotic behavior of our GLS (7.12), (7.27), (7.28) and GLS-ML (7.25), (7.26) estimators. In particular, we prove their efficiency by showing that asymptotically as $m_c \to \infty; \forall_c$ they yield exactly the same estimation errors as the MLE (7.10), which is known to be efficient. In addition, we get a closed-form expression for the CRB for subarray sampling. The approach follows [2].

Let ϕ denote a vector composed of all the real free parameters:

$$\phi \stackrel{\text{def}}{=} (\theta^T, \text{vec}^T(\bar{\mathbf{P}}), \sigma^2)^T \tag{7E.1}$$

where $\bar{\mathbf{P}}$ is a real matrix formed from the free real parameters of the Hermitian matrix $\mathbf{P}$ in the following way: the elements on or above the diagonal of $\mathbf{P}$ are replaced by their real parts, while the elements below the diagonal are replaced by their imaginary parts.

Let ϕ_o denote the true value of ϕ.

Using $\mathbf{T}_c = \sqrt[4]{m_c}\hat{\mathbf{R}}_c^{-1/2}$, (7.28), (7.8), and (7.9), the GLS and the ML estimators are the values $\hat{\phi}_{GLS}, \hat{\phi}_{ML}$ minimizing, respectively, the following functions:

$$L_{GLS}(\phi) = \frac{1}{2}\sum_c m_c \operatorname{tr}(\mathbf{I} - \mathbf{R}_c\hat{\mathbf{R}}_c^{-1})^2 \tag{7E.2}$$

$$L_{ML}(\phi) = \sum_c m_c \left[(-\log|\mathbf{R}_c^{-1}\hat{\mathbf{R}}_c| + \operatorname{tr}(\mathbf{R}_c^{-1}\hat{\mathbf{R}}_c) - p_c\right] \tag{7E.3}$$

where we used the following identity (valid for any matrix $\mathbf{A}$) $\|\mathbf{A}\|_F^2 = \operatorname{tr}(\mathbf{A}\mathbf{A}^H)$, and multiplied L_{GLS} by a factor of $\frac{1}{2}$ (without affecting $\hat{\phi}_{GLS}$), and added non-parameter-dependent terms to L_{ML}.

The derivative of L_{GLS} with respect to ϕ_i, the ith component of ϕ, is given by

$$\frac{\partial L_{GLS}(\phi)}{\partial \phi_i} = -\sum_c m_c \operatorname{tr}\left[\dot{\mathbf{R}}_c\hat{\mathbf{R}}_c^{-1}(\mathbf{I} - \mathbf{R}_c\hat{\mathbf{R}}_c^{-1})\right]$$

$$= -\sum_c m_c \operatorname{tr}\left[\hat{\mathbf{R}}_c^{-1}\hat{\mathbf{R}}_c\hat{\mathbf{R}}_c^{-1}\dot{\mathbf{R}}_c - \hat{\mathbf{R}}_c^{-1}\mathbf{R}_c\hat{\mathbf{R}}_c^{-1}\dot{\mathbf{R}}_c\right] \tag{7E.4}$$

$$= -\sum_c m_c \operatorname{vec}^H(\dot{\mathbf{R}}_c)\left[\hat{\mathbf{R}}_c^{-T} \otimes \hat{\mathbf{R}}_c^{-1}\right]\operatorname{vec}(\hat{\mathbf{R}}_c - \mathbf{R}_c)$$

where we have used the notation $\dot{\mathbf{R}} \stackrel{\text{def}}{=} \dfrac{\partial \mathbf{R}}{\partial \phi_i}$ and the result[2]

$$\operatorname{tr}(\mathbf{ABCD}) = \operatorname{vec}^T(\mathbf{D}^T)(\mathbf{C}^T \otimes \mathbf{A})\operatorname{vec}(\mathbf{B})$$

Thus

$$\frac{\partial L_{GLS}(\phi)}{\partial \phi_i} = -\sum_c m_c \left(\frac{\partial \operatorname{vec}(\mathbf{R}_c)}{\partial \phi}\right)^H \left[\hat{\mathbf{R}}_c^{-T} \otimes \hat{\mathbf{R}}_c^{-1}\right]\operatorname{vec}(\hat{\mathbf{R}}_c - \mathbf{R}_c) \tag{7E.5}$$

where the derivative of a scalar y with respect to a vector $\mathbf{x} \in \mathbf{R}^{n\times 1}$ is defined by $\dfrac{\partial y}{\partial \mathbf{x}} \stackrel{\text{def}}{=} \left(\dfrac{\partial y}{\partial x_1}, \ldots, \dfrac{\partial y}{\partial x_n}\right)^T$, and the derivative of a vector $\mathbf{y} \in \mathbf{R}^{m\times 1}$ with respect to a vector $\mathbf{x} \in \mathbf{R}^{n\times 1}$ is defined by $\dfrac{\partial \mathbf{y}}{\partial \mathbf{x}} \stackrel{\text{def}}{=} \left(\dfrac{\partial y_1}{\partial \mathbf{x}}, \ldots, \dfrac{\partial y_m}{\partial \mathbf{x}}\right)^T = \left(\dfrac{\partial \mathbf{y}}{\partial x_1}, \ldots, \dfrac{\partial \mathbf{y}}{\partial x_n}\right)$.

Similarly, the derivative of L_{ML} with respect to ϕ_i is given by

$$\frac{\partial L_{ML}(\phi)}{\partial \phi_i} = \sum_c m_c \left[\operatorname{tr}(\mathbf{R}_c^{-1}\dot{\mathbf{R}}_c) - \operatorname{tr}(\hat{\mathbf{R}}_c\mathbf{R}_c^{-1}\dot{\mathbf{R}}_c\mathbf{R}_c^{-1})\right] \tag{7E.6}$$

where we used $\dfrac{\partial \mathbf{R}_c^{-1}}{\partial \phi_i} = -\mathbf{R}_c^{-1}\dot{\mathbf{R}}_c\mathbf{R}_c^{-1}$ and $\dfrac{\partial \log|\mathbf{R}_c|}{\partial \phi_i} = \operatorname{tr}(\mathbf{R}_c^{-1}\dot{\mathbf{R}}_c)$. Repeating previous steps, we arrive at

$$\frac{\partial L_{ML}(\phi)}{\partial \phi_i} = -\sum_c m_c \left(\frac{\partial \operatorname{vec}(\mathbf{R}_c)}{\partial \phi_i}\right)^H \left[\mathbf{R}_c^{-T} \otimes \mathbf{R}_c^{-1}\right]\operatorname{vec}(\hat{\mathbf{R}}_c - \mathbf{R}_c) \tag{7E.7}$$

Notice that $\dfrac{\partial L_{ML}(\phi)}{\partial \phi}$ and $\dfrac{\partial L_{GLS}(\phi)}{\partial \phi}$ are asymptotically identical at $\phi = \phi_o$.

Assuming that $\mathbf{a}(\theta)$ is twice continuously differentiable in the neighborhood of θ_o, it follows that the matrices of second derivatives of L_{GLS} and L_{ML} are continuous and both have the same following probability limit at $\phi = \phi_o$:

$$\operatorname*{plim}_{m\to\infty}\left(\frac{1}{m}\frac{\partial^2 L_{GLS}(\phi)}{\partial \phi^2}\right)_{\phi=\phi_o} = \operatorname*{plim}_{m\to\infty}\left(\frac{1}{m}\frac{\partial^2 L_{ML}(\phi)}{\partial \phi^2}\right)_{\phi=\phi_o} \stackrel{\text{def}}{=} \mathbf{J}(\phi_o) \tag{7E.8}$$

where $\mathbf{J}(\phi_o)$ is given by

$$\mathbf{J}(\phi_o) = \sum_c \mu_c \mathbf{F}_c^H(\phi_o)\left[\mathbf{R}_c^{-T} \otimes \mathbf{R}_c^{-1}\right]\mathbf{F}_c(\phi_o) \tag{7E.9}$$

[2] This identity is easily obtained by applying $\operatorname{tr}(\mathbf{ED}) = \operatorname{vec}^T(\mathbf{D}^T)\operatorname{vec}(\mathbf{E})$ to the left side with $\mathbf{E} = \mathbf{ABC}$, and then using $\operatorname{vec}(\mathbf{ABC}) = (\mathbf{C}^T \otimes \mathbf{A})\operatorname{vec}(\mathbf{B})$.

with $\mu_c \overset{\text{def}}{=} \lim_{m\to\infty} \frac{m_c}{m}$, $\forall c$ and

$$\mathbf{F}_c(\phi) \overset{\text{def}}{=} \frac{\partial\, \text{vec}(\mathbf{R}_c)}{\partial \phi} \tag{7E.10}$$

Notice that $m\mathbf{J}(\phi_o)$ is in fact the Fisher Information Matrix (FIM).

Now, expanding the first derivative expressions using the fact that at the estimated values $\hat{\phi}_{GLS}, \hat{\phi}_{ML}$ these derivatives are zero, we get

$$\begin{aligned} 0 = \left(\frac{\partial L_{GLS}(\phi)}{\partial \phi}\right)_{\phi=\hat{\phi}_{GLS}} &= \left(\frac{\partial L_{GLS}(\phi)}{\partial \phi}\right)_{\phi_o} \\ &+ \left(\frac{\partial^2 L_{GLS}(\phi)}{\partial \phi^2}\right)_{\bar{\phi}_{GLS}} (\hat{\phi}_{GLS} - \phi_o) \end{aligned} \tag{7E.11}$$

$$\begin{aligned} 0 = \left(\frac{\partial L_{ML}(\phi)}{\partial \phi}\right)_{\phi=\hat{\phi}_{ML}} &= \left(\frac{\partial L_{ML}(\phi)}{\partial \phi}\right)_{\phi_o} \\ &+ \left(\frac{\partial^2 L_{ML}(\phi)}{\partial \phi^2}\right)_{\bar{\phi}_{ML}} (\hat{\phi}_{ML} - \phi_o) \end{aligned} \tag{7E.12}$$

where $\bar{\phi}$ denotes a midpoint on the line segment joining the true value and the estimated value.

Using (7E.5), (7E.7), (7E.8), and (7E.9), and assuming that $\mathbf{J}(\phi_o)$ is full rank, we asymptotically get

$$\begin{aligned} \hat{\phi}_{ML} - \phi_o &= \hat{\phi}_{GLS} - \phi_o \\ &= \mathbf{J}^{-1}(\phi_o) \sum_c \mu_c \mathbf{F}_c^H(\phi_o) \left[\mathbf{R}_c^{-T} \otimes \mathbf{R}_c^{-1}\right] \text{vec}\,(\hat{\mathbf{R}}_c - \mathbf{R}_c) \end{aligned} \tag{7E.13}$$

Thus, the asymptotic errors of the ML and the GLS estimators are identical. The GLS-ML estimator will also have the same error asymptotically, since it uses the MLE for some of the components of ϕ, and the GLS estimator for the rest.

Notice that from (7E.13) it follows that the asymptotical error decrease rate (as $m_c \to \infty; \forall c$) is the same as that of the elements of $\{\hat{\mathbf{R}}_c - \mathbf{R}_c\}$, since the errors are linear combinations of these elements.

E.1 THE CRAMÉR-RAO BOUND

The CRB is given by the inverse of the FIM:

$$\text{CRB} = \mathbf{J}^{-1}(\phi_o)/m \tag{7E.14}$$

where $\mathbf{J}(\phi_o)$ is given by (7E.9). We proceed now to get a closed-form expression for $\mathbf{J}(\phi_o)$. Using (7E.4) we can write $\text{vec}(\mathbf{R}_c)$ in terms of $\text{vec}(\bar{\mathbf{P}})$ as

$$\begin{aligned} \text{vec}(\mathbf{R}_c) &= \text{vec}(\mathbf{A}_c(\theta)\mathbf{P}\mathbf{A}_c^H(\theta)) + \sigma^2 \text{vec}(\mathbf{I}) \\ &= \left[\mathbf{A}_c^*(\theta) \otimes \mathbf{A}_c(\theta)\right] \text{vec}(\mathbf{P}) + \sigma^2 \text{vec}(\mathbf{I}) \\ &= \left[\mathbf{A}_c^*(\theta) \otimes \mathbf{A}_c(\theta)\right] \mathbf{U}\, \text{vec}(\bar{\mathbf{P}}) + \sigma^2 \text{vec}(\mathbf{I}) \end{aligned} \tag{7E.15}$$

where $\mathbf{U}$ is a matrix whose elements are $0, 1, +j$, or $-j$ defined by

$$\text{vec}(\mathbf{P}) \stackrel{\text{def}}{=} \mathbf{U}\,\text{vec}(\bar{\mathbf{P}}) \tag{7E.16}$$

The construction of $\mathbf{U}$ using this definition is straightforward. For example, for $q = 2$, that is, two sources

$$\text{vec}(\mathbf{P}) = (P_{11}, P_{21}, P_{12}, P_{22})^T$$
$$\text{vec}(\bar{\mathbf{P}}) = (P_{11}, P_{21I}, P_{12R}, P_{22})^T$$

and therefore

$$\mathbf{U} = \begin{bmatrix} 1 & 0 & 0 & 0 \\ 0 & j & 1 & 0 \\ 0 & -j & 1 & 0 \\ 0 & 0 & 0 & 1 \end{bmatrix}$$

Now, using (7E.10) and (7E.1) we have

$$\mathbf{F}_c(\phi) \stackrel{\text{def}}{=} \left[\frac{\partial\,\text{vec}(\mathbf{R}_c)}{\partial\boldsymbol{\theta}}\ \frac{\partial\,\text{vec}(\mathbf{R}_c)}{\partial\,\text{vec}(\bar{\mathbf{P}})}\ \frac{\partial\,\text{vec}(\mathbf{R}_c)}{\partial\sigma^2}\right] \tag{7E.17}$$

where the three different blocks on the right side can be computed from (7E.15):

$$\frac{\partial\,\text{vec}(\mathbf{R}_c)}{\partial\sigma^2} = \text{vec}(\mathbf{I}) \tag{7E.18}$$

$$\frac{\partial\,\text{vec}(\mathbf{R}_c)}{\partial\,\text{vec}(\bar{\mathbf{P}})} = \left[\mathbf{A}_c^*(\boldsymbol{\theta}) \otimes \mathbf{A}_c(\boldsymbol{\theta})\right]\mathbf{U} \tag{7E.19}$$

and the i-th column of the first block is given by

$$\frac{\partial\,\text{vec}(\mathbf{R}_c)}{\partial\theta_i} = \left[\left(\frac{\partial\mathbf{A}_c(\boldsymbol{\theta})}{\partial\theta_i}\right)^* \otimes \mathbf{A}_c(\boldsymbol{\theta}) + \mathbf{A}_c^*(\boldsymbol{\theta}) \otimes \left(\frac{\partial\mathbf{A}_c(\boldsymbol{\theta})}{\partial\theta_i}\right)\right] \text{vec}(\mathbf{P}) \tag{7E.20}$$

where

$$\frac{\partial\mathbf{A}_c(\boldsymbol{\theta})}{\partial\theta_i} = \left[0, \ldots, 0, \frac{\partial\mathbf{a}_c(\theta_i)}{\partial\theta_i}, 0, \ldots, 0\right]$$

is a matrix whose columns are all zero except its i-th column which is $\dfrac{\partial\mathbf{a}_c(\theta_i)}{\partial\theta_i}$.

Inserting these blocks into (7E.17) and then (7E.17) into (7E.9), we finally get the FIM. The upper left corner of the inverse of the FIM is the CRB for the DOA errors.

In special cases where some of the parameters are known a priori, the rows and the columns corresponding to these parameters should be omitted from the FIM, and the CRB is obtained as the inverse of this reduced FIM. Thus, for instance, when the signals are known to be uncorrelated, the columns corresponding to the nondiagonal elements of $\mathbf{P}$ should be omitted from $\mathbf{F}_c(\phi)$.

APPENDIX 7F ESTIMATION OF HERMITIAN $\hat{\mathbf{P}}$ AND REAL $\hat{\sigma}^2$

In this appendix we derive a GLS estimator that constrains $\hat{\mathbf{P}}$ to be Hermitian and $\hat{\sigma}^2$ to be real-valued. This GLS estimator is obtained by solving a real LS problem, where the unknown parameters are the free real parameters vec($\bar{\mathbf{P}}$) and σ^2; see (7E.1). This is in contrast to the GLS estimates derived in section 7.4 where $\hat{\mathbf{P}}$ and $\hat{\sigma}^2$ were not constrained to be Hermitian and real, respectively, and thus had twice as many real parameters. Due to the parameter reduction, the estimates derived here are bound to be more accurate and computationally simpler.

In addition, in the covariance equation (7.11):

$$\hat{\mathbf{R}}_c = \mathbf{R}_c + \mathbf{E}_c; \quad c = 1, \dots, K \tag{7F.1}$$

which is the basis for the GLS estimator, all the matrices (i.e., $\hat{\mathbf{R}}_c$, $\mathbf{R}_c$ and therefore $\mathbf{E}_c$) are Hermitian, and therefore the equations corresponding to the upper triangular elements are sufficient, giving rise to a further reduction in the computational complexity.

Substituting vec($\mathbf{P}$) $= \mathbf{U}$ vec($\hat{\mathbf{P}}$) into (7.18) [where $\mathbf{U}$ is a transformation matrix whose elements are 0, 1, or $\pm j$, see (7E.16)] and equating the real parts and the imaginary parts of both sides, we get the following real expression:

$$\begin{bmatrix} \mathbf{r} \\ 0 \end{bmatrix} = \begin{bmatrix} \mathbf{B}_R \\ \mathbf{B}_I \end{bmatrix} \begin{bmatrix} \text{vec}(\bar{\mathbf{P}}) \\ \sigma^2 \end{bmatrix} + \begin{bmatrix} \mathbf{e}_R \\ \mathbf{e}_I \end{bmatrix} \tag{7F.2}$$

where the subscripts R and I denote the real and imaginary parts, respectively, and where we have used the fact that $\mathbf{r}$ given in (7.23) is a real vector. The matrix $\mathbf{B}$ is defined by

$$\mathbf{B} \stackrel{\text{def}}{=} \begin{bmatrix} (\tilde{\mathbf{A}}_1^*(\boldsymbol{\theta}) \otimes \tilde{\mathbf{A}}_1(\boldsymbol{\theta}))\mathbf{U} & \sqrt{m_1}\,\text{vec}\,(\hat{\mathbf{R}}_1^{-1}) \\ (\tilde{\mathbf{A}}_2^*(\boldsymbol{\theta}) \otimes \tilde{\mathbf{A}}_2(\boldsymbol{\theta}))\mathbf{U} & \sqrt{m_2}\,\text{vec}\,(\hat{\mathbf{R}}_2^{-1}) \\ \vdots & \vdots \\ (\tilde{\mathbf{A}}_K^*(\boldsymbol{\theta}) \otimes \tilde{\mathbf{A}}_K(\boldsymbol{\theta}))\mathbf{U} & \sqrt{m_K}\,\text{vec}\,(\hat{\mathbf{R}}_K^{-1}) \end{bmatrix} \tag{7F.3}$$

with $\tilde{\mathbf{A}}_c(\boldsymbol{\theta}) = \sqrt[4]{m_c}\hat{\mathbf{R}}_c^{-1/2}\mathbf{A}_c(\boldsymbol{\theta})$, where we used the expression given (7.23) for $\mathcal{A}(\boldsymbol{\theta})$.

Now, dropping the equations that correspond to the strictly lower triangles of the covariances, as well as the equations that correspond to the (zero) imaginary parts of their diagonals, we finally get

$$\begin{bmatrix} \mathbf{r} \\ 0 \end{bmatrix}_{(h)} = \begin{bmatrix} \mathbf{B}_R \\ \mathbf{B}_I \end{bmatrix}_{(h)} \begin{bmatrix} \text{vec}(\bar{\mathbf{P}}) \\ \sigma^2 \end{bmatrix} + \begin{bmatrix} \mathbf{e}_R \\ \mathbf{e}_I \end{bmatrix}_{(h)} \tag{7F.4}$$

where the subscript (h) denotes an operator that removes those corresponding to the lower triangular half. Now, using the results of Appendix 7B, it can be easily shown that the components of

$$\begin{bmatrix} \mathbf{e}_R \\ \mathbf{e}_I \end{bmatrix}_{(h)}$$

are uncorrelated, with components corresponding to the diagonals of the covariances having variance 1, and the rest of the components having variance $\frac{1}{2}$:

$$E\left[\begin{pmatrix}\mathbf{e}_R\\ \mathbf{e}_I\end{pmatrix}_{(h)}\begin{pmatrix}\mathbf{e}_R\\ \mathbf{e}_I\end{pmatrix}_{(h)}^T\right]\stackrel{\text{def}}{=}\mathbf{W} \tag{7F.5}$$

where $\mathbf{W}$ is diagonal matrix with 1 or $\frac{1}{2}$ as its diagonal elements.

Therefore, the GLS solution for $\tilde{\mathbf{P}}$ and σ^2 is given by

$$\begin{pmatrix}\text{vec}(\hat{\tilde{\mathbf{P}}})\\ \sigma^2\end{pmatrix}=\left[\begin{pmatrix}\mathbf{B}_R\\ \mathbf{B}_I\end{pmatrix}_{(h)}^T\mathbf{W}^{-1}\begin{pmatrix}\mathbf{B}_R\\ \mathbf{B}_I\end{pmatrix}_{(h)}\right]^{-1}\begin{pmatrix}\mathbf{B}_R\\ \mathbf{B}_I\end{pmatrix}_{(h)}^T\mathbf{W}^{-1}\begin{pmatrix}\mathbf{r}\\ 0\end{pmatrix}_{(h)} \tag{7F.6}$$

The corresponding GLS estimator for the DOAs is given by

$$\hat{\theta}=\arg\min_{\theta}\left\{\begin{pmatrix}\mathbf{r}\\ 0\end{pmatrix}_{(h)}^T\mathbf{W}^{-1/2}\mathbf{P}_{\mathbf{B}}^{\perp}\mathbf{W}^{-1/2}\begin{pmatrix}\mathbf{r}\\ 0\end{pmatrix}_{(h)}\right\} \tag{7F.7}$$

where $\mathbf{P}_{\mathbf{B}}^{\perp}$ is defined by

$$\mathbf{P}_{\mathbf{B}}^{\perp}\stackrel{\text{def}}{=}\mathbf{W}^{-1/2}\begin{pmatrix}\mathbf{B}_R\\ \mathbf{B}_I\end{pmatrix}_{(h)}\left[\begin{pmatrix}\mathbf{B}_R\\ \mathbf{B}_I\end{pmatrix}_{(h)}^T\mathbf{W}^{-1}\begin{pmatrix}\mathbf{B}_R\\ \mathbf{B}_I\end{pmatrix}_{(h)}\right]^{-1}\begin{pmatrix}\mathbf{B}_R\\ \mathbf{B}_I\end{pmatrix}_{(h)}^T\mathbf{W}^{-1/2}$$

REFERENCES

1. T. W. Anderson. *An Introduction to Multivariate Statistical Analysis*, New York: John Wiley & Sons, 1984.
2. T. W. Anderson. "Linear Latent Variable Models and Covariance Structures," Technical Report 27, Econometric Workshop, Stanford University, Stanford, California, December 1987.
3. J. F. Bohme. "Estimation of Spectral Parameters of Correlated Signals in Wavefields," *Signal Processing*, 11: 329–37, 1986.
4. M. W. Browne. *Covariance Structures*, pp. 72–141. Cambridge: Cambridge University Press, 1982.
5. G. H. Golub and C. H. Van Loan. *Matrix Computations*, 2nd ed., Baltimore: The John Hopkins University Press, 1989.
6. A. Graham. *Kronecker Products and Matrix Calculus with Applications*, Chichester, UK: Ellis Horwood Ltd., 1981.
7. P. Hall and C. C. Heyde. *Martingale Limit Theory and Its Application*, New York: Academic Press, 1980.
8. S. Haykin. *Advances in Spectrum Analysis and Array Processing*, vol. 2, Englewood Cliffs, NJ: Prentice Hall, 1991.
9. J. E. Hudson. *Adaptive Array Principles*, Stevenage, UK: Peter Peregrinus Ltd., 1981.
10. A. G. Jaffer. "Maximum Lilelihood Direction Finding of Stochastic Sources: A Separable Solution," in *Proceedings of the IEEE International Conference on Acoustics, Speech, and Signal Processing (ICASSP)*, New York, pp. 2893–96, April 1988.

11. H. H. Jenkins. *Small-Aperture Direction Finding*, Norwood, MA: Artech House, 1991.
12. D. H. Johnson and D. E. Dudgeon. *Array Signal Processing*, Englewood Cliffs, NJ: Prentice Hall, 1991.
13. J. Rissanen. "Modeling by Shortest Data Description," *Automatica*, 14: 465–71, 1978.
14. R. O. Schmidt, *A Signal Subspace Approach to Multiple Emitter Location and Spectral Estimation*, Ph.D. thesis, Stanford University, 1981.
15. M. Viberg and B. Ottersten. "Sensor Array Processing Based on Subspace Fitting," *IEEE Trans. on ASSP*, 39: 1110–21, July 1991.
16. M. Wax and T. Kailath. "Detection of Signals by Information Theoretic Criteria," *IEEE Transactions on ASSP*, 33: 387–92, 1985.
17. L. C. Zhao, P. R. Krishnaia, and Z. D. Bai. "On Detection of the Number of Signals in Presence of White Noise," *Multivariate Analysis*, 20: 1–25, 1986.
18. I. Ziskind and M. Wax. "Maximum Likelihood Localization of Multiple Sources by Alternating Projections," *IEEE Trans. on ASSP*, 10: 1533–36, 1988.

8

Task-Specific Criteria for Adaptive Beamforming with Slow Fading Signals

Alfred O. Hero III and Ronald A. De Lap

8.1 INTRODUCTION

In this chapter we develop an approach to adaptive beamforming which seeks to optimize the best achievable signal detection or parameter estimation performance at the output of the beamsummer. The overall philosophy behind our approach is that the adaptation criterion for adaptive beamforming weights should be designed to optimize achievable performance for the primary task of interest, whether that may be signal detection, interference cancellation, rangefinding, or direction-of-arrival (DOA) estimation. The methodology behind our approach is the use of weight-dependent detection criteria and estimation theoretic lower bounds to specify adaptation criteria appropriate to the specific task of interest. In this way we obtain a criterion which is optimized precisely for those weights that minimize the achievable mean-square-error of any unbiased parameter estimator constructed on the beamformer output statistic. Here we focus on the tasks of designing unconstrained adaptive beamsummers for estimation of parameters of a spatially invariant constant-modulus signal amplitude and for DOA estimation. The case of constrained beamforming would be a simple generalization of these results. To show the effectiveness of our approach we present simulation results for several estimators operating on the optimal adaptive beamsummer output. The adaptive beamsummers are developed for sensor arrays operating in an environment that is characteristic of slow Rayleigh fading signals. In this environment, signal components in successive snapshots are uncorrelated but the signal amplitudes remain coherent over the array.

If the joint statistical distributions of the raw multiple-sensor measurements are available, maximum likelihood (ML) or other estimation criteria can be used to estimate the parameters of interest directly from the sensor data, for example, methods of [1–6]. While such direct methods can have excellent performance properties, in particular the asymptotically minimum variance and unbiased property of the ML estimator, they frequently suffer from high implementation complexity and may be sensitive to mismodeling errors. It therefore makes sense to constrain estimator complexity, for example, constrain to a beamsummer statistic, and then to optimally design the free parameters in the structure, such as to adapt the beamsummer weights, for best possible estimation performance.

The CR bound on the minimum achievable MSE of an unbiased estimator applied after beamsumming depends on the beamsummer weights and other signal- and noise-related parameters. Since large values of the CR bound preclude the existence of unbiased estimators with low mean-squared error (MSE), it makes sense to minimize the CR bound over the beamsummer weights. Even though the CR bound generally depends on unknown parameters, it is sometimes possible to identify a function of the weights which depends only on moments of measurable quantities and whose maximization also minimizes the CR bound. By implementing empirical moments in place of ensemble moments, we obtain a weight adaptation criterion which successively approximates the bound minimizing weights and converges to these weights in the limit of a large number of independent snapshots.

In recent years many methods have been proposed for adapting beamsummer weight vectors for direction finding, signal estimation, and interference nulling in antenna arrays [7–12]. In each of these methods, as new snapshots become available the weight vector is updated according to some empirical error criterion. For example, the Applebaum beamformer [13] attempts to maximize mean beamsummer output, the nulling beamformer of [14] and the partially adaptive beamformer of [15] attempt to minimize output power, while the CMA beamformer [16] attempts to extract a constant modulus signal. The common strategy underlying the development of these criteria has been to attempt to maximize some empirical "signal quality index" (SQI) formed from the snapshots of the beamsummer outputs. However, maximizing the SQI may be only indirectly related to the task of primary interest. For example, the Applebaum beamformer uses the modulus sample-mean of the beamsummer output to estimate the ensemble mean. The set of beamsummer weights is then adapted by successively maximizing this sample-mean SQI as snapshots are acquired. The DOA can be subsequently extracted from the beamsummer statistic, for example, by locating the global peak of the DFT of the adapted weights. However, maximizing the magnitude sample mean does not guarantee that the beamsummer statistic is in any way optimal for DOA estimation. Furthermore, when the impinging signals have random amplitudes, maximizing the variance or the r.m.s. beamsummer output may be more appropriate. The optimal design approach we adopt here attempts to get the most out of the beamsummer structure by designing the weights to minimize the Cramér-Rao lower bound on estimator variance.

For estimation of a constant-modulus signal parameter, we show that the beamsummer CR bound reduces to a monotone decreasing function of a compound beamsummer output signal-to-noise ratio (SNR): a convex combination of the squared-mean-signal to noise-variance (SM-SNR) and the mean-squared-signal to noise-variance (MS-SNR) ratios. This convex combination emphasizes the SM-SNR for very narrowband (little amplitude variation) signals while it emphasizes the MS-SNR for very broadband (much amplitude variation) signals. Maximizing this compound SNR gives a weight vector which achieves an optimal trade-off between the SM-SNR and MS-SNR for this parametric estimation task. Several weight adaptation algorithms are given which yield weight sequences that asymptotically converge to the CR bound minimizing weight vector in the limit of a large number of snapshots.

For the DOA estimation problem, we obtain a closed form for the optimal weight vector minimizing the single-signal beamsummer CR bound. The optimal weight vector is a linear combination of two components: the signal steering vector which maximizes beamsummer output power at the true signal DOA; and a vector orthogonal to the steering vector which minimizes the beamsummer output power at the true signal DOA. Thus, the optimal beamsummer consists of a combined matched filter and signal nuller. An adaptive weight update algorithm is given with an adaptation criterion which depends on the first and second moments of the sums and differences of a pair of coupled beamsummer outputs. This update algorithm yields a weight sequence which asymptotically converges to the optimal CR bound minimizing weight vector. We present simulations comparing several beampattern-based detection and estimation algorithms to a generalized likelihood ratio (GLR) signal detector and a ML signal DOA estimator implemented with Akaike's signal selection criterion. The GLR detector and the ML estimator are implemented under the following ideal conditions: (1) raw multiple-sensor data is available; (2) the signal and noise powers are known exactly. Even with such unfair disadvantages, our beamsummer-based algorithms perform remarkably well in comparison, especially in the near-narrowband large-number-of-snapshots regime.

Our approach is related to the techniques of optimal sequential design in the statistical design of experiments. The general principles of optimal design date back to R. A. Fisher's early work described in [17] and more recently described in [18, 19]. Theoretical justification for the use of the inverse Fisher information as a design criterion has been given in [20, 21]. The optimal design approach has been successfully applied to problems in many areas including: system identification and model selection [22–25]; identification of compartmental models for pharmacokinetics in biological systems [26, 27]; optimum scheduling of biometric measurements [28, 29]; and optimizing data acquisition strategies in radio-pharmaceutical cardiac imaging [30].

This chapter is organized as follows. An outline of notational conventions is given in section 8.2. In section 8.3 we review the regimes of narrowband and broadband array operation based on the spatio-temporal correlation coefficient. In section 8.4 we derive the deflection index for optimal detection performance and show that it is proportional to the square-mean, variance, and mean-square beamsummer output under various narrowband and broadband signal conditions. In sections 8.5–8.7, we derive the beamsummer-dependent CR bounds for the constant modulus and DOA parameter estimation problems. In these latter sections we obtain solutions for the optimum bound minimizing weight vectors for each of these problems and we specify adaptive beamsummer updates using empirical statistics. In section 8.8 we present results of simulations.

8.2 NOTATION

Throughout the chapter we will use the following notational conventions:

1. Bold uppercase letters will denote matrices, for example, **A, B, C, V, I**.
2. Vectors will always be column vectors, and will be denoted by bold lowercase letters, for example, **a, b, c, v, i**. The vector Hermitian transpose will be denoted by superscript H, for example, $\mathbf{s}^H$.
3. The complex conjugate of a scalar, vector, or matrix will be denoted by superscript *, for example, $\mathbf{x}^*$.
4. When referring to waveforms, for example, spatio-temporal functions, the following conventions will be used:
   - Caligraphic letters, such as $\zeta, \mathcal{N}, \chi$, will denote complex waveforms.
   - Italic letters, such as s, n, x, will denote complex baseband equivalent waveforms.

8.3 ARRAY SIGNAL AND NOISE MODELS

Consider a spatially and temporally varying field whose amplitude at a particular time and a particular spatial location can be measured from the current induced in a field-sensitive sensor. For example, in the case of a radio wave, an electromagnetic field induces a current at the output of an antenna element, while in the case of an acoustical wave a pressure field induces a current at the output of an acoustic transducer. Denote by $\chi(\mathbf{z}, t)$ the output current of such a sensor at time t placed at position $\mathbf{z}$ in space IR^3. It will be convenient to use the complex analytic signal representation for $\chi(\mathbf{z}, t)$:

$$\chi(\mathbf{z}, t) = x(\mathbf{z}, t)e^{jw_o(\mathbf{z})t} \tag{8.1}$$

where $x(\mathbf{z}, t)$ is the complex-valued baseband equivalent waveform.

Throughout this work we assume a general random process setting for signal and noise processes. In this setting a purely deterministic field $\chi(\mathbf{z}, t)$ is simply a random field with zero covariance. By the distribution $F(x)$ of $\chi(\mathbf{z}, t)$ is meant the set of finite dimensional joint distributions of spatio-temporal samples $\{\chi(\mathbf{z}_i, t_i), \mathbf{z}_i \in \mathrm{IR}^3, t_i \in \mathrm{IR}, i = 1, 2, \ldots\}$. Throughout this chapter it will be assumed that the random fields are Gaussian. Let $E[y]$ denote the expectation of any function $y = g(x)$ of x:

$$E[y] = \int g(x) dF(x)$$

The ensemble mean and covariance of $x(\mathbf{z}, t)$ are:

$$E[x(\mathbf{z}, t)] = \mu_x(\mathbf{z}, t) e^{j w_o(\mathbf{z}) t} \tag{8.2}$$

$$\mathrm{cov}\,[x(\mathbf{z}_1, t_1), x(\mathbf{z}_2, t_2)] = R_x(\mathbf{z}_1, \mathbf{z}_2, t_1, t_2) e^{j[w_o(\mathbf{z}_1)t_1 - w_o(\mathbf{z}_2)t_2]} \tag{8.3}$$

where μ_x and R_x are the mean and covariance functions of the baseband waveform $x(\mathbf{z}, t)$:

$$\mu_x(\mathbf{z}, t) = E[x(\mathbf{z}, t)] \tag{8.4}$$

$$\begin{aligned} R_x(\mathbf{z}_1, \mathbf{z}_2, t_1, t_2 &= \mathrm{cov}\,[x(\mathbf{z}_1, t_1), x(\mathbf{z}_2, t_2)] \\ &= E\{[x(\mathbf{z}_1, t_1) - \mu_x(\mathbf{z}_1, t_1)]\,[x^*(\mathbf{z}_2, t_2) - \mu_x^*(\mathbf{z}_2, t_2)]\} \end{aligned} \tag{8.5}$$

Wide-Sense-Stationarity and Spherical Symmetry

When the complex baseband random process $x(\mathbf{z}, t)$ is a wide-sense-stationary space-time process, its mean is constant:

$$\mu_x(\mathbf{z}, t) = \mu_x \tag{8.6}$$

and its covariance function collapses to a function of $\tau = t_1 - t_2$:

$$R_x(\mathbf{z}_1, \mathbf{z}_2, t_1, t_2) = R_x(\mathbf{z}_1 - \mathbf{z}_2, t_1 - t_2), \forall \mathbf{z}_1, \mathbf{z}_2, t_1, t_2 \tag{8.7}$$

Furthermore, when $x(\mathbf{z}, t)$ is wide-stationary spherically symmetric, R_x is a function of $\mathbf{z}_1, \mathbf{z}_2, t_1$, and t_2 only through the distance $\zeta = \|\mathbf{z}_1 - \mathbf{z}_2\| = \sqrt{(\mathbf{z}_1 - \mathbf{z}_2)^T(\mathbf{z}_1 - \mathbf{z}_2)}$:

$$R_x(\mathbf{z}_1 - \mathbf{z}_2, t_1 - t_2) = R_x(\|\mathbf{z}_1 - \mathbf{z}_2\|, t_1 - t_2), \forall \mathbf{z}_1, \mathbf{z}_2, t_1, t_2 \tag{8.8}$$

For a wide-sense-stationary spherically symmetric space-time process $x(\mathbf{z}, t)$ we define the temporal width W_t and the spatial width $W_{\mathbf{z}}$ of the covariance function as the real scalar quantities:

$$W_t = \frac{\int_0^\infty \tau^2 |R_x(0, \tau)| d\tau}{\int_0^\infty |R_x(0, \tau)| d\tau} \tag{8.9}$$

and

$$W_{\mathbf{z}} = \frac{\int_0^\infty \zeta^2 |R_x(\zeta, 0)| d\zeta}{\int_0^\infty |R_x(\zeta, 0)| d\zeta} \tag{8.10}$$

These quantities are inversely proportional to spectral bandwidth. W_t and $W_{\mathbf{z}}$ can be interpreted as the length of the temporal and spatial decorrelation intervals, respectively, in the sense that $x(\mathbf{z}_1, t_1)$ and $x(\mathbf{z}_2, t_2)$ are approximately uncorrelated for $\|\mathbf{z}_1 - \mathbf{z}_2\| > W_{\mathbf{z}}$ and for $|t_1 - t_2| > W_t$. We will generally assume all ambient noise processes to be wide-sense-stationary, spherically symmetric, and broadband relative to the signal.

Signal Propagation

Let $\varsigma_o(t) = s_o(t)e^{jw_o t}$ be a temporal signal omnidirectionally emitted from a fixed point $\mathbf{z} = \mathbf{z}_o$ in space. Assuming that the baseband signal s_o is wide sense stationary we have:

$$E\left[s_o(t)\right] = \mu_o \tag{8.11}$$

$$\text{cov}\left[s_o(t_1), s_o(t_2)\right] = R_o(t_1 - t_2) \tag{8.12}$$

where μ_o is the temporal mean and R_o is the temporal autocovariance function of the baseband equivalent signal $s_o(t)$. The time decorrelation interval of the signal $s_o(t)$ is:

$$W_t^o = \frac{\int_0^\infty \tau^2 |R_o(\tau)| d\tau}{\int_0^\infty |R_o(\tau)| d\tau} \tag{8.13}$$

Spherical Wavefronts. The spatial wavefronts of the propagating signal are the continuous closed contours in space $\mathbf{z}$ over which the complex amplitude of the signal field $\varsigma(\mathbf{z}, t)$ is constant at any given time t. The propagation medium is said to be isotropic if the velocity of signal propagation is constant in all directions for a sinusoidal signal at a given frequency, and the medium is said to be nondispersive if this propagation velocity does not depend on the sinusoidal frequency. Under the assumptions that the medium is isotropic and nondispersive, the signal propagates radially outward from $\mathbf{z}_o$ along spherical wavefronts $W(r)$ (see Figure 8.1), that is, for any fixed time t, and any fixed distance r, the signal field $\varsigma(\mathbf{z}, t)$ is constant over the set $\{\mathbf{z} : \|\mathbf{z} - \mathbf{z}_o\| = r\}$. Under these assumptions the signal field at location $\mathbf{z}_1$ is simply an attenuated and delayed version of the signal emitted at location $\mathbf{z}_o$, with the delay proportional to the radial distance $\Delta_{o1} \stackrel{\text{def}}{=} \|\mathbf{z}_1 - \mathbf{z}_o\|$. Under the above conditions, the signal field at spatial location $\mathbf{z}_1$ is simply:

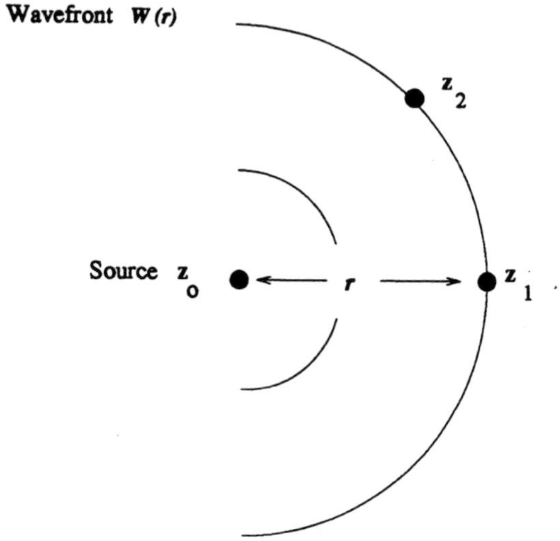

Figure 8.1 Spherical wavefronts.

$$\begin{aligned}\varsigma(\mathbf{z}_1, t) &= \varsigma(\mathbf{z}_o, t - \Delta_{o1}/\upsilon) \\ &= \varsigma_o(t - \Delta_{o1}/\upsilon)\end{aligned} \tag{8.14}$$

where υ is the radial speed of propagation associated with the medium and we have normalized the attenuation scale factor to unity for simplicity. Using $\varsigma_o(t) = s_o(t)e^{jw_o t}$ we have:

$$\begin{aligned}\varsigma(\mathbf{z}_1, t) &= s_o(t - \Delta_{o1}/\upsilon)e^{jw_o\Delta_{o1}/\upsilon}e^{jw_o t} \\ &= s(\mathbf{z}_1, t)e^{jw_o t}\end{aligned} \tag{8.15}$$

where we have identified $s(\mathbf{z}_1, t)$ as the baseband component of $\varsigma(\mathbf{z}_1, t)$,

$$s(\mathbf{z}_1, t) = s_o(t - \Delta_{o1}/\upsilon)e^{-jw_o\Delta_{o1}/\upsilon} \tag{8.16}$$

It also follows that for any spatial coordinates $\mathbf{z}_1$and $\mathbf{z}_2$ the signal field at $\mathbf{z}_2$ is a time-delayed version of the signal field at $\mathbf{z}_1$ where the delay is proportional to the radial distance between the spherical wavefront contours passing through points $\mathbf{z}_1$ and $\mathbf{z}_2$. This implies the important relation between baseband sensor components:

$$s(\mathbf{z}_2, t) = s(\mathbf{z}_1, t - \frac{\Delta_{o2} - \Delta_{o1}}{\upsilon})e^{-jw_o(\Delta_{o2}-\Delta_{o1})/\upsilon} \tag{8.17}$$

where $\Delta_{o1} = \|\mathbf{z}_1 - \mathbf{z}_o\|$ and $\Delta_{o2} = \|\mathbf{z}_2 - \mathbf{z}_o\|$.

For wide-sense-stationary $s_o(t)$ the mean and covariance functions of the baseband signal field $s(\mathbf{z}, t)$ are:

$$\begin{aligned}\mu_s(\mathbf{z}_1, t) &\stackrel{\text{def}}{=} E\left[s(\mathbf{z}_1, t)\right] \\ &= \mu_o e^{-jw_o\Delta_{o1}/\upsilon}\end{aligned} \tag{8.18}$$

$$\begin{aligned}R_s(\mathbf{z}_1, \mathbf{z}_2, t_1, t_2) &\stackrel{\text{def}}{=} \text{cov}\left[s(\mathbf{z}_1, t_1), s(\mathbf{z}_2, t_2)\right] \\ &= R_o\left(t_1 - t_2 - \frac{\Delta_{o1} - \Delta_{o2}}{\upsilon}\right)e^{-jw_o(\Delta_{o1}-\Delta_{o2})/\upsilon}\end{aligned} \tag{8.19}$$

Observe that as a consequence of the temporal wide-sense-stationarity of $s_o(t)$, the baseband signal field $s(\mathbf{z}, t)$ is itself wide-sense-stationary in t. However, since $\Delta_{01} - \Delta_{02} \neq \Delta_{12}$, $s(\mathbf{z}, t)$ is not wide-sense-stationary in $\mathbf{z}$.

Multi-Element Spatial Arrays

An M-element planar array consists of M sensors or elements located at positions $\mathbf{z}_1, \ldots, \mathbf{z}_M$ (see Fig. 8.2). Define $\mathbf{z}_c = \frac{1}{M}\sum_{l=1}^{M}\mathbf{z}_l$ the phase center of the array and $\mathbf{v}_A$ a direction vector perpendicular to the array plane. Let $\mathcal{N}(\mathbf{z}, t)$ be a zero-mean wide-sense-stationary spherically symmetric ambient noise field. A spherically propagating signal produces the array output at time t:

$$y(t) = \varsigma(t) + \mathcal{N}(t) \tag{8.20}$$

where

$$\begin{aligned}\varsigma(t) &= [\varsigma(\mathbf{z}_c, t - \tau_{1c}), \ldots, \varsigma(\mathbf{z}_c, t - \tau_{Mc})]^T \\ \mathcal{N}(t) &= [\mathcal{N}(\mathbf{z}_1, t), \ldots, \mathcal{N}(\mathbf{z}_M, t)]^T\end{aligned}$$

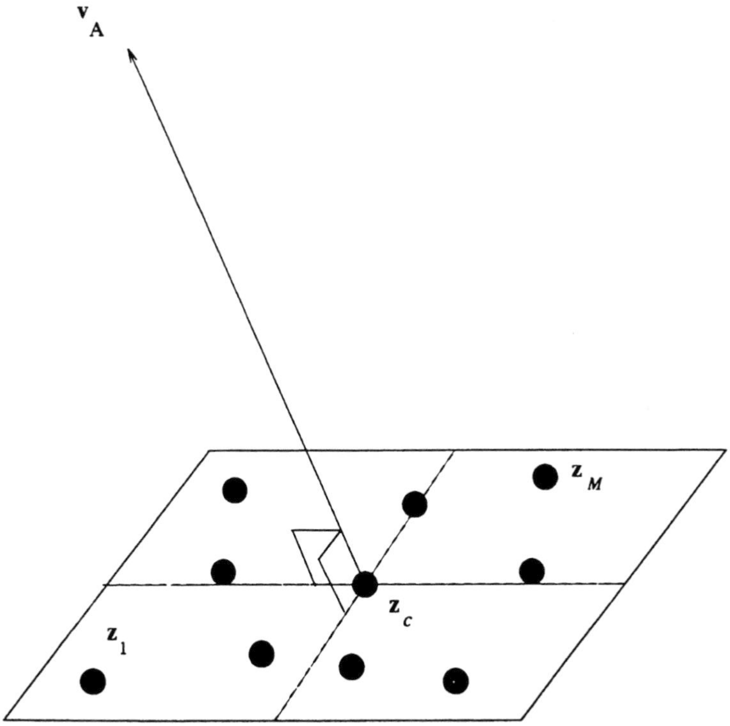

Figure 8.2 Two-dimensional (planar) sensor array.

and $\tau_{lc} = \dfrac{\Delta_{ol} - \Delta_{oc}}{v}$ is the signal propagation delay between the phase center of the array and the l-th sensor.

The baseband equivalent array output is:

$$\mathbf{y}(t) = \text{diag}_l(e^{-jw_o\tau_{lc}})\mathbf{s}(t) + \mathbf{n}(t) \tag{8.21}$$

where $\text{diag}_l(e^{-jw_o\tau_{lc}})$ is a diagonal $M \times M$ matrix with $e^{-jw_o\tau_{1c}}, \ldots, e^{-jw_o\tau_{Mc}}$ along the diagonal

$$\mathbf{s}(t) = [s(\mathbf{z}_c, t - \tau_{1c}), \ldots s(\mathbf{z}_c, t - \tau_{Mc})]^T \tag{8.22}$$

is the baseband equivalent signal referenced to phase center, and

$$\mathbf{n}(t) = [n(\mathbf{z}_1, t) \ \ldots \ n(\mathbf{z}_M, t)]^T$$

is the baseband equivalent ambient noise.

The baseband array output $\mathbf{y}(t)$ is measured at N distinct time instants $t = t_1, \ldots, t_N$. Each time sample is called a snapshot and $\Delta t_i = t_i - t_{i-1}$ is called the i-th snapshot interval. It will be assumed that the sensor spacings and snapshot intervals are sufficiently large so as to decorrelate the ambient noise field, that is, $n(\mathbf{z}_l, t_k), l = 1, \ldots, M, k = 1, \ldots, N$, defines a zero mean array of uncorrelated random variables with equal variance.

Uniformly Spaced Line Arrays. Consider the uniformly spaced line array of length $L = \|\mathbf{z}_M - \mathbf{z}_1\|$ with an even number M of elements spaced apart by D (Fig. 8.3). Under the far-field assumption, valid when the distance between the source and the array phase center satisfies $\|\mathbf{z}_c - \mathbf{z}_o\| \gg 2L^2/\lambda$, $\lambda = \upsilon/w_o$, the wavefronts as seen by the line array are planar (Fig. 8.4). It follows that in (8.21) $\tau_{lc} = \dfrac{(2l - M - 1)}{2}\dfrac{D \sin\theta}{\upsilon}$, $l = 1, \ldots, M$, where $\theta = \dfrac{\pi}{2} - \cos^{-1}\left(\dfrac{(\mathbf{z}_c - \mathbf{z}_1)^H(\mathbf{z}_c - \mathbf{z}_o)}{\|\mathbf{z}_c - \mathbf{z}_1\| \cdot \|\mathbf{z}_c - \mathbf{z}_o\|}\right)$ is the signal direction-of-arrival (DOA) relative to broadside, $\theta \in [-\pi/2, \pi/2]$.

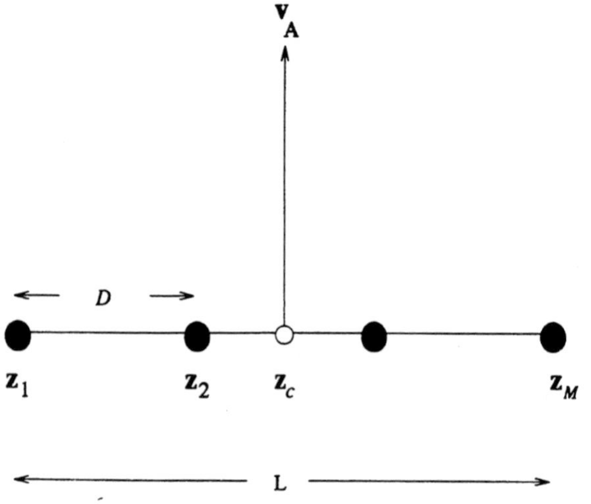

Figure 8.3 A uniformly spaced M element line array.

The Spatio-Temporal Correlation Coefficient: ρ. We use the spatio-temporal correlation coefficient to define regions of narrowband versus broadband operation of the array. For any spatial array of sensors, consider the line array formed by two sensors at locations $\mathbf{z}_1$ and $\mathbf{z}_2$. Let the variable d be the distance between these two sensors, and let the variable Δt be the time between two snapshots. Define the spatio-temporal correlation coefficient between the signal field amplitudes at spatial locations $\mathbf{z}_1$ and $\mathbf{z}_2$ and at times t and $t + \Delta t$:

$$\rho(d, \Delta t) \stackrel{\text{def}}{=} \left|\frac{\text{cov}\left[s(\mathbf{z}_1, t), s(\mathbf{z}_2, t + \Delta t)\right]}{\sqrt{\text{var}\left[s(\mathbf{z}_1, t)\right] \cdot \text{var}\left[s(\mathbf{z}_2, t + \Delta t)\right]}}\right|$$

$$= \left|\frac{\text{cov}\left[s_o(t - \frac{\Delta_{o1}}{\upsilon}), s_o(t + \Delta t - \frac{\Delta_{o2}}{\upsilon})\right]}{\sqrt{\text{var}\left[s_o(t - \frac{\Delta_{o1}}{\upsilon})\right] \cdot \text{var}\left[s_o(t + \Delta t - \frac{\Delta_{o2}}{\upsilon})\right]}}\right|.$$

Using the wide-sense-stationarity of $s_o(t)$ and the relation $(\Delta_{o2} - \Delta_{o1}) = d\sin(\theta)$, where θ is the angle of arrival relative to the baseline $\dfrac{\mathbf{z}_1 - \mathbf{z}_2}{\|\mathbf{z}_1 - \mathbf{z}_2\|}$ described by the two sensors, we obtain:

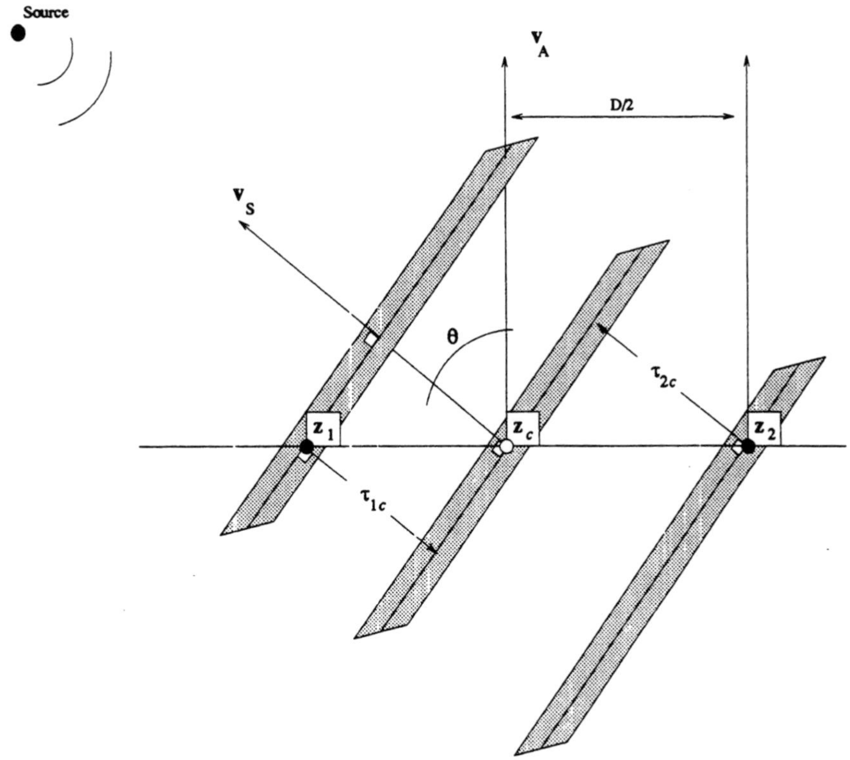

Figure 8.4 A plane wave signal propagates across two sensors at z_1, z_2 at angel θ. The propagation delays from the two elements to phase center z_C are respectively: $\tau_{1c} = \frac{1}{2}\frac{D \sin\theta}{v}$ and $\tau_{2c} = \frac{1}{2}\frac{D \sin\theta}{v}$.

$$\rho(d, \Delta t) = \frac{\left| R_o\left(\Delta t - \frac{d \sin(\theta)}{v}\right) \right|}{R_o(0)} \tag{8.23}$$

The correlation coefficient $\rho(d, \Delta t)$ is a function of the sensor spacing d, the time interval Δt, the DOA θ, and the shape of the temporal correlation function R_o.

We observe the following important property which directly follows from elementary properties of correlation functions and the assumed wide-sense-stationarity of $s_o(t)$:

Property 8.1. For fixed θ and $\Delta_{02} - \Delta_{01} = d \sin(\theta)$, the spatial-temporal correlation coefficient $\rho(d, \Delta t)$ equals one if and only if $s_o\left(t - \frac{\Delta_{01}}{v}\right) = s_o\left(t + \Delta t - \frac{\Delta_{02}}{v}\right)$ for all t with probability one, that is, the signal amplitudes are perfectly coherent across the array.

For a fixed DOA θ, the constant contours of $\rho(d, \Delta t)$ over $(d, \Delta t)$ are described by the equation of a line with slope $\frac{\sin(\theta)}{v}$ and intercept K:

$$\Delta t = \left(\frac{\sin(\theta)}{v}\right) d + K \tag{8.24}$$

K is a constant which indexes the contour lines of $\rho(d, \Delta t)$ for the signal correlation function $R_o(\tau)$ displayed in Figure 8.5.

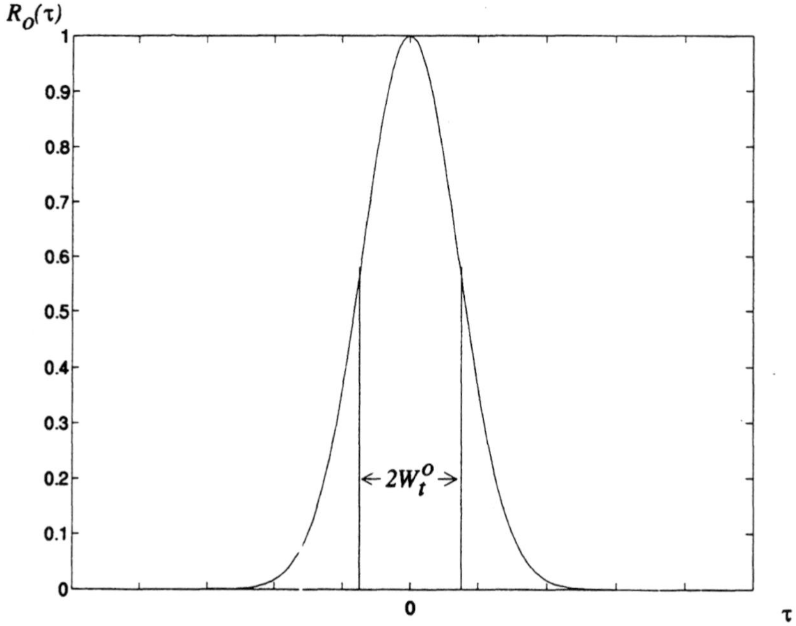

Figure 8.5 Typical signal correlation function with decorrelation with W_t^o.

The correlation coefficient goes to zero as K increases beyond the signal decorrelation time W_t^o. The darkly shaded strip in Figure 8.6 represents pairs $(d, \Delta t)$ where $\rho \approx 1$. In this region the signal field measured at the array is perfectly coherent over time and space. The lightly shaded region in Figure 8.6 represents pairs $(d, \Delta t)$ where $K \leq W_t^o$ and ρ is less than 1 but ρ can be significantly greater than 0. In this region the measured signal field is only partially coherent over time and space. Finally, the unshaded region in Figure 8.6 represent pairs $(d, \Delta t)$ for which $K > W_t^o$ and $\rho \approx 0$. In this region the measured signal field is incoherent over time and space.

For a particular array, let two sensors be separated by distance D_{lm} and let two snapshots be separated by T_s. The spatial correlation between the two sensor outputs for one of the snapshots is the value of the surface $\rho(d, \Delta t)$ at the point $(D_{lm}, 0)$, shown lying on the d axis in Figure 8.6. The temporal correlation between the two snapshots for any single sensor output is the value of the surface $\rho(d, \Delta t)$ at the point $(0, T_s)$, shown lying on the Δt axis in Figure 8.6. Finally, the spatio-temporal correlation between the two snapshots and two sensor outputs is the value of the surface $\rho(d, \Delta t)$ at the point (D_{lm}, T_s). Let the maximum interelement spacing be given by $D_o \stackrel{\text{def}}{=} \max_{lm}(D_{lm})$, which is the maximum spatial extent of the array. Let the total observation time be given by $T_o = t_N - t_1$.

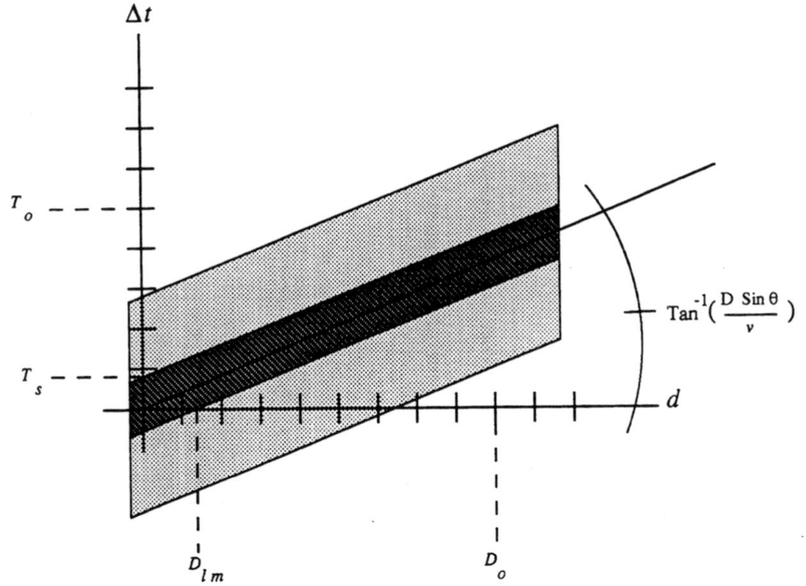

Figure 8.6 Constant contours of ρ over $(d, \Delta t)$. $\rho(\alpha, \Delta t) = 1$ over darkly shaded region, $0 \ll |\rho(\alpha, \Delta t| \ll 1$ over lightly shaded region, $\rho(\alpha, \Delta t) \approx 0$ over unshaded region.

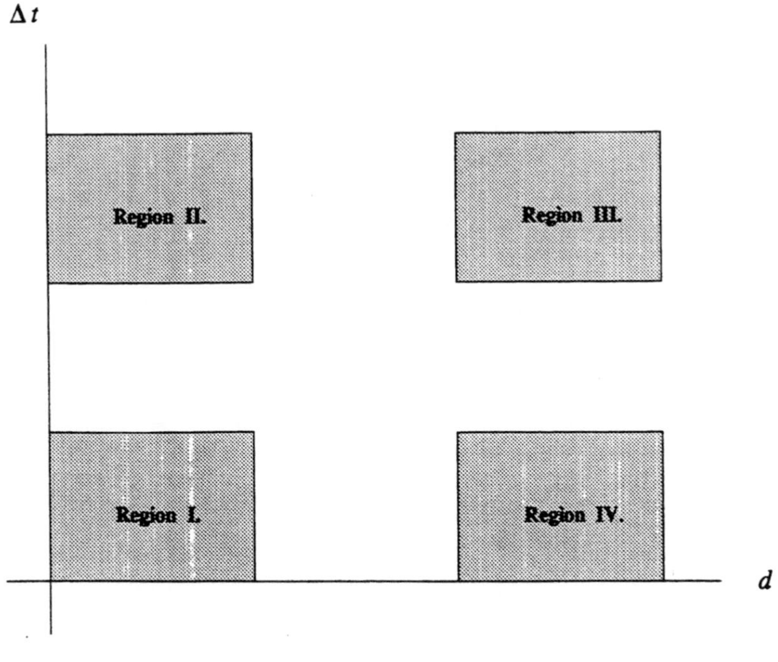

Figure 8.7 Spatio-temporal array operating regimes.

Regions of Array Operation. We separate the operation of the array into four $(d, \Delta t)$ regions of interest (Figs. 8.7 and 8.8): Region I (small Δt small d), Region II (large Δt small d), Region III (large Δt large d), and Region IV (small Δt large d). We call these regimes of array operation the spatially narrowband/temporally narrowband, spatially narrowband/temporally wideband, spatially wideband/temporally wideband, and the spatially wideband/temporally narrow-

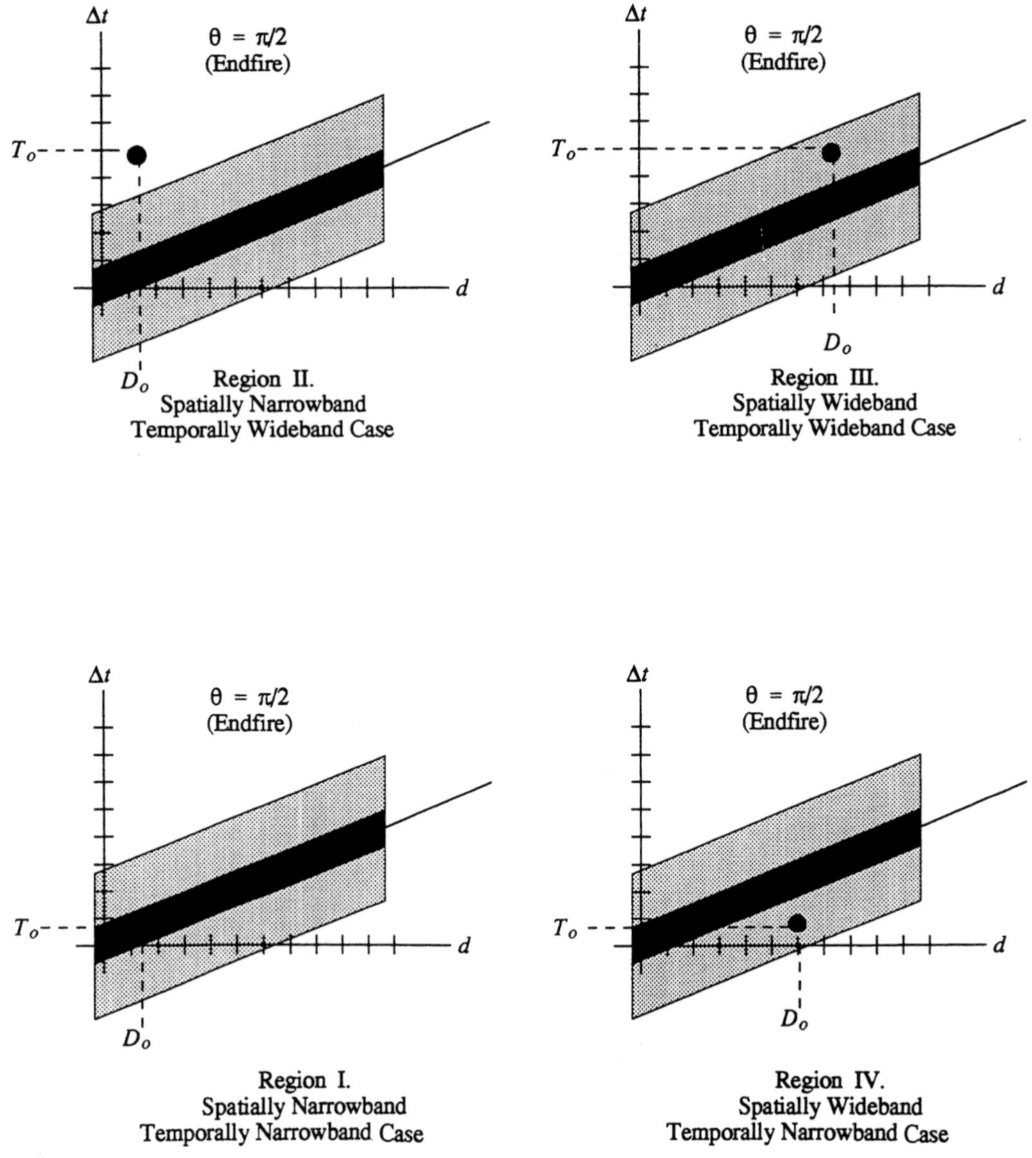

Figure 8.8 Constant contours in ρ for the four spatio-temporal array operating regimes.

band regimes, respectively. Region I is the standard narrowband regime where the measured signal amplitudes are perfectly coherent over snapshots and across the array, while Regions II–IV are wideband regimes where measured complex signal amplitudes may be incoherent or partially coherent.

Region I: Spatially Narrowband/Temporally Narrowband Regime. Referring to Eq. (8.23), assume that the total observation time T_o and the array spatial extent D_o are such that for all $d \in [0, D_o]$ and all $\Delta t \in [0, T_o] : |\Delta t - \frac{d\sin(\theta)}{v}| << W_t^o$. Then for all $d \in [0, D_o]$ and $\Delta t \in [0, T_o] : \rho(d, \Delta t) \approx 1$, and the measured signal amplitudes $s_o\left(t - \frac{\Delta_{ol}}{v}\right)$ are perfectly coherent over all sensors $l = 1, \ldots, M$ and all snapshots $t = t_1, \ldots, t_N$. If this holds for all DOAs $\theta \in [-\pi/2, \pi/2]$ the measured signal fields $s(\mathbf{z}_l, t_k), l = 1, \ldots, M$ and $k = 1, \ldots, N$, are well approximated by:

$$s(\mathbf{z}_l, t_k) = s_I \cdot e^{-jw_o\tau_{lc}} \tag{8.25}$$

where $s_I \overset{\text{def}}{=} s(\mathbf{z}_c, t_1)$ is a complex signal amplitude independent of l and k.

Region II: Spatially Narrowband/Temporally Wideband Regime. In this regime the snapshot interval $T_s \gg W_t^o$ and $|\frac{D_o\sin(\theta)}{v}| \ll W_t^o$ so that $\rho(d, 0) \approx 1, d \in [0, D_o]$, while $\rho(0, T_s) \approx 0$. Here, the source signal amplitudes are perfectly coherent across the array but are incoherent over the snapshots. In this case, the measured signal fields $s(\mathbf{z}_l, t_k), l = 1, \ldots, M$ and $k = 1, \ldots, N$, are well approximated by

$$s(\mathbf{z}_l, t_k) = s_{II}(k) \cdot e^{-jw_o\tau_{lc}} \tag{8.26}$$

where $s_{II}(k) \overset{\text{def}}{=} s(\mathbf{z}_c, t_k)$ is an uncorrelated sequence over k.

Region III: Spatially Wideband/Temporally Wideband Regime. In this regime the snapshot interval $T_s \gg W_t^o$ but $|\frac{D_o\sin(\theta)}{v}| \not\ll W_t^o$, so that $\rho(0, T_s) \approx 0$ and $0 < \rho(D_o, 0) \ll 1$. Here the measured signals are incoherent over the set of snapshots but are partially coherent across the sensor array. In this case, the measured signal fields $s(\mathbf{z}_l, t_k), l = 1, \ldots, M$ and $k = 1, \ldots, N$, are well approximated by

$$s(\mathbf{z}_l, t_k) = s_{III}(l, k) \cdot e^{-jw_o\tau_{lc}} \tag{8.27}$$

where $s_{III}(l, k) \overset{\text{def}}{=} s(\mathbf{z}_c, t_k - \tau_{lc})$ is a spatio-temporal process uncorrelated over k but correlated over l.

Region IV: Spatially Wideband/Temporally Narrowband Regime. In this regime $T_o \ll W_t^o$ so that $\rho(0, T_o) \approx 1$ while $0 < \rho(D_o, 0) \ll 1$. Thus, while the source signal amplitudes may vary over the M array elements, at any particular array element the snapshots are nearly identical. Here, the source signal amplitudes are only partially coherent across the array, while they are perfectly coherent over the set of snapshots. The measured signal fields $s(\mathbf{z}_l, t_k), l = 1, \ldots, M$ and $k = 1, \ldots, N$, are then well approximated by:

$$s(\mathbf{z}_l, t_k) = s_{IV}(l) \cdot e^{-jw_o\tau_{lc}} \tag{8.28}$$

where $s_{IV}(l) \overset{\text{def}}{=} s(\mathbf{z}_c, t_1 - \tau_{lc})$ is a correlated sequence over l.

Array Model for Slow Fading Signals

In many transmission channels Rayleigh scattering produces amplitude variations in the received signal, giving rise to Rayleigh fading. This scattering phenomenon is characteristic of optical and radio frequency tropospheric and ionospheric propagation [31], land, sea, and free space radar backscatter [32], multipath in mobile radio [33], and volume and surface reverberation in underwater acoustic propagation [34]. In slow Rayleigh fading [35] the received signal amplitudes are perfectly coherent over the array for any particular snapshot but vary randomly for different snapshots.

For a single signal impinging on a uniformly spaced line array from direction θ, we will use the following slow Rayleigh fading model for the baseband outputs of the array at times $t = t_1, \ldots, t_N$:

$$\mathbf{y}(t_k) = \mathbf{v}(\theta) \cdot s(t_k) + \mathbf{n}(t_k), \ k = 1, \ldots, N \tag{8.29}$$

where $s(t_k)$ is an i.i.d. sequence of Gaussian random variables with mean μ_s and variance σ_s^2, $\mathbf{n}(t_k)$ is an i.i.d. sequence of zero-mean Gaussian random vectors with covariance matrix $\sigma_n^2\mathbf{I}_M$, $\mathbf{I}_M$ is the $M \times M$ identity matrix, and $\mathbf{v}(\theta)$ is the steering vector:

$$\mathbf{v}(\theta) = \left[e^{-jw_o\tau_{1c}(\theta)}, \ldots, e^{-jw_o\tau_{Mc}(\theta)}\right]^T \tag{8.30}$$

The model (8.29) is simply a vector representation of the temporally wideband/spatially narrowband model (8.26). Note that when the signal variance σ_s^2 approaches 0, the signal amplitudes $s(t_k)$ become constant over time and the temporally wideband/spatially narrowband regime becomes equivalent to the temporally narrowband/spatially narrowband regime. Thus, the model (8.29) covers both of the regimes described by regions I and II in Figure 8.7.

For the case of p correlated signals $\varsigma_1, \ldots, \varsigma_p$ impinging on the array from directions $\theta = \left[\theta_1, \ldots, \theta_p\right]^T$ the baseband slow Rayleigh fading model (8.29) generalizes to:

$$\mathbf{y}(t_k) = \mathbf{V}(\theta) \cdot \mathbf{s}(t_k) + \mathbf{n}(t_k), \qquad k = 1, \ldots, N \tag{8.31}$$

where $\mathbf{s}(t_k) = \left[s_1(t_k), \ldots, s_p(t_k)\right]^T$, is an i.i.d. sequence of p-element Gaussian random vectors with p-element mean vector $\boldsymbol{\mu}_s$, $p \times p$ covariance matrix $\boldsymbol{\Sigma}_s$, and $\mathbf{V}(\theta) = \left[\mathbf{v}(\theta_1), \ldots, \mathbf{v}(\theta_p)\right]$ is an $M \times p$ matrix of steering vectors associated with the p signal directions.

8.4 TASK-SPECIFIC ADAPTIVE BEAMFORMING METHODS

If a specific estimation or detection task is of interest, for theoretically best performance an optimal estimator or detector should be constructed directly on the raw multiple-sensor outputs $\mathbf{y}$. Many methods, for example, maximum likelihood estimation and generalized likelihood ratio tests [1, 2, 5], have been studied for a wide variety of estimation and detection problems. However, these direct methods frequently suffer from high implementation complexity, analytical intractability, or sensitivity to statistical model assumptions. The alternative, studied here, is to constrain the complexity of the estimator statistic by specializing to a simple beamformer structure, and then to optimize the beamformer structure for the task of interest according to some suitable performance criterion.

Beamforming was developed as a method of linearly combining the elements of the array to produce a directionally sensitive *antenna gain pattern* which can enhance signal components of interest while suppressing noise and interference. In nonadaptive beamforming the coefficients, or weights, of the linear combiner are computed in advance and do not depend on the measured array output. In adaptive beamforming the weights are computed on the fly as a function of the measurements, and in this case they can adapt to unknown signal and noise environments such as unknown signal DOAs, unknown signal amplitudes, and unknown interference. While the results to be described in the following sections can be extended to wideband beamsummers [36–38], in this presentation we restrict attention to the narrowband beamsummer described below.

Narrowband Beamsummer

Define the weight vector $\mathbf{w} = [w_1 \ldots w_M]^T$. A narrowband beamsummer is simply a spatial FIR filter with filter coefficients $\mathbf{w}^H = [w_1^*, \ldots, w_M^*]$ shown in Figure 8.9. Samples are taken of the beamsummer output $x(t)$ at times $t = t_1, \ldots, t_N$. The baseband output of the beamsummer is the scalar sequence:

$$x(t_k) = \mathbf{w}^H \mathbf{y}(t_k), \quad k = 1, \ldots, N \tag{8.32}$$

where $\mathbf{y}(t)$ is the baseband M-sensor array output (8.31).

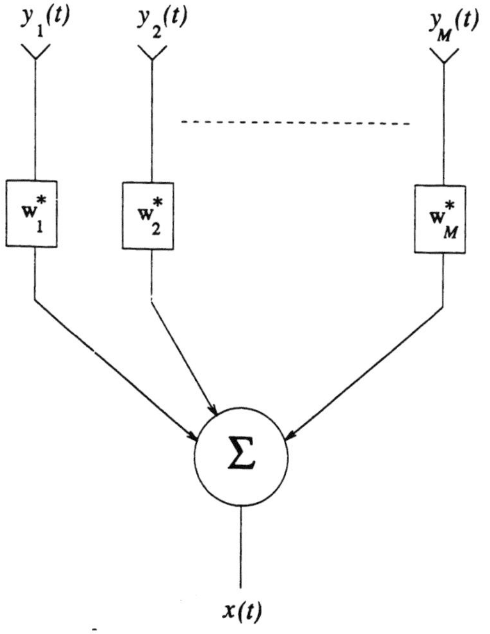

Figure 8.9 The narrowband beamsummer array.

The beamsummer gain pattern is defined as the noiseless single-signal response of the beamsummer as a function of incident signal angle θ:

$$|\mathbf{w}^H \mathbf{v}(\theta)| = \left| \sum_{l=1}^{M} w_l^* e^{-jw_o \tau_{lc}(\theta)} \right| \tag{8.33}$$

Signal Detection Criteria for Weight Selection

For a given estimation or detection task, let FOM be a nonrandom scalar function of the beamsummer weight vector. The value FOM($\mathbf{w}$) will be called the *figure of merit* (FOM) of the beamsummer for a particular weight vector $\mathbf{w}$. The FOM will be a function of ensemble statistics, for example, lower order moments, of the beamsummer output $x(t)$ and will measure the average performance of the beamsummer in fulfilling the aims of a given task. The weight vector which maximizes the FOM will be called the optimal weight vector for this task.

Consider the problem of testing the between hypotheses: $H_1 : \mathbf{y} = \mathbf{V}(\theta)\mathbf{s} + \mathbf{n}$ (signals present) versus $H_0 : \mathbf{y} = \mathbf{n}$ (signals absent) based on the beamsummer outputs $\mathbf{x} = [x(t_1), \ldots, x(t_N)]^T$. Let $h(\mathbf{x})$ be a detection statistic which is compared to a detection threshold to decide between H_0 and H_1. When the conditional p.d.f.s $f(\mathbf{x}|H_0)$ and $f(\mathbf{x}|H_1)$ of $\mathbf{x}$ are known exactly under H_0 and H_1, the optimal detection statistic is the log-likelihood ratio $h(\mathbf{x}) = \ln f(\mathbf{x}|H_1)/f(\mathbf{x}|H_0)$ since, when used with a suitable detection threshold, it gives the Neyman-Pearson detector [39, 40].

The deflection index d associated with a signal detection statistic $h(\mathbf{x})$ was introduced by Uhlenbeck and Lawson [41] as a convenient measure of detection performance:

$$d = \frac{E_{H_1}[h(\mathbf{x})] - E_{H_0}[h(\mathbf{x})]}{\sqrt{\operatorname{var}_{H_0}[h(\mathbf{x})]}}, \tag{8.34}$$

where $E_{H_1}[h(\mathbf{x})]$ is the mean under $f(\mathbf{x}|H_1)$, while $E_{H_0}[h(\mathbf{x})]$ and $\operatorname{var}_{H_0}[h(\mathbf{x})]$ are the mean and variance under $f(\mathbf{x}|H_0)$. While the exact detection probabilities of the detector may depend on additional factors, very large (small) values of the deflection d generally indicate that the test statistic $h(\mathbf{x})$ is capable (incapable) of accurately discriminating between H_0 and H_1 [35, 42]. Since the deflection depends on the weights $\mathbf{w}$ through the beamsummer output $\mathbf{x}$ the deflection $d = d(\mathbf{w})$ can be used to gauge the best achievable detection performance associated with a given set of beamsummer weights.

Using the temporally wideband model (8.31) for the baseband array output $\mathbf{y}$, the mean $E[x(t)] = \mu_x$ and variance $\operatorname{var}[x(t)] = \sigma_x^2$ of the beamsummer output at time t are easily seen to be:

$$\mu_x = \mathbf{w}^H \mathbf{V}(\theta)\boldsymbol{\mu}_s, \tag{8.35}$$

$$\sigma_x^2 = \mathbf{w}^H \left[\mathbf{V}(\theta)\boldsymbol{\Sigma}_s\mathbf{V}^H(\theta) + \sigma_n^2\mathbf{I}_M\right]\mathbf{w} \tag{8.36}$$

where $\boldsymbol{\mu}_s = [\mu_1, \ldots, \mu_p]^T$ is the vector of mean signal amplitudes, and $\boldsymbol{\Sigma}_s$ is the $p \times p$ signal amplitude covariance matrix.

In Appendix 8A it is shown that under the slow Rayleigh fading model (8.31) for **y**, the deflection for the optimal Neyman-Pearson signal detector is equal to:

$$\text{FOM}_D \stackrel{\text{def}}{=} d = \frac{\alpha^2(\alpha^2+\gamma^2)+2\gamma^2}{\sqrt{\alpha^4+\gamma^2}}\sqrt{\frac{N}{2}} \tag{8.37}$$

where γ^2 is the beamsummer output squared-mean signal-to-noise ratio (SM-SNR):

$$\gamma^2(\mathbf{w}) = \frac{|E_{H_1}[x(t)]|^2}{\text{var}_{H_0}[x(t)]} \tag{8.38}$$

$$= \frac{\mathbf{w}^H\left[\mathbf{V}(\boldsymbol{\theta})\boldsymbol{\mu}_s\boldsymbol{\mu}_s^H\mathbf{V}^H(\boldsymbol{\theta})\right]\mathbf{w}}{\mathbf{w}^H\mathbf{w}}\cdot\frac{1}{\sigma_n^2} \tag{8.39}$$

and α^2 is the variance signal-to-noise ratio (V-SNR):

$$\alpha^2(\mathbf{w}) = \frac{\text{var}_{H_1}[x(t)] - \text{var}_{H_0}[x(t)]}{\text{var}_{H_0}[x(t)]} \tag{8.40}$$

$$= \frac{\mathbf{w}^H\left[\mathbf{V}(\boldsymbol{\theta})\boldsymbol{\Sigma}_s\mathbf{V}^H(\boldsymbol{\theta})\right]\mathbf{w}}{\mathbf{w}^H\mathbf{w}}\cdot\frac{1}{\sigma_n^2} \tag{8.41}$$

The deflection index d can be used as a detection figure of merit, FOM_D, in that maximization of FOM_D over the beamsummer weights will maximize the capability of the beamsummer to detect signals arriving at angles $\theta_1, \ldots, \theta_p$.

Narrowband Signal FOM: Squared Mean Beamsummer Output. When $\boldsymbol{\Sigma}_s = 0$ (zero variance narrowband signals) then $\alpha^2 = 0$ and the deflection (8.37) is proportional to the square-root square-mean SNR (SM-SNR):

$$d = \sqrt{\gamma^2}\sqrt{2N}$$

In this case d is monotone increasing in the normalized squared-mean output of the beamsummer:

$$\text{FOM}_{SM}(\mathbf{w}) = \frac{\mathbf{w}^H\left[\mathbf{V}(\boldsymbol{\theta})\boldsymbol{\mu}_s\boldsymbol{\mu}_s^H\mathbf{V}^H(\boldsymbol{\theta})\right]\mathbf{w}}{\mathbf{w}^H\mathbf{w}} \tag{8.42}$$

Using Rayleigh's theorem [43] we know that, up to an arbitrary scale factor, the optimal weight vector maximizing FOM_{SM} is the eigenvector of the matrix $\mathbf{A} = \mathbf{V}(\boldsymbol{\theta})\boldsymbol{\mu}_s\boldsymbol{\mu}_s^H\mathbf{V}^H(\boldsymbol{\theta})$ corresponding to the maximum eigenvalue of $\mathbf{A}$.

Broadband Signal FOM: Variance at Beamsummer Output. If $\boldsymbol{\mu}_s = 0$ (zero mean broadband signals) then $\gamma^2 = 0$ and the deflection (8.37) is proportional to the variance SNR (V-SNR):

$$d = \alpha^2\sqrt{\frac{N}{2}}$$

In this case d is monotone increasing in the normalized signal variance at the output of the beamsummer:

$$\text{FOM}_V(\mathbf{w}) = \frac{\mathbf{w}^H\left[\mathbf{V}(\boldsymbol{\theta})\boldsymbol{\Sigma}_s\mathbf{V}^H(\boldsymbol{\theta})\right]\mathbf{w}}{\mathbf{w}^H\mathbf{w}} \tag{8.43}$$

Up to an arbitrary scale factor, the optimal weight vector maximizing FOM_V is the eigenvector of the matrix $\mathbf{B} = \mathbf{V}(\boldsymbol{\theta})\boldsymbol{\Sigma}_s\mathbf{V}^H(\boldsymbol{\theta})$ corresponding to the maximum eigenvalue of $\mathbf{B}$.

Broadband Signal FOM: Mean Squared at Beamsummer Output. When $\boldsymbol{\mu}_s \neq 0, \boldsymbol{\Sigma}_s \neq 0$ and $\gamma^2/\alpha^2 \ll 1$ (variance dominated broadband signals) the deflection (8.37) is approximately proportional to the mean-square SNR (MS-SNR):

$$d = \frac{\alpha^2}{\sqrt{\alpha^4+\gamma^2}}\left[\alpha^2+\gamma^2+2\frac{\gamma^2}{\alpha^2}\right]\sqrt{\frac{N}{2}}$$
$$\approx (\alpha^2+\gamma^2)\sqrt{\frac{N}{2}}$$

and is a monotone increasing function of the normalized mean-square of the beamsummer output $\text{FOM}_{SM}(\mathbf{w}) + \text{FOM}_V(\mathbf{w})$:

$$\text{FOM}_{MS}(\mathbf{w}) = \frac{\mathbf{w}^H\left[\mathbf{V}(\boldsymbol{\theta})\left[\boldsymbol{\Sigma}_s+\boldsymbol{\mu}_s\boldsymbol{\mu}_s^H\right]\mathbf{V}^H(\boldsymbol{\theta})\right]\mathbf{w}}{\mathbf{w}^H\mathbf{w}} \tag{8.44}$$

Up to an arbitrary scale factor, the optimal weight vector maximizing FOM_V is the eigenvector of the matrix $\mathbf{C} = \mathbf{V}(\boldsymbol{\theta})\left[\boldsymbol{\Sigma}_s+\boldsymbol{\mu}_s\boldsymbol{\mu}_s^H\right]\mathbf{V}^H(\boldsymbol{\theta})$ corresponding to the maximum eigenvalue of $\mathbf{C}$.

In the case of one signal ($p = 1$), $\mathbf{V}(\boldsymbol{\theta}) = \mathbf{v}(\boldsymbol{\theta})$, $\boldsymbol{\mu}_s$ and $\boldsymbol{\Sigma}_s$ are scalars and it is easily shown that the weight vector

$$\mathbf{w}_{opt} = b\mathbf{v}(\theta_1) \tag{8.45}$$

simultaneously maximizes FOM_{SM}, FOM_V, and FOM_{MS} where b is an arbitrary complex scale factor. The gain pattern for the detection optimal weight vector $\mathbf{w} = \mathbf{v}(\theta)$ and a 6 element line array is shown in Figure 8.10 for a single signal at direction $\theta = -.22$ radians. For this choice of weights the maximum beamsummer gain occurs in the direction θ of the signal.

Adaptive Beamsumming via Empirical Signal Quality Indices

Even for $p = 1$, when the signal DOA θ is unknown, the optimal weight vector (8.45) cannot be implemented. We define a *signal quality index* (SQI) as a function of $\mathbf{w}$ which depends only on empirical statistics computed from the beamsummer outputs $x(t), t = t_1, \ldots, t_N$. A SQI can frequently be defined which asymptotically converges to an appropriate FOM as the number of snapshots becomes large. Let $\mathbf{w}^{(N)}$ be the weight vector which maximizes the SQI when N snapshots are available. If the SQI converges as $N \to \infty$ to a FOM, then the sequence of weight vectors $\mathbf{w}^{(1)}, \mathbf{w}^{(2)}, \ldots,$ adapt to the unknown signal parameters in the sense that they converge to the optimal weight vector maximizing the FOM. In this case the weight sequence $\mathbf{w}^{(1)}, \mathbf{w}^{(2)}, \ldots$ is said to be adaptive.

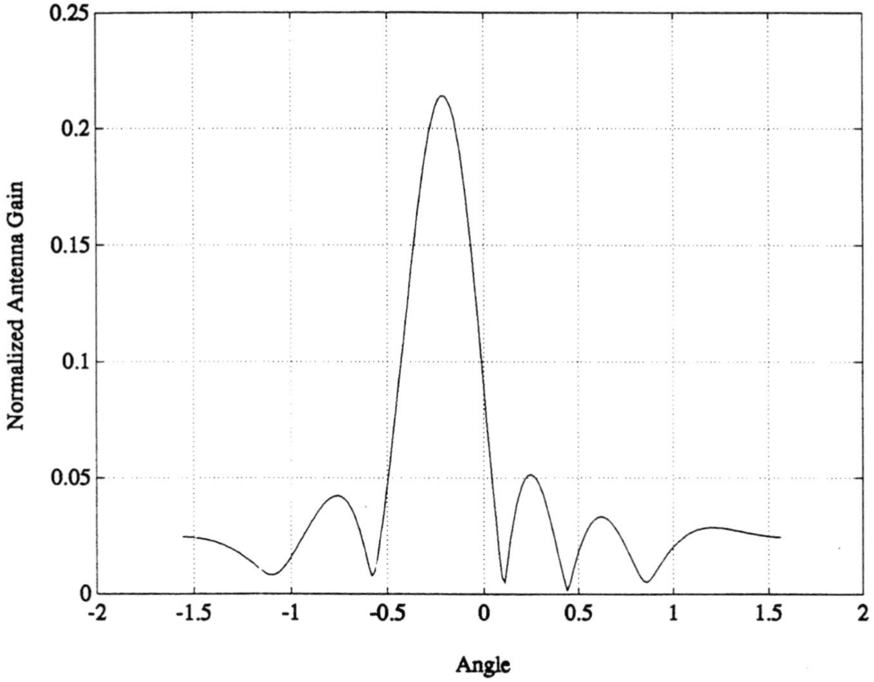

Figure 8.10 Beamsummer gain pattern as a function of signal direction angle θ for the detection optimal weight vector $\mathbf{w} = \mathbf{v}$ for one signal.

For $N = 2, 3, \ldots$ snapshots consider the three detection SQI's:

$$|\overline{x(N)}|^2 \stackrel{\text{def}}{=} \frac{1}{\mathbf{w}^H\mathbf{w}} \cdot \left| \frac{1}{N} \sum_{k=1}^{N} x(t_k) \right|^2 \tag{8.46}$$

$$\overline{|x(N)|^2} \stackrel{\text{def}}{=} \frac{1}{\mathbf{w}^H\mathbf{w}} \cdot \frac{1}{N} \sum_{k=1}^{N} |x(t_k)|^2 \tag{8.47}$$

$$\overline{|x(N) - \overline{x(N)}|^2} \stackrel{\text{def}}{=} \frac{1}{\mathbf{w}^H\mathbf{w}} \cdot \frac{1}{N-1} \sum_{k=1}^{N} \left| x(t_k) - \frac{1}{n} \sum_{k=1}^{N} x(t_k) \right|^2 \tag{8.48}$$

Under the assumption of uncorrelated snapshots $|\overline{x(N)}|^2$, $\overline{|x(N)|^2}$, and $\overline{|x(N) - \overline{x(N)}|^2}$ converge to FOM_{SM}, $\text{FOM}_{MS} + \sigma_n^2$, and $\text{FOM}_V + \sigma_n^2$, respectively, as the number of snapshots N goes to infinity. Thus, the sequences of adaptive weight vectors obtained by respectively maximizing $|\overline{x(N)}|^2$, $\overline{|x(N)|^2}$, or $\overline{|x(N) - \overline{x(N)}|^2}$ are asymptotically optimal for detection under the SM-SNR, MS-SNR, and V-SNR criteria.

8.5 OPTIMAL DESIGN USING CR BOUND

Here we introduce the methodology of optimal weight design using Fisher information and the Cramér-Rao (CR) bound.

Let ϕ be an unknown parameter of interest. For fixed beamsummer weights $\mathbf{w}$ let the snapshots of the beamsummer outputs $\mathbf{x} = [x(t_1), \ldots, x(t_N)]^T$ have joint probability density function $f_{\mathbf{x}}(\mathbf{x}; \phi)$ parameterized by the unknown scalar parameter ϕ. Define the Fisher information:

$$J_\phi = E_\phi \left[\frac{\partial}{\partial \phi} \ln f(\mathbf{x}; \phi) \frac{\partial}{\partial \phi} \ln f(\mathbf{x}; \phi) \right] \tag{8.49}$$

where, for any integrable function h, $E_\phi[h(\mathbf{x})] = \int h(\mathbf{x}) f(\mathbf{x}; \phi) d\mathbf{x}$ is the expectation of $h(\mathbf{x})$ when the true value of the underlying parameter is ϕ. Note that, while only functional dependence on ϕ is explicitly indicated in (8.49), the Fisher information is a function of both ϕ and $\mathbf{w}$.

Under appropriate regularity conditions [44, 45] any unbiased estimator $\hat{\phi} = \hat{\phi}(\mathbf{x})$ of ϕ has variance which satisfies the CR lower bound:

$$\text{var}\left[\hat{\phi}\right] \geq \frac{1}{J_\phi} \tag{8.50}$$

The methodology of optimal design of experiments [18, 19] attempts to maximize the Fisher information $J_\phi = J_\phi(\mathbf{w})$ over $\mathbf{w}$ in order to provide data $\mathbf{x}$ which allows the most accurate parameter estimation performance.

There are two well-known problems which make practical application of the optimal design methodology to estimation of ϕ difficult.

1. The weight vector which minimizes the CR bound on a parameter generally depends on the unknown parameters.
2. While the optimal weight vector may provide the most informative beamsummer output data, the form of a post-beamsummer optimal parameter estimator is unknown.

The first issue will be addressed by using empirically derived quantities to approximate the unknowns in the CR bound. The second issue is more difficult and will require additional work. In the sequel we will use heuristic estimation methods to extract parameter estimates from the optimal beamsummer output data.

Note that if other parameters besides ϕ are also unknown, the bound (8.50) may not be the tightest lower bound. For the case of multiple unknown parameters, a tighter CR bound exists but requires inversion of a Fisher information matrix [46]. While in principle generalizable to the tighter bound, here we will use the weaker version (8.50) of the CR bound.

Under the assumed Gaussian model (8.31) for the the array output $\mathbf{y}$, the Fisher information is given by the formula [47]:

$$J_\phi = \text{tr}\left\{ \frac{d\boldsymbol{\Gamma}_{\mathbf{x}}}{d\phi} \boldsymbol{\Gamma}_{\mathbf{x}}^{-1} \frac{d\boldsymbol{\Gamma}_{\mathbf{x}}}{d\phi} \boldsymbol{\Gamma}_{\mathbf{x}}^{-1} \right\} + 2\text{Re}\left\{ \frac{d\boldsymbol{\mu}_{\mathbf{x}}^H}{d\phi} \boldsymbol{\Gamma}_{\mathbf{x}}^{-1} \frac{d\boldsymbol{\mu}_{\mathbf{x}}}{d\phi} \right\} \tag{8.51}$$

where $\mu_x = E_\phi[\mathbf{x}]$ and $\Gamma_x = E_\phi[(\mathbf{x} - \mu_x)^T(\mathbf{x} - \mu_x)]$ are the mean vector and covariance matrix of the vector of beamsummer snapshots $\mathbf{x}$.

8.6 OPTIMAL DESIGN FOR CONSTANT MODULUS PARAMETERS

Consider the following time varying modification of the p-signal array output model (8.29) for equally spaced snapshot times:

$$\mathbf{y}(kT_s) = g(kT_s, \phi) \cdot \mathbf{V}(\boldsymbol{\theta})\mathbf{s}(kT_s) + \mathbf{n}(kT_s), \quad k = 1, \ldots, N \tag{8.52}$$

where the scalar sequence, $g(kT_s, \phi)$, modulates the p signals' amplitudes and depends on the unknown ϕ. We impose two restrictions on the function g: (1) g is constant modulus: $|g(t, \phi)|^2 = 1$, for all t and all ϕ; (2) g is differentiable in ϕ. We will also assume that $T_s \gg \max W_t^0$ the maximum decorrelation time of the p signal autocorrelation functions.

Our main motivation for the model (8.52) is the following heuristic: a beamsummer which is optimized for estimation of a hypothetical constant-power multiplicative factor $g(t, \phi)$ should also provide a good statistic for detection and estimation of the signal compound $\mathbf{V}(\boldsymbol{\theta}) \cdot \mathbf{s}(t)$. On the other hand, the model (8.52) may be of interest in its own right, for example, for decoding of phase or frequency-modulated communications signals when ϕ is encoded by spatially invariant modulation of the signal.

The Fisher information associated with measurement $\mathbf{x}$ is derived from the formula (8.51) in Appendix 8B. Specializing the result (8B.2) of Appendix 8B to the case of uncorrelated snapshots ($P = 1$):

$$J_\phi = \left(\alpha^2 \cdot \frac{\alpha^2}{1 + \alpha^2} + \gamma^2\right) \cdot 2\sum_{k=1}^{N} |g'(kT_s)|^2 \tag{8.53}$$

where

$$g'(t) = \frac{\partial g(t, \phi)}{\partial \phi}$$

γ^2 is the $\mathbf{w}$-dependent beamsummer SM-SNR defined in (8.38), and α^2 is the $\mathbf{w}$-dependent beamsummer V-SNR defined in (8.40).

We make the following observations concerning the Fisher information (8.53).

- The Fisher information (8.53) can be written in equivalent form:

$$J_\phi = \left[(1 - \varepsilon)\gamma^2 + \varepsilon(\gamma^2 + \alpha^2)\right] \cdot 2\sum_{k=1}^{N} |g'(kT_s)|^2$$

where

$$\varepsilon = \frac{\alpha^2}{1 + \alpha^2}$$

Since $\varepsilon \in [0, 1]$, the Fisher information is a convex combination of the beamsummer SM-SNR (γ^2) and MS-SNR ($\gamma^2 + \alpha^2$) criteria for beamsummer performance. The convex proportionality constant ε is a monotonic function of the beamsummer V-SNR α^2. For $\alpha^2 \approx 0$ (narrowband signal regime), we see that J_ϕ becomes proportional to γ^2 while for $\alpha^2 \approx 0$ (broadband signal regime) J_ϕ becomes proportional to α^2. Therefore, ε "adapts" the set of weights, obtained by maximizing $J_\phi(\mathbf{w})$, to the deflection index (8.37) governing the narrowband (8.42) and broadband (8.44) signal scenarios.

- Using (8.42), (8.44), and (8.43) J_ϕ (8.53) is seen to be equivalent to:

$$J_\phi = \text{FOM}_{CM}(\mathbf{w}) \cdot \frac{2\sum_{k=1}^{N} |g'(kT_s)|^2}{\sigma_n^2} \tag{8.54}$$

where

$$\text{FOM}_{CM} = \text{FOM}_V \cdot \frac{\text{FOM}_V}{\sigma_n^2 + \text{FOM}_V} + \text{FOM}_{SM} \tag{8.55}$$

and the factor $2\sum_{k=1}^{N} |g'(kT_s)|^2/\sigma_n^2$ does not depend on $\mathbf{w}$. FOM_{CM} is a universal figure of merit for the beamsummer in the sense that maximizing FOM_{CM} minimizes the CR bound (8.53) for *arbitrary* ϕ-differentiable functions $g(t, \phi)$ satisfying the constant modulus constraint. It is this universal aspect of FOM_{CM} that allows us to interpret FOM_{CM} as a reasonable beamsummer signal detection criterion.

- Note that $\text{FOM}_V \dfrac{\text{FOM}_V}{\sigma_n^2 + \text{FOM}_V}$ is a monotone increasing function of FOM_V. Recall that for the special case of a single signal ($p = 1$) the weight vector $\mathbf{w} = \mathbf{v}(\theta)$ simultaneously maximizes FOM_{SM} and FOM_V. Therefore, it follows that for $p = 1$ the weight vector $\mathbf{w} = \mathbf{v}(\theta)$ also maximizes FOM_{CM} and thus for this case $\text{FOM}_{SM}, \text{FOM}_{MS}, \text{FOM}_V$, and FOM_{CM} are equivalent figures of merit for optimal beamsummer design. For the case of $p > 1$, however, maximizing these different figures of merit can lead to significantly different optimal beamsummer weights.

Adaptive Beamsummer Algorithms for Constant Modulus Parameter Estimation

To implement an adaptive beamsummer based on FOM_{CM}, we must be able to express the figure of merit in terms of moments of measurable quantities $x(t)$ or $\mathbf{y}(t)$. If the noise variance σ_n^2 is known then sample moment approximations (8.46) and (8.48) to FOM_{SM} and FOM_V can be implemented in (8.55) to yield an SQI maximization criterion for adapting the weights. Otherwise we can either estimate σ_n^2, for example, using eigendecomposition methods on the covariance matrix cov $(\mathbf{y})$, or we can directly estimate α^2 by taking an extra correlated snapshot at each sampling time, as described below.

Let Δ be a positive constant which is smaller than the decorrelation time widths of the signal correlation functions associated with signals $1, \ldots, p$. Let an extra snapshot be taken at each of the sampling times $t + \Delta, t = T_s, \ldots, NT_s$. The Fisher information for this case is given by specializing (8.85) in Appendix 8B to $P = 2$, which as in the $P = 1$ case depends on $\mathbf{w}$ only through FOM_{CM} (8.55). It is easily shown that:

$$|E[x(t)]|^2 = \text{FOM}_{SM} \cdot \|\mathbf{w}\|^2 \tag{8.56}$$

$$|\text{cov}[x(t), x(t+\Delta)]| = \text{FOM}_V \cdot \|\mathbf{w}\|^2 \tag{8.57}$$

$$\frac{|\text{cov}[x(t), x(t+\Delta)]|}{\sqrt{\text{var}[x(t)]\,\text{var}[x(t+\Delta)]}} = \frac{\text{FOM}_V}{\sigma_n^2 + \text{FOM}_V} \tag{8.58}$$

Hence, FOM_{CM} has the form:

$$\text{FOM}_{CM} = \left(\text{cov}[x(t), x(t+\Delta)]\frac{|\text{cov}[x(t), x(t+\Delta)]|}{\sqrt{\text{var}[x(t)]\,\text{var}[x(t+\Delta)]}} + |E[x(t)]|^2\right)\frac{1}{\|\mathbf{w}\|^2} \tag{8.59}$$

Direct maximization of $\text{FOM}_{CM}(\mathbf{w})$ can be accomplished via hill-climbing or other iterative optimization strategies. We describe two strategies below.

From the definitions (8.42) and (8.43) of FOM_{SM} and FOM_V, the function FOM_{CM} can be written as:

$$\text{FOM}_{CM} = \mathbf{w}^H\left[\frac{\alpha^2}{1+\alpha^2}\cdot \mathbf{V}\boldsymbol{\Sigma}_s\mathbf{V}^H + \mathbf{V}\boldsymbol{\mu}_s\boldsymbol{\mu}_s^H\mathbf{V}^H\right]\mathbf{w}\frac{1}{\|\mathbf{w}\|^2} \tag{8.60}$$

where we have suppressed the argument $\boldsymbol{\theta}$ in $\mathbf{V} = \mathbf{V}(\boldsymbol{\theta})$. If $\dfrac{\alpha^2}{1+\alpha^2}$ were independent of $\mathbf{w}$ then FOM_{CM} would be maximized by taking $\mathbf{w}$ equal to the eigenvector $\max_{e-vector}\mathbf{A}$ of the matrix $\mathbf{A} = \dfrac{\alpha^2}{1+\alpha^2}\cdot \mathbf{V}\boldsymbol{\Sigma}_s\mathbf{V}^H + \mathbf{V}\boldsymbol{\mu}_s\boldsymbol{\mu}_s^H\mathbf{V}^H$ associated with its largest eigenvalue. Since $\dfrac{\alpha^2}{1+\alpha^2}$ does depend on $\mathbf{w}$ through $\alpha^2 = \alpha^2(\mathbf{w})$ we propose the following iterative algorithm:

Iterative Eigenmaximization Algorithm

{`Initialize:` $\mathbf{w}^{(0)}$ = initial weights, ν = tolerance, $k = 1$.
    `while` $\|\mathbf{w}^{(k)} - \mathbf{w}^{(k-1)}\| > \nu$
    {

$$\mathbf{w}^{(k)} = \max_{e-vector}\left[\frac{\alpha^2(\mathbf{w}^{(k-1)})}{1+\alpha^2(\mathbf{w}^{(k-1)})}\cdot \mathbf{V}\boldsymbol{\Sigma}_s\mathbf{V}^H + \mathbf{V}\boldsymbol{\mu}_s\boldsymbol{\mu}_s^H\mathbf{V}^H\right]$$

        $k = k + 1$
    }
}

An alternative algorithm is the following:

Steepest Descent Algorithm

{`Initialize:` $\mathbf{w}^{(0)}$ = initial weights, ν = tolerance, $k = 1$.
    `while` $\|\mathbf{w}^{(k)} - \mathbf{w}^{(k-1)}\| > \nu$
    {
        $\mathbf{w}^{(k+1)} = \mathbf{w}^{(k)} + \kappa\nabla_{\mathbf{w}^{(k)}}\text{FOM}_{CM}(\mathbf{w}^{(k)})$
        $k = k + 1$
    }
}

In the steepest descent algorithm, κ is a relaxation parameter controlling speed and radius of convergence of the sequence $\{\mathbf{w}^{(k)}\}_k$. The gradient in this algorithm has the explicit form:

$$\nabla_{\mathbf{w}} \mathrm{FOM}_{CM}(\mathbf{w}) = 2\frac{1}{\mathbf{w}^H\mathbf{w}}\left[\mathbf{Q} - \|\mathbf{Q}\| \cdot \mathbf{I}_M\right] \cdot \mathbf{w} \tag{8.61}$$

where $\mathbf{I}_M$ is the $M \times M$ identity matrix, $\mathbf{Q}$ is the $\mathbf{w}$-dependent $M \times M$ matrix

$$\mathbf{Q} = \beta \cdot \mathbf{V}\boldsymbol{\Sigma}_s\mathbf{V}^H + \mathbf{V}\boldsymbol{\mu}_s\boldsymbol{\mu}_s^H\mathbf{V}^H \tag{8.62}$$

$\|\mathbf{Q}\|$ is the normalized quadratic form:

$$\|\mathbf{Q}\| = \frac{\mathbf{w}^H\mathbf{Q}\mathbf{w}}{\mathbf{w}^H\mathbf{w}}$$

and β is the (non-negative) $\mathbf{w}$-dependent scalar:

$$\begin{aligned}\beta &= 2\frac{\alpha^2}{1+\alpha^2} - \left(\frac{\alpha^2}{1+\alpha^2}\right)^2 \\ &= \frac{\alpha^2(2+\alpha^2)}{(1+\alpha^2)^2}\end{aligned} \tag{8.63}$$

Note that the gradient (8.61) is equal to zero when $\mathbf{w}$ is any eigenvector of the matrix $\mathbf{Q}$ and, therefore, these eigenvectors are local maxima of $\mathrm{FOM}_{CM}(\mathbf{w})$. Since the steepest descent algorithm stops updating when the gradient equals zero, the algorithm only guarantees convergence to one of these local maxima. While the iterative eigenmaximization algorithm does not have this problem, it is more computationally demanding requiring the computation of the maximizing eigenvector of an $M \times M$ matrix at each iteration.

Adaptive Algorithm Implementations via Empirical Moments. We have the identities:

$$E\left[\mathbf{y}(t)\right]E\left[\mathbf{y}^H(t)\right] = \mathbf{V}\boldsymbol{\mu}_s\boldsymbol{\mu}_s^H\mathbf{V}^H \tag{8.64}$$

$$\mathrm{cov}\left[\mathbf{y}(t), \mathbf{y}(t+\Delta)\right] = \mathbf{V}\boldsymbol{\Sigma}_s\mathbf{V}^H \tag{8.65}$$

By using sample moments in place of ensemble moments in the eigenmaximization and steepest descent algorithms, we obtain two implementable algorithms for generating an adaptive weight sequence $\mathbf{w}^{(1)}, \mathbf{w}^{(2)}, \ldots$, based on snapshots taken at times $kT_s, kT_s + \Delta, k = 1, 2, \ldots$.

The following sample moments can be used in the k-th iteration of these two algorithms:

$$\mathbf{V}\boldsymbol{\mu}_s\boldsymbol{\mu}_s^H\mathbf{V}^H \to \left(\frac{\hat{E}^{(k)}\left[\mathbf{y}(t)\right] + \hat{E}^{(k)}\left[\mathbf{y}(t+\Delta)\right]}{2}\right)\left(\frac{\hat{E}^{(k)}\left[\mathbf{y}(t)\right] + \hat{E}^{(k)}\left[\mathbf{y}(t+\Delta)\right]}{2}\right)^H$$

$$\mathbf{V}\boldsymbol{\Sigma}_s\mathbf{V}^H \to \widehat{\mathrm{cov}}^{(k)}\left[\mathbf{y}(t), \mathbf{y}(t+\Delta)\right]$$

$$\frac{\alpha^2}{1+\alpha^2} \to \frac{\left|\widehat{\mathrm{cov}}^{(k)}\left[x(t), x(t+\Delta)\right]\right|}{\sqrt{\widehat{\mathrm{var}}^{(k)}\left[x(t)\right]\widehat{\mathrm{var}}^{(k)}\left[x(t+\Delta)\right]}}$$

where the sample moments for x and $\mathbf{y}$ are obtained iteratively by updates of the form:

$$\hat{E}^{(k+1)}[z(t)] = \frac{k}{k+1}\hat{E}^{(k)}[z(t)] + \frac{1}{k+1}[z(kT_s)] \tag{8.66}$$

$$\begin{aligned}\widehat{\text{cov}}^{(k+1)}[z_1(t), z_2(t)] = & \frac{k-1}{k}\widehat{\text{cov}}^{(k)}[z_1(t), z_2(t)] \\ & + \frac{1}{2k}\left[z_1(kT_s) - \hat{E}^{(k)}[z_1(t)]\right]\left[z_2(kT_2) - \hat{E}^{(k)}[z_2(t)]\right]^H \\ & + \frac{1}{2k}\left[z_2(kT_s) - \hat{E}^{(k)}[z_2(t)]\right]\left[z_1(kT_2) - \hat{E}^{(k)}[z_1(t)]\right]^H\end{aligned} \tag{8.67}$$

For example, taking $z_1(t) = \mathbf{y}(t)$ and $z_2(t) = \mathbf{y}(t+\Delta)$ the recursion (8.67) yields an unbiased and consistent estimate of the matrix $\mathbf{V\Sigma}_s\mathbf{V}^H$.

8.7 OPTIMAL DESIGN FOR DOA ESTIMATION

For DOA estimation we specialize to a single incident signal, a uniform line array with M elements spaced by D, and N equally spaced uncorrelated snapshots $t_k = kT_s$, $T_s \gg W_t^0$ the decorrelation time of the signal. Here $\phi = \theta$ is the scalar DOA of the signal.

The Fisher information for the signal direction θ can be reduced to the form (Appendix 8C):

$$J_\theta = \left(\left(\frac{\tilde{\mathbf{w}}^H\left[\mathbf{v}_p\mathbf{v}^H + \mathbf{v}\mathbf{v}_p^H\right]\tilde{\mathbf{w}}}{\|\tilde{\mathbf{w}}\|^2}\right)^2 \cdot \frac{1}{M^2}\delta_0^2 + 2\frac{\tilde{\mathbf{w}}^H\left[\mathbf{v}_p\mathbf{v}_p^H\right]\tilde{\mathbf{w}}}{\|\tilde{\mathbf{w}}\|^2} \cdot \frac{1}{M}\gamma_0^2\right) \cdot Nu^2\cos^2\theta \tag{8.68}$$

where $u = Dw_0/v = D/\lambda$

$$\mathbf{v} = \left[e^{j\frac{2l-M-1}{2}u\sin\theta}\right]_{l=1}^{M} \tag{8.69}$$

$$\mathbf{v}_p = \mathbf{Bv} \tag{8.70}$$

$$\mathbf{B} = \text{diag}\left(j\frac{2l-M-1}{2}\right)_{l=1,\ldots,M} \tag{8.71}$$

$$\tilde{\mathbf{w}} = \left[\mathbf{v}\mathbf{v}^H\sigma_s^2 + \mathbf{I}_M\sigma_n^2\right]^{\frac{1}{2}}\mathbf{w} \tag{8.72}$$

and

$$\gamma_0^2 = \frac{M|\mu_s|^2}{\sigma_n^2}$$

$$\delta_0^2 = \alpha_0^2\frac{\alpha_0^2}{1+\alpha_0^2}$$

$$\alpha_0^2 = \frac{M\sigma_s^2}{\sigma_n^2}$$

A number of features of the Fisher information provide insight into the achievable performance of an estimator of signal angle θ based on beamsummer outputs $x(T_s), \ldots, x(NT_s)$ implemented with weights $\mathbf{w}$.

- The parameters $\gamma_o^2 = \max_{\mathbf{w}} \gamma^2(\mathbf{w}), \alpha_o^2 = \max_{\mathbf{w}} \alpha^2(\mathbf{w})$ are the maximum beamsummer SM-SNR defined in (8.38) and beamsummer V-SNR defined in (8.40). Furthermore, since $\delta^2 = \alpha^2 \dfrac{\alpha^2}{1+\alpha^2}$ is an increasing function of α^2, $\delta_o^2 = \max_{\mathbf{w}} \delta^2$. Comparing (8.53) and (8.68), it is interesting that the Fisher information is parameterized by a linear combination of δ^2 and γ^2 for the constant modulus parameter estimation problem while it is parameterized by a linear combination of the corresponding δ_o^2 and γ_o^2 for the DOA estimation problem.
- The weight vector $\mathbf{w}$ can be chosen to emphasize either the V-SNR related term δ_o^2 or the SM-SNR term γ_o^2 in the Fisher information. By minimizing $J_\theta(\mathbf{w})$ with respect to $\mathbf{w}$, we achieve an optimal compromise between these two SNRs for estimating signal DOA.
- It is easily verified that the vector $\mathbf{v}_p$ is orthogonal to the steering vector $\mathbf{v}$. It is also evident from (8.72) that when $\mathbf{w}$ is chosen as a scaled version of $\mathbf{v}$, $\tilde{\mathbf{w}}$ is also a scaled version of $\mathbf{v}$. Using the above two facts, it can easily be seen from (8.68) that the SM-SNR-optimal weight vector (8.45) $\mathbf{w} = \mathbf{v}$ gives zero Fisher information, that is, infinite estimator variance! This establishes that no unbiased estimator for θ exists if the steering vector weights for a signal at angle θ just happen to be implemented.

Optimal Weight Vector Design for DOA Estimation. The DOA Fisher information (8.68) is maximized by the weight vector $\mathbf{w}_{opt}$ which maximizes the FOM:

$$\mathrm{FOM}_{DOA}(\mathbf{w}) = \left(\frac{\tilde{\mathbf{w}}^H \left[\mathbf{v}_p \mathbf{v}^H + \mathbf{v}\mathbf{v}_p^H \right] \tilde{\mathbf{w}}}{\|\tilde{\mathbf{w}}\|^2} \right)^2 \cdot \frac{1}{M^2} \delta_o^2 + 2 \frac{\tilde{\mathbf{w}}^H \left[\mathbf{v}_p \mathbf{v}_p^H \right] \tilde{\mathbf{w}}}{\|\tilde{\mathbf{w}}\|^2} \cdot \frac{1}{M^2} \gamma_o^2 \tag{8.73}$$

Define $\tilde{\mathbf{w}}_{opt}$ as a vector $\tilde{\mathbf{w}}$ which maximizes FOM_{DOA}. The optimal weight vector $\mathbf{w}_{opt}$ can be recovered from $\tilde{\mathbf{w}}_{opt}$ through relation (8.72).

Lemma 8.1. *Any vector* $\tilde{\mathbf{w}} = \tilde{\mathbf{w}}_{\mathrm{opt}}$ *which maximizes* $\mathrm{FOM}_{\mathrm{DOA}}$ *(8.73) has the representation:*

$$\tilde{\mathbf{w}} = b \left[\tilde{\mathbf{v}}_p + c \tilde{\mathbf{v}} \right]$$

where c is a complex constant, yet to be determined, b is an arbitrary complex scale factor which does not affect the value of $\mathrm{FOM}_{\mathrm{DOA}}$, $\tilde{\mathbf{v}} = \mathbf{v}/\|\mathbf{v}\|$, $\tilde{\mathbf{v}}_{\mathrm{p}} = \mathbf{v}_{\mathrm{p}}/\|\mathbf{v}_{\mathrm{p}}\|$ *are normalized versions of the vectors* $\mathbf{v}$ *and* $\mathbf{v}_{\mathrm{p}}$, *respectively.*

Proof. The vector $\tilde{\mathbf{w}}$ has the decomposition $\tilde{\mathbf{w}} = \tilde{\mathbf{w}}_\parallel + \tilde{\mathbf{w}}_\perp$ where $\tilde{\mathbf{w}}_\parallel$ lies in the linear span, span$\{\mathbf{v}, \mathbf{v}_p\}$, and $\tilde{\mathbf{w}}_\perp$ is orthogonal to span$\{\mathbf{v}, \mathbf{v}_p\}$. Now $\tilde{\mathbf{w}}_\perp$ is in the nullspace of the symmetric rank 2 matrix $\left[\mathbf{v}_p \mathbf{v}^H + \mathbf{v}\mathbf{v}_p^H\right]$ and the symmetric rank 1 matrix $\mathbf{v}_p \mathbf{v}_p^H$. Since $\tilde{\mathbf{w}}^H \tilde{\mathbf{w}} \geq \tilde{\mathbf{w}}_\parallel^H \tilde{\mathbf{w}}_\parallel$, this implies that $\mathrm{FOM}_{DOA}(\tilde{\mathbf{w}}) \leq \mathrm{FOM}_{DOA}(\tilde{\mathbf{w}}_\parallel)$ and therefore $\tilde{\mathbf{w}}_{opt} \in$ span$\{\mathbf{v}, \mathbf{v}_p\}$. This establishes that $\tilde{\mathbf{w}}_{opt} = b\left[\tilde{\mathbf{v}}_p + c\tilde{\mathbf{v}}\right]$ for some scale factors b, c. Since FOM_{DOA} is invariant to scaling of $\tilde{\mathbf{w}}$, b is arbitrary and the lemma follows. ■

Lemma 8.1 thus reduces the problem of finding an optimum vector $\tilde{\mathbf{w}}_{opt}$ to finding an optimum complex constant c.

Lemma 8.2. *The scale factor c in $\tilde{\mathbf{w}} = b(\tilde{\mathbf{v}}_\text{p} + c\tilde{\mathbf{v}})$ which maximizes* FOM_DOA *is given by c_opt:*

$$c_{opt} = \begin{cases} \pm\sqrt{\dfrac{2\delta_o^2 - \gamma_o^2}{2\delta_o^2 + \gamma_o^2}}, & \delta_o^2 > \gamma_o^2/2 \\ 0, & \delta_o^2 \le \gamma_o^2/2 \end{cases}$$

Proof. Taking $b = 1$ in $\tilde{\mathbf{w}}$ we have the identities:

$$\tilde{\mathbf{w}}^H\tilde{\mathbf{w}} = 1 + |c|^2$$

$$\tilde{\mathbf{w}}^H\left[\mathbf{v}_p^H\mathbf{v}_p\right]\tilde{\mathbf{w}} = \|\mathbf{v}_p\|^2$$

$$\tilde{\mathbf{w}}^H\left[\mathbf{v}_p\mathbf{v}^H + \mathbf{v}\mathbf{v}_p^H\right]\tilde{\mathbf{w}} = (c + c^*)\|\mathbf{v}\|\ \|\mathbf{v}_p\|$$

Substitution of these identities into the expression (8.73) yields:

$$\begin{aligned} \text{FOM}_{DOA}(\tilde{\mathbf{w}}) &= \frac{c_R^2}{1 + c_R^2 + c_I^2}\beta_1 + \frac{1}{1 + c_R^2 + c_I^2}\beta_2 \\ &= \frac{\beta_2 + c_R^2(\beta_1 + \beta_2) + c_I^2\beta_2^2}{1 + c_R^2 + c_I^2} \end{aligned} \tag{8.74}$$

where c_R and c_I are the real and imaginary parts of c, $\beta_1 = 4\|\mathbf{v}\|^2\|\mathbf{v}_p\|^2\dfrac{\delta_o^2}{M}$, and $\beta_2 = 2\|\mathbf{v}_p\|^2\dfrac{\gamma_o^2}{M}$.

Now, by considering the constant contours in (c_R, c_I) of the functions $\beta_2 + c_R^2(\beta_1 + \beta_2) + c_I^2\beta_2$ and $1 + c_R^2 + c_I^2$ (Fig. 8.11), and noting that β_1 and β_2 are non-negative, it is seen that $\text{FOM}_{DOA}(\tilde{\mathbf{v}}_p + c\tilde{\mathbf{v}}) \le \text{FOM}_{DOA}(\tilde{\mathbf{v}}_p + c_R\tilde{\mathbf{v}})$. Hence, c_I can be taken as zero and we have from (8.74):

$$\max_c \text{FOM}_{DOA}(\tilde{\mathbf{w}}) = \max_{c_R} \frac{\beta_1 + c_R^2(\beta_1 + \beta_2)}{(1 + c_R^2)^2}$$

The function $\dfrac{\beta_1 + c_R^2(\beta_1 + \beta_2)}{(1 + c_R^2)^2}$ of c_R takes its maximum at the point: $c_R = \pm\sqrt{\dfrac{\beta_1 - \beta_2}{\beta_1 + \beta_2}}$ if $\beta_1 - \beta_2 > 0$ and at $c_R = 0$ if $\beta_1 - \beta_2 \le 0$. Lemma 8.2 follows by noting that $\beta_1 - \beta_2 = \|\mathbf{v}_p\|^2\left[2\delta_o^2 - \gamma_o^2\right]\frac{2}{M}$ and $\beta_1 + \beta_2 = \|\mathbf{v}_p\|^2\left[2\delta_o^2 + \gamma_o^2\right]\frac{2}{M}$ ■

Using Lemma 8.2 and (8.72) we obtain the principal result of this section.

Proposition 8.1. The optimal weight vector that maximizes the Fisher information on DOA is given up to a complex scale factor by:

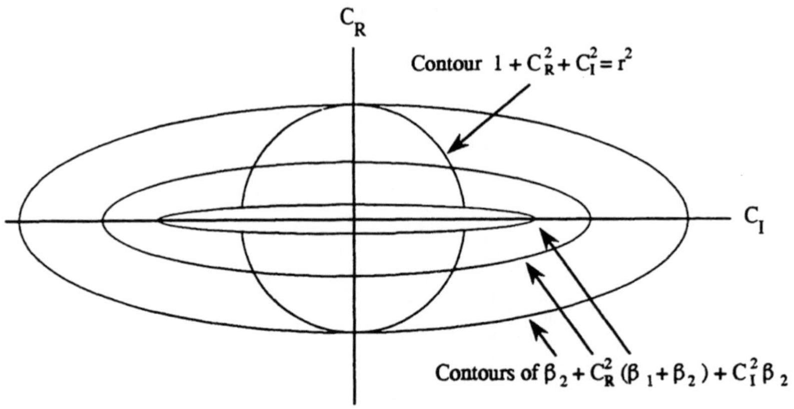

Figure 8.11 Constant contours over domain $c = c_R + jc_I$ of the numerator and denominator of the surface $\dfrac{\beta_2 + c_R^2(\beta_1+\beta_2) + c_I^2\beta_2^2}{1 + c_R^2 + c_I^2}$.

$$\mathbf{w}_{opt} = \begin{cases} \tilde{\mathbf{v}}_p + a\tilde{\mathbf{v}}, & \delta_o^2 > \gamma_o^2/2 \\ \tilde{\mathbf{v}}_p & \delta_o^2 \le \gamma_o^2/2 \end{cases} \tag{8.75}$$

where

$$a = \pm\frac{1}{1+\alpha_o}\sqrt{\frac{2\delta_o^2 - \gamma_o^2}{2\delta_o^2 + \gamma_o^2}} \tag{8.76}$$

The corresponding maximum Fisher information is:

$$\max_{\mathbf{w}} J_\theta(\mathbf{w}) = \begin{cases} \dfrac{M^2-1}{12}\delta_o^2\left(1 + \dfrac{\gamma_o^2/2}{\delta_o^2}\right)^2 \cdot Nu^2\cos^2\theta, & \delta_o^2 > \gamma_o^2/2 \\ \dfrac{M^2-1}{6}\gamma_o^2 \cdot Nu^2\cos^2\theta, & \delta_o^2 \le \gamma_o^2/2 \end{cases} \tag{8.77}$$

The following interpretations follow directly from Proposition 8.1.

- The optimal weight vector consists of the sum of the two unity normalized orthogonal components $\tilde{\mathbf{v}} = \mathbf{v}/\|\mathbf{v}\|$ and $\tilde{\mathbf{v}}_p = \mathbf{v}_p/\|\mathbf{v}_p\|$. We have seen that $\mathbf{w} = \mathbf{v}$ gives the narrowband "detection optimal" beamsummer — it maximizes mean array gain in the direction of the signal. On the other hand, $\mathbf{w} = \mathbf{v}_p$ gives a beamsummer which is a "signal nuller" — it minimizes mean array gain in the direction of the signal. A block diagram of the DOA-optimal beamsummer is given in Figure 8.12.
- In can be verified that the scale factor a is a convex function of α_o^2 over the range $\gamma_o^2/2 \le \delta_o^2 < \infty$, taking the value zero at the endpoints and a maximum value somewhere in between (see Fig. 8.13). Interestingly, the sign of a is immaterial to maximization of Fisher information.

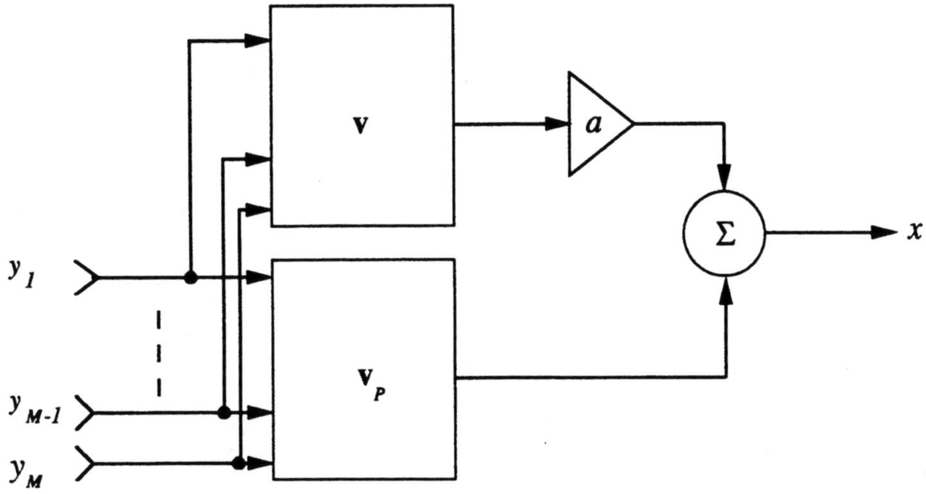

Figure 8.12 Block diagram for the DOA optimal beamsummer.

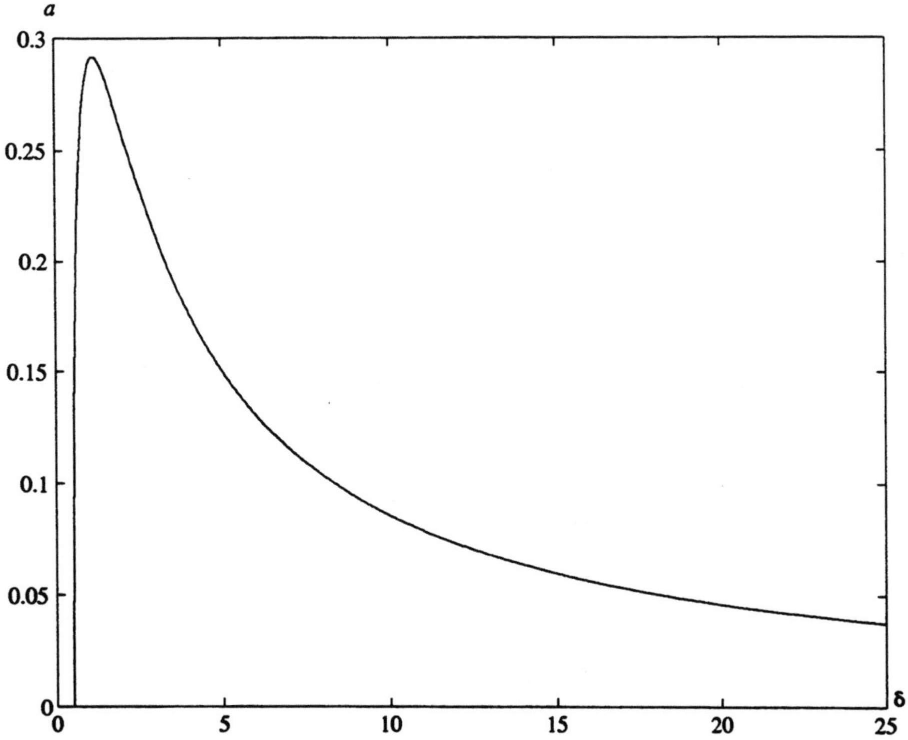

Figure 8.13 Plot of the scale factor a in $\mathbf{W}_{opt} = \tilde{\mathbf{v}} \pm a\tilde{\mathbf{v}}_p$.

- For $\delta_o^2 \leq \gamma_o^2/2$ then, up to an arbitrary scale factor, $\mathbf{w}_{opt} = \mathbf{v}_p$, that is, the optimal weight vector places a null exactly at the signal angle θ. This suggests that for this case, the best θ-estimator performance is achieved by trying to null out the signal, for example, by minimizing beamsummer response, rather than by trying to maximize beamsummer response to the signal. Such a strategy has been proposed in "super-resolution" DOA estimation methods [14, 48] and methods proposed for nulling of multiple interferers [49, 13, 50, 9, 51]. The intuition behind this is that a beamformer is better able to generate sharp nulls than sharp peaks in its gain pattern (compare Fig. 8.10 to Fig. 8.14). Hence, for a strong constant signal amplitude ($\gamma_o^2 \gg \alpha_o^2$), there is much greater variation in output power as θ varies about a null (a deep valley in beampattern) than as θ varies about the main beam (a relatively flat peak in beampattern), thereby leading to a greater ability to discriminate between small changes in θ. On the other hand, when the signal energy is concentrated in the signal variance ($\alpha_o^2 \gg \gamma_o^2$) the Fisher information indicates that the optimal weights must pass some signal energy — adding or subtracting $a\tilde{\mathbf{v}}$ from $\tilde{\mathbf{v}}_p$ accomplishes this.
- In Figures 8.15 and 8.16, the minimum beamsummer CR bound for DOA estimation variance is compared to the CR bound [38] for DOA estimation

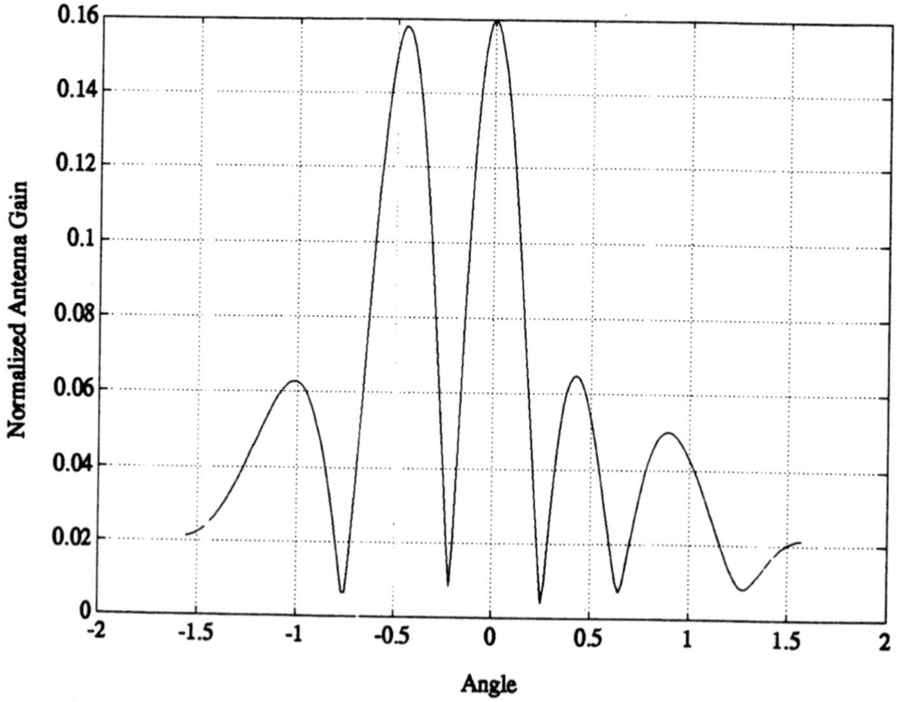

Figure 8.14 Directional beampattern of DOA optimal weight vector for signal at −0.22 radians and a 6-element uniform line array.

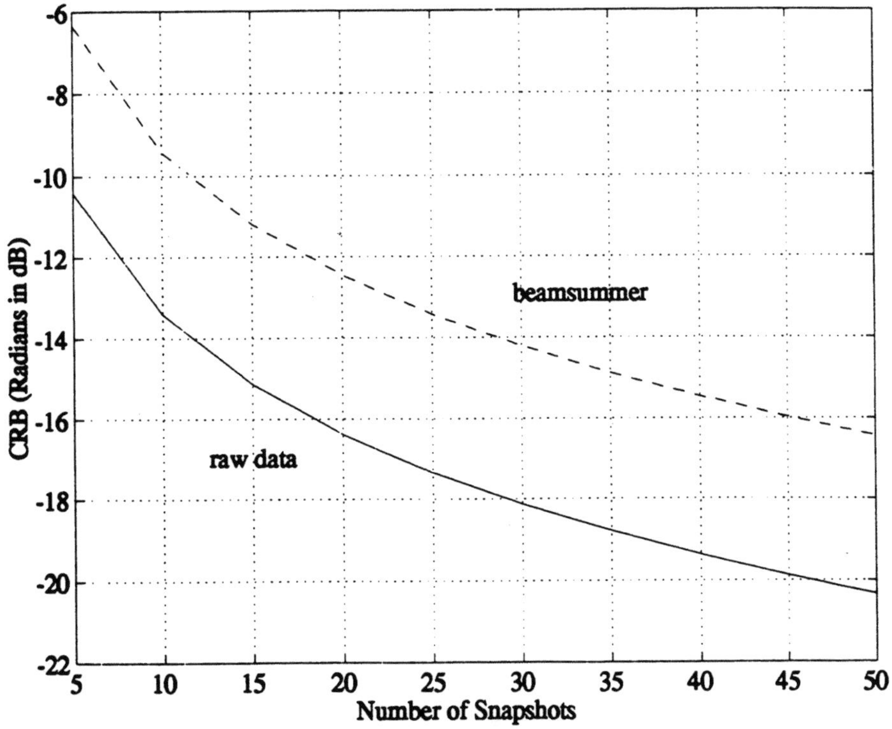

Figure 8.15 Plot of raw sensor data CR bound and beamsummer CR bound on DOA estimation implemented with bound minimizing beamsummer weights $\mathbf{w}_{opt} = \tilde{\mathbf{v}} \pm a\mathbf{v}_p$. $M = 6, \gamma^2 = 1, \alpha^2 = 10, \theta = -0.22$ radians.

variance computed using the raw sensor measurements $\mathbf{y}$ for a six-element array and a random signal arriving at -0.22 radians. The figures indicate that a greater loss in estimation performance occurs for high signal variance ($\gamma_o^2 = 6, \alpha_o^2 = 60$ in Fig. 8.15) than for low signal variance ($\gamma_o^2 = 60, \alpha_o^2 = 6$ in Fig. 8.16) when the beamsummer output data is used rather than the raw multiple-sensor data.

While $\mathbf{w}_{opt}$ given in Proposition 8.1 is of analytical closed form, it is unimplementable since it requires knowledge of θ and the SNR-dependent constant a. An eigendecomposition of the sample correlation matrix could probably provide reasonable estimates of $\mathbf{v}$ and $\mathbf{v}_p$. Here we develop a more direct approach.

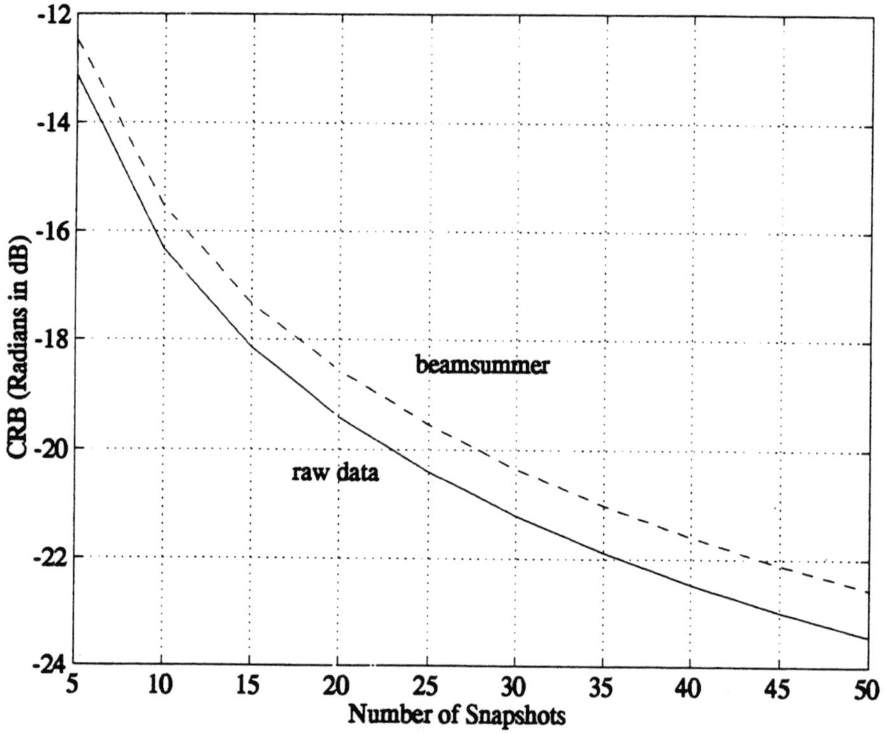

Figure 8.16 Plot of raw sensor data CR bound and beamsummer CR bound on DOA estimation implemented with bound minimizing weights $\mathbf{w}_{opt} = \tilde{\mathbf{v}} \pm a\mathbf{v}_p$. $M = 6, \gamma^2 = 10, \alpha^2 = 1, \theta = 0.22$ radians.

An Adaptive Implementation of the Optimal Beamsummer for DOA

By manipulation of pairs of beamsummer means and variances it is possible to express the function FOM_{DOA} in terms of first and second moments of the beamsummer output $x(t)$. Define the two weight vectors:

$$\mathbf{w}_1 = (\mathbf{B}_1 + \mathbf{B}_2)^H \mathbf{w} \tag{8.78}$$

$$\mathbf{w}_2 = (\mathbf{B}_1 + \mathbf{B}_2)^H \mathbf{w} \tag{8.79}$$

where $\mathbf{B}_1$ and $\mathbf{B}_2$ are the following $M \times M$ complex matrices:

$$\mathbf{B}_1 = \text{diag}_{l=1,\dots,M}\left(j\frac{(2l - M - 1)/2}{\sqrt{M(M^2 - 1)/6}} \right) \tag{8.80}$$

$$\mathbf{B}_2 = \text{diag}\left(\frac{1}{\sqrt{2M}} \right) \tag{8.81}$$

Let $x_1(t) = \mathbf{w}_1^H \mathbf{y}(t)$ and $x_2(t) = \mathbf{w}_2^H \mathbf{y}(t)$ be outputs of the two beamsummers. Then it can be shown that:

$$\mathbf{w}^H \mathbf{v}_p \cdot \mu_s = \frac{1}{\sqrt{2}} \|\mathbf{v}_p\| E\left[x_1(t) + x_2(t)\right]$$

$$\mathbf{w}^H \left[\mathbf{v}\mathbf{v}^H \sigma_s^2 + \sigma_n^2 \mathbf{I}_M\right] \mathbf{w} = \frac{1}{2} \|\mathbf{v}\|^2 \text{var}\left[x_1(t) - x_2(t)\right]$$

$$\mathbf{w}^H \left[\mathbf{v}_p \mathbf{v}^H + \mathbf{v}\mathbf{v}_p^H\right] \mathbf{w} = \|\mathbf{v}\| \|\mathbf{v}_p\| (\text{var}\left[x_1(t)\right] - \text{var}\left[x_2(t)\right])$$

From these identities we obtain an equivalent expression for Fisher information (8.68):

Proposition 8.2. The Fisher information matrix for the DOA θ of a single signal impinging on a uniform M-element linear array is given by:

$$J_\theta = \left[\left(\frac{\text{var}\left[x_1\right] - \text{var}\left[x_2\right]}{\frac{1}{2}\text{var}\left[x_1 - x_2\right]}\right)^2 + \frac{E\left[|x_1 + x_2|^2\right]}{\frac{1}{2}\text{var}\left[x_1 - x_2\right]}\right] \cdot \frac{\|\mathbf{v}_p\|^2}{\|\mathbf{v}\|^2} N u^2 \cos^2(\theta)$$

where $x_1 = x_1(t)$ and $x_2 = x_2(t)$ are outputs of beamsummers $\mathbf{w}_1$ and $\mathbf{w}_2$ defined in (8.78) and (8.79).

Maximization of $J_\theta(\mathbf{w})$ is equivalent to maximization of the following function of the means and variances beamsummer outputs:

$$\text{FOM}_{DOA}(\mathbf{w}) = \left(\frac{\text{var}\left[x_1\right] - \text{var}\left[x_2\right]}{\frac{1}{2}\text{var}\left[x_1 - x_2\right]}\right)^2 + \frac{E\left[|x_1 + x_2|^2\right]}{\frac{1}{2}\text{var}\left[x_1 - x_2\right]} \tag{8.82}$$

As in the case of FOM_D, hill-climbing methods can be applied to FOM_{DOA} to generate a sequence of weights $\mathbf{w}^{(1)}, \mathbf{w}^{(2)}, \ldots$ which converges to $\mathbf{w}_{opt}$. By replacing the ensemble moments in (8.82) with sample moments, such as obtained from the recursions (8.66) and (8.67) implemented with $z = x_1, x_2, x_1 - x_2$, we obtain an empirical SQI which can be maximized online to adapt the weights over snapshots $N = 2, 3, \ldots$. A block diagram of the adaptive weight computer based on iterative maximization of this empirical estimate of FOM_{DOA} is shown in Figure 8.17.

The function FOM_{DOA} in (8.82) can be expressed in the equivalent form:

$$\begin{aligned}
&\text{FOM}_{DOA}(\mathbf{w}) \\
&= \left(\frac{\mathbf{w}^H \left[\mathbf{B}_2 \text{cov}\left[\mathbf{y}\right] \mathbf{B}_1^H + \mathbf{B}_1 \text{cov}\left[\mathbf{y}\right] \mathbf{B}_2^H\right] \mathbf{w}}{\mathbf{w}^H \left[\mathbf{B}_2 \text{cov}\left[\mathbf{y}\right] \mathbf{B}_2^H\right] \mathbf{w}}\right)^2 + 2\frac{\mathbf{w}^H \left[\mathbf{B}_1 E\left[\mathbf{y}\right] E\left[\mathbf{y}^H\right] \mathbf{B}_1^H\right] \mathbf{w}}{\mathbf{w}^H \left[\mathbf{B}_2 \text{cov}\left[\mathbf{y}\right] \mathbf{B}_2^H\right] \mathbf{w}} \\
&= \left(\frac{\tilde{\mathbf{w}}^H \mathbf{G}_1 \tilde{\mathbf{w}}}{\tilde{\mathbf{w}}^H \tilde{\mathbf{w}}}\right)^2 + \frac{\tilde{\mathbf{w}}^H \mathbf{G}_2 \tilde{\mathbf{w}}}{\tilde{\mathbf{w}}^H \tilde{\mathbf{w}}} \qquad (8.83) \\
&= \frac{\tilde{\mathbf{w}}^H \left[\mathbf{G}_1 K(\tilde{\mathbf{w}}) + \mathbf{G}_2\right] \tilde{\mathbf{w}}}{\tilde{\mathbf{w}}^H \tilde{\mathbf{w}}}
\end{aligned}$$

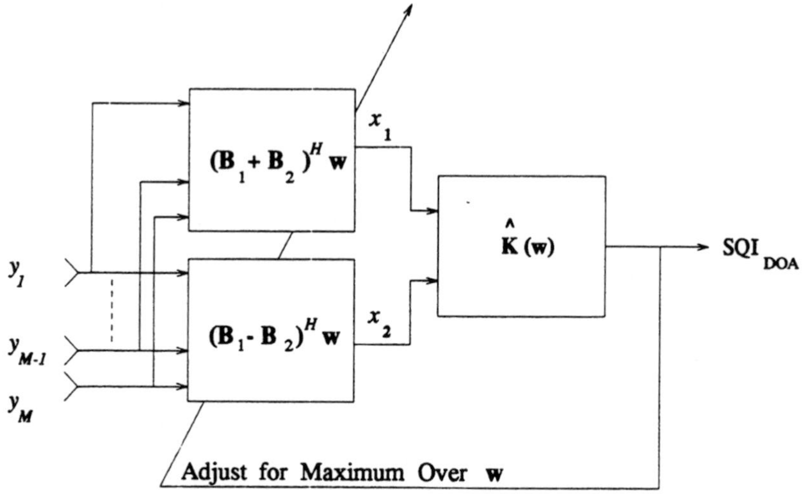

Figure 8.17 The optimal weights **w** are found by maximization of an empirical estimate of FOM_{DOA}.

where

$$\tilde{\mathbf{w}} = \text{cov}^{\frac{1}{2}}[\mathbf{y}]\,\mathbf{w}$$

$$\mathbf{G}_1 = \text{cov}^{\frac{1}{2}}[\mathbf{y}]\,\mathbf{B}_1^H \text{cov}^{-\frac{1}{2}}[\mathbf{y}] + \text{cov}^{-\frac{1}{2}}[\mathbf{y}]\,\mathbf{B}_1 \text{cov}^{\frac{1}{2}}[\mathbf{y}]$$

$$\mathbf{G}_2 = 4M \cdot \text{cov}^{-\frac{1}{2}}[\mathbf{y}]\,\mathbf{B}_1 E[\mathbf{y}]\,E\left[\mathbf{y}^H\right]\mathbf{B}_1^H \text{cov}^{-\frac{1}{2}}[\mathbf{y}]$$

$$K(\tilde{\mathbf{w}}) = \frac{\tilde{\mathbf{w}}^H \mathbf{G}_1 \tilde{\mathbf{w}}}{\tilde{\mathbf{w}}^H \tilde{\mathbf{w}}}$$

The form (8.83) suggests the following iterative algorithm:

Iterative Eigenmaximization Algorithm for DOA-Optimal Weights

{`Initialize:` $\tilde{\mathbf{w}}^{(0)} =$ initial weights, $\nu =$ tolerance, $k = 1$.

  `while` $\|\mathbf{w}^{(k)} - \mathbf{w}^{(k-1)}\| > \nu$

  {

    $\tilde{\mathbf{w}}^{(k)} = \max_{e-vector}\left[\mathbf{G}_1 K(\tilde{\mathbf{w}}^{(k-1)}) + \mathbf{G}_2\right]$

    $k = k + 1$

  }

  $\mathbf{w} = \text{cov}^{-\frac{1}{2}}[\mathbf{y}]\,\tilde{\mathbf{w}}^{(k+1)}$

}

This algorithm requires finding the square root factors $\text{cov}^{\frac{1}{2}}[\mathbf{y}]$ and $\text{cov}^{-\frac{1}{2}}[\mathbf{y}]$, for example, by Cholesky decomposition, and finding the eigenvector associated with the largest eigenvalue of the matrix $\mathbf{G}_1 K(\tilde{\mathbf{w}}) + \mathbf{G}_2$. A steepest descent algorithm can also be derived which reduces the computational load but only guarantees convergence to a local maximum of $\text{FOM}_{DOA}(\mathbf{w})$.

8.8 ESTIMATION AND DETECTION PERFORMANCE COMPARISONS

While the adaptive weights obtained from the $\text{FOM}_{CM}(\mathbf{w})$ (8.59) and $\text{FOM}_{DOA}(\mathbf{w})$ (8.82) maximizations are asymptotically optimal, in the sense of minimizing the respective CR bounds, and the adaptive weights obtained from the $\text{FOM}_D(\mathbf{w})$ are optimal in the sense of maximizing the detectability index, a method for signal detection and/or extraction of estimates from the beamsummer outputs $x(t_1), x(t_2), \ldots$ remains to be specified. The purpose of this section is show that significant performance gains can be achieved even by use of simple ad hoc estimation and detection algorithms applied to the optimized beamsummer output statistics.

In this study we used the empirical beampatterns computed from the magnitude of the discrete time Fourier transform (DTFT) of the adaptive weight vectors $\mathbf{w}^{(k)}$. Four adaptive beamsummers were implemented by adapting the weights to maximize empirical estimates of $\text{FOM}_{SM}(\mathbf{w})$, $\text{FOM}_D(\mathbf{w})$, $\text{FOM}_{CM}(\mathbf{w})$, and $\text{FOM}_{DOA}(\mathbf{w})$, which we call the classical adaptive beamsummer, the D-optimal adaptive beamsummer, the CM-optimal adaptive beamsummer, and the DOA-optimal adaptive beamsummer, respectively.

Signal Detection Algorithms. For detection we performed peak selection on the beampatterns of the classical, D-optimal, and CM-optimal adaptive beamsummers. Define a miss as the event that the detector declares fewer than the p signals present and a false alarm as the event that the detector falsely declares more than p signals present. Signal detection was performed by examining the highest peaks, declaring k signals to be present if k peaks exceeded a threshold chosen to achieve a specified proportion of false alarms (10%) over all trials. An ideal generalized likelihood ratio (GLR) detector was also implemented that has the relative advantage of operating directly on the raw sensor data $\mathbf{y}$ and of knowing the signal and noise powers exactly [38]. The family of signal-present/signal-absent log-likelihood ratios for the Gaussian raw multiple-sensor measurements were computed as a function of the unknown number p of signals and the unknown signal DOAs $\theta_1, \ldots, \theta_p$. The Akaike information criterion ($2p$) was added to penalize for overestimation of the number of signals, and the maximum of the resultant penalized log-likelihood ratio was computed via table lookup, over the range $p = 1, 2, 3$, and $\theta_1, \theta_2, \theta_3 \in [-\pi/2, \pi/2]$, and compared to a threshold determined so that the false alarm probability equals 0.1. Since raw sensor measurements and errorless signal and noise power estimates are available to the GLR detector, this detector was expected to show superior detection performance than the beamsummer-restricted detectors introduced in this work.

DOA Estimation Algorithms. For DOA estimation we implemented peak localization for the classical, D-optimal, and CM-optimal adaptive beamsummers while we implemented null localization for the DOA-optimal adaptive beamsummers. For comparison the clairvoyant (known signal and noise powers) maximum likelihood estimator was also implemented which operates directly on the raw sensor data $\mathbf{y}$. From the classical, D-optimal, and CM-optimal beampatterns we find the angles

$\theta_1, \ldots, \theta_k \in [-\pi/2, \pi/2]$ at which the k highest peaks of the beampattern are located, while from the DOA-optimal beampattern we find the location of the nulls detected within one beamwidth of each of the k-peaks detected by thresholding the D-optimal beampattern. We call these three techniques, respectively, the classical peak picking method, the D-optimal peak picking method, the CM-optimal peak picking method, and the null picking method of DOA estimation. For each estimator, the MSE was computed only for those cases for which correct detection occurred.

Simulation Results

An i.i.d. Gaussian sequence of zero mean noise variates was generated with variance $\sigma_n^2 = 1$, as was a pair of independent i.i.d. Gaussian sequences of equal power signal variates with prescribed means $\mu_{s_1} = \mu_{s_2}$ and variances $\sigma_{s_1}^2 = \sigma_{s_2}^2$. From these three sequences, 1000 realizations of groups of 100 snapshots of the output of a six-element uniform linear array were synthesized for: (1) a single coherent signal arriving from angle -0.22 radians in spatially incoherent noise; and (2) a pair of uncorrelated equal power signals arriving from angles -0.22 and $+0.22$ radians imbedded in spatially incoherent noise.

Single Signal. In Figures 8.18 and 8.19, typical sequences of beampatterns are shown for a single signal and varying numbers of snapshots for classical, D-optimal, CM-optimal, and DOA-optimal beamsummers. Also shown is the ensemble average beampattern. Note that, as expected, the beampatterns stabilize as N increases. The classical, D-optimal, and CM-optimal beampatterns exhibit broad peaks, with the respective global maxima near the signal DOA (dashed line), while the DOA-optimal beampattern exhibits a sharp null near the signal DOA. This null can be detected as the sharpest and deepest null within one beamwidth of the peak of the D-optimal beampattern. Figures 8.20 and 8.21 display the average miss probability for the peak picking detector based on the classical, D-optimal, and CM-optimal beampatterns, and for the clairvoyant GLR detector based on the raw sensor data $\mathbf{y}$ with the false alarm probability constraint set at 0.1. Figures 8.22 and 8.23 show the average false alarm probability for these four detectors when the miss probability is set at 0.1. Observe that the clairvoyant GLR detector has the best error performance, followed by the D-optimal peak picking detector, the CM-optimal peak picking detector, and finally the classical peak picking detector.

Figures 8.24 and 8.25 display the MSE of each of the classical peak picking, D-optimal peak picking, CM-optimal peak picking, null picking, and clairvoyant ML estimation methods for $\gamma_o^2 = 10, \alpha_o^2 = 1$ (narrowband regime) and $\gamma_o^2 = 1, \gamma_o^2 = 10$ (wideband regime), respectively. It was observed that all five of these methods gave unbiased estimates of DOA. Also shown for comparison is the minimum beamsummer CR bound, that is the inverse of (8.77), which is virtually attained by the null picking method as N increases. Note that the clairvoyant ML estimator MSE falls somewhat below the minimum beamsummer CR bound since it has access to the raw sensor measurements. However, the difference between the MSE of the null picking method and the clairvoyant ML method becomes very small for the case $\alpha_o^2 = 1$ as N increases.

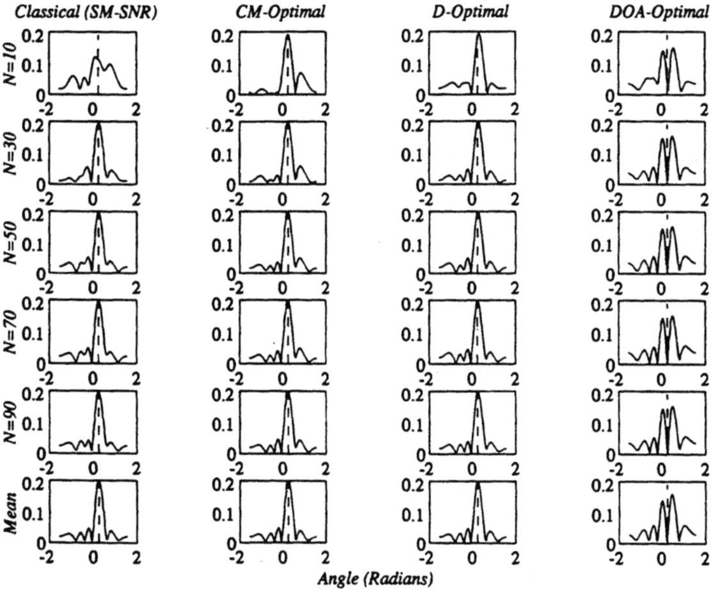

Figure 8.18 A single realization of the beampattern sequence for one signal, $\gamma^2 = 10, \alpha_o^2 = 1, N = 10, 30, 50, 70, 90$ snapshots and classical, D-optimal, CM optimal, and DOA-optimal beamsummers. At bottom are shown the ensemble mean beampatterns. Dashed line indicates the location $\theta = 0.22$ of the signal DOA.

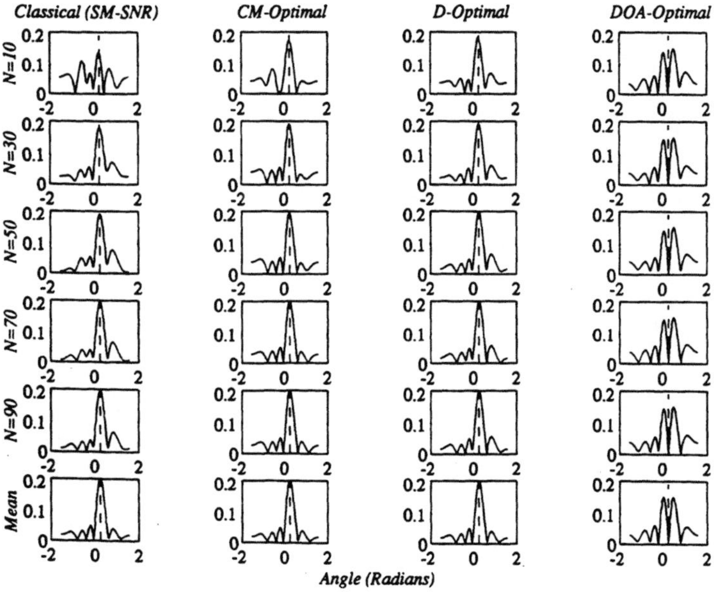

Figure 8.19 Same as previous figure except that $\gamma_o^2 = 1$ and $\alpha_o^2 = 10$.

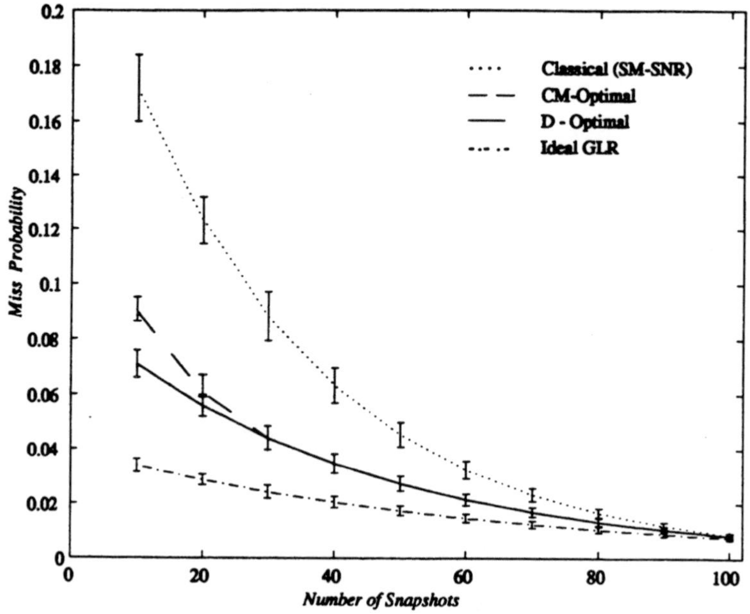

Figure 8.20 The miss probability for peak picking detectors based on classical, D-optimal, and CM-optimal beampatterns, and the ideal GLR detector for one signal. $\gamma_o^2 = 10$, $\alpha_o^2 = 1$. All detectors have false alarm probability fixed at 0.1. Vertical lines are 95% confidence intervals.

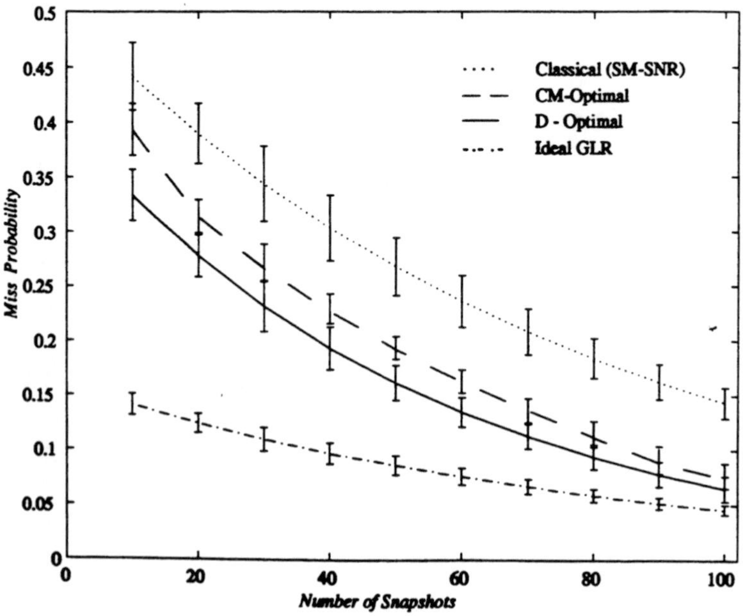

Figure 8.21 Same as previous figure except that $\gamma_o^2 = 1$ and $\alpha_o^2 = 10$.

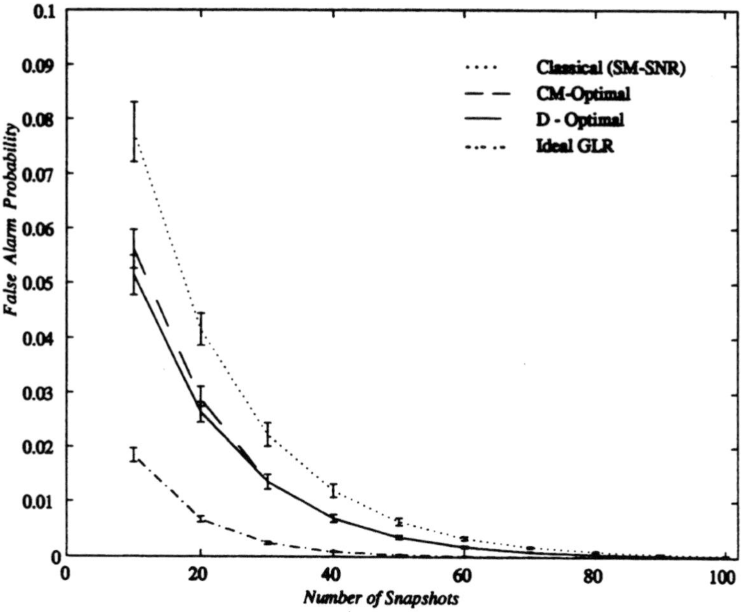

Figure 8.22 The false alarm probability for peak picking detector based on classical, D-optimal, and CM-optimal beampatterns, and the ideal GLR detector for one signal. $\gamma_o^2 = 10, \alpha_o^2 = 1$. All detectors have miss probability fixed at 0.1. Vertical lines are 95% confidence intervals.

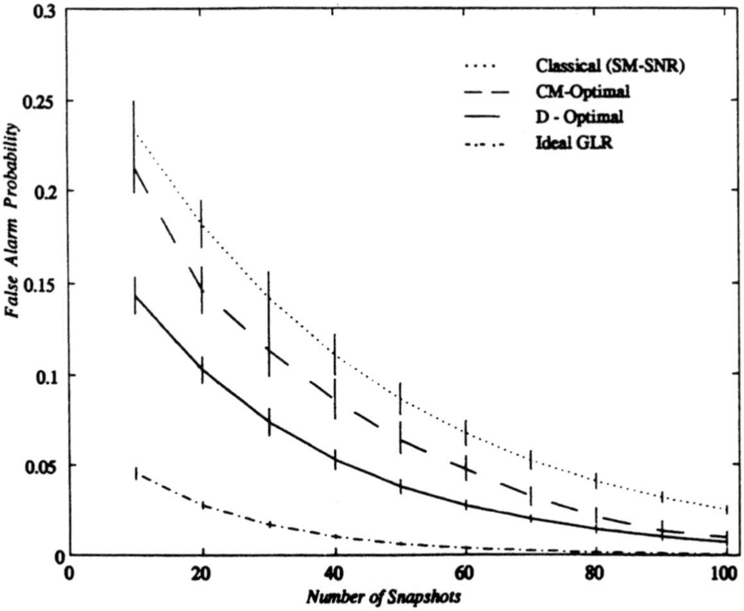

Figure 8.23 Same as previous figure except $\gamma_o^2 = 1, \alpha_o^2 = 1$.

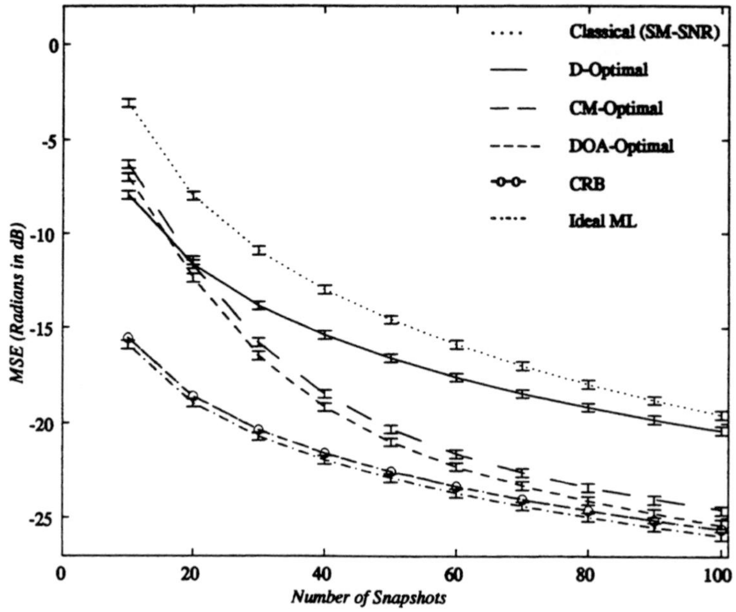

Figure 8.24 MSE of DOA estimates obtained from classical, D-optimal, and CM-optimal peak picking, DOA-optimal null picking, and ideal ML estimation methods for one signal, $\gamma_o^2 = 10, \alpha_o^2 = 1$. Also shown is the beamsummer CR bound. Vertical lines are 95% confidence intervals.

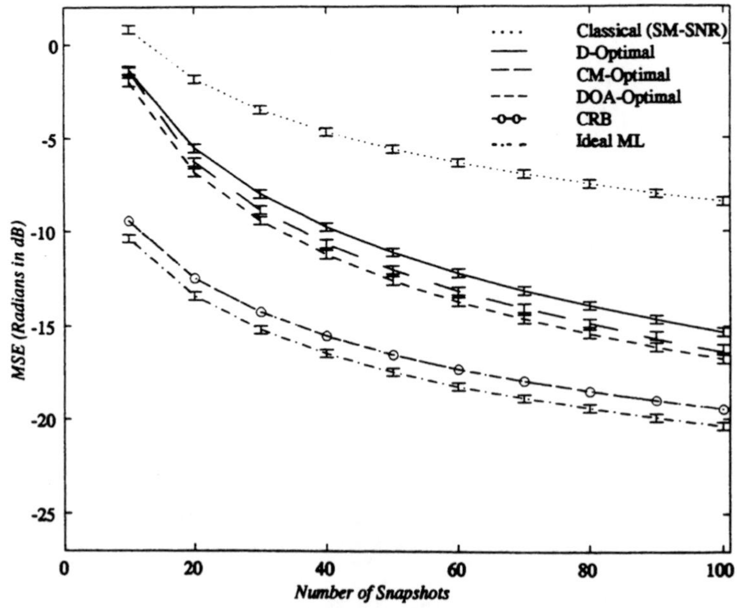

Figure 8.25 Same as previous figure except that $\gamma_o^2 = 1$, and $\alpha_o^2 = 10$.

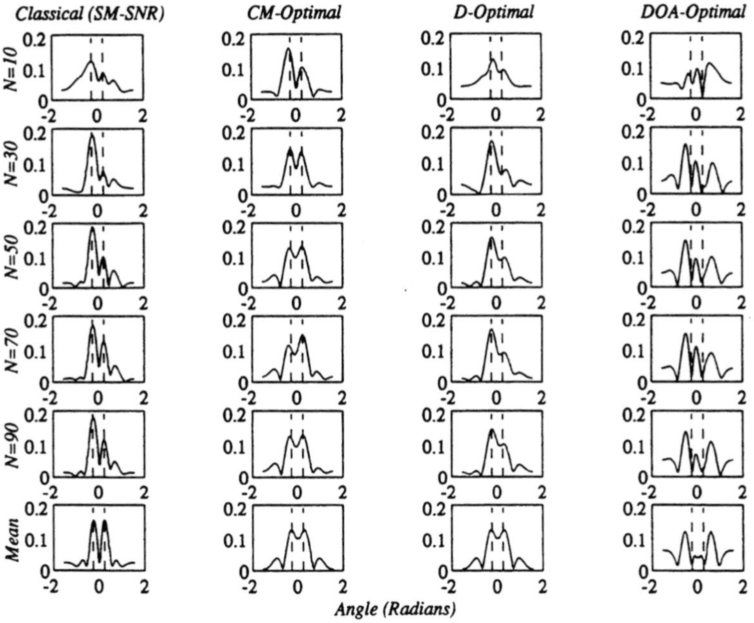

Figure 8.26 A single realization of the beampattern sequence for two uncorrelated equal power signals, $\gamma_o^2 = 10, \alpha_o^2 = 1, N = 10, 30, 50, 70, 90$ snapshots and classical D-optimal, CM optimal, and DOA-optimal beamsummers. At botton are shown the ensemble mean beampatterns. Dashed line indicates the location $\theta_1 = 0.22, \theta_2 = -0.22$, of the signal DOA's.

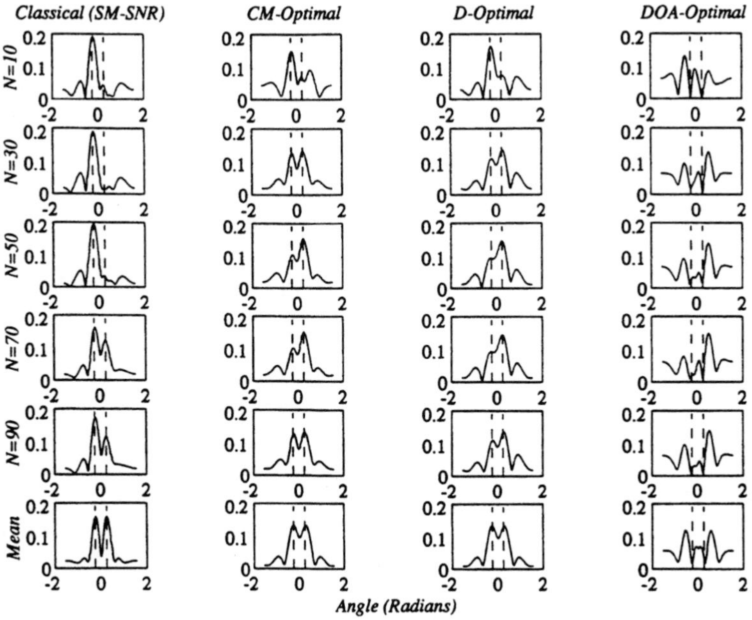

Figure 8.27 Same as previous figure except that $\gamma_o^2 = 1$ and $\alpha_o^2 = 10$.

Two Signals. In Figures 8.26 and 8.27, typical sequences of beampatterns are shown for two equal power signals and varying numbers of snapshots for classical, D-optimal, CM-optimal, and DOA-optimal beamsummers. Observe that due to the close spacing of the signal DOAs the classical, D-optimal, and CM-optimal beampatterns frequently have trouble resolving the two signal components, especially in the most broadband regime $\alpha_o^2 = 10$. On the other hand, the DOA-optimal beampattern more consistently indicates the presence of these signals by two sharp nulls near the signal DOAs within the main lobe of the D-optimal beampattern.

In Figures 8.28 and 8.29, the probability of miss is plotted as a function of N for peak detectors based on the classical, D-optimal, and CM-optimal beampatterns, and the clairvoyant GLR detector based on the raw sensor measurements. As in the one signal example, for all detectors the false alarm probability is fixed at 0.1. In Figures 8.30 and 8.31 the probability of false alarm is plotted as a function of N for the same detectors, for probability of miss fixed at 0.1. It is noted that in all cases the detection performance of the D-optimal peak picking method rapidly approaches the performance of the GLR detector as N increases.

Figures 8.32 and 8.33 show the MSE of the θ_1-estimator for the classical peak picking, D-optimal peak picking, CM-optimal peak picking, null picking, and clairvoyant ML estimation methods, for two uncorrelated equal power signals and $\gamma_o^2 = 10, \alpha_o^2 = 1$ (narrowband regime) and $\gamma_o^2 = 1, \alpha_o^2 = 10$ (wideband regime), respectively. It is interesting that even though the DOA-optimal null picking estimator is based on maximizing Fisher information for the erroneous one-signal model, it outperforms the D-optimal peak peaking estimator which is optimal, although not for DOA estimation, for any number of signals. It is believed that there still exists a margin for improvement of the DOA-optimal strategy in the multiple signal case. However, to achieve this improvement, one must study the minimization of the DOA CR bound information for the more difficult case of multiple signals.

8.9 CONCLUSION

We have presented a methodology for beamsummer optimization for particular estimation tasks based on maximization of the weight-dependent beamsummer detectability index and minimization of the weight-dependent CR bound. If a figure of merit can be identified that is a known function of the moments of the beamsummer output, and whose maximization minimizes the beamsummer CR bound, then an optimal weight adaptation criterion can be specified. Here we have shown that this can be done for the tasks of estimation of a spatially invariant constant-modulus signal component, and estimation of signal DOA. Simulations have shown that accurate parameter estimates can be extracted from the optimal beamsummer output even using very simple heuristic methods. If tractable, more sophisticated methods, such as post-beamsummer maximum likelihood, could be expected to improve the parameter estimation performance especially in the broadband small-number-of-snapshots regime. If a suitable figure of merit can be identified, extension of the DOA results to the tighter multiple-parameter matrix CR bound could also provide some beamsummer improvements. Finally, an

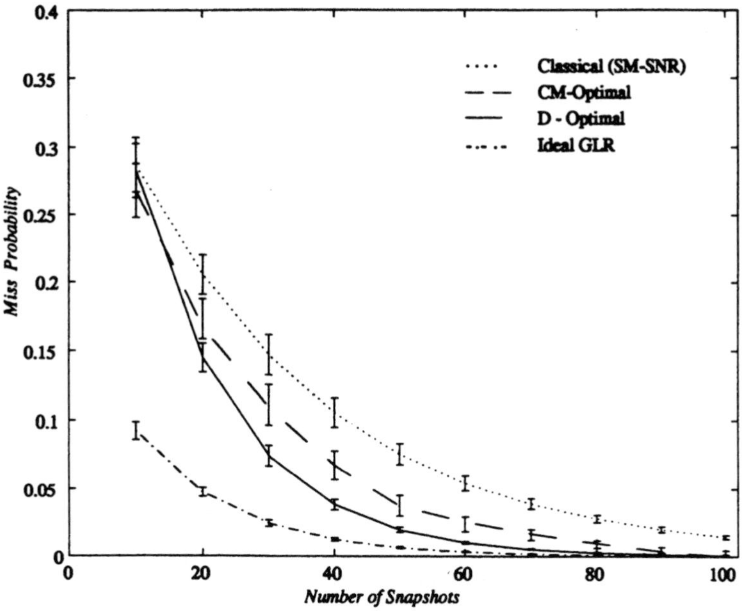

Figure 8.28 The miss probability for peak picking detectors based on classical, D-optimal, and CM-optimal beampatterns, and the ideal GLR detector for two uncorrelated equal power signals. $\gamma_o^2 = 10, \alpha_o^2 = 1$. All detectors have false alarm probability fixed at 0.1. Vertical lines are 95% confidence intervals.

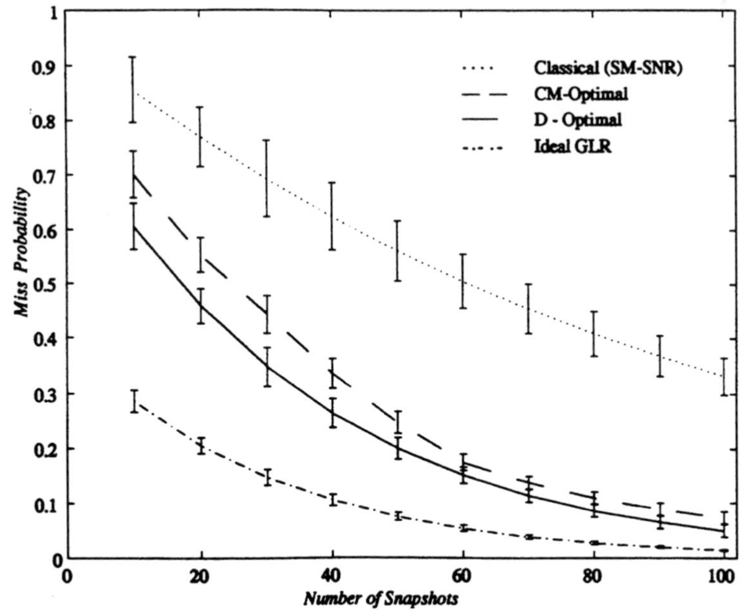

Figure 8.29 Same as in previous figure except that $\gamma_o^2 = 1, \alpha_o^2 = 10$.

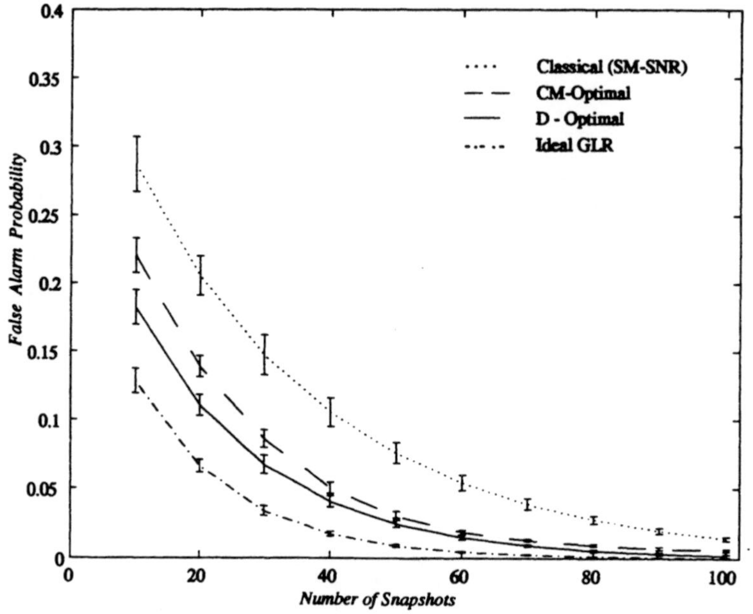

Figure 8.30 The false alarm probability for peak picking detectors based on classical, D-optimal, and CM-optimal beampatterns, and the ideal GLR detector for two uncorrelated equal power signals. $\gamma_o^2 = 10$, $\alpha_o^2 = 1$. All detectors have miss probability fixed at 0.1. Vertical lines are 96% confidence intervals.

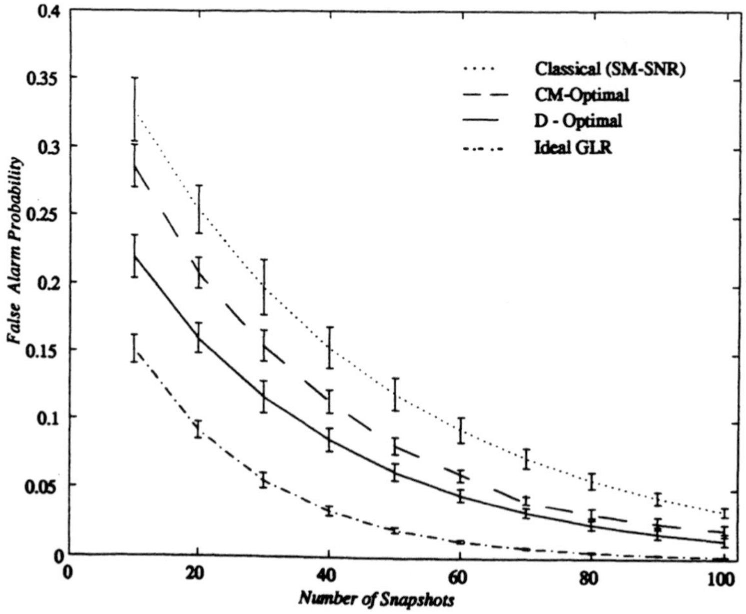

Figure 8.31 Same as previous figure except that $\gamma_o^2 = 1$, $\alpha_o^2 = 10$.

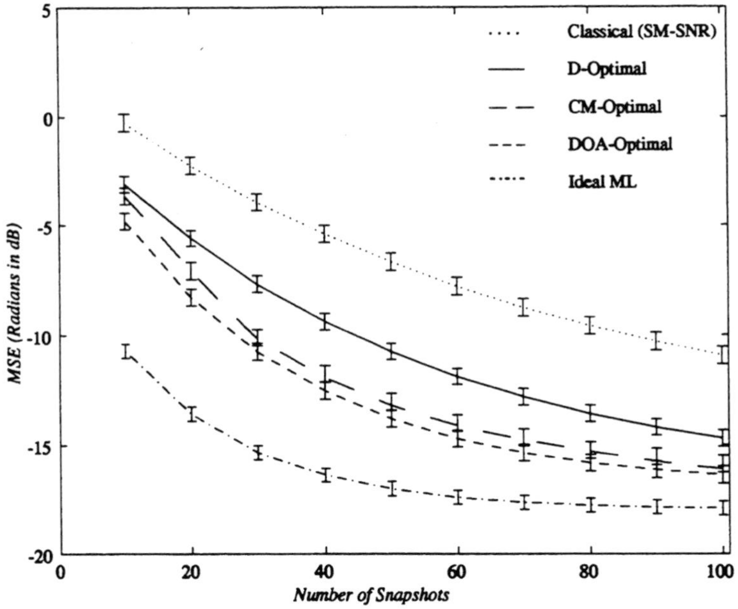

Figure 8.32 MSE of DOA estimates obtained from classical, D-optimal, and CM-optimal peak picking, DOA-optimal null picking, and ideal ML estimation methods for two uncorrelated equal power signals. $\gamma_o^2 = 10$, $\alpha_o^2 = 1$. Vertical lines are 95% confidence intervals.

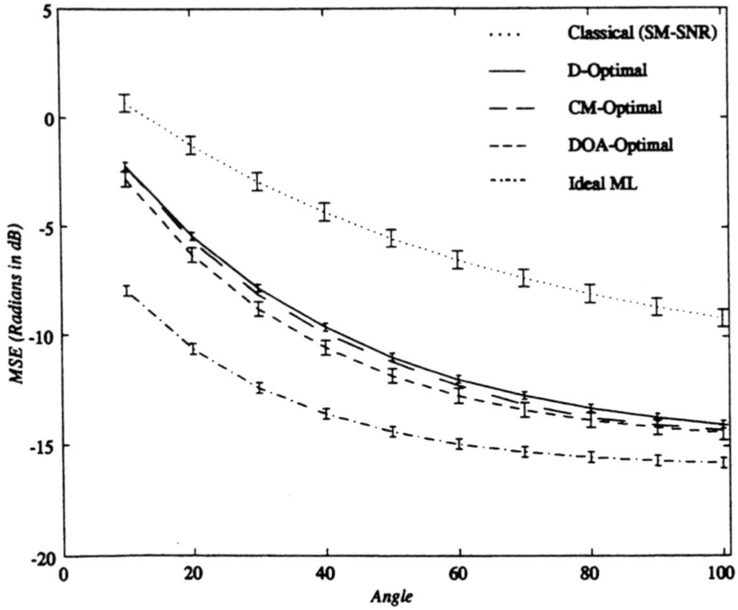

Figure 8.33 Same as previous figure except that $\gamma_o^2 = 1$, $\alpha_o^2 = 10$.

extension to biased estimators can be accomplished by investigating the CR-type bound derived in [52] for estimators satisfying a uniform bias constraint.

A beamsummer may frequently be used for multiple tasks, for example, simultaneous signal DOA estimation and interference nulling. One simple way of extending our methodology to this case is to consider constrained minimization of the CR bound, for example, requiring the weights to have null response in the direction of the interferers. An alternative would be to combine many different figures of merit and use multi-objective optimization techniques to adapt the weights.

APPENDIX 8A DEFLECTION INDEX FOR OPTIMAL DETECTION

Under the model (8.31) for **y**, the vector **x** of N snapshots of the beamsummer output are i.i.d. Gaussian with mean $E_{H_1}[x] = \mu_x$ and variance $\mathrm{var}_{H_1}[x] = \sigma_x^2$ given by (8.35) and (8.36). Therefore, the log-likelihood ratio is:

$$\ln \frac{f(\mathbf{x}|H_1)}{f(\mathbf{x}|H_0)} = \sum_{k=1}^{N} \left[\frac{|x(t_k)|^2}{\sigma_o^2} - \frac{|x(t_k) - \mu_x|^2}{(1+\alpha^2)\sigma_o^2} \right]$$

$$= \sum_{k=1}^{N} \left(|x(t_k)|^2 \frac{1}{\sigma_o^2} \frac{\alpha^2}{1+\alpha^2} + \frac{x^*(t_k)\mu_x + x(t_k)\mu_x^*}{(1+\alpha^2)\sigma_o^2} - \frac{\gamma^2}{1+\alpha} \right)$$

where $\sigma_o^2 = \mathrm{var}_{H_o}[x(t_k)]$. Now, since under H_0 the random variable $x = x(t_k)$ is zero-mean symmetric complex-Gaussian: $E_{H_0}[|x|^2 x] = E_{H_0}[x^2] = 0$, and $\mathrm{var}_{H_0}[|x|^2] = 2\,\mathrm{var}^2[x]$ [40]. Thus, $|x(t_k)|^2$ and $x^*(t_k)\mu_x + x(t_k)\mu_x^*$ are uncorrelated under H_0 and:

$$\mathrm{var}_{H_0}\left[|x(t_k)|^2\right] = 2\sigma_o^4$$

$$\mathrm{var}_{H_0}\left[x^*(t_k)\mu + x(t_k)\mu^*\right] = 2\gamma^2\sigma_o^4$$

Therefore, since $x(t_k), k = 1, \ldots, N$, are i.i.d.:

$$\mathrm{var}_{H_0}\left[\ln \frac{f(\mathbf{x}|H_1)}{f(\mathbf{x}|H_0)}\right] = 2\left(\frac{1}{1+\alpha^2}\right)^2 (\alpha^4 + \gamma^2)N$$

Furthermore, since $E_{H_i}\left[|x+c|^2\right] = \mathrm{var}_{H_i}[x] + |E_{H_i}[x] + c|^2$, we have:

$$E_{H_0}\left[\ln \frac{f(\mathbf{x}|H_1)}{f(\mathbf{x}|H_0)}\right] = \left[1 - \frac{1+\gamma^2}{1+\alpha^2}\right]N$$

$$E_{H_1}\left[\ln \frac{f(\mathbf{x}|H_1)}{f(\mathbf{x}|H_0)}\right] = \left[\alpha^2 + \gamma^2\right]N$$

Plugging the above into the formula $d = (E_{H_1}[h(\mathbf{x})] - (E_{H_0}[h(\mathbf{x})])/\sqrt{\mathrm{var}_{H_0}[h(\mathbf{x})]}$ for the deflection coefficient associated with the likelihood ratio $h(\mathbf{x})$, we obtain the expression (8.37).

APPENDIX 8B CR BOUND FOR CONSTANT MODULUS PARAMETERS

We consider a more general model than (8.31) which incorporates $P-1$ additional *correlated* time samples $kT_s+\Delta, kT_s+2\Delta, \ldots, kT_s+(P-1)\Delta, k=1,\ldots,N$, where $T_s \gg \max W_t^o$, the maximum decorrelation time and $\Delta \ll \min W_t^o$ the minimum decorrelation time of the p signal autocorrelation functions. Then the observations have the equivalent form:

$$\mathbf{x}(t) = \mathbf{g}(t)\cdot \mathbf{w}^H \mathbf{V}\mathbf{s}(t) + \mathbf{n}_x(t), \quad t = T_s, \ldots, NT_s$$

where $\mathbf{g}(t) = [g(t,\phi), \ldots, g(t+(P-1)\Delta,\phi)]^T$, $\mathbf{s}(t) = \left[s_1(t), \ldots, s_p(t)\right]^T$, is an i.i.d. Gaussian random vector with mean $\boldsymbol{\mu}_s$ covariance $\boldsymbol{\Sigma}_s$, and $\mathbf{n}_x = \left[\mathbf{w}^H\mathbf{n}(t), \ldots, \mathbf{w}^H\mathbf{n}(t+(P-1)\Delta)\right]^T$ is a zero mean i.i.d. Gaussian random vector with covariance $\sigma_n^2\mathbf{I}_M\|\mathbf{w}\|^2$. We will assume that the parameter ϕ does not affect the power of $\mathbf{g}(t), t = T_s, \ldots, NT_s$:

$$\|\mathbf{g}(t)\|^2 = \sum_{i=0}^{P-1} |g(t+i\Delta)|^2 = 1 \tag{8B.1}$$

The mean and covariance of $\mathbf{x}(t)$ are simply:

$$\begin{aligned}\boldsymbol{\mu}_x(t) &= \mathbf{g}(t)\cdot \mathbf{w}^H\mathbf{V}\boldsymbol{\mu}_s \\ \boldsymbol{\Gamma}_x(t) &= \mathbf{g}(t)\mathbf{g}^H(t)\cdot \mathbf{w}^H\left[\mathbf{V}\boldsymbol{\Sigma}_s\mathbf{V}^H + \sigma_n^2\mathbf{I}_M\right]\mathbf{w}\end{aligned}$$

Using the Woodbury identity and the fact that $\|\mathbf{g}(t)\|^2 = 1$ we have:

$$\boldsymbol{\Gamma}_x^{-1}(t) = \frac{1}{\sigma_n^2}\left(\mathbf{I}_M - \frac{\alpha^2}{1+\alpha^2}\mathbf{g}(t)\mathbf{g}^H(t)\right)$$

Furthermore:

$$\begin{aligned}\frac{d\boldsymbol{\mu}_x(t)}{d\phi} &= \mathbf{g}'(t)\mathbf{w}^H\mathbf{V}\boldsymbol{\mu}_s \\ \frac{d}{d\phi}\boldsymbol{\Gamma}_x(t) &= (\mathbf{g}'(t)\mathbf{g}^H(t) + \mathbf{g}(t)\left[\mathbf{g}'(t)\right]^H)\mathbf{w}^H\mathbf{V}\boldsymbol{\Sigma}_s\mathbf{V}^H\mathbf{w}\end{aligned}$$

where $\mathbf{g}'(t) = d\mathbf{g}(t)/d\phi$. Now, since $\|\mathbf{g}(t)\|^2 = 1$, we have $\mathbf{g}^H(t)\mathbf{g}'(t) = 0$. Therefore:

$$\begin{aligned}2\frac{d\boldsymbol{\mu}_x^H(t)}{d\phi}\boldsymbol{\Gamma}_x^{-1}(t)\frac{d\boldsymbol{\mu}_x(t)}{d\phi} &= 2\frac{|\mathbf{w}^H\mathbf{V}\boldsymbol{\mu}_s|^2}{\sigma_n^2}\cdot\left[\mathbf{g}'(t)\right]^H\left(\mathbf{I}_M - \frac{\alpha^2}{1+\alpha^2}\mathbf{g}(t)\mathbf{g}^H(t)\right)\mathbf{g}'(t) \\ &= 2\gamma^2\|\mathbf{g}'(t)\|^2\end{aligned}$$

and

$$\begin{aligned}\boldsymbol{\Gamma}_x^{-1}(t)\frac{d}{d\phi}\boldsymbol{\Gamma}_x(t)\boldsymbol{\Gamma}_x^{-1}(t)\frac{d}{d\phi}\boldsymbol{\Gamma}_x(t) &= \alpha^4\left(\mathbf{g}'(t)\mathbf{g}^H(t) + \mathbf{g}(t)\left[\mathbf{g}'(t)\right]^H\frac{1}{1+\alpha^2}\right)^2 \\ &= 2\frac{\alpha^4}{1+\alpha^2}\cdot\mathbf{g}'(t)\left[\mathbf{g}'(t)\right]^H\end{aligned}$$

From the formula (8.51) for the Fisher information we therefore obtain:

$$J_\theta = \left(\alpha^2 \frac{\alpha^2}{1+\alpha^2} + \gamma^2\right) \cdot 2 \sum_{k=1}^{N} \|\mathbf{g}'(kT_s)\|^2 \tag{8B.2}$$

APPENDIX 8C CR BOUND FOR DOA ESTIMATION

For $p = 1$ signal it is easily seen from (8.35), (8.36), and $\mathbf{v}(\theta) = \left[e^{j\frac{2l-M-1}{2}\frac{D\sin\theta}{v}}\right]_{l=1,\dots,M}$ that:

$$\frac{\partial E\left[x(t)\right]}{\partial \theta} = \mathbf{w}^H \mathbf{v}_p(\theta) \cdot \mu_s u \cos\theta$$

$$\frac{\partial \operatorname{var}\left[x(t)\right]}{\partial \theta} = \mathbf{w}^H \left[\mathbf{v}_p(\theta)\mathbf{v}^H(\theta) + \mathbf{v}(\theta)\mathbf{v}_p^H(\theta)\right] \mathbf{w} \cdot \sigma_s^2 u \cos\theta \tag{8C.1}$$

Thus, since $x(T_s), \dots, x(NT_s)$ are i.i.d. Gaussian, we have from the Fisher information formula (8.51):

$$\begin{aligned} J_\theta &= \left(\left[\frac{\mathbf{w}^H\left[\mathbf{v}_p\mathbf{v} + \mathbf{v}\mathbf{v}_p^H\right]\mathbf{w}}{\mathbf{w}^H\left[\mathbf{v}\mathbf{v}^H\sigma_s^2 + \sigma_n^2\mathbf{I}_M\right]\mathbf{w}}\right]^2 \sigma_s^4 + 2\frac{|\mathbf{w}^H\mathbf{v}_p|^2}{\mathbf{w}^H\left[\mathbf{v}\mathbf{v}^H\sigma_s^2 + \sigma_n^2\mathbf{I}_M\right]\mathbf{w}}|\mu_s|^2\right) \cdot Nu^2\cos^2\theta \\ &= \left(\left[\frac{\tilde{\mathbf{w}}^H\mathbf{Q}_1\tilde{\mathbf{w}}}{\tilde{\mathbf{w}}^H\tilde{\mathbf{w}}}\right]^2 \sigma_s^4 + 2\frac{\tilde{\mathbf{w}}^H\mathbf{Q}_2\tilde{\mathbf{w}}}{\tilde{\mathbf{w}}^H\tilde{\mathbf{w}}}|\mu_s|^2\right) \cdot Nu^2\cos^2\theta \end{aligned} \tag{8C.2}$$

where

$$\tilde{\mathbf{w}} = \left[\mathbf{v}\mathbf{v}^H\sigma_s^2 + \sigma_n^2\mathbf{I}_M\right]^{\frac{1}{2}}\mathbf{w}$$

$$\mathbf{Q}_1 = \left[\mathbf{v}\mathbf{v}^H\sigma_s^2 + \sigma_n^2\mathbf{I}_M\right]^{-\frac{1}{2}}\left[\mathbf{v}\mathbf{v}_p^H + \mathbf{v}_p\mathbf{v}^H\right]\left[\mathbf{v}\mathbf{v}^H\sigma_s^2 + \sigma_n^2\mathbf{I}_M\right]^{-\frac{1}{2}}$$

$$\mathbf{Q}_2 = \left[\mathbf{v}\mathbf{v}^H\sigma_s^2 + \sigma_n^2\mathbf{I}_M\right]^{-\frac{1}{2}}\mathbf{v}_p\mathbf{v}_p^H\left[\mathbf{v}\mathbf{v}^H\sigma_s^2 + \sigma_n^2\mathbf{I}_M\right]^{-\frac{1}{2}}$$

Now it can be shown that $\mathbf{v}\mathbf{v}^H\sigma_s^2 + \sigma_n^2\mathbf{I}_M$ and $\mathbf{v}\mathbf{v}_p^H + \mathbf{v}_p\mathbf{v}^H$ have the eigendecompositions:

$$\mathbf{v}\mathbf{v}^H\sigma_s^2 + \sigma_n^2\mathbf{I}_M = \sum_{k=1}^{M}\lambda_k\mathbf{r}_k\mathbf{r}_k^H$$

$$\mathbf{v}\mathbf{v}_p^H + \mathbf{v}_p\mathbf{v}^H = \sum_{k=1}^{M}\eta_k\mathbf{q}_k\mathbf{q}_k^H$$

where

$$\mathbf{r}_1 = \frac{1}{\|\mathbf{v}\|}\mathbf{v}, \quad \lambda_1 = M\sigma_s^2 + \sigma_n^2$$

$$\mathbf{r}_2 = \frac{1}{\|\mathbf{v}_p\|}\mathbf{v}_p, \quad \lambda_2 = \sigma_n^2$$

$\mathbf{r}_3, \ldots, \mathbf{r}_M$ are obtained by completion of $\mathbf{r}_1, \mathbf{r}_2$ in $\mathbb{C}^M$ and $\lambda_3 = \ldots = \lambda_M = \sigma_n^2$

$$\mathbf{q}_1 = \frac{1}{\sqrt{2}} \left[\frac{\mathbf{v}_p}{\|\mathbf{v}_p\|} + \frac{\mathbf{v}}{\|\mathbf{v}\|} \right], \ \eta_1 = \|\mathbf{v}\| \ \|\mathbf{v}_p\|$$

$$\mathbf{q}_2 = \frac{1}{\sqrt{2}} \left[\frac{\mathbf{v}_p}{\|\mathbf{v}_p\|} - \frac{\mathbf{v}}{\|\mathbf{v}\|} \right], \ \eta_2 = -\|\mathbf{v}\| \ \|\mathbf{v}_p\|$$

$\mathbf{q}_3, \ldots, \mathbf{q}_M$ are obtained by completion of $\mathbf{q}_1, \mathbf{q}_2$ in $\mathbb{C}^M$ and $\eta_3 = \ldots = \eta_M = 0$.

Since $\mathbf{r}_1, \mathbf{r}_2$ lie in the linear span of $\mathbf{q}_1, \mathbf{q}_2$ we obtain after some algebraic manipulation:

$$\begin{aligned} \mathbf{Q}_1 &= \sum_{k=1}^{2} \eta_k \mathbf{q}_k \mathbf{q}_k^H \cdot \frac{1}{\sqrt{\lambda_1 \lambda_2}} \\ &= \left[\mathbf{v}\mathbf{v}_p^H + \mathbf{v}_p \mathbf{v}^H \right] \cdot \frac{1}{\sigma_n^2 (M\sigma_s^2 + \sigma_n^2)} \end{aligned}$$

and

$$\mathbf{Q}_2 = \mathbf{v}_p \mathbf{v}_p^H \cdot \frac{1}{\sigma_n^2}$$

Plugging the above expressions for $\mathbf{Q}_1$ and $\mathbf{Q}_2$ into the expression (8C.2), we obtain the result (8.68).

REFERENCES

1. J. E. Evans, J. Johnson, and D. Sun. "Application of Advanced Signal Processing Techniques to Angle of Arrival Estimation in ATC Navigation and Surveillance Systems," Technical Report 582, M.I.T. Lincoln Laboratory, Lexington, MA, June 1982.
2. I. Ziskind and M. Wax. "Maximum Likelihood Localization of Multiple Sources by Alternating Projections," *IEEE Trans. Acoust., Speech, and Sig. Proc.*, vol. ASSP-37, pp. 1553–60, Oct. 1988.
3. A. Paulraj and T. Kailath. "Eigenstructure Methods for Direction-of-Arrival Estimation in the Presence of Unkown Noise Fields," *IEEE Trans. Acoust., Speech, and Sig. Proc.*, vol. ASSP-34, pp. 13–20, 1986.
4. J. P. LeCadre. "Parametric Methods for Spatial Signal Processing in the Presence of Unknown Colored Noise Fields," *IEEE Trans. Acoust., Speech, and Sig. Proc.*, vol. ASSP-37, pp. 965–83, July 1989.
5. R. L. Kashyap. "Maximum Likelihood Estimation for Direction of Arrival with Narrowband and Wideband Signals," *Proc. of Conference on Information Science and Systems*, pp. 187–92, Princeton, NJ, March 1991.
6. H. Wang and M. Kaveh. "Coherent Signal Subspace Processing for the Detection and Estimation of Angles of Arrival of Multiple Wideband Sources," *IEEE Trans. Acoust., Speech, and Sig. Proc.*, vol. ASSP-33, pp. 823–31, Aug. 1985.
7. R. A. Monzingo and T. W. Miller. *Introduction to Adaptive Arrays*, New York: John Wiley & Sons, 1980.

8. J. E. Hudson. *Adaptive Array Principles*, Peter Peregrinus Ltd., 1981.
9. B. Widrow and S. D. Stearns. *Adaptive Signal Processing*, Englewood Cliffs, NJ: Prentice Hall, 1985.
10. S. Haykin. *Array Signal Processing*, Englewood Cliffs, NJ: Prentice Hall, 1985.
11. R. T. Compton. *Adaptive Antennas: Concepts and Performance*, Englewood Cliffs, NJ: Prentice Hall, 1988.
12. D. H. Johnson and D. E. Dudgeon. *Array Signal Processing*, Englewood Cliffs, NJ: Prentice Hall, 1993.
13. S. P. Applebaum. "Adaptive Arrays," *IEEE Trans. Antennas and Propagation*, vol. AP-24, pp. 585–98, Sept. 1976.
14. W. F. Gabriel. "Spectral Analysis and Adaptive Array Superresolution Techniques," *IEEE Proceedings*, vol. 68, pp. 654–56, June 1980.
15. B. D. VanVeen and R. A. Roberts. "Partially Adaptive Beamformer Design via Output Power Minimization," *IEEE Trans. Acoust., Speech, and Sig. Proc.* vol. ASSP-35, pp. 1524–32, Nov. 1987.
16. J. R. Treichler and B. G. Agee. "A New Approach to Multipath Correction of Constant Modulus Signals," *IEEE Trans. Acoust., Speech, and Sig. Proc.*, vol. ASSP-31, pp. 459–72, April 1983.
17. R. Fisher. *The Design of Experiments*, Edinburgh: Oliver and Boyd, 1935.
18. V. V. Federov. *Theory of Optimal Experiments*, New York: Academic Press, 1972.
19. S. D. Silvey. *Optimal Design*, London: Chapman and Hall, 1980.
20. H. Chernoff. "Locally Optimal Designs for Estimating Parameters," *Ann. Math. Statist.*, vol. 24, pp. 586–602, 1953.
21. P. Whittle. "Some General Points in the Theory of Optimal Experimental Design," *J. Royal Stat. Soc., Ser. B,* vol. 35, pp. 123–30, 1973.
22. J. Kiefer and J. Wolfowitz. "Optimum Design in Regression Problems," *Ann. Math. Statist.*, vol. 30, pp. 271–94, 1959.
23. A. C. Atkinson and D. R. Cox. "Planning Experiments for Discriminating Between Models," *J. Royal Stat. Soc., Ser. B,* vol. 36, pp. 321–48, 1974.
24. G. C. Goodwin and R. L. Paynes. *Dynamic System Identification: Experiment Design and Data Analysis*, New York: Academic Press, 1977.
25. M. B. Zarrop. *Optimal Experimental Design for Dynamic System Identification*, New York: Springer-Verlag, 1979.
26. J. A. Jacquez. *Compartmental Analysis in Biology and Medicine*, 2nd ed., Ann Arbor: University of Michigan Press, 1988.
27. E. M. Landaw. "Optimal Multicompartmental Sampling Designs for Parameter Estimation: Practical Aspects of the Identification Problem," *Math. and Computers in Simulation*, vol. 24, pp. 525–30, 1982.
28. F. Mori and J. J. DiStefano. "Optimal Nonuniform Sampling Interval and Test Design for Identification of Physiological Systems from Very Limited Data," *IEEE Trans. Automatic Control*, vol. AC-24, pp. 893–900, 1979.
29. D. Z. D'Argenio. "Optimal Sampling Times for Pharmacokinetic Experiments," *J. Pharmacokinetics and Biopharmaceutics*, vol. 9, pp. 739–56, 1981.
30. P. C. Chiao. "Parameter Estimation Strategies for Dynamic Cardiac Studies using Emission Computed Tomography," Ph.D. thesis, Bioengineering Program, The University of Michigan, Ann Arbor, 1991.

31. E. A. Lee and D. G. Messerschmitt. *Digital Communications*, Boston, MA: Kluwer, 1988.

32. F. E. Nathanson, J. P. Reilly, and M. N. Cohen. *Radar Design Principles*, New York: McGraw-Hill, 1990.

33. W. C. Jakes. *Microwave Mobile Communications*, New York: John Wiley & Sons, 1974.

34. W. S. Burdic. *Underwater Acoustic System Analysis*, Englewood Cliffs, NJ: Prentice Hall, 1984.

35. J. I. Marcum and P. Swerling. "Studies of Target Detection by Pulsed Radar," *IRE Trans. on Inform. Theory*, vol. IT-6, pp. 59–308, April 1960.

36. R. A. DeLap and A. O. Hero. "Adaptive Beamforming for Slow Raleigh Fading Signals," *IEEE Workshop on Statistical Signal and Array Proc.*, Quebec City, Canada, 1994.

37. R. A. DeLap and A. O. Hero. "A New Method for Adaptive Wideband Beamforming," *IEEE Conf. Acoust., Speech, and Sig. Proc.*, Minneapolis, MN, 1993.

38. R. A. DeLap. "Adept: Task Specific Adaptive Beamforming," Ph.D. thesis, Electrical Engineering Program, The University of Michigan, Ann Arbor, 1994.

39. H. L. Van-Trees. *Detection, Estimation, and Modulation Theory: Part I*, New York: John Wiley & Sons, 1968.

40. A. D. Whalen. *Detection of Signals in Noise*, Orlando: Academic Press, 1971.

41. G. E. Uhlenbeck and J. Lawson. *Threshold Signals*, New York: McGraw-Hill, 1948.

42. D. O. North. "Analysis of the Factors which Determine Signal/Noise Discrimination in Radar," *Proc. IRE*, vol. 51, pp. 1016–28, July 1963.

43. G. H. Golub and C. F. Van Loan. *Matrix Computations*, 2nd ed., Baltimore: Johns Hopkins University Press, 1989.

44. J. D. Gorman and A. O. Hero. "Lower Bounds for Parametric Estimation with Constraints," *IEEE Trans. on Inform. Theory*, vol. IT-36, pp. 1285–1301, Nov. 1990.

45. I. A. Ibraginov and R. Z. Has'minskii. *Statistical Estimation: Asymptotic Theory*, New York: Springer-Verlag, 1981.

46. B. Z. Bobrovsky, E. Mayer-Wolf, and M. Zakai. "Some Classes of Global Cramér-Rao Bounds," *Annals of Statistics*, vol. 15, no. 4, pp. 1421–38, 1987.

47. W. J. Bangs. "Array Processing with Generalized Beamformers," Ph.D. thesis, Yale University, New Haven, CT, 1971.

48. R. O. Schmidt. "Multiple Emitter Location and Signal Parameter Estimation," *IEEE Trans. Antennas and Propagation*, vol. AP-34, pp. 276–80, March, 1986.

49. R. L. Riegler and R. Compton. "An Adaptive Array for Interference Rejection," *IEEE Proceedings*, vol. 61, pp. 748–58, June 1973.

50. R. N. Adams, L. L. Horowitz, and K. D. Senne. "Adaptive Main Beam Nulling for Narrow Band Antenna Arrays," *IEEE Trans. on Aerosp. Electron, and Systems,* vol. AES-16, pp. 509–51, July 1980.

51. M. G. Amin. "Concurrent Nulling and Localization of Multiple Interferences in Adaptive Antenna Arrays," *IEEE Trans. on Signal Processing,* vol. 40, pp. 2658–68, Nov. 1992.

52. A. O. Hero. "A Cramér-Rao Type Lower Bound for Essentially Unbiased Parameter Estimation," Technical Report 890, M.I.T. Lincoln Laboratory, Lexington, MA, Jan. 1992. DTIC AD-A246666.

9

Cumulants and Array Processing: A Unified Approach

Mithat C. Doğan and Jerry M. Mendel

9.1 INTRODUCTION

The main problems of interest in array processing are: detection of far-field sources, estimation of their parameters, such as angles of arrival (azimuth and elevation), power levels, correlation structure of sources, polarization properties, and estimation of the source waveforms. All these problems fall into two (overlapping) categories: detection or estimation of signals, and their parameters from multichannel information provided by an array of antennas. In some applications, perfect knowledge of the antenna responses over their field of view is required, whereas in other applications only partial information or some contraints on array geometry are sufficient.

In this chapter, we focus on narrowband array processing. For narrowband array processing applications, the following generic model is used to estimate the parameters of interest:

$$\mathbf{r}(t) = \sum_{l=1}^{P} \mathbf{a}(\theta_l)s_l(t) + \mathbf{n}(t) \tag{9.1}$$

where $\mathbf{r}(t)$ is the vector of complex envelopes of sensor measurements, and $\mathbf{n}(t)$ is the contribution of noise to the measurements. The complex envelope of the waveform emitted from the lth source is denoted as $s_l(t)$. The vector $\mathbf{a}(\theta)$ is referred to as the *steering vector*, which represents the complex envelope of the voltage induced on the array antennas when a unit amplitude wavefront illuminates the array from the direction represented by the angle θ. In general, the steering vector may depend on multiple parameters, for example, azimuth and elevation angles, polarization, and center frequency. If there are P sources and M sensors, $\mathbf{a}(\theta)$, $\mathbf{n}(t)$, and $\mathbf{r}(t)$ are M-vectors. The steering vectors can be stacked in an $M \times P$ matrix $\mathbf{A}$, which is know as the steering matrix, so that we have

$$\mathbf{r}(t) = \mathbf{A}(\theta)s(t) + \mathbf{n}(t) \tag{9.2}$$

where $\mathbf{s}(t)$ is a P-vector that contains the source waveforms. The *parameter space* Ω is defined as the set in which θ is allowed to vary. The collection of steering vectors over the parameter space is defined as the *array manifold*, $\mathcal{A}$

$$\mathcal{A} = \{\mathbf{a}(\theta) \mid \theta \in \Omega\} \tag{9.3}$$

The definition of array manifold was introduced by Schmidt [53]. An important point to note is that array manifold is not a subspace, it is only a set, that is, it is not necessarily closed under vector addition. Direction-finding algorithms described in the literature are based on the following property of the array manifold: the set of any $P(P < M)$ steering vectors (with distinct parameters) is assumed to be linearly independent.

The performance of high-resolution direction-finding algorithms is critically dependent on the knowledge of the array manifold. Small errors in the array manifold information yield considerable performance loss in direction-finding and signal recovery performance.

In general, array processing algorithms utilize the array covariance matrix for direction-finding purposes [45, 53]. Second-order statistics are sufficient if the received signals can be characterized by using their first two moments; however, this is rarely the case. Most signals that are used in the communication and telemetry systems are non-Gaussian, with their higher-than-second-order comulants providing extra information about the signals. Therefore, an important research direction is to investigate the extra information supplied by computing higher-order statistics in array processing applications. In section 9.2, after a brief introduction to cumulants and their properties as applied to array processing, we show that an appropriate treatment of cumulant statistcs increases the directional information when compared with second-order statistics, and enables us to estimate the parameters of more sources than sensors. More specifically, we introduce the concept of virtual sensors whose locations are determined by the locations of the actual sensors, and propose a way to compute the second-order statistics among virtual and actual sensors using the cumulants of actual sensor measurements. The second-order statistics computed in this way can be processed by any direction-finding algorithm to estimate the source parameters.

After identifying the extra information provided by cumulants in array processing applications, we address the problem of joint array calibration and parameter estimation in section 9.3. In this problem the goal is to estimate the source parameters (directions and waveforms) using minimal information necessary about the array manifold. We investigate this sufficient information about the array manifold. We determine that two sensors with identical but arbitrary gain pattern, with a know displacement vector between them, is necessary to solve the problem of joint array calibration and parameter estimation using cumulants. The two sensors are called "guiding sensors," and cumulants are used to compute the cross-correlations between the actual sensors and a virtual copy of the original array displaced by the vector between the guiding sensors. These cross-correlations are used as if the ESPRIT data model were available [45], in order to estimate source bearings, steering vectors, and source waveforms. We name this approach a "virtual-ESPRIT algorithm" (VESPA), since the required cross-correlations are computed virtually by using cumulants.

Based on the advantages of cumulants, it is of interest to learn how many virtual sensors can be created, given the number and locations of actual sensors. We derive bounds on the number of virtual sensors in section 9.4. We also propose minimum-redundancy array designs for use in cumulant-based direction-finding applications in which the number of effective sensors (sum of actual and virtual sensors) grows with the fourth power of the number of actual sensors. Our designs cover both linear and two-dimensional arrays.

Cumulants have been promoted in the literature mainly because of their capability to suppress additive Gaussian noise of unknown covariance. In section 9.5, we propose methods to suppress additive non-Gaussian noise by using the extra directional information provided by cumulants. We also determine non-Gaussian caapabilities of VESPA and propose a method of combining second- and fourth-order cumulants for better noise suppression.

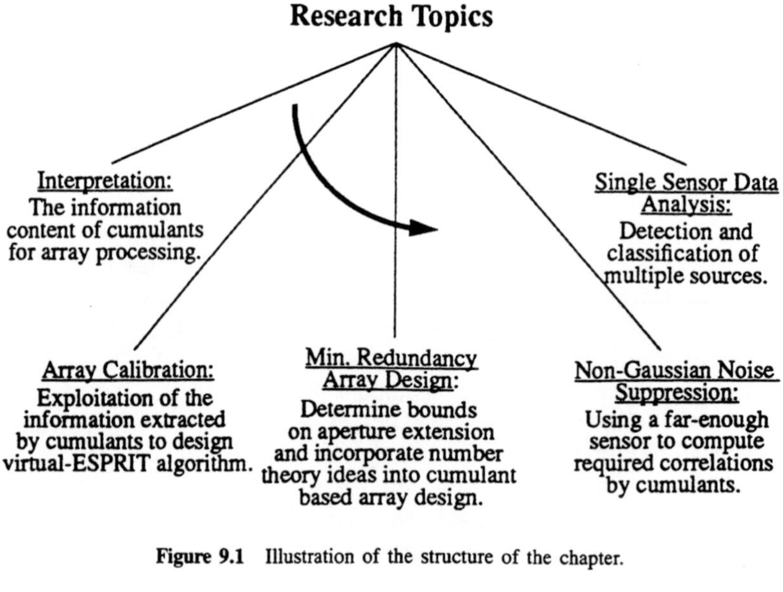

Figure 9.1 Illustration of the structure of the chapter.

The last topic that we investigate in this chapter is the application of array processing techniques to the analysis of cumulant statistics. We investigate single-sensor detection and classification of multiple sources in section 9.6. The motivation of using higher-order statistics in this problem is the observation that extra information in cumulants is provided in their multiple arguments, that is, cumulants have an array of arguments unlike second-order statistics. Using the multiple arguments of cumulants, we can construct cumulant matrices which resemble the array covariance matrix and then apply array processing methods to detect and classify the sources. As an extension, we describe the application of the single-sensor processing method to time-delay estimation problems, for multiple sources using two sensors.

Section 9.7 concludes our work with final observation and research directions. Figure 9.1 illustrates the structure of the chapter.

9.2 AN INTERPRETATION OF CUMULANTS FOR ARRAY PROCESSING

In this section, we propose an interpretation for the use of fourth-order cumulants in narrowband array processing problems. We start with the definitions and properties of cumulants for complex random processes. Using these properties, we show how fourth-order cumulants of multichannel observations increase the directional information available compared with second-order statistics. Based on our interpretation, we explain how cumulants can increase the effective aperture of antenna arrays. We give several examples to illustrate our interpretation. In addition, we extend our results to third-order cumulants. Finally, we describe the role of our interpretation in the following sections of the chapter.

Cumulants-Definitions and Properties

In our work, we consider the signals of interest to be non-Gaussian (possibly complex) narrowband random processes. Measurement noise components, being the outcome of many unknown, independent factors, are then assumed to be Gaussian random processes with unknown covariances. Not knowing the noise covariance reflects our lack of knowledge about the spatial propagation of noise. Spatial correlation of noise creates problems in covariance-based processing (e.g., biased direction-of-arrival—DOA—estimates), but with HOS this will not be an issue since higher-than second-order cumulants are blind to additive Gaussian noise. However, Gaussian noise suppression is only one benefit of using cumulants and it is not the major issue that we address here. In addition, we note that use of cumulants constrains the range of applicability to non-Gaussian signals.

To handle symmetric probability density functions from the sources of interest (e.g., communication signals), we shall use fourth-order cumulants of sensor outputs. Fourth-order (zero-lag) cumulants of zero-mean stationary measurements are defined in a balanced way as follows [42]:

$$\begin{aligned}\mu_{i,j}^{k,l} &\triangleq \operatorname{cum}\left\{r_i^*(t), r_j(t), r_k^*(t), r_l(t)\right\} \\ &= E\left\{r_i^*(t)r_j(t)r_k^*(t)r_l(t)\right\} - E\left\{r_i^*(t)r_j(t)\right\}E\left\{r_k^*(t)r_l(t)\right\} \\ &\quad - E\left\{r_i^*(t)r_k^*(t)\right\}E\left\{r_j(t)r_l(t)\right\} - E\left\{r_i^*(t)r_l(t)\right\}E\left\{r_j(t)r_k^*(t)\right\}\end{aligned} \tag{9.4}$$

for $(i, j, k, l) \in \{1, \ldots, M\}$, where $\{r_k(t)\}_{k=1}^{M}$ denotes the received signal vector from an array of M sensors. This definition is in keeping with the definition of cross-covariance, which can be expressed as $\mu_{i,j} \triangleq \{r_i^*(t)r_j(t)\}$, and which has only two arguments i and j. The richness of cumulants in terms of arguments will prove to be an interesting feature.

If the signal of interest also possesses a nonzero third-order cumulant, then we may also consider third-order cumulants. However, third-order cumulants cannot be defined in a balanced way

$$\mu_{i,j}^{k} \triangleq \text{cum}\left\{r_i^*(t), r_j(t), r_k(t)\right\} \triangleq E\left\{r_i^*(t)r_j(t)r_k(t)\right\} \tag{9.5}$$

since the number of arguments is odd.

Before giving an interpretation for the additional information provided by the higher-order statistics in (9.4), we list properties[1] of cumulants that are useful to us in the sequel [31]:

[CP1] If $\{\alpha_i\}_{i=1}^{n}$ are constants and $\{x_i\}_{i=1}^{n}$ are random variables, then

$$\text{cum}\,(\alpha_1 x_1, \alpha_2 x_2, \ldots, \alpha_n x_n) = \left(\prod_{i=1}^{n} \alpha_i\right) \text{cum}\,(x_1, x_2, \ldots, x_n) \tag{9.6}$$

[CP2] If a subset of random variables $\{x_i\}_{i=1}^{n}$ are independent of the rest, then

$$\text{cum}\,(x_1, x_2, \ldots, x_n) = 0 \tag{9.7}$$

[CP3] Cumulants are additive in their arguments;

$$\text{cum}\,(x_1 + y_1, x_2, \ldots, x_n) = \text{cum}\,(x_1, x_2, \ldots, x_n) + \text{cum}\,(y_1, x_2, \ldots, x_n) \tag{9.8}$$

[CP4] If the random variables $\{x_i\}_{i=1}^{n}$ are independent of the random variables $\{y_i\}_{i=1}^{n}$, then

$$\text{cum}\,(x_1 + y_1, x_2 + y_2, \ldots, x_n + y_n) = \text{cum}\,(x_1, x_2, \ldots, x_n) + \text{cum}\,(y_1, y_2, \ldots, y_n) \tag{9.9}$$

[CP5] The permutation of the random variables does not change the value of the cumulant.

[CP6] Cumulants suppress Gaussian noise of arbitrary covariance, that is, if $\{z_i\}_{i=1}^{n}$ are Gaussian random variables independent of $\{x_i\}_{i=1}^{n}$ and $n > 2$, we have

$$\text{cum}\,(x_1 + z_1, x_2 + z_2, \ldots, x_n + z_n) = \text{cum}\,(x_1, x_2, \ldots, x_n) \tag{9.10}$$

[CP7] If α_o is a constant, then

$$\text{cum}\,(\alpha_o + x_1, x_2, \ldots, x_n) = \text{cum}\,(x_1, x_2, \ldots, x_n) \tag{9.11}$$

An Interpretation for Array Processing

Conventional array processing techniques utilize only the second-order statistics of received signals. Second-order statistics are sufficient whenever the signals can be completely characterized by knowledge of the first two moments, as in the Gaussian

---

[1] These properties are given for the general case of nth-order ($n > 2$) cumulants.

case; however, in real applications, far-field sources emit non-Gaussian signals as in a communications scenario.

Whenever second-order statistics cannot completely characterize all of the statistical properties of underlying signals, it is beneficial to consider information embedded in higher than second-order moments. Higher-order statistics (HOS) prove to be rewarding alternatives to second-order statistics; there are many signal processing problems that are not solvable without access to HOS [31]. In this section, we provide an interpretation for the use of fourth-order cumulants in array processing problems based on the definition and properties of higher-order statistics summarized in the previous section.

High-resolution direction-finding methods such as MUSIC and ESPRIT, which use second-order statistics of array measurements, have been developed for the model

$$\mathbf{r}(t) = \mathbf{A}\mathbf{s}(t) + \mathbf{n}(t) \tag{9.12}$$

where $\mathbf{A}$ is the full-rank steering matrix, $\mathbf{s}(t)$ denotes the source waveforms, and $\mathbf{n}(t)$ is the noise contribution. If there are M sensors and P sources, then $\mathbf{r}(t)$ and $\mathbf{n}(t)$ are M-vectors, $\mathbf{s}(t)$ is a P-vector, and $\mathbf{A}$ is an $M \times P$ matrix. If the noise is spatially white, then the covariance matrix of $\mathbf{r}(t)$ takes the form

$$\mathbf{R} \triangleq E\left\{\mathbf{r}(t)\mathbf{r}^H(t)\right\} = \mathbf{A}\mathbf{\Sigma}_{ss}\mathbf{A}^H + \sigma^2\mathbf{I} \tag{9.13}$$

where $\mathbf{\Sigma}_{ss} \triangleq E\{\mathbf{s}(t)\mathbf{s}^H(t)\}$. If there are P sources, and they are all incoherent, then the source covariance matrix $\mathbf{\Sigma}_{ss}$ is diagonal, and $\mathbf{R}$ can be reexpressed as

$$\mathbf{R} = \mathbf{A}\mathbf{\Sigma}_{ss}\mathbf{A}^H + \sigma^2\mathbf{I} = \sum_{k=1}^{P} \sigma_k^2 \mathbf{a}_k\mathbf{a}_k^H + \sigma^2\mathbf{I} \tag{9.14}$$

where σ_k^2 and $\mathbf{a}_k$ denote the power and the steering vector for the kth source, respectively. If at another data collection time, the power of the kth source is scaled by β_k, then the array covariance matrix $\tilde{\mathbf{R}}$ for this scenario takes the form

$$\tilde{\mathbf{R}} = \sum_{k=1}^{P} \beta_k\sigma_k^2 \mathbf{a}_k\mathbf{a}_k^H + \sigma^2\mathbf{I} \tag{9.15}$$

High-resolution direction-finding methods use the structure of (9.14) to eliminate the noise component ($\sigma^2\mathbf{I}$), and then search for the vectors in the array manifold that lie in the range space of $\mathbf{A}\mathbf{\Sigma}_{ss}\mathbf{A}^H$. The presence of scale factors (β_k's) in (9.15) does not cause a problem for direction-finding algorithms unless the scale factors are zero, because the signal subspaces of $\mathbf{R}$ and $\tilde{\mathbf{R}}$ are the same. If the noise is spatially colored, elimination of noise in the array covariance matrix is not possible unless one knows the noise covariance matrix and whitens the receveid signals. If the noise is Gaussian, then cumulants can be used to suppress its effects.

In order to provide an interpretation for the use of second- and higher-order statistics in array processing applications, we illustrate an array set-up in Figure 9.2. For convenience, the elements of the array (represented by circles) are assumed to be isotropic, and the narrowband sources that illuminate the array are assumed to be statistically independent. In this case, we can further assume the presence of a single stationary source $s(t)$ (without loss of generality due

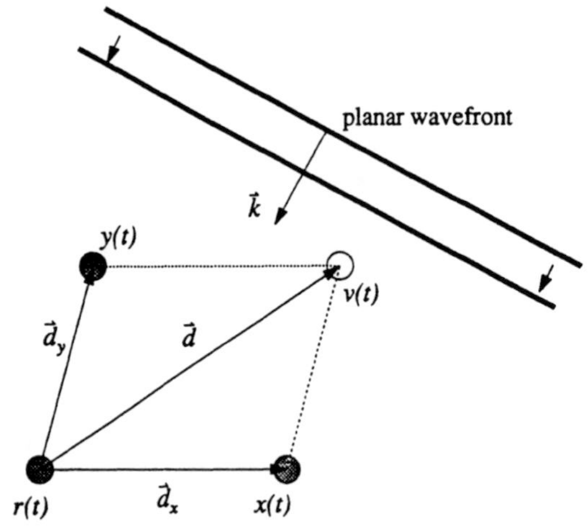

Figure 9.2 Mechanism of second- and higher-order cumulants for array processing: The sensors that measure $r(t), x(t), y(t)$ are actual sensors, whereas $v(t)$ is a virtual process measured by a virtual sensor. From the array geometry ($\vec{d} = \vec{d}_x + \vec{d}_y$), if $r(t) = s(t)$, then $x(t) = s(t)\exp(-j\vec{k}\cdot\vec{d}_x)$, $y(t) = s(t)\exp(-j\vec{k}\cdot d_y)$ and $v(t) = s(t)\exp(jk\cdot d)$.

do [CP4]), with propagation vector $\vec{k}$, where $\vec{k} = k_x\hat{a}_x + k_y\hat{a}_y(\hat{a}_x$ and $\hat{a}_y$ denote the unit vectors along the x and y axis, respectively), power σ_s^2 and fourth-order cumulant $\gamma_{4,s}$.

Suppose we compute the cross-correlation (ignoring noise effects) between the measured signal $r(t)$ and virtual signal $v(t), E\{r^*(t)v(t)\}$, assuming the reference point to be the position of the sensor that records $r(t)$. We call such a cross-correlation a "virtual" cross-correlation. Because $r(t) = s(t)$, and the virtual signal $v(t) = s(t)\exp(-j\vec{k}\cdot\vec{d})$, it follows that the directional information provided by the correlation operation is embedded in the dot product, $\vec{k}\cdot\vec{d}$, that is

$$\mu_{r,v} \triangleq E\{r^*(t)v(t)\} = \sigma_s^2 \exp(-j\vec{k}\cdot\vec{d}) \tag{9.16}$$

The source power σ_s^2 does not provide any directional information, and the vector $\vec{k}$ is common to all terms; hence, the information recovered by *cross-correlation* of two sensor outputs can be represented by the *vector* extending *from* the conjugated sensor, *to* the unconjugated one, that is, cross-correlation is a *vector* in the geometrical sense; in other words, $\mu_{r,v} \equiv \vec{d}$. Here, equivalence "$\equiv$" means "to carry the same directional information."

After providing such an interpretation for cross-correlation, the problem is how to interpret fourth-order cross-cumulants that have four arguments. Consider the cumulant

$$\begin{aligned}\mu_{r,x}^{r,y} &\triangleq \text{cum}\,(r^*(t), x(t), r^*(t), y(t)) \\ &= \text{cum}\,(s^*(t), s(t)\exp(-j\vec{k}\cdot\vec{d}_x)s^*(t), s(t)\exp(-j\vec{k}\cdot\vec{d}_y))\end{aligned} \tag{9.17}$$

and use **[CP1]** to obtain:

$$\text{cum}\,(r^*(t), x(t), r^*(t), y(t)) = \gamma_{4,s}\exp(-j\vec{k}\cdot\vec{d}_x)\exp(-j\vec{k}\cdot\vec{d}_y) = \gamma_{4,s}\exp(-j\vec{k}\cdot\vec{d}) \tag{9.18}$$

Finally, by comparing (9.16) and (9.18), we observe the following:

$$E\{r^*(t)\upsilon(t)\} = \frac{\sigma_s^2}{\gamma_{4,s}} \text{cum}(r^*(t), x(t), r^*(t), y(t)) \tag{9.19}$$

Equation (9.19), which relates a fourth-order statistic to a second-order statistic, demontrates that it is possible to recover directional information which is provided by channels $r(t)$ and $\upsilon(t)$ without actually using a real sensor to measure $\upsilon(t)$. We refer to (9.19) as a "virtual cross-correlation computation" (**VC**^{3}). Note that the cumulant computation in (9.19) carries directional information only for the non-Gaussian components of the measurements, and the cumulant computation in (9.19) suppresses the Gaussian components.

From the development of (9.19), we see that fourth-order cumulants can be interpreted as *addition* of *two vectors* each extending *from* a conjugated channel *to* an unconjugated one, that is, $\mu_{r,x}^{r,y} \equiv \vec{d}_x + \vec{d}_y = \vec{d} \equiv \mu_{r,\upsilon}$. It can be shown that (9.19) holds for both: multiple independent sources (due to [**CP4**]) and in the presence of additive colored Gaussian noise (due to [**CP6**]). Furthemore, [**CP5**], which indicates that cumulants are invariant with respect to a permutation of the random variables, is a restatement of the fact that addition of two vectors is a *commutative* operation.

To investigate the limits on **VC**^{3}, we redraw the three-element array of Figure 9.2 in Figure 9.3, and indicate the *lattice* structure defined by the vectors $\vec{d}_x, \vec{d}_y, \vec{d}_z$. We only have three actual sensors available: the ones labeled as $r(t), x(t)$, and $y(t)$. The intersections of the lines in the lattice determine the candidate locations for virtual sensors. To implement a covariance-like subspace algorithm, we need to compute the cross-correlation of all sensor outputs, actual or virtual. In other words, we need to connect the sensor to be used, with a single vector. With fourth-order cumulants, we have the opportunity of using two vectors for connection purposes. These connecting vectors must be selected from the set of vectors that define the lattice. In Figure 9.3 we have indicated a group of sensors (not a unique selection) that consist of four virtual and three actual sensors which can communicate by two jumps (vector additions). It is possible to form a 7×7 matrix in which we use cumulants to compute the required covariance (see Fig. 9.4). Example of cross-statistics are

$$\begin{aligned}
&\text{(between actual sensor)} \quad & E\{r^*(t)y(t)\} &= \frac{\sigma_s^2}{\gamma_{4,s}} \text{cum}(r^*(t), y(t), r^*(t), r(t)) \\
&\text{(actual and virtual sensors)} \quad & E\{r^*(t)\upsilon_1(t)\} &= \frac{\sigma_s^2}{\gamma_{4,s}} \text{cum}(r^*(t), x(t), r^*(t), y(t)) \\
&\text{(between virtual sensors)} \quad & E\left\{\upsilon_1^*(t)\upsilon_3(t)\right\} &= \frac{\sigma_s^2}{\gamma_{4,s}} \text{cum}(y^*(t), x(t), y^*(t), r(t))
\end{aligned} \tag{9.20}$$

It is not possible to create more than four virtual channels with this array configuration using fourth-order cumulants.

Observe, from (9.20), that the computed cumulants are scaled versions of the cross-correlations. If in forming the array covariance matrix, we replace every cross-correlation with its cumulant counterpart that carries the same directional information (as in Fig. 9.4), then the contribution of the source to the resulting covariance matrix

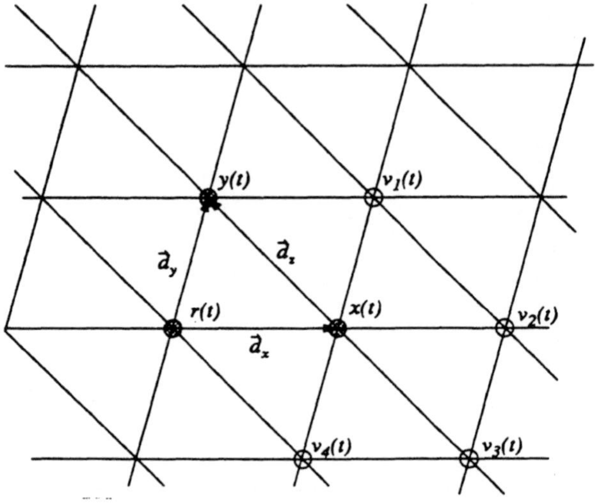

Figure 9.3 Virtual aperture extension by cumulants: lattice indicates the possible locations for virtual sensors. To be used in a DOA algorithm all sensors must be connected by at most two vectors. One such group (not unique) is illustrated in the figure.

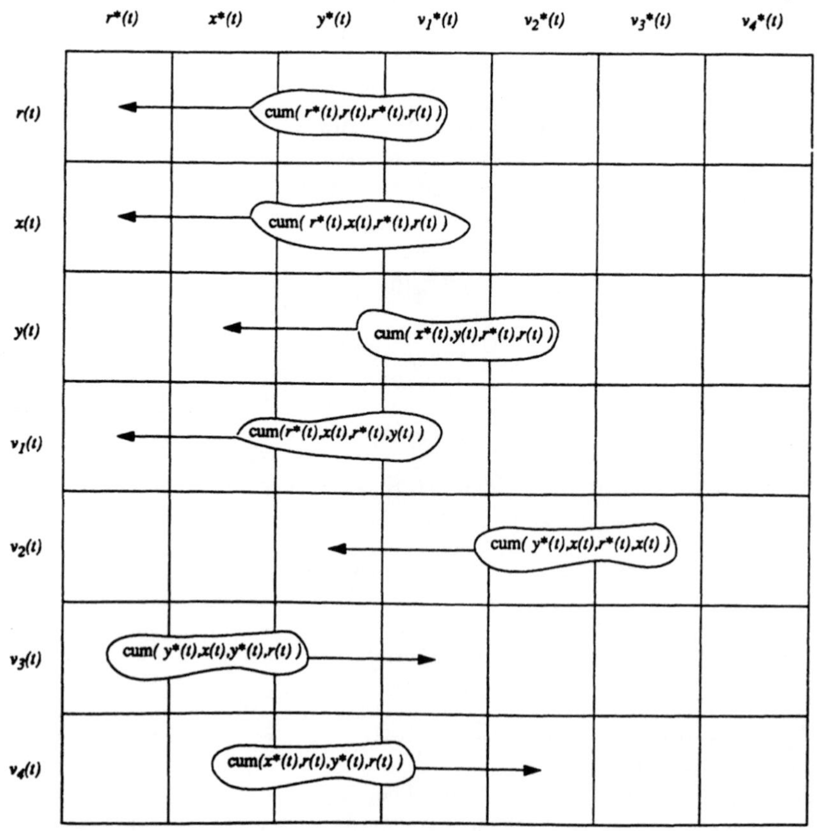

Figure 9.4 Construction of a matrix (**R**) to be used in the direction-finding processor using cumulants.

$\tilde{\mathbf{R}}$ will be scaled by an unknown scale factor $\beta_k = \gamma_{4,k}/\sigma_k^2$; therefore, $\tilde{\mathbf{R}}$ will take the form of the covariance matrix as in (9.15), in which source powers are scaled by β_k :

$$\tilde{\mathbf{R}} = \sum_{k=1}^{P} \beta_k \sigma_k^2 \mathbf{a}_k \mathbf{a}_k^H = \sum_{k=1}^{P} \gamma_{4,k} \mathbf{a}_k \mathbf{a}_k^H = \mathbf{A}\mathbf{\Gamma}\mathbf{A}^H \tag{9.21}$$

Note also that, because $\tilde{\mathbf{R}}$ is computed using cumulants, $\sigma^2 = 0$ (cumulants suppress additive Gaussian noise), but $\beta_k \neq 0$ since the sources are non-Gaussian. In (9.21), the steering vectors ($\mathbf{a}_k$) are 7×7, that is, they fully represent the delays encountered by the wavefronts as if we have seven actual elements at the locations selected in Figure 9.3. The diagonal matrix $\mathbf{\Gamma}$ consists of fourth-order cumulants (which can be negative[2]), of the non-Gaussian sources; therefore, the cumulant matrix $\tilde{\mathbf{R}}$ is not necessarily positive-definite, unlike the covariance matrix of sensor measurements. This difference does not pose a problem in the direction-finding processor, when the eigenvalues are sorted with respect to their magnitude.

Our geometric interpretation suggests why cumulants have a great potential for array processing: cumulants can be used to form virtual covariance matrices (e.g., Eq. 9.21) whose signal subspace can be characterized by an extended aperture. For the example in Figure 9.3, a covariance-based algorithm can estimate the bearings of two sources, whereas the cumulant-based approach can estimate the parameters of six sources (one less than the number of elements that form the extended, or effective, aperture). Although all the elements that form the effective aperture are not physically present, the covariance matrix can be computed (to within scale factors for each source) and used in a direction-finding algorithm; hence, we refer to the cumulant-based method as a "virtual aperture extension" approach. In addition, the cumulant-based approach can survive in the presence of colored noise due to **[CP6]**. In section 9.3, we provide more examples and simulations using different array configurations.

Note that there are various ways to compute the *cross-correlation between the actual sensors* using cumulants, because cross-correlation is a vector, and any vector can be expressed as the addition of the zero vector to itself. Therefore, by using two of the arguments of the fourth-order cumulant as required by correlation, and then using the remaining two arguments by repeating one of the sensor measurements in the cumulant expression, we obtain the required cross-correlation to within a scale factor, for example

$$\frac{\gamma_{4,s}}{\sigma_s^2} E\left\{x^*(t)y(t)\right\} = \mu_{x,y}^{x,x} = \mu_{x,y}^{y,y} = \mu_{x,y}^{r,r} \tag{9.22}$$

From (9.22), we can deduce that there exist M theoretically identical ways of computing cross-correlations between actual sensors using cumulants.[3] The advantage of computing $E\{x^*(t)y(t)\}$ by (9.22) is that additive Gaussian noise can be suppressed by the cumulant calculation. If $E\{x^*(t)y(t)\}$ were computed directly, it would be severely affected by additive Gaussian noise. Equation (9.22) was the approach taken in the initial attempts to incorporate higher-order statistics into direction-

[2] For example, the fourth-order cumulant for 4-QAM signals with variance σ_p^2 is $-(\sigma_p^2)^2$.

[3] Here we do not count the cumulants with symmetrical arguments, i.e., $\mu_{x,y}^{x,x} = \mu_{x,x}^{x,y}$ due to **[CP5]**.

finding algorithms for Gaussian noise suppression purposes [17, 37]; however, our various options for computing a cross-correlation provide a new way of smoothing to decrease the variance. For example, we can average all the options for the cross-correlation in (9.22), and use this average in place of the required correlation.

Similar analysis indicates that there are $M(M+1)/2$ ways to compute the *autocorrelation at a sensor*, for example

$$\frac{\gamma_{4,s}}{\sigma_s^2}\{x^*(t)x(t)\} = \mu_{x,x}^{x,x} = \mu_{x,y}^{y,x} = \mu_{x,r}^{r,x} = \mu_{y,y}^{y,y,} = \mu_{y,r}^{r,y} = \mu_{r,r}^{r,r} \tag{9.23}$$

The *cross-correlation between an actual and a virtual sensor* can be represented either in only one way, or in M theoretically identical ways using cumulants, depending on the relative locations of the sensors involved. For example (see Fig. 9.3), $E\{x^*(t)\upsilon_1(t)\} = E\{r^*(t)y(t)\}$ can be implemented in M ways, as in (9.22); whereas, $E\{r^*(t)\upsilon_1(t)\}$ has the unique implementation $\mu_{r,x}^{r,y}$.

Finally, we note that implementation of *cross-correlation between virtual sensors* is either identical to cross-correlation between actual sensors, or there is only one way to compute it (e.g., $E\{\upsilon_1^*(t)\upsilon_4(t)\}$). *Autocorrelation at virtual sensors* is identical to that at actual sensors and can be computed as in (9.23).

From the above discussion, we have determined that the extra information provided by cumulants is actually in the cross-correlation terms that can be uniquely represented by cumulants of actual measurements. Averaging (whenever possible) is proposed as a way to smooth the cumulants against estimation errors which are zero-mean, and Gaussian [42]. We have shown, however, that cumulants have more important properties than just Gaussian noise suppression. In section 9.5 we address the issue of non-Gaussian noise suppression.

Examples of Aperture Extension

We illustrate three different array geometries: circular, linear, and rectangular.

Circular Array. In Figure 9.5a, we illustrate a circular array with a sensor at the center. Such a configuration is very suitable for linear-prediction direction-finding when the sensor at the center is used as the reference [19]. However, some sensors are redundant if we use cumulants; the empty sensors in Figure 9.5b indicate the virtual elements whose second-order statistics can be computed by higher-order statistics.

A more careful investigation (see Fig. 9.6) of the circular array reveals that with the actual sensors of Fig. 9.5b, it is possible to extend the aperture so that the effective aperture is even larger than that of the original array in Figure 9.5a. This result indicates that cumulants are very promising in the design of two-dimensional *minimum-redundancy* arrays. Section 9.4 investigates this issue.

Linear Array. Consider the fully redundant linear array of N isotropic sensors in Figure 9.7a. In [55] it is proved that the aperture can be extended to $2N-1$ elements. We now provide a very simple and geometric proof for this fact: the most distant point from the origin that we can reach by two vector additions is $(2N-2)$, (since the

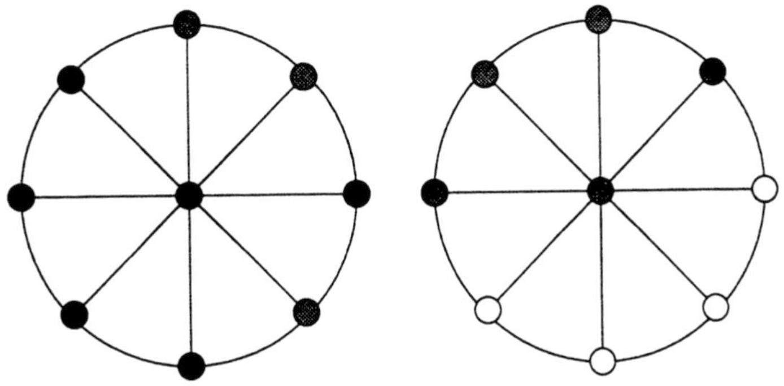

(a) (b)

Figure 9.5 Circular arrays and element reduction by cumulants: (a) circular array for linear-prediction direction finding, (b) empty sensors become redundant with the use of cumulants.

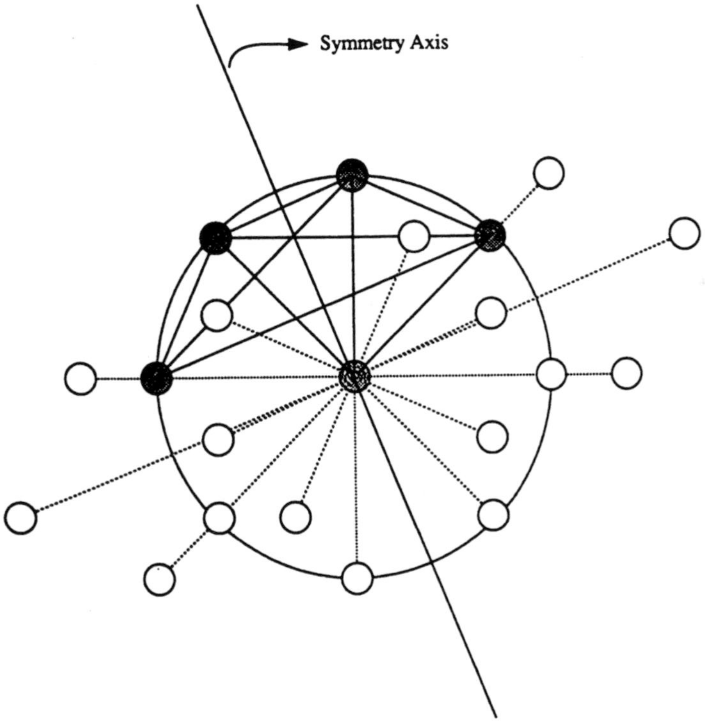

Figure 9.6 It is possible to generate 12 more virtual sensors by using cumulants. These new locations are obtained by picking one of the 6 possible vectors among the 4 actual sensors ($4!/2!2! = 6$) on the circle (excluding the sensor at the center), and attaching this vector to the sensor at the center. There are two ways to do this; hence, we have 12 new locations. Since these 12 new locations communicate with the center sensor with only one vector, they communicate with the sensors of previous design (virtual sensors on the circle) with at most two jumps, implying that the aperture can be extended to 21 elements using cumulants.

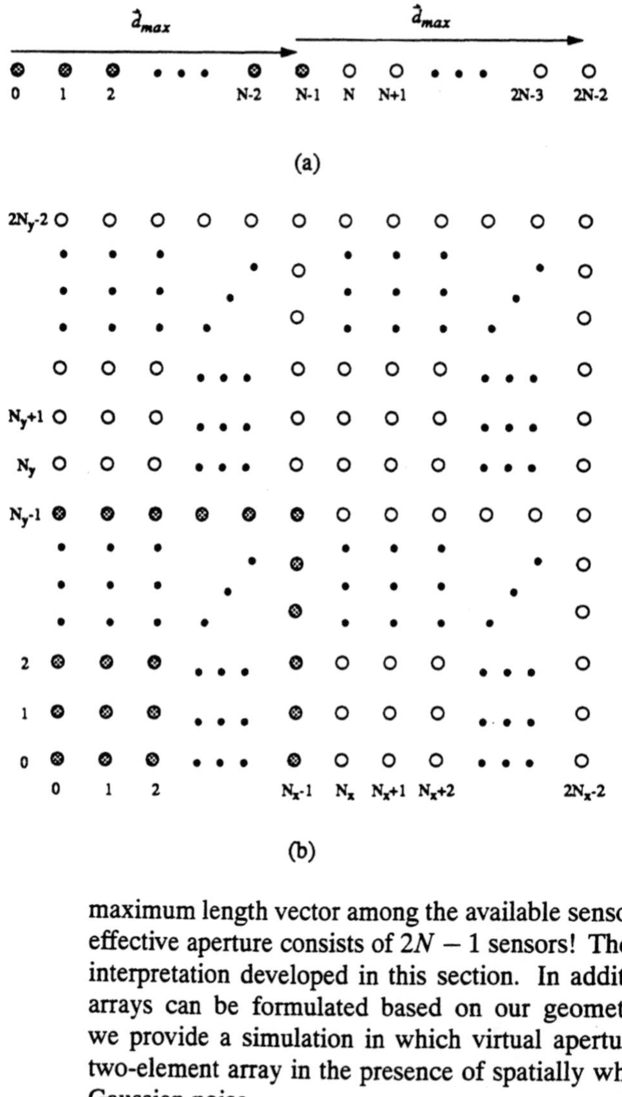

Figure 9.7 Linear and rectangular array aperture extension: (a) fully redundant linear array (N_x filled sensors), and virtual sensors that can be generated by cumulants ($N_x - 1$ empty sensors); (b) fully redundant rectangular array (N_x, N_y filled sensors), and virtual sensors that can be generated by cumulants. Effective aperture consists of $(2N_x - 1)(2N_y - 1)$ sensors.

maximum length vector among the available sensors is $(N-1)$ units long), that is, the effective aperture consists of $2N - 1$ sensors! The proof is this simple because of the interpretation developed in this section. In addition, design of linear nonredundant arrays can be formulated based on our geometric interpretation. In section 9.5, we provide a simulation in which virtual aperture extension is accomplished for a two-element array in the presence of spatially white but spatially nonstationary non-Gaussian noise.

Rectangular Array. We can now analyze the more general case of the fully redundant rectangular array in Figure 9.7b. We have an array of $N_x \cdot N_y$ sensors. With similar reasoning to the linear case, the effective aperture consists of $(2N_x - 1)(2N_y - 1)$ sensors.

Third-Order Cumulants

Third-order cumulants are used less frequently in array processing applications than fourth-order cumulants [14, 17, 38, 60]. This is due to the nonsymmetric nature of the definition of the third-order moments for complex random processes.

Another important reason is that third-order cumulants of almost all manmade signals are identically zero (communication signals, telemetry signals); however, there are physical processes that have nonzero third-order statistics, such as signals associated with rotating machinery, radar returns, and voiced speech signals. For localization of these processes, it is advantageous to use third-order cumulants, since they are more computationally attractive than fourth-order cumulants, their data length requirements are less when compared with fourth-order statistics, and they can suppress symmetrically distributed ambient noise.

An important observation about third-order moments is that they are *not spatially stationary* even in the case of independent sources. This is a result of the asymmetry in the third-order moments with respect to conjugation. For example, refer to Figure 9.2, and consider

$$\text{cum}\,(r(t), r^*(t), r(t)) = \gamma_{3,s} \neq \text{cum}\,(x(t), x^*(t), x(t)) = \gamma_{3,s} \exp(-j\vec{k} \cdot \vec{d}_x) \tag{9.24}$$

This indicates that third-order cumulants have a *reference point problem*, unlike second- and fourth-order statistics. They are *floating* in this sense. However, this can be cured very easily by allowing the third-order cumulant $\gamma_{3,s}$ to be an unknown complex quantity so as to include the multiplication by an unknown phase term due to the floating reference point. This is possible since the statistic of the source waveform does not provide information about the location of the source as pointed earlier.

Aperture extension (for identical sensors) starts by computing the autocorrelation as

$$E\,\{r(t)r^*(t)\} = E\,\{x(t)x^*(t)\} = E\,\{y(t)y^*(t)\} = \frac{\sigma_s^2}{\gamma_{3,s}}\,\text{cum}\,(r(t), r^*(t), r(t)) \tag{9.25}$$

Next, we must compute the cross-correlation between actual sensors, which can be done as follows (direct extension of the fourth-order cumulant results):

$$E\,\{r^*(t)x(t)\} = \frac{\sigma_s^2}{\gamma_{3,s}}\,\text{cum}\,(r^*(t), x(t), r(t)) \tag{9.26}$$

$$E\,\{r^*(t)y(t)\} = \frac{\sigma_s^2}{\gamma_{3,s}}\,\text{cum}\,(r^*(t), y(t), r(t)) \tag{9.27}$$

$$E\,\{x^*(t)y(t)\} = \frac{\sigma_s^2}{\gamma_{3,s}}\,\text{cum}\,(x^*(t), y(t), r(t)) \tag{9.28}$$

Observe that we always fix the last component of the cumulant in (9.26)–(9.28) to be $r(t)$. We call $r(t)$ the "common element." Now we shall complete the description of virtual aperture extension by demonstrating the way to compute the cross-correlations between real elements and the lone virtual sensor (since we have only one virtual location, we cannot calculate cross-correlations between two virtual sensor measurements), that is,

$$E\,\{r^*(t)v(t)\} = \frac{\sigma_s^2}{\gamma_{3,s}}\,\text{cum}\,(r^*(t), x(t), y(t)) \tag{9.29}$$

Equation (9.29) implies that for aperture extension purposes, the third-order cumulant can be considered as *addition* of *two vectors*, which are *constrained* to originate from the same sensor (the conjugated one). Therefore, aperture extension is possible with third-order statistics, but it is not as powerful as using fourth-order cumulants. For example, in Figure 9.3, we can take the set $\{r(t), x(t), y(t), \upsilon_1(t)\}$ as an extended aperture with third-order cumulants, whereas with fourth-order cumulants the aperture can be extended to seven elements (see Fig. 9.3). A third-order 4×4 cumulant matrix can be set up as described in the fourth-order cumulants case (Fig. 9.4).

In the circular array example (Fig. 9.5) using third-order cumulants, it is possible to extend the aperture to nine elements (show as an exercise) as in Figure 9.5b; however, this is less than the result obtained by using fourth-order cumulants (see Fig. 9.6), since fourth-order cumulants can be viewed as unconstrained vector addition.

Overview

We have proposed an interpretation for describing the potential of cumulants in array processing applications. It has led to the virtual cross-correlation computer[4] (**VC**^{3}). The **VC**^{3} can be used to: calibrate arbitrary arrays and jointly estimate the direction parameters of far-field sources, design minimun-redundancy arrays, and suppress undesired non-Gaussian signals. Figure 9.8 summarizes these applications and their treatment in this chapter.

The most surprising result of this section is the explanation of how cumulants can increase the *effective aperture* of an arbitrary array. This fact is reasonable when one considers that forming the covariance matrix is a *data reduction* technique (rather than storing multichannel snapshots, we compute the sample covariance matrix). It is a well-known fact of information theory that we always lose information by data reduction. Forming a cumulant matrix can be considered as an alternate data reduction technique, but it is much better than using only second-order statistics, since it is possible to recover more information from the cumulant matrix about the sources illuminating the array. The interpretation provided for cumulants can be extended for cyclostationarity-based array processing applications [52, 54, 72].

9.3 ARRAY CALIBRATION ISSUES: VIRTUAL-ESPRIT ALGORITHM

In section 9.2 we proposed a novel interpretation for the use of cumulants in narrowband array processing problems. Based on this interpretation, we may now investigate the *amount* of partial information necessary to *jointly* calibrate an arbitrary array and estimate the directions of far-field sources in this section. In so doing, we prove that the presence of a doublet and use of fourth-order cumulants is sufficient to accomplish this task. Our approach is computationally efficient and more general than covariance-based algorithms that have addressed this problem under constraints. A class of beamforming techniques is proposed to recover the source waveforms.

[4] A patent for **VC**^{3} has been filed for by the University of Southern California.

APPLICATION	PROBLEM	SOLUTION	DESCRIPTION
Aperture Extension	To extend the effective aperture of an array without adding real sensors.	Compute all the required second-order statistics virtually, by fourth-order cumulants.	Section 2, 4.
Array Calibration	An identical copy of the array at a known displacement is required to compute cross-correlations. This requires extra hardware and precision.	Compute the required cross-correlations virtually by using a single doublet and fourth-order cumulants.	Section 3.
Antenna Array Design	Minimum redundancy array design is limited to linear arrays. Linear arrays have ambiguity problems.	Use the interpretation of cumulants to design 2-D minimum redundancy arrays.	Section 5.
Noise and Interference Suppression	To decorrelate the undesired components, another array which is far-away from the existing one is required. This means doubling the hardware.	A single sensor which is far-away from the existing array is sufficient. Compute the required statistics virtually, by using fourth-order cumulants.	Section 6.

Figure 9.8 An overview of the possible applications of our interpretation.

All of the developed estimation procedures are based on cumulants, which bring insensitivity to the spatial correlation structure of additive Gaussian measurement noise. Simulations indicate excellent results by the proposed algorithms.

The Array Calibration Problem

During the 1980s, there were revolutionary advances in high-resolution direction-of arrival estimation problems. Among the algorithms proposed in the literature, the so-called subspace methods that are based on the eigendecomposition of the sample covariance matrix possess very appealing features: they have modest computational requirements when compared with the maximum-likelihood method [74]; there exists solid work on the detection of sources problem [67]; and they provide asymptotically exact values for the parameters of interest.

Among this class of subspace methods, the popular MUSIC algorithm [53] stands out due to its applicability to arrays of arbitrary orientation and response. In addition, it can estimate a multitude of parameters for each far-field source, such as azimuth, elevation angles, and polarization. This generality results in a major drawback: array response must be measured and stored for every possible combination of source parameters. This procedure, known as array calibration, is very undesirable, since it requires an enormous amount of memory to store the array manifold, especially in the case of multiple parameters. In addition, the MUSIC algorithm is very sensitive to calibration errors [18]. The direction-finding (DF) step of the MUSIC algorithm is also computationally expensive except for some specific array configurations. These problems limit the applicability of the MUSIC-like subspace algorithms.

In this section, we address the problem of *joint* array calibration and direction-of-arrival estimation (DOA) with arbitrary arrays. Our goal is to determine the minimal information necessary about the array structure to accomplish this task and develop an algorithm that utilizies this *sufficient* information. The problem resembles the blind equalization problem in data communications, where the data symbols are distorted by finite-memory channels. In blind equalization, the goal is to "open the eye" so that it is possible to jointly estimate the impulse response of the channel and recover the symbols. In the array problem, the aim is to "open the eye of the processor," so that it can "see" the far-field sources (DOA estimation), and "listen to" each of them (waveform recovery).

The blind equalization problem is known to be unsolvable for nonminimun phase systems if processing is limited to the power spectrum. Higher-than-second-order statistics (cumulants) have been shown to be invaluable for solving this problem, since it is possible to recover phase information [31] with cumulants. In array processing, we obtain phase information by cross-correlating channel measurements. It is this phase information that makes eigenstructure-based high-resolution spatial-spectrum estimation possible. In the array processing context, the *motivation* for using cumulants is to recover *more* phase terms than is possible by using only second-order statistics. This goal is accomplished in section 9.2.

Given an arbitrary array, *joint* calibration and DOA estimation problems can be solved, if we have an identical copy of the array displaced in space, by using the ESPRIT algorithm [45]. In this way, the problems associated with the array calibration procedure are alleviated by incorporating a specific type of *redundancy* into the array configuration. The ESPRIT algorithm can blindly identify the steering vectors and DOA's of sources whose number is limited by the subarray size; hence, waveforms of the sources whose directions are identified can be estimated.

The special array geometry required by ESPRIT is not available in general, so it is difficult to calibrate arbitrary arrays in practice. Rockah and Schultheiss [43] did pioneering work on the conditions required for calibrating isotropic sensors of arbitrary arrays. They proved that if there are three non-colinear spectrally/temporally disjoint sources with unknown bearings, the calibration errors tend to zero as the signal-to-noise ratio (SNR) of the sources tends to infinity. In addition, if the direction of one sensor to the reference sensor is known, then DOA estimation can be done by taking the known direction vector as a reference. These results give us hope that not all of the redundancy required by ESPRIT is necessary, and that one may do well by using a single doublet rather than having all the sensors occur in pairs.

In this aspect, the benefits of incorporating redundancy in an array structure are analogous to the benefits of channel coding for communication systems which incorporates redundancy into the actual data. In the channel coding problem, the natural question is "How much redundancy is needed to transmit information with sufficient protection?" For high-rate transmission objectives, the task of the communication engineer is to minimize the required redundancy while maintaining the specifications for error correction. Similarly, the task of the antenna signal processing engineer is to minimize the redundancy required by the ESPRIT algorithm, while maintaining the capability of DOA estimation, array calibration, and waveform estimation so that the algorithm can be applied to a wider class of array processing scenarios. In this section, 9.3, we propose a solution to the redundancy minimization problem so that it will be possible to calibrate an arbitrary array by using a single doublet and fourth-order cumulants. Our results are in agreement with the results of Rockah and Schultheiss, although their paper takes the approach of evaluating Cramér-Rao bounds for the unknown parameters and assuming Gaussian processes, whereas ours uses cumulants which are blind to Gaussian processes. The results of Rockah and Schultheiss [43], about the conditions required for calibration and estimation, play the role of the Shannon Theorem for the capacity of a communication channel; they provide the conditions necessary for calibration but not the way to achieve it without further a priori information. (The analogy between the array processing problem and the communication problem was emphasized in the review paper [21]). Our approach is totally different: we propose a method for calibration and then address its consistency.

Following the excellent work of Rockah and Schultheiss [43], work on direction finding in the presence of sensor uncertainties continued [44], based mainly on the maximum-likelihood approach [68], or utilizing calibration sources whose bearings are known [30–36]. Problems with the existing methods arise from iterative nonlinear optimization procedures that require good nominal knowledge about array geometry, unrealistic constraints about sensor responses (e.g., all sensors are assumed to be isotropic), computational complexity of optimization techniques, local convergence problems, and sensitivity to noise spatial correlation structure. The method that is proposed in this section is nonrecursive, depends on the eigendecomposition of cumulant matrices obtained from the observed data, and does not involve any computationally intensive optimization or search procedure. In this way, local convergence problems and noise effects are avoided. Later in this section, 3.3, we provide simulation experiments to illustrate the effectiveness of our method. We conclude the section with some final observations and extensions.

Joint Calibration and Parameter Estimation

In this subsection, we propose the use of $\mathbf{VC}^3$ for calibrating arrays which are illuminated by multiple incoherent far-field sources from unknown directions. The calibration problem can be summarized as estimation of the directions of far-field sources, with an array of unknown array manifold (i.e, sensor locations and responses) Contrary to [43], we: allow sources to overlap in time and frequency, consider sensors with arbitrary responses, and do not assume a nominal knowledge for the array geometry and spatial correlation of measurement noise. Clearly, there is need for some information about the array; here we investigate what that sufficient information is.

Given an arbitrary array, joint array calibration and source parameter estimation can easily be solved if we have an identical copy of the array displaced in space with a known displacement vector $\vec{\Delta}$ so that we can apply the ESPRIT algorithm [45]. Given such an array configuration, ESPRIT can determine the source bearings within a cone of ambiguity, for which the axis of the cone coincides with the known displacement vector. This ambiguity can be removed if one uses multiple-invariance ESPRIT to estimate both azimuth and elevation angles [62]. In this subsection, we consider bearing estimation within a cone of ambiguity.

The main questions we answer herein are: given an array of arbitrary geometry and sensor responses: (1) Is it necessary to have a full copy of the array for calibration? (2) If not, how much redundancy is necessary? (3) How can such an algorithm be implemented? We provide the answers based on the results of section 9.2.

Consider the arbitrary array and its copy in Figure 9.9. In order to use the ESPRIT algorithm to jointly estimate the DOA parameters of multiple sources and the associated steering vectors, we need to compute the cross-correlations between subarrays. For example [see (9.16) and Fig. 9.9]

$$E\{r_1^*(t)v_M(t)\} = \sigma_s^2 a_1^* a_M \exp(-j\vec{k}\cdots\vec{d}_{1M})\exp(j\vec{k}\cdot\vec{\Delta}) \tag{9.30}$$

where a_M denotes the response of the Mth sensor to the wavefront from the source. Unfortunately, we cannot compute (9.30) since we do not have access to the virtual signal $v_M(t)$. Next, consider

$$E\{r_1^*(t)r_M(t)\}=\sigma_s^2 a_1^* a_M \exp(-j\vec{k}\cdots\vec{d}_{1M})=E\{r_1^*(t)v_M(t)\}\exp(-j\vec{k}\cdot\vec{\Delta}) \tag{9.31}$$

which follows from (9.30). Note that $E\{r_1^*(t)r_M(t)\}$ is computable because $r_1(t)$ and $r_M(t)$ are actual signals. If we knew $e^{j\cdot\vec{k}\cdot\vec{\Delta}}$ then we could solve (9.31) for $E\{\{r_1^*(t)v_M(t)\}$; however, this is not possible since we do not know the propagation vector $\vec{k}$.

From Figure 9.9 we observe that all vectors joining two sensors in separate subarrays can be decomposed as addition of two vectors, one between the sensors of the main array, and the other one being the displacement vector $\vec{\Delta}$, for example, $\vec{d}'_{12} = \vec{d}'_{12} + \vec{\Delta}$. All the computable correlations (9.31) lack the common term $\exp(-j\vec{k}\cdot\vec{\Delta})$. It is necessary to form a bridge between the main array and its copy to recover this phase term.

From our results in section 9.2, we know that by using fourth-order cumulants, and assuming that only one doublet $\{r_1(t), v_1(t)\}$ is available, we can compute the cross-correlations between subarray elements, as:

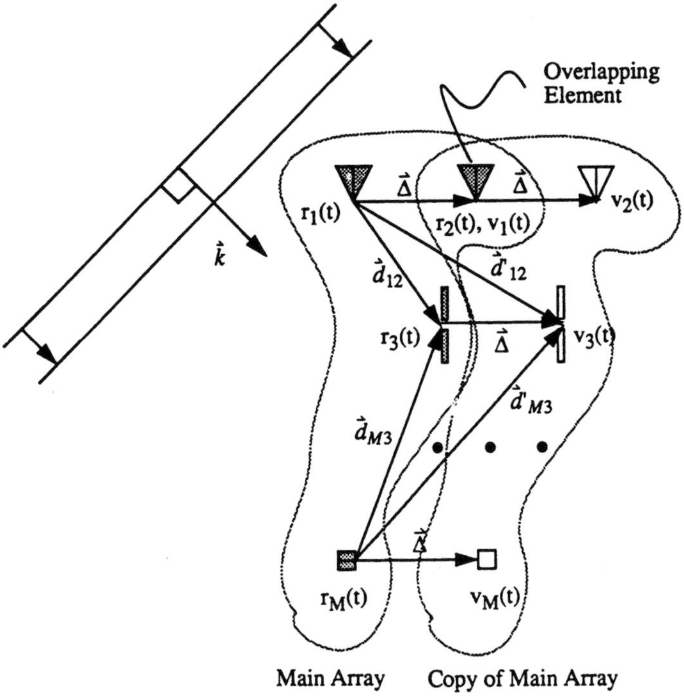

Figure 9.9 Virtual-ESPRIT algorithm: If the virtual sensor measurements $\{v_k(t)\}_{k=1}^{M}$ denote the required data from a displaced copy of the main array which measures $\{r_k(t)\}_{k=1}^{M}$, then it is possible to reach any virtual sensor location from an actual sensor location by addition of two vectors between actual sensors when the doublet $\{r_1(t), r_2(t)\}$ is available. In other words, cross-correlation between actual and virtual sensor elements can be calculated by using cumulants, since cumulants can be interpreted as vector addition, whereas cross-covariance can be interpreted as a single vector. From the geometry, the copy of the first sensor $v_1(t)$ overlaps the second sensor that measures $r_2(t)$. If we use the actual sensors as a subarray in the ESPRIT algorithm and virtual sensors as another subarray, then these two subarrays share an element, namely the second guiding sensor. The number of elements in a subarray equals the number of actual elements M, whereas the total aperture consists of $M-1$ sensors.

(between actual sensors)

$$E\{r_k^*(t)r_l(t)\} = \frac{\sigma_s^2}{\gamma_{4,s}|a_1|^2}\,\text{cum}\,(r_1^*(t), r_1(t), r_k^*(t), r_l(t)) \tag{9.32}$$

(actual and virtual sensors)

$$E\{r_k^*(t)v_l(t)\} = \frac{\sigma_s^2}{\gamma_{4,s}|a_1|^2}\,\text{cum}\,(r_1^*(t), r_2(t), r_k^*(t), r_l(t)) \tag{9.33}$$

Cross-correlation between virtual sensors is identical to that between actual sensors, that is $\mu_{v_k,v_l} = \mu_{r_k,r_l}$.

Equations (9.32) and (9.33) can therefore be used to form the covariance matrix required by ESPRIT [45]. The presence of $|a_1|^2$ in the scale factor is not important since this term is common to all cumulant terms; it makes the effective fourth-order source cumulant be $\gamma_{4,s}|a_1|^2$.

Two ways exist (which is the number of guiding sensors) to implement the required cross-correlations between actual elements, that is

$$E\{r_k^*(t)r_l(t)\} = \frac{\sigma_s^2}{\gamma_{4,s}|a_1|^2}\mu_{r_1,r_1}^{r_k,r_l} = \frac{\sigma_s^2}{\gamma_{4,s}|a_1|^2}\mu_{r_2,r_2}^{r_k,r_l} \tag{9.34}$$

because it is known that the response of the guiding sensors that measure $r_1(t)$ and $r_2(t)$ are identical (i.e., $a_1 = a_2$). Since the sensors of the main array are not otherwise identical, we cannot substitute the guiding sensor measurements with the measurement of another sensor, as we did in (9.22), that is

$$E\{r_k^*(t)r_l(t)\} = \frac{\sigma_s^2}{\gamma_{4,s}|a_1|^2}\mu_{r_1,r_1}^{r_k,r_l} \neq \frac{\sigma_s^2}{\gamma_{4,s}|a_1|^2}\mu_{r_m,r_m}^{r_k,r_l}, \quad 3 \le m \le M \tag{9.35}$$

because $\mu_{r_m,r_m}^{r_k,r_l} = |a_m|^2/|a_1|^2\mu_{r_1,r_1}^{r_k,r_l}$ and it is unknown whether $|a_1| = |a_m|$. Similar reasoning indicates that there is only a single way to compute the cross-correlations between actual and virtual sensors.

For obvious reasons, we call the single pair of sensors that form the doublet "guiding sensors" and the resulting algorithm "virtual-ESPRIT algorithm (VESPA)."

Note that the VESPA requires only a single doublet rather than a full copy of the array, resulting in enormous hardware reductions. VESPA also alleviates the problems resulting from the perfect sampling synchronization requirements of the covariance-ESPRIT for the two subarrays. In VESPA, synchronization must be maintained only between the elements of the single doublet. A block diagram for VESPA is provided in Figure 9.10.

In applications, we do not have the true cumulants; they are replaced by their *consistent* estimates which converge to true values as the data length grows to infinity. The results in [34] indicate that convergence to the true values is rapid at high SNR.

The requirement of a single doublet with known orientation $\vec{\Delta}$ is in fact the necessary requirement in [43]. Rockah and Schultheiss [43] derive this result assuming isotropic sensors and Gaussian processes that are disjoint in time or frequency. Levi and Messer [28] relaxed the Gaussian assumption on the sources for isotropic sensors. VESPA utilizes fourth-order cumulants which are blind to Gaussian processes, yet we obtain the same requirement about the necessary information for identifiability. In addition, the consistency results for our approach and the method in [43] require high SNR and longer data lengths when compared with methods that utilize full knowledge of the array manifold. On the other hand, the requirements of our approach are very mild when compared to those in [43] and its extensions: (1) We allow multiple sources sharing[5] the same frequency band due to **[CP4]**, (2) Our approach is applicable to arbitrary arrays (the isotropic sensor arrays of [28, 43] are a subclass); (3) Our approach is also applicable to nominally linear arrays, for which the bearing estimates are within a cone of ambiguity, and the cone axis coincides with the displacement vector between guiding sensors. This ambiguity

[5] Although Levi and Messer [28] claim this condition can also be relaxed by their purely geometrical analysis, it is not possible to estimate the underlying phase factors by their approach when there are multiple cochannel signals and there is uncertainty about sensor locations.

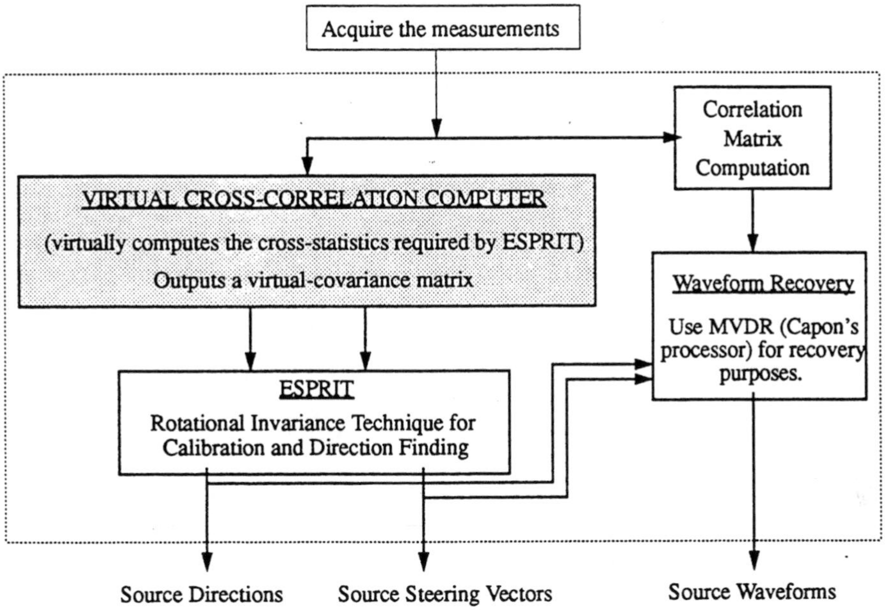

Figure 9.10 The role of virtual cross-correlation computer (**VC**^{3}) in the joint array calibration and DOA estimation problem. **VC**^{3} provides ESPRIT the required covariance matrix without the need for a full copy of the sensor array. The matrices provided by **VC**^{3} are also insensitive to colored Gaussian noise.

can be resolved by adding a third guiding sensor which is not colinear with the other two guiding sensors, in order to create another virtual copy of the original array and to then apply the multiple-invariance ESPRIT algorithm [62]; (4) From an implementation point of view, our approach is noniterative, and eliminates parameter search by using ESPRIT, unlike the method in [68] which requires iterations, and very good nominal knowledge about the array to start a multidimensional search; (5) The cumulant-based approach does not require information about noise spatial correlation, unlike the covariance-based algorithm; (6) In the presence of white observation noise, it is possible to initialize the maximum-likelihood approach proposed in [68] by the results of VESPA to get even better estimates; (7) In the presence of colored Gaussian noise, a trispectral maximum-likelihood approach can be designed along the lines of [17] initialized by VESPA to calibrate arbitrary arrays in a maximum-likelihood fashion without the knowledge of noise color; and, finally (8) Section 9.5 sets the stage for removing the effects of non-Gaussian colored noise, which can be followed by a ML approach.

Finally, we address an alternative approach which is designed to apply ESPRIT to arbitrary arrays [69]. In [69], array measurements are processed by transformation matrices in order to approximate the measurements of a uniform linear array, so that the ESPRIT algorithm can be applied to estimate the source parameters. In other words, this method first interpolates the measurements to estimate the received signals as though they were from a linear array under the same far-field conditions.

The design of the transformation matrices requires perfect knowledge of the original array manifold; hence the method described in [69] can be thought of as a DOA estimation method rather than a calibration method. On the other hand, VESPA does not assume perfect knowledge of the complete array manifold; it requires knowledge about a doublet, and interpolates fourth-order cumulants of measurements to implement the missing cross-correlations required by the ESPRIT algorithm.

To recover the waveforms associated with the far-field sources, we first estimate the steering vectors by subspace rotation [45]. The measured signals are represented as

$$\mathbf{r}(t) = [r_1(t), r_2(t), \ldots, r_M(t)]^T \tag{9.36}$$

in which the first two channels are the two guiding sensor measurements. Let $\mathbf{a}_d$ be the $M \times 1$ steering vector of the signal of interest (SOI) $s_d(t)$ with estimated bearing θ_d, and let the augmented steering matrix be decomposed as $\mathbf{A} = [\mathbf{a}_d, \mathbf{A_J}]$, so that

$$\mathbf{r}(t) = \mathbf{a}_d s_d(t) + \mathbf{A_J j}(t) + \mathbf{n}(t) \tag{9.37}$$

where $\mathbf{j}(t)$ contains the co-channel signals [except $s_d(t)$] and $\mathbf{n}(t)$ is the measurement noise.

We propose two approaches for signal recovery

Minimum-Variance Distortionless Response Beamformer (MVDR). This beamformer estimates the desired waveform in the mean-square sense, as

$$\hat{s}_d(t) = \mathbf{w}_1^H \mathbf{r}(t) = (\mathbf{R}^{-1}\mathbf{a}_d)^H \mathbf{r}(t) \tag{9.38}$$

where $\mathbf{R} = E\{\mathbf{r}(t)\mathbf{r}^H(t)\}$. In (9.38), we ignored a scale factor since it does not effect the output signal-to-interference plus noise ratio (SINR).

MVDR with Perfect Nulling (null-MVDR). This beamformer estimates the SOI waveform in the mean-square sense while putting perfect nulls on the interferers, that is

$$\hat{s}_d(t) = \mathbf{w}_2^H \mathbf{r}(t) \tag{9.39}$$

where the weight vector $\mathbf{w}_2$ is the solution of the linearly constrained minimum variance problem:

$$\mathbf{w}_2 = \min_{\mathbf{w}} \mathbf{w}^H \mathbf{R} \mathbf{w} \quad \text{subject to} \quad \left[\mathbf{a}_d,\ \mathbf{A_J}\right]^H \ \mathbf{w} = \mathbf{f} = [1, 0, \ldots, 0]^T \tag{9.40}$$

which has the solution

$$\mathbf{w}_2 = \mathbf{R}^{-1}\mathbf{A}(\mathbf{A}^H\mathbf{R}^{-1}\mathbf{A})^{-1}\mathbf{f} \tag{9.41}$$

Both of the above beamformers do not require knowledge of the measurement noise covariance matrix. Derivations of (9.38) and (9.41) can be found in [33].

Simulations

In this subsection, we present simulation experiments to demonstrate the performance of the joint parameter estimation and calibration algorithms proposed herein.

Direction-Finding using VC³ and MUSIC. Here we provide an experiment to illustrate the use of **VC³** with the MUSIC algorithm in order to estimate the bearings of six far-field sources which illuminate the array from[6] $\{50°, 70°, 90°, 100°, 120°, 160°\}$. The sources are assumed to be of equal power and assumed to broadcast binary phase shift keyed (BPSK) waveforms. We use a three-element array of isotropic sensors which is depicted in Figure 9.11. This estimation problem is not solvable with second-order statistics even when one uses minimum-redundancy array concepts (see section 9.4); however, using **VC³** it is possible to extend the effective aperture to seven sensors.

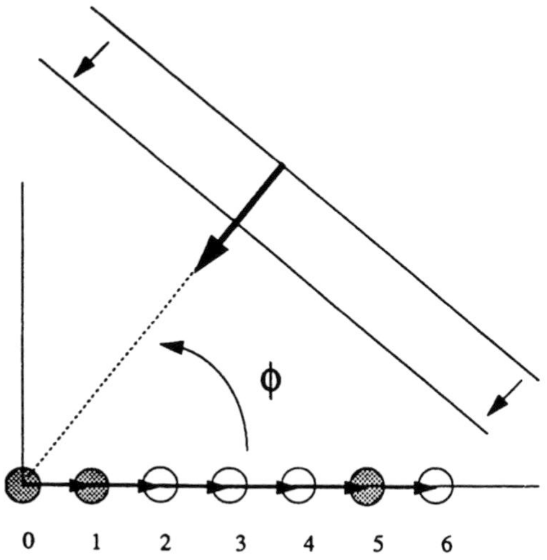

Figure 9.11 Using 3 actual sensors (filled sensors) and **VC³**, it is possible to generate the cross-correlations from a 7-element virtual array. The sensor separation is half-wavelength.

Using cumulants to compute cross-correlations, we form a 7×7 virtual covariance matrix. In forming this matrix, we calculated all the alternatives for computing a cross-correlation and averaged them before subsituting for the cross-correlation. We then applied the MUSIC algorithm to the virtual covariance matrix. Figure 9.12 indicates results from 10 independent realizations at 20 dB SNR, for two different data lengths in spatially white noise. It is clear that the **VC³**-based MUSIC can resolve all the sources in both cases. In addition, the variation of peak locations decreases with increasing data length.

Joint Calibration and Parameter Estimation. We next investigate direction-finding and waveform estimation capability of VESPA using an array of a dipoles. The response of a dipole takes the form $\cos(\phi)$, and is illustrated in Figure 9.13. If the dipole is rotated counterclockwise by α, then its response becomes $\cos(\phi - \alpha)$.

We consider an array of four dipoles with locations $\{(0,0), (0, -\lambda/2), (-\lambda/2, -\lambda), (\lambda/2 - \lambda)\}$ on the x-y plane, with orientations $\{90°, 95°, 87°, 100°\}$. Sensor locations are depicted in Figure 9.14. The steering vector for this ar-

[6] Note the angle measurement convention in Figure 9.11

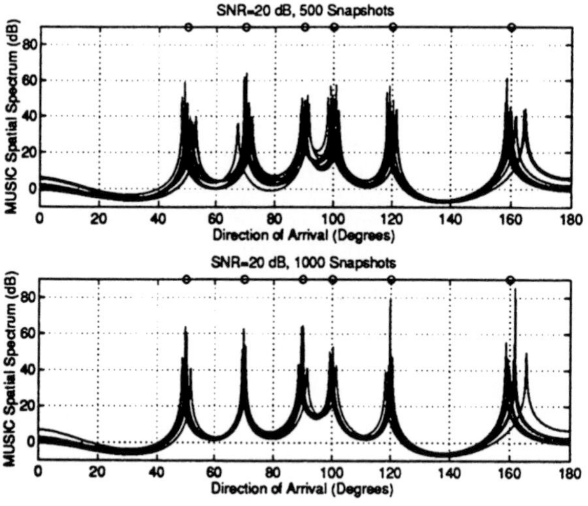

Figure 9.12 Direction finding for more sources than sensors: MUSIC spatial spectrum resolves all of the sources in all 10 realizations. Increase in data length decreases cumulant estimation errors and decreases the variation in peak locations.

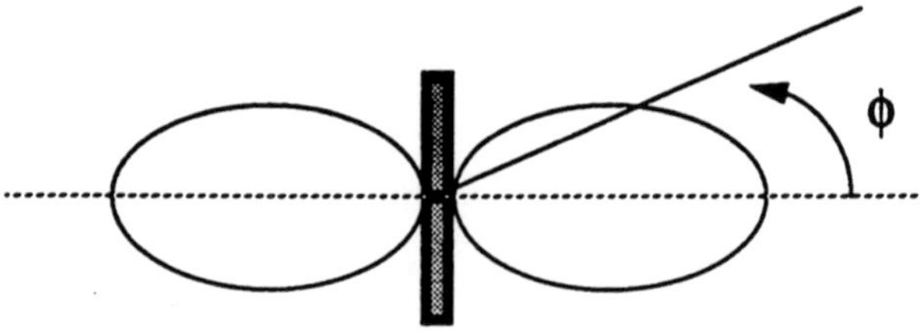

Figure 9.13 Response of the dipole antenna used in the experiment.

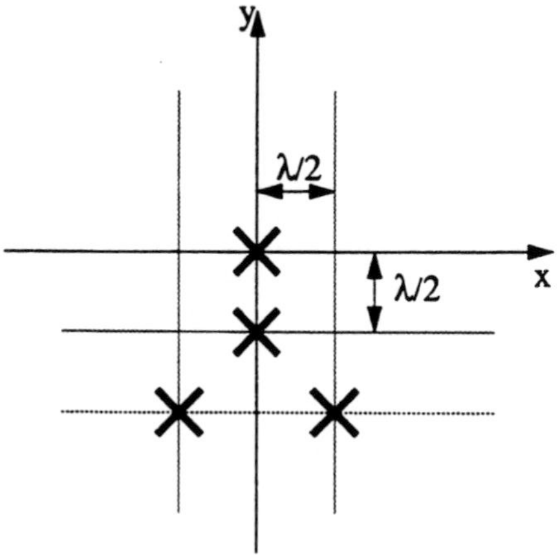

Figure 9.14 Sensor locations for the four-element dipole array.

ray takes the form:

$$\mathbf{a}(\phi) = [\sin(\phi), \sin(\phi - 5^\circ)e^{-j\pi\sin(\phi)}, \sin(\phi + 3^\circ)e^{-j\pi(2\sin(\phi)+\cos(\phi))}$$
$$\sin(\phi - 10^\circ)e^{-j\pi(2\sin(\phi)-\cos(\phi))}]^T \tag{9.42}$$

Direction Finding. We assume that three equal power BPSK sources illuminate the array with directions $\{\phi_1 = 85^\circ, \phi_2 = 90^\circ, \phi_3 = 95^\circ\}$. The bearings are measured counterclockwise as indicated in Figure 9.13. The resulting 4×3 steering matrix has singular values {3.3959, 0.5347, 0.0337}, which indicates that sources are very hard to resolve with the array, that is, the ratio between the extreme singular values is 1/100.

To implement VESPA, we need a doublet that consists of sensors that have identical responses separated with a known displacement vector. To maintain this requirement, we provide a fifth sensor at location $(\lambda/2, 0)$ on the x-axis, with orientation 90°, that is, a twin of the first element in the original four-element array. In this scenario, VESPA uses five sensors, therefore it is called VESPA(5). The subarray size of VESPA(5) is five elements and one sensor is shared between subarrays. Figure 9.15a illustrates the set-up for VESPA(5).

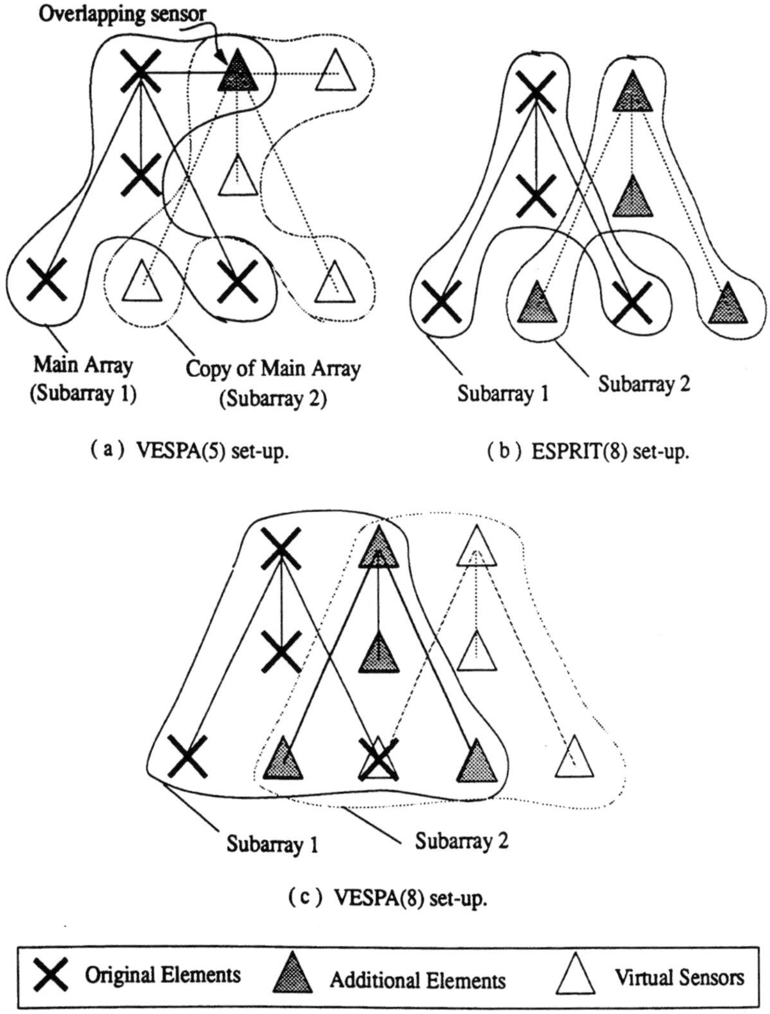

Figure 9.15 Array configurations for (a) VESPA(5): (b) ESPRIT(8): and (c) VESPA(8).

To compare VESPA with the original ESPRIT algorithm, we provided a full copy of the original array to the ESPRIT algorithm. This copy is dispaced $(\lambda/2)\hat{a}_x$ from the original array. ESPRIT uses eight sensors, hence it is called ESPRIT(8). The subarray size of ESPRIT(8) is four elements with no overlap between subarrays. Figure 9.15.b illustrates the set-up for ESPRIT(8).

Finally, to make a reasonable comparison between VESPA and ESPRIT, we let VESPA use all eight channels which are available to ESPRIT(8). We label this scenario as VESPA(8). The only information available to VESPA(8) is the presence of the doublet as in VESPA(5), that is, VESPA(8) does not use the fact that array measurements conform to the ESPRIT data model. Figure 9.15c illustrates the set-up for VESPA(8). The subarray size in VESPA(8) is eight elements, and five elements are common to both subarrays. Since VESPA(8) does not know the fact that the ESPRIT model is available, it does not know the degree of subarray overlap.

White Noise: We first test the three methods in additive, spatially white, circular Gaussian noise. We define the power of a source as the variance of its wavefront measured by an isotropic sensor with unity gain.

Figure 9.16 illustrates the estimates from the three methods for 200 independent realizations using 100 snapshots. It can be observed that ESPRIT(8) failed in general, whereas VESPA(5) and VESPA(8) resolved the sources and provided acceptable estimates.

To investigate the performance of cumulant-based methods, we performed additional experiments by changing both the SNR and the data length. Table 9.1 reports results for 200 and 500 snapshots, with SNR levels of 0, 10, 20, and 30 dB. Each mean and standard deviation (Std) pair in this table is obtained from 500 independent realizations. We note that low-SNR performance (0–20 dB SNR) of VESPA(5) and VESPA(8) are significantly better than that of ESPRIT(8). At 30 dB SNR, ESPRIT(8) performed slightly better than the cumulant-based methods: at low-SNR's noise is the primary factor for separation of signal and noise subspaces for close signals, and at high-SNR's the performance depends on the presence of cross-terms in the estimation of cumulant and covariance matrices. For cumulant-based methods, these unwanted cross-terms decay with increasing data length at a slower rate than covariance-based methods. Similar experiments were carried out for data lengths of 1000 and 2000 snapshots, and are reported in Table 9.2. We observe improvement with increasing data length when SNR $>$ 20 dB, and a dominant behavior of cumulant-based methods in the low-SNR region. Increase in data length does not provide as much improvement at low-SNR (SNR $=$ 0 dB) as it does for high-SNR scenarios.

Colored Noise: We now investigate the perfomance of VESPA(8) in spatially colored noise and compare its performance with that of ESPRIT(8).

Due to the presence of colored noise, ESPRIT(8) will provide biased results. One alternative to overcome this may be to provide the perfect knowledge of noise-covariance matrix to ESPRIT(8) to enable prewhitening of the measurements; however, the prewhitened measurements do not conform to the ESPRIT data model.

The colored noise is generated so that its principal component overlaps with the spatial spectra of the far-field signals. The noise spatial spectrum is illustrated

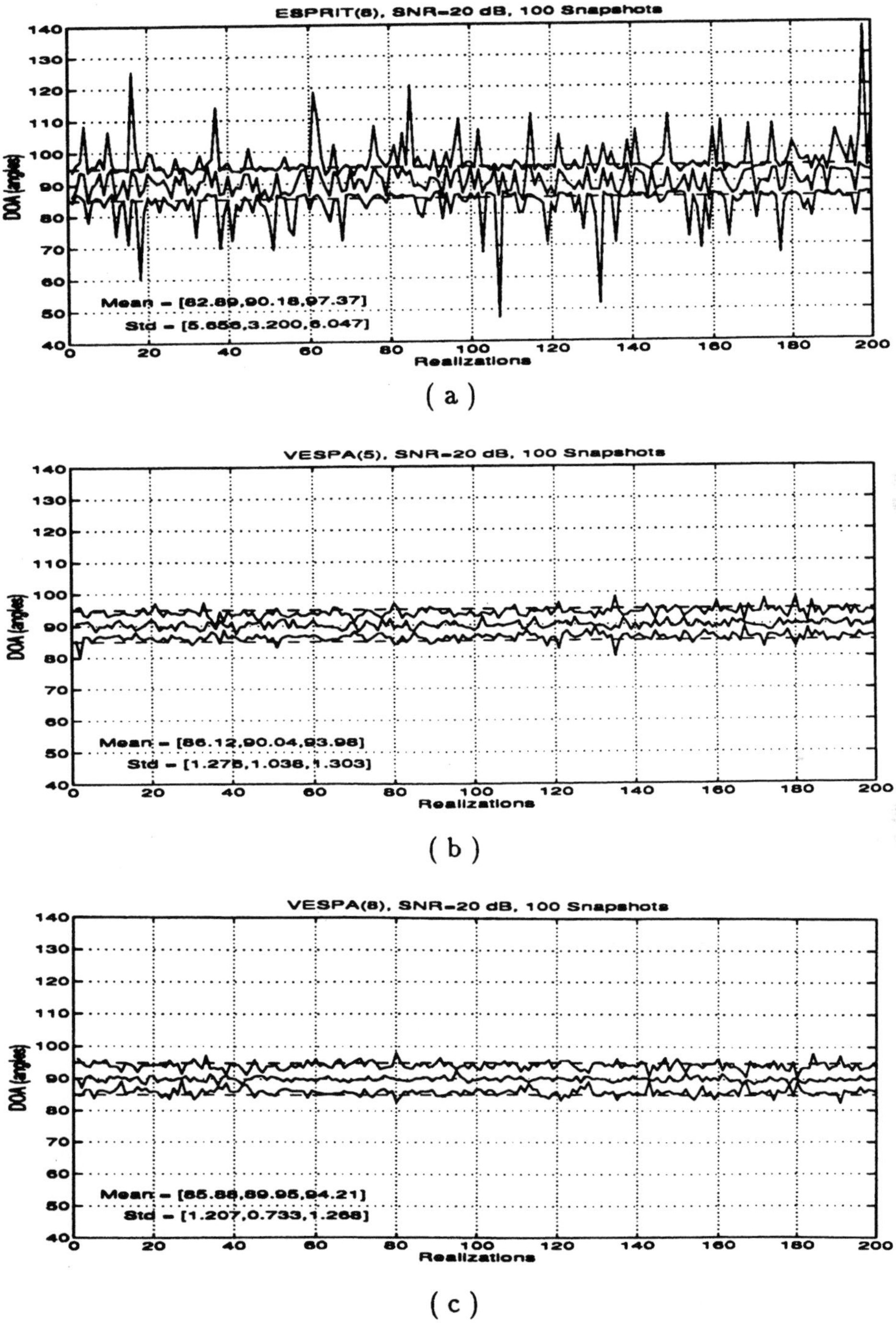

Figure 9.16 Estimates at SNR = 20 dB, $N = 100$; (a) ESPRIT(8) estimates are not acceptable; (b) VESPA(5) resolved sources; (c) VESPA(8) provided estimates with less standard derivation.

TABLE 9.1 Results from 500 Monte-Carlo Runs for Three BSPK Signals in White Noise with 200 and 500 Snapshots

DOA	Method		200 Snapshots SNR (dB)				500 Snapshots SNR (dB)			
			0	10	20	30	0	10	20	30
$\phi_1 = 85°$	VESPA(5)	Mean	83.4099	85.8595	85.8701	84.7064	84.3271	85.9369	85.4293	84.8976
		Std.	4.9298	1.0028	0.9921	0.9439	3.6020	0.7631	0.6560	0.5539
	ESPRIT(8)	Mean	71.9885	71.6384	93.9761	84.9883	70.4335	74.9728	84.7067	84.9994
		Std.	19.7574	20.9615	3.4550	0.5334	21.7166	17.0000	1.4659	0.3394
	VESPA(8)	Mean	84.8958	85.9855	85.0155	84.7630	85.3630	85.9726	84.9238	84.9212
		Std.	2.6868	0.9121	0.9108	0.7787	1.9784	0.5547	0.5435	0.5162
$\phi_2 = 90°$	VESPA(5)	Mean	90.0494	90.0698	89.9968	90.0242	89.9729	89.8865	89.9860	90.0118
		Std.	3.2003	1.8163	1.0002	0.5203	2.9186	1.6900	0.8112	0.3318
	ESPRIT(8)	Mean	89.9651	89.7413	90.0499	89.8953	89.9428	90.1959	90.0041	90.0273
		Std.	4.6123	3.8651	2.9914	1.0534	4.3559	3.9474	2.3730	0.6763
	VESPA(8)	Mean	90.3493	90.0557	89.9363	90.0227	90.0700	89.9152	90.183	90.0104
		Std.	2.5198	1.2815	0.5238	0.4971	2.3795	1.1464	0.3457	0.3286
$\phi_3 = 95°$	VESPA(5)	Mean	96.2618	94.1182	94.2291	95.2839	95.6538	94.1415	94.5853	95.0918
		Std.	4.9411	1.0771	1.0123	0.9703	3.2144	0.7577	0.6264	0.5398
	ESPRIT(8)	Mean	108.7386	105.1270	96.2200	95.0167	107.2382	106.5508	95.3995	95.0057
		Std.	20.3948	17.7241	3.2253	0.5389	20.0022	18.4354	1.4766	0.3259
	VESPA(8)	Mean	95.1700	93.9623	95.0619	95.2237	94.8661	94.0596	95.0844	95.0671
		Std.	2.8559	1.0172	0.9514	0.8386	2.0728	0.5875	0.5139	0.5073

TABLE 9.2 Results from 500 Monte-Carlo Runs for Three BSPK Signals in White Noise with 1000 and 2000 Snapshots

DOA	Method		1000 Snapshots SNR (dB)				2000 Snapshots SNR (dB)			
			0	10	20	30	0	10	20	30
$\phi_1 = 85°$	VESPA(5)	Mean	84.5509	85.9303	85.2081	84.9471	84.7314	85.8948	85.0985	84.9699
		Std.	3.2603	0.6019	0.4462	0.3768	3.3044	0.4658	0.4223	0.2715
	ESPRIT(8)	Mean	73.0389	76.0120	84.8508	85.0048	73.1729	77.8325	85.0204	84.9992
		Std.	19.4379	16.6112	0.9694	0.2444	19.9394	14.1232	0.5994	0.1684
	VESPA(8)	Mean	85.3913	85.9225	84.9432	84.9521	85.5524	85.8856	84.9730	84.9726
		Std.	1.6176	0.4338	0.3872	0.3494	1.4340	0.3042	0.2917	0.2569
$\phi_2 = 90°$	VESPA(5)	Mean	89.8669	90.1814	90.0363	90.0103	90.1357	89.8951	90.0449	89.9919
		Std.	2.7809	1.7362	0.6549	0.2283	2.7833	1.7299	0.4729	0.1635
	ESPRIT(8)	Mean	90.1068	89.9544	90.0619	89.9875	90.1757	89.7641	89.9586	90.0099
		Std.	4.0431	3.7572	1.9355	0.4749	4.0295	3.7143	1.4276	0.3164
	VESPA(8)	Mean	89.8812	90.1058	89.9911	90.0062	89.9852	89.9054	89.9964	89.9918
		Std.	2.4352	1.1438	0.2407	0.2209	2.3509	1.0271	0.1906	0.1599
$\phi_3 = 95°$	VESPA(5)	Mean	95.3994	94.1593	94.8092	95.0506	95.3629	94.1575	94.9347	95.0373
		Std.	3.3417	0.5867	0.5202	0.3749	2.6707	0.4140	0.3885	0.2689
	ESPRIT(8)	Mean	109.9478	103.8032	95.0330	95.0128	107.5084	101.2070	95.0225	95.0025
		Std.	22.6759	15.9755	0.9627	0.2522	19.1691	13.4493	0.3715	0.1653
	VESPA(8)	Mean	94.5979	94.0813	95.0563	95.0436	94.5748	94.1598	95.0329	95.0344
		Std.	1.4775	0.4135	0.3844	0.3500	1.4713	0.3008	0.2848	0.2551

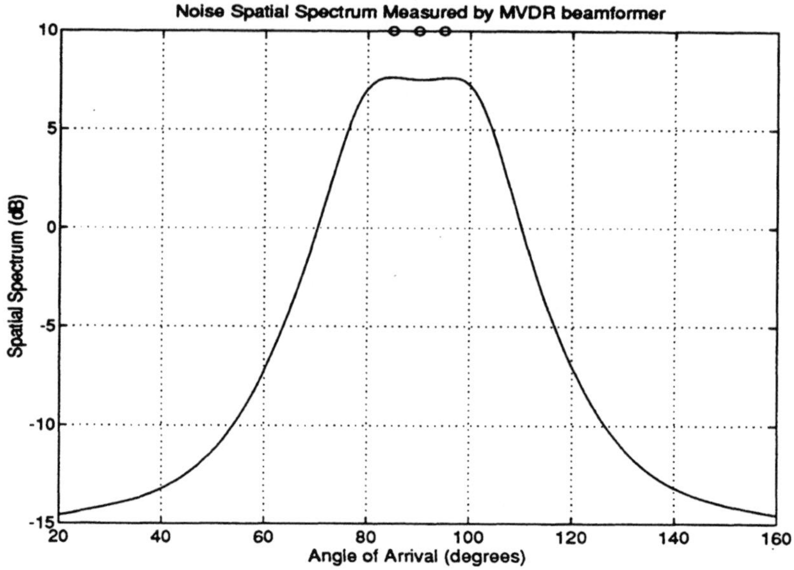

Figure 9.17 Spatial spectrum of measurement noise computed using the MVDR beamformer: we plot $E\{|\omega^H(\theta)\mathbf{r}(t)^2|\}$, in which $\omega(\theta) = \mathbf{R}_n^{-1}a(\theta)/a^H(\theta)\mathbf{R}_n^{-1}a(\theta)$ Here, $a(\theta)$ is an 8-vector that corresponds to the ESPRIT(8) configuration. The signal bearings $\{85°, 90°, 95°\}$ are indicated by circles.

in Figure 9.17. The corresponding noise covariance matrix $\mathbf{R}_n$ is formed as the scaled sum of two covariance matrices $\mathbf{R}_d$ and $\mathbf{R}_w$. $\mathbf{R}_d$ represents the directional-noise component and $\mathbf{R}_w$ represents the white-noise component. More specifically

$$\mathbf{R}_d = \beta_1 \sum_{\theta=60°}^{120°} \alpha((\theta - 90°)/30°)\mathbf{a}(\theta)\mathbf{a}^H(\theta) \tag{9.43}$$

where $\mathbf{a}(\theta)$ corresponds to the steering vector for the ESPRIT(8) configuration, and $\alpha(\theta)$ is a triangular function which is centered at 0° (i.e., maximum value of 1 at 0°) and extends from $-1°$ to $1°$. The constant β_1 is provided to make the trace of $\mathbf{R}_d$ equal to 8 (number of elements).

The white-noise component $\mathbf{R}_w$ carries one-fifth of the power of the directional-component, that is, $\mathbf{R}_w = 0.2\mathbf{I}_8$. The noise covariance matrix is obtained as

$$\mathbf{R}_n = \beta_2(\mathbf{R}_d + \mathbf{R}_w) \tag{9.44}$$

where β_2 is present to make the trace of $\mathbf{R}_n$ equal to 8. We then performed a Cholesky decomposition of $\mathbf{R}_n$, so that $\mathbf{R}_n = \mathbf{L}\mathbf{L}^H$, to generate the measurements

$$\mathbf{r}(t) = \mathbf{A}\mathbf{s}(t) + 10^{-(\text{SNR}/20)}\mathbf{L}\mathbf{n}(t) \tag{9.45}$$

in which $\mathbf{n}(t)$ is complex circular Gaussian noise with zero-mean and covariance matrix $\mathbf{I}_8$.

The data generated using (9.45) is used in VESPA(8). Since ESPRIT(8) yields biased results in colored noise, and does not apply to prewhitened data, we replaced

$\mathbf{L}$ by $\mathbf{I}_8$ in (9.45) for generating the data for ESPRIT(8), so we compare colored noise performance of VESPA(8) with white noise performance of ESPRIT(8).

We performed 200 independent experiments for VESPA(8) in colored noise and ESPRIT(8) in white noise. We let the SNR be 10 dB (see Eq. 9.45), and collected 1000 snapshots. The noise covariance matrices for colored and white noises have the same trace, that is, total noise power introduced to ESPRIT and VESPA are the same. The mean and standard deviation of ESPRIT(8) estimates for this scenario have already been displayed in Table 9.2, for a different set of 500 independent realizations. Figure 9.18 displays the results for the present 200 realizations; it indicates that at this SNR level ESPRIT(8) provides unreliable estimates for data corrupted by only white noise, whereas VESPA(8) provides very good estimates for data corrupted by colored noise.

Waveform Recovery. In this experiment, we compare the performance of the cumulant-based signal recovery techniques and Capon's MVDR beamformer with perfectly known steering vector in terms of the signal-to-interference-plus-noise (SINR) at the output of the processor. We also compare SINR performances of ESPRIT and VESPA that use the same number of sensors.

Capon's MVDR beamformer is a very sensitive processor to mismatches in the steering vector of the source whose waveform is to be estimated. In this experiment, we assume that perfect knowledge of the steering vector is available to the Capon processor to implement (9.38). Even in this case, Capon's MVDR has problems due to the inversion of sample covariance matrix in (9.38).

There are two major techniques to address the mismatches created by inversion of the sample covariance matrix: (1) Robust adaptive beamforming technique [9] which contrains the norm of the weight vector of the MVDR processor to lie in a hypersphere, and (2) Steering vector projection approach of [16] which, as its name implies, first projects the known steering vector on the signal subspace of the sample covariance matrix, and then computes (9.38). Mathematically, if $\mathbf{E}_s$ denotes the $M \times P$ eigenvector matrix for the signal subspace of the sample covariance matrix, and $\mathbf{\Omega}_{ss}$ is the corresponding eigenvalue matrix, then the weight vector of the linear combiner (instead of Eq. 9.38) is computed as:

$$\mathbf{w}_1 = \mathbf{E}_s \mathbf{\Omega}_{ss}^{-1} \mathbf{E}_s^H \mathbf{a} \tag{9.46}$$

in which $\mathbf{a}$ is the known or estimated steering vector of the source of interest.

We decided to investigate the performance improvement offered by the projection method since the hypersphere constraint in the robust beamforming approach prevents directional interference from being *sufficiently* removed from the output.

We consider the array manifold in (9.42) and assume the presence of two BPSK sources from 85° and 90° at an SNR of 30 dB. We are interested in the waveform of the latter source, and regard all other contributions to the measurements as interference.

In the first part of this experiment, we compare a beamformer that uses steering vector estimates provided by VESPA(5) with an MVDR beamformer that uses the true steering vector for the source of interest. We call this processor CUM1. The

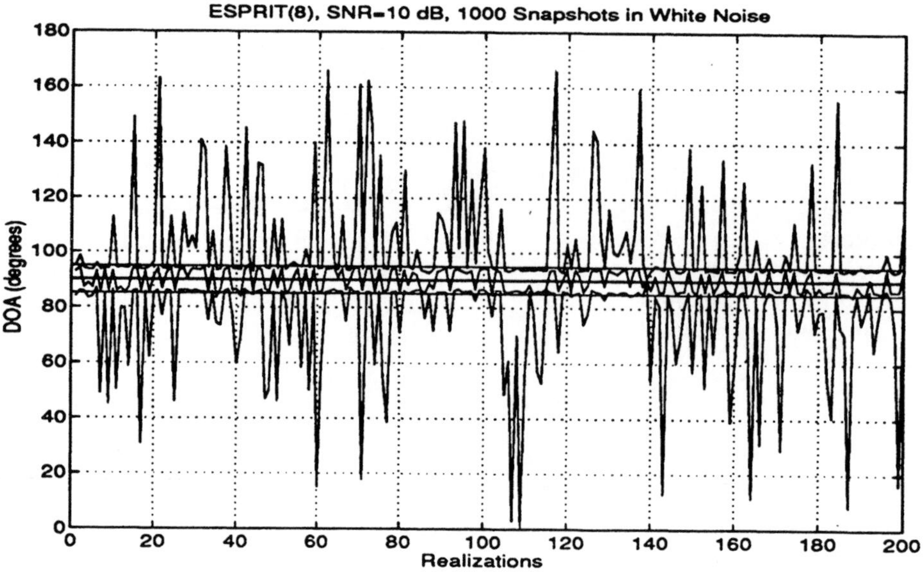

(a)

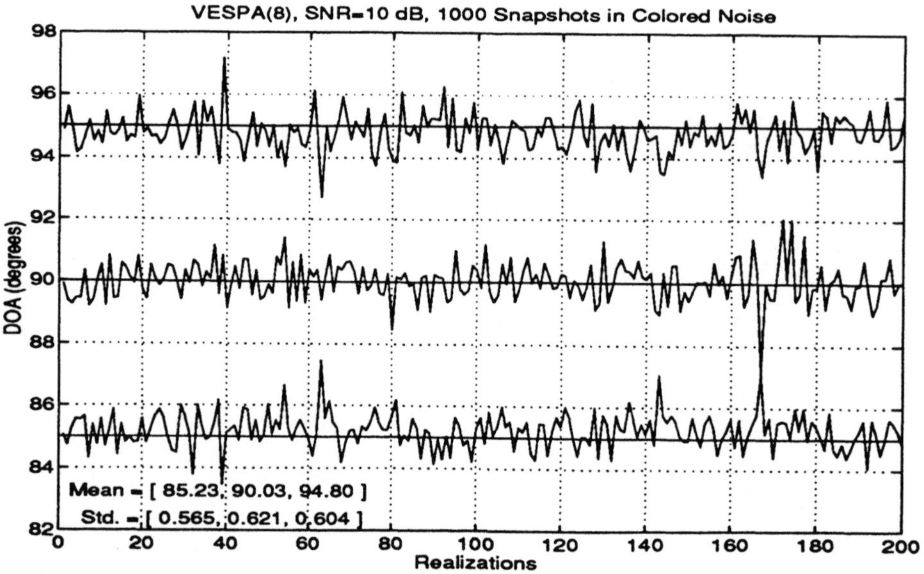

(b)

Figure 9.18 Estimates obtained from (a) ESPRIT(8) in white noise, and (b) VESPA(8) in colored noise. ESPRIT(8) breaks down many times, yielding unreliable estimates.

MVDR beamformer which processes the same measurements as CUM1, but uses the true steering vector for the source of interest, is called MVDR1. We then project the estimated steering vector from VESPA(5) onto the signal subspace as in (9.46), and compute the weight vector of the processor; this is CUM2. Finally, we project the true steering vector onto the signal subspace and compute the weight vector for the processor; this is MVDR2. The maximum possible output-SINR using the five channels is 23.02 dB.

We varied the data length from 50 to 1000 snapshots in 50 sample increments. We performed 500 independent realizations for each data length, computed the output-SINR for each of the four beamformers for each realization, and averaged the results from the realizations. The results are given in Figure 9.19. It is observed that cumulant-based beamformers are superior to MVDR beamformers that utilize perfectly known steering vectors.

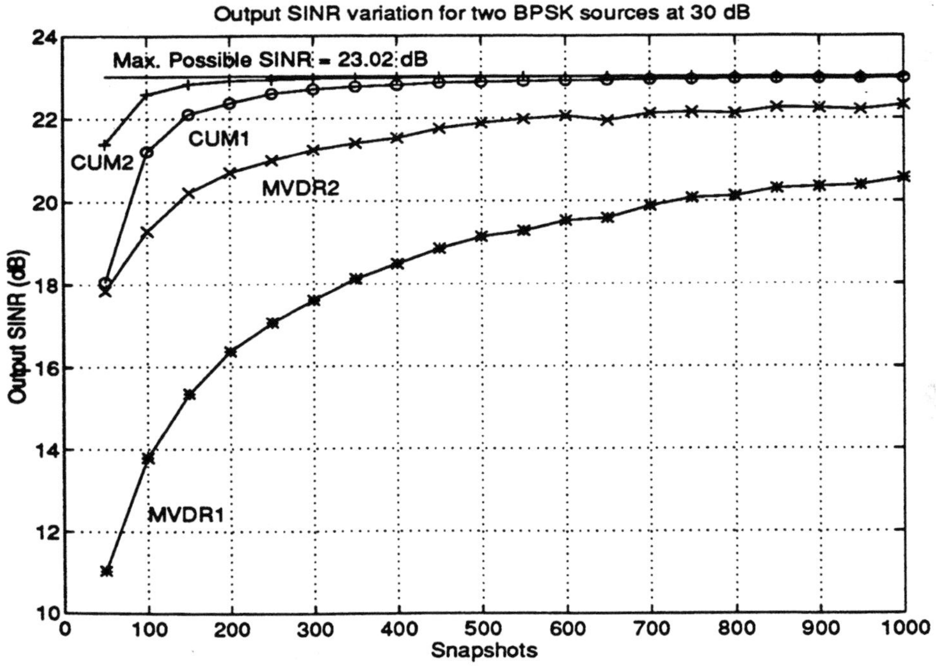

Figure 9.19 Performance comparison for the beamformers with five sensors and two sources illuminating the array with SNR = 30 dB. Cumulant-based beamformers are superior to the MVDR beamformers that use the true steering vector. The advantage is more evident for small number of snapshots.

Next, we provided a full copy of the array so that we have an array of eight elements. Doing this increases the output SINR to 26.37 dB. In this setting, we can also use ESPRIT(8) to estimate the source steering vectors. The beamformer COV1 uses the steering vector estimated by ESPRIT(8) in (9.38) to compute the beamformer weights. COV2 uses the projection method on the estimated steering vector from ESPRIT(8) to compute the beamformer weights. We compare COV1 and COV2 with the four beamformers defined above. The results are given in Figure 9.20. We observe that waveform recovery using ESPRIT(8) is the worst alternative and that cumulant-based beamformers are superior to all others except for the case of 50 snapshots, where CUM1 is inferior to COV2 and MVDR2. We note that, in practice, it is not possible to implement MVDR1 and MVDR2 because we do not know the steering vectors for the sources of interest. Then, as Figure 9.20 illustrates, COV1 and COV2 provide good approximations to MVDR1 and MVDR2, as long as measurements conform to the ESPRIT data model.

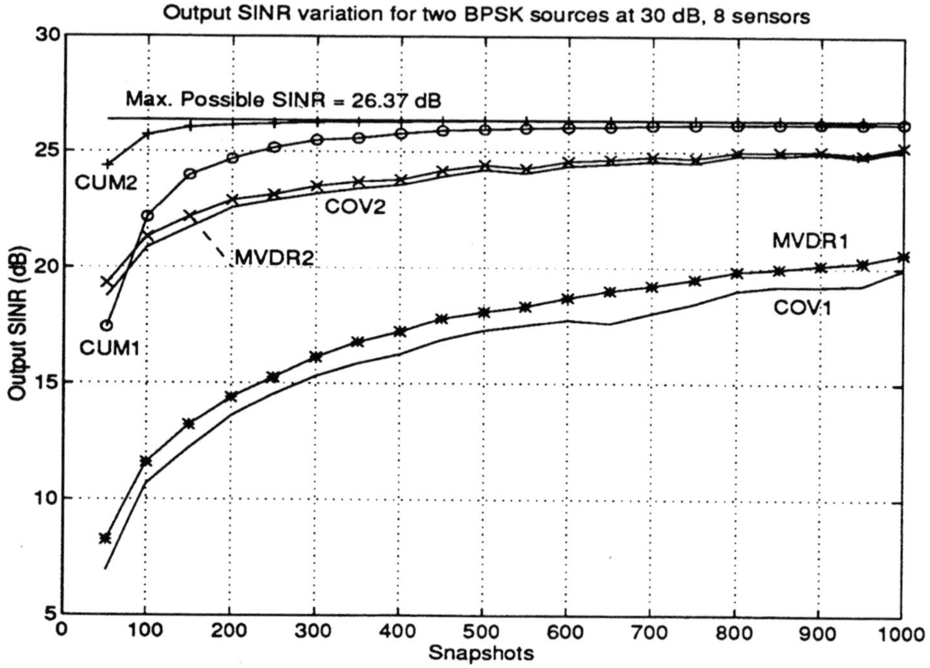

Figure 9.20 Performance comparison for the beamformers with 8 sensors and 2 sources illuminating the array with SNR = 30 dB. Cumulant-based beamformers are superior to the MVDR beamformers that use the true steering vector. The advantage is more evident for small number of snapshots.

Finally, we investigate the advantages of estimating the steering vectors of all sources on the signal recovery performance in an environment where sources may change their power levels. If the steering vectors of jammers are known or they can be reliably estimated, then it is possible to form a null-MVDR beamformer (9.41) that puts perfect nulls on jammers. Such a beamformer will suppress the jammers, even when their powers increase due to the perfect nulling condition. The MVDR beamformer in (9.38) will not be able to null the jammers if their powers increase after the weights are computed. With VESPA, we can first estimate the steering vectors of all sources and then put nulls on jammers. We assume the VESPA(5) configuration (see Fig. 9.15.a), and assume one continuous wave (CW) and one BPSK source at 20 dB SNR, with bearings of 85° and 90° respectively. We estimate the bearings and steering vectors using VESPA(5) with 1000 snapshots and select the source with bearing 90° as the signal of interest. We compute the weights for MVDR and null-MVDR based on steering vector estimates from VESPA(5). Then we collect 1000 more snapshots, during which the power of the CW jammer increased to 50 dB, and process the measurements using the beamformers formed by using the first 1000 snapshots. The beamformer outputs are shown in Figure 9.21 together with the received measurement from the first sensor at the equal SNR case. It is observed that the jammer leaks through the MVDR beamformer and distorts the message in the MVDR beamformer, but null-MVDR already has a perfect null on the jammer, so that it is not very much affected by this change in the experiment condition. The increase in the size of the signal "cloud" in null-MVDR is due the fact that one degree of freedom is used to put a perfect null on the jammer which results in increased white noise gain.

In summary, cumulant-based blind signal recovery offers significant advantages over beamformers that use a perfectly known steering vector for the source of interest. The reason for this is mismatches due to the use of the sample covariance matrix in place of true statistics. In addition, knowledge of the steering vectors of all sources provides additional advantages such as putting perfect nulls on jammers that are robust to changes in jammer power levels.

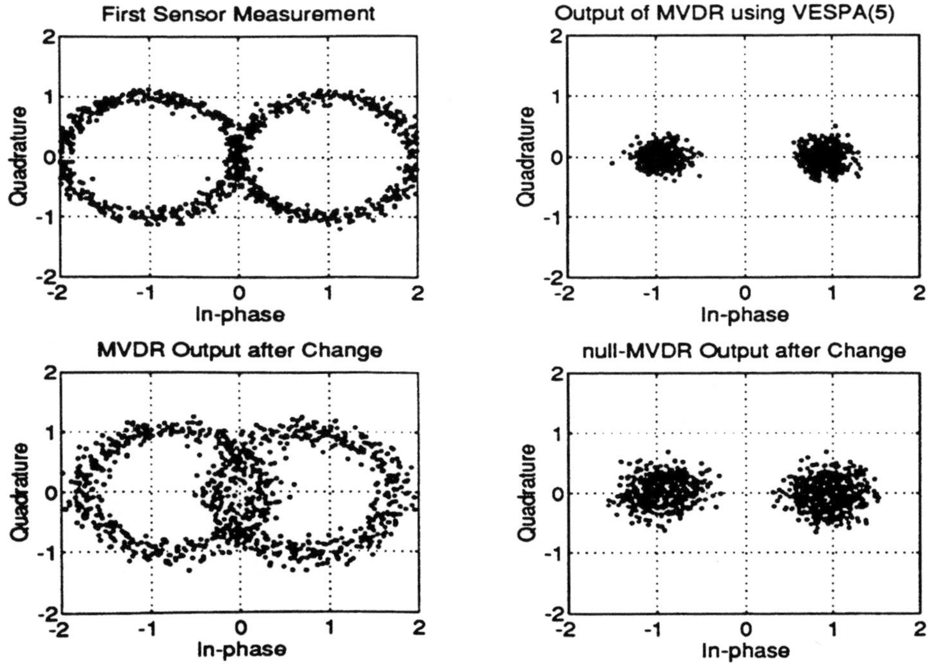

Figure 9.21 CW and BPSK sources illuminate the array: (top left) received signal at the reference sensor; (top right) MVDR output which nulls the CW component; (bottom left) CW jammer increases its power and penetrates through the MVDR processor destroying the BPSK message: (bottom right) null-MVDR nulls the interference with the precomputed weights and recovers the message.

Overview

In this section, we applied virtual cross-correlation computation (section 9.2) to the joint array calibration and direction-finding problem. We established several relationships between the proposed method and exsting work.

The algorithms of this section (DOA estimation, array calibration, and waveform recovery) are asymptotically insensitive to the spatial correlation structure of additive Gaussian sensor noise. Furthemore, our cumulant-based algorithms are *computationally simpler* than their covariance-based counterparts, which require multidimensional search and suffer from local convergence problems [68]. The consistency requirements of cumulant- and covariance-based approaches are found to be similar; but the latter are applicable to a very limited class of scenarios, that is, isotropic sensor arrays [43], whereas the former can calibrate arbitrary arrays using only a doublet of unknown characteristics.

When compared with existing cumulant-based signal recovery algorithms [4, 7, 11, 64, 47], virtual-ESPRIT provides both source bearings in addition to source waveforms with minimal information on the array manifold. We illustrated experiments in which the virtual-ESPRIT algorithm provides bearing estimates that are more accurate than those from the ESPRIT algorithm that uses the same number of sensors. We compared signal recovery performance of cumulant-based beamformers with that of the MVDR beamformer, which uses a perfectly known steering vector, and demonstrated that the cumulant-based signal recovery method which estimates the steering vectors from the statistics of the measurements outperforms the MVDR beamformer, especially when the data lengths are short. Extensions of virtual-ESPRIT are provided in the following discussion.

Extensions

In this subsection, we extend the application of the virtual-ESPRIT algorithm to different scenarios. We start with the use of third-order cumulants for the joint array calibration and direction-finding problem, which in turn leads to a new class of algorithms suitable for fourth-order cumulants. We also extend the virtual-ESPRIT algorithm to wideband signals and near-field source localization problems.

Virtual-ESPRIT with Third-Order Cumulants. In section 9.2, we showed how to compute cross-correlations between sensors by using third-order cumulants. We now extend our results to compute the statistics required by the ESPRIT algorithm by using third-order cumulants and a doublet of guiding sensors.

Let $\{g_1(t), g_2(t)\}$ denote the guiding sensor measurements and consider the location of the first guiding sensor as the reference point. The cross-correlations among the rest of the array measurements $\{r_1(t), r_2(t), \ldots, r_M(t)\}$ can be computed by using our previous results (9.26–9.28) as:

$$E\{r_p^*(t)r_q(t)\} = \frac{\sigma_s^2}{\gamma_{3,s}}\text{cum}(r_p^*(t), r_q(t), g_1(t)) \quad 1 \le p, q \le M \tag{9.47}$$

Our goal is to compute the cross-correlation matrix between the actual sensor measurements and their virtual counterparts denoted as $\{v_1(t), v_2(t), \ldots, v_M(t)\}$. Before proceeding we note that

$$v_p(t) = r_p(t)\exp(-j\vec{k}\cdot\vec{\Delta}) \quad 1 \le p \le M \tag{9.48}$$

and similarly,

$$g_2(t) = g_1(t)\exp(-j\vec{k}\cdot\vec{\Delta}) \tag{9.49}$$

which implies that we can compute the required statistics by using cumulants as follows:

$$\begin{aligned}
E\{r_p^*(t)v_q(t)\} &= E\{r_p^*(t)r_q(t)\}\exp(-j\vec{k}\cdot\vec{\Delta}) \\
&= \frac{\sigma_s^2}{\gamma_{3,s}}\operatorname{cum}(r_p^*(t), r_q(t), g_1(t))\exp(-j\vec{k}\cdot\vec{\Delta}) \\
&= \frac{\sigma_s^2}{\gamma_{3,s}}\operatorname{cum}(r_p^*(t), r_q(t), g_1(t)\exp(-j\vec{k}\cdot\vec{\Delta})) \\
&= \frac{\sigma_s^2}{\gamma_{3,s}}\operatorname{cum}(r_p^*(t), r_q(t), g_2(t))
\end{aligned} \tag{9.50}$$

These expressions indicate that three arguments in a cumulant are sufficient to generate the statistics required by the ESPRIT algorithm. The interpretation of (9.47) and (9.50) is that if we change the location of the so-called common element ($g_1(t)$ in Eq. 9.47) by $\vec{\Delta}$, then we obtain the cross-correlation matrix of the actual array measurements and their virtual counterparts which are dispaced by $\vec{\Delta}$, due to the spatial-nonstationarity of the third-order cumulants. This result indicates that one of the arguments of the fourth-order version of the virtual-ESPRIT may be redundant and it can be used in alternate ways, as we shall see in the next section.

Modifications of Virtual-ESPRIT. We indicated previously that three arguments in a cumulant expression are sufficient to implement the virtual-ESPRIT algorithm; hence, the question is "how can the fourth-argument be utilized when fourth-order cumulants are employed?" The first two parts of the material presented herein are devoted to this issue. The third part describes use of multiple guiding sensors.

Consider an array of M elements that measures $\{r_1(t), r_2(t), \ldots, r_M(t)\}$, and the two guiding sensor measurements, $\{g_1(t), g_2(t)\}$ which are actually obtained from the first two sensors [i.e., $g_1(t) = r_1(t)$ and $g_2(t) = r_2(t)$]. Let $a(t)$ be a linear function of the array measurements (i.e., $a(t) = \mathbf{w}^H\mathbf{r}(t)$)]. Assume there is only one source $s(t)$, and the signal part of $a(t)$ is equal to $\beta s(t)$. The fourth-order cumulant obtained by fixing one of the arguments to be $a(t)$ is functionally identical to the third-order cumulant for direction-finding purposes, that is

$$\operatorname{cum}(a^*(t), g_1(t), r_p^*(t), r_q(t)) = \frac{\gamma_{4,s}\beta^*}{\gamma_{3,s}}\operatorname{cum}(r_p^*(t), r_q(t), g_1(t)) \tag{9.51}$$

because we can view the constant term in (9.51) as a scaling of the new third-order cumulant of the source which does affect the direction-finding performance. Covariances computed by using cumulants as in (9.51) are used to form the autocorrelation matrix required by the ESPRIT algorithm. Similarly

$$\mathrm{cum}\,(a^*(t), g_2(t), r_p^*(t), r_q(t)) = \frac{\gamma_{4,s}\beta^*}{\gamma_{3,s}}\,\mathrm{cum}\,(r_p^*(t), r_q(t), g_2(t)) \tag{9.52}$$

can be used to compute the cross-correlation matrix required by the ESPRIT algorithm. The way to obtain $a(t)$ is important, and it will be discussed in the following.

Increasing Dimensionality. The process $a(t)$ can be selected as one of the M sensor measurements. Let

$$[\mathbf{C}_1(l)]_{p,q} = \mathrm{cum}\,(a^*(t), g_1(t), r_p(t), r_q^*(t))\big|_{a(t)=r_l(t)} \tag{9.53}$$

$$[\mathbf{C}_2(l)]_{p,q} = \mathrm{cum}\,(a^*(t), g_2(t), r_p(t), r_q^*(t))\big|_{a(t)=r_l(t)} \tag{9.54}$$

In this way, there are M selections for $a(t)$. In the original development of virtual-ESPRIT, we let $a(t) = g_1(t)$, which is in turn only a small subset of all possible choices. Of course, the reason behind that selection was to illustrate the operation of the virtual-ESPRIT more clearly. With each selection of $a(t)$ in (9.53) and (9.54), we obtain a third-order virtual-ESPRIT problem. All M of these problems can be combined to provide the direction estimates, if we construct the following $M^2 \times M$ matrices:

$$\mathbf{T}_1 = \begin{bmatrix} \mathbf{C}_1(1) \\ \mathbf{C}_1(2) \\ \vdots \\ \mathbf{C}_1(M) \end{bmatrix} \quad \mathbf{T}_2 = \begin{bmatrix} \mathbf{C}_2(1) \\ \mathbf{C}_2(2) \\ \vdots \\ \mathbf{C}_2(M) \end{bmatrix} \tag{9.55}$$

which take the form

$$\mathbf{T}_1 = \mathbf{B}\mathbf{\Gamma}_{4,s}\mathbf{A}^H \quad \mathbf{T}_2 = \mathbf{B}\mathbf{\Phi}\mathbf{\Gamma}_{4,s}\mathbf{A}^H \tag{9.56}$$

where $\mathbf{B}$ can be viewed as an effective steering matrix with the kth column $\mathbf{b}_k = \mathbf{a}_k^* \otimes \mathbf{a}_k$. $\mathbf{\Gamma}_{4,s}$ is a diagonal matrix that contains the fourth-order cumulants of the far-field sources. The derivation of (9.56) is possible by (9.51) and (9.52). The diagonal matrix $\mathbf{\Phi}$ contains the directional information in the ESPRIT data model. We can use the ESPRIT algorithm to solve for the elements of $\mathbf{\Phi}$ which contain the direction information and the effective steering matrix $\mathbf{B}$, which consists of the effective steering vectors of sources. After the columns of $\mathbf{B}$ are determined, then the M^2-vectors ($\mathbf{b}_k$'s) can be reconfigured in a Hermitian matrix which is rank one (unvec $(\mathbf{b}, M, M) = \mathbf{aa}^H$). Then, the steering vectors can be determined by taking the principal components of these reconfigured matrices. This provides additional smoothing for the estimation of the steering vectors.

Beamforming and Virtual-ESPRIT. In the previous section, we indicated a computationally demanding extension to the original virtual-ESPRIT. An alternate approach which can parallelize the computations is possible by the selection of

$a(t)$ to be spatial-filtered array measurements, that is, $a(t) = \mathbf{w}^H\mathbf{r}(t)$. The weight-vector can be determined (using the MVDR) to suppress all but one of the sources illuminating the array by using the initial estimates of the steering vectors provided by virtual-ESPRIT. Performing this procedure for all of the P sources, we generate P separate, theoretically rank-one virtual-ESPRIT problems in which the cross-correlation matrices between the subarrays can be computed as:

$$\text{cum}\,(a^*(t), g_2(t), r_p(t), r_q^*(t)) \tag{9.57}$$

Generating rank-one virtual-ESPRIT problems is motivated to alleviate the effects of finite-sample estimates of cumulants by suppressing the residual cross-terms between multiple signals which only decay to zero asymptotically. These problems can be solved in parallel for each source of interest.

This technique can be further improved by reducing the dimensionality of the main array by putting null-constraints on all but one of the sources. If $\mathbf{A}$ is an $M \times (P-1)$ matrix that contains the estimates of steering vectors of the undesired sources, and $\mathbf{a}_1$ is the estimate of the steering vector for the desired source, then the modified array measurements can be obtained by the transformation

$$\tilde{\mathbf{r}}(t) = \mathbf{E}^H\mathbf{r}(t) \tag{9.58}$$

where the columns of $\mathbf{E}$ constitute an orthonormal basis for the left-nullspace of $\mathbf{A}$. This transformation cannot suppress the desired signal, since $\mathbf{a}_1$ cannot be represented as a linear combination of the columns of $\mathbf{A}$. Due to the transformation in (9.58), $\tilde{\mathbf{r}}(t)$ is an $(M-P+1)$ vector. The virtual-ESPRIT subarray cross-correlation matrix can be computed as

$$\text{cum}\,(a^*(t), g_2(t), \tilde{r}_p(t), \tilde{r}_q^*(t)) \tag{9.59}$$

Multiple Guiding Sensors. We now investigate how to use multiple guiding sensors in the virtual-ESPRIT algorithm. One approach is to lay the guiding sensors as a minimum-redundancy array so as to generate as many copies of the main array as possible. This is motived by the 2-D MRA design procedure of section 9.4. Figure 9.22 illustrates an example in which four guiding sensors are used to generate six virtual copies of an arbitrary array. The seven subbarrays can be grouped as two overlapping super-subarrays such that each super-subarray contains four subarrays (other grouping options are also possible). The ESPRIT algorithm can be applied with the effective subarray size being equal to the super-subarray size. An alternative approach is to perform beamforming on the virtual covariance matrices as suggested in the previous section (on the received signals). Here we do not have the virtual signals, so we do the processing on the statistics.

In some applications, both azimuth and elevation information is necessary. Modifications of the ESPRIT algorithm (see the introduction of [62] for an extensive summary of algorithms) can be applied if the array contains displaced copies of a subarray for every dimension. We can compute the virtual statistics with $\mathbf{VC}^3$ if we have three guiding sensors. Figure 9.23 illustrates the guiding sensor configuration for this application. For example

$$\text{cum}\,(g_2^*(t), g_3(t), r_k(t), r_l^*(t) = \frac{\gamma_{4,s}}{\sigma_s^2} E\{y_k(t)z_l^*(t)\} \tag{9.60}$$

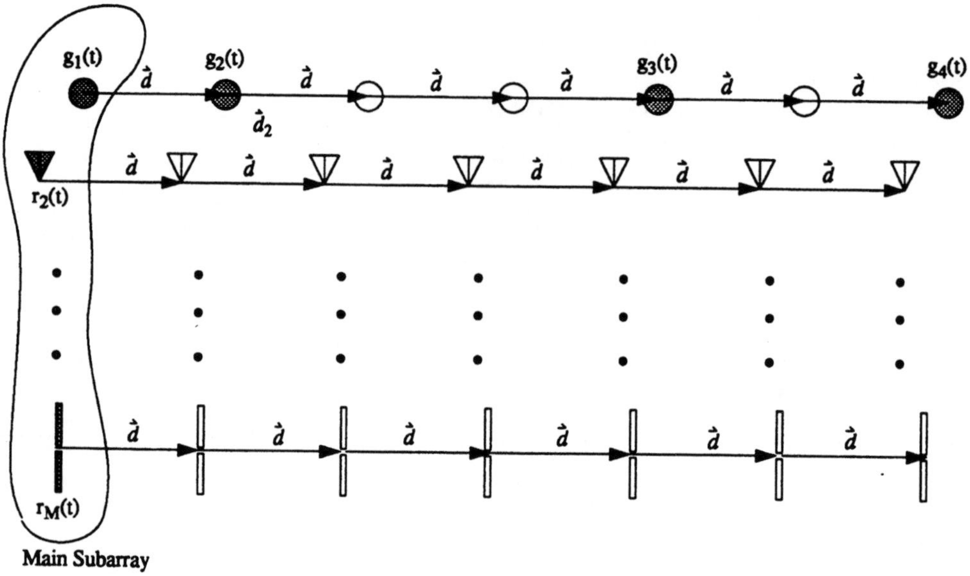

Figure 9.22 A minimum redundacy array of our guiding sensors is used to create six copies of the main array. The filled sensors indicate the actual sensors.

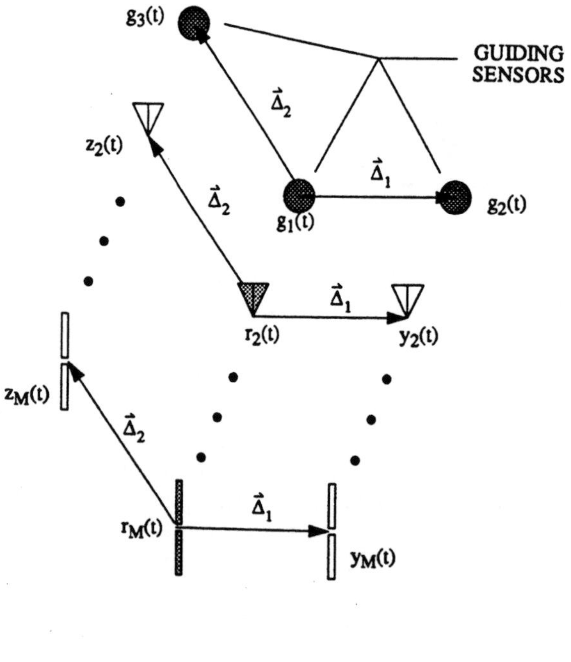

Figure 9.23 Azimuth elevation direction finding is possible if we use $\mathbf{VC}^3$ and three guiding sensors.

Wideband Signals. We next extend the virtual-ESPRIT algorithm to wideband signals. The spectrum of interest can be divided into L narrowbands around the center frequency w_c. Let $r_k(t, \omega_j)$ denote the measurement from the kth sensor at frequency ω_j, and $\{g_1(t, w_c), g_2(t, w_c)\}$ denote the guiding sensor measurements at the center frequency ω_c. Let

$$\begin{aligned} \left[\mathbf{C}_1(\omega_j)\right]_{k,l} &= \text{cum}\,(g_1^*(t, \omega_c), g_1(t, \omega_c), r_k(t, \omega_j), r_l^*(t, \omega_j)) \\ &= \frac{\gamma_{4,s}}{\sigma_s^2} E\{r_k(t, \omega_j), r_l^*(t, \omega_j)\} \end{aligned} \tag{9.61}$$

and

$$\begin{aligned} \left[\mathbf{C}_2(\omega_j)\right]_{k,l} &= \text{cum}\,(g_1^*(t, \omega_c), g_2(t, \omega_c), r_k(t, \omega_j), r_l^*(t, \omega_j)) \\ &= \frac{\gamma_{4,s}}{\sigma_s^2} E\{\upsilon_k(t, \omega_j), r_l^*(t, \omega_j)\} \end{aligned} \tag{9.62}$$

where $\upsilon_k(t)$ is the process recorded by a virtual sensor, which is located at $\vec{d}$ away from the sensor that measures $r_k(t)[\vec{d}$ is the vector from $g_1(t)$ to $g_2(t)]$.

If $\mathbf{A}(\omega_j$ denotes the steering matrix at frequency ω_j, we have

$$\mathbf{C}_1(\omega_j) = \mathbf{A}(\omega_j)\mathbf{\Gamma}_{4,s}\mathbf{A}^H(\omega_j) \tag{9.63}$$

where $\mathbf{\Gamma}_{4,s} \triangleq \text{diag}(\gamma_{4,1}|g_{1,1}|^2, \ldots, \gamma_{4,P}|g_{1,P}|^2)$, in which $g_{1,k}$ represents the response of the first guiding sensor to the kth source at the center frequency. Similarly

$$\mathbf{C}_2(\omega_j) = \mathbf{A}(\omega_j)\mathbf{\Phi}\mathbf{\Gamma}_{4,s}\mathbf{A}^H(\omega_j) \tag{9.64}$$

where $\mathbf{\Phi}$ is the diagonal matrix that contains the direction information (at the center frequency) as in the ESPRIT algorithm [45]. Using all the frequency bins (L of them), we obtain two matrices:

$$\mathbf{T}_1 = \begin{bmatrix} \mathbf{C}_1(\omega_1) \\ \mathbf{C}_1(\omega_2) \\ \vdots \\ \mathbf{C}_1(\omega_L) \end{bmatrix} = \underbrace{\begin{bmatrix} \mathbf{A}(\omega_1) \\ \mathbf{A}(\omega_2) \\ \vdots \\ \mathbf{A}(\omega_L) \end{bmatrix}}_{\triangleq \mathbf{B}} \mathbf{\Gamma}_{4,s}\mathbf{B}^H \tag{9.65}$$

$$\mathbf{T}_2 = \begin{bmatrix} \mathbf{C}_2(\omega_1) \\ \mathbf{C}_2(\omega_2) \\ \vdots \\ \mathbf{C}_2(\omega_L) \end{bmatrix} = B\mathbf{\Phi}\mathbf{\Gamma}_{4,s}\mathbf{B}^H \tag{9.66}$$

Now, we can use the ESPRIT algorithm to solve for the elements of $\mathbf{\Phi}$, which contain the direction information, and the effective steering matrix $\mathbf{B}$, which consists of the steering vectors for individual narrowbands over the spectral band of interest. The dimensionality of the cumulant matrices $\mathbf{T}_1$ and $\mathbf{T}_2$ can be increased along the lines of the previous section. Section 9.6 further investigates the use of higher-order statistics for wideband signals.

Near-Field Direction Finding. Next we extend the virtual-ESPRIT algorithm to the case when the guiding sensors lie in the *Fresnel* region of the sources. This region is intermediate between the true near-field case where the wavefronts are spherical and the true far-field case where the wavefronts are very well approximated as planar.

Let the ith source be located at an unknown range $\mathbf{R}_{i,1}$ from the first guiding sensor (reference point), with an unknown bearing θ_i with respect to the vector joining the two guiding sensors. Figure 9.24 illustrates an M element uniform linear array for the near-field problem, with interelement spacing Δ. The distance from the ith source to the kth sensor is given by applying the law-of-cosines:

$$R_{i,k} = \sqrt{R_{i,1}^2 + (k-1)^2\Delta^2 - 2(k-1)\Delta R_{i,1}\sin\theta_i} \tag{9.67}$$

For sources in the far-field, $R_{i,1} \gg \Delta$, and $R_{i,k}$ is approximated by taking only the first two terms of the binomial expansion of (9.67):

$$R_{i,k} \simeq R_{i,1} - (k-1)\Delta\sin\theta_i \tag{9.68}$$

In many applications, the distances from the array to the emitters are on the order of only a few apertures, and the plane-wave approximation (9.68) is not valid. Retaining an additional term from the binomial expansion of (9.67) leads to an expansion that is quadratic in k:

$$R_{i,k} \simeq R_{i,1} - (k-1)\Delta\sin\theta_i + \frac{((k-1)\Delta\cos\theta_i)^2}{2R_{i,1}} \tag{9.69}$$

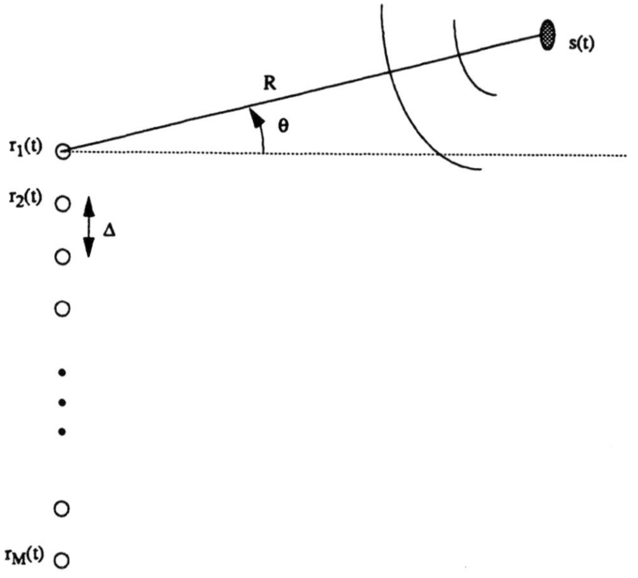

Figure 9.24 A uniform linear array in the near-field localization problem.

Equation (9.69) is called the *Fresnel* approximation; it models the spherical wavefronts as quadratic surfaces in the vicinity of the array; therefore, the received signal model for the linear array can be approximated as:

$$r_k(t) = \sum_{i=1}^{P} s_i(t)\alpha_i^{(k-1)}\beta_i^{(k-1)^2} + n_k(t) \tag{9.70}$$

where

$$\alpha_i(\theta_i) = \exp(-j(2\pi/\lambda)\Delta \sin\theta_i) \tag{9.71}$$

and

$$\beta_i(\theta_i, R_{i,1}) = \exp\left(j(2\pi/\lambda)\frac{(\Delta\cos\theta_i)^2}{2R_{i,1}}\right) \tag{9.72}$$

In this case, the steering vector for the uniform linear array takes the form:

$$\mathbf{a}(\theta_i, R_{i,1}) = \left[1, \alpha_i\beta_i\alpha_i^2\beta_i^4, \ldots, \alpha_i^{(M-1)}\beta_i^{(M-1)^2}\right]^T \tag{9.73}$$

It is interesting to note that the array manifold for the uniform linear array *does not* conform directly to the displacement invariance structure required by the ESPRIT algorithm since the Vandermonde structure of the steering vector is lost. Therefore, even for this simple array configuration a search over the array manifold is necessary. Staser and Nehorai [58] proposed a path-following algorithm for localization of sources using a uniform linear array and parellelized computations. Swindlehurst and Kailath [59] proposed a spatial Wigner-Ville analysis (a heavy computational load) which is followed by the ESPRIT algorithm.

We now investigate how to calibrate arbitrary arrays with the virtual-ESPRIT algorithm using a guiding doublet that measures $\{g_1(t), g_2(t)\}$. If the location of the first guiding sensor is chosen as reference, the response of the doublet can be expressed as

$$[1, \alpha_i\beta_i] \tag{9.74}$$

which follows from viewing the guiding sensors as a uniform linear array with only two elements and using (9.73). If there is only one signal $s(t)$, then the autocovariance matrix required by ESPRIT can be computed as before by using cumulants.

$$\text{cum}\,(g_1^*(t), g_1(t), r_k(t), r_l^*(t)) = \frac{\gamma_{4,s}}{\sigma_s^2}E\{r_k(t), r_l^*(t)\} \tag{9.75}$$

A difference appears in the cross-correlation matrix required by ESPRIT, that is

$$\text{cum}\,(g_1^*(t), g_2(t), r_k(t), r_l^*(t)) = \text{cum}\,(g_1^*(t), g_1(t), r_k(t), r_l^*(t))\,\underbrace{e^{-j(2\pi/\lambda)\Delta\sin\theta_i}}_{\alpha}\,\beta \tag{9.76}$$

Due to near-field effects the additional term β appears in (9.76) instead of α alone. This determines the generalized eigenvalue for the ith source provided by virtual-ESPRIT to be $\lambda_i = \alpha_i\beta_i$ instead of α_i alone; hence, the DOA cannot be estimated as simply as in the far-field case; but the steering vector estimates corresponding to the sources are not affected by this modification (since we assumed no modeling for them), which implies that signal recovery can be done without any difference from the far-field case. Using the definition of λ_i, the direction of the ith source can be found as the minimizer of

$$\hat{\theta}_i = \arg\min_{R,\theta} \| \exp(-j(2\pi/\lambda)\Delta \sin\theta_i) \exp\left(j(2\pi/\lambda) \frac{(\Delta \cos\theta_i)^2}{2R_{i,1}} \right) - \lambda_i \| \tag{9.77}$$

The minimization is implemented for each source separately, and this parallelization provides fast throughput.

Our approach based on virtual-ESPRIT works even when the sources are very close in the near field in which case the *Fresnel* approximation is not valid for the entire array but only for the guiding sensor part of the aperture that consist of only elements.

9.4 MINIMUM REDUNDANCY ARRAY DESIGN FOR CUMULANT-BASED DIRECTION FINDING

Minimum-redundancy arrays (MRA) have long been used to increase the effective aperture of uniform linear arrays [32, 40]. These arrays have been designed with tools from number theory and numerical search algorithms. MRA design exploits the redundant structure of the uniform linear array for independent sources. MRA designs provide an effective aperture proportional to the square of the number of actual sensors, but the resulting arrays are constrained to be uniform and linear.

We proposed a cumulant-based approach to increase the effective aperture of antenna arrays in section 9.2. We showed how cumulants can be substituted for the cross-correlantions between actual and so-called virtual sensors, to form a virtual covariance matrix which resembles the array covariance matrix, as if the measurements from the virtual sensors were available. In this way, the effective aperture of antenna arrays can be increased without geometrical constraints, unlike MRA designs. This effective aperture increase method is used to calibrate arbitrary antenna arrays using a doublet, and the resulting algorithm is called the virtual-ESPRIT algorithm in section 9.3. In section 9.4, we established upper and lower bounds for the size of effective aperture.

The MRA design methods and the cumulant-based approach of section 9.2 increase the effective aperture of an array; therefore, it is a natural expectation that combining the two viewpoints should lead to considerable improvements over the results offered by these viewpoints individually. Toward this objective, we first describe MRA design concepts based on second-order statistics. Then, we propose cumulant-based minimum-redundancy array designs. We show that it is possible to design both single and two-dimensional MRA's using cumulants with apertures proportional to the fourth power of the number of actual sensors. We describe several design methods and compare them with existing results. The section concludes our work on array design with final observation and comparisons.

Bounds on Aperture Extension

We derive lower and upper bounds for cumulant-based virtual-aperture extension as described in section 9.2. We do this to enumerate the maximum number of independent sources whose directions can be identified by an array of identical sensors at arbitrary locations. Our interest in identical sensor arrays stems from their important role in the design of minimum-redundancy antenna systems. The bounds are derived by letting the sensors have no area in the physical space, that is, by representing them by dots; however, all the results apply to the real case, where the sensors consume a volume in three-dimensional space because the virtual aperture created by using actual sensor outputs and cumulants does not occupy any area either.

Lower Bound

Definition 9.1. Consider an antenna array of M identical sensors, where the set of vectors $\{\mathbf{r}_k\}_{k=1}^{M}$ denote sensor locations. The *diameter* d of such an array is defined as the maximum distance between any pair of sensors, that is

$$d \triangleq \max_{1 \le i,j \le M} \|\vec{r}_i - \vec{r}_j\| F \tag{9.78}$$

Definition 9.2. A pair of sensors is called *limit-point sensors* if the distance between them is identical to the diameter (d) of the array. For a given array, there may be more than one pair of limit-point sensors.

Fact 9.1. For a given antenna array configuration, let $(\vec{r}_1, \vec{r}_2)$ denote the locations of a limit-point pair. The rest of the sensors must lie in the intersection of the following two spheres:

$$\begin{aligned} &\mathbf{S}_1 : \text{centered at } \vec{r}_1, \text{ with radius } d \\ &\mathbf{S}_2 : \text{centered at } \vec{r}_2, \text{ with radius } d \end{aligned} \tag{9.79}$$

Proof. If there is a single sensor that does not lie in the intersection of spheres defined above, then its distance to at least one of the limit-point sensors must be larger than d. But this implies that the diameter of the array must be larger than d, which is a contradiction. Figure 9.25 gives an illustration. The intersection of spheres $\mathbf{S}_1, \mathbf{S}_2$ that contain all the actual sensors is defined as the *region of support*. ■

Theorem 9.1. Using the measurements from a limit-point pair, it is possible to extend the effective aperture form M to at least $2M - 1$, regardless of array geometry. The uniform linear array satisfies this lower bound.

Proof. Suppose that the limit-point sensor measurements are used as in the virtual-ESPRIT algorithm and let the vector from the first limit-point sensor to the second one be $\vec{d}$. Then, it is possible to generate a virtual array which is the shifted version of the actual array using $\mathbf{VC}^3$. The virtual sensors do not coincide with the actual sensors except for one of the limit-point sensor positions, due to Fact 9.1;

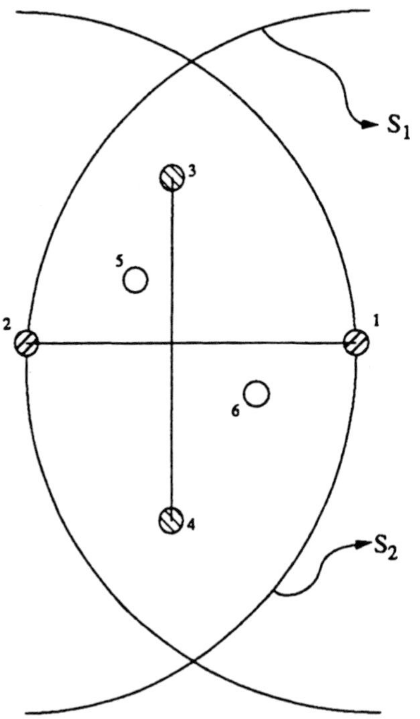

Figure 9.25 An antenna array of identical elements: the pairs (1 2) and (3,4) constitute the limit-point sensors. All sensors must lie within the intersection of the two spheres. This intersection is defined as the region of support.

hence, the effective aperture can be extended to $2M - 1$ sensors; see Figure 9.26 for an illustration. ■

Consider the fully redundant linear of N isotropic sensors in Figure 9.27. The most distant point from the origin that we can reach by two vector additions is $(2N - 2)$, (since the maximum length vector among the available sensors in $(N - 1)$ units long), that is, the effective aperture consists of $2N - 1$ sensors.

Comment 9.1. The aperture extension result for the uniform linear array was reported in [55] by using Carathedory's theorem. This result can be easily obtained by using $\mathbf{VC}^3$.

Comment 9.2. Theorem 9.1 indicates that we can extend the aperture to at least $2M - 1$ sensors when we know which sensors are the limit-point sensors. This raises the issue of whether it is possible to accomplish this extension without this knowledge. We answer this question affirmatively in the next subsection, in the context of a direction-finding method proposed by Porat and Friedlander [42] (an equivalent method is described by Cardoso [5]).

Upper Bound. Consider an M element array that measures the signals $\{r_1(t), r_2(t), \ldots, r_M(t)\}$, and consider cumulant matrices defined as

$$[\mathbf{C}(a, b)]_{k,l} = \text{cum}\,(\underbrace{r_a^*(t), r_b(t)}_{\text{guiding sensors}}, r_k(t), r_l^*(t)) \quad 1 \leq a, b, k, l, \leq M \tag{9.80}$$

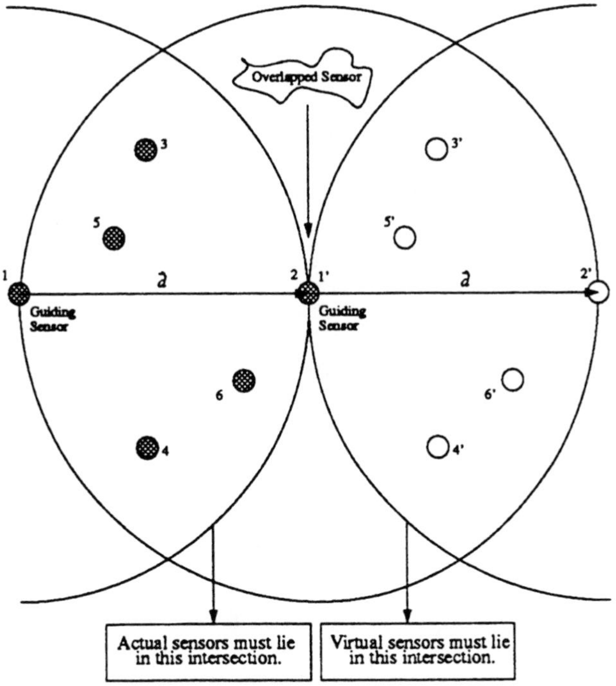

Figure 9.26 Aperture extension by using the limit-point sensors (sensors 1 and 2) as guiding sensors as in virtual-ESPRIT. Actual and virtual sensors must lie in distinct support regions due to Fact 9.1, and the two support regions intersect only at the guiding sensor 2. Each of the virtual sensors (e.g., 5′) is obtained by shifting an actual sensor (e.g., 5) by the vector d.

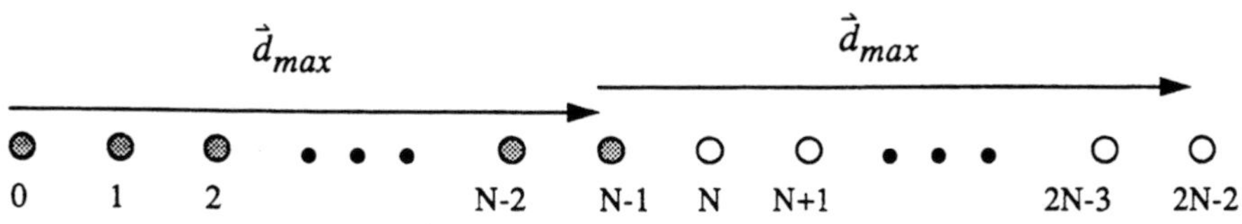

Figure 9.27 A fully redundant linear array (N filled sensors), and virtual sensors that can be generated by cumulants ($N - 1$ empty sensors).

Hence, $\mathbf{C}(a, b)$ is equivalent to the cross-correlation matrix of the actual array and its virtual copy which is shifted by the vector from $r_a(t)$ to $r_b(t)$. For example, in Figure 9.26, if we select $a = 1$ and $b = 2$, then $\mathbf{C}(a, b)$ is the cross-correlation matrix between the displayed actual array and its virtual copy. Similarly, $\mathbf{C}(a, a)$ is equivalent to the autocorrelation matrix.

Porat and Friedlander [42] propose to form the following $M^2 \times M^2$ matrix:

$$\mathbf{C} = \begin{bmatrix} \mathbf{C}(1,1) & \mathbf{C}(1,2) & \dots & \mathbf{C}(1,M) \\ \mathbf{C}(2,1) & \mathbf{C}(2,2) & \dots & \mathbf{C}(2,M) \\ \vdots & \vdots & \ddots & \vdots \\ \mathbf{C}(M,1) & \mathbf{C}(M,2) & \dots & \mathbf{C}(M,M) \end{bmatrix} \tag{9.81}$$

If a_k denotes the steering vector of the kth source, then $\mathbf{C}$ can be decomposed as

$$\mathbf{C} = \sum_{k=1}^{P} \gamma_{4,k}(\mathbf{a}_k^* \otimes \mathbf{a}_k)(\mathbf{a}_k^* \otimes \mathbf{a}_k)^H \tag{9.82}$$

where $\gamma_{4,k}$ denotes the fourth-order cumulant of the kth source.

The following interpretation of the matrix $\mathbf{C}$ is important for determining the upper bound. We fix one sensor (e.g., $r_1(t)$ in Figure 9.28) to be one of the guiding sensors, and pick one of the remaining four sensors [say $r_4(t)$] as the first guiding sensor, and extend the aperture as in virtual-ESPRIT using the matrix $\mathbf{C}(4,1)$ [which shifts the actual array by the vector from $r_4(t)$ to $r_1(t)$]. Then, pick another sensor from the remaining three sensors [say $r_2(t)$], and use it as the second guiding sensor by computing $\mathbf{C}(2,1)$. Continuing this we obtain the first block-column of $\mathbf{C}$ in (9.81). Of course, cross-correlations between these two sets of virtual sensors must be computed. For example, the cross-correlation matrix between these two sets of virtual sensors is computed by estimating $\mathbf{C}(2,4)$ and using the fact that $\mathbf{C}(4,2) = \mathbf{C}^H(2,4)$. This explains the remaining block-components of $\mathbf{C}$ in (9.81).

Fact 9.2. The effective aperture provided by $\mathbf{C}$ does not depend on the choice of the reference sensor. This is obvious from the structure of the effective steering vectors, which take the form:

$$\mathbf{a}^*\otimes\mathbf{a} = \left[a_1^*a_1, a_1^*a_2, \dots, a_1^*a_M, a_2^*a_1, a_2^*a_2, \dots, a_2^*a_M, \dots, a_M^*a_1, a_M^*a_2, \dots, a_M^*a_M\right]^T \tag{9.83}$$

Suppose that we change the labels on the first and second sensors, that is, the second sensor is named as the first, etc. Then, the effective steering vector for this labeling takes the form:

$$\tilde{\mathbf{a}}^*\otimes\tilde{\mathbf{a}} = \left[a_2^*a_2, a_2^*a_1, \dots, a_2^*a_M, a_1^*a_2, a_1^*a_1, \dots, a_1^*a_M, \dots, a_M^*a_2, a_M^*a_1, \dots, a_M^*a_M\right]^T \tag{9.84}$$

The elements of the vector in (9.84) are a permutation of the elements of the effective steering vector in (9.83), which means a relabeling of the actual sensors result in a relabeling of the virtual sensors.

Theorem 9.2. The effective aperture provided by $\mathbf{C}$ is at least $2M - 1$ elements. This result is valid regardless of the selection of the reference sensor among the M actual sensors.

Proof. Fact 9.2 provides a way to prove Theorem 9.2. The choice of the reference point does not affect the effective aperture; hence, we can choose one of the limit points at the reference sensor. When the other limit-point sensor is chosen

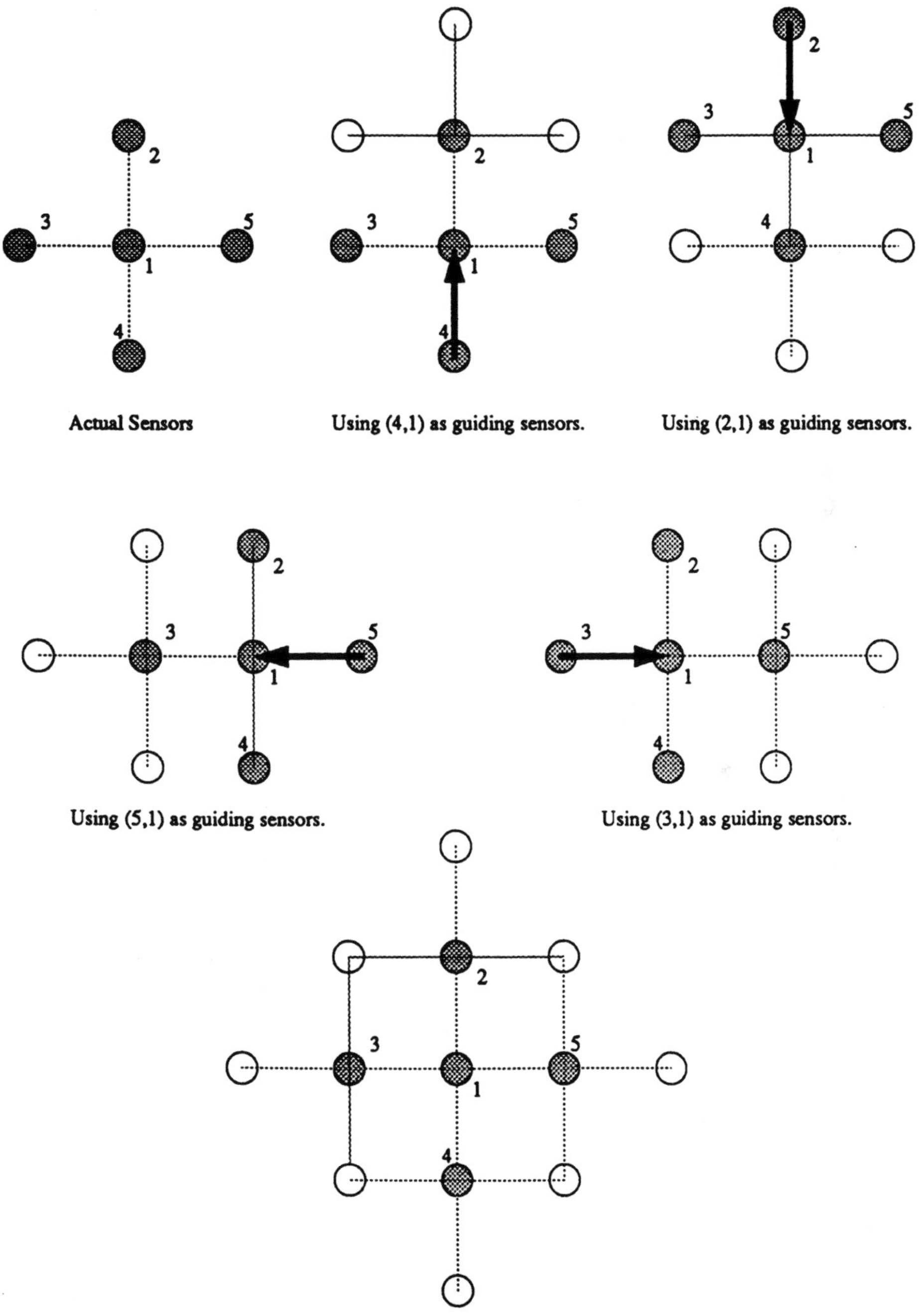

Extended Aperture Using Cumulants

Figure 9.28 An interpretation of the matrix construction proposed in (9.81). We select 1 as the reference sensor, and shift the actual array by the vectors from the other actual sensors to the reference sensor to obtain the effective aperture. Note that there exist virtual sensors at all actual sensor locations and this redundancy decreases the capacity to extend the effective aperture.

as a guiding sensor to shift the array, the number of the resulting virtual sensors that do not overlap with the actual ones will be $M-1$, as stated in Theorem 9.1, and as displayed in Figure 9.26. Then, the effective aperture consists of at least $2M-1$ elements. An alternate proof based on array geometry and on the interpretation of cumulants is provided in [12]. ■

Theorem 9.3. The effective aperture provided by **C** can be at most M^2-M+1 elements.

Proof. After choosing the reference sensor (this choice does not affect the result, due to Fact 9.2), we shift the array as in virtual-ESPRIT by using the reference sensor as a guiding sensor and picking another sensor [say $r_a(t)$] from the remaining $M-1$ sensors as the other guiding sensor. With each such shift, there is at least one overlap between the virtual sensor locations and the actual ones [the reference sensor position coincides with the virtual sensor location created by the shift of $r_a(t)$]. Hence, each shift adds at most $M-1$ virtual sensors. We can make $M-1$ such shifts, resulting in a total of $(M-1)^2 = M^2-2M+1$ virtual sensors. Adding the number of actual sensors, M, to the number of virtual sensors, we obtain the upper bound for the number of effective sensors as M^2-M+1.

The number of the effective sensors clearly depends on the array geometry, but it is always lower bounded by $2M-1$. The result reported in Theorem 9.3 is also claimed in [5] without proof. The remaining question is whether there exist a class of arrays which always achieve the upper bound. We shall provide an affirmative answer in the context of two-dimensional array design later in the section. ■

Minimum Redundancy Arrays (MRA)

The structure of the array covariance matrix in the case of incoherent sources illuminating a uniformly spaced linear array of identical sensors has led to algorithms that can estimate the directions of more sources than sensors. This section is devoted to reviewing this concept within the framework developed in the previous sections for the role of cumulants in array processing.

Consider a uniformly spaced antenna array of M identical sensors which is illuminated by P incoherent sources with waveforms $\{s_1, \ldots, s_P(t)\}$, where

$$x_k(t) = \sum_{l=1}^{P} s_k(t)\exp(-j(k-1)\pi\sin(\theta_l)) + n_k(t) \tag{9.85}$$

where d is the sensor spacing, and the noise components $\{n_1(t), \ldots, n_M(t)\}$ are uncorrelated and have the same variance σ^2. The restrictions on the noise color are necessary for covariance-based processing. In (9.85), we assume the sensor separation is a half-wavelength to eliminate additional parameters. We also assume that noise components are independent of signals. After these assumptions, the cross-covariance between sensor outputs can be expressed as

$$E\{x_m(t)x_n^*(t)\} = \sum_{k=1}^{P} \sigma_k^2 \exp(-j(m-n)\pi\sin(\theta_l)) + \sigma^2\delta(m-n) \tag{9.86}$$

where σ_k^2 is the power of the kth far-field source, and $\delta(m-n)$ is the Kronecker delta function, which is unity if and only if $m=n$. Clearly, (9.86) indicates that the *cross-covariance* between two sensors can be interpreted as an *integer* which is the difference of their locations, that is, $E\{x_m(t)x_n^*(t)\} = r_{m-n}$. This is also obvious from our previous results: covariance can be interpreted as a vector between two sensors, and when the sensors are constrained to lie on a line, the sign and the magnitude of this vector is sufficient to represent cross-covariance. We can extend this interpretation to fourth-order cumulants: cumulants are addition of two "covariance" integers.

The steering vectors for the uniform linear array take the form

$$\mathbf{a}(\theta) = [1, \exp(-j\pi \sin(\theta)), \ldots, \exp(-j(M-1)\pi \sin(\theta))]^T \tag{9.87}$$

and as a result, the steering matrix possesses a Vandermonde structure. Furthermore, since signals are independent and because of this structure of the steering vectors, the array covariance matrix $\mathbf{R}$ is Toeplitz, that is

$$\mathbf{R} = \begin{bmatrix} r_0 & r_1^* & r_2^* & \cdots & r_{M-1}^* \\ r_1 & r_0 & r_1^* & \cdots & r_{M-2}^* \\ r_2 & r_1 & r_0 & \ddots & \vdots \\ \vdots & \vdots & \vdots & \ddots & \vdots \\ r_{M-1} & r_{M-2} & r_{M-3} & \cdots & r_0 \end{bmatrix} \tag{9.88}$$

which implies that if we can compute the set of covariances $\{r_0, r_1, \ldots, r_{M-1}\}$, then we can reconstruct $\mathbf{R}$ due to the Toeplitz property. An arbitrary $M \times M$ covariance matrix has $M(M+1)/2$ parameters due to the fact that $\mathbf{R} = \mathbf{R}^H$. When the covariance matrix is constrained to be Toeplitz, the number of free parameters reduces to M. It is possible to remove some of the M sensors, and still be able to compute $\mathbf{R}$ from the second-order statistics of the remaining sensor measurements. An example is given in Figure 9.29, in which a seven-element uniform linear array is reduced to four elements. The covariance matrix can still be computed using the measurements of the remaining four sensors; therefore, there is redundancy in the uniform linear array, that is, we do not need all the sensors in order to compute the set of necessary covariances. The problem is to eliminate the redundancy by removing the maximum number of sensors while still being able to compute the covariance matrix of the original array.

The problem of eliminating redundancy by exploiting the special signal model in (9.85) and the array covariance matrix (9.88) can be restated as the following combinatorial problem [27]: "Represent $1, 2, \ldots, N$ by differences of $M (M < N)$ non-negative integers drawn from non-negative natural numbers up to N." In this problem, N corresponds to the number of sources that can be resolved. The mathematical description of this problem is as follows:

Find a set of integers $\{d_1, d_2, \ldots, d_M\}$ such that every positive integer $k (0 < k \leq N)$ can be represented in the form $k = d_i - d_j$ with the constraint

$$0 = d_1 < d_2 < \cdots < d_M = N \tag{9.89}$$

The set of integers that satisfies (9.89) is called the *restricted difference basis*

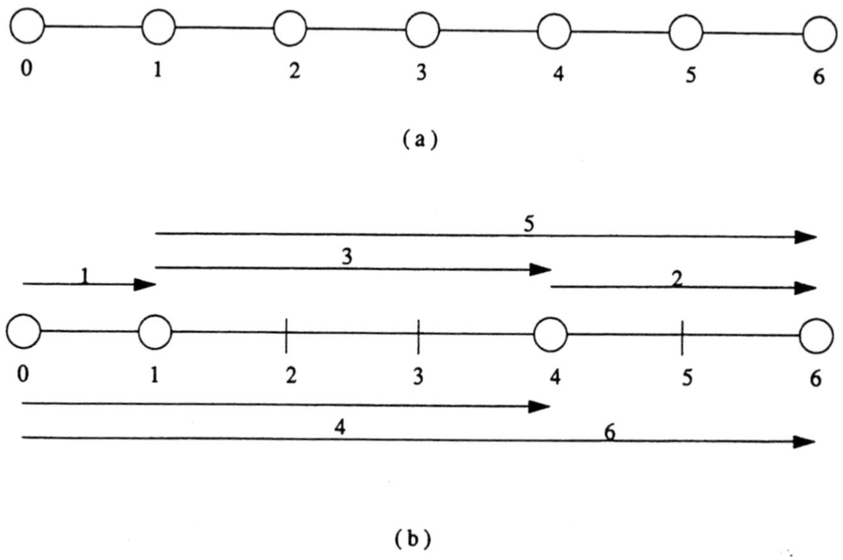

Figure 9.29 Elimination of redundant sensors in a uniform linear array where identical sensors are separated by $\lambda/2$: (a) Redundant array with 7 sensors; (b) nonredundant array obtained by removing 3 sensors; as shown, it is possible to generate all integers from zero to six, which means that the $7 \times$ covariance matrix of (a) can be constructed using only 4 sensors.

with respect to N [27]. The minimum number of integers used to construct a restricted difference basis with respect to N is denoted as M_N. In the array processing context, M_N corresponds to the minimum number of sensors to estimate the directions of N sources, and the difference basis indicates the sensor locations. In engineering applications, the convese problem is stated more frequently as follows: given M sensors, design an array to achieve N_M which denotes the length of the longest restricted difference basis, or the maximum number of sources whose bearing can be estimated by the array. Figure 9.29 gives an example for the case with $N = 6$ in which the restricted difference basis $\{0, 1, 4, 6\}$ is used. Table 9.3 illustrates the optimal locations of the sensors for minimum redundancy up to $M = 17$ [39]. These locations were obtained by exhaustive search methods whose running time grows exponentially with the size of the array.

Finding the minimum number of elements to construct restricted sets and their locations has been of interest for a long time and results are published in different contexts [15, 21, 23, 25–27, 29, 32, 40, 63, 66]. The first important result is that given M integers, it is possible to generate at most $\binom{M}{2} = M(M-1)/2$ integers that satisfy (9.89) which are not necessarily unique; hence, an upper bound for N_M is:

$$N_M \leq \frac{M(M-1)}{2} \tag{9.90}$$

Bracewell [1] showed that the strict inequality holds for $M > 4$. Leech [27] provided the asymptotic result

TABLE 9.3 Interelement Spacing for Optimal Restricted Difference Basis Determined by Exhaustive Search [39] (M is the Number of Sensors and N_M is the Array Length). Note that the First Sensor is Always Located at the Origin ($d_1 = 0$)

M	N_M	Interelement Spacing	M^2/N_M
3	3	•1•2•	3.0
4	6	•1•3•2•	2.667
5	9	•1•3•3•2•	2.778
6	13	•1•1•4•4•3•	2.769
7	17	•1•1•4•4•4•3•	2.882
8	23	•1•3•6•6•2•3•2•	2.783
9	29	•1•4•4•7•7•3•2•1•	2.793
10	36	•1•4•4•7•7•7•3•2•1•	2.778
11	43	•1•4•4•7•7•7•7•3•2•1•	2.814
12	50	•1•4•4•7•7•7•7•7•3•2•1•	2.88
13	58	•1•4•3•4•9•9•9•9•5•1•2•2•	2.914
14	68	•1•1•6•6•6•11•11•11•5•5•3•1•1•	2.882
15	79	•1•1•6•6•6•11•11•11•11•5•5•3•1•1•	2.848
16	90	•1•1•6•6•6•11•11•11•11•11•5•5•3•1•1•	2.844
17	101	•1•1•6•6•6•11•11•11•11•11•11•5•5•3•1•1•	2.861

$$2.434\ldots \leq \lim_{N\to\infty} \frac{M_N^2}{N} \leq \frac{375}{112} = 3.348\ldots \tag{9.91}$$

More important, Pearson et al. [39] proved the following result by providing a *constructive method* to generate a restricted difference basis: for any given $M > 3$ (M is the number of sensors), it is always possible to chose "a" restricted difference basis such that

$$M^2/N < 3 \tag{9.92}$$

which implies $M^2/N_M < 3$. Note that this is not just an asymptotic result. For $M = 3, M^2/N_M = 3$ (see Table 9.3). We refer the reader to [39, or 41, chap. 2] for the proof of (9.92). It is important to note that asymptotically it is not possible to find configurations that use 10 percent fewer elements than the design procedure of [41, chap. 2]. This follows from the lower bound (9.91) of Leech [27].

There are also improved search procedures for finding the optimal configuration of antennas. A recent approach [46] uses a numerical annealing method for the search and provides identical results with the exhaustive search method up to $M = 17$. Additional results are also given for $M > 17$, but there is no proof that these designs are optimal.

The minimum redundancy array concept is only applicable to linear arrays, which can only provide the azimuth of the far-field sources. In many applications, both azimuth and elevation information is necessary; this requires at least a two-dimensional array configuration. Bracewell [1] has noted that there is no two-dimensional analog to the minimum redundancy arrays except for the case of a four-element "T" shaped array. Because of the lack of a solution, two-dimensional resolution can only be obtained by linear minimum redundancy arrays if the array is physically rotated. For example, in astronomy the *earth rotation synthesis* technique of [48] exploits the rotation of the earth with respect to the far-field source, which in turn changes the orientation of the linear array. The collection times are very large, and the performance depends on the actual direction of the source. In addition, during the rotation time of the array, the source is assumed to be stationary. These assumptions will never hold in small-scale applications such as radar and sonar.

Next we provide methods to design two-dimensional minimum redundancy arrays based on the interpretation of cumulants detailed in section 9.2, and already existing design procedures for linear minimum redundancy arrays (see Table 9.3). We also provide methods to design cumulant-based minimum redundancy linear arrays which can significantly outperform the existing covariance-based designs.

Cumulant-Based MRA Design

In this subsection, we first design two-dimensional minimum redundancy arrays based on the existing one-dimensional covariance-based designs. We then concentrate on cumulant-based designs for linear arrays.

Two-Dimensional Arrays

Suppose we wish to design a rectangular array so that both azimuth and elevation estimation is possible through the use of subspace rotation techniques (ESPRIT). Specifically, if we have $M_x M_y$ elements, then it is possible to generate a rectangular grid of length M_x and width M_y and put the sensors in the intersection points. For example, if we have 49 sensors, then a square array can be generated by using the intersections of the grids in Figure 9.30. To implement a direction-finding algorithm, all the concections between the sensors must be completed by a single vector between actual sensors. Any vector in the rectangle can be decomposed into two parts: x-component and a y-component. If one uses fourth-order cumulants, then it is possible to add two vectors between actual sensors to obtain virtual cross-correlations; hence, the 2-D design problem reduces to designing two linear minimum redundancy arrays which can cover the axes of the rectangle. This design was described in the previous section (the restricted difference basis). Figure 9.30 illustrates the design procedure with four actual sensors per axis, where one sensor is common to the two axes. It is possible to extend the aperture to 49 sensors by using 7 actual elements. Note that this design breaks the bound of the existing linear minimum redundancy arrays (see Bracewell's bound in (9.90), which upper bounds the number of resolvable sources to be 21).

Our approach provides both two-dimensional resolution and more virtual sensors than covariance-based designs. Given a number of actual sensors, the best placement is the one in which the extended aperture will be close to a square. The reason is that the perimeter of a rectangle is proportional to the number of actual elements, and the area of a rectangle is proportional to the number of sensors in the extended aperture. The minimum redundancy problem requires us to maximize the aperture (area) while the number of actual sensors (perimeter) is kept constant. The solution is the special rectangle, namely, the square. If the number of actual sensors M is odd, then we can use $(M+1)/2$ sensors per axis to form a linear minimum redundancy array, whose length is lower bounded by $(M+1)^2/12$, which

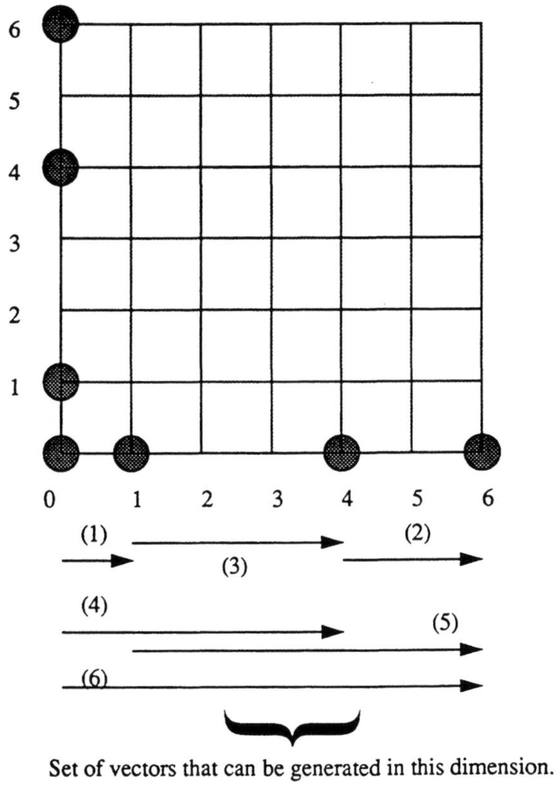

Figure 9.30 2-D minimum redundancy array design: combining two 1-D arrays and using $\mathbf{VC}^3$ allows the computation of cross-correlations among all the 49 grid points, using the data available just from the 7 real sensors (filled circles).

comes from the constructive method of Pillai [41]. The total number of elements in the square aperture is the square of this number; hence, the effective aperture is lower bounded by $(M+1)^4/144$. We describe the general *design steps* for the cumulant-based two-dimensional, rectangular MRA placement method (CUM-REC) given the number of sensors (M):

1. If M is even, let $M_x = (M+2)/2$ and $M_y = M/2$. If M is odd, let $M_x = M_y = (M+1)/2$.
2. Decompose the two-dimensional design problem into two separate linear array design problems, with number of sensors equal to M_x and M_y, respectively.
3. Use Table 9.3, or the extended results reported in [46], to find the solutions to the linear array design problems. Let $N[M_x]$ denote the effective number of sensors with M_x actual sensors, and define $N[M_y]$ similarly.
4. Put the array with M_x actual elements on the x-axis with the locations obtained from the previous step. Put the array with M_y actual elements on the y-axis such that the first actual elements of both arrays coincide.
5. It is now possible to compute the cross-correlations between any pair of sensors located at the grid points of the rectangular structure of area $N[M_x]N[M_y]$ by using the $\mathbf{VC}^3$. An example was given for the $M=7$ case in Figure 9.30.

Based on the design procedure just described, we compare the covariance-based linear MRA (COV-1D) and the cumulant-based two-dimensional MRA (CUM-REC) in Table 9.4 for a small number of actual sensors. In this table, CUM-ALL indicates the effective aperture generated by the sensors designed for CUM-REC, but whose measurements are processed by forming the matrix **C** in (9.81). An example with four actual sensors is given in Figure 9.31 for CUM-ALL. In Table 9.4, COV-1D refers to the effective sensors from the one-dimensional covariance-based MRA design (Table 9.3). Observe that when the number of available sensors in less than 7, CUM-REC provides less sensors than CUM-ALL, and does not provide a "dramatic" improvement over COV-1D. CUM-REC results in an array whose effective number of sensors is less than the upper bound $(M^2 - M + 1)$ without the MRA design concept, if $M < 7$, for example, with six actual sensors the effective number of sensors from CUM-ALL and CUM-REC are 31 and 28, respectively. The reason for this is that the partitioning of the available elements reduces the number of sensors per axis to less than four, which in turn causes an inefficiency by insisting on a rectangular-shaped array. If we can guarantee that there exists a class of array configurations that meet the upper bound $M^2 - M + 1$, then we can use the matrix

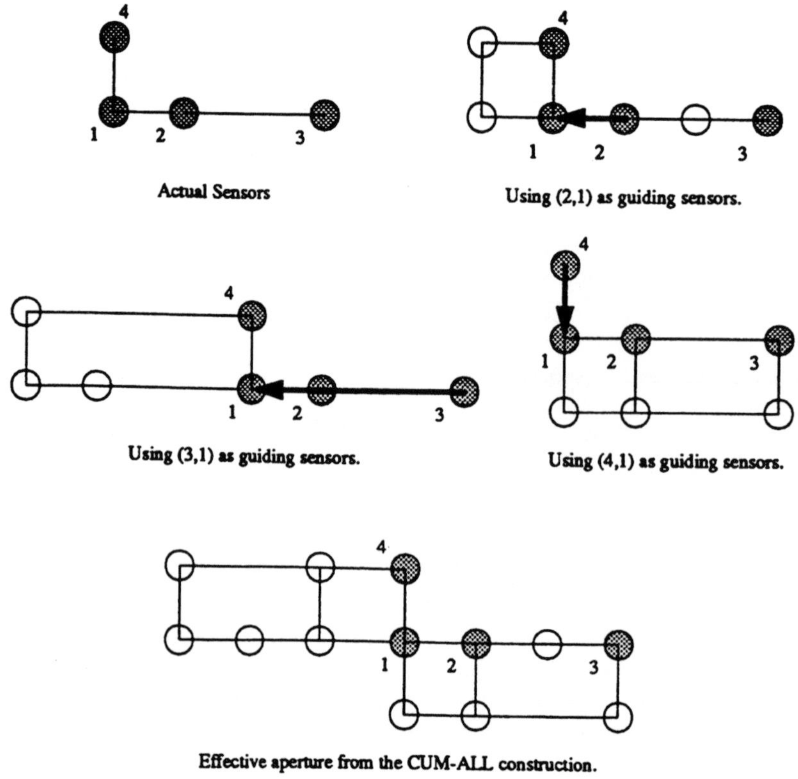

Figure 9.31 An example for aperture extension with CUM-ALL: there are 4 actual sensors: $M_x = 3$ and $M_y = 2$ from the CUM-REC procedure. Sensor 1 is selected as the reference sensor.

TABLE 9.4 Comparison of Aperture Extension for Small Number of Actual Sensors

Actual Sensors	3	4	5	6	7
CUM-REC	4	8	16	28	49
CUM-ALL	7	13	21	31	43
COV-1D	4	7	10	14	18

C defined in (9.81) to increase the aperture for $M < 7$. We now show that Costas arrays achieve the upper bound.

Definition 9.3. For each positive integer M, construct an $M \times M$ permutation matrix with the property that for all possible $x - y$ shift combinations at most one pair of ones coincide with the unshifted matrix [8]. The resulting structure is called a *Costas array*. An alternate definition is to construct an $M \times M$ permutation matrix with the property that the $\binom{M}{2}$ vectors connecting two 1's of the matrix are all distinct vectors [20]. An example is given for the case $M = 6$ in Figure 9.32.

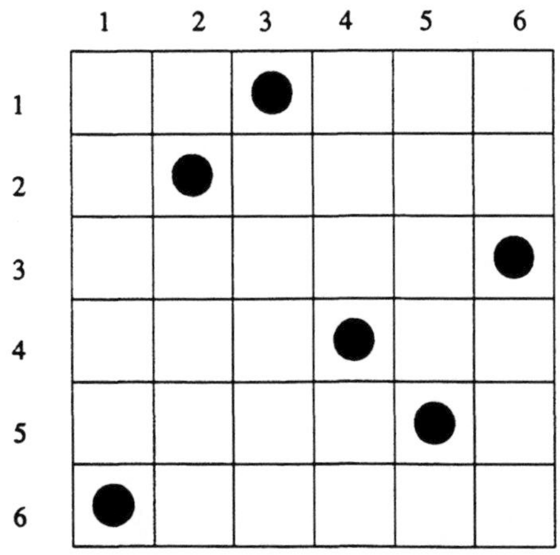

Figure 9.32 An example of a Costas array of order six.

Theorem 9.4. If the M sensors are placed at locations determined by a Costas array of order M, then the effective aperture consists of sensors $M^2 - M + 1$ for the direction-finding method based on processing **C** in (9.81).

Proof. The virtual array is obtained by shifting the actual array by vectors between the actual sensors. In Theorem 9.3, we indicated that in each shift operation, one of the virtual sensors coincides with the reference sensor, regardless of the sensor that is chosen to shift the array. Therefore, by definition of the Costas arrays, there can be no other overlaps, that is, each shift creates $M - 1$ more virtual sensors. There are $(M - 1)$ possible shifts, hence $(M - 1)^2$ virtual sensors that do not overlap with each other and the actual sensors. When we add to $(M - 1)^2$, the number of actual sensors M, we obtain $M^2 - M + 1$. ■

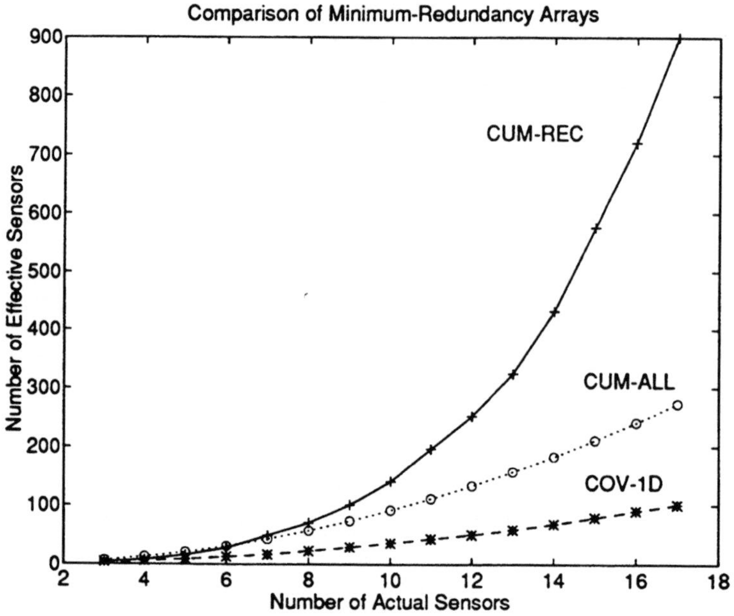

Figure 9.33 CUM-REC outperforms other designs and provides two-dimensional resolution. CUM-ALL also represents the upper bound of section 4.1.2.

An extended comparison of CUM-ALL, CUM-REC, and COV-1D is provided in Figure 9.33.

Linear Arrays

An initial attempt to design linear MRA's based on cumulants is to take an existing design from Table 9.3, and recall that cumulants are addition of two integers, and can be used to double the aperture. Therefore, to obtain an effective aperture of N units long, we need to look for a covariance-based design for an aperture $N/2$ units long; hence, replacing N by $N/2$ in (9.92), we determine the upper bound for this simple cumulant-based procedure as

$$\frac{M^2}{N} < 3 \cdot \frac{1}{2} \tag{9.93}$$

Equation (9.93) implies that the effective aperture (N) is still proportional to the square of the number of actual sensors. Cardoso [5] also claimed that linear MRA's can be designed with an effective aperture of $M^2 - M + 1$ sensors, which is slightly better than (9.93). Cardoso did not prove whether or not the resulting effective aperture has "holes," that is, whether it constitutes sensors configured as a uniform linear array with no missing sensors.

Of course, these results are somewhat primitive because we already showed that it is possible to design two-dimensional MRA's (a harder problem than the linear array problem) whose effective aperture is proportional to the fourth power of the number of the actual sensors.

For illustrative purposes, we start with a simple method which generates similar primitive results without using existing covariance-based MRA results.

We know that cumulants involve the addition of two integers in the linear array problem; hence, if we have all the odd numbers, then we can generate even numbers by adding existing odd numbers. Let us place our sensors with this idea in mind: let the first sensor be the reference element; put the second sensor one unit to the right of the first one; put the third sensor three units to the right of the second one; put the fourth sensor five units to the right of the third sensor; etc. With six sensors, this placing method results in the following array:

$$\bullet\ 1 \bullet 3 \bullet 5 \bullet 7 \bullet 9 \bullet \tag{9.94}$$

with the sensor locations $\{0, 1, 4, 9, 16, 25\}$. This method actually puts the sensors at the squares of integers. This observation can be proved as follows: if M actual sensors are available, then the location of the last sensor is

$$\sum_{k=1}^{M-1}(2k-1) = (M-1)^2, \quad 1 \leq M \tag{9.95}$$

The next thing to show is that the effective aperture that can be created using cumulants has *no holes* in it, at least for the range $0 < k < (M-1)^2$ (end points are guaranteed to be sensor locations, by design). This can be proved by induction as follows: for $M = 1$ or $M = 2$, it can be proved by observing (9.94). Then, for any M, we observe that the last two elements of the M element array are located at $\{(M-2)^2, (M-1)^2\}$, and these elements are separated by 2M-3 units (just subtract the numbers). By construction [see (9.94)], we have the odd numbers $\{1, \ldots, 2M-2\}$ available between the actual sensors. Now, all the even (odd) numbers between $\{(M-2)^2, (M-1)^2\}$ can be generated by adding the appropriate number from the set of odd numbers to $(M-2)^2$ if M is odd (even). Similarly, all the odd (even) numbers between $\{(M-2)^2, (M-1)^2\}$ can be obtained by subtracting the appropriate number from the odd numbers set from $(M-1)^2$ if M is odd (even). For example, if $M = 6$ (even), the last two sensors are at 16 and 25. An odd number in this range can be generated by adding an integer between the actual sensors [see (9.94)] to 16, for example, $21 = 16+5$. Similarly, for the even numbers, we subtract from 25, for example, $22 = 25 - 3$. This completes the proof that the effective aperture has no holes in it for $0 < k < (M-1)^2$.

Actually, this simple design can extend the aperture to more than $(M-1)^2$ sensors. For example, using a four-element design with sensor locations $\{0, 1, 4, 9\}$, it is possible to create effective sensors at the locations $\{0-14, 16-18\}$, with a sensor missing at 15 (a total of 18 sensors). Our aim was not to find the exact number of effective sensors for this design, but to show how to reach claims made at the beginning of this section.

We now propose a cumulant-based linear MRA design method (CUM-LIN) that uses results from covariance-based designs, and achieves an effective aperture proportional to the fourth power of the number of actual sensors. The proposed method is actually a mapping of our two-dimensional array design method to a single dimension. The CUM-LIN method is described below:

1. Given M actual sensors, divide them into two groups: the first group has M_x sensors and the second group has M_y sensors. If M is even, then let $M_x = (M+2)/2$ and $M_y = M/2$. If M is odd, then let $M_x = M_y = (M+1)/2$. This step is identical to the first step of the two-dimensional method; however, the ordering $M_y \leq M_x$ is important in the linear MRA design problem and is explained in the last item of this list.
2. Given M_x actual sensors, design a one-dimensional MRA based on the results presented in Table 9.3. Let $N[M_x]$ denote the effective length of the array (one less than the number of sensors) using second-order statistics.
3. By adding two vectors between the M_x actual sensors located in the previous step (using cumulants of measurements), it is possible to extend (double) the aperture to have a length of $2N[M_x]$. Therefore, if we put new sensors at multiples of $2N[M_x]+1$, then we can generate all integers using the M_x element array of Step 2 and these new sensors. For example, if $k = l(2N[M_x]+1)+m$, where $0 \leq k$ and $0 \leq m \leq 2N[M_x]$, then k can be represented by the addition of two vectors between actual sensors; if $m \leq M_x$, then m can be obtained as the difference between the locations of the M_x element array of Step 2; if $M_x < m \leq 2N[M_x]$, then we rewrite $k = (l+1)(2N[M_x]+1) - (2N[M_x]+1-m)$, where the second term $(2N[M_x]+1-m)$ is not larger than $N[M_x]$ and hence can be obtained as the difference between the locations of the M_x element array of Step 2.
4. The next step is to minimize the number of sensors which are separated by $2N[M_x]+1$ units, but maintain the maximum length possible. This is no different from a covariance-based MRA design problem with M_y sensors, but the separation between each sensor is a *superunit*, which is defined as $2N[M_x]+1$ units. The first element of this array coincides with the first element of the M_x array. The design can be done by using Table 9.3. Let $N[M_y]$ be the length of the MRA array from Table 9.3.
5. Clearly, it is possible to obtain an array of length $N[M_y](2N[M_x]+1)$ by using cumulants and the actual sensors deployed as stated in the previous steps. We can also generate integers from $N[M_y](2N[M_x]+1)$ to $N[M_y](2N[M_x]+1)+N[M_x]$ by adding vectors from the M_x element array to the last element of the M_y element array which is located at $N[M_y](2N[M_x]+1)$. Finally, the difference between the last element of the M_x element array (with location $N[M_x]$) and the second element of the M_y element array (located at $2N[M_x]+1$) is $N[M_x]+1$, which can be added to the last element of the M_y element array to obtain an effective aperture of length $L[M_x, M_y] = N[M_x](2N[M_y]+1)+N[M_y]+1$ without holes. The expression for $L([M_x, M_y]$ explains our selection for $M_y \leq M_x$ in Step 1.

As an example, consider the $M = 7$ case. We let $M_x = M_y = 4$ and find the locations of the M_x array (from Table 9.3) as $\{0, 1, 4, 6\}$, which indicates that $N[M_x] = 6$. Now, a superunit is 13 units (i.e., $2 \cdot 6 + 1$). We design the

M_y array by using Table 9.3, and multiplying the results by a superunit, that is, locations are $13 \cdot \{0, 1, 4, 6\} = \{0, 13, 52, 78\}$; hence, the location of all seven sensors are $\{0, 1, 4, 6, 13, 52, 78\}$. Integers from 79 to 84 can be obtained by adding the differences between the M_x array to 78. Finally, 85 can be obtained as $(78 - 0) + (13 - 6)$. The aperture is 85 units long, or it consists of 86 sensors. For comparison, CUM-REC provided 49 sensors, and COV-1D provided 18 sensors. In Table 9.5 we provide designs for $3 \leq M \leq 17$. In Table 9.6 we compare the length of the effective aperture with that of COV-1D and previously described two-dimensional designs. The results from Table 9.3 can be used to design CUM-LIN and CUM-REC arrays for up to $M = 34$. For $M > 34$ the results from [46] can be used.

TABLE 9.5 Cumulant-Based Linear MRA Design (CUM-LIN) for $3 \leq M \leq 17$

Actual Sensors M	M_x ARRAY Sensor Locations	M_y ARRAY Sensor Locations
3	$\{0, 1\}$	$\{0, 3\}$
4	$\{0, 1, 3\}$	$\{0, 7\}$
5	$\{0, 1, 3\}$	$\{0, 7, 21\}$
6	$\{0, 1, 4, 6\}$	$\{0, 13, 39\}$
7	$\{0, 1, 4, 6\}$	$\{0, 13, 52, 78\}$
8	$\{0, 1, 4, 7, 9\}$	$\{0, 19, 76, 114\}$
9	$\{0, 1, 4, 7, 9\}$	$\{0, 19, 76, 133, 171\}$
10	$\{0, 1, 2, 6, 10, 13\}$	$\{0, 27, 108, 189, 243\}$
11	$\{0, 1, 2, 6, 10, 13\}$	$\{0, 27, 54, 162, 270, 351\}$
12	$\{0, 1, 2, 6, 10, 14, 17\}$	$\{0, 35, 70, 210, 350, 455\}$
13	$\{0, 1, 2, 6, 10, 14, 17\}$	$\{0, 35, 70, 210, 350, 490, 595\}$
14	$\{0, 1, 4, 10, 16, 18, 21, 23\}$	$\{0, 47, 94, 282, 470, 658, 799\}$
15	$\{0, 1, 4, 10, 16, 18, 21, 23\}$	$\{0, 47, 188, 470, 752, 846, 987, 1081\}$
16	$\{0, 1, 5, 9, 16, 23, 26, 28, 29\}$	$\{0, 59, 236, 590, 944, 1062, 1239, 1357\}$
17	$\{0, 1, 5, 9, 16, 23, 26, 28, 29\}$	$\{0, 59, 295, 531, 944, 1357, 1534, 1652, 1711\}$

From the results of Table 9.6, we observe that for large M, CUM-LIN provides twice the number of effective sensors that CUM-REC can provide. This is due to the ease of designing linear arrays as compared to two-dimensional arrays. This observation can be proved as follows: let M be odd, so that $M_x = M_y = (M+1)/2$. Then $N[M_x] = N[M_y]$ are lower bounded by $(M+1)^2/12$ [due to (9.92)], which implies the effective length $L[M_x, M_y]$ is lower bounded by $(M+1)^4/72$, for large M, which is twice the lower bound for CUM-REC.

We now present a final linear MRA design for small $M (M < 7)$ which does not employ covariance-based MRA methods and is competitive with CUM-LIN (actually slightly better). It has the very important property that whenever a new sensor is available, the locations of the sensors from the previous design remain the same, so that calibration problems are not repeated every time the designer can afford a new sensor. This new cumulant-based design procedure for small linear arrays (called CUM-SL) starts with the observation that given an existing linear array, cumulants can be used to double the aperture. Therefore, given a previous design, we put the new element as far as possible from the reference element with the constraint that all

TABLE 9.6 Comparison of CUM-LIN, CUM-REC, CUM-ALL ($M^2 - M + 1$) and COV-1D Designs for Total Number of Effective Sensors

M	CUM-LIN	CUM-REC	CUM-ALL	COV-1D
3	6	4	7	4
4	12	8	13	7
5	26	16	21	10
6	47	28	31	14
7	86	49	43	18
8	125	70	57	24
9	182	100	73	30
10	258	140	91	37
11	366	196	111	44
12	474	252	133	51
13	614	324	157	59
14	824	432	183	69
15	1106	576	211	80
16	1388	720	241	91
17	1742	900	273	102

TABLE 9.7 Sensor Locations for CUM-SL Method

M	New Sensor Location	Effective Sensor Locations
2	$\{1\}$	$\{0-2\}$
3	$\{5\}$	$\{0-6, 8-10\}$
4	$\{13\}$	$\{0-14, 16-18, 20, 21, 24-26\}$
5	$\{28\}$	$\{0-33, 35, 36, 38-43, 46, 50, 51, 54-56\}$
6	$\{57\}$	$\{0-62, 64, 65, 67-73, 75, 79-81, 83-86,$ $88, 96, 100, 101, 104, 108, 109, 112-114\}$

integers from 0 to the location of the new element can be produced by addition of two integers between the actual sensors. The procedure starts by putting the referene sensor at the origin ($M = 1$). When a second sensor is available, it can be put at 1. When the third sensor is available, it can be put 5 units to the right of the reference sensor, since all the integers $\{0, 1, \ldots, 5\}$ can be generated from the set of integers between the three actual sensors by addition or subtraction of two elements. If we put the third element more than 5 units to right of the second sensor, holes appear, so we fix the location of the third element to be 5. The continuation of this search produces the results of Table 9.7, where effective sensor locations are the integers that can be obtained by adding/subtracting two integers between the actual sensors. CUM-SL is always better than CUM-LIN in this range, and by construction it guarantees an effective aperture without holes whose length is lower bounded by the location of the last actual element. In addition, whenever a new sensor is available we do not have to alter the locations of the previously available sensors as in the other methods CUM-LIN, CUM-REC, COV1D. Limited results are given for the CUM-SL method, since it involves a search procedure that becomes complicated when the number of actual sensors is large.

Overview

In this section we determined lower and upper bounds on aperture extension by using the geometric interpretation of cumulants for array processing applications; cross-correlation can be viewed as a vector between sensors, and fourth-order cumulants can be viewed as an operator that processes two correlation vectors. We proved that cumulants can be used to at least double the effective aperture with an upper bound of $M^2 - M + 1$ sensors where M is the number of actual sensors. These bounds do not assume that we have the option to design the array geometry.

We showed ways to exceed the upper bound by designing arrays for cumulant-based processing. We started with two-dimensional MRA arrays and proposed the CUM-REC algorithm. We showed that the CUM-REC algorithm can significantly outperform the covariance-based optimum design (COV-1D), particularly when the number of actual sensors is large. After investigating the issues related to small number of sensors in the two-dimensional MRA design problem, we returned to the linear MRA design, and proposed the CUM-LIN method which is inspired by the CUM-REC method. CUM-LIN provides twice as many effective sensors when compared to CUM-REC for large M. Finally, we described the CUM-SL method, which addresses the problem of linear MRA design in which availability of a new sensor does not affect the locations of the previous sensors, thus less calibration efforts are required. The effective sensors provided by all of the investigated methods are summarized in Figure 9.34. We finally note that all linear array methods (CUM-LIN, CUM-SL, COV-1D) suffer from a cone of ambiguity, whereas CUM-2D

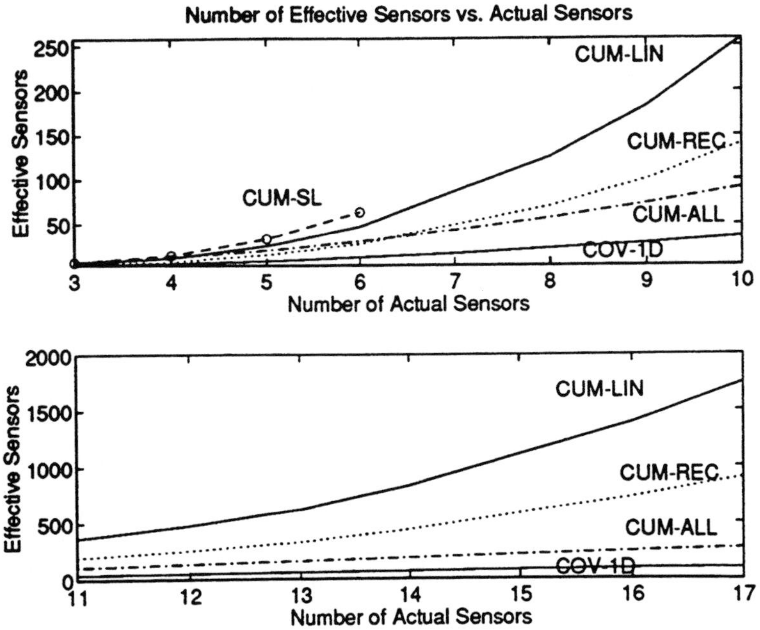

Figure 9.34 Comparison of minimum redundancy arrays.

provides an effective cross-correlation matrix that yields both azimuth and elevation estimates, and removes the cone of ambiguity.

Non-Gaussian Noise Suppression

The main motivation of using higher-order statistics in signal processing applications has been their insensitivity to additive colored Gaussian noise. The main objection to those methods is their possible vulnerability to non-Gaussian noise. In this section, we investigate the effects of non-Gaussian ambient noise on cumulant-based direction-finding systems using the interpretation for the information provided by cumulants for array processing applications described in section 9.2. We first demonstrate the suppression of uncorrelated non-Gaussian noise that has spatially varying statistics. Then, we indicate methods to suppress spatially colored non-Gaussian noise using cumulants and an additional sensor whose measurement noise component is independent of the noise components of the original array measurements. In addition, we propose a method that combines second- and fourt-order statistics together in order to suppress spatially colored non-Gaussian noise. We also illustrate how to suppress spatially colored non-Gaussian noise when the additional sensor measurement is not available. We finally indicate the noise suppression properties of the virtual-ESPRIT algorithm proposed in section 9.3. Simulations are presented to verify our results.

Non-Gaussian Noise Suppression (Uncorrelated Noise)

Theorem 9.5. *Consider an array of isotropic sensors, which is illuminated by statistically independent non-Gaussian sources. Furthermore, assume that measurements are contaminated by additive non-Gaussian sensor noise, which is independent from sensor to sensor, and whose noise components can have varying power and kurtosis over the aperture. If one uses cumulants, it is possible to:*

1. *identify the signal subspace, although noise statistics vary from sensor to sensor; this implies the directions of far-field sources can be estimated using subspace techniques; and*
2. *extend the aperture regardless of the sensor noise.*

Proof. Since the far-field sources are assumed to be independent, we can consider the presence of a single source without loss of generality. Consider Figure 9.2, which illustrates an array of three sensors. Since noise components are independent from sensor to sensor, statistical expressions such as $E\{r^*(t)x(t)\}$ or $\text{cum}(r^*(t), x(t), r^*(t), r(t))$ are not affected by the noise. Noise affects the computation of variance at a sensor, for example, if $r(t) = s(t) + n_r(t)$, then $E\{r^*(t)r(t)\} = \sigma_s^2 + \sigma_{n_r}^2 \neq \sigma_s^2$, whereas $E\{r^*(t)x(t)\} = \sigma_s^2 \exp(-j\vec{k} \cdot \vec{d}_x)$.

When noise power changes from sensor to sensor in an unknown way, it is not possible to remove its effects by an eigenanalysis of the sample covariance matrix, since the diagonal terms of the covariance matrix are corrupted by unknown (not necessarily identical) positive numbers; however, if one uses cumulants to compute correlations, then it is possible to exploit the sensor-to-sensor independence of noise, that is

$$\begin{aligned}\frac{\sigma_s^2}{\gamma_{4,s}} \operatorname{cum}(r^*(t), x(t), x^*(t), r(t))\,|_{\text{with non-Gaussian noise}} &= E\{s^*(t)s(t)\} \\ &= E\{r^*(t)r(t)\}\,|_{\text{no noise}}\end{aligned} \tag{9.96}$$

The left-hand side of (9.96) (to within the scale factor $\sigma_s^2/\gamma_{4,s}$) is computed in the actual scenario where additive non-Gaussian noise is present. To derive (9.96), let $r(t) = s(t) + n_r(t)$ and $x(t) = s(t)\exp(-j\vec{k}\cdot\vec{d}_x) + n_x(t)$. Then, because the noise components $n_r(t)$ and $n_x(t)$ are independent of the signal component $s(t)$, it follows that

$$\begin{aligned}\operatorname{cum}(r^*(t), x(t), x^*(t), r(t)) &= \underbrace{\operatorname{cum}(s^*(t), s(t)e^{-j\vec{k}\cdot\vec{d}_x}, s^*(t)e^{j\vec{k}\cdot\vec{d}_x}, s(t))}_{\gamma_{4,s}} \\ &\quad + \operatorname{cum}(n_r^*(t), n_x(t), n_x^*(t), n_r(t))\end{aligned} \tag{9.97}$$

Since the noise components are independent of each other, the second term $\operatorname{cum}(n_r^*(t), n_x(t), n_x^*(t), n_r(t))$ is equal to zero ([**CP2**]). Scaling (9.97) by $\sigma_s^2/\gamma_{4,s}$ gives the left equality in (9.96), because $E\{s^*(t)s(t)\} = \sigma_s^2$. If there is no noise, that is, $n_r(t) = 0$, then $r(t) = s(t)$, which results in the right equality in (9.96).

The right-hand side of (9.96) can only be computed in the hypothetical case where there is no measurement noise, in which case $r(t) = s(t)$; however, when noise is present, $E\{r^*(t)r(t)\} \neq E\{s^*(t)s(t)\}$, but $E\{s^*(t)s(t)\}$ is still equal to $\operatorname{cum}(r^*(t), x(t)r^*(t), r(t))$ to within the scale factor $\beta_s \triangleq \sigma_s^2/\gamma_{4,s}$ since the noise contributions in $r(t)$ and $x(t)$ are independent. The scale factor $\sigma_s^2/\gamma_{4,s}$ does not cause a problem, because if all the required covariances are computed through cumulants, then the resulting covariance matrix will correspond to the case in which source powers are scaled by these unknown (but nonzero) factors, which preserves the signal subspace. We refer the reader to section 9.2 for further discussion on scaling. For example, [with proof similar to that of (9.96)]

$$\operatorname{cum}(r^*(t), x(t), r^*(t), r(t))\,|_{\text{with non-Gaussian noise}} = \frac{\gamma_{4,s}}{\sigma_s^2} E\{r^*(t)x(t)\}\,|_{\text{no noise}} \tag{9.98}$$

In the case of P non-Gaussian sources with cumulants $\{\gamma_{4,k}\}_{k=1}^P$, and powers $\{\sigma_k^2\}_{k=1}^P$, if one constructs a matrix of covariances, computed by using cumulants, through relations such as (9.96) and (9.98), and ignores the scale factors $\beta_k = \gamma_{4,k}/\sigma_k^2$, then the resulting matrix will be identical to the covariance matrix in which the source powers (σ_k^2's) are scaled by β_k's. This matrix takes the form $\mathbf{A}\boldsymbol{\Gamma}_{ss}\mathbf{A}^H$, where $\boldsymbol{\Gamma}_{ss}$ is a diagonal matrix that contains the fourth-order cumulants of sources, and has a rank that is equal to the number of sources, that is, the noise subspace will be spanned by the eigenvectors that have zero eigenvalue; therefore, the signal subspace can be identified as the eigenvectors of this cumulant matrix that have nonzero (but, perhaps negative, since scale factors may be negative) eigenvalues. This proves part 1 of Theorem 9.1.

Virtual aperture extension is the term coined in section 9.2 to explain how cumulants increase the aperture of antenna arrays. Aperture extension is accomplished by using the cumulants of received signals to compute the cross-correlation between actual and virtual elements (e.g., see Fig. 9.2, where $\vec{d} = \vec{d}_x + \vec{d}_y$). From our interpretation, this can be viewed as reaching a virtual location by adding two nonzero vectors (otherwise we cannot reach a virtual location) that extend between actual array elements. A nonzero vector implies that its tail and head do not coincide, that is, in the cumulant expression to compute the virtual corss-correlation, at least one of the four components must be different than the other components; for example (see Fig. 9.2)

$$\text{cum}\,(r^*(t), x(t), r^*(t), y(t))\,|_{\text{with non-Gaussian noise}} = \frac{\gamma_{4,s}}{\sigma_s^2} E\{r^*(t)v(t)\}\,|_{\text{no noise}} \qquad (9.99)$$

The derivation of (9.99) can be done by observing similar results in section 9.2. $E\{r^*(t)v(t)\}$ is not computable since we do not have access to $v(t)$ (a virtual sensor); however, we have $r(t)$ and $x(t)$, and the noise in these two channels is independent; hence, equality in (9.99). We have therefore shown that $E\{r^*(t)v(t)\}$ (virtual statistic) can be computed using cumulants by processing the measure signals $r(t)$ and $x(t)$, even in the presence of non-Gaussian noise. This proves the second part of Theorem 9.1. ■

COMMENTS

- The convention established in (9.96) will be used throughout this section. It is important to note that (9.96) is valid only for ensemble averages. With finite samples, the standard deviations of the two sides will be different. In addition, there may be a bias due to finite sample size.
- The geometric interpretation of (9.96) is: with cumulants, we *move* from one sensor to another one (which has non-Gaussian but independent noise), and come back to the starting point using the same path. This approach is in fact an interpretation of the technique proposed by Cardoso [5] for accomplishing non-Gaussian noise insensitivity by removing the diagonal elements of quadricovariance steering matrices.
- The primary limitation of the proposed approach comes from the assumption about the sensor-to-sensor independence of the non-Gaussian noise. In addition, we used the far-field assumption and independence of sources.

Non-Gaussian Noise Suppression (Correlated Noises)

Theorem 9.6. *Consider an array of arbitrary sensors which is illuminated by linearly correlated non-Gaussian sources. Assume that array measurements are contaminated by non-Gaussian sensor noise of arbitrary cross-statistics. Then, it is possible to identify the signal subspace to estimate the DOA parameters by subspace techniques if there is a single sensor whose measurements are contaminated by non-Gaussian noise which is independent of the noise component of other sensors. Furthermore, there is no need to store the spatial response of that sensor.*

Proof. To begin, we assume that the sources are independent. Later we consider linearly correlated sources. Consider Figure 9.35, in which there is an array

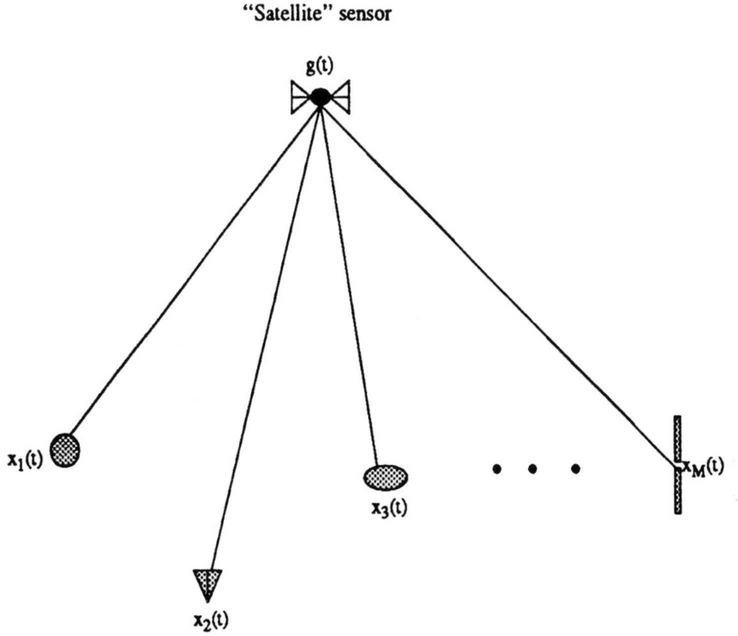

Figure 9.35 Non-Gaussian noise suppression for correlated noises in the main array that measures $\{x_K(t)\}_{K2}^{M}$. An arbitrary array of M sensors whose measurements are corrupted by colored non-Gaussian noise can still be used for direction finding if the noise in $g(t)$ is independent of the noise present in the rest of the array elements. Such a unit can be imposed in the field to correct the performance of existing systems that suffer from colored noise.

of M sensors which measure $\{x_k(t)\}_{k=1}^{M}$ and there is another sensor that measures $g(t)$ where

$$g(t) = \mathbf{g}^T\mathbf{s}(t) + n_g(t) \tag{9.100}$$

whereas the main array measurements take the form

$$\mathbf{x}(t) = \mathbf{A}\mathbf{s}(t) + \mathbf{n}_\mathbf{x}(t) \tag{9.101}$$

We assume the noise component $n_g(t)$ in $g(t)$ is independent of the noise component $\mathbf{n}_\mathbf{x}(t)$ in the main array. The satellite sensor, $g(t)$, can be used to compute the second-order statistics by using cumulants (assume a single source for the moment), because:

$$\text{cum}\,(x_j^*(t), g(t), g^*(t), x_k(t))\,|_{\text{with non-Gaussian noise}}$$
$$= \frac{\gamma_{4,s}|g_s|^2}{\sigma_s^2} E\{x_j^*(t)x_k(t)\}\,|_{\text{no noise}} \tag{9.102}$$

where g_s is the response of the satellite sensor to the source wavefront. The derivation of (9.102) is similar to that of (9.96). In this way, the following noise-free array covariance matrix can be constructed using cumulants to replace the second-order statistics [using (9.102)], where we assume multiple independent sources and that superposition holds:

$$\mathbf{C} = \mathbf{A}\mathbf{\Gamma}\mathbf{A}^H \tag{9.103}$$

where $\mathbf{A}$ is the steering matrix for M sensors (except the satellite sensor), $\mathbf{\Gamma}$ is a diagonal matrix whose kth diagonal entry is $\gamma_{4,k}|g_k|^2$, and g_k is the response of the satellite sensor to the kth source (the vector $\mathbf{g}$ is the collection of such responses). Equation (9.103) follows from the fact that in the absence of noise the array covariance matrix takes the form $\mathbf{A}\mathbf{\Sigma}_{ss}\mathbf{A}^H$, where $\mathbf{\Sigma}_{ss}$ is a diagonal matrix that contains source powers. When cumulants are used to compute the cross-correlations, we obtain a covariance matrix for the case in which source powers are scaled by unknown constants, $\beta_k = \gamma_{4,k}|g_k|^2/\sigma_k^2$, which yields (9.103).

Any subspace method can be applied to $\mathbf{C}$ in (9.103), whose elements are computed using (9.102). There is no need to know the response of the satellite sensor to the far-field sources [i.e., $\mathbf{g}$ in (9.100)]; but, the elements of $\mathbf{g}$ must be nonzero in order to make $\mathbf{\Gamma}$ a nonsingular matrix. We just need the time series recorded by the satellite sensor to actually compute the left-hand side of (9.102).

Next, consider the source signals $\mathbf{s}(t)$ correlated in the following way:

$$\mathbf{s}(t) = \mathbf{Q}\mathbf{u}(t) \tag{9.104}$$

where $\mathbf{Q}$ is nonsingular (but arbitrary), and components of $\mathbf{u}(t)$ are independent. Then the observation equations (9.100) and (9.101) change to

$$\begin{aligned} g(t) &= \mathbf{h}^T\mathbf{u}(t) + n_g(t) \\ \mathbf{x}(t) &= \mathbf{B}\mathbf{u}(t) + \mathbf{n}_\mathbf{x}(t) \end{aligned} \tag{9.105}$$

where $\mathbf{B} \triangleq \mathbf{A}\mathbf{Q}$ and $\mathbf{h}^T \triangleq \mathbf{g}^T\mathbf{Q}$. Since the components of $\mathbf{u}(t)$ are independent, they can be viewed as the actual source waveforms of (9.100) and (9.101) with an effective steering matrix $\mathbf{B}$, and a response vector $\mathbf{h}$. The cumulant matrix $\mathbf{C}$, computed as described in (9.102) for the independent sources scenario, now takes the form

$$\mathbf{C} = \mathbf{B}\tilde{\mathbf{\Gamma}}\mathbf{B}^H \tag{9.106}$$

which was obtained by substituting $\mathbf{B}$ for $\mathbf{A}$ and $\mathbf{h}$ for $\mathbf{g}$ in (9.103). In (9.106), $\tilde{\mathbf{\Gamma}}$ is defined as the diagonal matrix whose kth diagonal entry is $\tilde{\gamma}_{4,k}|h_k|^2$ and $\tilde{\gamma}_{4,k}$ is the fourth-order cumulant of $u_k(t)$. Note that $\mathbf{Q}\tilde{\mathbf{\Gamma}}\mathbf{Q}^H$ is full-rank, so that $\mathbf{C}$, expressed as

$$\mathbf{C} = \mathbf{A}(\mathbf{Q}\tilde{\mathbf{\Gamma}}\mathbf{Q}^H)\mathbf{A}^H \tag{9.107}$$

maintains all the requirements for subspace algorithms like MUSIC and ESPRIT for direction finding, even in the presence of correlated sources, correlated non-Gaussian noise, and arbitrary array characteristics. It is also important to note that we do not need to know the response of the satellite sensor to the waveforms (i.e., $\mathbf{g}$), as long as the components of $\mathbf{h}(\mathbf{h}^T = \mathbf{g}^T\mathbf{Q})$ are nonzero. This completes the proof of Theorem 9.2. ■

COMMENTS

- This method can be interpreted as follows: consider a totally different problem in which the sensors $\{x_k\}_{k=1}^{M}$ are viewed as mobile communication antennas which suffer from interference effects, so that they cannot communicate directly. It is necessary to use a satellite transponder ($g(t)$) to maintain communications among sensors. From the results in section 9.2, we know that communication between sensors means implementing cross-covariance. Here we cannot do that because of non-Gaussian sensor noise of arbitrary statistics; however, the satellite sensor, $g(t)$, can be used to make that communication possible: to implement $E\{x_j^*(t)x_k(t)\}$ first move from $x_j(t)$ to the satellite sensor $g(t)$, then let the satellite distribuite the message; that is, move from $g(t)$ to $x_k(t)$.
- A similar technique was developed in [71] as a covariance-based approach; however, it requires a second array of sensors whose noise component is independent of the noise in the existing array. Consequently, [71] ends up doubling the number of sensors for direction finding. We have accomplished noise reduction by using only one extra sensor. This gain on the number of required sensors is similar to the gain observed in the virtual-ESPRIT algorithm (VESPA) of section 9.3 as compared with covariance-ESPRIT. The reason for this difference is that cumulants, unlike cross-correlation, have an array of arguments. Based on this observation, we have also developed algorithms for single sensor detection/classification of multiple sources, and two-sensor multiple source nonparametric time-delay estimation in section 9.6.
- If the original array is linear and consists of uniformly spaced sensors of identical response, then it is possible to apply the spatial-smoothing algorithm of [56] to the covariance matrix in (9.107) to estimate the parameters of *coherent sources* (i.e., when **Q** is singular). Simulations presented later in the section investigate the coherent sources in non-Gaussian noise scenario. In addiiton, an approach that applies spatial smoothing to the generalized steering vectors estimated by VESPA is proposed in [13]. This approach can estimate more sources than sensors even in the coherent sources case, and it can utilize sensors that are nonuniformly spaced with different responses.
- Virtual aperture extension using Theorem 9.2 is also possible: it requires fixing one of the four arguments of the cumulant to be $g(t)$. Consequently, this problem is similar to aperture extension using third-order cumulants, since we now only have three free cumulant arguments with which to extend the aperture.

Virtual-ESPRIT and Non-Gaussian Noise

Theorem 9.7. *Assume independent non-Gaussian sources illuminate an array of arbitrary sensors whose measurements are corrupted by non-Gaussian noise of unknown statistics. Joint array calibration and direction finding is possible, if we have a doublet and at least one of the doublet element's measurement noise component is independent of the noise components measured by the rest of the sensors.*

Proof. We apply Theorem 9.2 to VESPA (see Figure 9.9). Let us assume that the noise component of the second sensor measurement ($r_2(t)$) is independent of the rest of the sensors, and consider a single wavefront. Let a_s denote the response of this sensor to the single wavefront. We choose the second sensor as the "satellite" sensor, that is, $g(t) = r_2(t)$. Since the responses of the first two sensors are identical, the response of the first sensor to the wavefront is also a_s.

ESPRIT autocorrelations for the measurements $\{r_k(t)\}_{k=1}^{M}$ can now be generated using cumulants as

$$\text{cum}\,(g^*(t), g(t), r_k^*(t), r_l(t))\,|_{\text{non-Gaussian noise}}$$
$$= \frac{\gamma_{4,s}|a_s|^2}{\sigma_s^2} E\{r_k^*(t) r_l(t)\}\,|_{\text{no noise}} \tag{9.108}$$

where $1 \leq k, l \leq M$. This is in fact the idea presented for Theorem 9.2: we first move to the satellite sensor $g(t)$ and then come back. A slight modification of this idea can be used to compute the ESPRIT cross-correlations virtually, as:

$$\text{cum}\,(r_1^*(t), g(t), r_k^*(t), r_l(t))\,|_{\text{non-Gaussian noise}}$$
$$= \frac{\gamma_{4,s}|a_s|^2}{\sigma_s^2} E\{r_k^*(t) v_l(t)\}\,|_{\text{no noise}} \tag{9.109}$$

although $v_l(t)$ is not physically available. For multiple independent sources, the superposition property of cumulants holds.

The cross-correlations between virtual sensors are identical to those between actual sensors (e.g., $E\{v_k^*(t)v_l(t)\} = E\{r_k^*(t)r_l(t)\}$); therefore, (9.108) can also be used to compute cross-correlations between virtual sensors. This completes the proof of Theorem 9.3. ■

Combining Second- and Fouth-Order Statistics

We have shown several ways to use higher-order statistics to suppress non-Gaussian noise. In this subsection, we investigate possible use of second-order statistics along with fourth-order cumulants. We show that the results from Theorem 9.2 can be improved by using second-order statistics. We also propose a cumulant-based method to generate a vector that can be used to estimate source bearings when an additional sensor is not available.

Consider the cross-correlation vector **d**, defined as

$$\mathbf{d} \triangleq E\{\mathbf{x}(t) g^*(t)\} \tag{9.110}$$

where $\mathbf{x}(t)$ denotes the measurements of the main array and $g(t)$ is the measurement of the satellite sensor (see Fig. 9.35). Since the noise component of $\mathbf{x}(t)$ is independent of the noise component in $g(t)$, **d** is free of the effects of noise (only when the ensemble average is considered). If **A** is the steering matrix for the main array, and the sources are linearly correlated [$\mathbf{s}(t) = \mathbf{Q}\mathbf{u}(t)$; see (9.104)], then, since the noise components $n_g(t)$ and $\mathbf{n}_\mathbf{x}(t)$ are independent, we consider only the signal components of measurements to obtain

$$\mathbf{d} = E\{\mathbf{x}(t)g(t)\} = E\{\mathbf{AQu}(t)\mathbf{u}^H(t)\mathbf{Q}^H\mathbf{g}^*\} = \mathbf{AQ}\boldsymbol{\Sigma}_{uu}\mathbf{Q}^H\mathbf{g}^* \triangleq \mathbf{Az} \tag{9.111}$$

where $\boldsymbol{\Sigma}_{uu} \triangleq E\{\mathbf{u}(t)\mathbf{u}^H(t)\}$. If none of the components of $\mathbf{z}$ are zero, then $\mathbf{d}$ is a superposition of steering vectors from the sources, and leads to an algorithm that we describe next to estimate the directions-of-arrival.

With finite samples, we estimate the noise-free vector $\mathbf{d}$ as

$$\hat{\mathbf{d}}_N = \frac{1}{N}\sum_{t=1}^{N}\mathbf{x}(t)g^*(t) \tag{9.112}$$

If the received signal vectors are independent and identically distributed, with finite moments up to order eight, then $\hat{\mathbf{d}}_N$ is an asymptotically normal sequence of random vector [10], that is,

$$\sqrt{N}(\hat{\mathbf{d}}_N - \mathbf{d}) \xrightarrow{\mathcal{L}} \mathcal{N}(\mathbf{0}, \boldsymbol{\Sigma}) \tag{9.113}$$

where the $(m, n)th$ entry of $\boldsymbol{\Sigma}$ can be expressed as

$$\begin{aligned}\Sigma_{m,n} &\triangleq \lim_{N\to+\infty} N(E\{(\hat{d}_N(m) - d(m))(\hat{d}_N(n) - d(n))^*\}\\ &= E\{|g(t)|^2 r_m(t)r_n^*(t)\} - d(m)d^*(n)\end{aligned} \tag{9.114}$$

which is derived in the Appendix section.

Since the estimation errors are asymptotically Gaussian, we can use the following cost function to estimate the parameters of interest [74]:

$$\hat{\theta} = \arg\min_{\theta,z} J(\theta) \triangleq \arg\min_{\theta,z}\left\|\boldsymbol{\Sigma}^{-1/2}\left(\hat{\mathbf{d}}_N - \mathbf{A}(\theta)\mathbf{z}\right)\right\|^2 \tag{9.115}$$

It is possible to eliminate $\mathbf{z}$ in the optimization procedure, since given the optimal estimates for θ, namely $\hat{\theta}$, $\hat{\mathbf{z}}$ can be estimated as

$$\hat{\mathbf{z}} = (\mathbf{A}^H(\hat{\theta})\boldsymbol{\Sigma}^{-1}\mathbf{A}(\hat{\theta}))^{-1}\mathbf{A}^H(\hat{\theta})\boldsymbol{\Sigma}^{-1/2}\hat{\mathbf{d}}_N \tag{9.116}$$

Substituting (9.116) into (9.115), we obtain

$$\hat{\theta} = \arg\min_{\theta}\left\|\left[\mathbf{I} - \mathbf{A}(\theta)(\mathbf{A}^H(\theta)\boldsymbol{\Sigma}^{-1}\mathbf{A}(\theta))^{-1}\mathbf{A}^H(\theta)\right]\boldsymbol{\Sigma}^{-1/2}\hat{\mathbf{d}}_N\right\|^2 \tag{9.117}$$

Since the true covariance matrix $\boldsymbol{\Sigma}$ is not available, we may estimate it from the data using fourth-order statistics of the received signals. Alternatively, we may take a suboptimal approach by letting $\boldsymbol{\Sigma} = \mathbf{I}$ in which case, the contents of $\hat{\theta}$ are the least-squares estimates of source bearings.

The direction estimation from (9.117) requires a P-dimensional search procedure (P is the number of sources). This search is quite complex unless we have good initial estimates. We use the estimates provided by the method described in Theorem 9.2 for initialization. The minimization in (9.117) can be performed by the alternating projection (AP) method, as suggested by Ziskind and Wax for the cost function associated with the deterministic maximum likelihood method for direction finding [74]. We refer the reader to [74] for the implementation of the AP algorithm.

Since this approach of suppressing non-Gaussian noise uses second- and fourth-order statistics together, we call it the SFS method. Simulations presented in section 9.5 indicate that the SFS method can decrease the variance of the estimates from the cumulant-based approach, which would now only be used for initialization purposes.

If a satellite sensor is not available, cumulants can still be used to obtain a cumulant vector in which the contribution of noise is significantly reduced; then, this vector can be used to estimate source bearings. Specifically, consider the following simple scenario in which there are only two identically distributed and independent non-Gaussian processes, $\{u_1(t), u_2(t)\}$, with variance σ^2 and fourth-order cumulant γ_4, that is

$$\mathbf{r}(t) = \mathbf{a}_1 u_1(t) + \mathbf{a}_2 u_2(t) + \mathbf{n}(t) \tag{9.118}$$

where $\mathbf{n}(t)$ is Gaussian noise, and the second non-Gaussian component, $u_2(t)$, is also treated as noise. Suppose that we preprocess the measurements with a weight-vector $\mathbf{w}$ to obtain $g(t) = \mathbf{w}^H \mathbf{r}(t)$. The selection of $\mathbf{w}$ depends on the *a priori* information on the source bearings, that is, $\mathbf{w}$ is selected in a way to not only pass the signal $u_1(t)$ relatively undistorted, but to also considerably suppress the second non-Gaussian component, $u_2(t)$. Let the beamformer weights be designed so that $\mathbf{w}^H \mathbf{a}_1 = g_1 (|g_1| \simeq 1)$ and $\mathbf{w}^H \mathbf{a}_2 = g_1 \beta$ where $|\beta| < 1$. Then, the cumulant vector

$$(\mathbf{c})_k \triangleq \text{cum}(g^*(t), g(t), g^*(t), r_k(t)) \quad 1 \le k \le M \tag{9.119}$$

can be expressed as

$$\mathbf{c} = \gamma_4 |g_1|^2 g_1^* \mathbf{a}_1 + \gamma_4 |g_1 \beta|^2 g_1^* \beta^* \mathbf{a}_2 \tag{9.120}$$

This implies that if the second source is scaled by a factor β during the beamforming step (that uses $\mathbf{w}$) to form the satellite sensor measurement, then its contribution to $\mathbf{c}$ is scaled by $|\beta|^2 \beta^*$; therefore, the ratio of contributions of the non-Gaussian noise to non-Gaussian desired source in the expression for $\mathbf{c}$ will be $|\beta|^2 \beta^*$, which is small enough to ignore provided that $|\beta|$ is small enough. If we can ignore the non-Gaussian noise component in $\mathbf{c}$ due to this scaling, we can use $\mathbf{c}$ in place of the second-order statistics vector $\mathbf{d}$ for DOA estimation. For the multiple sources/multiple non-Gaussian noises scenario, the superposition property of cumulants (**[CP4]**) applies to yield a similar result.

Simulations

In this section, we provide simulations that demonstrate our proposed non-Gaussian noise-insensitive direction-finding methods. Our first simulation experiment illustrates virtual aperture extension in the presence of spatially nonstationary but independent non-Gaussian sensor noise. Our second simulation compares cumulant and covariance-based algorithms in the presence of correlated non-Gaussian noise. Our third simulation investigates direction finding for coherent sources in non-Gaussian colored noise; it also investigates the performance improvement obtained by using both second- and fourth-order statistics. Our final simulation illustrates the non-Gaussian noise suppression properties of VESPA.

Experiment 9.1: Virtual Aperture Extension in Non-Gaussian Noise. In this experiment, we consider a two-element linear array illuminated by two equal-power, independent non-Gaussian sources from 70 and 110 degrees measured from the array axis (90° is the broadside). We assume that the sensors are isotropic and separated by a half-wavelength. The signal model is as follows:

$$\begin{bmatrix} r_1(t) \\ r_2(t) \end{bmatrix} = \begin{bmatrix} 1 & 1 \\ e^{j\pi\cos(70^\circ)} & e^{j\pi\cos(110^\circ)} \end{bmatrix} \begin{bmatrix} s_1(t) \\ s_2(t) \end{bmatrix} \beta + \begin{bmatrix} 1 & 0 \\ 0 & 2 \end{bmatrix} \begin{bmatrix} n_1(t) \\ n_2(t) \end{bmatrix} \tag{9.121}$$

The statistically independent signal components $\{s_1(t), s_2(t)\}$ are zero-mean and non-Gaussian with unit variance. The noise components are generated as a mixture of non-Gaussian and Gaussian components as follows:

$$n_k(t) = \frac{1}{\sqrt{2}}(n_{k,1}(t) + n_{k,2}(t)) \quad k = 1, 2 \tag{9.122}$$

where $n_{k,1}(t)$ is circular Gaussian and $n_{k,2}(t)$ is non-Gaussian and represents the contribution of 4-QAM communication signals. The noise components $\{n_{1,1}(t), n_{1,2}(t), n_{2,1}(t), n_{2,2}(t)\}$ are zero-mean, have unit variance, and are statistically independent; therefore, $n_1(t)$ and $n_2(t)$ are statistically independent as well. The SNR is defined as $20\log_{10}(\beta)$.

Using cumulants, it is possible to extend the aperture to three sensors. Since the noise components in the actual sensor measurements, $\{n_1(t), n_2(t)\}$, are independent, it is possible to apply Theorem 9.1 to construct a 3×3 matrix which is blind to the presence of non-Gaussian noise, as data length grows to infinity. It is therefore possible to estimate the autocorrelation at a sensor uniquely, and the cross-correlation between the two actual sensors in two different ways, using cumulants (see section 9.2 for additional discussion on this issue). The cumulant statistics are placed in a 3×3 matrix that plays the role of the array covariance matrix, as if there were three actual sensors. The MUSIC algorithm was then applied to this 3×3 matrix to estimate the bearings of two sources. If we are constrained to use only second-order statistics, we cannot identify the bearings of the sources, since the number of actual sensors (two) is not larger than the number of sources (two).

We present the mean and standard deviation of the estimates for various data lengths and SNR levels, for two different source distributions: for the first case we let the sources be BPSK and for the second case we let the sources be 4-QAM. The first case has a greater cumulant-to-power ratio;[7] hence, we expect the results for BPSK sources to be better than those for 4-QAM sources. Figure 9.36 depicts the means obtained from 100 snapshots for BPSK and 4-QAM sources for varying SNR's. BPSK sources yielded slightly better results in terms of mean. Figure 9.37 illustrates the standard deviation versus SNR for data lengths of 100, 500, 1000, and 2000 snapshots. Observe that the bearing estimates for BPSK sources have a smaller estimation error than that of 4-QAM sources. At high SNR's, performance is limited by the presence of cross-terms in the 3×3 cumulant matrix, between the independent waveforms in the cumulant expressions (these components should converge to zero

[7] For a BPSK signal with variance σ_p^2, the fourth-order cumulant is $-2(\sigma_p^2)^2$. For a 4-QAM signal with variance σ_p^2, the fourth-order cumulant is $-(\sigma_p^2)^2$.

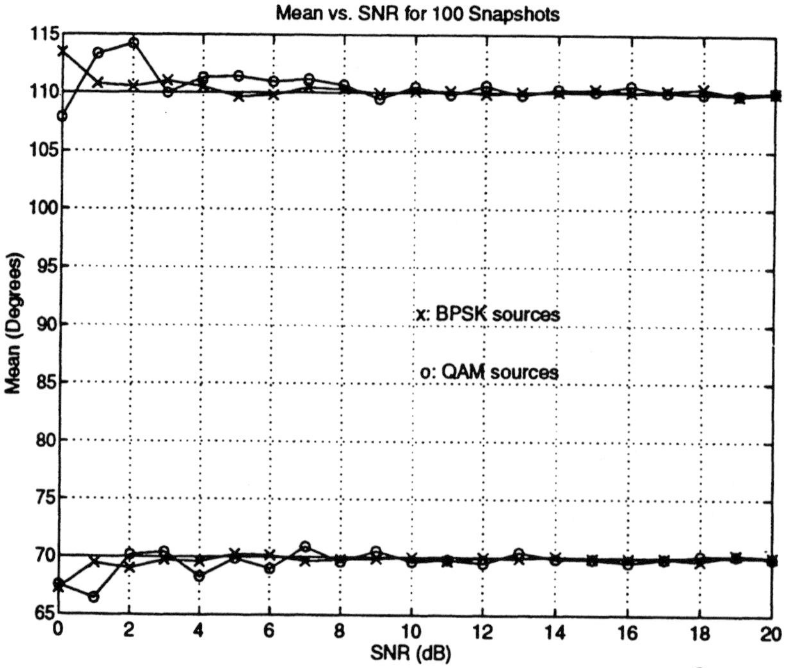

Figure 9.36 The mean of estimates for BPSK and 4-QAM sources for 100 snapshots. True bearing are 70 and 110 degrees. Each estimate is obtained from 100 independent realizations. Means of estimates are satisfactory except in the low-SNR region.

in theory). At low-SNR, performance is limited by the estimation errors due to the presence of high levels of noise which dominate the presence of cross-terms.

Experiment 9.2: Incoherent Sources in Non-Gaussian Noise. Here we estimate the bearings of two far-field sources which illuminate a uniformly spaced linear array of five sensors from $\{85^\circ, 95^\circ\}$. The bearings are measured from the axis of the linear array. Both sources broadcast BPSK waveforms of unity variance. The signal model for the experiment is as follows:

$$\begin{bmatrix} r_1(t) \\ r_2(t) \\ \vdots \\ r_5(t) \end{bmatrix} = \begin{bmatrix} 1 & 1 \\ e^{j\pi\cos(85^\circ)} & e^{j\pi\cos(95^\circ)} \\ \vdots & \vdots \\ e^{j\pi 4\cos(85^\circ)} & e^{j\pi 4\cos(95^\circ)} \end{bmatrix} \begin{bmatrix} s_1(t) \\ s_2(t) \end{bmatrix} \beta + \mathbf{L} \begin{bmatrix} n_1(t) \\ n_2(t) \\ \vdots \\ n_5(t) \end{bmatrix} \tag{9.123}$$

The signal components $\{s_1(t), s_2(t)\}$ are zero-mean, non-Gaussian with unit variance. The independent noise components $\{n_k(t)\}_{k=1}^{5}$ are generated as a mixture of circular Gaussian and 4-QAM non-Gaussian processes as in (9.122), and they have unit variance. The SNR is defined as $20\log_{10}(\beta)$.

The matrix $\mathbf{L}$ represents the spatial color of the noise in the linear array.

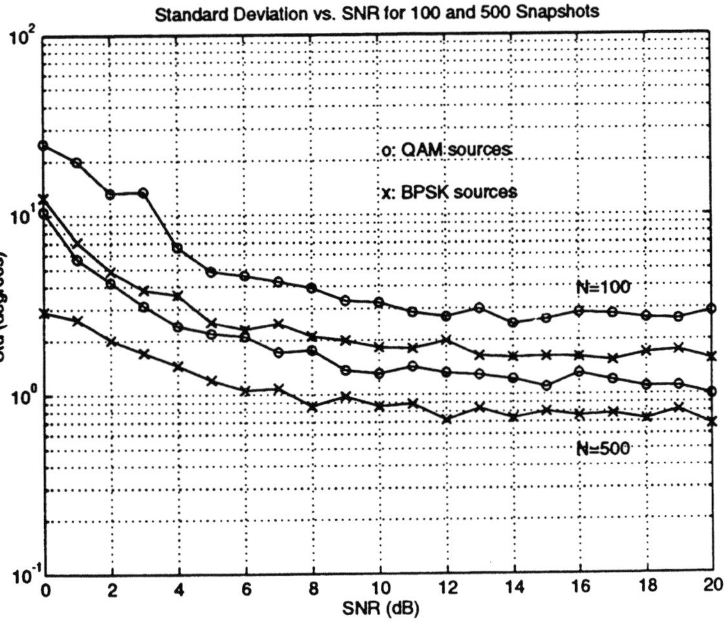

(a)

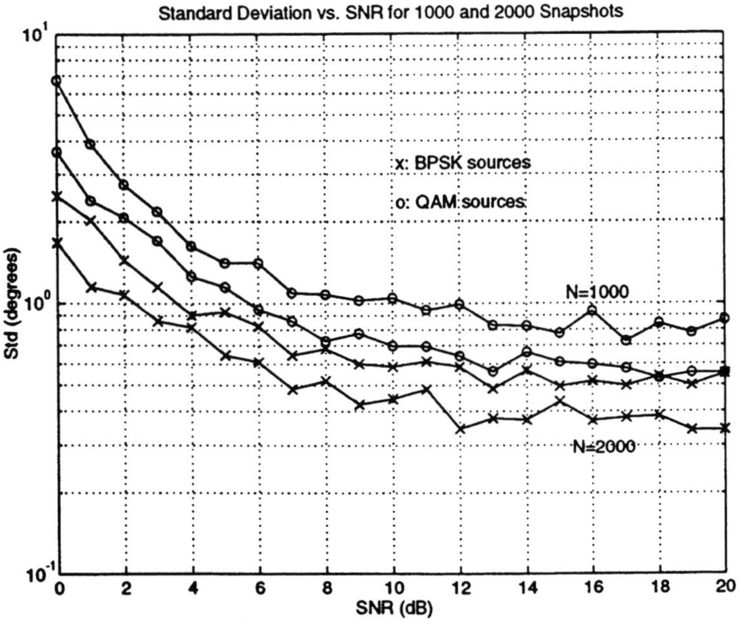

(b)

Figure 9.37 Standard deviation of estimate versus SNR for different data lengths and source distributions: (a) 100 and 500 snapshots; (b) 1000 and 2000 snapshots. Each estimate is obtained from 100 independent realizations. Standard deviations of bearing estimates for BPSK sources are always lower than those for 4-QAM sources. The results are given for the source from 70°.

It is obtained as the Cholesky decomposition of a noise covariance matrix that corresponds to major noise contribution from the range [50°, 70°]. More specifically, we summed the rank-one matrices $\mathbf{a}(\theta)\mathbf{a}^H(\theta)$ in the range [50°, 70°] at increments of 1°, and scaled the result by α_1 so that it has a trace equal to 5 (the number of elements), to obtain $\alpha_1 \sum_{\theta=50^\circ}^{70^\circ} \mathbf{a}(\theta)\mathbf{a}^H(\theta)$, and then to this we added the identity matrix that represents the spatially white part of the measurement noise, scaled by 0.2, so that the directional part is stronger in power than the white part. The resulting noise covariance matrix is

$$\mathbf{R_n} = \frac{1}{1.2}\left(\alpha_1 \sum_{\theta=50^\circ}^{70^\circ} \mathbf{a}(\theta)\mathbf{a}^H(\theta) + 0.2\mathbf{I}\right) \tag{9.124}$$

where the factor 1/1.2 is included to make the trace of $\mathbf{R_n}$ equal to 5. Then, we determined $\mathbf{L}$ so that $\mathbf{R_n} = \mathbf{L}\mathbf{L}^H$. The spatial spectrum of this measurement noise, as measured by an MVDR beamformer, is illustrated in Figure 9.38, that is, for each source bearing, we constructed an MVDR beamformer that passes the noise component from that particular direction undistorted, and minimizes the contributions of noise from other directions.

In this experiment, we do not assume the availability of a satellite sensor. Instead, we process the measurements to obtain one such signal without an additional sensor. Using the *a priori* information that the sources are close to broadside (90°),

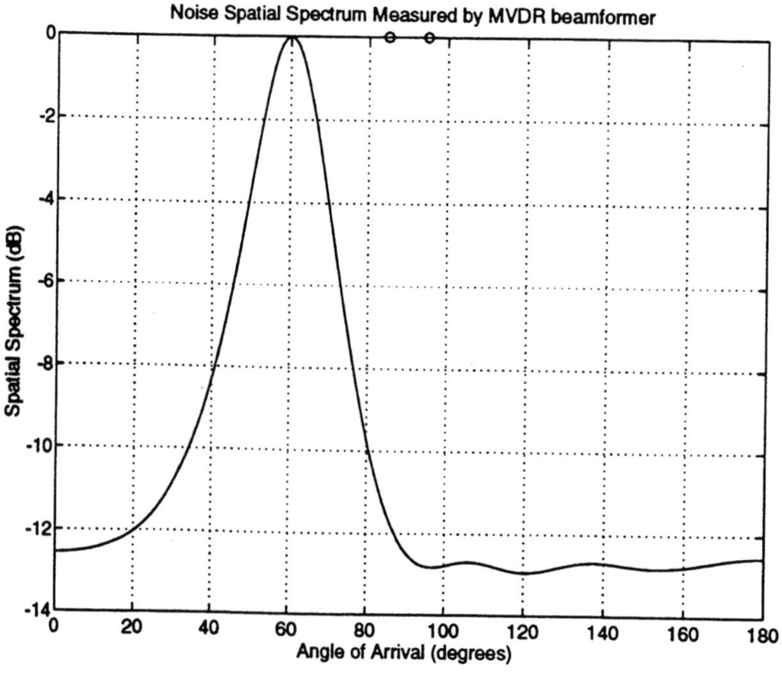

Figure 9.38 Spatial spectrum of the noise as measured by an MVDR beamformer. The two circles indicate the source bearings.

we utilize the conventional beamformer with a look direction of 90°, to obtain $g(t)$ as

$$g(t) = \mathbf{a}^H(90°)\mathbf{r}(t) \tag{9.125}$$

Since the far-field sources are close to 90°, they pass through the conventional beamformer almost undistorted, whereas the noise, with its principal component in the range [50°, 70°], is attenuated severely.

We construct the cumulant matrix

$$(\mathbf{C})_{m,n} = \text{cum}(g_*(t), g(t), r_m(t), r_n^*(t)) \tag{9.126}$$

which takes the form:

$$\mathbf{C} = \sum_{k=1}^{2} \gamma_{4,k} |\mathbf{a}^H(90°)\mathbf{a}(\theta_k)|^2 \mathbf{a}(\theta_k)\mathbf{a}^H(\theta_k) + \sum_{k=1}^{5} \gamma_{4,n_k} |\mathbf{a}^H(90°)\mathbf{l}_k|^2 \mathbf{l}_k \mathbf{l}_k^H \tag{9.127}$$

where γ_{4,n_k} is the fourth-order cumulant of the noise component $n_k(t)$ in (9.123), and $\mathbf{l}_k$ is the kth column of $\mathbf{L}$. To derive (9.127), we first assume a single source, so that $\mathbf{C} = \gamma_4 |g(\theta)|^2 \mathbf{a}(\theta)\mathbf{a}^H(\theta)$, where $g(\theta) \triangleq \mathbf{a}^H(90°)\mathbf{a}(\theta_k)$. We consider the columns of $\mathbf{L}$ as steering vectors for the non-Gaussian noise sources and use the superposition property of cumulants to obtain the final result. Since the directional noise illuminates the array from directions that correspond to sidelobes of the conventional beamformer weight vector, we have $|\mathbf{a}^H(90°)\mathbf{a}(\theta_k)|^2 \gg |\mathbf{a}^H(90°)\mathbf{l}_k|^2$, which implies noise suppression, that is, we can ignore the second sum $\sum_{k=1}^{5}(\cdot)$ in (9.127). Algorithms such as MUSIC or ESPRIT can be used to process $\mathbf{C}$ to estimate source bearings.

We performed experiments to compare the proposed method that uses ESPRIT on the cumulant matrix $\mathbf{C}$ [defined in (9.126)] versus covariance-based ESPRIT that uses the array covariance matrix. We varied SNR [defined after (9.123)] from -10 dB to 20dB in 1dB steps, and estimated the directions of arrival for data lengths of 500, 1000, 2000, and 5000 snapshots. For each experiment (with data length and SNR fixed) we performed 100 independent trials to estimate the directions of arrival. Since the bearing estimates from cumulant and covariance-based methods are expected to be biased due to the presence of colored non-Gaussian noise (the cumulant-based method does not use a true satellite sensor measurement whose noise component is statistically independent of the noise component in the main array measurements), we used the following performance criterion (RMSE) to compare the results;

$$\text{RMSE} \triangleq \sqrt{\frac{1}{N_e}\sum_{k=1}^{N_e}\left(\hat{\theta}_k - \theta_{\text{true}}\right)^2}$$

in which N_e is the number of independent realizations, and $\hat{\theta}_k$ is the estimate provided from the kth realization.

The RMSE results are shown in Figure 9.39 for the two sources. At low-SNR, the performances of both methods are similar: the covariance-ESPRIT is biased by the colored noise (see Fig. 9.40 for the mean of estimates), and the cumulant-based

ESPRIT has high RMSE and bias due to high level of noise; however, the effects of colored noise decrease sharply for the cumulant-based approach; for example, for 500 snapshots, the RMSE drops to an acceptable level of 1°, at an SNR of 0 dB, due still to estimation errors in cumulant estimates, but not due to bias (see also Fig. 9.40). On the other hand, for the same data length, covariance-ESPRIT achieves the same RMSE performance at an SNR of 11 dB. As the data length increases, the gap between cumulant and covariance-based results increases, since the covariance-based ESPRIT is limited by the bias in covariance estimates (not the variance of estimates) and its performance does not vary with data length (see Fig. 9.40). For this reason, increasing data length does not affect the RMSE from covariance-based ESPRIT for SNR's below 10 dB, since the algorithm is SNR limited in this range. For an SNR larger than 10 dB, bias gradually decreases and estimation errors in finite-sample covariance estimates become visible (i.e., observe in Fig. 9.39 that the performance shifts down slightly with increasing data length); however, the performance of covariance-based ESPRIT is always worse than that of cumulant-ESPRIT (even when the former uses 5000 snapshots and the latter uses 500 snapshots) for all SNR levels.

Experiment 9.3: Coherent Sources in Non-Gaussian Noise. In this experiment, we show how to incorporate a satellite sensor measurement into an array processing scenario which requires spatial smoothing, that is, into a linear array. To compare the performance of the covariance- and cumulant-based algorithms for the case of coherent sources, we used a uniform linear array with eight sensors in order to improve resolution and enable spatial smoothing [56]. The uniform spacing between sensors is half-wavelength.

First we investigate the case of spatially white non-Gaussian noise. We consider a BPSK source which illuminates the array from 85°, and, due to multipath, a perfectly coherent equal-power replica illuminates the array from 95°. We use an additional sensor in order to apply the method described in Theorem 9.2. This sensor is located 10 wavelentgths to the left of the first element of the linear array. A covariance-based MUSIC algorithm cannot incorporate this measurement in the present case of coherent sources, because the resulting array manifold does not possess a VanderMonde structure which is required by the spatial smoothing algorithm [56] to decorrelate the sources.

The noise components are assumed to originate from 4-QAM communications equipment, that is, they are not a mixture of Gaussian and non-Gaussian components as we have considered in the previous experiments. This is done to introduce more problems to the cumulant-based approach: from cumulant properties, it is well known that cumulants can suppress additive Gaussian noise (in theory), but not non-Gaussian noise. We assume the noise power is identical (unit variance) at all sensors including the satellite sensor, and that the signal power is equal to the noise power (0 dB). We used the MUSIC algorithm (cov) after one level of spatial smoothing to obtain a 7×7 matrix from the original 8×8 array covariance matrix [56], to investigate the performance of signal coherence on second-order statistics-based direction finding. Similarly, we applied the MUSIC algorithm to the spatially smoothed array cumulant matrix defined in (9.102) and (9.103) to

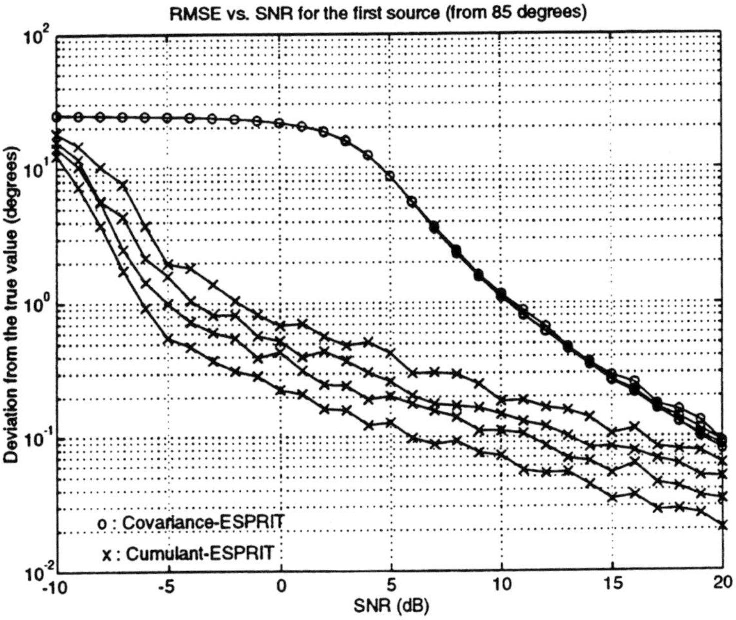

(a)

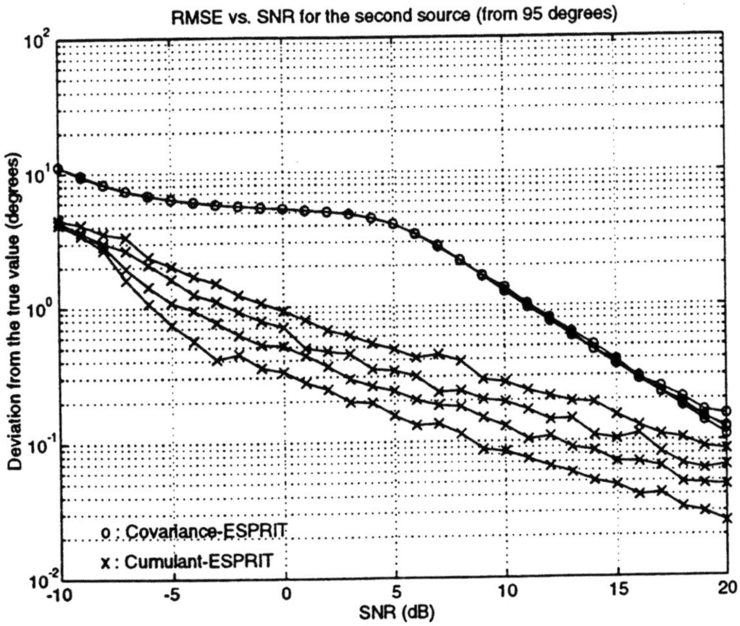

(b)

Figure 9.39 Performance comparisons for covariance-ESPRIT and cumulant-based ESPRIT that uses a preprocessed sensor measurement for noise suppresion: (a) RMSE for the first source; (b) RMSE for the second source. For each SNR level, we used four different data lengths of 500, 1000, 2000, 5000 snapshots, and performed each experiment 100 times to obtain the results. Performance of the cumulant-based approach improves with increasing data length for all SNR's whereas only high-SNR results improve for the covariance-ESPRIT.

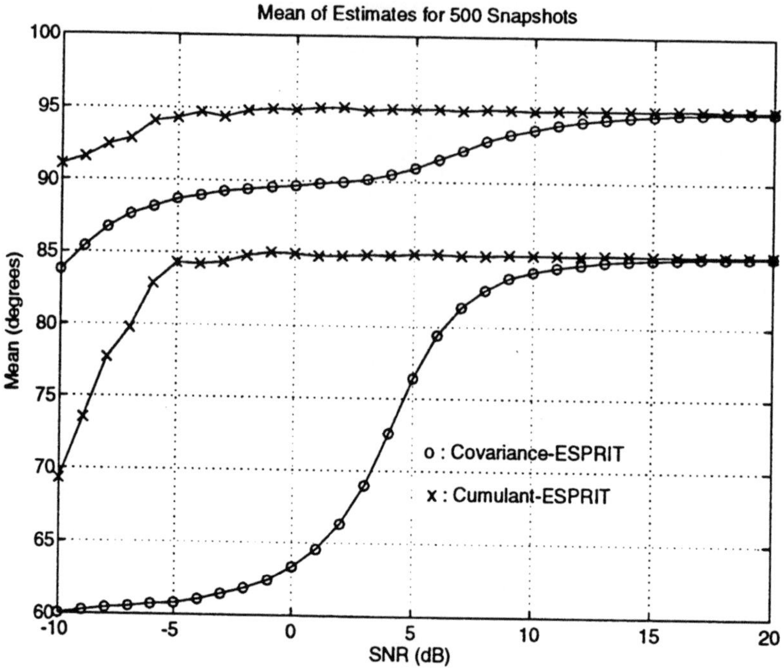

Figure 9.40 Mean of estimates for 500 snapshots: cumulant-based ESPRIT recovers from colored noise effects much faster than does the covariance-ESPRIT. The actual sources are located at 85 and 95 degrees.

investigate the performance of the cumulant-based MUSIC algorithm (CBMA). We used 1000 snapshots to estimate the required statistics and display spatial spectra in Figure 9.41 for 50 independent realizations. Observe that, in many realizations, cov is unable to resolve the sources in a satisfactory way. In addition, the estimates are biased whenever cov can resolve the sources (indicated by the two vertical lines). In this case, CBMA is able to resolve the sources, as illustrated in Figure 9.42.

Next, we investigate the effects of colored non-Gaussian noise on the direction-finding methods. To make matters even worse, we assume the two coherent wavefronts are closer to each other: the bearings are now $\{87.5°, 92.5°\}$. The noise covariance matrix for the main array takes the form: $\mathbf{R_n} = \mathbf{a}(90°)\mathbf{a}^H(90°) + 0.01\mathbf{I}$ where $\mathbf{a}(90°)$ is the steering vector that corresponds to 90°, that is, $\mathbf{a}(90°) = [1, 1, \ldots, 1]^T$. $\mathbf{R_n}$ represents an ambient noise structure whose major component illuminates the array from 90° (broadside) and shadows the presence of sources. The noise power at the satellite sensor remains at unity, and the signal power remains at 0 dB. Figure 9.43 illustrates the results from the covariance-based MUSIC algorithm in this scenario: sources are never resolved since the processor confuses the noise with a signal that arrives from the noise direction of 90°. On the other hand, the cumulant-based MUSIC algorithm (see Fig. 9.44) successfully resolves the two sources and suppresses the noise; however, CBMA estimates are slightly biased, because the sample size (1000 snapshots) is not large enough to suppress the effects

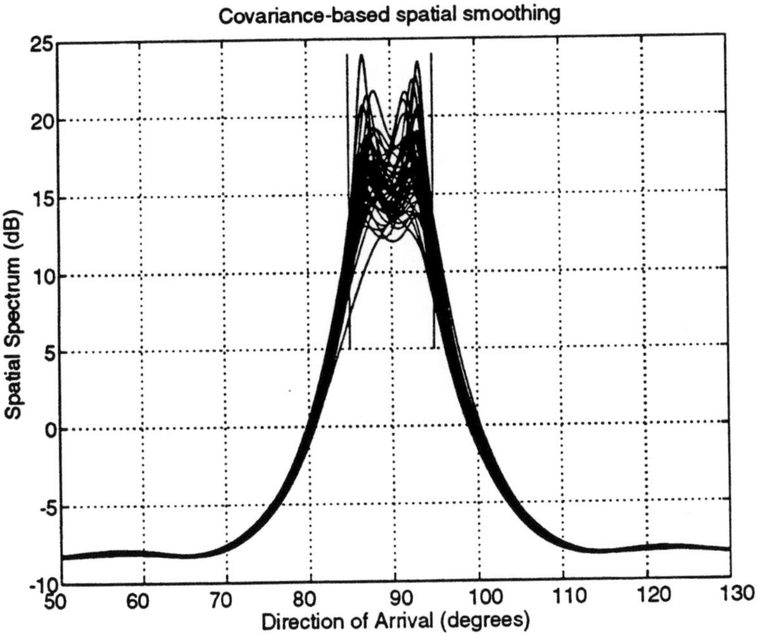

Figure 9.41 Covariance-based MUSIC algorithm is unable to resolve the sources in general. Even when resolution is possible, the estimates are biased. The vertical lines indicate true source locations.

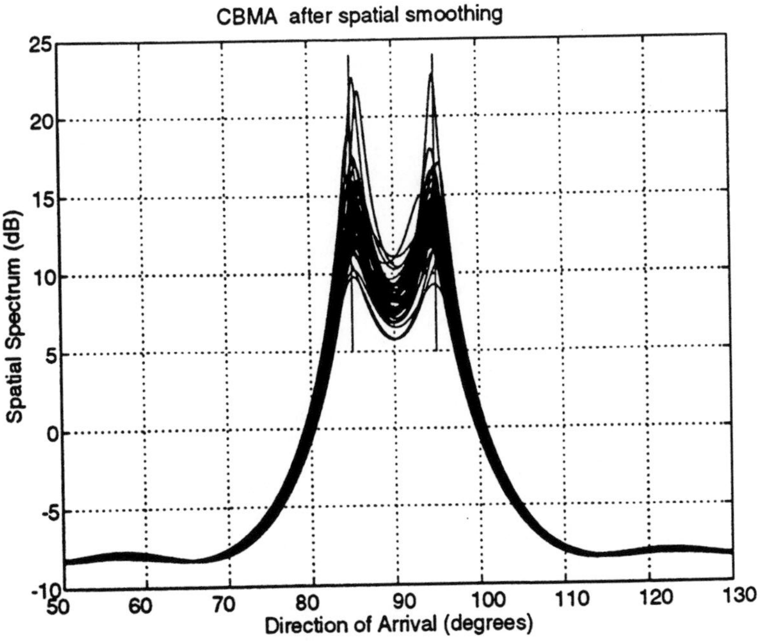

Figure 9.42 Cumulant-based algorithm which uses an extra (satellite) sensor successfully resolves the sources and estimates the directions without bias. The vertical lines indicate true source locations.

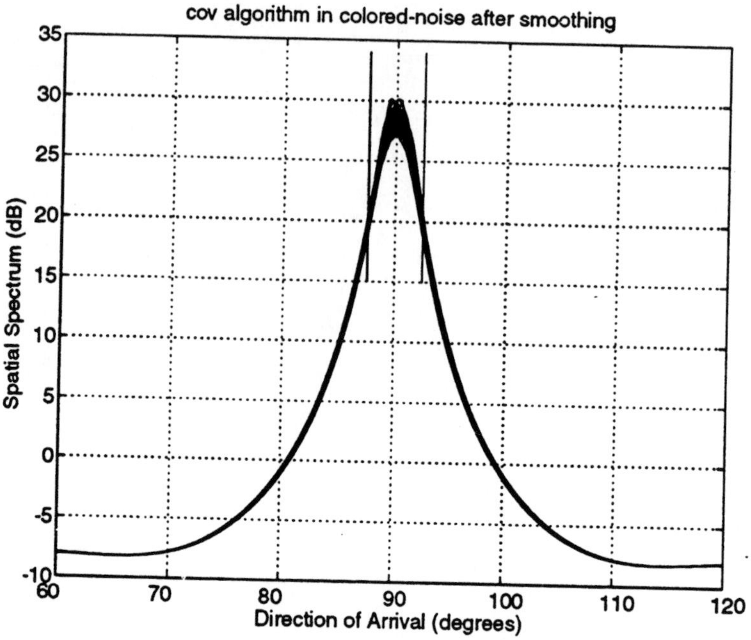

Figure 9.43 Covariance-based MUSIC algorithm (cov) is unable to resolve the sources when the non-Gaussian noise is colored. The vertical lines indicate true source locations.

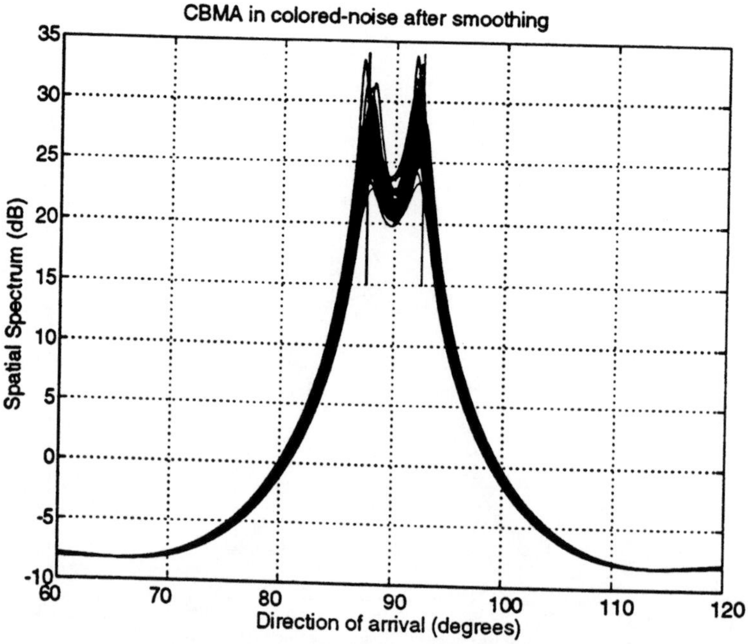

Figure 9.44 Cumulant-based algorithm which uses an extra (satellite) sensor successfully resolves the sources is colored non-Gaussian noise. The estimates can be fine-tuned by the SFS algorithm which uses CBMA estimate for initialization. The vertical lines indicate true source locations.

of the high-power noise source from 0° which leaks into the spatial smoothing algorithm and pulls the estimates towards 0°. This observation is in accordance with the results of Xu and Buckley [72], who indicate that as the correlation increases between closely separated sources, bias plays an increasingly important role.

Finally, we illustrate the improvement provided by the SFS algorithm. We initialized the search required by SFS using the results of CBMA. We display the estimates provided by CBMA and the suboptimal SFS algorithm [in which Σ is replaced by $\mathbf{I}$ in (9.117)] for 50 trials in Figure 9.45. Observe that suboptimal-SFS significantly reduces the variation and bias in the estimates.

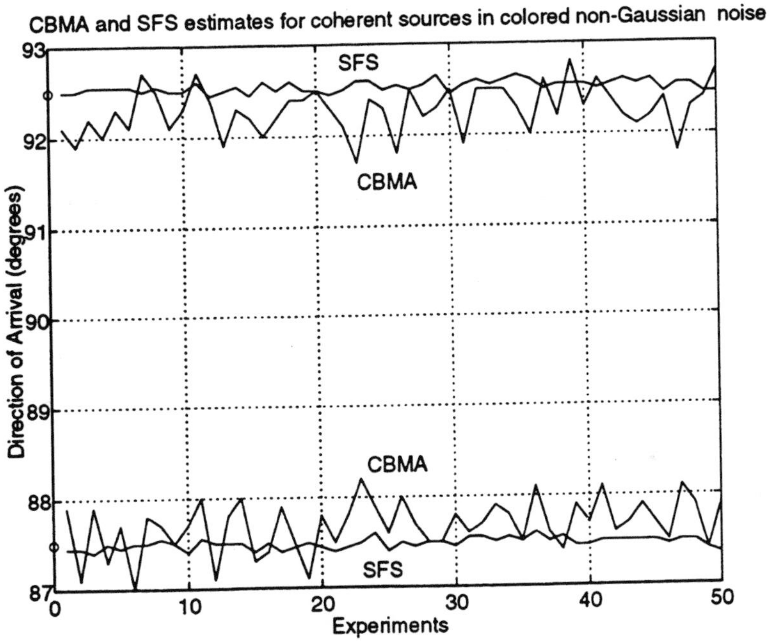

Figure 9.45 SFS and CBMA performance comparison: SFS decreases the variation and the bias (due to finite number of samples) of CBMA estimates, since it uses second-order statistics for estimation and fourth-order statistics for initialization. The mean of CBMA estimates are {87.680, 92.280} whereas the mean of SFS estimates are {87.482, 92.536}.

Experiment 9.4: Virtual-ESPRIT and Non-Gaussian Noise. In this experiment, we used the 8-element linear array of Experiment 9.3 with the same noise correlation structure and strength. Two equal power, independent signals illuminate the array from {87.5°, 92.5°}. Non-Gaussian noise suppression can be achieved by VESPA in two ways: (1) Use one new sensor that is a copy of an existing sensor, but whose additive noise in independent of noise in the other sensors; or (2) Use a doublet located sufficiently far away from the original array so that the noise contribution of the doublet measurements are independent of the noise in the original array. The first method applies only when one of the responses of the main array elements is known. This is the major reason for using the second approach. In addition, the second method can be made insensitive to the noise correlation structure between the two guiding sensors if we only create a copy of the original array; that is, an 8×8 cross-correlation matrix between the original array and its virtual copy, rather than a 10×10 matrix (see the description of VESPA in section 9.3). Consequently, we used two guiding sensors separated by $\lambda/2$. The guiding sensors are located on the axis of the main array, the first one of which is 10λ to the left of the leftmost element of the main array. The noise power at the guiding sensors is identical to the noise level at the satellite sensor of the third experiment (unity power). The signals are at 0 dB with respect to noise at the guiding sensor.

We performed 1000 independent experiments to estimate source directions using virtual-ESPRIT. The distribution of estimates is given in Figure 9.46. VESPA resolves the sources as does the cumulant-based MUSIC algorithm (CBMA) described in Experiment 9.3. The bias in the estimates is less than that of the CBMA algorithm, since sources are independent for this experiment.

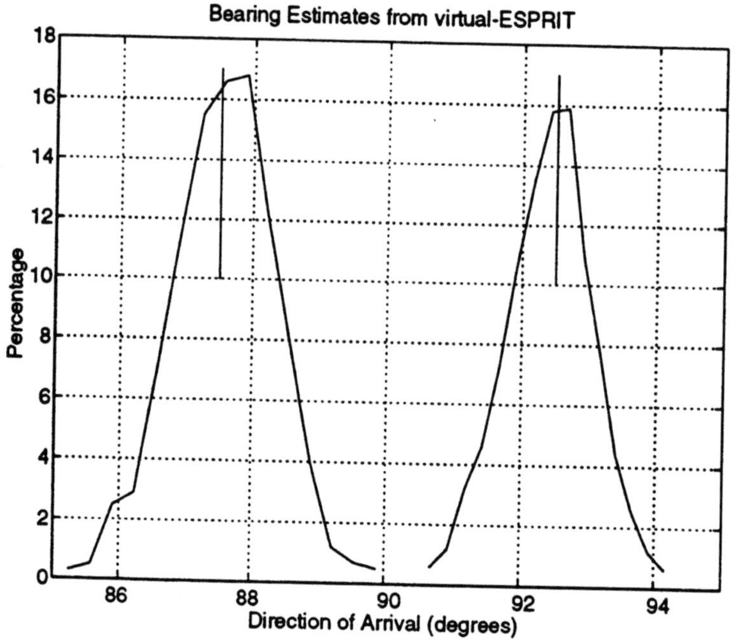

Figure 9.46 Virtual-ESPRIT algorithm can estimate the source bearings in the presence of non-Gaussian noise. The graph indicates that the sources are resolved successfully. The mean of the estimates are {87.5852, 92.4206}.

Overview

We have developed algorithms which are capable of suppressing the effects of non-Gaussian noise in array processing problems. We accomplished this by using the geometric interpretation of cumulants for array processing problems developed in section 9.2. We first showed how to suppress statistically independent non-Gaussian noise that has different statistics at each sensor. Then, we generalized this method to suppress correlated non-Gaussian noise by using an additional sensor which is remotely located to the main array and whose noise component is uncorrelated with that of the main array. We also showed that it is possible to improve cumulant-based results by using second-order statistics. Our simulations indicated that doing this significantly improves the bias and standard deviation in the estimates over a cumulant-based algorithm. In addition, we demonstrated noise suppression capabilities of the virtual-ESPRIT algorithm.

Our overall conclusions are:

- The richness of fourth-order cumulants over second-order statistics in terms of arguments provides ways to increase the effective aperture of antenna arrays and reduce the adverse effects of additive correlated non-Gaussian noise.
- Combining second- and higher-order statistics provides better results than the results obtained using only cumulants.
- Suppressing correlated measurement noise requires the use of an additional sensor that is located far from the main array, so that its noise component is not correlated with those of the main array; however, it is also possible to suppress non-Gaussian noise by using the received signals from the main array using cumulants and a priori information about the sources of interest, without using an additional sensor.
- ESPRIT can be implemented in a practical manner using cumulants. Doing this saves hardware costs and achieves non-Gaussian as well as Gaussian noise suppression. This implementation is extended to the case of coherent sources in [13].

Appendix

Here we derive the asymptotic covariance, given in (9.114), of the estimation error associated with $\hat{\mathbf{d}}_N$, defined in (9.112). From (9.112)

$$\hat{d}_N(m) = \frac{1}{N}\sum_{t=1}^{N} g^*(t) r_m(t) \tag{9.128}$$

hence

$$\begin{aligned}\Sigma_{m,n} &\triangleq \lim_{N\to+\infty} NE\left\{(\hat{d}_N(m) - d(m))(\hat{d}_N(n) - d(n))^*\right\} \\ &= \lim_{N\to\infty} E\left\{\frac{1}{N}\sum_{t_1,t_2=1}^{N} g^*(t_1)r_m(t_1)g(t_2)r_n^*(t_2)\right\} \\ &\quad - E\{d^*(n)\sum_{t_1=1}^{N} g^*(t_1)r_m(t_1)\}E\{d(m)\sum_{t_1=1}^{N} g(t_2)r_n^*(t_2)\} + Nd(m)d^*(n)\end{aligned} \tag{9.129}$$

Using the definition of $\mathbf{d}$ in (9.110), we observe $E\{g^*(t_1)r_m(t_1)\} = d(m)$, and we are able to reexpress (9.129) as

$$\Sigma_{m,n} = \lim_{N\to\infty} \frac{1}{N} \sum_{t_1=t_2=t=1}^{N} E\{|g(t)|^2 r_m(t) r_n^*(t)\} + \frac{1}{N} \sum_{t_1 \neq t_2=1}^{N} E\{g^*(t_1) r_m(t_1)\} E\{g(t_2) r_n^*(t_2)\} - N d(m) d^*(n) \tag{9.130}$$

The first summation has N identical terms, and the second summation has $(N^2 - N)$ identical terms. Using the definition of $d(m)$, we can express (9.130) as

$$\Sigma_{m,n} = \lim_{N\to\infty} E\{|g(t)|^2 r_m(t) r_n^*(t)\} - d(m)d^*(n) = E\{|g(t)|^2 r_m(t) r_n^*(t)\} - d(m)d^*(n) \tag{9.131}$$

which is the result stated in (9.114).

To gain more insight into the asymptotic covariance matrix of the vector $\hat{\mathbf{d}}_N$, we first reexpress (9.131) in terms of cumulants as

$$\Sigma_{m,n} = \text{cum}\,(g^*(t), g(t), r_n^*(t), r_m(t)) + E\{|g(t)|^2\} E\{r_m(t) r_n^*(t)\} + E\{g_1(t) r_m(t)\} E\{g_1^*(t) r_n^*(t)\} \tag{9.132}$$

The last term vanishes for measurements that are circularly symmetric. Next, assume that there exist P far-field sources with powers $\{\sigma_k\}_{k=1}^P$, fouth-order cumulants $\{\gamma_{4,k}\}_{k=1}^P$, and steering vectors $\{\mathbf{a}_k\}_{k=1}^P$. If the noise covariance matrix for the main array is denoted by $\mathbf{R_n}$, the response of the satellite sensor to the kth source as g_k, and the variance of noise in $g(t)$ is $\sigma_{n,v}^2$, then the matrix form of (9.132) is

$$\boldsymbol{\Sigma} = \sum_{k=1}^{P} \gamma_{4,k} |g_k|^2 \mathbf{a}_k \mathbf{a}_k^H + (\sigma_{n,v}^2 + \sum_{k=1}^{P} \sigma_k^2 |g_k|^2)(\sum_{k=1}^{P} \sigma_k^2 \mathbf{a}_k \mathbf{a}_k^H + \mathbf{R_n}) \tag{9.133}$$

which can be simplified to

$$\boldsymbol{\Sigma} = \sum_{k=1}^{P} (\gamma_{4,k} |g_k|^2 + \alpha \sigma_k^2) \mathbf{a}_k \mathbf{a}_k^H + \alpha \mathbf{R_n} \tag{9.134}$$

where $\alpha \triangleq (\sigma_{n,v}^2 + \sum_{k=1}^{P} \sigma_k^2 |g_k|^2)$. Observe that $\boldsymbol{\Sigma}$ (covariance matrix of a second-order statistics vector) depends only on the second-order statistics of the noise components; it does not depend on the higher-order statistics of the noise component.

9.5 SINGLE-SENSOR DETECTION AND CLASSIFICATION OF MULTIPLE SOURCES

In many problems in signal processing, observations can be modeled as a superposition of an unknown number of signals corrupted by additive noise. An important issue is to detect the number of sources that emit the waveforms and classify them using a priori information about their statistical characteristics.

Existing approaches to the *multiple* source detection problem employ *multichannel* data and utilize information-theoretic criteria for model selection, as introduced by Akaike (AIC) or by Schwartz and Rissanen (MDL) [67]. The number of signals is determined as the value for which one of these criteria is optimized. If, however, multichannel data are not available (e.g., when observations are limited to data from a *single* sensor), these approaches are not applicable. This the problem we address in this section. In addition, after estimating the number of sources, it is important to classify them. We propose a classification method that utilizes a priori knowledge of the shape of the spectrum of the sources.

We first formulate our problem, state the limitation of second-order statistics, and describe the relevant work in the area of higher-order spectral analysis. Then we establish an analogy with array processing that results in an algorithm for detection and classification of multiple sources. Next, we present simulation experiments, and conclude the section indicating possible extensions of the proposed methods.

Formulation of the Problem

We first address the problem of detecting the number of sources that emit non-Gaussian signals, where we have access to only the superposition of the waveforms, and this observation may be further corrupted by additive Gaussian noise of unknown covariance. Mathematically, we have the measurements

$$x(t) = \sum_{k=1}^{P} x_k(t) + n(t) \tag{9.135}$$

where $n(t)$ represents the Gaussian noise with spectrum $S_n(w)$ and $\{x_k(t)\}_{k=1}^{P}$ are the waveforms from sources, which can in turn be modeled as

$$x_k(t) = h_k(t) \star u_k(t) \quad k = 1, 2, \ldots, P \tag{9.136}$$

where $\{u_k(t)\}_{k=1}^{P}$ are real, stationary, white, non-Gaussian excitation sequences which are statically independent among themselves, with variance σ_k^2 and fourth-order cumulant $\gamma_{4,k}$, and the filter $h_k(t)$ models the waveform generation process of the kth source, with a frequency response $H_k(w)$. The signal model is illustrated in Figure 9.47. This linear process model is used extensively in the cumulant-based time-delay estimation problem [65, and the references therein], which has significant underwater military applications. The algorithms to be proposed here for source detection and classification will assume this model. In addition, we provide an extension to the time-delay estimation for a source with known spectrum in additive non-Gaussian colored noise.

The detection problem is to determine the number of sources, P, whereas the classification problem is to sort the signals into specific categories based on some characteristics of the emitted waveforms. In an underwater military application, the detection problem can be to determine the number of submarines in a specific zone with only a single sensor. In this scenario, the classification problem will be to identify the submarines as friendly/hostile, or as conventional/nuclear. In the speaker verification problem of speech processing, it is desirable to identify the presence of the true speaker in the presence of noise, interference, or an imitating speaker.

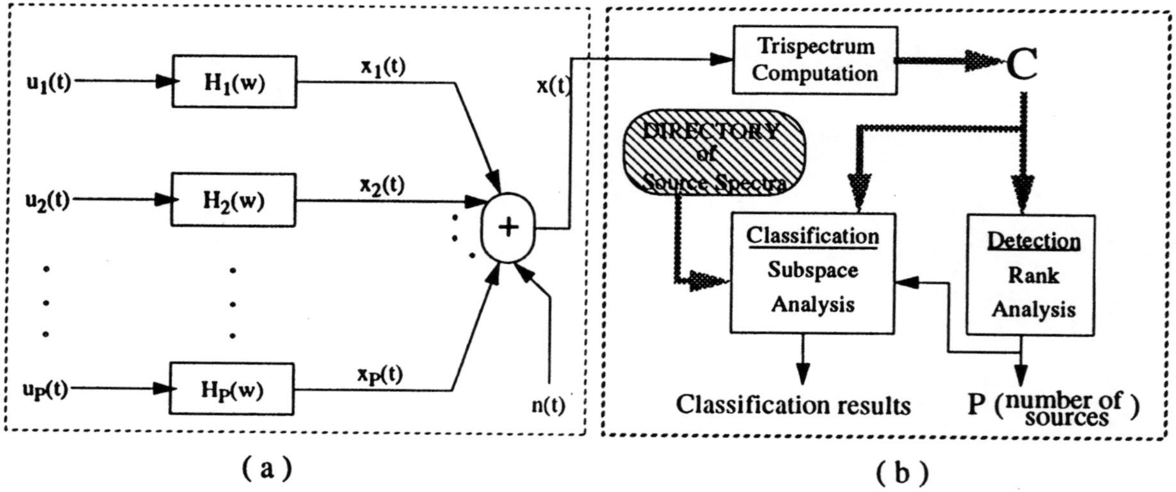

Figure 9.47 (a) Signal generation process, (b) Proposed system.

For purposes of classification, there exists a need for *templates*. In this section, we consider the availability of spectrum shape information of sources, that is, we have in our directory

$$S_k(w) = H_k(w)H_k^*(w) \tag{9.137}$$

for all sources that we want to classify. If $x(t)$ contains signals from sources with spectral shape that do not exist in our directory, the corresponding sources will be classified as "unknown."

Neither the detection nor the estimation problem can be solved by second-order statistics of the observation, since the spectrum of the observed signal $x(t)$, $S_x(w)$, can be expressed as

$$S_x(w) = \sum_{k=1}^{P} \sigma_k^2 S_k(w) + S_n(w) \tag{9.138}$$

for all frequencies of interest.

To demonstrate the inadequacy of output spectral information more clearly, let us consider the following vector formulation of the problem: the received signal spectrum evaluated at discrete frequencies (FFT bins), and stacked in a vector $\mathbf{s}_x \triangleq [S_x(w_1), S_x(w_2), \ldots, S_x(w_M)]^T$, $w_k = 2\pi(k-1)/M$, can be represented as a superposition of *source spectrum* vectors

$$\mathbf{s}_x = \sum_{k=1}^{P} \sigma_k^2 \mathbf{s}_k + \sigma_n^2 \mathbf{s}_n \tag{9.139}$$

where $\mathbf{s}_k$ and $\mathbf{s}_n$ are defined analogously to $\mathbf{s}_x$ (all are $M \times 1$ vectors). If all contributing sources (including noise) have their spectrum vectors in our directory, that is, the spectrum vectors are known and linearly independent, then it is possible to identify the sources, since (9.139) can be uniquely expressed as a linear combination of spectrum vectors; but, with unknown contributions, which may also have linearly independent spectrum vectors, it is not possible to express (9.139) as a linear combination of known spectrum vectors. Even when all the observed sources are registered in our directory and the noise spectrum is of known shape (e.g., white), then the detection and classification task will require an exhaustive search procedure over our directory of spectral information, which will stop only after

$$\sum_{l=1}^{P+1} \binom{d}{l} \tag{9.140}$$

iterations where d denotes the total number of sources registered in the directory. The exhaustive search procedure stops when there is no improvement in approximating $\mathbf{s}_x$ with the spectra of $P+1$ sources. As a numerical example, in order to detect and classify five sources which contribute to the received data, and which are registred in a ten-source directory, we need 847 iterations. Each iteration requires the construction of a projection matrix that spans the columns of selected source spectrum vectors; therefore, even in this simplified case of known noise spectrum and all registred sources, computations become very excessive. Clearly, the *observation dimensionality* provided by second-order statistics is *inadequate* to solve our problem.

Recently, higher-order statistics have been proposed to increase the *effective* dimensionality of an array [5]; however, the single-sensor problem has received very little attention except for [49], which is an excellent paper in which the problem of separating the spectrum for the sum of two time series is treated. That paper utilizes a particular submanifold of the trispectrum of the observed signal, $T_x(w_i, w_k, w_j)$, for which $w_k = -w_i$. For the multiple sources case, this statistic can be expressed as[8]

$$T_x(w_i, -w_i, w_j) = \sum_{k=1}^{P} \gamma_{4,k} \cdot S_k(w_i) S_k(w_j) 1 \le i, j \le M, w_i = 2\pi(i-1)/M \tag{9.141}$$

Unfortunately, the approach in [49] does not handle measurement noise (since it requires the spectrum of received signals) and is limited by the assumption that it requires one of the sources to have a null in its spectrum when the other source must have a finite value in its spectrum. The authors claim that their method can be extended to the case where there are more than two time series, but this makes the assumption about the spectra of the sources even less reasonable.

In our work, we assume the presence of an unknown number of sources. The detection algorithm to be presented next estimates the number of sources. For the classification problem, the information about the shape of the source spectra plays the role of a steering vector in an array processing scenario; this duality will be utilized to construct a subspace-based approach for the classification problem which is described later.

Analogy with Array Processing

In this section, we construct an analogy between our problem and a narrowband array processing problem, in which the goal is to detect the number of far-field sources and estimate their directions-of-arrival (DOA).

Let us consider a narrowband array processing scenario, where there are M sensors and P far-field sources ($M > P$) with steering vectors $\mathbf{a}_k$. The measured $M \times 1$ signal vector $\mathbf{y}(t)$ can be expressed as

$$\mathbf{y}(t) = \sum_{k=1}^{P} \mathbf{a}_k y_k(t) + \mathbf{n}(t) \tag{9.142}$$

where $\mathbf{n}(t)$ represents the effects of spatially white measurement noise with power σ^2, and the source waveforms, $y_k(t)$, are assumed to be not fully correlated (coherent) among themselves [53]. Then the covariance matrix of measurements takes the form

$$\mathbf{R} = \mathbf{A}\mathbf{R}_s\mathbf{A}^H + \sigma^2\mathbf{I} \tag{9.143}$$

in which $M \times P$ matrix $\mathbf{A}$ is the steering matrix and $\mathbf{R}_s$ is the positive-definite covariance matrix of sources.

[8] For more background on trispectrum, refer to [3, 24].

Now, let us return to our single-channel problem. We can form an $M \times M$ trispectrum matrix $\mathbf{C}(M > P)$ that contains samples of the trispectrum of the received signal $x(t)$, as

$$c_{ij} = T(w_i, -w_i, w_j) = 1 \leq i, j \leq M, w_i = 2\pi(i-1)/M \tag{9.144}$$

Then using (9.141) and the definition of source spectrum vectors, we obtain

$$\mathbf{C} = \sum_{k=1}^{P} \gamma_{4,k}\mathbf{s}_k\mathbf{s}_k^T = \mathbf{S\Gamma S}^T \tag{9.145}$$

where $\mathbf{S}$, refered to as a *source spectrum matrix*, has columns that are the source spectrum vectors (hence, $\mathbf{S}^T = \mathbf{S}^H$), and the $P \times P$ diagonal matrix $\mathbf{\Gamma}$ consists of the fourth-order cumulants $\gamma_{4,k}$ of the sources (hence, $\gamma_{4,k}$ are real). The trispectrum matrix $\mathbf{C}$ has the following properties:

- It is real, since all of its terms are the product of real factors (9.141).
- It is symmetric, that is, $\mathbf{C} = \mathbf{C}^T = \mathbf{C}^H$. This follows from (9.145).
- Therefore, $\mathbf{C}$ can be viewed as a pseudo-covariance matrix; but, it is *indefinite*, since fouth-order cumulants of driving sources are not necessarily positive.

The analogy between the array problem and the single-sensor problem is summarized in Table 9.8. Based on this analogy, we can utilize the detection and DOA algorithms already formulated for array processing for the single-sensor detection and classification problem.

TABLE 9.8 Analogy Between Two Problems

Array Problem	Single-Sensor Problem
Steering Vector: $\mathbf{a}_k$	Spectrum Vector: $\mathbf{s}_k$
Steering Matrix: $\mathbf{A}$	Spectrum Matrix: $\mathbf{S}$
Source Covariance Matrix: $\mathbf{R}_s$	Source Cumulant Matrix: $\mathbf{\Gamma}$
Noise Covariance Matrix: $\sigma^2\mathbf{I}$	Noise Cumulant Matrix: $\mathbf{0}$
Array Covariance Matrix: $\mathbf{R}$	Trispectrum Matrix: $\mathbf{C}$

Detection of the Number of Sources. It is a common assumption in array processing to have the steering vectors ($M \times 1$) corresponding to the sources illuminating the array be linearly independent. The equivalent assumption for the single-sensor trispectral analysis scheme is that all the source spectrum vectors corresponding to the sources in the field must be linearly independent. Clearly, when this assumption is violated, a solution becomes impossible even with the exhaustive search scheme described in section 9.6; hence, in this section we assume that this is a valid assumption. Note that this is a very mild assumption, and is less restrictive than the one used in [49].

If the source spectrum vectors are linearly independent (there are P of them), then the rank of the matrix $\mathbf{C}$ must be P (since $M > P$). This follows from the above assumption and (9.145) and is motivated by the similar use of covariance matrix $\mathbf{R}$ in array processing [53]; hence, the number of sources can be detected by computing the rank of $\mathbf{C}$.

Classification of Sources. Using the MUSIC algorithm, the DOA's of sources in array processing are determined by a search procedure [53]. A vector from the array manifold is selected and its distance from the so-called noise subspace is computed. If the vector is in the signal subspace, then this implies that we have an arrival from a source with this particular steering vector.

The MUSIC algorithm can be used to classify the sources in the single-sensor problem as follows:

1. Compute the rank of $\mathbf{C}$ to reveal the number of sources P. This is the detection algorithm.
2. Form the $M \times (M - P)$ matrix $\mathbf{E}_n$ containing the eigenvectors of $\mathbf{C}$, associated with its zero eigenvalues, as its columns.
3. Pick a source spectrum vector $\mathbf{s} \in \{\mathbf{s}_j\}_{j=1}^{d}$ from our directory that contains spectral shape information about the kth source, and compute
$$f(k) = \frac{\mathbf{s}_k^T \mathbf{s}_k}{\mathbf{s}_k^T \mathbf{E}_n \mathbf{E}_n^T \mathbf{s}_k} \quad \text{for } k = 1, 2, \ldots, d \tag{9.146}$$
The numerator is included to provide normalization. After $f(\cdot)$ is computed for all the sources in the directory, further normalization can be applied to force this function to have a maximum of unity.
4. The higher the value of $f(k)$, the higher the possibility of existence of the kth source in the field.
5. If only K sources $(K < P)$ from our directory are classified to be in the field, then there must be $(P - K)$ sources of unknown spectra.

The proposed system is illustrated in Figure 9.47b.

Simulations

We tested our method with a directory of size eight. The available spectral information about eight sources is illustrated in Figure 9.48. The last source of the directory is one of the two non-Gaussian sources in the field.

The signal generation process is described in Figure 9.49. The non-Gaussian processes are obtained by passing a Gaussian process through a cubic nonlinearity. The measurement noise is colored due to the filter $h_n(l)$. The spectra of non-Gaussian sources, noise and the received signal are illustrated in Figure 9.50. We observe that the received signal spectrum alone cannot help to detect/classify the sources in the field (see the directory in Fig. 9.48).

For each realization, we collected 16,384 samples $x(l)$ and computed the fourth-order cumulant function $C_{4,x}(\tau_1, \tau_2, \tau_3)$ for all $|\tau_1|, |\tau_2|, |\tau_3| \leq 7$. We obtain the trispectrum matrix $\mathbf{C}$ by using Discrete Fourier Transform. We generated 60

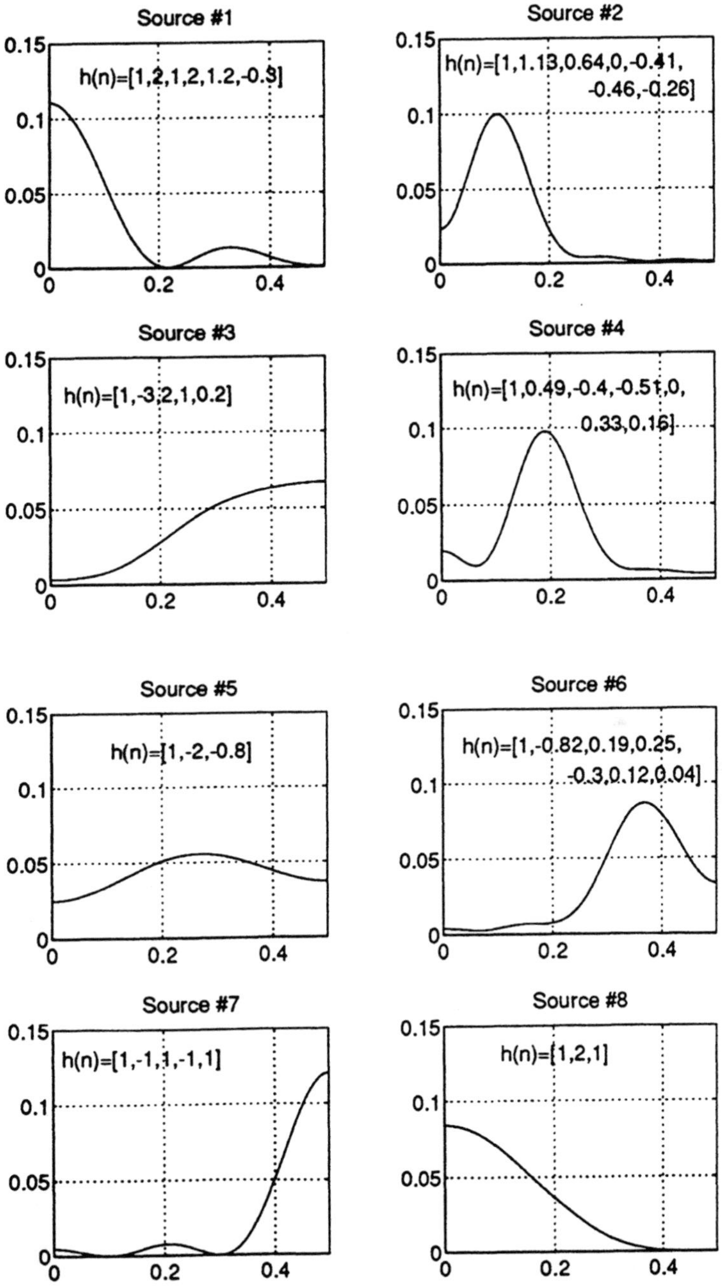

Figure 9.48 Spectral information on the eight sources of our directory. The impulse responses are also given for each source, although this is not required for detection or classification. The last registered source is actually in the field.

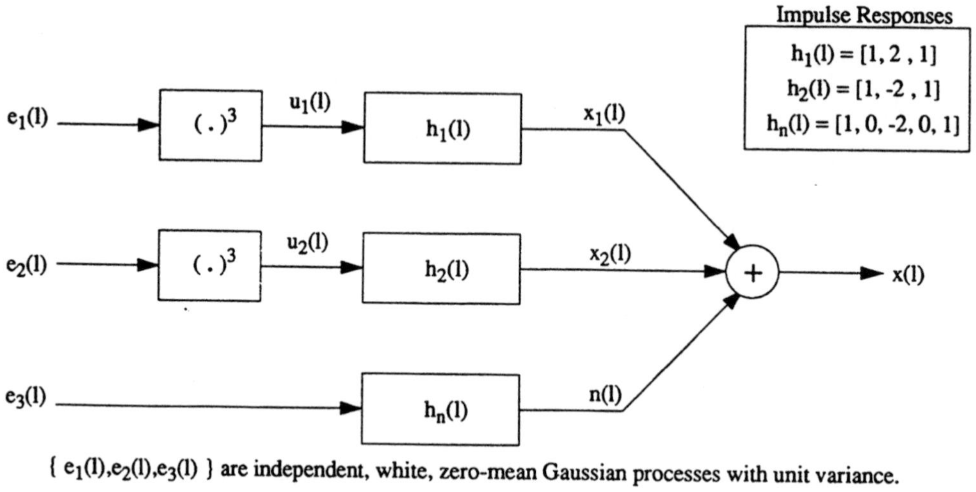

Figure 9.49 Data generation for the experiment. There are two non-Gaussian sources and measurements are corrupted by colored Gaussian noise. The source that contributes $x_1(t)$ to the measurements is the last registred source in the directory. The second non-Gaussian source and the spectral shape of the additive Gaussian noise are unknown to the processor.

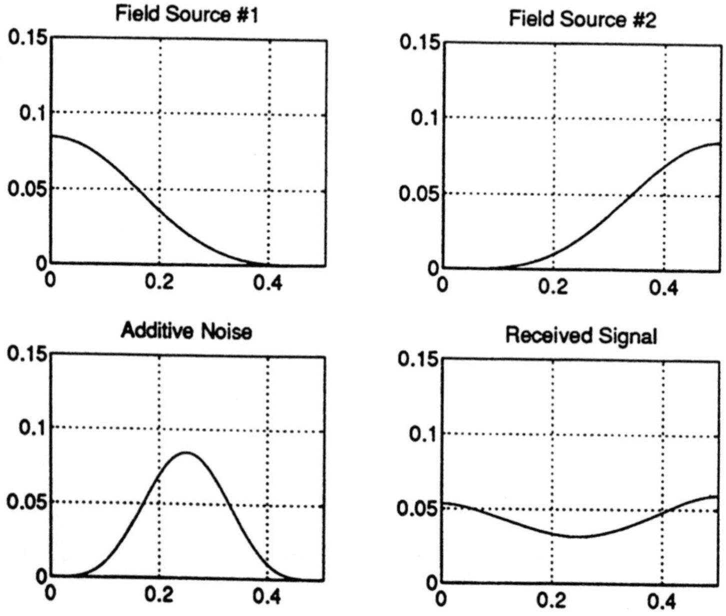

Figure 9.50 Spectra of the sources and noise in the field, and the spectrum of the received data estimated from a single realization.

independent realizations for the experiment. Table 9.9 presents the statistics of the eigenvalues of the trispectrum matrix indicating two principal eigenvalues, that is, $P = 2$.

TABLE 9.9 Statistics of the Eigenvalues of the Trispectrum Matrix Over 60 Realizations

Eigenvalue Number	1	2	3	4	5	6	7	8
Mean	0.5929	0.3509	0.0238	0.0130	0.0091	0.0059	0.0033	0.0012
Standard Deviation	0.0880	0.0792	0.0115	0.0044	0.0031	0.0018	0.0014	0.0011

After determining the presence of two non-Gaussian sources, we use our directory of spectral information (Fig. 9.48) to classify the sources. Table 9.10 illustrates the variation of the membership functions over 60 realizations for the eight sources. For each realization, we normalized the membership function to have a maximum of unity. We observe that there is one significant peak in the memberschip functions which corresponds to the last sources of the directory (which is the first field source). The standard deviation for directory source 8 is zero, indicating that the membership function peaks at 8 for all realizations. The membership function for the third directory source is slightly larger than that of the directory sources because of the small angle between the spectrum vector of source 3 and that of the unknown source (see Figures 9.48 and 9.50).

TABLE 9.10 Variation of Membership Functions Over 60 Realizations. The Last Source (the Only One in the Field From the Directory) Yields a Significant Peak When Compared with the Other Sources of the Directory.

Souce Number	1	2	3	4	5	6	7	8
Mean	0.0537	0.0388	0.1464	0.0148	0.0493	0.0499	0.0367	**1.0000**
Standard Deviation	0.0541	0.0542	0.1589	0.0153	0.0532	0.0512	0.0323	0

Overview and Extensions

We have established a framework for the analysis of single-channel, multicomponent data using higher-order statistics. We have shown that by using only second-order statistics, it is impossible to detect and classify multiple sources from single-channel data, and that, by using higher than second-order statistics, it is possible to form an analogy with the direction-of-arrival estimation problem in array processing, and accomplish single-channel detection and classification for multiple sources.

Generally, cumulants are used in place of covariance information within popular algorithms such as ESPRIT, with the hope of suppressing the effects of Gaussian noise of arbitrary covariance. In this section, we suggest another viewpoint for the use of the extra information supplied by cumulants, namely: cumulants (or, equivalent, polyspectra) possess an *array of lags* unlike second-order statistics. When properly organized, signal processing tasks that are impossible to accomplish with just second-order statistics alone are possible with higher-order statistics. Furthermore, another important interpretation is: since cumulants possess an array of lags, it is possible to extract extra information provided by cumulants in a better way if HOS information is processed using array processing techniques.

We conclude this section by providing several extensions of the methods developed for single-sensor data analysis.

Single-Source Scenario. In the single-source case, the trispectral matrix defined in section 9.6 is rank one, and its principal eigenvector is the spectrum vector of the non-Gaussian source. It is possible to develop a nonparametric detection test based on the observation that the trispectral matrix is null under the no source hypothesis. When "the source is present" hypothesis is true, the principal eigenvector will be a smoothed estimate of the spectrum of the non-Gaussian signal over trispectral slices.

Blind Spectrum Estimation for Multiple Sources. If we resort to higher-than-fourth-order cumulants, it is also possible to estimate the spectral shape of sources blindly by using subspace rotation direction finding. Before illustrating how to do this by higher-than-fourth-order statistics, we first describe how fourth-order cumulants can be used to solve the problem under some special conditions.

Consider two sets of data, collected at different times from the available single channel. We assume the excitation source powers (and hence cumulants) are scaled during the interval between the two collections. For the first set, we have

$$\mathbf{C}_1 = \mathbf{S}\boldsymbol{\Gamma}\mathbf{S}^T \tag{9.147}$$

In the second collection, excitation sequence is scaled (channels are excited with different power), which yields the trispectrum matrix:

$$\mathbf{C}_2 = \mathbf{S}\mathbf{D}\boldsymbol{\Gamma}\mathbf{S}^T \tag{9.148}$$

where $\mathbf{D}$ is a diagonal matrix that contains the scaling factors. We can consider the matrices in (9.147) and (9.148) as the covariance matrices required by the ESPRIT algorithm, and recover the spectral shapes of the sources.

The use of higher-than-fourth-order cumulants enables us to generate the matrices in (9.147) and (9.148) without requiring two collection times. For example, if $F_x(\omega_1, \omega_2, \omega_3, \omega_4)$ denotes the tetraspectrum (Fourier Transform of the fifth-order cumulant function) of $x(t)$, then the following submanifold

$$F_x(\omega_i, -\omega_i, \omega_i, -\omega_j, 0) = \sum_{k=1}^{P} \gamma_{5,k} H_k(\omega_i) H_k(-\omega_i) H_k(\omega_j) H_k(-\omega_j)(0) \tag{9.149}$$

can be expressed as

$$F_x(\omega_i, -\omega_i, \omega_i, -\omega_j, 0) = \sum_{k=1}^{P} \gamma_{5,k} H_k(0) S_k(\omega_i) S_k(\omega_j) \tag{9.150}$$

An alternative expression is

$$F_x(\omega_i, -\omega_i, \omega_j, -\omega_j, 0) = \sum_{k=1}^{P} \frac{\gamma_{5,k} H_k(0)}{\gamma_{4,k}} \gamma_{4,k} S_k(\omega_i) S_k(\omega_j) \tag{9.151}$$

We can construct an $M \times M$ matrix $\mathbf{C}_2$ from the samples of the tetraspectrum slice on the manifold (9.151) by varying ω_i and ω_j. Comparing (9.151) with (9.141), we observe that $\mathbf{C}_2$ can be decomposed as

$$\mathbf{C}_2 = \mathbf{SD\Gamma S}^T \tag{9.152}$$

where the trispectrum matrix was decomposed as $\mathbf{S\Gamma S}^T$. $\mathbf{D}$ is a diagonal matrix $\mathbf{D} = \text{diag}\left(\frac{\gamma_{5,1} H_1(0)}{\gamma_{4,1}}, \ldots, \frac{\gamma_{5,P} H_P(0)}{\gamma_{4,P}}\right)$. Now, the ESPRIT algorithm can be used to analyze $\mathbf{C}$ and $\mathbf{C}_2$, for estimating the spectral shapes contained in $\mathbf{S}$.

Time-Delay Estimation for Multiple Sources. Time-delay estimation using higher-order statistics is motivated to suppress the effects of additive spatially colored Gaussian noise [22, 35, 50, 51, 65, 73]. We now propose the use of cumulants in the case of multiple sources instead of a single non-Gaussian source. Let us assume that $\{r_1(t), r_2(t)\}$ are two sensor measurements satisfying

$$\begin{aligned} r_1(t) &= \sum_{k=1}^{P} x_k(t) + n_1(t) \\ r_2(t) &= \sum_{k=1}^{P} x_k(t - D_k) + n_2(t) \end{aligned} \tag{9.153}$$

where $\{x_k(t)\}_{k=1}^{P}$ are non-Gaussian, stationary, linear processes. The additive Gaussian noise sources $\{n_1(t), n_2(t)\}$ may have unknown spectra.

Let us consider the following fourth-order cross-cumulant function for two zero-mean, stationary random processes, $a(t)$ and $b(t)$:

$$f_{a,b}(\tau_1, \tau_2, \tau_3) \triangleq \text{cum}\,(a(t), a(t+\tau_1), b(t+\tau_2), b(t+\tau_3)) \tag{9.154}$$

The corresponding cross-trispectrum is the three-dimensional Fourier transform:

$$T_{a,b}(\omega_1, \omega_2, \omega_3) \triangleq \sum_{\tau_1,\tau_2,\tau_3=-\infty}^{\infty} f_{a,b}(\tau_1, \tau_2, \tau_3) \exp(-j(\omega_1\tau_1 + \omega_2\tau_2 + \omega_3\tau_3)) \tag{9.155}$$

Following the signal model in (9.153), the following holds:

$$T_{r_1,r_1}(\omega_1, \omega_2, \omega_3) = \sum_{k=1}^{P} \gamma_{4,k} H_k(\omega_1) H_k(\omega_2) H_k(\omega_3) H_k(-(\omega_1 + \omega_2 + \omega_3)) \tag{9.156}$$

$$\begin{aligned} &T_{r_1,r_2}(w_1, w_2, w_3) \\ &= \sum_{k=1}^{P} \gamma_{4,k} H_k(\omega_1) H_k(\omega_2) H_k(\omega_3) H_k(-(\omega_1 + \omega_2 + \omega_3)) e^{-j(\omega_2+\omega_3)D_k} \end{aligned} \tag{9.157}$$

In practical applications, polyspectra are computed only at discrete points in the frequency domain which are located at multiples of $2\pi/M$, in which M corresponds to the Fourier analysis window length. The frequency variable w takes on the values in the range $\{2\pi(k-1)/M\}_{k=1}^{M}$. Consider the functions

$$g(\omega_1, \omega_2) \triangleq T_{r_1,r_1}(\omega_1, \omega_2, 2\pi/M - \omega_2) \tag{9.158}$$

$$h(\omega_1, \omega_2) \triangleq T_{r_1,r_2}(\omega_1, \omega_2, 2\pi/M - \omega_2) \tag{9.159}$$

which can be expressed as

$$g(\omega_1, \omega_2) = \sum_{k=1}^{P} \gamma_{k,4} H_k(\omega_1) H_k(\omega_2) H_k(2\pi/M - \omega_2) H_k(-(\omega_1 + 2\pi/M)) \tag{9.160}$$

$$h(\omega_1, \omega_2) = \sum_{k=1}^{P} \gamma_{k,4} H_k(\omega_1) H_k(\omega_2) H_k(2\pi/M - \omega_2) H_k(-(\omega_1 + 2\pi/M)) \cdot e^{-j2\pi D_k/M} \tag{9.161}$$

Let us consider the vectors $\{\mathbf{s}_k, \mathbf{v}_k\}_{k=1}^{P}$ defined as follows:

$$\mathbf{s}_k(l) \triangleq H_k(\omega_1) H_k(-(\omega_1 + 2\pi/M))|_{\omega_1 = 2\pi(l-1)/M} \tag{9.162}$$

$$\mathbf{v}_k(l) \triangleq H_k(\omega_2) H_k(2\pi/M - \omega_2)|_{\omega_2 = 2\pi(l-1)/M} \tag{9.163}$$

Now we form two matrices, $\mathbf{G}$ and $\mathbf{H}$, where $g_{ij} \triangleq g(2\pi(i-1)/M, 2\pi(j-1)/M)$ and $h_{ij} \triangleq h(2\pi(i-1)/M, 2\pi(j-1)/M)$. Then, it is possible to express these matrices as

$$\mathbf{G} = \sum_{k=1}^{P} \gamma_{4,k} \mathbf{s}_k \mathbf{v}_k^T \tag{9.164}$$

$$\mathbf{H} = \sum_{k=1}^{P} \gamma_{4,k} \mathbf{s}_k \exp(-j\beta_k) \mathbf{v}_k^T \tag{9.165}$$

in which $\beta_k \triangleq 2\pi D_k/M$. In matrix form, if $\mathbf{S} \triangleq [\mathbf{s}_1, \mathbf{s}_2, \ldots, \mathbf{s}_P]$ and $\mathbf{V} \triangleq [\mathbf{v}_1, \mathbf{v}_2, \ldots, \mathbf{v}_P]$ with $\mathbf{\Gamma} \triangleq \mathrm{diag}(\gamma_{4,1}, \ldots, \gamma_{4,P})$ and $\mathbf{D} \triangleq \mathrm{diag}(\exp(-j\beta_1), \ldots, \exp(-j\beta_P))$ we have

$$\mathbf{G} = \mathbf{S\Gamma V}^T \tag{9.166}$$

$$\mathbf{H} = \mathbf{SD\Gamma V}^T \tag{9.167}$$

which implies

$$\begin{bmatrix} \mathbf{G} \\ \mathbf{H} \end{bmatrix} = \begin{bmatrix} \mathbf{S} \\ \mathbf{SD} \end{bmatrix} \mathbf{\Gamma V}^T \tag{9.168}$$

We observe that (9.168) is identical in form to the signal subspace provided by the ESPRIT data model [45] provided that the matrices $\mathbf{S}$ and $\mathbf{V}$ are full-rank. In the ESPRIT data model, the diagonal matrix $\mathbf{D}$ contains the DOA parameters and plays the same role as $\mathbf{D}$ does in (9.168). Therefore, the ESPRIT algorithm can be used to determine the components of the diagonal matrix that correspond to the time delays encountered by the sources.

9.6 CONCLUSIONS

In this chapter, we have shown that the use of higher-order statistics can provide solutions to difficult array signal processing problems. The multiple arguments of cumulants provide additional information about the phase of source wavefronts. In this respect, the use of cumulants instead of correlation seems analogous to the use of antenna arrays rather than a single sensor; when one uses an array it is possible to overcome the sensitivity limitations of a single sensor; when one uses higher-order statistics, it is possible to eliminate the phase recovery limitations of cross-spectrum. In the single-channel problem, spectrum cannot provide phase information, whereas polyspectra can.

Computation of cumulants (especially fouth-order cumulants) is more expensive than computation of second-order statistics; however, the former can be parallelized. In array processing applications, the main computational burden that causes processing delay is the eigenanalysis of covariance and cumulant matrices. Provided that these matrices are of the same size, the major computational costs for their eigenanalyses are identical.

In section 9.2, we provided an interpretation for the cumulants of narrowband sensor measurements. We showed that fourth-order cumulants *increase* the phase information available from the sensor measurements; fourth-order cumulants are interpreted as two "covariance vectors." We introduced the concept of virtual sensors, and explained how cumulants can be used to compute cross-correlations among actual and virtual sensors to increase the effective aperture of the array.

In section 9.3, we addressed the joint array calibration and direction-finding problem using an arbitrary antenna array. In this problem, an identical copy of the array was needed so that the ESPRIT algorithm might be applicable. We considered extra sensors required by the ESPRIT algorithm as virtual sensors and proved that, using a single doublet, it is possible to compute all the cross-correlations required by the ESPRIT algorithm by means of fouth-order cumulants. The resulting algorithm was called the *virtual*-ESPRIT algorithm (VESPA). It is also possible to estimate the directions, steering vectors, and the waveforms of the non-Gaussian sources using cumulants and a doublet. We provided extensions to indicate how VESPA can be applied in wideband senarios, near-field sources, and using third-order cumulants.

In section 9.4, we determined bounds on effective aperture extension by using cumulants. We proved that given an array of M identical sensors, fourth-order cumulants can at least double the aperture $2M - 1$ regardless of the location of the sensors, with an upper bound of $M^2 - M + 1$. These bounds are derived for existing cumulant-based algorithms, and therefore the upper bound was exceeded when we used minimum redundancy array design concepts together with cumulants. We proposed designs for both linear and two-dimensional arrays. The effective number of sensors provided by our designs is proportional to M^4.

Cumulants have long been promoted in signal processing applications for their ability to suppress additive Gaussian observation noise. In section 9.5, we showed that it is possible to suppress additive *non*-Gaussian noise if there exists a sensor far enough from the main array whose noise component can be non-Gaussian, but is independent from the noise components in the main array sensors.

We proved that our approach of computing cross-correlations using cumulants suppresses the effects of additive noise even when the sources of interest are correlated. We also indicated the noise suppression capabilities of the virtual-ESPRIT algorithm.

Finally, in section 9.6, we addressed the problem of single-sensor detection and classification of multiple linear non-Gaussian processes. This problem was solved by exploiting the fact that polyspectra possess an *array* of arguments, unlike spectrum. We showed how to form a trispectral matrix that has the same structure as the array covariance matrix in the narrowband array processing problem. Techniques that are well known in array processing were then applied for source detection and classification. This approach completed the "circle" in our array processing and cumulants research, by showing how array processing techniques can be used to extract information from cumulants, instead of using cumulants to extract extra information for array processing problems. We indicated extensions of this approach for time-delay estimation and blind spectrum estimation problems.

The use of cumulants for array processing applications is an important research area, and several issues remain to be explored

- Asymptotical performance analysis of the virtual-ESPRIT algorithm for direction finding and signal recovery purposes. In [34], a performance analysis is carried out to evaluate the behavior of a cumulant-based direction-finding algorithm. In [6], it is proved that signal recovery by a cumulant-based method provides better results than an informed beamformer that uses the steering vector of the source of interest. This fact is also observed in [11]. Although algorithm development is easier with Kronecker products [2], performance analysis is simpler with tensors [6].
- Extension of the virtual-ESPRIT algorithm to correlated and/or coherent sources. Use of multiple guiding sensors seems a good starting point to address this problem.
- Development of algorithms for randomly perturbed arrays, for which sensor locations change randomly over the observation interval [57]. Investigation of the possibility of "virtually stationarizing" the array by a doublet which is stationary is an important issue.
- Development of adaptive virtual subspace rotation algorithms for tracking moving non-Gaussian sources without array calibration.
- Development of computationally efficient, cumulant-based algorithms for azimuth/elevation and polarization estimation. This was briefly discussed in section 9.3.
- Extending virtual-ESPRIT algorithms to specific array geometries such as rotationally invariant arrays [61], and arrays that satisfy original ESPRIT [45] requirements.
- Blind recovery of weak signals under strong jamming using multiple arguments of cumulants.
- Development and assessment of new methods for time-delay estimation using the framework developed for single-sensor data analysis.

Although we showed how cumulants provide ways to handle important problems in array processing, we certainly do not wish to suggest that second-order statistics should be abandoned. Rather, whenever applicable, the use of second-order statistics can further improve the estimation results, as indicated by the SFS method of section 9.5.

REFERENCES

1. R. N. Bracewell. "Radio Astronomy Techniques," in *Handbuch der Physik*, vol. 54, pp. 42–129, Berlin: Springer, 1962.
2. J. W. Brewer. "Kronecker Products and Matrix Calculus in System Theory," *IEEE Trans. Circuits Systems*, vol. 25, no. 9, pp. 772–81, September 1978.
3. D. R. Brillinger and M. Rosenblatt. "Asymptotic Theory of Estimates of *k*th-order Spectra," in *Spectral Analysis of Time Series*, B. Harris, ed., New York: John Wiley & Sons, pp. 189–232, 1967.
4. J. F. Cardoso. "Blind Identification of Independent Components with Higher-order Statistics," *Proc. Vail Workshop on Nigher-Order Spectral Analysis*, pp. 157–62, June 1989.
5. J. F. Cardoso. "Higher-order Narrowband Array Processing," *International Signal Processing Workshop on Higher Order Statistics*, pp. 121-30, Chamrousse-France, July 10–12, 1991.
6. J. F. Cardoso and A. Souloumiac. "An Efficient Technique for the Blind Separation of Complex Sources," *Third International Workshop on Higher-Order Spectral Analysis*, pp. 275–79, Lake Tahoe, June 1993.
7. P. Comon. "Separation of Stochastic Processes," *Proc. Vail Workshop on Higher-Order Spectral Analysis*, pp. 174–79, June 1989.
8. J. P. Costas. "A Study of a Class of Detection Waveforms Having Nearly Ideal Range-Doppler Ambiguity Properties," *Proc. IEEE*, vol. 72, no. 8, pp. 996–1009, August 1984.
9. H. Cox, H. M. Ziskind and M. M. Owen. "Robust Adaptive Beamforming," *IEEE Trans. Acoust., Speech, Signal Processing*, vol. 35, no. 10, pp. 1365–76, October 1987.
10. P. McCullagh. *Tensor Methods in Statistics.* Monographs on Statistics and Applied Probability, Chapman and Hall, 1987.
11. M. C. Doğan and J. M. Mendel. "Cumulant-based Blind Optimum Beamforming," *IEEE Trans. Aerospace and Electronic Systems*, April 1994.
12. M. C. Doğan. "Cumulants and Array Processing." Ph.D. dissertation, Signal and Image Processing Institute, USC, Los Angeles, CA, May 1994.
13. M. C. Doğan, E. Gönen, and J. M. Mendel. "Cumulant-based Approach for Coherent Source Direction-finding," in preparation.
14. T. S. Durrani, A. R. Leyman, and J. J. Soraghan. "New Algorithms for Array Processing Using Higher-order Statistics," *Proc. ICASSP-93*, vol. 4, pp. 500–03, 1993.
15. P. Erdös and I. S. Gál. "On the representation of $1, 2, \ldots, N$ by Differences," *Nederl. Akad. Wetensch. Proc.*, vol. 51, pp. 1155–48, 1948.
16. D. Feldman and L. J. Griffiths. "A Constraint Projection Approach for Robust Adaptive Beamforming," *Proc. IEEE Intl. Conf. Acoust., Speech, Signal Processing*, vol. 2, pp. 1381–84, May 1991.

17. P. FORSTER and C. L. NIKIAS. "Bearing Estimation in the Bispectrum Domain," *IEEE Trans. Acoust., Speech, Signal Processing*, vol. 39, no. 9, pp. 1994–2006, September 1991.

18. B. FRIEDLANDER. "A Sensitivity Analysis of the MUSIC Algorithm," *IEEE Trans. Acoust., Speech, Signal Processing*, vol. 38, no. 10, pp. 1740–51, October 1990.

19. W. GABRIEL. "Spectral Analysis and Adaptive Array Superresolution Techniques," *Proc. of IEEE*, vol. 68, no. 6, pp. 654–66, June 1980.

20. S. W. GOLOMB. "Algebraic Constructions for Costas Arrays," *J. Combinatorial Theory*, Series A, vol. 37, no. 1, pp. 13–21, July 1984.

21. S. HAYKIN, J. P. REILLY, V. KEZYS and E. VERTATSCHITSCH. "Some Aspects of Array Signal Processing," *IEEE Proceedings-F*, vol. 139, no. 1, pp. 1–26, February 1992.

22. M. J. HINICH and G. R. WILSON. "Time Delay Estimation Using the Cross–Bispectrum," *IEEE Trans. Signal Processing*, vol. 40, no. 1, pp. 106–13, January 1992.

23. X. P. HUANG, J. P. REILLY and K. M. WONG. "Optimal Design of Linear Arrays of Sensors," *Proc. ICASSP-91*, vol. 2, pp. 1405-08, May 1991.

24. P. T. KIM. "Consistent Estimation of Fourth-order Cumulant Spectral Density," *J. Time Series Analysis*, vol. 12, no. 1, pp. 63–71, January 1991.

25. S. W. LANG, G. L. DUCKWORTH and J. H. MCCLELLAN. "Array Design for MEM and MLM Array Processing," *Proc. ICASSP-81*, vol. 1, pp. 145–48, March 1981.

26. R. E. LEAHY and B. D. JEFFS. "On the Design of Maximally Sparse Beamforming Arrays," *IEEE Trans. Antennas and Propagation*, vol. 39, no. 8, pp. 1178–87, August 1991.

27. J. LEECH. "On the Representation of $1, 2, \ldots, n$ by Differences," *J. London Math. Soc.*, vol. 31, pp. 160–69, 1956.

28. M. LEVI and H. MESSER. "Sufficient Conditions for Array Calibration Using Sources of Mixed Types," *ICASSP-90*, pp. 2943–46, New Mexico, April 1990.

29. D. A. LINEBARGER, I. H. SUDBOROUGH, and I. G. TOLLIS. "Difference Bases and Sparse Sensor Arrays," *IEEE Trans. Information Theory*, vol. 39, no. 2, pp. 716–21, March 1993.

30. Y. T. LO and S. L. MARPLE Jr. "Observability Conditions Multiple Signal Direction Finding and Array Sensor Localization," *IEEE Trans. Signal Processing*, vol. 40, no. 11, pp. 2641–50, November 1992.

31. J. M. MENDEL. "Tutorial on Higher-order Statistics (Spectra) in Signal Processing and System Theory: Theoretical Results and Some Applications," *Proc. IEEE*, vol. 79, no. 3, pp. 278–305, March 1991.

32. A. T. MOFFET. "Minimum-Redundancy Linear Arrays," *IEEE Trans. Antennas and Propagation*, vol. 16, no. 2, pp. 172–75, March 1968.

33. R. A. MONZINGO and T. W. MILLER. *Introduction to Adaptive Arrays*. New York: John Wiley & Sons, 1980.

34. E. MOULINES and J. F. CARDOSO, "Direction-finding Algorithms Using Fourth-order Statistics: Asymptotic Performance Analysis," *Proc. ICASSP-92*, vol. 2, pp. 437–40, March 1992.

35. C. L. NIKIAS and R. PAN. "Time-delay Estimation in Unknown Gaussian Spatially Correlated Noise," *IEEE Trans. Signal Processing*, vol. 36, no. 11, pp. 1706–14, November 1988.

36. B. OTTERSTEN, M. VIBERG, and B. WAHLBERG. "Robust Source Localization Based on Array Response Modeling," *Proc. ICASSP-92*, vol. 2, pp. 441–44, March 1992.

37. R. Pan and C. L. Nikias. "Harmonic Decomposition Methods in Cumulant Domains," *Proc. ICASSP'88*, pp. 2356–59, April 1988.

38. H. Parthasarathy, S. Prasad, and S. D. Joshi. "An ESPRIT-like Algorithm for the Estimation of Quadratic Phase Coupling," Sixth SSAP Workshop on Statistical Signal and Array Processing, pp. 189–92, Victoria, British Columbia, Canada, October 1992.

39. D. Pearson, S. Pillai, and Y. Lee. "An Algorithm for Near Optimal Placement of Sensor Elements," *IEEE Trans. Information Theory*, vol. 36, no. 6, pp. 1280–84, November 1990.

40. S. U. Pillai, Y. Bar-Ness, and F. Haber. "A New Approach to Array Geometry for Improved Spatial Spectrum Estimation," *Proc. IEEE*, vol. 73, no. 10, pp. 1522–24, October 1985.

41. S. U. Pillai. *Array Signal Processing*. New York: Springer-Verlag, 1989.

42. B. Porat and B. Friedlander. "Direction Finding Algorithms Based on High-order Statistics," *IEEE Trans. Acoust., Speech, Signal Processing*, vol. 39, no. 9, pp. 2016–24, September 1991.

43. Y. Rockah and P. M. Schultheiss. "Array Shape Calibration Using Sources in Unknown Locations–Part I: Far-field Sources," *IEEE Trans. Acoust., Speech, Signal Processing*, vol. 35, no. 3, pp. 286–99, March 1987.

44. Y. Rockah and P. M. Schultheiss. "Array Shape Calibration Using Sources in Unknown Locations–Part II: Near-field Sources and Estimator Implementation," *IEEE Trans. Acoust., Speech, Signal Processing*, vol. 35, no. 6, pp. 724–35, March 1987.

45. R. Roy and T. Kailath. "ESPRIT–Estimation of Signal Parameters via Rotational Invariance Techniques," *Optical Engineering*, vol. 29, no. 4, pp. 296–313, April 1990.

46. C. S. Ruf. "Numerical Annealing of Low-redundancy Linear Arrays," *IEEE Trans. Antennas and Propagation*, vol. 41, no. 1, pp. 85–90, January 1993.

47. P. Ruiz and J. L. Lacoume. "Extraction of Independent Sources from Correlated Sources: A Solution Based on Cumulants," *Proc. Vail Workshop on Higher-Order Spectral Analysis*, pp. 146–51, June 1989.

48. M. Rytle, B. Elsmore, and A. C. Neville. "Observations of Radio Galaxies with the One Mile Telescope at Cambridge," *Nature*, vol. 207, pp. 1024–27, September 1965.

49. F. Sakaguchi and H. Sakai. "A Spectrum Separation Method for the Sum of two Non-Gaussian Stationary Time Series Using Higher-order Periodograms," *IEEE J. Oceanic Engineering*, vol. 12, no. 1, pp. 80–89, January 1987.

50. K. Sasaki, T. Sato, and Y. Nakamura. "Holographic Passive Sonar," *IEEE Trans. Sonics Ultrasonics*, vol. 24, no. 3, pp. 193–200, May 1977.

51. T. Sato and K. Sasaki. "Bispectral Holography," *J. Acoustical Society of America*, vol. 62, no. 2, pp. 404–08, August 1977.

52. S. V. Schell and W. A. Gardner. "Progress on Signal Selective Direction Finding," *Fifth ASSP Workshop on Spectrum Estimation and Modelling*, pp. 144–48, October 1990.

53. R. O. Schmidt. "Multiple Emitter Location and Signal Parameter Estimation," *IEEE Trans. Antennas and Propagation*, vol. 34, no. 3, pp. 276–80, March 1986.

54. S. Shamsunder and G. Giannakis. "Non-Gaussian Source Localization via Exploitation of Higher-order Cyclostationarity," *Sixth Signal Processing Workshop on Statistical Signal and Array Processing*, pp. 193–96, Victoria, Canada, October 1992.

55. S. Shamsunder and G. Giannakis. "Modeling of non-Gaussian Array Data Using Cumulants: DOA Estimation of More Sources with Less Sensors," *Signal Processing*, vol. 30, no. 3, pp. 279–97, February 1993.

56. T. SHAN and T. KAILATH. "Adaptive Beamforming for Coherent Signals and Interference," *IEEE Trans. Acoust., Speech, Signal Processing*, vol. 33, no. 3, pp. 527–36, June 1985.

57. P. M. SCHULTHEISS and E. ASHOK. "Localization with Arrays Subject to Sensor Motion," *Proc-ICASSP-83*, vol. 1, pp. 371–74, May 1983.

58. D. STARER and A. NEHORAI. "Path following Algorithm for Passive Localization of Near-Field Sources," Report #9008, Center of Systems Science, Electrical Engineering, Yale University, New Haven, CT, 1990.

59. A. L. SWINDLEHURST and T. KAILATH. "Near Field Source Parameter Estimation Using a Spatial Wigner Distribution Approach," *Proc. 4th ASSP Workshop on Spectrum Estimation and Modeling*, pp. 123–28, Minneapolis, August 1988.

60. A. L. SWINDLEHURST and T. KAILATH. "Detection and Estimation Using the Third-moment Matrix," *Proc. ICASSP-89*, vol. 4, pp. 2325–28, 1989.

61. A. L. SWINDLEHURST. "DOA Identifiability for Rotationally Invariant Arrays," *IEEE Trans. Signal Processing*, vol. 40, no. 7, pp. 1825–28, July 1992.

62. A. L. SWINDLEHURST and T. KAILATH. "Azimuth/Elevation Direction Finding Using Regular Array Geometries," *IEEE Trans. on Aerospace and Electronic Systems*, vol. AES-29, no. 1, pp. 145–56, January 1993.

63. H. TAYLOR and S. W. GOLOMB. *Rulers Part 1.* CSI Technical Report 85-05-01, University of Southern California, 1985.

64. L. TONG, Y. INOUYE, and R. LIU. "Waveform Preserving Blind Estimation of Multiple Independent Sources," *IEEE Trans. Signal Processing*, vol. 41, no. 7, pp. 2461–70, July 1993.

65. J. K. TUGNAIT. "Time-delay Estimation with Unknown Correlated Gaussian Noise Using Fourth-order Cumulants and Cross-cumulants," *IEEE Trans. Signal Processing*, vol. 39, no. 6, pp. 1258–67, June 1991.

66. E. VERTATSCHITSCH and S. HAYKIN. "Impact of Linear Array Geometry on Direction of Arrival Estimation in a Multipath Environment," *IEEE Trans. Antennas and Propagation*, vol. 39, no. 5, pp. 576–84, May 1991.

67. M. WAX and I. ZISKIND. "On Unique Localization of Multiple Sources by Passive Sensor Arrays," *IEEE Trans. Acoust., Speech, Signal Processing*, vol. 37, no. 7, pp. 996–1000, July 1989.

68. A. J. WEISS and B. FRIEDLANDER. "Array Shape Calibration Using Sources in Unknown Locations—A Maximum-likelihood Approach," *IEEE Trans. Acoust., Speech, Signal Processing*, vol. 37, no. 12, pp. 1958–66, December 1989.

69. A. WEISS and M. GAVISH. "Direction-finding Using ESPRIT with Interpolated Arrays," *IEEE Trans. Acoust., Speech, Signal Processing*, vol. ASSP-39, no. 7, pp. 1589–1603, June 1991.

70. B. WIDROW, J. R. GLOVER, J. M. MCCOOL, J. KAUNITZ, C. S. WILLIAMS, R. H. HEARN, J. R. ZEIDLER, E. D. DONG, and R. C. GOODLIN. "Adaptive Noise Cancelling: Principles and Applications," *Proc. IEEE*, vol. 63, no. 12, pp. 1692–1716, December 1975.

71. Q. WU, K. M. WONG, and J. P. REILLY. "Maximum-likelihood Estimation for Array Processing in Unknown Noise Environments," *Proc. IEEE Intl. Conf. Acoust., Speech, Signal Processing*, pp. 241–44, vol. 5, San Francisco, March 1992.

72. X. L. XU and K. M. BUCKLEY. "Bias Analysis of the MUSIC Location Estimator," *IEEE Trans. Signal Processing*, vol. SP-40, no. 10, pp. 2559–69, October 1992.

73. W. ZHANG and M. RAGHUVEER. "Nonparametric Bispectrum-based Time-delay Estimators for Multiple Sensor Data," *IEEE Trans. Signal Processing*, vol. 39, no. 3, pp. 770–74, March 1991.

74. I. ZISKIND and M. WAX. "Maximum Likelihood Localization of Multiple Sources by Alternating Projection," *IEEE Trans. Acoust., Speech, Signal Processing*, vol. 36, no. 10, pp. 1553–60, October 1988.

10

Array Processing Using Radial-Basis Function Neural Network

Henry Leung, and Titus Lo

10.1 INTRODUCTION

Traditionally, the DOA problem has been dealt with using time series modeling methods such as autoregressive (AR) and moving-average (MA) models [1]. To improve the resolution of these approaches, high-resolution methods such as the MUSIC [2] algorithm and the maximum likelihood (ML) [3], offering asymptotically unbiased estimates of the direction of the radiating sources, have been proposed. However, these techniques are computationally expensive and real-time implementation seems to be very difficult to realize. For instance, the time series methods involve the inversion of covariance matrix requires a $O(M^3)$ multiply/accumulate operations, and the MUSIC algorithm typically requires $O(M^4)$ multiply/accumulate operations for the eigendecomposition of the covariance matrix, where M is the number of sensors in an array. The ML is an optimal estimator, but it requires the minimization of a highly nonlinear, multimodal, and multidimensional cost function. When a coarse initial estimate is available, the ML cost function can be minimized using the Gauss-Newton algorithm. However, in the absence of a reliable estimate, computationally intensive heuristic algorithms are needed to minimize the ML cost function. Although some methods to decrease the computational complexity of the ML criterion have been proposed, it is still difficult to deliver the desired real-time performance. To handle the computational problem, neural networks have been proposed recently as one possible solution to the computation problem [4–6].

There is a resurgence of interest in the use of neural networks to solve some of the problems arising in signal processing applications. Neural networks are useful in many applications of signal/image processing and understanding. A number of methods of signal processing can also be used in neural networks to improve performance, to reduce complexity and cost of implementation, to reduce learning and recall times, to generate truly parallel architectures, to achieve better generalization, and to come up with new learning techniques.

There are two main trends in applying neural networks to signal processing problems. The first trend is the representation of the signal processing problem as one of optimizing an energy function that matches the energy function of a particular neural network [7]. Processing using neural networks leads to the solution of an optimization problem by minimizing the energy function. An exciting realization in this approach is that analog computations with binary outputs can be used to solve such problems in a manner similar to the way in which biological neural networks carry out their computations. This is one reason why the input and output signals in neural networks are often represented in binary codes. The second trend is the application of neural networks to signal recognition problems, especially speech/image recognition and vision [8], in which the information processing operation is an approximation of a mathematical mapping. A feedforward network which has a full capability of approximation is found to be an efficient model for this application.

The application of neural networks to the DOA estimation problem in the literature belongs to the first category. Following the neural network approach to combinatorial optimization proposed by Hopfield and Tank [9], different DOA cost functions including the MUSIC [10] and the ML [5] have been mapped onto the quadratic energy function of the Hopfield model network in a minimum mean-square sense. In this chapter we consider the DOA problem as a mapping $G : \mathbb{R}^K \rightarrow \mathbb{C}^M$ from the space of DOA, $\{\theta = [\theta_1, \theta_2, \cdots, \theta_K]\}$, to the space of sensor output, $\{\mathbf{s} = [s_1, s_2, \cdots, s_M]\}$, in a noiseless environment, where

$$s_m = \sum_{k=1}^{K} a_k e^{j(m\frac{\omega_0}{c} d \sin\theta_{k+\alpha_k})} \tag{10.1}$$

where K is the number of signals, M is the number of elements of a linear array, a_k represents the complex amplitude of the kth signal, α_k represents the corresponding initial phase, ω_0 is the center frequency of the narrowband signals, d is the sensor spacing, and c is the propagation velocity. A neural network, $F : \mathbb{C}^M \rightarrow \mathbb{R}^K$, is used to form an inverse mapping of G. When a signal $\mathbf{x} = \mathbf{s} + \mathbf{n}$ where $\mathbf{n}$ is the measurement noise is received from the array, the DOA of the received signal $\mathbf{x}$ is then estimated as the output of the neural net inverse mapping. In other words, a neural network F tries to associate a received signal $\mathbf{x}$ with an estimate $\hat{\theta}$ where the association is based on the ideal array signal (10.1).

One reason for using a neural network to form such an inverse associative mapping is that the DOA mapping given in Eq. (10.1) is not an onto function because $M \geq K$, and hence the inverse of (10.1) does not exist everywhere in $\mathbb{C}^M$. In addition, a closed-form inverse formula for (10.1) would be difficult to derive. When a neural network F is used to construct a DOA associative mapping, the neural network is first trained by samples $\{(\boldsymbol{\theta}(1), \mathbf{s}(1)), (\boldsymbol{\theta}(2), \mathbf{s}(2)), \cdots, (\boldsymbol{\theta}(N), \mathbf{s}(N))\}$ generated by (10.1). This memory produces the data $\mathbf{s}(i)$ in response to a memory address

$$\mathbf{s}(i) = G(\boldsymbol{\theta}(i)) \quad i = 1, 2, \cdots, N \tag{10.2}$$

An associative memory based on the neural network produces the address $\boldsymbol{\theta}(i)$ in response to the data $\mathbf{s}(i)$:

$$\boldsymbol{\theta}(i) = F(\mathbf{s}(i)) \tag{10.3}$$

The associative memory is identical to the functionality of an inverse mapping.

Since the DOA mapping G is highly nonlinear, a neural network with nonlinearity should be used to form an inverse mapping. In this chapter, the Radial Basis Function (RBF) neural network is used for this purpose. Recently, the RBF network has been studied intensively. Besides many applications and improvements, several theoretical results have also been obtained, for example, RBF network can be naturally derived from regularization theory [11], RBF network has the universal approximation ability [12], and the best approximation ability [13]. Furthermore, RBF network has been shown to relate to the Parzen Window Estimator for probability density [14] and fuzzy logic [15]. In particular, its universal approximation and regularization capability make it an attractive model for constructing an inverse DOA mapping.

10.2 DOA ESTIMATION USING ASSOCIATIVE MEMORY

Classical DOA estimation methods such as the ML technique are typically formulated as nonlinear optimization problems. Assuming the knowledge of the probability distribution of the measurement noise $\mathbf{n}$ where Gaussianity is usually employed, $\boldsymbol{\theta}$ is estimated by optimizing the likelihood when an actual sensor array output $\mathbf{x} = \mathbf{s} + \mathbf{n}$ is received. Although the ML method has an optimum performance in a statistical sense, the computational load of the optimization procedure is too heavy for real-time processing. In addition, the performance of the classical DOA methods degrades when the signal or noise model departs from the ideal statistical assumption.

As mentioned, we propose computing the DOA estimates from the sensor outputs by forming an inverse mapping of Eq. (10.1). Specifically, before an actual observation is obtained, many training cases are constructed. For each DOA vector $\boldsymbol{\theta}(i)$ in some selected finite set $\{\boldsymbol{\theta}(i) \mid i = 1, 2, \cdots, N\}$, the DOA mapping G is used to generate a corresponding vector $\mathbf{s}(i) = G(\boldsymbol{\theta}(i))$ of sensor outputs. A neural network F is then trained to associate the sensor output vectors $\mathbf{s}(1), \mathbf{s}(2), \cdots, \mathbf{s}(N)$ with the corresponding DOA vectors $\boldsymbol{\theta}(1), \boldsymbol{\theta}(2), \cdots, \boldsymbol{\theta}(N)$. When an actual sensor array output $\mathbf{x} = \mathbf{s} + \mathbf{n}$ is received, the associative memory processor produces a DOA estimate of $\mathbf{x}$ using the trained neural network, that is, $F(\mathbf{x})$. The crucial point is that the associative memory approach encodes the nonlinear inverse mapping between the sensor output $\mathbf{s}(i)$ and the DOA vector $\boldsymbol{\theta}(i)$ for each $\boldsymbol{\theta}(i)$ in the selected training grid $\boldsymbol{\theta}(1), \cdots, \boldsymbol{\theta}(N)$. The training procedure can be compared with the maximum likelihood method for associating the most "likely" DOA vector with any given sensor output.

Three types of associative memories have been distinguished in the literature [16]. They are:

1. *Heteroassociative memory*, which implements a mapping, F, of $\mathbf{x}$ to $\mathbf{y}$ such that $F(\mathbf{x}(i)) = \mathbf{y}(i)$, and if an arbitrary $\mathbf{x}$ is "closer" to $\mathbf{x}(i)$ than to any other $\mathbf{x}(j), j = 1, \cdots, N$, then $F(\mathbf{x}) = \mathbf{y}(i)$.
2. *Interpolative associative memory*, which implements a mapping, F, of $\mathbf{x}$ to $\mathbf{y}$ such that $F(\mathbf{x}(i)) = \mathbf{y}(i)$, but, if the input vector differs from one of the training samples by the vector $\mathbf{n}$, such that $\mathbf{x} = \mathbf{x}(i) + \mathbf{n}$, then the output of the memory also differs from one of the training vectors by some vector $\mathbf{e} : F(\mathbf{x}) = F(\mathbf{x}(i) + \mathbf{n}) = \mathbf{y}(i) + \mathbf{e}$.
3. *Autoassociative memory*, which assumes $\mathbf{y}(i) = \mathbf{x}(i)$ and implements a mapping, F, of $\mathbf{x}$ to $\mathbf{x}$ such that $F(\mathbf{x}(i)) = \mathbf{x}(i)$, and, if some arbitrary $\mathbf{x}$ is "closer" to $\mathbf{x}(i)$ than to any other $\mathbf{x}(j), j = 1, \cdots, N$, then $F(\mathbf{x}) = \mathbf{x}(i)$.

Since $\boldsymbol{\theta}$ and $\mathbf{s}$ are different quantities, the autoassociative memory is not suitable for this application. Either the heteroassociative memory or the interpolative associative memory can be used for DOA estimation. When the heteroassociative memory is used, only those DOA values $\boldsymbol{\theta}(i), i = 1, 2, \cdots, N$ can be provided by the neural network as an estimate for any received sensor output. In that case, the training vectors must be sampled very closely so that the neural network is capable of producing a reasonably accurate estimate. When the number of signals is large, this situation becomes worse and memorizing such a large training set is prohibitive.

In this chapter, we concentrate on the interpolative associative memory. It is our belief that the size of the training set can be reduced by interpolation. In addition, an interpolative mapping can produce a more accurate DOA estimate for those sensor outputs outside the training set. Building such a memory is not such a difficult task mathematically if we make the further restriction that the vectors, $\mathbf{s}(i)$, form an orthonormal set. To build a linear interpolative associative memory for DOA estimation, we consider

$$F(\mathbf{s}) = (\boldsymbol{\theta}(1)\mathbf{s}^T(1) + \boldsymbol{\theta}(2)\mathbf{s}^T(2) + \cdots + \boldsymbol{\theta}(N)\mathbf{s}^T(N))\mathbf{s} \tag{10.4}$$

If $\mathbf{s}(i)$ is the input vector, then $F(\mathbf{s}(i)) = \boldsymbol{\theta}(i)$, since the set of $\mathbf{s}$ vectors is orthonormal. That is

$$\begin{aligned} F(\mathbf{s}(i)) &= (\boldsymbol{\theta}(1)\mathbf{s}^T(1) + \boldsymbol{\theta}(2)\mathbf{s}^T(2) + \cdots + \boldsymbol{\theta}(N)\mathbf{s}^T(N))\mathbf{s}(i) \\ &= \boldsymbol{\theta}(1)\delta_{1i} + \boldsymbol{\theta}(2)\delta_{2i} + \cdots + \boldsymbol{\theta}(N)\delta_{Ni} \end{aligned} \tag{10.5}$$

where $\delta_{ij} = 1$ if $i = j$ and $\delta_{ij} = 0$ if $i \neq j$. The result is then a perfect recall of $\boldsymbol{\theta}(i)$. If the input vector is different from one of the training vectors, such that $\mathbf{x} = \mathbf{s}(i) + \mathbf{n}$, then the output is

$$F(\mathbf{x}) = F(\mathbf{s}(i) + \mathbf{n}) = \boldsymbol{\theta}(i) + \mathbf{e} \tag{10.6}$$

where

$$\mathbf{e} = (\boldsymbol{\theta}(1)\mathbf{s}^T(1) + \boldsymbol{\theta}(2)\mathbf{s}^T(2) + \cdots + \boldsymbol{\theta}(N)\mathbf{s}^T(N))\mathbf{n} \tag{10.7}$$

Note that $\mathbf{n}$ may be measurement noise, sampling errors, or the combination of both. Furthermore, there was no training involved in the definition of this linear associator. The function that mapped $\mathbf{s}$ into $\boldsymbol{\theta}$ was defined by the mathematical expression in Eq. (10.4).

A linear associative memory may be regarded as the lowest-order approach to the problem of associative transformations designed to find a mathematical mapping F, by means of which arbitrary output patterns can be obtained from arbitrary input patterns. At least when the number of patterns is greater than their dimensionality, the vectors must be linearly dependent and no exact solution exists. For this reason there arises a need to generalize the class of associative mappings [17]. Using nonlinear transformations, it is possible to implement selective associative mappings for patterns which, because of their linear dependence, would not be distinguishable from a linear transformation [17]. For the DOA estimation application, the need for a nonlinear associative mapping becomes essential. The DOA mapping G given by Eq. (10.1) is a higly nonlinear mapping, and hence the inverse mapping of G is also expected to be highly nonlinear. A nonlinear neural network F is therefore needed to form a DOA associative memory.

10.3 ASSOCIATIVE MEMORY USING THE RADIAL-BASIS FUNCTION NEURAL NETWORK

Given a set of data $\{(\mathbf{s}(i), \boldsymbol{\theta}(i)) \mid i = 1, \cdots, N\}$, the RBF corresponds to choosing the form of the interpolating function

$$\theta_k(j) = \sum_{i=1}^{N} w_i^k h(\|\mathbf{s}(j) - \mathbf{s}(i)\|^2) \quad k = 1, \cdots, K,\ j = 1, \cdots, N \tag{10.8}$$

where h is a smooth univariate function and $\|\cdot\|$ is a norm. This formula means that the RBF is expanded on a finite N-element basis that is given from the set of function h translated and centered at data points. Define the matrices $\boldsymbol{\Theta}$, $\mathbf{W}$, and $\mathbf{H}$ as follows:

$$\boldsymbol{\Theta} = \begin{bmatrix} \theta_1(1) & \theta_1(2) & \ldots & \theta_1(N) \\ \theta_2(1) & \theta_2(2) & \ldots & \theta_2(N) \\ \vdots & \vdots & \ddots & \vdots \\ \theta_K(1) & \theta_K(2) & \ldots & \theta_K(N) \end{bmatrix}, \quad \mathbf{W} = \begin{bmatrix} w_1^1 & w_2^1 & \ldots & w_N^1 \\ w_1^2 & w_2^2 & \ldots & w_N^2 \\ \vdots & \vdots & \ddots & \vdots \\ w_1^K & w_2^K & \ldots & w_N^K \end{bmatrix}, \quad \text{and} \tag{10.9}$$

$$\mathbf{H} = \begin{bmatrix} h(\|\mathbf{s}(1) - \mathbf{s}(1)\|^2) & h(\|\mathbf{s}(2) - \mathbf{s}(1)\|^2) & \ldots & h(\|\mathbf{s}(N) - \mathbf{s}(1)\|^2) \\ h(\|\mathbf{s}(1) - \mathbf{s}(2)\|^2) & h(\|\mathbf{s}(2) - \mathbf{s}(2)\|^2) & \ldots & h(\|\mathbf{s}(N) - \mathbf{s}(2)\|^2) \\ \vdots & \vdots & \ddots & \vdots \\ h(\|\mathbf{s}(1) - \mathbf{s}(N)\|^2) & h(\|\mathbf{s}(2) - \mathbf{s}(N)\|^2) & \ldots & h(\|\mathbf{s}(N) - \mathbf{s}(N)\|^2) \end{bmatrix}$$

We may then write

$$\Theta = \mathbf{WH} \tag{10.10}$$

One of the most widely used functions for h is the Gaussian function, that is

$$\theta_k(j) = \sum_{i=1}^{N} w_i^k e^{-\frac{\|\mathbf{s}(j)-\mathbf{s}(i)\|^2}{\sigma^2}} \tag{10.11}$$

which has been proved to provide an invertible $\mathbf{H}$ [18]. Morever, it has been shown that it corresponds to the solution of a class of ill-posed, inverse problems involving the reconstruction of a function from a sparse set of example data [11].

The Gaussian RBF is nonzero only in a small neighborhood of the center $\mathbf{s}(i)$, in other words, each hidden unit has its own receptive field. The hidden unit will be activated when input value $\mathbf{x}$ lies in the set $\{\mathbf{x} : \|\mathbf{x} - \mathbf{s}(i)\| < \delta\}$, and the size will be proportional to the variance σ^2. To build an RBF associative memory for the data $\{(\mathbf{s}(i), \theta(i)) \mid i = 1, \ldots, N\}$, we first follow the same strategy for building a linear interpolative associative memory, that is, $w_j^k = \theta_k(j)$. Accordingly, Eq. (10.11) gives the DOA estimate as

$$\hat{\theta}_k(j) = \sum_{i=1}^{N} \theta_k(i) e^{-\frac{\|\mathbf{s}(j)-\mathbf{s}(i)\|^2}{\sigma^2}} \tag{10.12}$$

That is

$$\begin{aligned}\hat{\theta}_k(j) &= \theta_k(j) + \theta_k(1) e^{-\frac{\|\mathbf{s}(j)-\mathbf{s}(1)\|^2}{\sigma^2}} + \cdots + \theta_k(j-1) e^{-\frac{\|\mathbf{s}(j)-\mathbf{s}(j-1)\|^2}{\sigma^2}} \\ &\quad + \theta_k(j+1) e^{-\frac{\|\mathbf{s}(j)-\mathbf{s}(j+1)\|^2}{\sigma^2}} + \cdots + \theta_k(N) e^{-\frac{\|\mathbf{s}(j)-\mathbf{s}(N)\|^2}{\sigma^2}}\end{aligned} \tag{10.13}$$

When σ is set to be very small, all the exponential terms in Eq. (10.13) approach zero. A perfect recall is then produced. Therefore, the RBF interpolative associative memory given by Eq. (10.12) produces perfect recall if the Gaussian function is chosen to be sufficiently localized. If the variance σ^2 is relatively large, then the above RBF associative memory will produce a weighted average of all the $\theta_k(j)$, $j = 1, 2, \cdots, N$ for $\mathbf{s}(j)$.

Although the above RBF associative memory works, at least in principle, it basically memorizes all the training vectors in its coefficients. In order for the RBF to learn the underlying rules of the inverse DOA mapping from the training samples, the weight matrix is usually obtained by taking the inverse of the matrix $\mathbf{H}$. That is

$$\hat{\mathbf{W}} = \Theta \mathbf{H}^{-1} \tag{10.14}$$

The invertibility of $\mathbf{H}$ depends on the choice of the function h, and the Gaussian function has been proven to have an inverse [18].

Substituting $\mathbf{s}(j)$ into this inverse matrix RBF for associative memory, it produces $\hat{\theta}(j)$ as the output:

$$\hat{\theta}^T(j) = \hat{\mathbf{W}} \left[e^{-\frac{\|\mathbf{s}(j)-\mathbf{s}(1)\|^2}{\sigma^2}}, \cdots, e^{-\frac{\|\mathbf{s}(j)-\mathbf{s}(N)\|^2}{\sigma^2}} \right]^T \tag{10.15}$$

Substituting (10.14) into (10.15), we have

$$\hat{\theta}^T(j) = \mathbf{\Theta}\mathbf{H}^{-1}\left[e^{-\frac{\|\mathbf{s}(j)-\mathbf{s}(1)\|^2}{\sigma^2}}, \cdots, e^{-\frac{\|\mathbf{s}(j)-\mathbf{s}(N)\|^2}{\sigma^2}}\right]^T \tag{10.16}$$

Since the vector on the right-hand side of Eq. (10.16) is just the jth column of $\mathbf{H}$, (10.16) becomes

$$\hat{\theta}(j) = \mathbf{\Theta}\,[0, \ldots, 1, 0, \ldots, 0,]^T \tag{10.17}$$

where the 1 is the jth position; hence, a perfect recall follows.

If the input value is different from one of the training vectors, say $\mathbf{x} = \mathbf{s}(j) + \mathbf{n}$, the RBF associative memory will try to associate $\mathbf{x}$ with a DOA estimate $\hat{\theta}$, where

$$\begin{aligned}\hat{\theta} &= \hat{W}\left[e^{-\frac{\|\mathbf{x}-\mathbf{s}(1)\|^2}{\sigma^2}}, \cdots, e^{-\frac{\|\mathbf{x}-\mathbf{s}(N)\|^2}{\sigma^2}}\right]^T \\ &= \mathbf{\Theta}\mathbf{H}^{-1}\left[e^{-\frac{\|\mathbf{s}(j)-\mathbf{s}(1)+\mathbf{n}\|^2}{\sigma^2}}, \cdots, e^{-\frac{\|\mathbf{s}(j)-\mathbf{s}(N)+\mathbf{n}\|^2}{\sigma^2}}\right]^T \\ &\approx \mathbf{\Theta}\mathbf{H}^{-1}e^{-\frac{\|\mathbf{n}(j)\|^2}{\sigma^2}}\left[e^{-\frac{\|\mathbf{s}(j)-\mathbf{s}(1)\|^2}{\sigma^2}}, \cdots, e^{-\frac{\|\mathbf{s}(j)-\mathbf{s}(N)\|^2}{\sigma^2}}\right]^T\end{aligned} \tag{10.18}$$

The last statement of Eq. (10.18) holds since in DOA estimation, the signal vector $\mathbf{s}$ and noise vector $\mathbf{n}$ are usually assumed to be independent, and hence their norm or dot product, that is, summation of cross-products, will be close to zero. Since the column vector on the right side of (10.18) is just the jth column of $\mathbf{H}$, we have

$$\hat{\theta}(j) = e^{-\frac{\|\mathbf{n}\|^2}{\sigma^2}}\mathbf{\Theta}\,[0, \ldots, 0, 1, 0, \ldots, 0,]^T = e^{-\frac{\|\mathbf{n}\|^2}{\sigma^2}}\theta(j) \tag{10.19}$$

The estimate $\hat{\theta}$ deviates from the ideal DOA θ by the exponential factor given in Eq. (10.19). When the power of $\mathbf{n}$ is close to zero, the exponential factor will go to one and the ideal estimate will be generated by the RBF associative memory. When the power of $\mathbf{n}$ is large, the exponential factor will approach zero. Note that the noise effect can be controlled by the parameter σ in the sense that a large value of σ can reduce the effect of $\|\mathbf{n}\|$.

In the above RBF formulation, the computation of the weights is the most time-consuming operation in the modeling process, and its complexity grows polynomially with N, since in order to find the coefficients of the expansions, an $N \times N$ matrix has to be inverted. However, once the coefficients have been computed, the time needed to evaluate the function at a point grows only linearly in N and this process can easily be parallelized. In the DOA application the number of training examples can be very large if the number of signals is large, and the inversion of such a huge matrix is not only formidable but could also be meaningless since the probability that a large matrix is ill-conditioned is high. In Eq. (10.8), the basis on which the function is expanded is given by the set of functions h translated and centered on a set of centers, coincident with the data. The interpolation conditions can be weakened if the number of centers is made lower than the number of data and their coordinates are allowed to be chosen arbitrarily [19]. In this case, denoting with $\mathbf{c}(1), \mathbf{c}(2), \cdots, \mathbf{c}(L)$ the coordinates of the L centers ($L < N$), the RBF network becomes

$$\mathbf{\Theta} = \mathbf{W}\mathbf{H} \tag{10.20}$$

where Θ, $\mathbf{W}$, and $\mathbf{H}$ are $K \times N$, $K \times L$, and $L \times N$ matrices, respectively, defined as

$$\mathbf{\Theta} = \begin{bmatrix} \theta_1(1) & \theta_1(2) & \dots & \theta_1(N) \\ \theta_2(1) & \theta_2(2) & \dots & \theta_2(N) \\ \vdots & \vdots & \ddots & \vdots \\ \theta_K(1) & \theta_K(2) & \dots & \theta_K(N) \end{bmatrix}, \quad \mathbf{W} = \begin{bmatrix} w_1^1 & w_2^1 & \dots & w_L^1 \\ w_1^2 & w_2^2 & \dots & w_L^2 \\ \vdots & \vdots & \ddots & \vdots \\ w_1^K & w_2^K & \dots & w_L^K \end{bmatrix}, \text{ and}$$

$$\mathbf{H} = \begin{bmatrix} h(\|\mathbf{s}(1) - \mathbf{c}(1)\|^2) & h(\|\mathbf{s}(2) - \mathbf{c}(1)\|^2) & \dots & h(\|\mathbf{s}(N) - \mathbf{c}(1)\|^2) \\ h(\|\mathbf{s}(1) - \mathbf{c}(2)\|^2) & h(\|\mathbf{s}(2) - \mathbf{c}(2)\|^2) & \dots & h(\|\mathbf{s}(N) - \mathbf{c}(2)\|^2) \\ \vdots & \vdots & \ddots & \vdots \\ h(\|\mathbf{s}(1) - \mathbf{c}(L)\|^2) & h(\|\mathbf{s}(2) - \mathbf{c}(L)\|^2) & \dots & h(\|\mathbf{s}(N) - \mathbf{c}(L)\|^2) \end{bmatrix} \tag{10.21}$$

A least squares (LS) approach can then be adopted and the optimal solution can be written as

$$\hat{\mathbf{W}} = \mathbf{\Theta}\mathbf{H}^+ \tag{10.22}$$

where $\mathbf{H}^+$ is the pseudo-inverse of $\mathbf{H}$ given as

$$\mathbf{H}^+ = \mathbf{H}^{\mathrm{T}}(\mathbf{H}\mathbf{H}^{\mathrm{T}})^{-1} \tag{10.23}$$

This formulation makes sense if the matrix $\mathbf{H}\mathbf{H}^T$ is nonsingular, and the existence of the inverse of this matrix has been proven [18]. Another advantage of using a rectangular matrix is that an adaptive algorithm such as the Least Mean Squares (LMS) algorithm [20] and the Recursive Least Squares (RLS) algorithm [20] can be used to estimate the weights. This further reduces the computational time and the hardware complexity for using this model in signal processing applications.

Using this LS-RBF network for associative memory, we consider

$$\hat{\mathbf{\Theta}} = \hat{\mathbf{W}}\mathbf{H} = \mathbf{\Theta}\mathbf{H}^{\mathrm{T}}(\mathbf{H}\mathbf{H}^{\mathrm{T}})^{-1}\mathbf{H} \tag{10.24}$$

The difference between the estimation DOA and the actual DOA is then the LS fitting error

$$\begin{aligned} \|\mathbf{\Theta} - \hat{\mathbf{\Theta}}\| &= \|\mathbf{\Theta}(\mathbf{I} - \mathbf{H}(\mathbf{H}\mathbf{H}^{\mathrm{T}})^{-1}\mathbf{H})\| \\ &\le \|\mathbf{\Theta}\|\,\|\mathbf{I} - \mathbf{H}(\mathbf{H}\mathbf{H}^{\mathrm{T}})^{-1}\mathbf{H}\| \\ &= \min(N - L, 1)\|\mathbf{\Theta}\| \end{aligned} \tag{10.25}$$

The properties of this orthogonal projection can be found in [21]. The magnitude of the estimation error depends on the subspace spanned by the data matrix $\mathbf{H}$ is bounded, and the error of DOA estimates is bounded by the norm of the corresponding actual DOA vector [22] as shown in the above equation.

10.4 DOA ESTIMATION USING THE RBF ASSOCIATIVE MEMORY

As mentioned previously, the implementation of the RBF associative memory for DOA estimation consists of two phases: learning and estimation. In the DOA estimation problem, the initial phases α_k in Eq. (10.1) are of no interest. In order to reduce the dimensionality of the signal parameter space in which the network learns, instead of using the ideal signal **s** directly as the input to the network, a preprocessing procedure is used to eliminate α_k from the observation **s**. As shown in Figure 10.1(a), the learning phase consists of the generation of the array output vectors, transformation of the array vectors into the input vectors in the preprocess procedure, and applying the least squares method to obtain a set of optimal weights. The estimation phase consists of transforming the sensor output vector into an input vector and producing the DOA estimate by retrieving it from memory, as shown in Figure 10.1(b). The RBF network for DOA estimation is shown in Figure 10.1(c). It consists of three layers: an input layer, a hidden layer, and output layer.

The input layer consists of a number of input units, which distribute the inputs to the next layer. In the learning phase, the covariance matrix **R** is first formed in the preprocessing procedure. Assuming that there is no noise and taking the expectation, the elements of **R** are expressed as

$$R_{mm'} = \sum_{k=1}^{K} p_k e^{j(m-m')\frac{\omega_0}{c} d \sin\theta_k} + \delta R_{mm'} \tag{10.26}$$

where $p_k = E[a_k^2]$ and $\delta R_{mm'}$ contains all the cross-correlated terms between signals, which becomes zero for uncorrelated signals. When $m = m'$, we have $R_{mm'} = \sum_{k=1}^{K} p_k$ which do not carry information on the DOA and are irrelevant for the associated memory. We may then rearrange the rest of the elements into a new input vector, **b**, given as

$$\mathbf{b} = \left[R_{21}, \cdots, R_{M1}, R_{12}, \cdots, R_{M2}, \cdots, R_{m(m-1)}\right]^T \tag{10.27}$$

The number of input units is determined by the dimension of **b**, that is, $M(M-1)$.

The hidden layer consists of L Gaussian basis functions. As shown in the previous section, L can be set to equal N, the number of training vectors, and this allows perfect recall. However, if N is large, ill-conditioning of the matrix **H** may result in a poor solution. Therefore, the least squares associative memory is recommended. In the next section, we demonstrate some effects of L on learning and estimation. For $L < N$, we have to determine the L centers $\{\mathbf{c}_i, I = 1, \ldots, L\}$ in Eq. (10.21). In our application, we choose these centers uniformly from the training set. To avoid nonlinear optimization, the value of σ must be decided a priori. As we have discussed above, the use of a large σ may reduce the effect of noise. However, a large σ will decrease the ability of a basis function to localize and will increase the chance that the matrix **H** becomes ill-conditioned. A compromise must therefore be made in the choice of σ.

The network is trained separately for a different number of signals. The required number of units in the output layer is equal to the number of signals. That is, the output layer consists of K units. In addition, for a particular number of signals, it requires a particular set of centers and the corresponding set of optimal weights. Thus, in this approach, the a priori information on the number of signals is required in both the training and estimation phases.

The term p_k in Eq. (10.26) is the absolute power of the kth signal. In practice, it

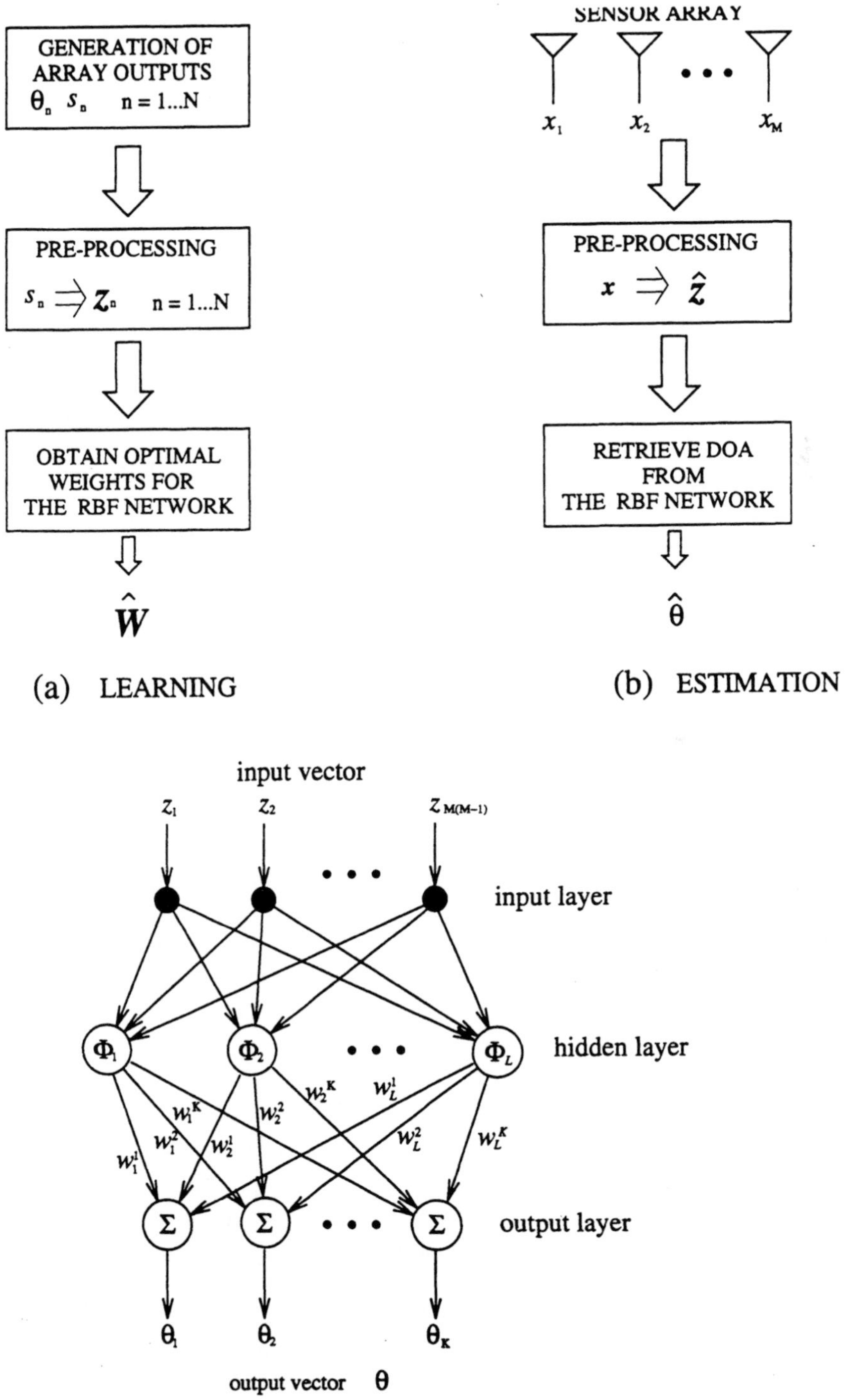

Figure 10.1 Implementation of the RBF associative memory for DOA estimation (a) learning phase; (b) estimation phase; (c) the RBF network.

can have an arbitrary value. In order to avoid training the network with different values of p_k we consider the relative power of the signal with respect to a reference which may be a fixed value. We may set $p_k = 1$ in training, that is, the mean of the relative signal power is considered to be unity. We then normalize the input vector by its norm in both the training and estimation phases, that is, the new input vector is given by

$$\mathbf{z} = \frac{\mathbf{b}}{\|\mathbf{b}\|} \tag{10.28}$$

The learning vectors and the corresponding output vectors are $\{(\boldsymbol{\theta}(i), \mathbf{z}(i)), i = 1, \cdots, N\}$. The DOA is uniformly sampled from a visible interval v where $v = (-90°, 90°]$ in our application. For the single-signal case, there are $N = N_L$ learning samples. For the two-signal case, $N = (N_L + 1)N_L/2$ learning samples are required, instead of $N_L \times N_L$ samples because of symmetry. We then define the ratio $r = v/N_L$ as the learning resolution. If r is sufficiently small, it may provide an adequate accuracy for the interpolative estimation. In Table 10.1, we summarize the learning procedure.

TABLE 10.1 Learning Procedure for DOA Processing Using RBF Network

1.	Generate array output vectors $\{\mathbf{s}(n), n = 1, \ldots, N\}$ using Eq. (10.1).
2.	Evaluate the covariance matrice $\{\mathbf{R}(n), n = 1, \ldots, N\}$.
3.	Form the vectors $\{\mathbf{b}(n), n = 1, \ldots N\}$ using Eq. (10.29).
4.	Normalize the input vectors $\{\mathbf{z}(n) = \mathbf{b}(n)/\|\mathbf{b}(n)\|, n = 1, \ldots N\}$.
5.	Compute the data matrix $\mathbf{H}$ using $h_l(\mathbf{z}(i)) = \exp(-\|\mathbf{z}(i) - \mathbf{c}_l\|^2/\sigma^2)$.
6.	Compute the least squares solution for $\hat{\mathbf{W}} = \left[\boldsymbol{\theta}(1) \ldots \boldsymbol{\theta}(N)\right]\mathbf{H}^+$.

In the estimation phase, an observation $\mathbf{x}$ is presented to the network, and the DOA vector $\boldsymbol{\theta}$ with a corresponding $\mathbf{z}$ close to $\mathbf{x}$ will be retrieved. If $\mathbf{x}$ is noise-free and is one of the input vectors in the training set, a perfect recall is possible for a network with the number of hidden units L equal to the number of training samples N. If $\mathbf{x}$ is not in the training set, the network will produce an estimation output through interpolation. The first-order perturbation in $b_{mm'}$ is given by

$$\delta b_{mm'} = (m - m')\frac{\omega_0}{c}d \sum_{k=1}^{K} \delta\theta_k \cos\theta_k e^{j((m-m')\frac{\omega_0}{c}d \sin\theta_k - \pi)} \tag{10.29}$$

It is clear that as $\{\theta_k, k = 1, \ldots K\}$ approaches zero, $\delta b_{mm'}$ will diminish and hence the DOA estimate $\hat{\boldsymbol{\theta}}$ approaches $\boldsymbol{\theta}(j)$. In addition, $\delta b_{mm'}$ is the combination of all $\{\delta\theta_k, k = 1, \ldots K\}$. This implies that in the case that all $\delta\theta_k$ are zero except one of them, $\delta b_{mm'}$ will not vanish and there will be small errors in $\hat{\boldsymbol{\theta}}$, that is, $\{\hat{\theta}_k \neq \theta_k(j), k = 1, \ldots, K\}$.

If the input vector is corrupted with noise, that is, $\mathbf{x} = \mathbf{s} + \mathbf{n}$, the estimate of $\mathbf{R}$ is evaluated as

$$\hat{\mathbf{R}} = \frac{1}{N_s} \sum_{i=1}^{N_s} \mathbf{x}(i)\mathbf{x}^+(i) = \mathbf{s}\mathbf{s}^+ + \boldsymbol{\varepsilon} \tag{10.30}$$

where N_s is the number of snapshots used in the estimation. The off-diagonal elements of ε are the statistical mean of all the uncorrelated terms. If N_s is sufficiently large, these terms are approximately Gaussian distributed with mean equal to zero. As N_s approaches infinity, $\varepsilon = \sigma_n^2\mathbf{I}$ where σ_n^2 is the noise power and $\mathbf{I}$ is the identity matrix. The noise-corrupted $\hat{\mathbf{b}}$ is expressed as

$$\begin{aligned}\hat{\mathbf{b}} &= \left[\hat{R}_{21}, \cdots, \hat{R}_{M1}, \cdots, \hat{R}_{M2}, \cdots, \hat{R}_{M(M-1)}\right]^T \\ &= \left[R_{21}, \cdots, R_{M1}, R_{12}, \cdots, R_{M2}, \cdots, R_{M(M-1)}\right]^T \\ &\quad + \left[\varepsilon_{21}, \cdots, \varepsilon_{M1}, \varepsilon_{12}, \cdots, \varepsilon_{M2}, \cdots, \varepsilon_{M(M-1)}\right]^T \\ &= \mathbf{b} + \varepsilon'\end{aligned} \tag{10.31}$$

That is, the noise in the input vector can be considered as an additive noise

$$\hat{\mathbf{z}} \simeq \mathbf{z} + \frac{\varepsilon'}{\|\hat{\mathbf{b}}\|} \tag{10.32}$$

This preprocessing procedure will not affect the associative memory function described in the previous section. It should also be noted that since the diagonal elements of $\mathbf{R}$ are dropped out, the noise power term σ_n^2 is removed. That is, $\hat{\mathbf{b}}$ approaches $\mathbf{b}$ as N_s goes to infinity. In Table 10.2, we summarize the DOA estimation procedures.

TABLE 10.2 Estimation Procedure for DOA Processing Using RBF Network

1. Evaluate the sample covariance matrice $\hat{\mathbf{R}}$ using the array outputs $\{\mathbf{x}(n), n = 1, \ldots, N_s\}$.
2. Form the vectors $\hat{\mathbf{b}}$ using Eq. (10.30).
3. Normalize the input vectors $\hat{\mathbf{z}} = \hat{\mathbf{b}}/\|\hat{\mathbf{b}}\|$.
4. Compute the vector $\mathbf{h}$ using $h_l(\hat{\mathbf{z}}) = \exp(-\|\hat{\mathbf{z}} - \mathbf{c}_l\|^2/\sigma^2)$.
5. Recall the corresponding DOA $\hat{\theta} = \hat{\mathbf{W}}\mathbf{h}$.

10.5 COMPUTER SIMULATIONS AND ANALYSIS

Simulations are now carried out to understand the efficiency of the RBF neural network approach. In the simulations, a linear array of eight elements ($M = 8$) is assumed. d is set to be a value such that $(\omega_0/c)d = \pi/2$. The variance of each basis function is set to be unity. The signal-to-noise (SNR) ratio is defined as the ratio of the total array power to the average noise power. For performance measure, we use the normalized mean squares error (MSE),

$$\mathcal{E} = \sum_{k=1}^{K} \frac{1}{N_t} \sum_{i=1}^{N_t} \left(\frac{\theta_k(i) - \hat{\theta}_k(i)}{BW} \right)^2 \tag{10.33}$$

where N_t is the number of ensembles to be averaged and BW is the 3-dB beamwidth of the array, which is given by $BW = 0.88\lambda/(M-1)d$ for a linear array.

The learning data are generated as described previously. In analyzing the effects of learning on the DOA estimatoin, we consider several factors including the number of learning samples N, the learning resolution r, the number of hidden units L, and the number of snapshots N_s for estimation. The simulation uses one signal only. We also set the value of σ to be one. For learning, we set $p_1 = 1$. For estimation, N_s is equal to 8. $\mathbf{n}$ is assumed to be white Gaussian, and the initial phase α_m takes random values which are uniformly distributed over $[-\pi, \pi]$. The MSE is plotted as a function of learning resolution r in Figure 10.2. For each value of r, the network is trained with $L = N$. The input vectors presented to the network for estimation are contained in the learning set so that a perfect recall is possible. For large values of r, the network gives extremely small MSE and we may say the network recalls perfectly. However, the MSE is comparatively large when r is small. The larger errors are due to ill-conditioned matrix $\mathbf{H}$. When the condition number of $\mathbf{H}$ is examined, as shown in Figure 10.3, it can be seen that for small values of r which correspond to a large value of N, the condition number is large, which indicates that $\mathbf{H}$ is ill-conditioned. To determine a suitable r, we also need to consider the network's capability for interpolative associative recalls. When the network is presented with input vectors that are not contained in the learning set, a different performance is obtained, which is shown in Figure 10.4. Small values of MSE are associated with small values of r. Thus, on the one hand, in order to keep the interpolative errors small, r should also be small. On the other hand, r should be large to avoid and ill-conditioned $\mathbf{H}$.

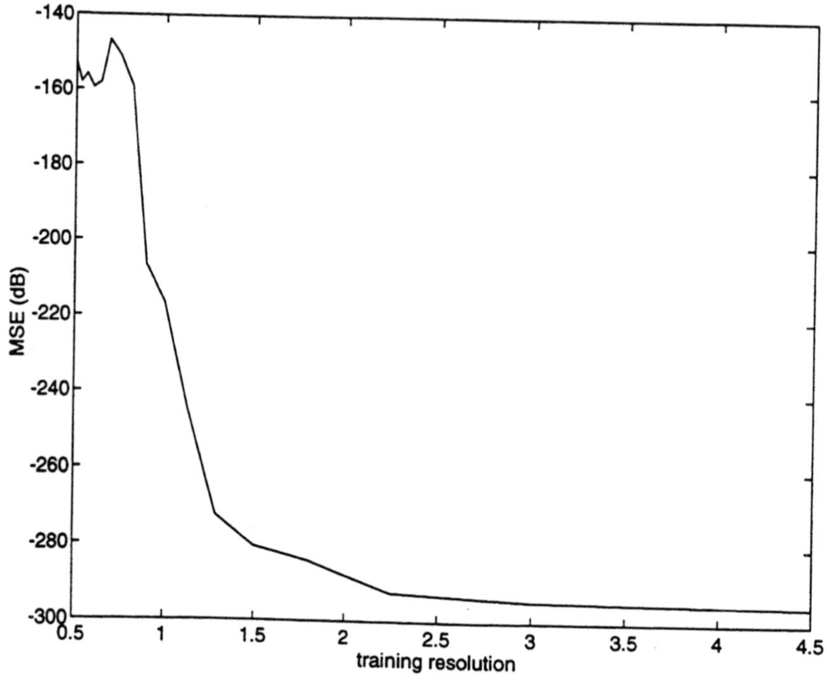

Figure 10.2 Learning MSE as a function of the learning resolution r for $L = N$.

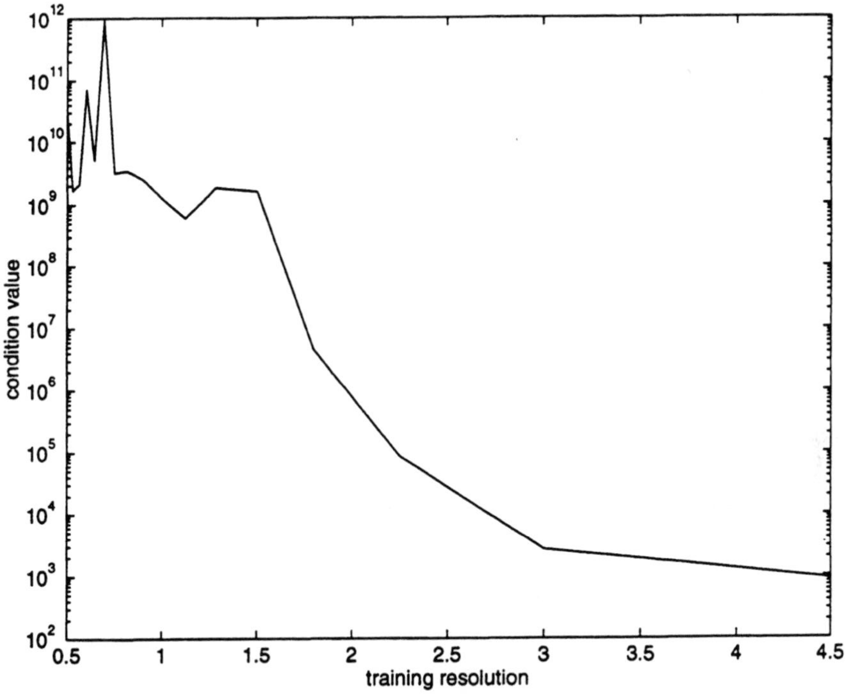

Figure 10.3 The matrix condition values as a function of the learning resolution r for the case shown in Figure 10.2.

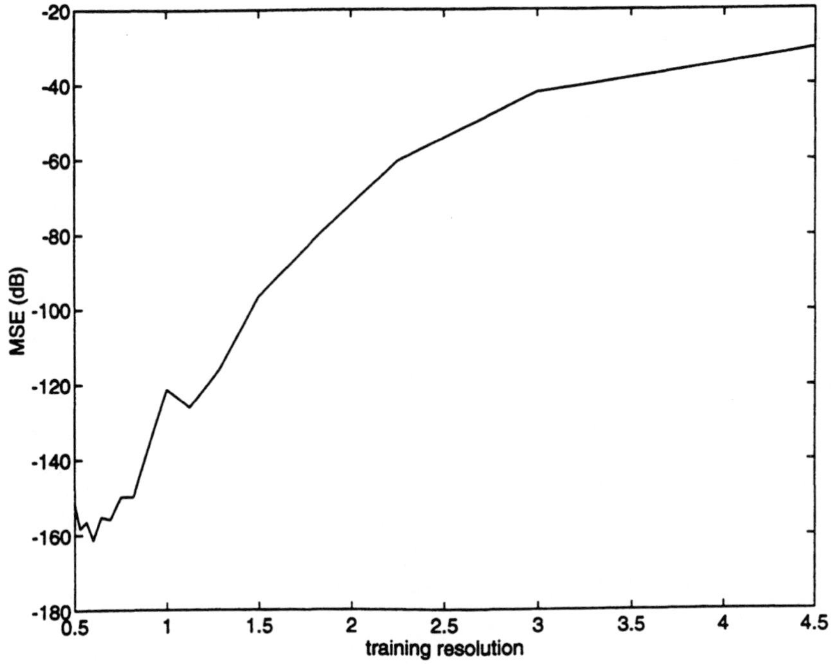

Figure 10.4 Interpolative MSE as a function of the learning resultion r for $L = N$.

Next, the effects of the number of hidden units, L, are examined. The relationship between the MSE and L is shown in Figure 10.5. The network is trained with $N = 180$. The input vectors presented to the network for associative recall are contained in the learning set. In the noiseless case, as L approaches N, the associative recall should approach a perfect condition. However, as L becomes larger and larger, the effect of the ill-condition of the matrix $\mathbf{H}$ becomes significant, that is, a perfect recall may not be obtainable, as shown by the dotted line in Figure 10.5. To understand this, we examine the condition nunber of the matrix $\mathbf{H}$, which is shown in Figure 10.6. The magnitudes of the values of the matrix condition are consistent with those of the MSE (dotted line in Figure 10.5). The effects of noise are also studied by adding Gaussian white noise to the input vectors. The dashed-dotted line in Figure 10.5 represents the MSE for SNR$=$80. As expected, the minimum level of MSE for recall is higher than in the noiseless case. Increasing the number of hidden units will not decrease the MSE of recall. It may even increase the MSE, as shown by the dashed line and solid line for SNR$=$30 and SNR$=$20, respectively. To investigate how well the network can perform the interpolative associative recall, the network is presented with noiseless input vectors that are not in the training set. The recall results are compared with those for input vectors inside the learning set, as shown in Figure 10.7. It is clear that there is no significant difference between the two cases. This is because a sufficiently large value of N is used.

The effectiveness of the RBF approach for DOA estimation is now demonstrated. In the simulation, we consider both the single-signal case and the two-signal case. In the single-signal case, the effects of Gaussian white noise and colored noise are studied. In the two-signal case, we consider both Gaussian and non-Gaussian signals. We also study the effects of using incorrect information on the number of signals for training and estimation. In addition , we consider the case of two correlated signals. The performance of the network will be compared with that of the MUSIC algorithm. In the simulation, p_k is set to be unity.

We start our analysis by considering the single-signal first. In this case, $L = 45$ and $N_L = 90$. For estimation, a_k is assumed to be Gaussian. The signal is located at 0°, which is not in the learning set. Each MSE value is obtained with $N_t = 200$. In Figure 10.8, the estimation performance MSE is plotted against N_s where SNR $= 20$ dB. For $N_s \geq 8$, the RBF network and the MUSIC method have a similar performance. As N_s increases, the error ε due to the input vector $\mathbf{z}$ in Eq. (10.30) decreases, and hence the MSE decreases. The MSE is shown in Figure 10.9 for $N_s = 8$. As shown in the figure, in the SNR range of 10 to 20 dB, the network shifts the threshold in the MSE curve to the left by at least 10 dB with respect to the MSE curve given by the MUSIC algorithm. For SNR$>$20 dB, the results are similar. It should be noted that in the MUSIC algorithm, the noise eigenvectors of the covariance matrix are used to form the subspace on which the steering vector is projected. At low SNR, the noise perturbation can cause the noise eigenvectors to fluctuate and hence affect the performance of the MUSIC algorthm [23]. The RBF network, on the other hand, does not involve any decomposition of the covariance matrix and is thus not affected by the behavior of the eigenvectors of the covariance matrix. However, the performance of the network depends on how much the input vector is corrupted by noise and on how well the network has learned. The learning

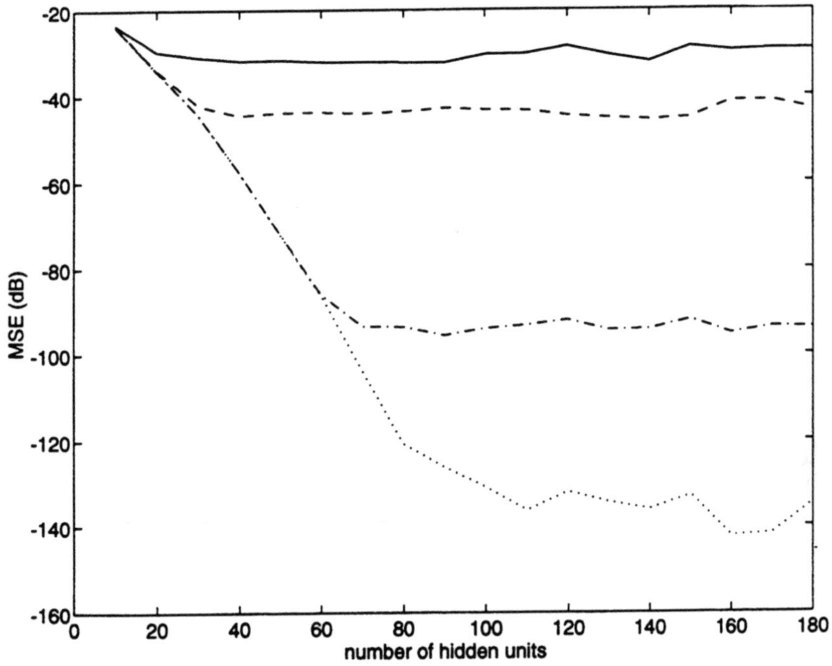

Figure 10.5 Learning MSE as a function of the number of hidden units: — SNR = 20 dB, -- SNR = 30 dB, -·- SNR = 80 dB, ··· SNR = ∞.

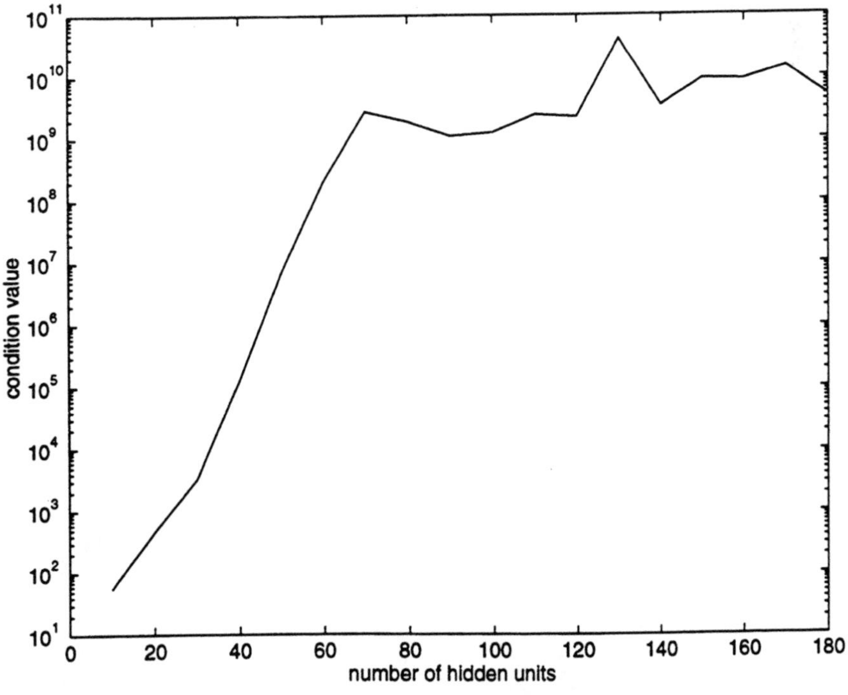

Figure 10.6 The matrix condition values as a function of the number of hidden units for the case shown in Figure 10.5.

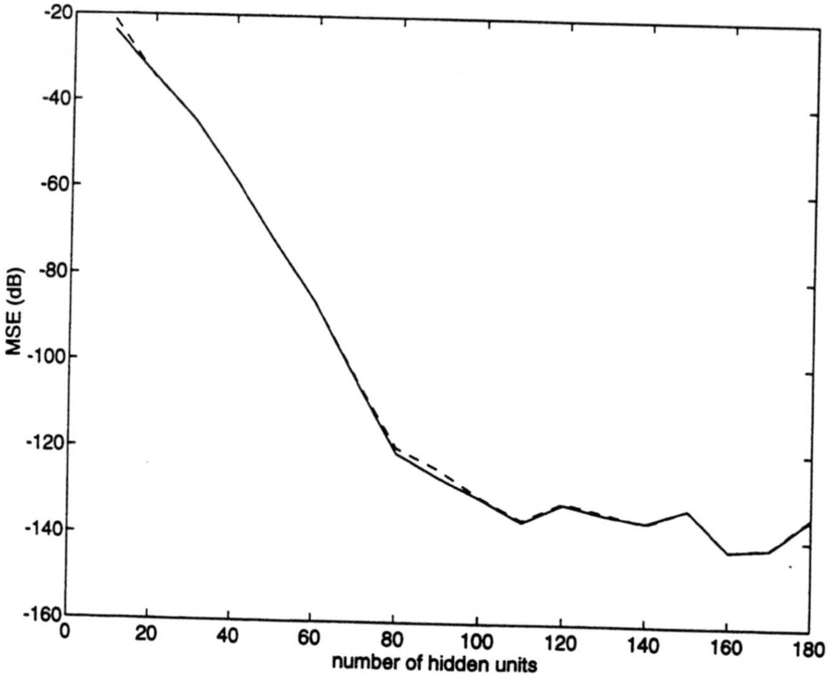

Figure 10.7 The comparison of learning MSE (—) and the interpolative MSE (--) for SNR $= \infty$.

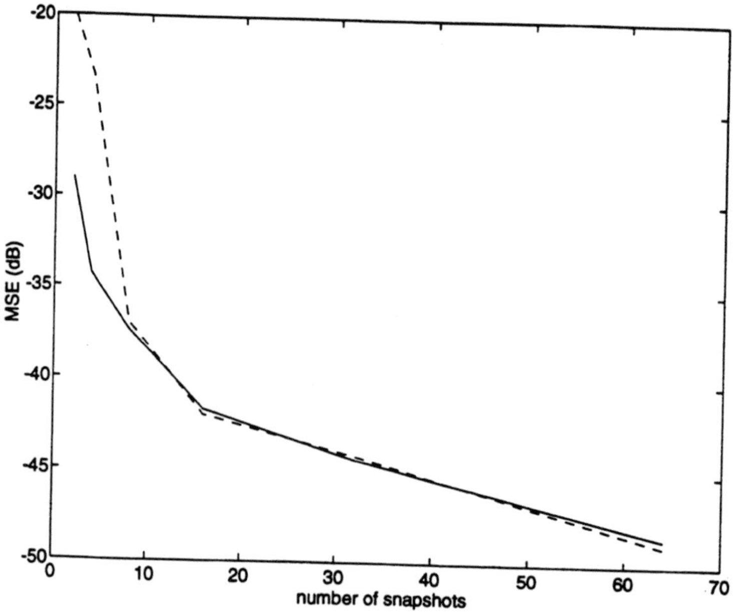

Figure 10.8 MSE as a function of the number of snapshots Ns for white Gaussian noise with SNR = 20 dB; — RBF, -- MUSIC.

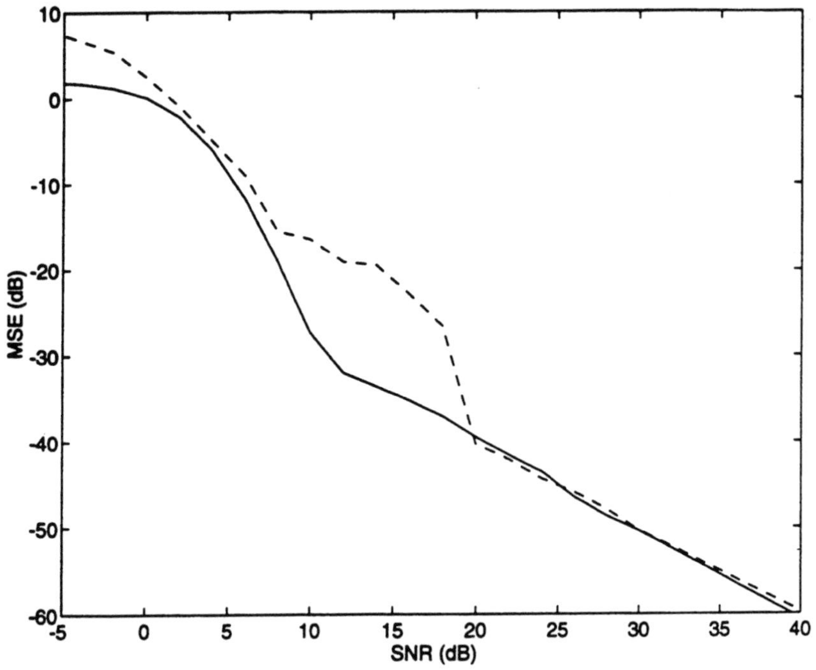

Figure 10.9 MSE as a function of SNR for white Gaussian noise with $Ns = 8$; — RBF, -- MUSIC.

error for the network is fixed once the centers and optimal weights are determined. Intuitively, the estimation errors are proportional to the SNR. When noise is dominant in the input vector, the RBF network will not be able to retrieve the correct DOA. Because the RBF network is a nonlinear estimator, it is also subject to a threshold effect. Qualitatively, the threshold is caused by outliers in the data [24]. Analysis of the threshold effect is complicated because of the nonlinear nature of the RBF network. However, Monte Carlo simulation does provide a means to investigate the threshold phenomenon.

In Figure 10.10, the estimation performance is shown when a Gaussian spatially colored noise is added to the signal. The colored noise is generated using a spatial correlation matrix, σ_s, the *ik*th element of which is given by

$$\sigma_s(ik) = \rho^{|i-k|} e^{j\phi_p(i-k)} \tag{10.34}$$

where the parameter ρ controls the relative height of the peak of the noise spatial spectrum and ϕ_p controls the angle at which the peak occurs. In our case, $\rho = 0.9$ and $\phi_p = \pi/8$. From the figure, we observe that the RBF performance is almost the same as in the Gaussian white noise case. When the noise is spatially colored, its correlation matrix is represented by $\boldsymbol{\varepsilon}$ in Eq. (10.30). As the input vector $\hat{\mathbf{z}}$ is formed by rearranging the elements of $\hat{\mathbf{R}}$, the correlation relation between the elements in $\boldsymbol{\varepsilon}$ is altered. In a sense, the spatially colored noise is whitened, and insofar as the associative memory is concerned it is no different from white noise. In contrast, the

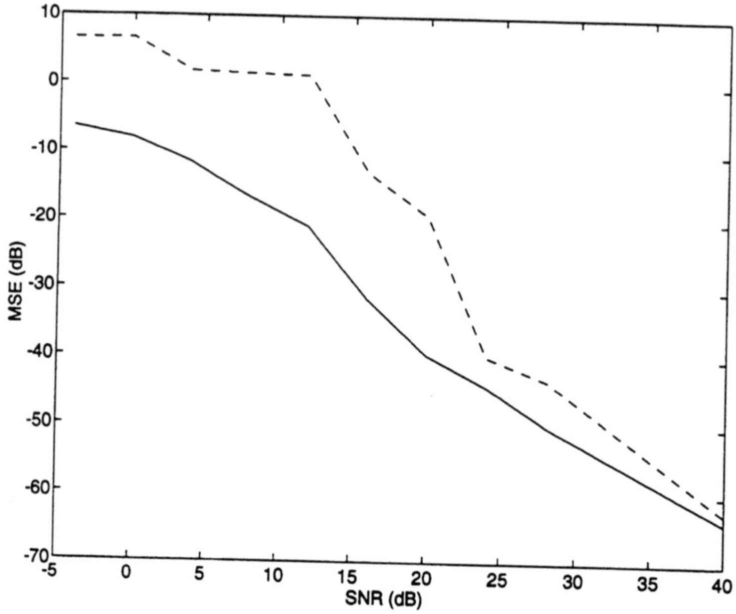

Figure 10.10 MSE as a function of SNR for colored Gaussian noise with $Ns = 8$; — RBF, -- MUSIC.

MUSIC algorithm depends on the eigendecomposition, which is in turn affected by ε. Thus, its performance is degraded with respect to the Gaussian white noise case.

We now consider the two-signal case. The network is trained with $\{p_k = 1, k = 1, 2\}, L = 90, N_L = 45$ is used for generating the learning samples for each of the signals. That is, the total number of learning samples is $N = 1035$. For estimation, two equal-strength signals are generated, one signal being located at 0° and the other at 5°. Both values are not in the training set. The angular separation of the two signals is about 35% beamwidth of the antenna.

In Figure 10.11, the performance of the DOA estimation for two uncorrelated signals is given. The amplitudes of the signals are assumed to be drawn from two independent Gaussian processes, and the initial phases are uniformly distributed. Gaussian white noise of different strengths is added to the signals to investigate the performance of the network. It can be seen that when the SNR is less than 20 dB, the MSE given by the RBF network is about 10 dB less than that given by the MUSIC algorithm. It is known that the angular separation of the signals affects the eigenvalue spread of the covariance matrix [25]. The smaller the angular separation is, the larger the eigenvalue spread is. The eigenvalue spread directly affects the performance of the MUSIC algorithm, in addition to the effects of the fluctuation of the noise eigenvectors of the covariance matrix at low SNR. On the other hand, the RBF network does not depend on the eigenvalue spread of the covariance matrix. However, at high SNR (> 30 dB), the RBF network has a greater MSE than the MUSIC algorithm. As discussed above, to keep the interpolative errors small, r should also be small. However, since N_L is set to be 45, the value of r is relatively

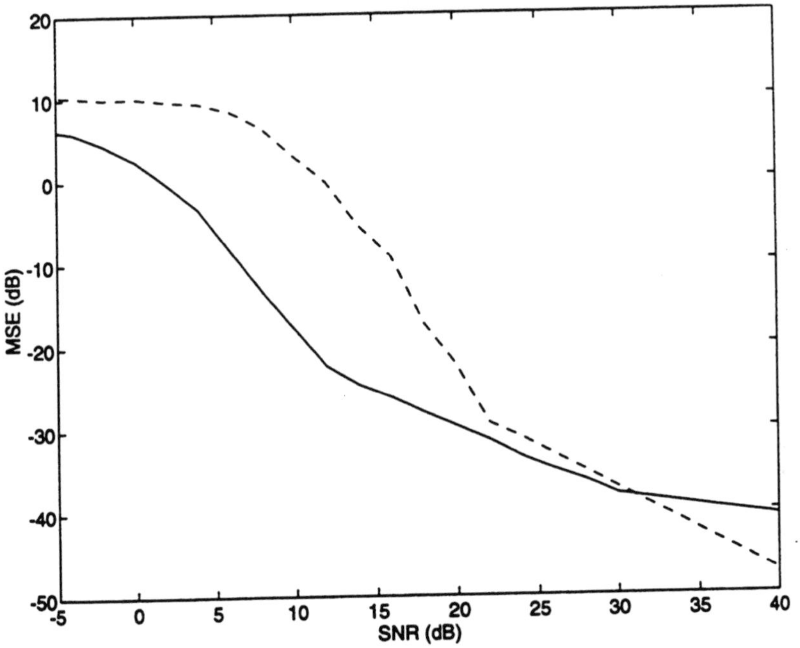

Figure 10.11 Two-signal MSE as a function of SNR for white Gaussian noise with $Ns = 8$; — RBF, -- MUSIC.

large. This leads to a relatively large interpolative errors, which is observable at high SNR.

In this chapter, we assume that the information on the number for signals is given. When this information is not correct, that is, a wrong signal model, the performance of the RBF may be degraded. A simulation is carried out to study the effect of using an incorrect model. In Figure 10.12, a RBF network trained by the two-signal model is used to estimate the DOA of a data set that contains only one signal. In this case, the SNR is set to 20 dB. Although there is some ambiguity of the two estimates provided by the trained RBF, the RBF network seems to be able to provide an acceptable value of the true DOA. Figure 10.13 shows another example of using an incorrect number of signals. The actual number of signals is three, that is, one more signal than the network is trained for. In this case, the network fails to provide the correct DOA. A detailed study of the effect of the number of signals is left for future research.

The performance shown in Figure 10.14 is for the same scenario as in Figure 10.11, except for changes in the amplitude of the incident signals. The amplitudes of the two signals are assumed to be two independent exponential distributed processes. If we compare the results so obtained with those in Figure 10.11, we see that the performance of the MUSIC algorithm is degraded; this is not surprising since the statistical model now departs from that used for the MUSIC algorithm. An interesting observation is that the performance of the RBF approach is also degraded, even though the degradation is much less compared with the MUSIC algorithm. A

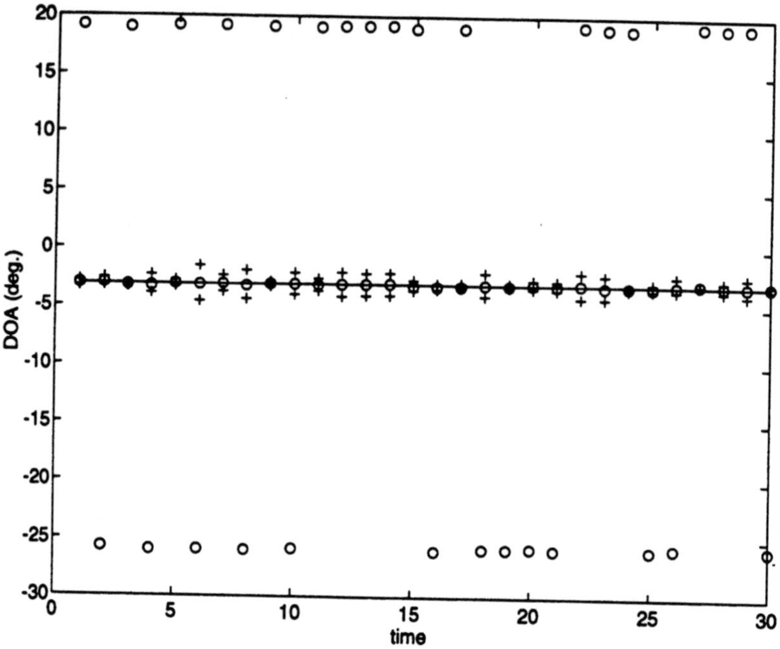

Figure 10.12 DOA estimates with incorrect information on number of signals. Actual number of signals is one and two signals are assumed for the estimation — actual DOA, + RBF, o MUSIC.

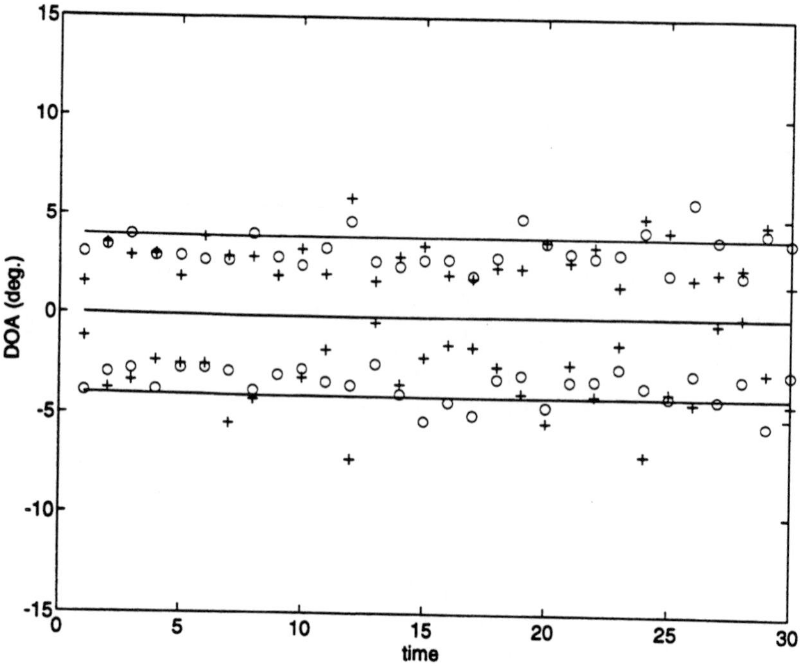

Figure 10.13 DOA estimates with incorrect information on number of signals. Actual number of signals is three and two signals are assumed for the estimation — actual DOA, + RBF, o MUSIC.

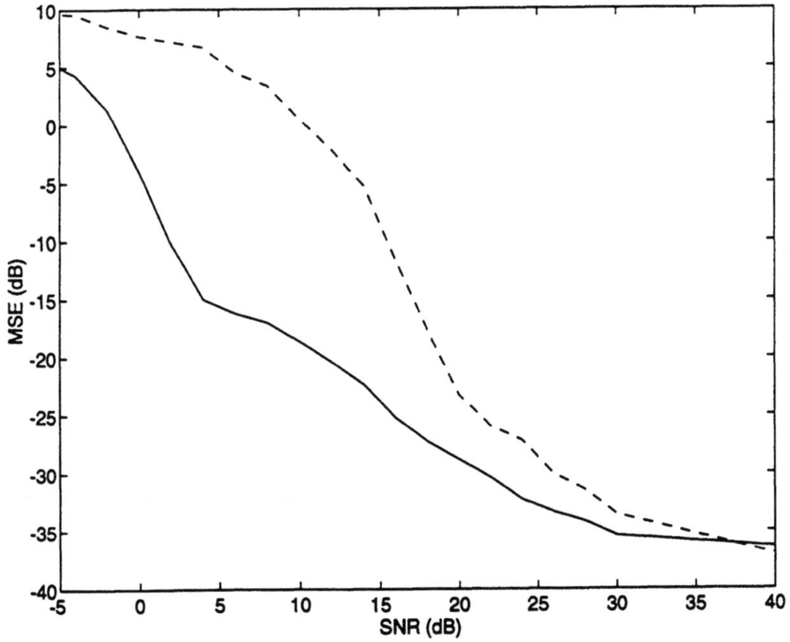

Figure 10.14 Two-signal MSE as a function of SNR for exponentially distributed amplitude with $Ns = 8$; — RBF, -- MUSIC.

lesser degradation is expected since the RBF net approach does not rely on statistical assumptions as strongly as the MUSIC algorithm.

The situation of correlated signals is considered next. The correlation coefficient of two signals is expressed as

$$C = \frac{cov\left[(a_1 e^{j\alpha_1})(a_2 e^{j\alpha_2})\right]}{\sqrt{var(a_1 e^{j\alpha_1}) var(a_2 e^{j\alpha_2})}} \tag{10.35}$$

where $cov[\cdot]$ denotes the covariance operator and $var(\cdot)$ denotes the variance operator. For correlated signals, the $\delta R_{mm'}$ in Eq. (10.26) has a finite value. Its magnitude depends on the correlation coefficient. Since the signals are correlated, a spatial smoothing technique has to be incorporated with the MUSIC algorithm. The subarray has a size of 7. We first use the same trained network as in the case shown in Figure 10.10 to carry out the estimation. The amplitude of each of the signals is assumed to be a Gaussian process, and the initial phase is uniformly distributed, with the correlation of the two signals set to be C. In Figure 10.15, the MSE is plotted as a function of the correlation coefficient C. In this case, the SNR is 25 dB. In the figure, the solid line represents the MSE given by the RBF network, which is consistently less by several dB than that of the MUSIC method with spatial smoothing, represented by the dashed line. Without spatial smoothing, the MSE of the MUSIC algorithm is even larger (dashed-dotted line). The results show that the network can handle correlated signals even though it is not trained specially for

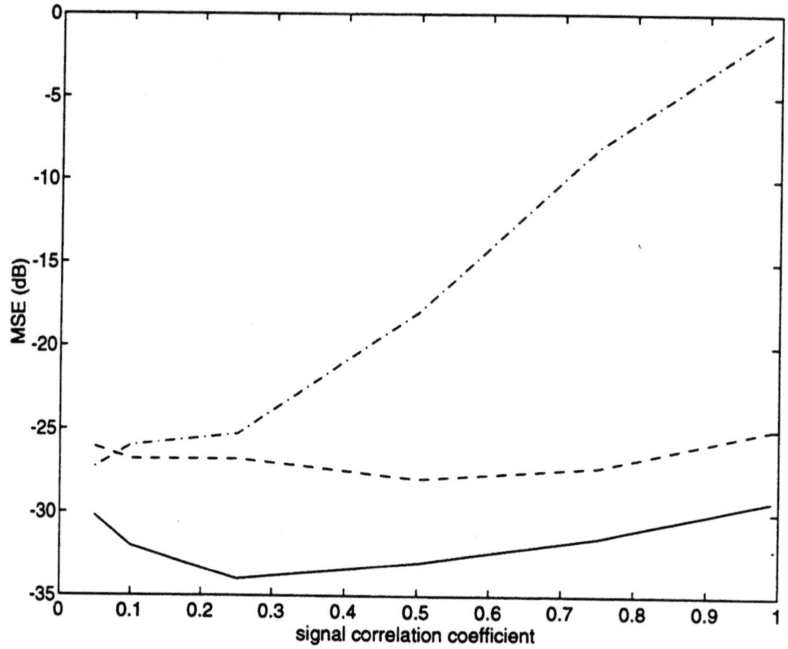

Figure 10.15 Two-signal MSE as a function of the signal correlation coefficient with $Ns = 8$; — RBF, -- MUSIC with spatiaal smoothing, -·- MUSIC without spatial smoothing.

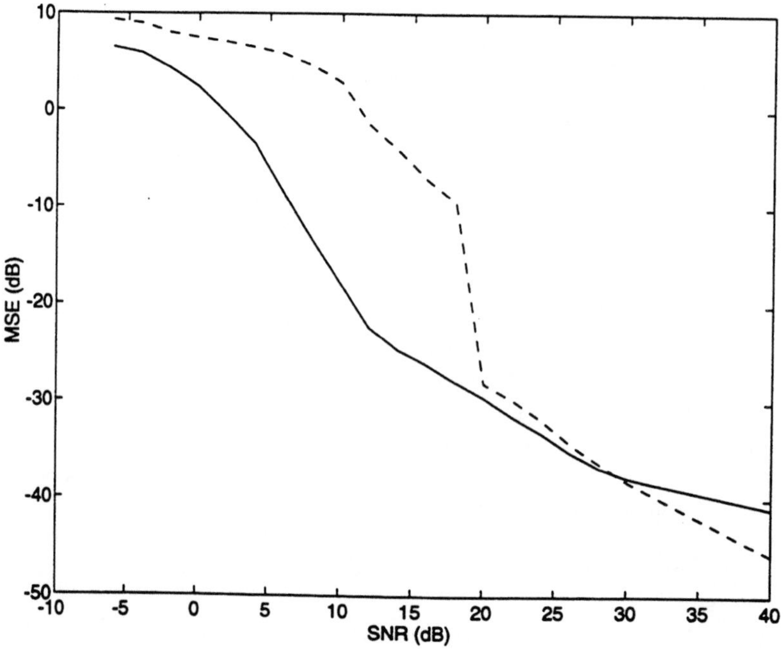

Figure 10.16 Two-signal MSE as a function of SNR for signal correlation coefficient = 0.99; — RBF, -- MUSIC.

correlated signals. This is because $\delta R_{mm'}$ is small compared with the first term on the right-hand side of Eq. (10.26) even though the signals are correlated. If we train the network using learning samples containing two correlated signals, we may expect some improvement. The training samples are generated with $\{p_k = 1, k = 1, 2\}$ and $C = 0.99$. For estimation, the signals also have a value of $C = 0.99$. The performance is shown in Figure 10.16. At SNR = 25 dB, the MSE given by the RBF network is about -34 dB, which suggests that about 4 dB improvement may be obtained if the network is trained with correlated signals. This is because the term $\delta R_{mm'}$ is included in the input vectors used for training. The information on $\delta R_{mm'}$ for different DOA is also memorized by the RBF network Thus, the network is able to retrieve the DOA more accurately. At a relatively low SNR, the performance of the MUSIC algorithm is worse for correlated signals than for uncorrelated ones (Figure 10.11), which is expected since the effective aperture of the array is reduced due to the use of smoothing. In addition, the use of a subarray of size 7 may not be able to totally decorrelate the correlation between the two signals.

10.6 APPLICATION TO AN EXPERIMENTAL DIRECTION-FINDING SYSTEM

Real data are now used to illustrate the effectiveness of the RBF associative memory approach. The measurement system is an experimental eight-element direction-finding system, which was developed at Defence Research Establishment Ottawa. The receiver consists of a radio frequency (RF), intermediate frequency (IF), and baseband data acquisition sections. The receiving array consists of eight dipole antenna elements operating at 60 MHz. The antennas are spaced by 2 m and mounted on a movable platform, which is made of fibreglass to prevent electromagnetic scattering and diffraction. In the front-end module of each element, the received signals are amplified, filtered, and down-converted to an intermediate frequency of 10.7 MHz. In order to improve the dynamic range, the IF bandwidth is chosen to be 200 KHz. The video processors use digitally controlled 5th-order switched capacitor low-pass filters. The channel bandwidth is further reduced to 300 Hz with a center frequency located at 200 Hz to avoid direct current drift.

The received baseband signals are digitally sampled at a rate of 1000 samples/second with a 16-bit precision. In a direction-finding system, both the in-phase and the quadrature components are usually required to form the complex signals. However, it is often found that there is an imbalance between the in-phase and the quadrature channels in the system. Instead of eliminating the imbalance, which is usually a difficult task, we sample only the in-phase component and derive the quadrature component by performing a digital Hilbert transform on the in-phase component [26].

Each dipole antenna has an independent receiver. Since the gain and phase of each receiver are different from the others, data calibration is required to correct for gain and phase differences. A simple but practical data calibration technique is chosen for our system, where the effect due to the signal environment is implicitly considered.

In order to find the gain and phase errors, which we call the calibration error vector, a source with a known direction θ_0 is used. The signal vector can then be expressed as

$$\mathbf{q} = a \cdot e^{j\alpha}\left[1, e^{j\phi}, \cdots, e^{j7\phi}\right]^T \tag{10.36}$$

where

$$\phi = \frac{\omega_0}{C} d \sin\theta_0$$

and α is the initial phase. The complex gains of the eight receivers can be expressed in vector form:

$$\mathbf{g}^c = \left[g_1^c, \cdots, g_8^c\right]^T \tag{10.37}$$

where

$$g_m^c = g_m e^{j\psi_m}$$

where g_m and ψ_m are the gain and phase of the mth channel, respectively. It follows that the outputs of the eight channels are

$$x^0 = \left[a g_1^c e^{j\alpha}, a g_2^c e^{j\alpha + j\phi}, \cdots, a g_8^c e^{j\alpha + j7\phi}\right]^T \tag{10.38}$$

The calibration error vector is obtained by normalizing $\mathbf{x}^0$ by the first element of $\mathbf{x}^0$, that is

$$\mathbf{x}^c = \left[1, \frac{g_2^c}{g_1^c}, \cdots, \frac{g_8^c}{g_1^c}\right]^T \tag{10.39}$$

Since the received signals are contaminated by receiver noise, statistical mean values are used as the estimate of the calibration error function. The receiver noise is assumed to be additive white Gaussian noise. The estimate of $\mathbf{x}^0(m)$ is given by

$$\hat{\mathbf{x}}_m^0 = \frac{1}{N_s}\sum_{n=1}^{N_s} x_m^0(n) \tag{10.40}$$

where N_s is the number of time samples used for temporal averaging, and $x_m^0(n)$ is the output of the mth element at time n. In our experiment, N_s is 2048, which is the maximum length of a data set. The estimate of the calibration function is given as

$$\hat{\mathbf{x}}^c = \left[1, \frac{\hat{x}^0(2)}{\hat{x}^0(1)}, \cdots, \frac{\hat{x}^0(8)}{\hat{x}^0(1)}\right]^T \tag{10.41}$$

Once this function is found, it can be applied to the raw received data to correct for the errors due to amplitude and phase differences between the receivers. Specifically, if the raw output from the mth receiver is x_m, the corrected output is $x_m/\hat{x}_m^c$.

The measurement trials were conducted during March 1993. A transmitter with a dipole antenna was used to simulate a radio source. The transmitting antenna was placed at predetermined locations, which were surveyed to determine the exact coordinates with respect to the array. These coordinates were later used to derive the bearing of the transmitted radio signals. Three sets of real data are used to evaluate the RBF technique for DOA estimation. Although the data have been

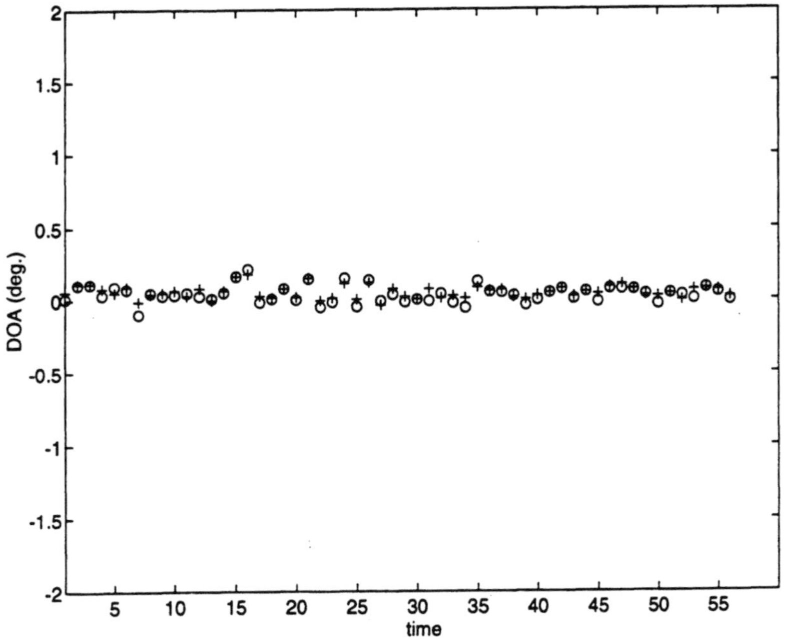

Figure 10.17 DOA estimates derived from real data. Measured DOA = 0 deg + RBF, o MUSIC.

calibrated, there are still errors contained in the data. These errors include (1) receiver noise, (2) phase and amplitude fluctuations due to motion of the dipole antenna caused by strong wind, (3) uneven spacing between antenna elements, and (4) residual calibration error. Among them, only the receiver noise may be described as white Gaussian noise. The others are usually correlated and non-Gaussian.

The trained network used for the simulation of the one-signal case is considered here. To form the normalized covariance matrix, $\hat{\mathbf{R}}$, $N_s = 32$ is used. That is, each DOA is derived using 32 snapshots. Three different datesets were used, for which the measured angular locations of the radio sources were 0°, 5.7°, and 11.3°. The DOA estimates as time progresses for the three data sets are shown in Figures 10.17–10.19, respectively. The plus symbol represents the RBF network estimates and the circle represents those for the MUSIC algorithm. It may not be clear from these figures which technique performs better. This is because the SNR of the real data are quite high. However, the comparison becomes clearer when we examine the statistics of the estimates. In Table 10.3, the mean and the MSE are calculated assuming the measured angles were the true angle. It is observed that the means estimated by both techniques are about the same; the RBF technique is however better than the MUSIC algorithm in the sense that the MSE is less by 0.3 to 3 dB.

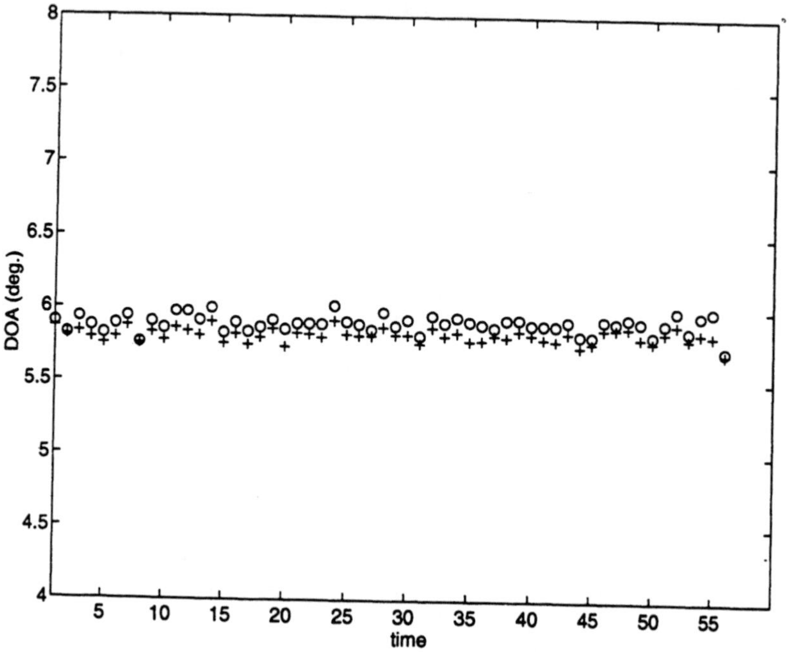

Figure 10.18 DOA estimates derived from real data. Measured DOA = 5.77 deg + RBF, o MUSIC.

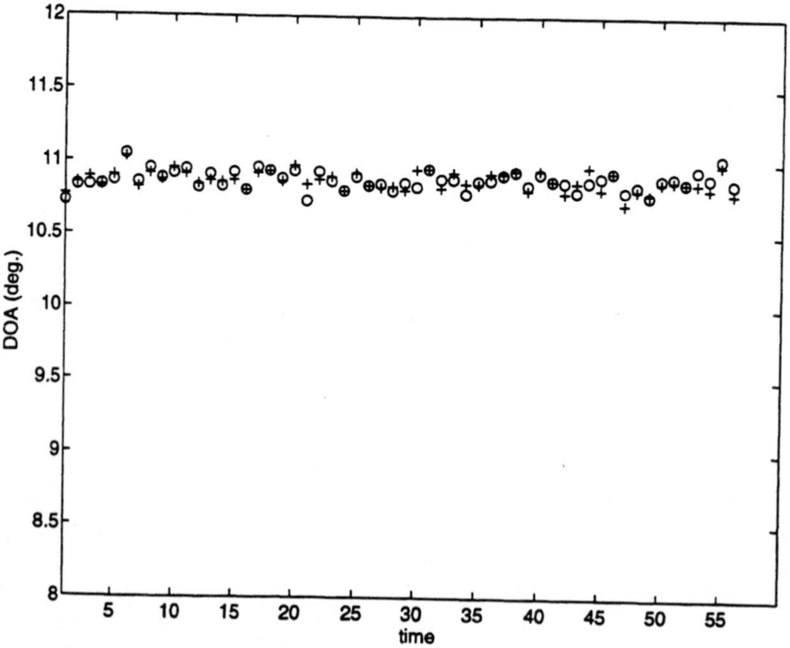

Figure 10.19 DOA estimates derived from real data. Measured DOA = 11.3 deg + RBF, o MUSIC.

TABLE 10.3 Summary of DOA Estimates Derived from Real Data

	Set 1	Set 2	Set 3
Actual:	0	5.7	11.3
RBF network:			
mean	0.05	5.77	10.8
MSE (dB)	−51	−51	−48
MUSIC:			
mean	0.04	6.0	10.9
MSE (dB)	−48	−48	−47.7

10.7 CONCLUSION

In this chapter, we have proposed an approach for DOA estimation using a nonlinear associative memory. In particular, the DOA problem is considered as a mapping from the space of DOA to the space of the sensor output. A nonlinear associative memory is used to from the inverse mapping from the space of sensor output to the space of DOA and this memory is realized using the radial-basis function neural network

Simulations have also been presented to understand the efficiency of the RBF neural network approach. In analyzing the effects of learning for DOA estimation, we have considered several parameters including the number of learning samples N, the learning resolution r, and the number of hidden units L. We have observed that a small r or a large N can increase accuracy of the DOA estimation through interpolation. In principle, a large L should also improve the accuracy. However, a large L may give rise to ill-conditioning in practice. In investigating the performance of the RBF network, we have considered both the single-signal case and the two-signal case. In the single-signal case, the effects of Gaussian white noise and colored noise have been considered. In the two-signal case, we have examined the performance for both Gaussian and non-Gaussian signals. We have also studied the effects when incorrect information on the number of signals is used for training and estimation. In addition, we have considered the case of correlated signals. The results of these simulations have shown that the RBF network performs reasonably well and is superior to the MUSIC algorithm. The results also show that the RBF approach is sensitive to the signal model used in the training. On the other hand, it is relatively insensitive to background noise. Moreover, real data have been used to illustrate the effectiveness of the RBF approach. The results showed that the RBF network also performs well in a practical environment.

In the RBF approach for DOA estimation, there is no inversion or decomposition of the covariance matrix required during the estimation phase. Therefore, its performance does not depend on the eigenvalue spread and the behavior of the eigenvectors of the covariance matrix of the input data. The associative recall process directly provides the DOA estimates. There is no search for peak values involved; as a result, the RBF approach is computationally efficient. Furthermore, parallel implementation of the RBF network makes it feasible to carry out real-time processing. Although a large learning set is used in the learning phase, the learning process itself can be carried out off-line. Thus, it should not diminish the possibility of real-time implementation. However, the network requires separate training for different signal environments such as different numbers of signal and degrees of signal correlation. Consequently, a priori information about the relevant signal environment is required.

REFERENCES

1. S. M. KAY. *Modern Spectral Estimation: Theory and Application*, Englewood Cliffs, NJ: Prentice Hall, 1988.
2. S. PILLAI. *Array Signal Processing*, New York: Springer-Verlag, 1989.
3. S. HAYKIN, ed. *Advances in Spectrum Analysis and Array Processing*, vol. I & II, Englewood Cliffs, NJ: Prentice Hall, 1992.
4. S. JHA and T. S. DURRANI. "Direction of Arrival Estimation Using Artificial Neural Networks," *IEEE Trans. Systems, Man and Cybernetics*, vol. 21, pp. 1192–1201, 1991.
5. L. FA-LONG and B. ZHENG. "Real-time Neural Computation of the Maximum Likelihood Criterion for Bearing Estimation Problems," *Neural Networks*, vol. 5, pp. 765–69, 1992.
6. G. MARTINELLI and R. PERFETTI. "Neural Network Approach to Spectral Estimation of Harmonic Processes," *IEE Proc.*, part G, vol. 140, pp. 95–100, 1993.
7. J. HERTZ, A. KROGH, and R. G. PALMER. *Introduction to the Theory of Neural Computation*, Redwood City: Addison-Wesley, 1991.
8. Y. PAO. *Adaptive Pattern Recognition and Neural Networks*, Redwood City: Addison-Wesley, 1989.
9. J. J. HOPFIELD and D. W. TANK. "Neural Computation of Decisions in Optimization Problems," *Biological Cybernetics*, vol. 52, pp. 141–52, 1985.
10. D. GORYN and M. KAVEH. "Neural Networks for Narrowband and Wideband Direction Findign," *Proc. ICASSP*, pp. 2164–67, 1988.
11. T. POGGIO and F. GIROSI. "Networks for Approximation and Learning," *Proc. IEEE*, vol. 78, pp. 1481–95, 1990.
12. J. PARK and I. W. SANDBERG. "Universal Approximation Using Radial-Basis Function Networks," *Neural Computation*, vol. 3, pp. 246–57, 1991
13. F. GIROSI and T. POGGIO. "Networks and the Best Approximation Property," *Biological Cybernetics*, vol. 63, pp. 169–76, 1990.
14. H. SCHIOLER and U. HARTMANN. "Mapping Neural Network Derived from the Parzen Window Estimator," *Neural Networks*, vol. 5, pp. 903–09, 1992.
15. J. S. R. JUNG and C. T. SUN. "Functional Equivalence Between Radial-Basis Function Networks and Fuzzy Inference Systems," *IEEE Trans. Neural Networks*, vol. 4, pp. 156–59, 1993.
16. J. A. FREEMAN and D. M. SKAPURA. *Neural Networks: Algorithms, Applications and Programming Techniques*, Redwood City: Addison-Wesley, 1992.
17. T. KOHONEN. *Self-Organization and Associative Memory*, 3rd ed., Springer-Verlag, 1989.
18. C. A. MICCHELLI. "Interpolation of Scattered Data: Distance Matrices and Conditionally Positive Definite Functions," *Constructive Approximation*, vol. 2, pp. 11–22, 1986.

19. D. S. Broomhead and D. Lowe. "Multivariable Functional Interpolation and Adaptive Networks," *Complex Systems*, vol. 2, pp. 321–55, 1988.

20. S. Haykin. *Adaptive Filter Theory*, 2nd ed. Englewood Cliffs, NJ: Prentice Hall, 1991.

21. L. L. Scharf. *Statistical Signal Processing—Detection, Estimation and Time Series Analysis*, Redwood City: Addison-Wesley, 1991.

22. G. H. Golub and C. F. Van Loan. *Matrix Computations*, 2nd ed., Baltimore: The Johns Hopkins University Press, 1983.

23. W. F. Gabriel. "Using Spectral Estimation Techniques in Adaptive Processing Antenna Systems," *IEEE Trans. Antennas and Propagation*, vol. 34, pp. 291–300, 1986.

24. D. C. Rife and R. R. Boorstyn. "Single-Tone Estimation from Discrete-Time Observations," *IEEE Trans. Information Theory*, vol. 20, pp. 591–98, 1974.

25. D. E. N. Davis, et al. "Array Signal Processing," in *The Handbook of Antenna Design*, by A. W. Rudge, et al., eds., Peter Peregrinus, 1986.

26. J. F. Claerbout. *Fundamentals of Geophysiccal Data Processing*, New York: McGraw-Hill, 1976

Index